Advances in Intelligent Systems and Computing

Volume 1092

The series "Advances in Intelligent Systems and Computing" contains publications on theory, applications, and design methods of Intelligent Systems and Intelligent Computing. Virtually all disciplines such as engineering, natural sciences, computer and information science, ICT, economics, business, e-commerce, environment, healthcare, life science are covered. The list of topics spans all the areas of modern intelligent systems and computing such as: computational intelligence, soft computing including neural networks, fuzzy systems, evolutionary computing and the fusion of these paradigms, social intelligence, ambient intelligence, computational neuroscience, artificial life, virtual worlds and society, cognitive science and systems, Perception and Vision, DNA and immune based systems, self-organizing and adaptive systems, e-Learning and teaching, human-centered and human-centric computing, recommender systems, intelligent control, robotics and mechatronics including human-machine teaming, knowledge-based paradigms, learning paradigms, machine ethics, intelligent data analysis, knowledge management, intelligent agents, intelligent decision making and support, intelligent network security, trust management, interactive entertainment, Web intelligence and multimedia.

The publications within "Advances in Intelligent Systems and Computing" are primarily proceedings of important conferences, symposia and congresses. They cover significant recent developments in the field, both of a foundational and applicable character. An important characteristic feature of the series is the short publication time and world-wide distribution. This permits a rapid and broad dissemination of research results.

**** Indexing: The books of this series are submitted to ISI Proceedings, EI-Compendex, DBLP, SCOPUS, Google Scholar and Springerlink ****

More information about this series at http://www.springer.com/series/11156

Manuel F. Silva · José Luís Lima ·
Luís Paulo Reis · Alberto Sanfeliu ·
Danilo Tardioli
Editors

Robot 2019: Fourth Iberian Robotics Conference

Advances in Robotics, Volume 1

Springer

Editors
Manuel F. Silva
School of Engineering
Polytechnic Institute of Porto
Porto, Portugal

José Luís Lima
Department of Electrical Engineering
Polytechnic Institute of Bragança
Bragança, Portugal

Luís Paulo Reis
Faculty of Engineering
University of Porto
Porto, Portugal

Alberto Sanfeliu
UPC
Universitat Politècnica de Catalunya
Barcelona, Spain

Danilo Tardioli
Centro Universitario de la Defensa (CUD)
Zaragoza, Spain

ISSN 2194-5357 ISSN 2194-5365 (electronic)
Advances in Intelligent Systems and Computing
ISBN 978-3-030-35989-8 ISBN 978-3-030-35990-4 (eBook)
https://doi.org/10.1007/978-3-030-35990-4

This Springer imprint is published by the registered company Springer Nature Switzerland AG
The registered company address is: Gewerbestrasse 11, 6330 Cham, Switzerland

Preface

This book contains a selection of papers accepted for presentation and discussion at Robot 2019—Fourth Iberian Robotics Conference—held in Porto, Portugal, November 20–22, 2019. Robot 2019 is part of a series of conferences that are a joint organization of SPR—Sociedade Portuguesa de Robótica/Portuguese Society for Robotics and SEIDROB—Sociedad Española para la Investigación y Desarrollo en Robótica/Spanish Society for Research and Development in Robotics. The conference organization had also the collaboration of several universities and research institutes, including School of Engineering of the Polytechnic Institute of Porto, Polytechnic Institute of Bragança, University of Porto, University Politécnica de Cataluña, University of Zaragoza/I3A, INESC TEC, Centro Universitario de la Defensa, CeDRI and LIACC.

Robot 2019 builds upon several previous successful events, including three biannual workshops (Zaragoza—2007, Barcelona–2009 and Sevilla–2011) and the three previous editions of the Iberian Robotics Conference held in Madrid in 2013, Lisbon in 2015 and Seville in 2017. The conference is focused on presenting the research and development of new applications, on the field of robotics, in the Iberian Peninsula, although open to research and delegates from other countries.

Robot 2019 featured five plenary talks on state-of-the-art subjects on robotics by Mirko Kovac, Director of the Aerial Robotics Laboratory, Reader in Aero-structures at Imperial College London and Royal Society Wolfson Fellow, UK, on "Soft Aerial Robotics for Digital Infrastructure Systems;" Gianni A. Di Caro, Associate Teaching Professor at the Department of Computer Science of the Carnegie Mellon University, Qatar, on "Robot Swarms and the Human-in-the-Loop;" Luis Merino, Associate Professor of Systems Engineering and Automation and Co-Principal Investigator of the Service Robotics Laboratory at the Universidad Pablo de Olavide, Spain, on "Human-Aware Decision Making and Navigation for Service Robots;" Nuno Lau, Assistant Professor at Aveiro University, Portugal, on "Optimization and Learning in Robotics;" and Elon Rimon, Professor in the Department of Mechanical Engineering at the Technion–Israel Institute of Technology, Israel, on "Perspectives on Minimalistic Robot Hand Design and a New Class of Caging-to-Grasping Algorithms."

Robot 2019 featured 16 special sessions, plus a main/general robotics track. The special sessions were about Aerial Robotics for Inspection and Maintenance; Agricultural Robotics and Field Automation; Autonomous Driving and Driver Assistance Systems; Autonomous Sailboats and Support Technologies; Collaborative Robots for Industry Applications; Core Concepts for an Ontology for Autonomous Robotics; Genealogy and Engineering Practice; Educational Robotics; Field Robotics In Challenging Environments; Future Industrial Robotics; Intelligent Perception and Manipulation; Machine Learning in Robotics; Mobile Robots for Industrial Environments; Radar-Based Applications for Robotics; Rehabilitation and Assistive Robotics; Simulation in Robotics; and the Workshop on Physical Agents.

In total, after a careful review process with at least three independent reviews for each paper, but in some cases 5 or 6 reviews, a total of 112 high-quality papers were selected for publication, with a total number of 468 authors, from 24 countries, including Aland Islands, Australia, Belgium, Brazil, Canada, Colombia, Croatia, Czechia, Ecuador, Estonia, France, Germany, Italy, Japan, Netherlands, Pakistan, Portugal, Puerto Rico, Spain, Sweden, United Arab Emirates, UK, USA and Venezuela.

We would like to thank all special sessions' organizers for their hard work on promoting their special session, inviting the Program Committee, organizing the special session review process and helping to promote Robot 2019 Conference. This acknowledgment goes especially to Adrià Colomé, Alberto Olivares Alarcos, Alejandro Mosteo, Angel Sappa, Armando Sousa, Artur Pereira, Arturo de la Escalera, Begoña Arrue, Benedita Malheiro, Brígida Mónica Faria, Bruno Ferreira, Cristina Manuela Peixoto Santos, Danilo Tardioli, Eurico Pedrosa, Filipe Neves dos Santos, Francisco Bellas, Francisco Curado, Francisco Rovira Más, Germano Veiga, Guillem Alenyà, Guillermo Heredia, Ismael García Varea, Jan Rosell, João Quintas, Jon Agirre Ibarbia, Jorge Cabrera Gámez, José Lima, Juan C Moreno, Julita Bermejo-Alonso, Luis Merino, Luis Piardi, Luis Riazuelo, Manuel Silva, Miguel Ángel Cazorla Quevedo, Miguel Oliveira, Nuno Cruz, Nuno Lau, Pablo Bustos García de Castro, Paulo Goncalves, Pedro Guedes, Pedro Neto, Raul Morais, Ricardo Sanz, Roemi Fernandez, Vicente Matellán Olivera and Vitor Santos.

We would also like to take the opportunity to thank the rest of the organization members (André Dias, Benedita Malheiro, Luís Lima, Nuno Dias, Paulo Ferreira, Pedro Costa, Pedro Guedes and Teresa Costa) for their hard and valuable work on the local arrangements, publicity, publication and financial issues. We also express our gratitude to the members of all the Program Committees and additional reviewers, as they were crucial for ensuring the high scientific quality of the event and to all the authors and delegates that with their research work and participation made this event a huge success. To the end of this preface, special thanks to our editors, Springer, that was in charge of this conference proceedings edition and, in particular, to Dr. Thomas Ditzinger.

October 2019

Manuel F. Silva
José Luís Lima
Luís Paulo Reis
Alberto Sanfeliu
Danilo Tardioli

Organization

Program Committee

Jon Agirre Ibarbia	Tecnalia Research and Innovation, Spain
António Pedro Aguiar	University of Porto, Portugal
Teodoro Aguilera	University of Extremadura, Spain
Eugenio Aguirre	University of Granada, Spain
Daniel Albuquerque	Polytechnic Institute of Viseu, Portugal
David Alejo	University Pablo de Olavide, Spain
Guillem Alenyà	Universitat Politècnica de Catalunya, Spain
Jorge Almeida	University of Aveiro, Portugal
Luis Almeida	University of Porto, Portugal
Fernando Alvarez	University of Extremadura, Spain
Angelos Amanatiadis	Democritus University of Thrace, Greece
Josep Amat	Universitat Politècnica de Catalunya, Spain
Rui Araújo	University of Coimbra, Portugal
Manuel Armada	Consejo Superior de Investigaciones Cientificas, Spain
Begoña C. Arrue	University of Seville, Spain
Helio Azevedo	University of São Paulo, Brazil
José Azevedo	University of Aveiro, Portugal
Mihail Babcinschi	University of Coimbra, Portugal
Pilar Bachiller	Universidad de Extremadura, Spain
Stephen Balakirsky	Georgia Institute of Technology, USA
Antonio Bandera	University of Malaga, Spain
Juan Pedro Bandera Rubio	Universidad de Málaga, Spain
Antonio Barrientos	Universidad Politécnica de Madrid, Spain
Richard Bearee	Arts et Métiers ParisTech, France
José Antonio Becerra Permuy	University of A Coruña, Spain
Daniel Beffler	University of Bremen, Germany
Francisco Bellas	University of A Coruña, Spain
Luis M. Bergasa	Universidad de Alcalá, Spain

Julita Bermejo-Alonso	Universidad Isabel I, Spain
Marko Bertogna	University of Modena, Italy
Mehul Bhatt	Örebro University, Sweden
Estela Bicho	University of Minho, Portugal
Ole Blaurock	Fachhochschule Lübeck, Germany
Thadeu Brito	Research Center in Digitalization and Intelligent Robotics, Portugal
Adrian Burlacu	The Gheorghe Asachi Technical University of Iasi, Romania
Pablo Bustos García de Castro	Universidad de Extremadura, Spain
Fernando Caballero Benítez	University of Seville, Spain
Jorge Cabrera	Universidad de Las Palmas de Gran Canaria, Spain
João Calado	Instituto Superior de Engenharia de Lisboa, Portugal
Rafael Caldeirinha	Polytechnic Institute of Leiria, Instituto de Telecomunicações, Portugal
Joel Luis Carbonera	Federal University of Rio Grande do Sul, Brazil
Angela Cardoso	INEGI, Portugal
Carlos Carreto	Polytechnic Institute of Guarda, Portugal
Alicia Casals	Institute for Bioengineering of Catalonia, Spain
Jose A. Castellanos	University of Zaragoza, Spain
Miguel Ángel Cazorla Quevedo	Universidad de Alicante, Spain
Jose Maria Cañas Plaza	Universidad Rey Juan Carlos, Spain
Marco Ceccarelli	University of Rome Tor Vergata, Italy
Abdelghani Chibani	Université Paris-Est Créteil, France
Benoit Clement	ENSTA Bretagne, France
Adrià Colomé	Universitat Politècnica de Catalunya, Spain
Luís Correia	University of Lisboa, Portugal
Paulo Costa	University of Porto, Portugal
Pedro Costa	University of Porto, Portugal
Hugo Costelha	Polytechnic Institute of Leiria, Portugal
Micael Couceiro	Ingeniarius, Ltd., Portugal
Vincent Creuze	University of Montpellier, France
Nuno Cruz	University of Porto, Portugal
Francisco Curado	University of Aveiro, Portugal
Arturo de la Escalera	Universidad Carlos III de Madrid, Spain
Félix de La Paz López	Universidad Nacional de Educación a Distancia, Spain
Antonio J. Del-Ama	National Hospital for Spinal Cord Injury, Spain
Mohammed Diab	Universitat Politècnica de Catalunya, Spain
Paulo Dias	University of Aveiro, Portugal

Antonio C. Domínguez-Brito	Universidad de Las Palmas de Gran Canaria, Spain
Fadi Dornaika	University of the Basque Country, Spain
Richard Duro	University of A Coruña, Spain
Luis Emmi	Centre National de la Recherche Scientifique, France
João Alberto Fabro	Federal Technological University of Parana, Brazil
Andres Faina	IT University of Copenhagen, Denmark
Brígida Mónica Faria	Polytechnic Institute of Porto, Portugal
Paulo Farias	Federal University of Bahia, Brazil
Vicente Feliu	Universidad de Castilla–La Mancha, Spain
Fernando Fernandez	Universidad Carlos III de Madrid, Spain
Roemi Fernandez	Center for Automation and Robotics CSIC-UPM, Spain
Camino Fernández Llamas	Universidad de León, Spain
Manuel Ferre	CAR UPM-CSIC, Spain
Bruno Ferreira	INESC TEC, Portugal
Hugo Ferreira	INESC TEC, Portugal
Paulo Ferreira	Polytechnic Institute of Porto, Portugal
Pedro Fonseca	University of Aveiro, Portugal
Oscar Fontenla-Romero	University of A Coruña, Spain
Anna Friebe	Åland University of Applied Sciences, Finland
Oren Gal	Technion, Israel
Nicolas Garcia-Aracil	Universidad Miguel Hernandez, Spain
Ismael Garcia-Varea	Universidad de Castilla–La Mancha, Spain
Fernando García	University Carlos III, Spain
Paulo Goncalves	Polytechnic Institute of Castelo Branco, Portugal
Pablo Gonzalez-De-Santos	Centre for Automation and Robotics CSIC-UPM, Spain
Pedro Guedes	Polytechnic Institute of Porto, Portugal
Tamás Haidegger	Óbuda University, Hungary
Florian Haug	Åland University of Applied Sciences, Finland
Guillermo Heredia	University of Seville, Spain
Carlos Hernández Corbato	Delft University of Technology, The Netherlands
Roberto Iglesias Rodríguez	University of Santiago de Compostela, Spain
Eduardo Iáñez	Miguel Hernández University of Elche, Spain
Nicolas Jouandeau	Paris8 University, France
Nuno Lau	University of Aveiro, Portugal
Fabrice Le Bars	ENSTA Bretagne, France
Agapito Ledezma	Universidad Carlos III de Madrid, Spain
Wilfried Lepuschitz	Practical Robotics Institute Austria, Austria
Frederic Lerasle	LAAS-CNRS, France
Howard Li	University of New Brunswick, Canada

José Luís Lima	Polytechnic Institute of Bragança, Portugal
Pedro U. Lima	University of Lisboa, Portugal
Alejandro Linares-Barranco	University of Seville, Spain
Vitor Lobo	Escola Naval–CINAV, Portugal
Ana Lopes	University of Coimbra, Portugal
Joaquín López Fernández	University of Vigo, Spain
Jose Manuel Lopez-Guede	Basque Country University, Spain
Pedro Machado	Nottingham Trent University, UK
Erik Maehle	University of Luebeck, Germany
Benedita Malheiro	Polytechnic Institute of Porto, Portugal
André Luís Marcato	Federal University of Juiz de Fora, Brazil
Yarik Marchukov	Universidad de Zaragoza, Spain
Yaroslav Marchukov	Universidad de Zaragoza, Spain
Lino Marques	University of Coimbra, Portugal
Humberto Martinez Barbera	Universidad de Murcia, Spain
Francisco Martín Rico	Universidad Rey Juan Carlos, Spain
Vicente Matellán Olivera	Universidad de León, Spain
Nuno Mendes	University of Coimbra, Portugal
Munir Merdan	Practical Robotics Institute Austria, Austria
Luis Merino	Pablo de Olavide University, Spain
Luis Montano	University of Zaragoza, Spain
Paulo Monteiro	University of Aveiro, Portugal
Hector Montes Franceschi	Universidad Tecnologica de Panama, Panama
Eduardo Montijano	Universidad de Zaragoza, Spain
Raul Morais	University of Trás-os-Montes e Alto Douro, Portugal
Antonio Morales	Universitat Jaume I, Spain
Antonio P. Moreira	University of Porto, Portugal
Juan C. Moreno	Cajal Institute, Spain
Luis Moreno	Universidad Carlos III de Madrid, Spain
Alejandro R. Mosteo	Centro Universitario de la Defensa de Zaragoza, Spain
Paulo Moura Oliveira	University of Trás-os-Montes e Alto Douro, Portugal
Ana C. Murillo	University of Zaragoza, Spain
Giovanni Muscato	University of Catania, Italy
Alberto Nakano	Federal Technological University of Paraná, Brazil
Hirenkumar Chandrakant Nakawala	University of Verona, Italy
Pedro Navarro	University Carlos III, Madrid, Spain
Pedro Neto	University of Coimbra, Portugal
António J. R. Neves	University of Aveiro, Portugal

Urbano Nunes	University of Coimbra, Portugal
Pedro Núñez	Universidad de Extremadura, Spain
Manuel Ocaña Miguel	Universidad de Alcalá de Henares, Spain
Alberto Olivares Alarcos	Universitat Politècnica de Catalunya, Spain
Alexandra Oliveira	Polytechnic Institute of Porto, Portugal
Andre Oliveira	Federal Technological University of Paraná, Brazil
Josenalde Oliveira	Federal University of Rio Grande do Norte, Brazil
Miguel Oliveira	University of Aveiro, Portugal
Joanna Isabelle Olszewska	University of West Scotland, UK
Jun Ota	The University of Tokyo, Japan
Dimitrios Paraforos	University of Hohenheim, Germany
Nieves Pavon-Pulido	Universidad Politécnica de Cartagena, Spain
Eurico Pedrosa	University of Aveiro, Portugal
Peter Peer	University of Ljubljana, Slovenia
Ana Pereira	Polytechnic Institute of Bragança, Portugal
Artur Pereira	University of Aveiro, Portugal
Marcelo Petry	INESC TEC, INESC P&D Brasil, Federal University of Santa Catarina, Brazil
Luis Piardi	Research Center in Digitalization and Intelligent Robotics, Portugal
Tatiana M. Pinho	INESC TEC, Portugal
Andry Pinto	INESC TEC, Portugal
Vítor Hugo Pinto	University of Porto, Portugal
Frédéric Plumet	Institut des Systemes Intelligents et de Robotique, France
Mihai Pomarlan	Universität Bremen, Germany
Cristiano Premebida	Loughborough University, UK
Edson Prestes	Federal University of Rio Grande do Sul, Brazil
Domeneç Puig Valls	Universitat Rovira i Virgili, Spain
Noé Pérez-Higueras	Pablo de Olavide University, Spain
Jonas Queiroz	University of Porto, Portugal
João Quintas	Instituto Pedro Nunes, Portugal
Sandro Rama Fiorini	IBM, USA
Francisco Ramos	University of Castilla–La Mancha, Spain
Pedro Jesús Reche López	Universidad de Jaén, Spain
Carlos V. Regueiro	Universidade da Coruña, Spain
Giulio Reina	University of Salento, Italy
Oscar Reinoso	Universidad Miguel Hernandez, Spain
Luis Paulo Reis	University of Porto, Portugal
Luis Riazuelo	Universidad de Zaragoza, Spain
Lluís Ribas-Xirgo	Universitat Autònoma de Barcelona, Spain
Isabel Ribeiro	University of Lisboa, Portugal
Eugénio A. M. Rocha	University of Aveiro, Portugal

Luis Rocha	INESC TEC, Portugal
Rui P. Rocha	University of Coimbra, Portugal
Nelson Rodrigues	University of Porto, Portugal
Francisco J. Rodríguez Lera	University of Luxembourg, Luxembourg
Ronnier Rohrich	Federal Technological University of Paraná, Brazil
Roseli Romero	University of São Paulo, Brazil
Cristina Romero González	Universidad de Castilla–La Mancha, Spain
Kostia Roncin	ENSTA Bretagne, France
Jan Rosell	Universitat Politècnica de Catalunya, Spain
Rosaldo Rossetti	University of Porto, Portugal
Francisco Rovira Más	Universitat Politècnica de València, Spain
José Rufino	Polytechnic Institute of Bragança, Portugal
Mohammad Safeea	University of Coimbra, Portugal
Veera Ragavan Sampath Kumar	Monash University, Malaysia
Cristina Manuela Peixoto Santos	University of Minho, Portugal
Filipe Neves Santos	INESC TEC, Portugal
Jose Santos	University of A Coruña, Spain
Vitor Santos	University of Aveiro, Portugal
Ricardo Sanz	Universidad Politécnica de Madrid, Spain
Rafael Sanz Dominguez	Universidad de Vigo, Spain
Angel Sappa	CVC Barcelona, Spain
Colin Sauze	Aberystwyth University, UK
Sophia Schillai	University of Southampton, UK
Alexander Schlaefer	Hamburg University of Technology, Germany
Craig Schlenoff	NIST, USA
Michael Schukat	NUI Galway, Ireland
André Scolari	Federal University of Bahia, Brazil
Joao Sequeira	University of Lisboa, Portugal
João Silva	Altran Portugal, Portugal
Manuel F. Silva	Polytechnic Institute of Porto, Portugal
Danijel Skocaj	University of Ljubljana, Slovenia
Filomena Soares	University of Minho, Portugal
Armando Sousa	University of Porto, Portugal
Bo Tan	Tampere University, Finland
Danilo Tardioli	University of Zaragoza, Spain
Ana Maria Tome	University of Aveiro, Portugal
Carme Torras	Institut de Robòtica i Informàtica Industrial CSIC-UPC, Spain
Elisa Tosello	University of Padua, Italy
Jesus Urena	University of Alcala, Spain
Alberto Vale	University of Lisboa, Portugal

Antonio Valente	University of Trás-os-Montes e Alto Douro, Portugal
Enrique Valero	Heriot Watt University, UK
Germano Veiga	INESC TEC, Portugal
Jose Vieira	University of Aveiro, Portugal
Antidio Viguria	Center for Advanced Aerospace Technologies, Spain
Blas M. Vinagre	University of Extremadura, Spain
David Vázquez Bermúdez	Computer Vision Center, Spain
Matias Waller	Åland University of Applied Sciences, Finland
Angel Garcia-Olaya	Universidad Carlos III de Madrid, Spain
Arturo Torres-González	Universidad de Sevilla, Spain
Augusto Gómez Eguíluz	Universidad de Sevilla, Spain
Bruno J. N. Guerreiro	University of Lisboa, Portugal
Fernando Gomez-Bravo	Universidad de Huelva, Spain
Francisco Javier Perez-Grau	CATEC, Spain
Ja Acosta	Universidad de Sevilla, Spain
Jesus Capitan	Universidad de Sevilla, Spain
Jose Joaquin Acevedo	Universidad de Sevilla, Spain
Julián Estévez	Universidad del País Vasco, Spain
Julio L. Paneque	Universidad de Sevilla, Spain
Marco Wehrmeister	Federal Technological University of Paraná, Brazil
P. J. Sanchez-Cuevas	Universidad de Sevilla, Spain
Pablo Ramon Soria	Universidad de Sevilla, Spain
Saeed Rafee Nekoo	Universidad de Sevilla, Spain
Shayok Mukhopadhyay	American University of Sharjah, Dubai
Valter Costa	INEGI, Portugal

Additional Reviewers

Alireza Asvadi	Université de Bretagne Occidentale, France
Ana Lopes	Polytechnic Institute of Tomar, Portugal
Anis Koubaa	Prince Sultan University, Saudi Arabia
Carlos Gómez-Huélamo	University of Alcalá, Spain
Daniele Di Vito	University of Cassino and Southern Lazio, Italy
Edmanuel Cruz	University of Alicante, Spain
Félix Escalona	University of Alicante, Spain
Francisco Gomez-Donoso	University of Alicante, Spain
Francisco J Rodríguez Lera	University of León, Spain
Gledson Melotti	Federal Institute of Espírito Santo, Brazil
Juan Monroy	University of A Coruña, Spain
Kai Li	Polytechnic Institute of Porto, Portugal
Luís Ramos Pinto	University of Porto, Portugal

Luka Čehovin Zajc	University of Ljubljana, Slovenia
Miguel Armando Riem De Oliveira	University of Aveiro, Portugal
Mihail Babcinschi	University of Coimbra, Portugal
Nara Doria	Federal Institute of Sergipe, Brazil
Pedro Miraldo	University of Lisboa, Portugal
Pedro Santos	University of Porto, Portugal
Ramon Fernandes	Federal University of Bahia, Brazil
Ricardo Sutana de Mello	Brazil
Richard Duro	University of A Coruña, Spain
Tiago Trindade Ribeiro	Federal University of Bahia, Brazil
Vítor Hugo Pinto	University of Porto, Portugal

Contents

Aerial Robotics for Inspection and Maintenance

Design Optimization of a Ducted-Drone to Perform Inspection Operations

João Vilaça[1], Alberto Vale[2(✉)], and Filipe Cunha[3]

[1] Centro de Investigação da Academia da Força Aérea, Força Aérea Portuguesa, Lisbon, Portugal
jlvilaca@emfa.pt

[2] Instituto de Plasmas e Fusão Nuclear (IPFN), Instituto Superior Técnico, Universidade de Lisboa, Lisbon, Portugal
avale@ipfn.tecnico.ulisboa.pt

[3] Associated Laboratory for Energy, Transports and Aeronautics (LAETA), Mechanical Engineering Institute (IDMEC), Instituto Superior Técnico, Universidade de Lisboa, Lisbon, Portugal
filipescunha@tecnico.ulisboa.pt

Abstract. The paper presents a study to design and implement a multi-copter with ducted propellers to perform radiological inspections, i.e., near to the radioactive sources and with the best flight time. It must operate in confined spaces or outdoor scenarios and be resilient to collisions. The design is optimized according to the number of rotors and components available in the market. A best combination is implemented and tested in laboratory and in a real outdoor scenarios. The improvements related to ducted propellers are evaluated and reported.

Keywords: Multi-copters · Remotely piloted aircraft system · Ducted propellers

1 Motivation to Perform Inspection Operations with UAV

The research and development with Unmanned Aerial Vehicles (UAV) is at full throttle due to the increasing interest in the airspace domination and the promising applications. UAV are able to perform the same activities as the manned aircraft with a significantly lower risk and at a reduced cost. Thus, UAVs are nowadays a valuable option to any kind of civil or even military mission.

UAV is an easily transportable tool to operate autonomously or controlled by a human operator and to acquire data in real-time, [1]. The usage of UAV is also a promising solution in radiological inspections, scalable to chemical, biological, radiological and nuclear (CBRN) defence missions. The time duration of the flight, as well as the loitering capability of UAV plays an important role in these type of inspection missions. The operations in confined scenarios, such as in indoor facilities, has an additional burden of mitigation the high risk of collisions.

M. F. Silva et al. (Eds.): ROBOT 2019, AISC 1092, pp. 3–15, 2020.
https://doi.org/10.1007/978-3-030-35990-4_1

The multi-copter Remotely Piloted Aircraft System (RPAS) is becoming a well established technology in industry, but the autonomy and the sharp behavior of the propellers while rotating are still some of the handicaps.

This paper presents a study to increase the autonomy and safety of a multi-copter based on ducted propellers, which is still an under-developed area.

2 Background of Aircrafts Using Ducted Propellers

In 1923, Georges Hamel patented an idea that describes a fixed wing RPAS with propellers inlaid in its wings. With this idea Hamel was aiming to obtain a system with good performance both in low speed horizontal flight and in vertical flight near the ground, [1].

In 1933, Stipa [2] investigated the effect of integrating a propeller into a hollow fuselage. The fuselage interior had the shape of a Venturi tube. He proved that the new configuration increased the thrust and reduced the power consumption when compared to the free propeller configuration.

In 1948, Platt [3] performed outdoors static experiments with two propellers configuration. One free and the other ducted. During those experiments he tested the length of the duct and the exit area influence by changing these variables. Platt was able to conclude that the thrust of the ducted propeller is the doubled when compared to the free one. He also concluded that the length of the duct and the exit area had no influence on performance.

In 1955, Parlett [4] tested the influence on lift, drag, and roll when a duct is being used. He also studied the duct's lip size influence on its performance. From this year on the number of aircrafts using ducted propellers technology increased. Nord 500 Cadet and Doak 16 VZ47 are some examples, as illustrated in Fig. 1. This technology was intended to enable those aircrafts to improve Short/Vertical Take-Off and Landing (S/VTOL) performance.

In 1970, the configuration Fan-in-Fin or Fenestron was presented by Mouille. This configuration boils down to the introduction of the helicopters tail rotor in a duct. This configuration improved the performance of the tail rotors reducing the power required to counter the torque of the rotor. It could also reduce the size of the tail and its operation safety.

The application of ducts propellers was the key to perform the first vertical take-off of a supersonic airplane. In 2001, X-35 took-off vertically.

In 2009, Zhao [5] studied the aerodynamic coefficients, and flow patterns at different velocities of a ducted propeller vertical take-Off and landing UAV. For this study Zhao used computational and wind-tunnel techniques. Also in 2009, Pereira [6] studied the dimensions of the duct for an optimal performance. This study is the starting point for the work herein presented.

More recently, some work has been done to study the influence of aerodynamics and disturbances of proximity effects in multirotors, [7,8]. In particular, the ground effect, [9], and ceiling effect, [20], and its influence in multirotor control.

Fig. 1. Examples of aircrafts using ducted propellers (from left to right): Nord 500 Cadet and Doak 16 VZ-47. Images from http://www.aviastar.org/.

2.1 Momentum Theory Applied to Free Rotors

The thrust provided by a RPAS propulsion system is evaluated using the Momentum Theory (MT). The thrust is considered to be a reaction to the required force used to accelerate an amount of air through a actuator disk, [11,18]. This theory is based on the mass, momentum, and energy conservation laws applied to a control volume, which contains the rotor and its wake, [11]. The left image of Fig. 2 sketches the control volume of a free rotor in hover. Hence, it is possible to calculate the thrust delivered and the power consumption by the diameter of the propeller and induced velocity on the plane of the propeller. The flow around the blades, its pitch and the rotation speed of the propeller are ignored here, [11].

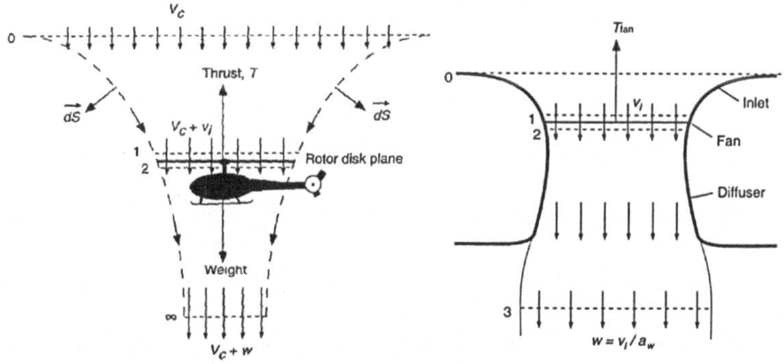

Fig. 2. Control volume for a hovering free rotor (left) and Momentum Theory applied to ducts (right), [15].

In a hover situation the vertical velocity is considered to be zero, $V_C = 0$. Thus, the induced velocity in the actuator disk is v_h and in the wake is w_h. Since the applied force to the control volume is equal to the time variation of the momentum, T is given by (1), where T is the thrust, $\dot{m}$ is the mass flow, ρ is the air density, A is the rotor's area, and v_h is the induced velocity in the actuator disk for a hover situation. Given the induced velocity value, it is possible to calculate the required power for hover by (2), where P_h is the power required for hover.

$$T = \dot{m}w = (\rho A v_h)(2v_h) \Rightarrow v_h = \sqrt{\frac{T}{2\rho A}} \tag{1}$$

$$P_h = Tv_h = \frac{T^{\frac{3}{2}}}{\sqrt{2\rho A}} \tag{2}$$

In an axial flight situation it is considered the rotor has a upwards/downwards velocity, and no horizontal velocity. It is also considered the axial flight has a constant speed being the acceleration and forces sum equal to zero, [11]. For the axial flight the power required is given by (3), where P_C is the power required for axial flight and V_C is the axial velocity. Solving (3) by (1), results (4).

$$P_C = T(V_C + v_h) = \frac{1}{2}\dot{m}(2V_C + w) \tag{3}$$

$$\dot{m}w(V_C + v_h) = V_C + \frac{w}{2} \Rightarrow w = 2v_i \tag{4}$$

Solving (3) with (4), the result is given by (5), where $''+''$ and $''-''$ are for upwards and downwards flight, respectively.

$$\frac{V_C}{v_h} = \left(\frac{V_C}{2v_h}\right) \pm \sqrt{\left(\frac{V_C}{2v_p}\right)^2 \pm 1} \tag{5}$$

The power required in axial flight is given by (6), where P_C is the power required for axial flight.

$$\frac{P_C}{P_h} = \left(\frac{V_C}{2v_h}\right) \pm \sqrt{\left(\frac{V_C}{2v_h}\right)^2 \pm 1} \tag{6}$$

The horizontal flight is studied taking into account the Rankine model, which considers that the total velocity, in the rotors level, may be split into a perpendicular and a parallel velocity to the rotor's plane.

Analyzing the momentum equation on the perpendicular direction to the rotor the result obtained is the same as the one for the axial and hover flight situations. The thrust delivered in the horizontal flight situation is given by (7), where V_∞ is the horizontal velocity in the plane of the rotor, and α is the angle of the rotor. The power required for horizontal flight is given by (8), where P_0 is the parasite power, where P_0 is the parasitic power due to the drag on the blades and must be calculated using a different model, [15].

$$T = 2\rho A v_i \sqrt{V_\infty^2 + 2v_\infty \sin\alpha + v_i^2} \tag{7}$$

$$P = TV_\infty \sin\alpha + Tv_i + P_0 \tag{8}$$

2.2 Momentum Theory Applied to Ducted Rotors

The MT may also be used for ducted rotors, see Fig. 2. By the mass conservation principle in (9), due to its design, the exit area of the duct is given by $A_e = a_w A$, where a_w is the flow contraction parameter. The results obtained in [15] is given by (10).

$$\dot{m} = \rho A v_i = \rho(a_w A)w \tag{9}$$

$$w = \frac{v_i}{a_w} \tag{10}$$

Using the Momentum Conservation Principle, the thrust delivered by the rotor is given by (11) and, hence, the induced velocity is given by (12).

$$T = T_{duct} + T_{rotor} = \dot{m}w = (\rho A v_i)w = \frac{\rho A v_i^2}{a_w} \tag{11}$$

$$v_i = \sqrt{\frac{a_w T}{\rho A}} \tag{12}$$

The relation between the total thrust delivered and the thrust delivered by the motor is given by (13). Since the power consumption of the rotor is given by $P_r = T_r v_i$ and solving (12) and (13), results in (14).

$$\frac{T_{rotor}}{T} = \frac{\frac{1}{2}\rho A w^2}{\rho A v_i w} = \frac{1}{a_w} \tag{13}$$

$$(P_i)_{rotor} = T_{rotor} v_i = \left(\frac{T}{2a_w}\right)\sqrt{\frac{a_w T}{\rho A}} = \frac{T^{\frac{3}{2}}}{\sqrt{a a_w \rho A}} \tag{14}$$

The relation between the power consumption between free and ducted propellers is given by:

$$\frac{(P_i)_{DR}}{(P_i)_{FR}} = \frac{1}{\sqrt{2a_w}}. \tag{15}$$

3 Proposed Design

3.1 Commercial Off-the-Shelf (COTS) Components

In [10] a comparison is made regarding different number of motors, from mono-copter to octa-copter, identifying the advantages and disadvantages of each configuration. This comparison allows to find the best RPAS for a specific type of missions. The hexa-copter configuration is proposed as the best one. Since the performance of a quad-copter is not so different from an hexa-copter and being its construction easier, a more detailed comparison between these two copters is performed. A cost function is defined to identify which configuration is adopted and implemented. The function takes into account all specifications of RPAS and a market survey is also performed to identify the available components.

Currently, the most used motors for RPAS are the brushless motors, which are typically characterized by a four digit number. Large dimensions of the stator allows the use of higher torque created by the motor. Other essential parameter to be considered is the KV, which is an approximation of the rotation speed, i.e., rotations per minute (RPM) increment imposed by a $1V$ increment to the no-load current. This is an estimation because when the propellers are mounted on the motor, the drag increases and the number of RPM decreases, [17].

The dimensions of the propellers are assessed according to the motors dimensions and KV, and with the RPAS dimensions. For motors with small dimensions, the KV is high, and small propellers must be used. For big motors, the KV is low, and hence, big propellers must be used, [14]. The efficiency of the propellers is an important key factor for the performance of the aircraft. From the experience of aeromodelling players, high number of blades in propeller decreases the efficiency of the propeller, [13], which is also influenced by the material used in its construction, being difficult to distinguish which material is better. Since herein the target is an efficient and stable aircraft, only propellers with large dimensions and with two blades are considered.

The Electronic Speed Controller (ESC) is the component that control the rotation speed of the motor, direction of rotation, and the braking system. The decision of which ESC is to be used have to take into account the electric current, size, and weight. Its current designates the maximum amperes supported by the controller without damaging the motor, [12]. The considered ESC are from $3A$ to $85A$ to assure that any type of assessed motors has a compatible ESC.

In RPAS the most relevant feature to take into account is its size. Usually, large RPAS requires large number of motors. Thus, aircraft with larger dimensions requires batteries with multiple cells to maximize the performance of the aircraft, [21]. The LiPo type of battery is the most used one, since it is high efficient for RPA given its low weight, high energy density, fast charging capability, and constant power delivery, [19]. The battery range, voltage, and discharge rate are important features to take into account when choosing a battery. The discharge rate is what sets the rotation speed of the propellers. For a stable aircraft with low horizontal speed, the discharge rate should be low, [19]. The assessed batteries had ranges between 2200 mAh and 20000 mAh, discharge C-rates between 10 and 50, and from 1 to 6 cells. If a single battery is not enough to accomplish the required endurance, the number of batteries is estimated as follows:

1. Knowing the total weight of RPAS with the assumed number of batteries, the required thrust, power for hover, horizontal and axial flight are calculated;
2. Knowing the required power for the different phases of the flight, the percentage of maximum power available is calculated;
3. The total power consumption is calculated by (16);
4. The number of batteries to use is calculated using (17).

$$P_{consumed} = \Delta t_h P_h + \Delta t_C P_C \Rightarrow \frac{5}{12} P_h + \frac{1}{12} P_C \tag{16}$$

$$N_{bat} = \left\lceil \frac{P_{consumed}}{P_{battery_{\mathbb{N}}}} \right\rceil \quad (17)$$

The procedure is repeated until the number of batteries is identified. In (17), the resulted is given by the upper boundary approximation, since the number of batteries must be an integer value. Thus the upper boundary approximation guarantees that the batteries provide the required energy.

The radio controller is used to control the aircraft and to exchange telemetry. The decision over this component takes into account the transmitter frequency, number of channels available and the price. No relevant issues were identified to select the radio controller, and thus, one of the most used is chosen, namely the *FlySky FS-I6S*.

The candidate components were organized and two lists contained all the possible combinations between all the type of components were created. One of the lists was related to the quad-copter configuration while the other was related with the hexa-copter configuration.

The total number of components provides approximately 2100 combinations. The resulted thrust, power consumption, weight, and purchase cost is computed for each combination. Figure 3 presents the following variables for all combination: $\frac{Thrust}{Power}$, $\frac{Thrust}{Weight}$, $\frac{Weight}{cost}$ and $\frac{Thrust}{Cost}$. The next goal is to find the best (maximum value) combinations.

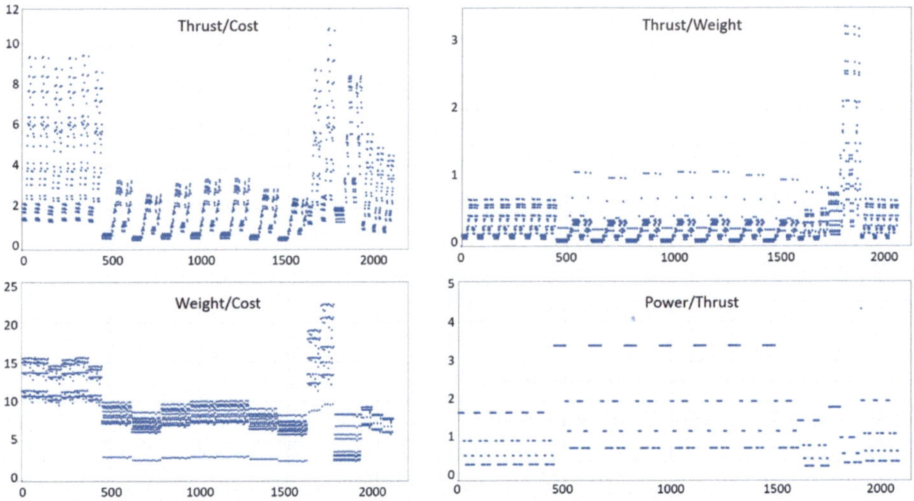

Fig. 3. Number of combinations and different variables calculated

3.2 Cost Function

To evaluate all combination the following cost function, CF, is considered:

$$CF = \frac{1}{0.6\frac{Thrust}{Power} + 0.3\frac{Thrust}{Weight} + 0.1\frac{Thrust}{Cost}} \tag{18}$$

The weights are assigned according to the mission. For an high autonomy, $\frac{Thrust}{Power}$ is the most important variable. The RPAS weight is also relevant because if $\frac{Thrust}{Weight}$ is lower than 1.3, than the RPAS flight is not profitable. The $\frac{Thrust}{Cost}$ has no impact on the performance of the RPAS and in a proof-of-concept the cost was not considered.

The five best combinations for each RPAS type were selected and compared in Fig. 4. The quad-copter configurations present better results when comparing the variable $\frac{Thrust}{Power}$. In relation to $\frac{Thrust}{Weight}$ the results with hexa-copter are slightly better. However, if considering $\frac{Thrust}{Cost}$ the hexa-copter is the winner. Since the last variable is almost irrelevant in this study and the simplicity of implementation, the quad-copter is considered as the adopted solution with the following configuration: motor NTM Prop-Drive 35-30, 1400 KV, 560 W; ESC Hobbyking Blueseries 50 A; battery TATTU LiPo 16000 mAh, 4 S, 15/30C; and propellers Gemfan 7 × 3.8 in.

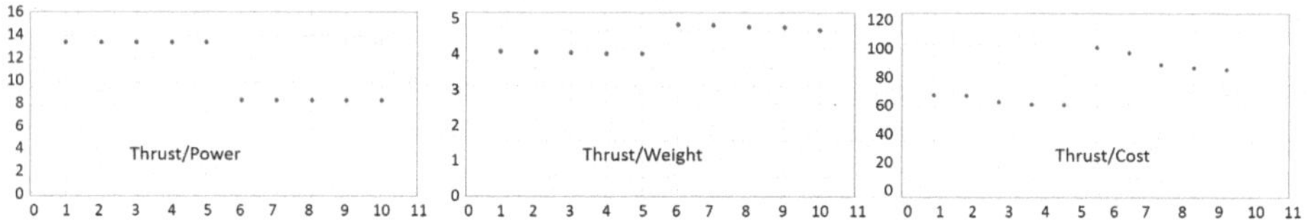

Fig. 4. Comparison of the five best combinations between quad- and hexa-copters (from left to right): $\frac{Thrust}{Power}$, $\frac{Thrust}{Weight}$ and $\frac{Thrust}{Cost}$. The first five values are related to the quad-copter configuration and the last five to the hexa-copter.

4 Implementation

For the aircraft take-off from any kind of terrain, the minimum distance between the propellers and the ground is evaluated to avoid the ground effect phenomenon, [7,9]. According to [15] the ground effect is minimum from a distance to the floor equals to a propeller diameter. Consequently, the used propeller has an height over 180 mm, without ground effect as confirmed after experimental tests.

The dimension of the ducts are in accordance with [6] to optimize the performance of the ducted propeller. Thereafter, the ducts were developed and implemented and tunned to reduce the weight. The first attempt was a $3D$ printed duct, as depicted in the left-top image of Fig. 5. The weight of a single duct is above 400g, which is relative heavy. The alternative solution was to produce a duct using a Computational Numeric Command (CNC) machine, resulting into a foam with the duct shape, as illustrated in the left-bottom image of Fig. 5. The foam was covered with fiber glass and, once dry, the foam was removed. The resulted duct has a low weight and with a good finish, as illustrated in the second image of Fig. 5. Finally, the outside part of the duct under the lips was removed to reduce even more the weight, as illustrated in the right image. The final shape has no influence in the performance of the ducted propeller based on fluid dynamics simulation performed in Flow Simulation software of Dassault Systémes.

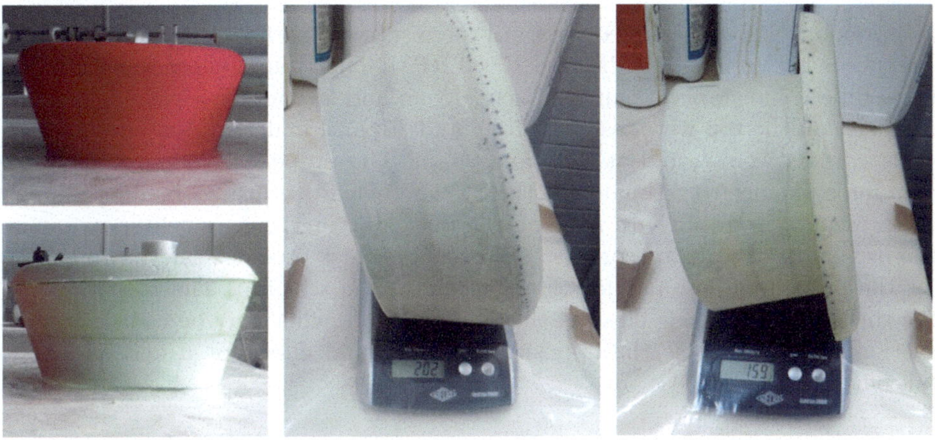

Fig. 5. The 3D printed duct and the CNC with hot wire shaping (left images)

The main frame was designed using aluminum bars, carbon fiber tubes and carbon fiber tripod boards. These materials were chosen after simulation bending tests, with a maximum deflection within an acceptable margin. The main frame was implemented with removable arms in order to be easier to transport and to have the capability to easily swap between free and ducted propellers. In order to have a stiff and suitable landing gear, carbon fiber was used with a quarter elliptical shape. The landing gear is attached to the central part of the main frame with a comfortable distance between its tips. The entire structure was designed in Solidworks, a CAD software of Dassault Systémes, and assembled in the lab, with the results presented in Fig. 6.

5 Experimental Results

Experimental tests with a single propeller (un-ducted and ducted) system were performed in laboratory to test: (i) the behavior of the propeller in free space, i.e., assuming no physical elements in the proximity of the propellers, (ii) the ground effect, (iii) the proximity of obstacles on the top of the propeller and

Fig. 6. CAD design (left image) and the implementation placed in the laboratory

(iv) proximity of walls and corners. All the tests were performed with five different speeds between 5000 RPM and 17000 RPM. The ground effect was tested at three different distances, namely 13 cm, 18 cm and 28 cm. The proximity of the obstacles on the top were tested at 5 cm and 15 cm. Finally the distances to the walls/corners were tested at 6 cm and 26 cm. The propeller system is installed on a force balance to estimate the thrust. The benefits of using the duct system are presented in Table 1. In general, the duct system brings benefits, except in the proximity of walls or corners. However, the duct provides safety protection for the propellers and also for the elements in the scenario. The high value of the thrust with obstacle close to the top of the propeller may be caused by the suction effect, which is the opposite in the ground effect for a fast and easy takeoff.

Table 1. Benefits of using the duct system in terms of thrust, power and $\frac{Thrust}{Power}$, comparing to a non-duct system with the same components.

	Thrust	Power	$\frac{Thrust}{Power}$
Free space (no elements in the proximity of the propeller)	7%	17%	20%
Ground effect	31%	22%	46%
Obstacles closer to the top of the propeller	48%	15%	n.a.
Proximity to walls or corners	−3%	8%	4%

After the static tests with a single propeller, the RPAS was totally assembled and the outdoors tests were performed in a real scenario, as illustrated in Fig. 7. The flight controller was calibrated, with and without ducted propellers, for different movements (pitch, roll, yaw, and throttle) using Ardupilot and a Pixhawk. During this process the RPAS was fixed to the floor using external weights. The tests were performed using a different battery from the selected one given delivery issues. In particular the used battery has a capacity of 5000 mAh and four cells (14.8 V).

The implemented duct drone was able to perform the common motion operations, as takeoff and landing, move forward/backward, rotation loiter or simply free roaming. It is possible to handle the RPAS by the ducts, which also provides safety against collision at low speed. In terms of performance with this less power battery, the theoretical flight time is 247 s and 267 s, without and with ducted propellers, respectively. The tests resulted in 235 s and 250 s, respectively, i.e., an improvement of 6% of using duct drones.

Fig. 7. Photos taken during the experimental tests performed in a real scenario. The person in the left image is holding a cable only for security reasons, since the tests were performed in military air force facility. Videos available here: https://www.ipfn.tecnico.ulisboa.pt/FRIENDS/research.html

6 Conclusions and Future Work

The paper presented a study to design and implement a multi-copter with ducted propellers. A cost function was created and more than 2100 combinations of COTS components were considered. The best combinations for quad-copter and hexa-copter were selected and compared. The adopted one, based on a quad-copter, was implemented and tested.

The experimental results concluded that a ducted propeller system provides an improvement on quad-copters in terms of thrust and power. This improvement occurs not only in free space, but also in the proximity of obstacles on top of the propellers and also during the ground effect. The last one is particular important for radiological inspections, where the UAV has to move slowly and near to the radiological sources. Ducted propellers are resilient to collisions at low speed during the flight, mainly in cluttered indoor scenarios. However, the proximity of obstacles on top of the propellers must be quantified in a future work based on similar studies presented in [8,20].

The extension of the flight time is not so substantial and further tests are required with the proper battery. Building the entire UAV body using a lightweight and strong type of material, such as carbon fiber, also reduces the total weight and increases the flight time.

In summary, ducted propellers requires additional material, but still improvements on the performance of the UAV and safety against collisions.

Acknowledgments. This work is a result of the project FRIENDS - Fleet of dRones for radIological inspEction, commuNication anD reScue, supported by the Fundação para a Ciência e Tecnologia (FCT), Compete 2020 and Lisboa 2020 under the Portugal 2020 Partnership Agreement, through the European Regional Development Fund (ERDF). IST activities also received financial support from FCT through project UID/FIS/50010/2019 and also through IDMEC under LAETA project UID/EMS/50022/2013.

References

1. Mueller, T., Gibson, R.: Aerodynamic measurements at low reynolds numbers for fixed wing micro-air vehicles. In: Development and Operation of UAVs for Military and Civil Aplications [Development et ulilisation des avions san pilote pour des applications civiles et militaires], University if Notre Dame, Department of Aerospace and Mechanical Engineering (1999)
2. Stipa, L.: Stipa monoplane with venturi fuselage. National Advisory Committee for Aeronautics, Number 753 (1933)
3. Platt, R.: Static tests of a shrouded and an unshrouded propeller. National Advisory Committee for Aeronautics, Number L7H25 (1948)
4. Parlett, L.: Aerodynamic characteristics of a Small-scale Shrouded Propeller at Angles of Attack from 0 deg to 90 deg. National Advisory Committee for Aeronautics, Number 55 (1955)
5. Zhao, H.: Development of a dynamic model of a ducted fan VTOL UAV. School of Aerospace, Mechanical and Manufacturing Engeneering, RMIT University (2009)
6. Pereira, J.: Hover and windtunnel testing of shrouded rotors for improved micro air cehicle design. In: Partial Fulfillement of the Requirements for the Degree of Doctor of Philosophy, University of Maryland (2008)
7. Powers, C., Mellinger, D., Kushleyev, A., Kothmann, B., Kumar, V.: Influence of aerodynamics and proximity effects in quadrotor flight. In: Experimental Robotics, Springer Tracts in Advanced Roboticsm, pp. 289–302 (2013)
8. Jain, K., Fortmuller, T., Byun, J., Makiharju, S., Mueller, M.: Modeling of aerodynamic disturbances for proximity flight of multirotors. In: International Conference on Unmanned Aircraft Systems, pp. 1261–1269 (2019)
9. Sanchez-Cuevas, P., Heredia, G., Ollero, A.: Characterization of the aerodynamic ground effect and its influence in multirotor control. Int. J. Aerosp. Eng. (2017)
10. Agrawal, K., Shrivastav, P.: Multi-rotors: a revolution in unmanned aerial vehicle. Int. J. Sci. Res. **4**, 1800–1804 (2015)
11. Cunha, F.: Teoria do Momento Linear. Instituto Superior Tècnico
12. Gualzaar: What to consider when buying a ESC for your multirotor (2018). https://fpvfrenzy.com/best-esc-for-quadcopters/
13. Guillaume: 2- vs 3- vs 4-blade propellers (2015). http://aerotrash.over-blog.com/2015/02/2-blade-vs-3-blade-and-4-blade-propellers.html
14. Kadamatt, V.: Chosing motor and propeller for multirotors (2017). http://www.droneybee.com/choosing-motors-and-propellers-for-multirotors
15. Leishman, J.: Principles of Helicopter Aerodynamics. Cambridge University Press, Cambridge (2006)

16. Liang, O.: What are ESC, UBEC and BEC (2015). https://oscarliang.com/what-is-esc-ubec-bec-quadcopter/
17. Reid, J.: Multirotor Motor Guide (2017). https://www.rotordronemag.com/guide-multirotor-motors/
18. Richards, R.: Principles of Helicopter Performance. U. S. Naval Test Pilot School, Number USNTPS-T-NO. 1 (1968)
19. Salt, J.: Understanding RC LiPo Batteries (2018). https://www.rchelicopterfun.com/rc-lipo-batteries.html
20. Sanchez-Cuevas, P., Heredia, G., Ollero, A.: Multirotor UAS for bridge inspection by contact using the ceiling effect. In: International Conference on Unmanned Aircraft Systems, pp. 767–774 (2017)
21. Liang, O.: How to choose battery Capacity for longer flight time (2014). https://oscarliang.com/how-to-choose-battery-for-quadcopter-multicopter

Positioning System for Pipe Inspection with Aerial Robots Using Time of Flight Sensors

Manuel Perez[(✉)], Alejandro Suarez, Guillermo Heredia, and Anibal Ollero

GRVC Robotics Labs Seville, School of Engineers of Seville, Seville, Spain
Manupl192@gmail.com

Abstract. This paper describes a positioning system for aerial robots consisting of a linear array of time-of-flight (ToF) sensors whose orientation angle is controlled with micro servo, allowing the detection and accurate localization of the contour of a pipe to be inspected. The system, integrated in a hexarotor vehicle, has three operation modes: searching, aligning, and tracking. In the first phase, a scan motion is conducted rotating the servo in a wide range until one of the four ToF sensors of the array detects a close obstacle. Then, the rotation angle is adjusted to align the array with the normal vector of the surface, tracking actively its contour while providing an estimation of the relative position and orientation. The paper details the design and implementation of the system, the control scheme, the position estimator, and its integration in an aerial robot. Experimental results carried out in test-bench show the performance of the system.

Keywords: Aerial robotic inspection · Time-of-flight sensor · Position sensor

1 Introduction

The ability of aerial robots to reach quickly high altitude or distant workspaces results of interest for a wide variety of inspection and maintenance operations in industrial facilities, conducted nowadays by human operators in risky conditions. Motivated by the convenience to reduce the time, cost and resources involved in these operations, the aerial manipulation field [1] proposes the integration of control, perception and planning capabilities in aerial platforms equipped with robotic arms. The installation and retrieval of inspection tools [2], the inspection by contact [3, 4], the inspection of power lines [5], or the cleaning of wind turbines [6] are some illustrative examples of applications that may benefit from this technology. Several prototypes of aerial robots have been developed in recent years, integrating single arm [7, 8] or dual arm [2, 9] manipulators in multirotor platforms, demonstrating the grasping objects [2] and control the contact forces [10].

In the execution of an aerial manipulation mission, it is possible to identify three phases: navigation through the environment, approaching to the workspace, and realization of the particular operation. Real-Time Kinematics (RTK) Global Positioning

M. F. Silva et al. (Eds.): ROBOT 2019, AISC 1092, pp. 16–27, 2020.
https://doi.org/10.1007/978-3-030-35990-4_2

System (GPS), vision-based [11] and range-only [12] Simultaneous Localization and Mapping (SLAM), or laser tracking systems [13] have been used for the position and trajectory control of multirotor vehicles in outdoors. However, higher positioning accuracies are needed once the aerial robot is close to the workspace and operating on flight, taking into account that the reach of the manipulator is relatively small (around 50 cm). ARUCo tags have been extensively used in indoors and outdoors [14], although the use of these markers and the computational resources may limit their application. A docking system is proposed in [15] for measuring the relative position of the aerial platform using a robotic arm attached to fixed point over a pipe. A cooperative virtual sensor is built in [16] exploiting the vision sensors on-board a group of quadrotors (observers) for estimating the position and velocity of a target quadrotor.

In our previous work we developed several prototypes of lightweight and compliant aerial manipulation robots [17] intended for inspection and maintenance of pipe structures in chemical plants [1]. The need of a high accuracy positioning system was evidenced in the grasping and installation of inspection devices on flight [2], taking into account that the effective reach of a human-size manipulator (w.r.t. the nominal operation position) is around 30 cm. Reference [9] analyzes the limitations in the performance of the manipulator due to the motion constraints and dynamic coupling with the aerial platform. The ARUCo markers were employed in [10] to measure the position of the aerial manipulator relative to the inspection point, although the detection of the marker was affected by shades and changes in the illumination. The long reach aerial manipulator presented in [18] introduced a time of flight sensor attached at the end effector to measure the distance to a pipe, enhancing the situational awareness of the human operator.

The main contribution of this work is the design and development of a positioning system based on Time of Flight (ToF) sensors to be used in the inspection of high altitude pipe structures with aerial manipulation robots. The system consists of two linear arrays of distance sensors whose orientation angle is controlled to track the contour of the pipe, providing an estimation of the relative position and orientation. The paper details the mechanical design and integration in a hexarotor platform, the electronics of the system, the operation modes for the detection and tracking of close surfaces, as well as the geometric relationships used for estimating the relative pose. Each linear array consist of four VL53L0 SATEL sensors manufactured by ST Microelectronics (2100 mm max distance, 20–50 Hz update rate, 3 g weight), using a micro servo to control its orientation in the pitch angle. Known this angle and the mean distance to the obstacle, it is possible to estimate the position in the two orthogonal axes (XZ). A second array is used to estimate the relative orientation, what can be useful for landing the aerial platform over the pipes. Experimental results conducted in an indoor test bench evaluate the performance of these sensors and the positioning system, paying special attention to the accuracy and update rate. The rest of the paper is organized as follows. Section 2 describes the positioning system, including the mechanical design, electronics, and its integration in a hexarotor platform. Section 3 explains the three operation modes (searching-aligning-tracking) of the sensor and the control of the orientation angle, whereas Sect. 4 describes the relative position estimator. The experimental results are presented in Sect. 5, summarizing the conclusions and future work in Sect. 6.

2 System Description

2.1 Mechanical Construction and Sensor Specifications

The prototype presented in this work and shown in Fig. 1 implements a laser tracking system capable to estimate the position and orientation of a multirotor with respect to a pipe or a flat surface at distances below 2 m. It consists of two moving arrays of four VL53L0X-SATEL time of flight sensor boards whose main specifications are summarized in Table 1. These devices, manufactured by ST Microelectronics, rely in the time taken by a laser pulse to hit the target and return to estimate the distance. The orientation of each array is controlled with a Pololu micro metal gear motor 250:1 and a DRV 8833 driver, measuring the rotation angle of the motor shaft with a Murata SV01A potentiometer. Known the distance measured by the array and its orientation angle, it is possible to obtain the position of an obstacle in the forward and vertical directions (XZ-axes). Combining the measurements provided by two arrays it is possible to obtain the relative orientation in the yaw angle with respect to the obstacle, as it will be seen in Sect. 4. This may be useful to control the position and orientation of the aerial robot relative to a pipe during the realization of the inspection operation or for landing the aerial platform.

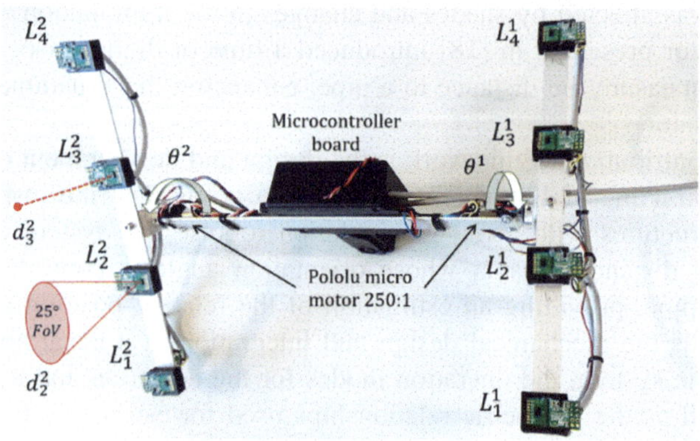

Fig. 1. Positioning system with two arrays of four time of flight sensors.

2.2 Electronics and Hardware Architecture

The positioning system is implemented in a STM32F103 microcontroller that features all the peripherals required for interfacing the sensors and for controlling the DC micro servos: two I2C buses (one per array) to read the VL53L0X sensors, one UART for sending the estimated pose to the main computer board, two ADC channels to get the rotation angle of the micro servos, and four PWM signals to control the two motor drivers. A picture of the components and the architecture of the positioning system is shown in Fig. 2. Each array of sensors is connected to an independent I2C bus for simplifying the wiring and increasing the read rate. A pair of Pololu micro metal gear

Table 1. Main specifications of the VL53L0X-SATEL time of flight sensor board.

Size/Weight	31 × 26,5 × 6,5 mm/318,2 g (total)
Ranging time	30–200 ms (depending on distance)
Max. ranging (17% grey)	80 cm (Indoor) – 50 cm (Outdoor)
Max. ranging (88% white)	+200 cm (Indoor) – 80 cm (Outdoor)
Ranging accuracy	4–7% (Indoor) – 6–12% (Outdoor)
Field of view	35° (Emitter) – 25° (Collector)

motors 250:1 are used to rotate the arrays in the pitch angle during the searching, aligning and tracking phases. The DRV 8833 driver allows the microcontroller to vary the rotational speed and the direction of the servos through the PWM signals, using the ADC to get position feedback. The whole system is powered by a 2S LiPo battery, exploiting the voltage regulator embedded in the microcontroller board. Communication with main computer is done using a custom protocol over full duplex UART.

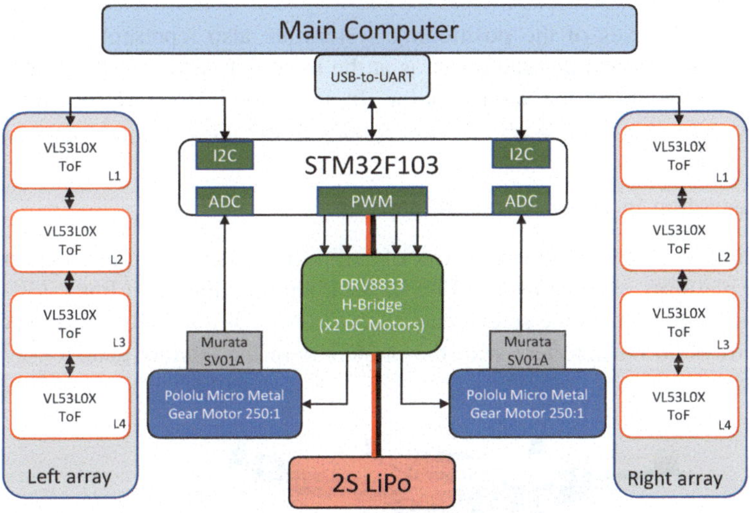

Fig. 2. Hardware architecture and components of the positioning system.

2.3 Geometric Model

The geometric model of a single array of sensors is illustrated in Fig. 3, considering its application for the relative localization of pipes. Here d_j^i is the distance measured by the j-th sensor of the i-th array, with $i = \{1, 2\}, j = \{1, 2, 3, 4\}$, θ^i is the rotation angle of the corresponding array, Δx is the separation distance between lasers whereas D_{pipe} is the diameter of the pipe to be inspected. It is imposed that $\Delta x \sim 0.8 \cdot D_{pipe}$ to ensure that the contour of the pipe is tracked by two laser sensors, otherwise the aligning phase described in Sect. 3 will fail.

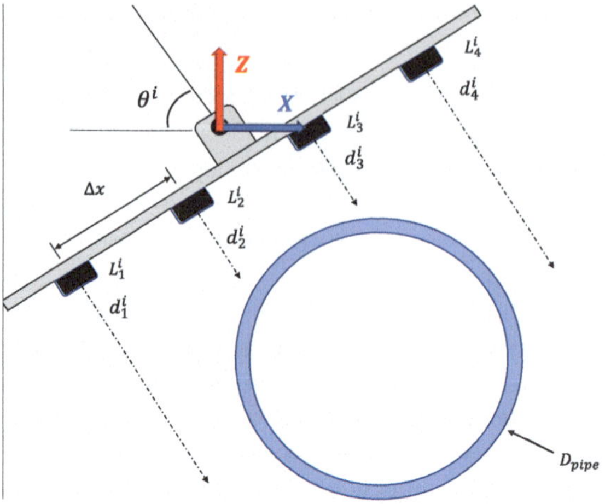

Fig. 3. Geometric model of an array of sensors tracking the contour of a pipe.

The local XZ-axes of the positioning system are also represented in Fig. 3. For simplicity, it is assumed that the origin is at the midpoint of the array, coinciding with the axis of rotation of the micro motor. In the experimental results presented in Sects. 5.2 and 5.3, the angle $\theta^i = 0$ corresponds to the array pointing downwards.

2.4 Integration in Hexarotor Platform

The positioning system with two arrays of ToF sensors has been integrated in a S550 hexarotor platform, as illustrated in Fig. 4. An L-shaped aluminum frame (30 × 2 mm section) is used to attach the sensor case with the base of the multirotor in such a way that the arrays do not interfere with the propellers or the landing gear.

Fig. 4. Positioning system integrated in a hexarotor platform.

3 Operation Modes and Control

The positioning system implements three operation modes in the microcontroller board according to the state machine represented in Fig. 5: searching, aligning, and tracking. A sequence of pictures illustrating the performance of the developed prototype can be seen in Sect. 5.2. In the following, it is assumed that the positioning system is integrated in a multirotor platform, which is approaching to a pipe structure to be inspected. The obstacles are initially far away, out of the range of the sensors, so the searching mode is firstly executed. In this mode, the two arrays perform a scan motion rotating the micro motors in a wide range (±100°) at constant speed (~60 °/s), trying to find close obstacles. If one of the external lasers (d_1^i or d_4^i) hits the incoming pipe, the system enters in the aligning mode in which the array is rotated quickly towards the direction of the obstacle until this is detected by the internal sensors (d_2^i or d_3^i). In this phase, the microcontroller acts over the motor to align the measurement of both sensors ($d_2^i = d_3^i$) in order to ensure that the array is oriented orthogonally to the normal vector of surface of the pipe (see Fig. 3). Note that the circular profile of the pipe facilitates the detection and tracking of its contour.

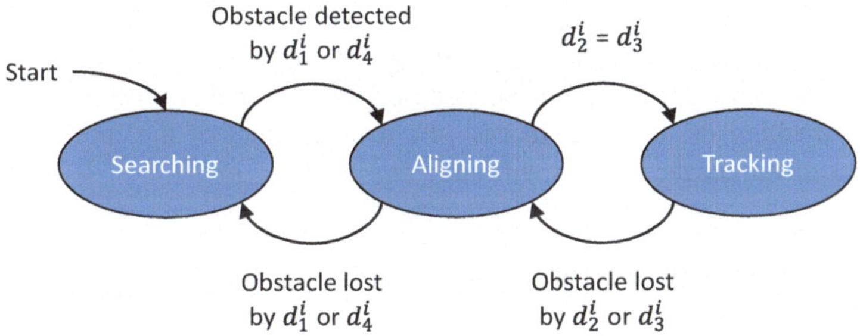

Fig. 5. State machine representing the three operation modes and transitions of the system.

The orientation of each array is controlled generating the following reference at 30 Hz in such a way that the measurements of the inner sensors are equal:

$$\theta_{ref}^i = \theta^i + k \cdot \left(d_2^i - d_3^i\right) \tag{1}$$

Here θ^i and θ_{ref}^i, are the current and reference orientation of the i-th array, and k is the proportional gain of the controller. This reference is taken as input by a low level PID position controller that acts over the PWM signal of the motor driver:

$$pwm^i = K_P\theta_e^i + K_I \int_0^t \theta_e^i d\tau + K_D\dot{\theta}_e^i \tag{2}$$

Here $pwm^i \in [-1, 1]$ is the normalized PWM signal, $\theta_e = \theta_{ref} - \theta$ is the angular error, whereas K_P, K_I and K_D are the proportional, integral and derivative gains of the position controller, respectively, which is executed at 200 Hz. Figure 6 shows the complete control scheme.

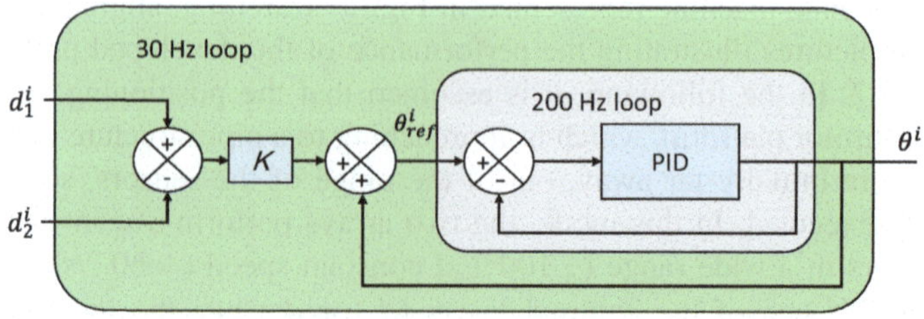

Fig. 6. Low level controller of the positioning system used in the tracking mode: inner PID position control loop and outer proportional controller used for aligning the internal sensors.

Once the aligning phase is complete $\left(d_2^i = d_3^i\right)$, the low level controller will keep active the tracking of the contour of the pipe while the relative position of the aerial platform w.r.t. the pipe is computed and sent to the main computer board at 30 Hz, as it will be described in next section. The position deviations in the multirotor due to wind perturbations or physical interactions will be compensated controlling the orientation angle of the array. If the measurement of one of the inner sensors (L_2 or L_3) goes suddenly out of range, the system will automatically switch to the aligning mode to recover the tracking of the pipe. In case the measurement of the four sensors of the array go out of range, the system will enter in searching mode to redetect the pipe.

4 Relative Position Estimation

As indicated in the introduction, the motivation of this work is the development of a high accuracy positioning system intended for an aerial robot operating close to pipes, where the distances are relatively small (below one meter). The proposed system uses ToF sensors to obtain the distance to the surface of the obstacle, and a potentiometer to measure the rotation angle of the array. Taking into account the diagram depicted in Fig. 3, the position of the obstacle relative to the rotation axis of the motor can be estimated as follows:

$$x^i \cong \frac{\left(d_2^i + d_3^i\right)}{2} \cdot sin(\theta) \tag{3}$$

$$z^i \cong \frac{\left(d_2^i + d_3^i\right)}{2} \cdot cos(\theta) \tag{4}$$

Note that the approximation in Eqs. (3) and (4) neglects the curvature of the pipe and considers the mean distance of the inner sensors. According to the datasheet of the VL53L0X sensor (see specifications in Table 1), the field of view of the collector is

25°, so the measured distance will correspond to closest point within the reception cone, in this case, the curved surface between the two inner sensors.

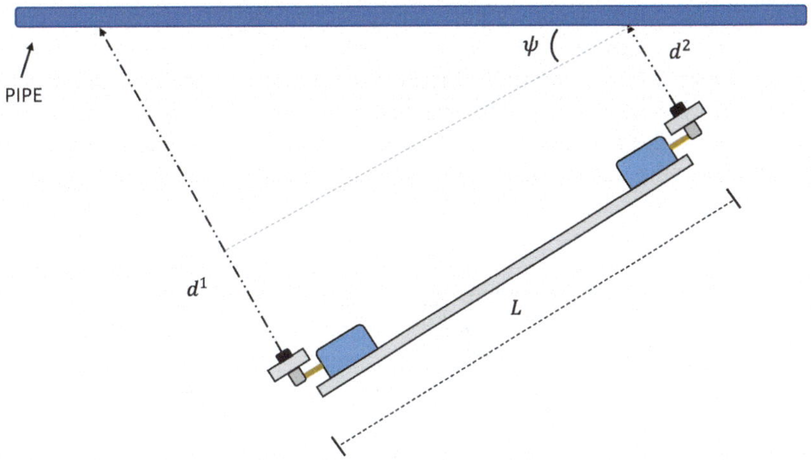

Fig. 7. Estimation of the orientation (yaw angle) of the positioning system relative to a pipe from the distance measured by both arrays.

Let us consider now the double-array positioning system depicted in Fig. 7, where L is the baseline between the left and right arrays, the orientation angle in the horizontal plane (yaw) relative to a target pipe is denoted by ψ, whereas d^i and θ^i represent the distance measured by each array and its corresponding rotation angle respectively, with, $i = \{1, 2\}$. The orientation of the baseline can be obtained geometrically from the projection in the XY-plane of the distances measured by each array as follows:

$$\psi = \tan^{-1}\left(\frac{d^1 sin(\theta^1) - d^2 sin(\theta^2)}{L}\right) \tag{5}$$

Here $d^i sin(\theta^i)$ represents the projection of the distance measured by the i-th array over the XY-plane, so the difference between both arrays corresponds to the length of the leg in the formed triangle. Note that the direction of the laser beam is orthogonal to the baseline of the pair of arrays.

It is necessary to highlight that the pose estimation given by Eqs. (3)–(5) is obtained assuming that the positioning system remains almost horizontal during the tracking phase, since the multirotor platform usually operates in hovering conditions.

5 Experimental Results

5.1 ToF Sensor Characterization

In order to provide a high accuracy positioning estimation, it is necessary to conduct a calibration process to determine the deviation in the position measurement given by the sensor with respect to a ground truth. Figure 8-left represents the measurements given by the VL53L0X sensor along with the real value obtained from a ruler, whereas Fig. 8-right shows the calibrated signal obtained applying a simple correction curve. Table 2 indicates the mean error and standard deviation for some representative distances.

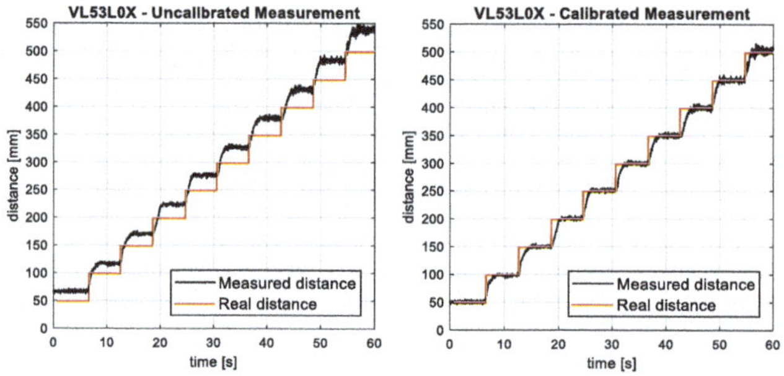

Fig. 8. Positioning accuracy of the VL53L0X-SATEL ToF sensor: uncalibrated (left) and calibrated (right) with offset correction. A 50 cm ruler is used as ground truth.

Table 2. Data obtained from sensor before and after calibration.

	Uncalibrated			Calibrated		
Ground truth	Measurement	Error	STD	Measurement	Error	STD
100	117,2	17,2	1,7	99,1	−0,9	1,6
200	223,3	23,3	1,8	200,8	0,8	1,8
300	326,6	26,6	2,8	299,8	−0,2	2,7
400	431,3	31,3	3,7	400,1	0,1	3,5
500	539	39	4,3	503,3	3,3	4,0

5.2 Detection and Position Estimation of Pipe Contour

The goal of this experiment is to evaluate the performance of the developed positioning system estimating the relative position of a pipe with respect to the local axes shown in Fig. 3. The experiment, conducted in an indoor testbed, consists of moving the baseline of the array along the X and Z axes following the contour of a 200 mm Ø PVC pipe while the system is in tracking mode. Figures 9 and 10 represent the estimated position given by Eqs. (3) and (4) along with the mean distance given by the inner sensors and

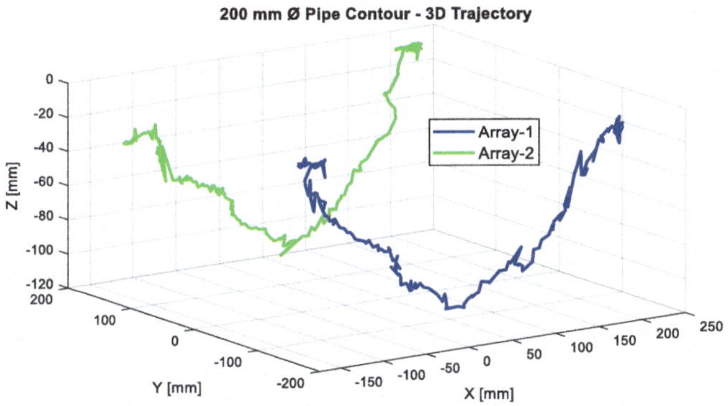

Fig. 9. 3D position estimation corresponding to the contour of the 200 mm Ø pipe.

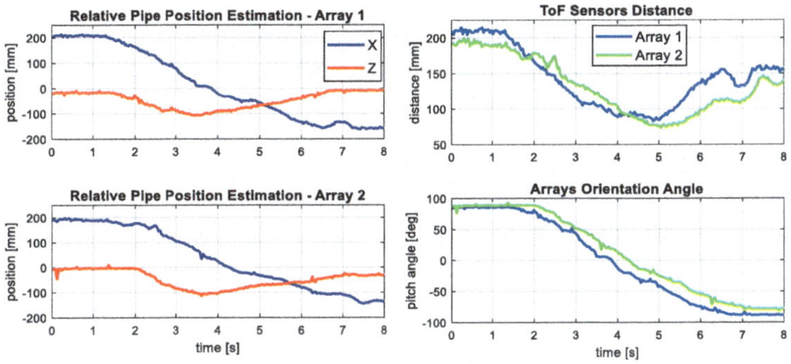

Fig. 10. Evolution of the relative position estimation provided by both arrays (left). Mean distance and orientation angle of the arrays (right).

the rotation angle of each array. The execution of the experiment can be followed in the sequence of pictures illustrated in Fig. 11. The estimation is obtained directly without applying any filter to the signal provided by the sensors. The mechanical clearance of the micro motors introduces additional noise in the estimation.

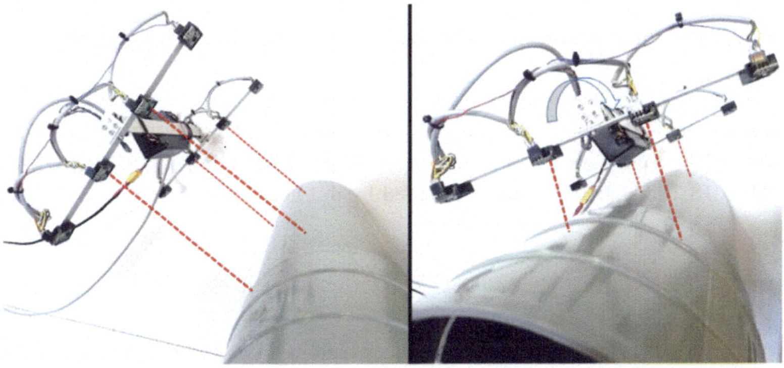

Fig. 11. Two poses of the positioning system while tracking the contour of the pipe.

5.3 Relative Orientation Estimation

In this experiment, the baseline of the positioning system is rotated in the yaw angle with respect to the pipe, maintaining fixed the relative position. Figure 12 represents the estimation given by Eq. (5) along with the distance and rotation angle of both arrays.

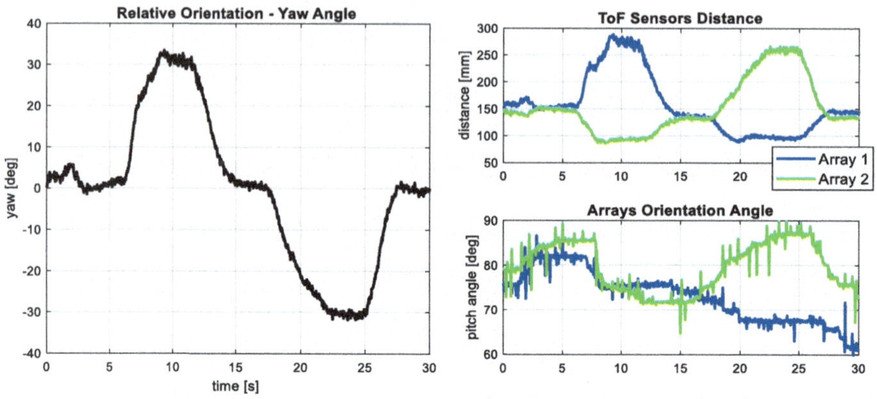

Fig. 12. Estimation of the relative orientation of the pipe in the yaw angle.

6 Conclusion

This paper described the design of a positioning system consisting of two orientable arrays of time of flight (ToF) sensors, used to estimate the relative position and orientation of a pipe to be inspected with an aerial robot. Each array counts with four sensors and a micro servo actuator that actively tracks the contour of the pipe, being able to detect and recover from losses. Besides their low weight, these devices provide highly accurate (<1 cm) distance measurements within a range around 2 m, and update rates above 30 Hz. Unlike vision sensors, which typically require a certain degree of texture to reconstruct the shape of a surface, the measurement obtained from the laser pulse does not depend on the target reflectance. However, the sunlight has a significant influence over the operation range and accuracy, what limits its application in outdoor scenarios.

Acknowledgments. This work was supported by the HYFLIERS (H2020-2018-779411) project, funded by the European Commission, and the ARM-EXTEND (DPI2017-89790-R) and ARTIC (RTI2018-10224-B-I00) projects, funded by the Spanish MINECO.

References

1. Ollero, A., et al.: The AEROARMS project: aerial robots with advanced manipulation capabilities for inspection and maintenance. IEEE Robot. Autom. Mag. **25**(4), 12–23 (2018)

2. Suarez, A., Heredia, G., Ollero, A.: Design of an anthropomorphic, compliant, and lightweight dual arm for aerial manipulation. IEEE Access **6**, 29173–29189 (2018)
3. Trujillo, M.Á., Martínez-de Dios, J.R., Martín, C., Viguria, A., Ollero, A.: Novel aerial manipulator for accurate and robust industrial NDT contact inspection: a new tool for the oil and gas inspection industry. Sens. (Basel, Switzerland) **19**(6), 1305 (2019)
4. Bartelds, T., Capra, A., Hamaza, S., Stramigioli, S., Fumagalli, M.: Compliant aerial manipulators: toward a new generation of aerial robotic workers. IEEE Rob. Autom. Lett. **1** (1), 477–483 (2016)
5. Deng, C., Wang, S., Huang, Z., Tan, Z., Liu, J.: Unmanned aerial vehicles for power line inspection: a cooperative way in platforms and communications. J. Commun. **9**, 687–692 (2014)
6. AERONES homepage. https://aerones.com/eng/wind_turbine_maintenance_drone/. Accessed 28 May 2019
7. Cano, R., Pérez, C., Pruano, F., Ollero, A., Heredia, G.: Mechanical design of a 6-DOF aerial manipulator for assembling structures using UAVs. In: 2nd RED-UAS Workshop on Research, Education and Development of Unmanned Aerial Systems (2013)
8. Bellicoso, C.D., Buonocore, L.R., Lippiello, V., Siciliano, B.: Design, modeling and control of a 5-DoF light-weight robot arm for aerial manipulation. In: 3rd Mediterranean Conference on Control and Automation (MED), Torremolinos, pp. 853–858 (2015)
9. Suarez, A., Jimenez-Cano, A.E., Vega, V., Heredia, G., Rodriguez-Castaño, A., Ollero, A.: Design of a lightweight dual arm system for aerial manipulation. Mechatronics **50**, 30–44 (2018)
10. Suarez, A., Heredia, G., Ollero, A.: Physical-virtual impedance control in ultralightweight and compliant dual-arm aerial manipulators. IEEE Rob. Autom. Lett. **3**(3), 2553–2560 (2018)
11. Artieda, J., Sebastian, J.M., Campoy, P., Correa, J.F., Mondragón, I.F., Martínez, C., Olivares, M.: Visual 3-d slam from UAVs. J. Intell. Rob. Syst. **55**(4–5), 299 (2009)
12. Torres-Gonzalez, A., Martinez-de-Dios, J.R., Ollero, A.: Range-only SLAM for robot-sensor network cooperation. Auton. Rob. **42**(3), 649–663 (2018)
13. Huh, S., Shim, D.H., Kim, J.: Integrated navigation system using camera and gimbaled laser scanner for indoor and outdoor autonomous flight of UAVs. In: IEEE/RSJ International Conference on Intelligent Robots and Systems, Tokyo, pp. 3158–3163 (2013)
14. Sani, M.F., Karimian, G.: Automatic navigation and landing of an indoor AR. drone quadrotor using ArUco marker and inertial sensors. In: International Conference on Computer and Drone Applications (IConDA), pp. 102–107 (2017)
15. Soria, P.R., Arrue, B., Ollero, A.: A 3D-printable docking system for aerial robots: controlling aerial manipulators in outdoor industrial applications. IEEE Robot. Autom. Mag. **26**(1), 44–53 (2019)
16. Suarez, A., Heredia, G., Ollero, A.: Cooperative virtual sensor for fault detection and identification in multi-UAV applications. J. Sens. **2018**, 19 (2018). Article ID 4515828
17. Suarez, A.: Compliant aerial manipulation. PhD thesis Dissertation, University of Seville (2019)
18. Suarez, A., Sanchez-Cuevas, P., Fernandez, M., Perez, M., Heredia, G., Ollero, A.: Lightweight and compliant long reach aerial manipulator for inspection operations. In: IEEE/RSJ International Conference on Intelligent Robots and Systems (IROS), Madrid, pp. 6746–6752 (2018)

Online Detection and Tracking of Pipes During UAV Flight in Industrial Environments

Augusto Gómez Eguíluz(✉), Julio Lopez Paneque, José Ramiro Martínez-de Dios, and Aníbal Ollero

GRVC-Robotics Laboratory, University of Seville, Seville, Spain
{ageguiluz,jlpaneque,jdedios,aollero}@us.es

Abstract. 3D LiDaR-based perception has been used in robotics for obtaining accurate representations of the robot surroundings. As pipes are one of the most common objects in industrial environments, cylinder detection systems provide valuable information for robot navigation in industrial scenarios. However, most cylinder detection approaches using 3D LiDaRs suffer from high computational requirements and, therefore, their on-line implementation on real robots is limited for certain applications. This work proposes a computationally-light probabilistic approach to pipe detection and tracking suitable for aerial robot mapping and navigation. The proposed method was tested in both simulation and real scenarios. Through a combination of previous estimates and the localisation of the robot, the proposed approach is capable of reducing the computational cost of the RANSAC algorithm while keeping high detection accuracy.

Keywords: Pipe detection · UAVs · LiDaR

1 Introduction

Robot perception in industrial environments is a challenging task, since the scenarios are typically highly complex, non-structured and dynamic. During autonomous navigation in an industrial scenario, a robot should deal with a number of perceptual issues such as highly changing illumination, reflections caused by metallic objects and poor GNSS reception. This work focuses on robotic inspection and maintenance activities using aerial robots that have been developed in the GRVC-Robotics Laboratory of the University of Seville, such as the H2020-AEROARMS and ERC-GRIFFIN projects.

Detecting common objects in industrial scenarios, such as pipes, would enable improving localisation and mapping accuracy with impact on robot navigation and perching, among others. A number of techniques have been recently developed for the detection of pipes using 3D light detection and ranging (LiDaR) sensors [1,3,5,6,8,10,11]. The work in [9] defined a taxonomy of the techniques used for point cloud segmentation categorising them as methods based on edges,

M. F. Silva et al. (Eds.): ROBOT 2019, AISC 1092, pp. 28–39, 2020.
https://doi.org/10.1007/978-3-030-35990-4_3

regions, models, attributes, and graphs. Edge and region-based approaches analyse local neighborhoods in the point cloud to find similar features. Model-based approaches find the points that fit a geometric model within an error threshold. Other methods rely on clustering regions according to some attributes or performing a classification procedure on a graph representing the point cloud. A detailed survey on these 3D segmentation techniques was presented in [9]. All the above techniques require intensive computational resources, which constrain their on-line execution onboard aerial robots. Generally, pipe detection is applied offline and many existing approaches reduce the computational cost through constraining the pipe orientations. For instance, methods such as [8] and [1] have focused on the detection of vertical and horizontal pipes.

This paper presents a method to detect and track pipelines in an online manner using an aerial robot equipped with a 3D LiDaR. The proposed method combines prior information with the robot localisation system to obtain a probabilistic model that is used to predict the pipe location in subsequent iterations. Thus, the computation cost is reduced in order to detect pipes in an online manner while keeping a robust estimation of the pipe geometric parameters. To the best of our knowledge this is the first LiDaR-based pipe detection and tracking approach that relies on the robot localisation for reducing computational cost, which enables its online execution onboard aerial robots. The proposed method has been implemented in a hexarotor robot and has been validated in outdoor experiments.

The rest of the paper is organised as follows. A literature review of pipe detection systems for industrial inspection and maintenance is presented in Sect. 2. Section 3 presents the methodology used for the proposed approach, including the probabilistic tracking and detection of pipes, the combination of prior estimations and the robot localisation, and the implementation details of the proposed algorithm. Section 4 shows results on pipe detection in both simulation and a real scenario. Finally, Sect. 5 presents conclusions and directions for future work.

2 Related Work

Recent research has explored the use of Unmanned Aerial Vehicles (UAVs) in industrial environments in order to endow robots with the necessary capabilities to navigate and interact with pipes. In a recent work, [13] developed a long reach aerial manipulator for pipe inspection. Assuming the location of the pipe structure was known and relying on contact force sensing, the authors presented a 2-DOF compliant arm attached at the tip of a pendulum bar to contact the non-accessible sides of the pipes. A motion planning algorithm that estimates the trajectory for a UAV to transport a long bar through a cluttered industrial scenario consisting of several pipe structures was presented in [2]. Although their experimental evaluation showed that their method is capable of successfully navigating in a complex environment, their work focused on the mechatronics and planning systems, whereas the location of the pipes was assumed to be known. An aerial repairing method for oil pipes was developed in [15]. The authors presented an aerial manipulator robot for sealing cracks and holes on damaged sections of pipes. Although their system was capable of compensating the pose offset of the UAV while achieving high precision at sealing the leaks, their perception relied on a VICON motion capture system and, therefore, a robust pipe/crack detection system is required to be used in other scenarios.

Despite the detection of pipes being an essential requirement for autonomous UAV navigation in many industrial environments, the online detection of pipes for UAV navigation is still an under researched area. An integral approach to inspection and maintenance of pipes in industrial scenarios was developed in the AEROARMS H2020 project [10]. The pipe detection approach used in AEROARMS [5] combined a visual line detection and robust estimation using 3D laser sensing to achieve accurate pipe detection in an online manner. The work in [6] proposed a UAV navigation approach for indoor environments that stabilises the LiDaR measurement plane using an Inertial Measurement Unit (IMU). Additionally, the authors propose a real-time classification method for pipelines. Their method evaluated a region of points surrounding a seed to determine if a cylinder representation of a given radius can be fitted. Their method does not perform pipe detection on the full point cloud but the classification of a neighbourhood representing a pipe, which was assumed to be known at every time step.

A number of approaches have explored cylinder detection from 3D data for non-robotic applications. The work in [3] proposed a two-step procedure to detect cylinders from 3D point clouds using a random sampling based algorithm. The authors used a Gaussian image to compute the direction of a cylinder candidate and, an extraction step was used to compute its size and location. To minimise the effects of individual random samples, the authors set the number of the required iterations to 500 iterations, which deeply constrains its online on-board execution. A method to automatically detect pipelines in point clouds was used in [8]. The authors simplify the problem of detecting 3D patterns to 2D (i.e. cylinders to circles) in order to deal with large scale data. However, their method is limited to horizontal a vertical pipelines, i.e. parallel and perpendicular to the ground.

Another pipe detection system was presented in [1]. Their approach splits the point cloud into a number of thin subsets, obtains circles of a given diameter using the Hough transform, and fits straight lines through the circle centers. Despite the experiments showing high accuracy on pipe detection, their approach is limited to vertical and horizontal pipes. The work in [11] detects and reconstructs pipes by projecting a point cloud onto a Gaussian sphere to detect circle patterns (i.e. to find the cylinder orientation), which are used to iteratively estimate cylinder candidates. Although their approach showed high recall on large datasets, the computation had to be performed offline involving 24 h in GPU. Besides, the results presented suffered from high variance.

Most of these works are either not suitable for UAV navigation (e.g. due to limitations on certain pipe orientations or computational expenses) or do not exploit the potential of the available robot localisation. The presented work contributes to UAV perception for industrial inspection and maintenance by proposing an approach for recursive detection and tracking of pipes that iteratively combines prior knowledge with information of the UAV odometry to obtain a probabilistic model that is used to predict the pipe location in subsequent iterations and hence drastically reduce its computational needs.

3 Methods

The scheme of the proposed approach is shown in Fig. 1. It refines the parameters of the pipe (location and orientation) following a recursive Prediction-Update approach. In the Prediction stage, the robot odometry –computed from 3D LiDaR scans as in [7]– is used to predict the pipe position and orientation in the point cloud. In the Update stage the Random Sample Consensus (RANSAC) algorithm is used for computing the pipe parameters that best fit within the point cloud. To reduce the computational cost of the RANSAC method, the LiDaR point cloud is filtered to eliminate the points that are not likely to lie on the surface of the pipe.

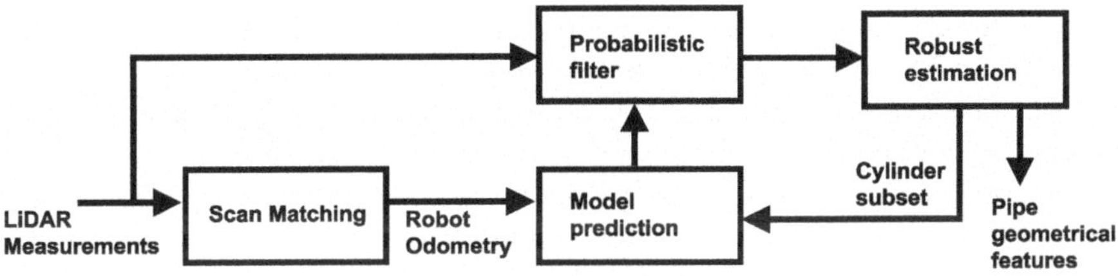

Fig. 1. Diagram of the proposed pipe detection and tracking method.

Section 3.1 details the proposed approach for detection of pipes using robust estimation. Section 3.2 describes the method used for probabilistic modelling, which is used to predict the points representing the pipe in subsequent iterations. Section 3.3 provides insights of the implementation of the proposed method.

3.1 Robust Estimation of Pipes

Random Sample Consensus [4] (RANSAC) is a robust estimation algorithm that fits a mathematical model to a dataset. The basic assumption is that the observed data can be explained by a set of model parameters and additional data that do not fit the model, i.e. outliers. The problem can be formulated as follows: given a set of data points $\mathbf{Z}$ and an error threshold ϵ, find the subgroup of points $\mathbf{X}$ that minimises the sum of squared perpendicular distances while ensuring none of the points in $\mathbf{X}$ deviates from a given model $m(\mathbf{X}, \boldsymbol{\theta})$ by more than ϵ units i.e. $\boldsymbol{\theta}$ are the pipe model parameters.

RANSAC starts by selecting a random subset of k LiDaR points and, then, the subset is used to compute an initial estimate for $\boldsymbol{\theta}$. The support of $\boldsymbol{\theta}$ is measured by the number of inliers computed as $||m(\mathbf{X}, \boldsymbol{\theta})|| \leq \epsilon$. Thus, the random selection is repeated n times and the best estimate of $\mathbf{X}$ is considered the robust fit. To ensure high probability of finding the true set of inliers, the algorithm requires the computation of a sufficient number of trials. The likelihood that all k random LiDaR points are inliers (i.e. in one trail) is p^k, where p is the probability that a point in $\mathbf{X}$ is inlier. The likelihood that n trials will not find a subset $\mathbf{X}$ in which all points are inliers is $1 - P = (1 - p^k)^n$, where P is the total probability of success after n trials. Hence, the required minimum number of trials is given by:

$$n = \frac{\log(1 - P)}{\log(1 - p^k)}. \tag{1}$$

In practice, the number of sample LiDaR points k is intended to be kept as low as possible when using RANSAC [14] since n, the number of RANSAC iterations, grows exponentially with k. However, 3D LiDaR scans provide a vast amount of points that entails the straightforward computation of RANSAC to be performed offline. The proposed approach performs an initial estimation and predicts the pipe data points that match the estimated cylinder. Thus, further estimations are performed evaluating only the sample points that fit the prediction within a given probability and, therefore, reducing the number of RANSAC iterations.

We denote $\mathbf{D}_t$ to the point cloud obtained from the LiDaR at time t. The subset of LiDaR points that fits the pipe is denoted as $\mathbf{C}_t$. A cylinder in 3D space can be defined as $||\mathbf{x} - \mathbf{p} - (\mathbf{x} - \mathbf{p}) \cdot \mathbf{v} \cdot \mathbf{v}||^2 = r^2$, where $\mathbf{p} = \{p_x, p_y, p_z\}$ is a point on its axis, $\mathbf{v} = \{v_x, v_y, v_z\}$ is the axis direction, and r is the radius of the cylinder. Therefore, the cylinder model was defined by seven coefficients $\mathbf{m}_c = \{p_x, p_y, p_z, v_x, v_y, v_z, r\}$ that RANSAC algorithm uses to find the best match in $\mathbf{D}_t$.

3.2 Model Prediction and Filtering

This section describes the method used for probabilistic modelling the pipeline. This model will be used to predict the points representing the pipe in subsequent times.

First, the center of the pipeline is predicted. At each time step, the average value $\boldsymbol{\mu}$ is computed from $\mathbf{C}_t$. However, $\boldsymbol{\mu}$ is an estimate of the average position of a pipe surface computed with the cylinder surface points visible at time t: it does not consider the occluded face of the cylinder. To cope with that, $\boldsymbol{\mu}$ is projected onto the cylinder axis direction vector $\mathbf{v}$. The projection is denoted $\bar{\boldsymbol{\mu}}$ and used as the average value for computing the covariance matrix of $\mathbf{C}_t$, which is denoted as $\boldsymbol{\Sigma}$. Thus, the pipe is modelled as a Normal distribution $\theta \sim N \in (\bar{\boldsymbol{\mu}}, \boldsymbol{\Sigma})$ and the likelihood function of its surface points is given by $p(\mathbf{d}_t^i|\theta)$, where $\mathbf{d}_t^i$ is the ith point of $\mathbf{D}_t$. In each iteration the algorithm performs as follows: (1) computes the posterior probability of each point lying at the surface of the pipe; (2) estimates the new parameters of the cylinder by applying RANSAC to the points with the highest probability; and (3) updates the value for $\bar{\boldsymbol{\mu}}$ and $\boldsymbol{\Sigma}$ using the new cylinder estimate $\mathbf{C}_t$.

It is worth noting that all LiDaR points in $\mathbf{D}_t$ and $\mathbf{C}_t$ are relative to the reference frame attached to the LiDaR. However, the LiDaR is assumed to be onboard an aerial robot and, therefore, its reference frame moves along with the robot. Assuming that the movement of the robot between t and $t+1$ is small enough to be neglected, the method described in Sect. 3.1 could be seen as a naïve solution to pipe detection. Instead, for higher accuracy in cases of non-negligible aerial robot motions, we opted for computing the transformation between the LiDaR reference frame L_t at times t and $t+1$. For simplicity, we will denote thereafter $L = L_t$ to the LiDaR reference frame at time t.

The pose of the robot in the global reference frame at every time t is assumed to be known (e.g. computed with accurate pose robot estimation methods) and it is denoted as $\boldsymbol{\phi}_t$. The proposed method uses $\boldsymbol{\phi}_t$ together with the Gaussian model obtained at $t-1$ in order to estimate the area in which the pipe is located at time t. This helps to filter the point cloud, both reducing the complexity and improving accuracy of the RANSAC-based robust pipe estimation performed at time t.

At each time the pipe model parameters, $\bar{\boldsymbol{\mu}}$ and $\boldsymbol{\Sigma}$, obtained at $t-1$ are transformed from L_{t-1} to L_t. The transformation of $\bar{\boldsymbol{\mu}}$ is computed in a straightforward manner using the homogeneous tranformation matrix. Differently, the covariance matrix is transformed from L_{t-1} to L_t by computing its rotation as $\hat{\boldsymbol{\Sigma}} = \boldsymbol{R}'\boldsymbol{\Sigma}\boldsymbol{R}$, where $\hat{\boldsymbol{\Sigma}} = {}_{L_t}\boldsymbol{\Sigma}$ and $\boldsymbol{R}$ is the rotation matrix from L_{t-1} to L_t.

3.3 Recursive Cylinder Detection and Tracking Algorithm

The proposed approach iteratively creates a probabilistic model of a pipe in the scene and, thus, achieves faster estimation in subsequent time-steps. The approach can be easily extended to all the pipes in the scene. This section is described assuming that there are N pipes in the scene. The pseudocode of the presented method is shown in Algorithm 1. The set of points lying at the surface of pipe j at time t is denoted as $\mathbf{C}_t^j$. The method is recursive. Hence, we assume $\mathbf{C}_0^j$, the initial pipe estimate for each pipe j, is known.

At time $t-1$ the method provided θ_{t-1}^j, the cylinder parameters for all pipes. At time t the predicted parameters $\hat{\theta}_t^j$ are computed using the transformations between L_{t-1} and L_t, the LiDaR frames at time $t-1$ and t. Next, the algorithm evaluates the likelihood of $\mathbf{D}_t$ to fit the predicted pipe models $\hat{\theta}_t^j$. The subset of the LiDaR points with greater probability to fit each model $\hat{\theta}_t^j$ is denoted $\mathbf{Z}_t^j$ and used for cylinder estimation using RANSAC. It is worth noting that a LiDaR point can fit multiple models, for instance, if two pipes intersect. To address these cases, a point fitting multiple models is considered as belonging to the subset $\mathbf{Z}_t^j$ that has the highest probability (i.e. normalised likelihood of all datapoints) of fitting the corresponding model. Therefore, the convergence properties of RANSAC are enhanced and $\mathbf{C}_t^j$ –the LiDaR points lying on the surface of cylinder j– are used to obtain the pipe models at time t, θ_t^j. Moreover, the rate of inliers in $\mathbf{Z}_t^j$ is used to estimate the maximum number of iterations n^j using Eq. (1), so that the method ensures the robust estimation of the pipes while keeping low computation cost.

Algorithm 1. Recursive cylinder detection and tracking algorithm

procedure RECURSIVE-CYLINDER-DETECTION-AND-TRACKING($\mathbf{D}_t$, ϕ_t)
    for $j = \{1, \ldots, N\}$ **do**
        $\hat{\theta} \leftarrow ModelTransformation(\theta^j_{t-1}, \phi_t)$
        $\mathbf{Z}^j_t \leftarrow FilterByProbability(\mathbf{D}_t, \hat{\theta})$
        $\mathbf{C}^j_t \leftarrow RANSAC(Z^j_t, n^j)$
        $\theta^j \leftarrow ComputeGaussian(\mathbf{C}^j_t)$
        $p^k \leftarrow \frac{Size(\mathbf{C}^j_t)}{Size(\mathbf{Z}^j_t)}$
        $n^j \leftarrow \frac{log(1-P)}{log(1-p^k)}$
    end for
end procedure

4 Experiments

The proposed method has been validated and evaluated in a number of experiments and simulations. The proposed approach aims at enhancing the detection speed to allow the algorithm to be used online while keeping high detection accuracy. Therefore, the maximum number of iterations n was set for robust pipe estimation at each time step and the angle deviation of the pipe axis was used as performance measure. In terms of computational performance, the main advantage of the proposed method is that it only evaluates the LiDaR points that have probability of fitting a model, which has been estimated in previous iterations. As detailed in Sect. 3.1, the number of outlier points to be evaluated is correlated to the number of iterations required for a robust estimation within a given error probability. In all the experiments, we set the total probability of successfully detect the pipe to 99.99%, i.e. $P = 0.9999$.

As mentioned in Sect. 3.3, the proposed approach assumes an initial estimation of the pipes in the scene. In all the experiments, the initial estimation of the pipes in the scene $\mathbf{C}^j_0$ was computed in an autonomous manner by filtering the point cloud at time $t = 0$ and using RANSAC. To filter the point cloud, the points beyond a distance threshold and the data representing the ground (i.e fitting a plane using RANSAC) were removed. It is worth noting that the proposed approach is not restricted for obtaining $\mathbf{C}^j_0$ and, although out of the scope of this paper, other initial estimation techniques can be used.

The first set of experiments was performed in a real scenario. The scene included an horizontal pipe with a diameter of, approximately, 70 cm supported in a metallic structure as shown in Fig. 2-left. The LiDaR sensor used for pipe detection was a Velodyne HDL-32E, embarked in the DARIUS hexarotor developed at the GRVC-Robotics Laboratory (see Fig. 2-right) with a MTOW (Maximum Take Off Weight) of 22 kg and tip-to-tip blade size of 1700 mm. The 3D LiDaR data was collected during both manual and GPS-assisted flights. The proposed method showed no differences between those two types of flights. Additionally, the robot was always controlled or supervised by an expert operator and the experiments

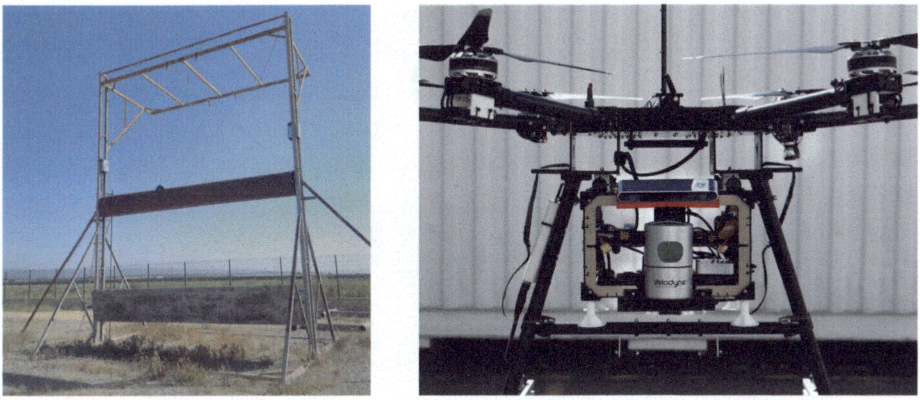

Fig. 2. Experimental scenario used for outdoor flights. On the left, the metallic structure supporting an horizontal pipe of diameter 70 cm. approximately. On the right, the aerial robot used in the experiments.

were performed in a secured area following the legal requirements. Moreover, the UAV was always kept at a safety distance from the pipe structure to avoid potential collisions.

The experiments performed in simulation were implemented using Gazebo and the abstraction layer for unmanned aerial vehicles (UAL) developed in [12]. The used scene has several pipes on vertical and horizontal orientations and diameters of 50 cm. We collected data from a simulated flight including measurements of the 3D LiDaR and the LiDaR-based robot odometry as described in Sect. 3. In both experiments, real scenario and simulation, the UAV performed a flight trajectory that moves around the pipes, maintaining the pipe structures (at least partially) within the range of the LiDaR.

4.1 Performance During Real Flight

The proposed approach was tested during >20 flights on the real scenario, i.e. more than one flight hour. To evaluate the method the estimated pipe axis, diameter, and the homogeneous transformation between the coordinate frame of the robot and the global reference frame were stored for further analysis. The robust estimation was set to find pipes with diameter in the range of [0.1 2] m.

Figure 3 shows the detection of the pipe during a flight of 50 s. Figure 3-top shows in red colour the points corresponding to the pipe (i.e. after applying robust estimation). The rest of the points are coloured in green. The Gaussian distribution $\hat{\theta}_t$ is represented with a white ellipsoid. Thus, the points lying inside the ellipsoid are approximately the subset that was evaluated for the pipe estimation in that time. Figure 3-bottom shows a map of the scene coloured in orange and the trajectory followed by the UAV represented by red arrows.

For each time step the pipe axis direction was transformed to the global reference frame, and the angular deviation $\alpha > 0$ of the pipe axis (w.r.t. the

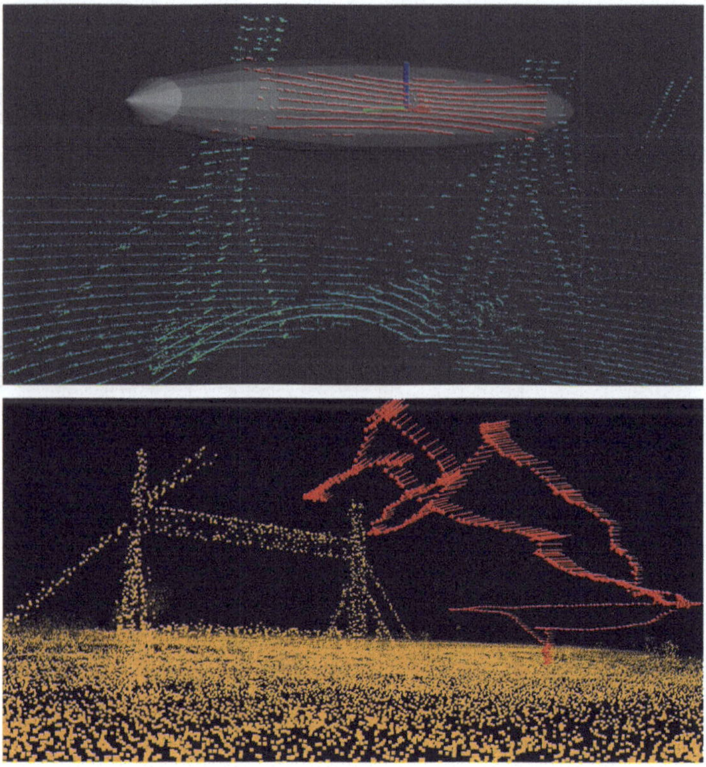

Fig. 3. (Top) Detection of the pipe at a given time in the method execution. The LiDaR scan is shown in green, the red points are detected as part of the pipe, and the white ellipsoid represents the model $\hat{\theta}$ used to predict the location of the pipe. (Bottom) Trajectory in one of the flights used for evaluating the method. The red arrows represent the drone odometry over time, and the orange points represent an obtained geometrical map of the environment.

average estimation) was computed. The average angle variation was $\mu_\alpha = 1.67°$ and standard deviation $\sigma_\alpha = 1.06°$ shows the cylinder axis was estimated with a maximum error of 4.85° 99.7% of the times, i.e. $\mu_\alpha \pm 3\sigma_\alpha$. It is worth noting that this error metric accumulates the inaccuracies of the robot localisation system since it was used to compute the homogeneous transformations between the coordinate frames.

Additionally, we registered the maximum number of iterations n used for robustly estimating the cylinders (i.e. pipes) using RANSAC at each time. It is worth noting that, as detailed in Sect. 4.2, the value of n was estimated at every time step using the rate of inliers of the previous time step. In this experiment, the probabilistic filter reduced the number of points to be used for robust estimation to 22.55% of the total. On average, 94.72% of the resultant points were subsequently detected as inliers. This allowed the pipe identification to be always

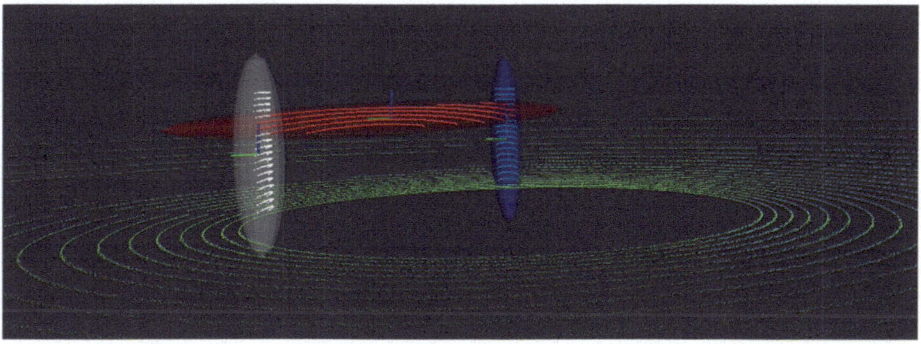

Fig. 4. LiDaR scan of the simulation multi-pipe scenario. The points lying at the surface of the three cylinders are coloured in white, red, and blue, while the points with low probability of being at any of the cylinders are coloured in green. Additionally, the same colour pattern is used for the ellipsoids representing the predicted location of the pipes.

performed in less than 3 iterations whereas processing the full point cloud would require on average 38 iterations. This result validates both the accuracy and the computational efficiency of the proposed method.

4.2 Simulation in a Complex Scenario

The performance of the proposed method was also evaluated in a scenario with multiple pipes. This section shows its performance for the simulated scenario shown in Fig. 4, which includes several pipes intersecting each other. The points corresponding to the three cylinders are annotated in red, white and blue. Similar colours are used for the ellipsoids representing to the Gaussian models in which the points were expected to lie. The rest of the input points are coloured in green.

The robust estimation parameters were set to find pipes with a diameter in the range of [0.20 0.60] m. It is worth noting that, generally, the smaller the pipe, the more difficult the detection is, due to the inliers-outliers ratio. The values of $\mathbf{D}_t$, $\mathbf{Z}_t^j$, $\mathbf{C}_t^j$ and ϕ_t^j were collected at each time step during 60 s. Despite the points near the intersection between the cylinders being candidates for more than one cylinder, we found that the algorithm does not mismatch the cylinders and always keep their track.

When we compare to the straightforward and alone RANSAC algorithm, we observed that only an average of 1.17%, 3.48%, 1.83% of the point cloud $\mathbf{D}_t$ were used for detecting the three cylinders. Additionally, we found that, on average, the 78.96%, 76.97% and 79.16% of the points in $\mathbf{Z}_t^j$ represent the corresponding pipe. Finally, the maximum number of iterations was set to $n = 6$, on average, for all pipes. According to Eq. (1), these results entail that the likelihood of RANSAC not finding the cylinder can be neglected. In contrast, the likelihood of each cylinder not being found in $n = 6$ when using RANSAC with the full point cloud $\mathbf{D}_t$ is 0.95%. According to Eq. (1), the straightforward version of RANSAC requires of 1114 iterations to detect the pipes with a success probability equivalent to the proposed approach, precluding the algorithm to run online and validating the proposed method.

5 Conclusions and Future Work

This paper presents a method for online detection and tracking of industrial pipes using a 3D LiDaR onboard an aerial robot. While state-of-the-art approaches generally assume that the sensor is static or their experimental scenarios are limited to certain pipe orientations, the proposed method takes advantage of the localisation system of the aerial robot to improve robustness and largely reduce the computational cost, enabling online execution. In short, the proposed method iteratively combines the prior probability distribution of the pipe with the localisation of the robot to filter the point cloud reducing the number of outliers, which improves detection accuracy and largely reduces the required computational effort. The proposed pipe detection system was extensively evaluated in both simulation and real flights. The results showed that the algorithm reduces the computation required for the robust estimation of cylinders and, thus, allows the robot to perform online detection and tracking of pipelines.

This work has been developed in the context of the ERC-GRIFFIN project. We plan to extend the proposed method to endow an ornithopter robot (i.e. robotic bird) with the necessary perception capabilities to both detect perching candidates and guide the robot during landing and perching. Additionally, we plan to extend the proposed approach in order to build semantic representation of the scenes that improve existing planning, localisation, and mapping algorithms. Thus, this work is a step towards the development of autonomous aerial robots that interact with industrial environments.

Acknowledgements. This work was supported by the European Research Council as part of GRIFFIN ERC Advanced Grant 2017 (Action 788247), the H2020 AEROARMS project under contract 644271, and the ARM-EXTEND project funded by the Spanish National R&D plan.

References

1. Ahmed, M.F., Haas, C.T., Haas, R.: Automatic detection of cylindrical objects in built facilities. J. Comput. Civil Eng. **28**(3), 04014009 (2014)
2. Caballero, A., Suarez, A., Real, F., Vega, V.M., Bejar, M., Rodriguez-Castaño, A., Ollero, A.: First experimental results on motion planning for transportation in aerial long-reach manipulators with two arms. In: 2018 IEEE/RSJ International Conference on Intelligent Robots and Systems (IROS), pp. 8471–8477. IEEE (2018)
3. Chaperon, T., Goulette, F.: Extracting cylinders in full 3D data using a random sampling method and the Gaussian image. In: Vision Modeling and Visualization Conference 2001 (VMV-01) (2001)

4. Fischler, M.A., Bolles, R.C.: Random sample consensus: a paradigm for model fitting with applications to image analysis and automated cartography. Commun. ACM **24**(6), 381–395 (1981)
5. Guerra, E., Munguía, R., Grau, A.: UAV visual and laser sensors fusion for detection and positioning in industrial applications. Sensors **18**(7), 2071 (2018)
6. Kumar, G.A., Patil, A., Patil, R., Park, S., Chai, Y.: A LiDAR and IMU integrated indoor navigation system for UAVs and its application in real-time pipeline classification. Sensors **17**(6), 1268 (2017)
7. Paneque, L., Martínez-de Dios, J., Ollero, A.: Multi-sensor 6-DoF localization for aerial robots in complex GNSS-denied environments. In: 2019 IEEE/RSJ International Conference on Intelligent Robots and Systems (IROS). IEEE (2019)
8. Liu, Y.J., Zhang, J.B., Hou, J.C., Ren, J.C., Tang, W.Q.: Cylinder detection in large-scale point cloud of pipeline plant. IEEE Trans. Visual Comput. Graphics **19**(10), 1700–1707 (2013)
9. Nguyen, A., Le, B.: 3D point cloud segmentation: a survey. In: 2013 6th IEEE Conference on Robotics, Automation and Mechatronics (RAM), pp. 225–230. IEEE (2013)
10. Ollero, A., Heredia, G., Franchi, A., Antonelli, G., Kondak, K., Sanfeliu, A., Viguria, A., Martinez-De Dios, J.R., Pierri, F., Cortés, J., et al.: The AEROARMS project: aerial robots with advanced manipulation capabilities for inspection and maintenance. IEEE Rob. Autom. Mag. (2018)
11. Qiu, R., Zhou, Q.Y., Neumann, U.: Pipe-run extraction and reconstruction from point clouds. In: European Conference on Computer Vision, pp. 17–30. Springer (2014)
12. Real, F., Torres-Gonzalez, A., Ramon Soria, P., Capitan, J., Ollero, A.: UAL: an abstraction layer for unmanned vehicles. In: 2nd International Symposium on Aerial Robotics (ISAR) (2018)
13. Suarez, A., Sanchez-Cuevas, P., Fernandez, M., Perez, M., Heredia, G., Ollero, A.: Lightweight and compliant long reach aerial manipulator for inspection operations. In: 2018 IEEE/RSJ International Conference on Intelligent Robots and Systems (IROS), pp. 6746–6752. IEEE (2018)
14. Szeliski, R.: Computer Vision: Algorithms and Applications. Springer Science & Business Media, London (2010)
15. Zhang, K., Kovac, M., Chermprayong, P., Xiao, F.: An autonomous aerial robot with integrated delta manipulator for aerial repair. IEEE Rob. Autom. Mag. (2019)

TCP Muscle Tensors: Theoretical Analysis and Potential Applications in Aerial Robotic Systems

Alejandro Ernesto Gomez-Tamm(✉), Pablo Ramon-Soria, B. C. Arrue, and Aníbal Ollero

Group of Robotics Vision and Control, University of Seville, Seville, Spain
{agtamm,prs,barrue,aollero}@us.es

Abstract. The use of aerial systems in a variety of real applications is increasing nowadays. These offer solutions to existing problems in ways that have never seen before thanks to their capability to perform perching, grasping or manipulating in inaccessible or dangerous places. Many of these applications require small-sized robots that can maneuver in narrow environments. However, these are required to have also strength enough to perform the desired tasks. This balance is sometimes unreachable due to the fact that traditional servomotors are too heavyweight for being carried by such small unmanned aerial systems (UAS). This paper, offers a innovative solution based on twisted and coiled polymers (TCP) muscles. These tensors have a high weight/strength ratio (up to 200 times) compared with traditional servos. In this work, the practical and modeling work done by the authors is presented. Then, a preliminary design of a bio-inspired claw for an unmanned aerial system (UAS) is shown. This claw has been developed using additive manufacturing techniques with different materials. Actuated with TCP, it is intrinsically compliant and offers a great force/weight ratio.

Keywords: Soft robotics · Bio-inspired robots · Aerial robots · Aerial manipulation

1 Introduction

The advances in unmanned aerial vehicles (UAV) over the years have promoted their use in manifold applications. However, there are still several limitations while working with these robots. These limitations affect mainly to three parameters: endurance, payload, and maneuverability. The balance between these parameters is essential during the design and development of UAV.

Due to that fact, researchers have centered their efforts in reducing the weight of these systems so that endurance and maneuverability could be optimized.

Our research, focuses on a particular type of UAS called Aerial Manipulators (AM). These aerial robots are equipped with limbs to extend their operation range. However, this equipment inevitably adds weight, and unbalances the robot.

M. F. Silva et al. (Eds.): ROBOT 2019, AISC 1092, pp. 40–51, 2020.
https://doi.org/10.1007/978-3-030-35990-4_4

Nowadays, the most used solution for actuating arms and legs are servo motors. The problem with this kind of devices is their weight. Authors in [16] developed a pair of novel lightweight arms with a total weight of 1.3 kg approximately. This design was extremely optimized, but it has a lower bound that cannot be overstepped due to the servos. Analyzing reviews in this matter [7,12], it is easy to realize that most of the approaches center their efforts in optimizing other parameters of the manipulators. Some approaches explore different options such as using additive manufacturing techniques [2]. However, the use of this technology has a consequent trade-off between weight and robustness.

Authors in [5] developed lightweight arms based in one conventional servo per arm, but using tensors to reduce the moments in the joints. The problem with this approach is that it has intrinsic limitations. The maximum weight to lift and the working volume is reduced due to the fact that all the movements are dependent on the same servo.

Some projects like [10] have centered efforts in developing optimized aerial manipulators. Others like [3] have tried to develop soft actuators for industrial inspection. However, drastically reducing weight while adding softness to manipulators are still unresolved challenges.

Replacing servos by twisted and coiled polymers (TCP) muscles fibers [8] could be one all-time solution for reducing the total weight of AM. Moreover, these actuators have some desirable properties that may be of interest in aerial manipulation. One of them is that these are intrinsically compliant and add softness to the system. Recent research [4] has proved that adding compliance to the manipulator, reduce the perturbations on the aerial platform and improves the maneuverability. However, this has only been achieved until now, by adding springs or elastic bands to conventional servos. Authors in [15] developed a compliant lightweight manipulator for aerial robots, but the weight is still limited by the use of servo motors.

TCP muscles are however still an immature technology. There is a strong need to settle the pillars to help new researchers to work with them. TCP works thanks to nylon elasto-plastic behavior [6]. One principal characteristic is their shape memory property, the ability of undergoing deformation at one temperature and then recover its original length.

One of the limitations of using these muscles nowadays is the lack of manufacturers. These, are currently produced in laboratories with handcrafted techniques.

This manufacturing process has been described in various articles [8,18]. In this article, we briefly cover these steps to settle a general manufacturing process. Two different types of Nylon threads are normally used. The first one, used since the first paper published in 2014 [8], is a common fishing line thread that can be found in different diameters. The more diameter, the more weight it can lift. These wires are strong and resistant but are difficult to heat, even more, when embedded on aerial systems. In the first years, these were heated using hot water or air. However, these methods are quite inefficient and impracticable when working with UAV.

To overcome this issue, authors in [14] started using copper wires twisted over the nylon threads, so that the threads could be electrically heated. Authors in [20], on the other hand, used silver covered nylon thread, thought for clothes manufacturing, to obtain better performances in terms of heating transmission. Even though this was a great step into this research, problems appeared due to the fact that these wires are not designed to lift weight, leading to unexpected breakages. They also had much less diameter than the normal fishing line wires, limiting the force they can perform.

In this paper, the combination of both types of threads is presented. This is a novel combination idea. The purpose is to extract the best properties of each thread to obtain the best possible muscle fiber.

Finally, it is worth mentioning that the work on this TCP done for these last 5 years have centered their efforts in developing models of the muscles behavior [1]. For proving their utility, several robotic arms have been developed [5,13,20], and also hybrid small robots like [19]. Despite this, all the models mentioned do not bring in a real application for them, offering only functional proof of concept. In this work, the conception will be different, centering the effort in developing a system for a real application in the field of aerial robotics. This has not been explored yet and could have huge potential. Particularly, the GRIFFIN project aims to develop multi-function ornithopters with manipulation capabilities, that can perform various tasks. This kind of AM have stricter payload limitations, making vital the research of new forms of actuation. This paper is a first step in making those system capable of performing perching and grasping in a bio-inspired compliant way and with much less payload requirements.

Following this idea, bio-inspired designs or the idea of developing systems based on the mother nature living beings is something researchers have tried to achieve in very different fields over the years. Some authors [9,11] inspired their design in insects. Nevertheless, they still use classic servo-actuators, and only the mechanical design is bio-inspired. Other approaches such as [17] try to solve that issue, developing passive actuators. However, this solution presents limited actuation capabilities. The goal of this work is trying to develop an innovative solution that could offer a bio-inspired design. That is, emulating the births muscle capabilities using tensors and mimicking biological muscle configuration for enhance performance. This innovative research could lead to replace servomotors in aerial systems, getting closer to a biological behavior, and optimizing their payload.

The remainder of the article is organized as follow. Section 2 introduces the developed model to better understand the behaviour of TCP muscle fibers. Section 3 focus on the novel TCP muscle configuration and potential. Section 4 shows the application proposed for aerial manipulators, an ultra light weight bio-inspired claw. At last, Sect. 5 present the conclusions and the future lines of work.

2 Muscle Models

This section introduces the models developed to better understand the behavior of TCP muscle fibers. At first, the muscle manufacturing process is introduced. After that, the system is modeled using different test-bench configurations. At the end, the model conclusion is presented.

2.1 Muscles Behavior

As mentioned in previous works such as [8], to grant that the manufacturing of the muscles is successful, it is required to have many details into consideration. This process is divided into the following steps: (1) Twisting; (2) Coiling; (3) Loading and (4) Annealing.

The process starts by spinning the nylon in the longitudinal direction. This is called twisting phase. The nylon starts to twist and generates an internal deformation. When the internal structure can not hold more torsion it reaches a collapse point, when the coiling starts. During the coiling, the nylon forms loops, or coils, over itself. Equation 1 is used to quantify the relation between the number of coils and weight for the coiling process according to the parameters of the fiber:

$$\begin{cases} N_{coils} = l_{thread}/\pi(d_{rod} + d_{nylon}) \\ m = (\pi(d_{nylon}/2)^2 T)/g \end{cases} \tag{1}$$

Where N_{coils} means the number of coils of the finished muscle fiber, d_{rod} is the diameter of the rod used for the coiling, if no rod is used (in the case of a self-coiled muscle fiber) this parameter is equal to zero, d_{nylon} is the diameter of the nylon thread, l_{thread} is the length of the nylon thread just before it starts to form the loops. m is the weight used for the coiling, T is the recommended load tension during coiling out of [8], and g is the gravity ($9.81\,\text{m/s}^2$).

The right selection of mass m is critical in the manufacturing process. If this mass is too high the fiber may break. Conversely, if it is not enough the fiber will bend instead of coil. In the case of this work, two different threads of diameter 0.2 mm and 0.7 mm where used. Using 17 MPa as the value for T, out of [8], weights of 54 and 667 g where obtained using Eq. 1. With experimental validations, it can be assert that this values are the lower limit for avoiding bending, there is a margin up to the upper limit, where the fibers brake.

As summary of the fabrication process of one muscle fiber, first step is to cut a desired length of the Nylon thread. Then, the thread is placed on a test-bench, and coiled with the recommended weight for applying tension. After a time, the thread will start coiling forming loops. When all the loops are formed, the coiling process must be stopped and the annealing can start. The first annealing can be done with any load, as its function is to eliminate residual tensions. Later on, a second annealing with more weight must be performed to create space between loops and allow compression. Finally, after this, the muscle fiber will be ready to use.

The direction of formation of the loops can be the same (homochiral) or the opposite (heterochiral). Homochiral refers to muscles fibers in which the direction of the coils are in the same spin than the internal twist deformation. In the heterochiral case, the spin of the coils occurs in the opposite direction of the internal torsion of the nylon. This difference makes that, when heated, the first type tends to compress and the second one tends to expand. Most of the works [14,18,20] focus on the use of homochiral muscles. On the other hand, heterochiral fibers are as best defined in the first paper published, [8], this expansive effect may be of interest for other specific applications.

The manufacturing process of heterochiral fibers is more complex due to the need of inverting the torsion in the coiling phase. This is typically done by manually coiling the muscle around a rod. Conversely, homochiral fibers can be created with and without a rod, because the coiling phase appears spontaneously when the nylon collapses (self-coiled fibers).

Using a rod for the manufacturing process has advantages and disadvantages. On one hand, manually rod-coiled muscles have a higher internal diameter (equal to the diameter of the rod). Enlarging this diameter allows the muscles to compress or expand more than the self-coiled ones (up to 50% compared with the 10–20% obtained by self-coiled). Also, this method allows to separate the loops. This fact is important, because it enables the muscle to compress even if it is not stressed (preloaded), due to the existing space between the loops. This behaviour is related with the elastic properties of the Nylon, such as elastic range and limit [6]. Conversely, such muscle fibers have less lifting force, being it inversely proportional to the diameter of the loops. Self-coiled muscle fibers have the minimum loop diameter and so the maximum force [8]. The problem with self-coiled fibers is that the loops do not have inter-distance, so that to be able to compress them, they should be previously stressed.

However, the experimentation made in this work has proven that self-coiled muscles fibers modify their length after a few annealing cycles, increasing their length notoriously compared with its initial length just after the coiling. This increment depends on the hanging weight. It means, that the muscle adjusts its length to the force applied on it when annealed. As a consequence, spaces between the loops of the spring are generated. Therefore, the muscle fiber will have an initial length between there loops, allowing it to contract even if not stressed. Only a minimum initial force must be applied to stretch the muscle fiber completely.

Nevertheless, the annealed state is reversible, i.e., if the muscle tensor is stored without any external force applied on it, it will eventually recover the original length, it had before the annealing, due to the shape memory property of the nylon. If that happens, the muscle can be retrained to recover the previous properties, using the same weight.

In order to obtain the model of the system, we have analyzed the behavior of the TCP muscles in two scenarios. At first, the muscles have been tested in a vertical test-bench, loaded with different weights and actuated with an electrical source. Then, a second set up has been prepared using a muscle fiber stressed by a spring. The idea behind this setup is to build a bio-inspired dual muscle system with antagonist actuator.

2.2 Test Bench with Free Weight

Figure 1 shows the basic set up. The muscle is held on one side, and a load is placed on the other side.

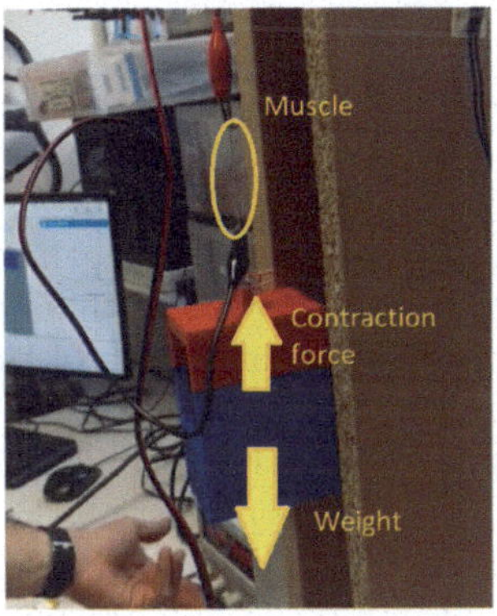

Fig. 1. Existing forces in the free-weight test-bench

The muscle behaves like a spring with no heating actuation. Thus, applying equilibrium of forces Eq. 2 is obtained.

$$K \cdot \Delta l = F_{ext} = m \cdot g \tag{2}$$

K means the static factor characteristic of the muscle, Δl is the increment of length of the muscle due to the load, m is the mass of the load and g is the gravity force. If the system is analyzed statically, it can be observed that at each instant, the equilibrium of forces is:

$$K \cdot \Delta l + F_{muscle} = m \cdot g \tag{3}$$

$$F_{muscle} = m \cdot g - K \cdot \Delta l \tag{4}$$

From the static point of view, the muscle acts with a force that is equal to the strength lost by the internal spring force but with the opposite direction. When heated, the Nylon tries to recover the untwisted state (shape memory), producing a force in the longitudinal direction and compressing the coils. This phenomena is modeled as a force F_{muscle} that compensates the decrease of Δl of the spring composed by muscle's coils.

Figure 2 shows the evaluation of the muscle against different loads given fixed energy from the power source.

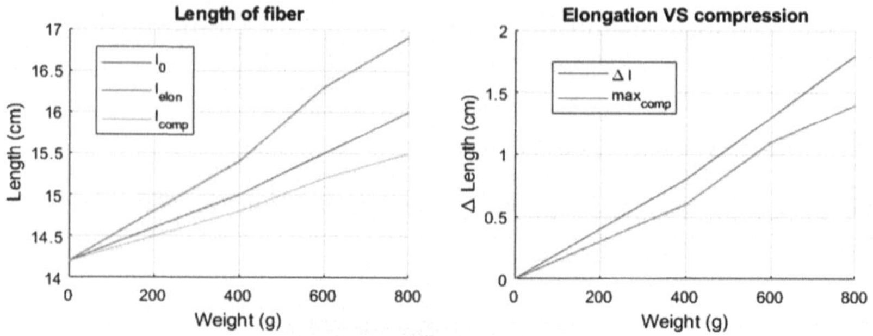

Fig. 2. Evaluation of muscle fiber against different fixed loads.

On the left image, it can be observed the length of the fiber against different weights in three different situations. l_0 refers to the initial static length of the unannealed fiber. Variable l_{elon} refers to the length of the fiber after the annealing. The last line, l_{compr}, is the maximum compression achieved with the fiber against the different loads. Right figure shows the absolute elongation and compression. This figure shows that the fiber recovers almost all the elongation when actuated.

When the muscle is heated, the force of the muscle grows and the length of the muscle decreases, maintaining the balance between forces. This assertion leads to the following conclusions:

- muscles need to have internal space to be able to compress and exert forces longitudinally.
- when fully compressed, the force is still present but can not be used to actuate.
- if the muscle is still actuated after reaching the maximum compression, it starts bending, trying to recover the untwisted shape.
- if this last state is reached, the muscle fiber will be near to collapse and break.

2.3 Bio-Inspired Muscle-Antagonist Configuration

From the previous experiments, we determined that the muscles can at least generate forces to lift fixed external forces. However, in this new set up, the muscle will stretch a spring, which means that the external force will increase as it self-compress. The reason for this new set up is to be able to reproduce a muscle antagonist pair, typical in animals. Consequently, when a muscle actuates, the second one, or antagonist, acts as a passive spring.

Figure 3 shows the proposed configuration. Equation 5 shows the proposed model for the equilibrium of forces for this set-up.

$$K_1 \Delta l_1(t) = K_2 \Delta l_2(t) + F_{muscle}(T, t) \tag{5}$$

Where F_{muscle} is dependant of the temperature. Temperature is, on the other hand dependant of the time, due to the fact that it increase with it when a tension is applied thanks to the Joule effect. l_1 and l_2 modules are the same, when one of

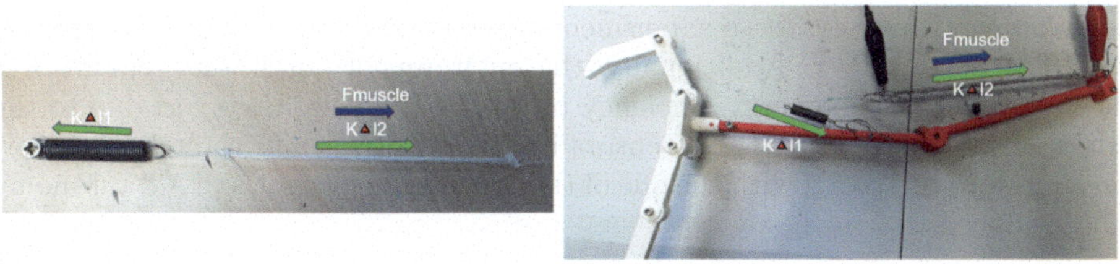

Fig. 3. On the right, spring-muscle configuration. On the left, early version of the claw with one finger actuated by a muscle fiber and a passive spring.

the terms decrease the other increases. K_1 is the stiffness coefficient of the spring and K_2 is the stiffness coefficient of the muscle fiber as seen in Eq. 2, being both of them constant. However, the increase of the F_{muscle} is able to compensate that decrease and increase and get to a new equilibrium. Doing regression to the experimental data, a quadratic regression is obtained for the model of the muscle fiber behaviour.

Figure 4 shows the compression of the muscle in the presented configuration. At first, the experiments show that the muscle overpass the force of the fixed spring, which is a necessary condition to validate the configuration.

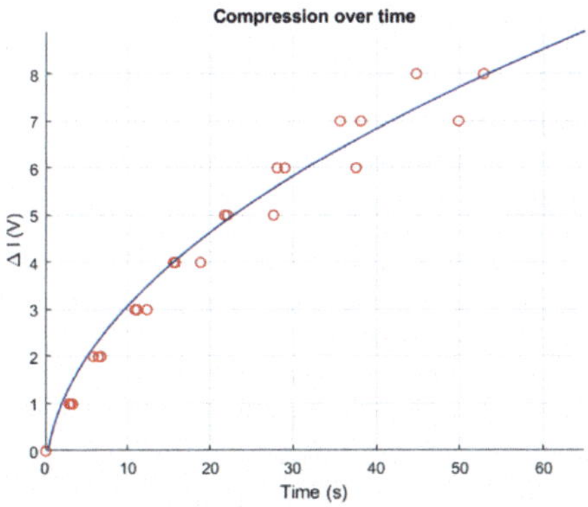

Fig. 4. Muscle fiber stressed by a spring, the force generated are shown

The muscle starts compressing as in the fixed-load experiments. However, due to the increment of Δl the force of the passive spring increases, opposing the muscle. As more the muscle compress, more is the force that the spring exert. Thus larger is the force that the muscle need to generate per unit of compression, or equivalently, smaller is the compression per unit of force. Figure 4, show the compression of the muscle against the spring over time. Contrary to Fig. 2 the trend is not linear.

According to that, the chosen model is an square root equation. For this model it is assumed that the muscle is heated by the Joule effect by applying a fixed voltage. Precisely, these experiments were carried out with 5 V, consuming around 10 W of power. Additionally, the environment conditions remain over the experiments. The energy efficiency of the TCP is low. However, due to the reduce of weight by using these tensors instead of servos and the relatively low consumption, compared with the one of the brushless motors, the UAV autonomy is not reduced.

$$\Delta l = a + b \cdot \sqrt{x} \tag{6}$$

3 Hybrid Self-coiled Muscles (HSM)

The design of the muscle used in this work tries to get the best performance and reliability using the existing technologies. For that purpose, a combination of Nylon wires is used. A non-conductive fishing line nylon thread of 0.7 mm diameter is coiled and used as core fiber for the muscle. This fiber can lift a large weight without breaking, but it does not have any conductive material for self-heating it. To overcome this issue, several muscle fibers of Shieldex 235/36 4-ply silver covered nylon wire are coiled with the previous fishing line wire. This muscle fibers, made of the Shieldex material, can lift up to 200 g each, and they can be actuated electrically. Figure 5 shows the resulting muscle hybrid after the first annealing phase.

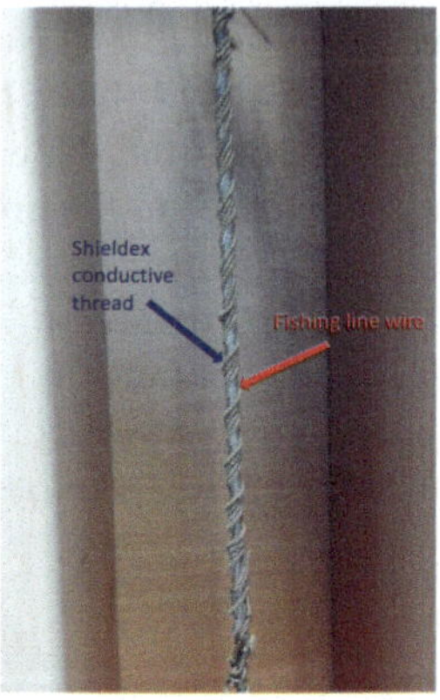

Fig. 5. An hybrid 6-ply muscle out of one fishing line (0.7 mm) muscle fiber and 5 Shieldex conductive (0.2 mm) silver threads

Particularly, for the experiments conducted in this article, five conductive fibers were coiled around the non conductive nylon thread. This muscle composition has been characterized for the model presented in Eq. 6 having the following parameters $(a, b) = (-0.6815, 1.188)$ with a confidence bound of 95%.

The resulting muscle, is capable of lifting up to 1 kg. With this solution, the system is capable of put up with real objects, lifting them without opening the limbs with the force generated by the weight. This is an important fact, because state of art muscles do have large limitations in terms of payload capabilities [13].

4 Application in Aerial Manipulation. Ultra Lightweight Claw

In this section, a preliminary version of an eagle-based claw with three fingers to be actuated with muscles fibers for ornithopters is presented. Taking inspiration in birds configuration, the muscles are placed on the legs as straight as possible, giving the muscle freedom to move and avoiding friction as much as possible.

At the tip of the muscles, normal fishing lines thread is placed. These act as tendons to transmit the force needed to the fingers. The idea of using fishing line as tendons is due to the property of nylon of not stretching and being resistant. All the designs are printed using additive manufacturing using Polylactic acid (PLA) and Filaflex to create flexible joints. Figure 6 shows the design of the claw. The distributions of the fingers mimics real eagle fingers.

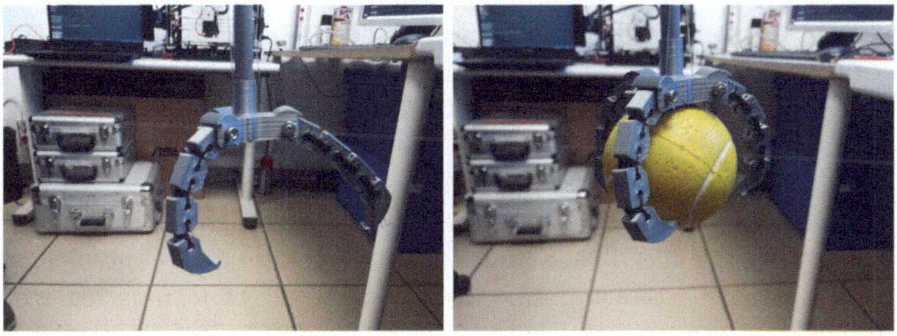

Fig. 6. Grasp of a plastic ball by the three finger of the claw, it can be observed that the claw can put up with the weight even looking downwards

For actuating all the phalanges with just one muscle fiber, the joints are made of Filaflex which, when not actuated, act as a spring, returning to their rest position. The resting state is the extended finger. When actuated, the Filaflex will apply some resistance to the movement, that will be enough for keeping the muscle stretched.

Also, this design provides smoothness to the movement of the claw, fact that we did not achieve using common axial joints. The fingers act as one articulation, composed by many, with Filaflex "springs", connected articulations (Fig. 6).

Fig. 7. Two claws on an ornithopter used for the project GRIFFIN an European ERC advanced grant

5 Conclusions

This paper has presented the foundations for working with Twisted and Coiled Polymers for aerial manipulators. At first, the manufacturing process has been wrapped up. This is one critical step in this new trend of AMs. Due to the absence of existing manufacturing is essential to establish standard procedures for the developers.

Then, a simple but yet effective model has been proposed to analyze the behaviour of the proposed muscles. These models have been validated experimentally in different experiments.

It is important to remark that the use of these actuators may greatly impact the field of aerial manipulators, providing ultra-lightweight actuators that overtake the capabilities of existing servomotors. Moreover, in the field rising field of ornithopters, which payload capabilities are extremely limited. Embedding this muscles will lead to a new generation of aerial robots.

Nevertheless, there is still a lot of intense research that needs to be developed in this field. Next steps will focus in developing a control system for this actuators. Also, exploiting the combination of TCP muscles with other bio-inspired/soft actuators may lead develop more efficient solutions in aerial robotics and replace traditional actuators.

Acknowledgments. We thank *Robotics, Vision and Control Group* for supporting us during this work. This work has been developed under the framework of the project GRIFFIN (General compliant aerial Robotic manipulation system Integrating Fixed and Flapping wings to INcrease range and safety) SI-1867/23/2018 ERC-ADG - Advanced Grant EU-funded project.

References

1. Arakawa, T., Takagi, K., Tahara, K., Asaka, K.: Position control of fishing line artificial muscles (coiled polymer actuators) from nylon thread. In: Electroactive Polymer Actuators and Devices (EAPAD) 2016, vol. 9798, p. 97982W. International Society for Optics and Photonics (2016)
2. Ramon-Soria, P., Gomez-Tamm, A.E., Garcia-Rubiales, F.J., Arrue, B.C., Ollero, A.: Hecatonquiros: open-source hardware for aerial manipulation applications. Int. J. Adv. Rob. Syst. **16**(3) (2019)
3. Ramon-Soria, P., Gomez-Tamm, A.E., Garcia-Rubiales, F.J., Arrue, B.C., Ollero, A.: Autonomous landing on pipes using soft gripper for inspection and maintenance in outdoor environments. In: 2019 IEEE/RSJ International Conference on Intelligent Robots and Systems (IROS), November 2019
4. Bartelds, T., Capra, A., Hamaza, S., Stramigioli, S., Fumagalli, M.: Compliant aerial manipulators: toward a new generation of aerial robotic workers. IEEE Rob. Autom. Lett. **1**(1), 477–483 (2016)
5. Bellicoso, C.D., Buonocore, L.R., Lippiello, V., Siciliano, B.: Design, modeling and control of a 5-DoF light-weight robot arm for aerial manipulation. In: 2015 23rd Mediterranean Conference on Control and Automation (MED), pp. 853–858. IEEE (2015)
6. Cherubini, A., Moretti, G., Vertechy, R., Fontana, M.: Experimental characterization of thermally-activated artificial muscles based on coiled nylon fishing lines. AIP Adva. **5**(6), 067158 (2015)

7. Ding, X., Guo, P., Xu, K., Yu, Y.: A review of aerial manipulation of small-scale rotorcraft unmanned robotic systems. Chin. J. Aeronaut. **32**(1), 200–214 (2019). http://www.sciencedirect.com/science/article/pii/S1000936118301894
8. Haines, C.S., Lima, M.D., Li, N., Spinks, G.M., Foroughi, J., Madden, J.D., Kim, S.H., Fang, S., de Andrade, M.J., Göktepe, F., et al.: Artificial muscles from fishing line and sewing thread. Science **343**(6173), 868–872 (2014)
9. Ignasov, J., Kapilavai, A., Filonenko, K., Larsen, J.C., Baird, E., Hallam, J., Büsse, S., Kovalev, A., Gorb, S.N., Duggen, L., et al.: Bio-inspired design and movement generation of dung beetle-like legs. Artif. Life Rob. **23**(4), 555–563 (2018)
10. Ollero, A., Heredia, G., Franchi, A., Antonelli, G., Kondak, K., Sanfeliu, A., Viguria, A., Martinez-De Dios, J.R., Pierri, F., Cortés, J., et al.: The aeroarms project: aerial robots with advanced manipulation capabilities for inspection and maintenance. IEEE Rob. Autom. Mag. (2018)
11. Oszwald, F., Wedler, A., Schiele, A.: Development of a bioinspired robotic insect leg, November 2004
12. Ruggiero, F., Lippiello, V., Ollero, A.: Aerial manipulation: a literature review. IEEE Rob. Autom. Lett. **3**(3), 1957–1964 (2018)
13. Saharan, L., de Andrade, M.J., Saleem, W., Baughman, R.H., Tadesse, Y.: iGrab: hand orthosis powered by twisted and coiled polymer muscles **26**(10), 105048 (2017). https://app.dimensions.ai/details/publication/pub.1091450247 and http://iopscience.iop.org/article/10.1088/1361-665x/aa8929/ampdf, exported from https://app.dimensions.ai. Accessed 28 May 2019
14. Semochkin, A.N.: A device for producing artificial muscles from nylon fishing line with a heater wire. In: 2016 IEEE International Symposium on Assembly and Manufacturing (ISAM), pp. 26–30. IEEE (2016)
15. Suarez, A., Heredia, G., Ollero, A.: Lightweight compliant arm for aerial manipulation. In: 2015 IEEE/RSJ International Conference on Intelligent Robots and Systems (IROS), pp. 1627–1632, September 2015
16. Suarez, A., Soria, P.R., Heredia, G., Arrue, B.C., Ollero, A.: Anthropomorphic, compliant and lightweight dual arm system for aerial manipulation. In: 2017 IEEE/RSJ International Conference on Intelligent Robots and Systems (IROS), pp. 992–997. IEEE (2017)
17. Sun, B., Jing, X.: A tracked robot with novel bio-inspired passive "legs". Rob. Biomimetics **4**(1), 18 (2017)
18. Wu, L., de Andrade, M.J., Saharan, L.K., Rome, R.S., Baughman, R.H., Tadesse, Y.: Compact and low-cost humanoid hand powered by nylon artificial muscles. Bioinspiration Biomimetics **12**(2), 026004 (2017)
19. Wu, L., Karami, F., Hamidi, A., Tadesse, Y.: Biorobotic systems design and development using TCP muscles. In: Electroactive Polymer Actuators and Devices (EAPAD) XX, vol. 10594, p. 1059417. International Society for Optics and Photonics (2018)
20. Yip, M.C., Niemeyer, G.: High-performance robotic muscles from conductive nylon sewing thread. In: 2015 IEEE International Conference on Robotics and Automation (ICRA), pp. 2313–2318. IEEE (2015)

Autonomous Drone-Based Powerline Insulator Inspection via Deep Learning

Anas Muhammad[1], Adnan Shahpurwala[1], Shayok Mukhopadhyay[1(✉)], and Ayman H. El-Hag[2]

[1] Department of Electrical Engineering, American University of Sharjah, P.O. Box 26666, Sharjah, UAE
{b00052374,b00055159}@alumni.aus.edu, smukhopadhyay@aus.edu

[2] Department of Electrical and Computer Engineering, University of Waterloo, 200 University Avenue W, Waterloo, ON N2L 3G1, Canada
ahalhaj@uwaterloo.ca

Abstract. Accumulation of pollutants on ceramic insulators is one of the major causes of dry band arcing, a predecessor to flashovers, which may further cause major outages of electricity. It is critical to know locations of polluted insulators to prevent flashovers to make the power-grid reliable. This paper proposes a solution to detect the location of polluted insulators along an overhead transmission line using a quadcopter. Once provided with the GPS locations of the electrical powerline transmission towers, the quadcopter autonomously hovers along the line. And while doing so, it sends a live video feed of the transmission line to the ground station. A pre-trained neural network on the ground station then detects insulators in the video and classifies the detected insulators as polluted or clean. Only if the insulator detected is polluted, its location is recorded and reported back to the ground station. The novelty of this work is the use of a drone to automate the process of insulator inspection via a deep learning based neural network approach. Experiments show that accurate inspection results are obtained. This work is an initial step in the direction of achieving completely autonomous drone-based powerline insulator inspection.

Keywords: Drone · Quadcopter · Overhead transmission line (OHTL) · Insulator · Pollution · Inspection · Autonomous · Deep learning · Neural network

1 Introduction

Dry band arcing is the phenomenon when electrical flashes occur between wet and dry spots over the polluted surface of ceramic insulators, which accelerates their deterioration. Often, this phenomenon leads to a flashover which is a major concern for electric power utility companies. One of the leading causes of dry band arcing is settling of pollutants on insulators, a polluted, and a clean insulator are shown in Fig. 1. Knowing the location of polluted insulators can aid in reducing chances of flashover and make the grid more reliable.

M. F. Silva et al. (Eds.): ROBOT 2019, AISC 1092, pp. 52–62, 2020.
https://doi.org/10.1007/978-3-030-35990-4_5

Fig. 1. A picture with two ceramic insulators, a 'polluted' insulator with visible pollutant coating its surface - on the left, and a clean insulator - on the right.

A lot of research exists on (1) Detecting insulator strings, (2) Detecting pollution on ceramic insulators, (3) Designing autonomous UAVs to survey overhead transmission lines (OHTL), but an autonomous system which combines all of the above is still not readily available, to the best of the authors' knowledge. For example, in [7,14], the authors use machine learning techniques to delineate insulator strings in images. However, this work does not focus on identifying any fault once the insulator string is identified. The authors in [8] used multi-source images of insulators to train a back propagation neural network (BPNN) to rate the level of pollution of insulators, on a scale of zero to four. However, details about acquiring the multi-source images and determination of the location of polluted insulators, are unavailable. The authors in [1,4,16] have developed techniques to calculate the level of pollution of insulators based on two factors, which are leakage current and partial discharge. One common disadvantage of the techniques in [1,4,16] is that the proposed techniques are contact methods, which makes them dangerous to use. The authors in [3] developed a quadcopter with a camera to be flown manually along an OHTL while autonomously detecting objects like towers, lines, and insulators. However, fault identification is not carried out once these objects are detected. A quadcopter, which flies autonomously and takes infrared and visible light images of insulators to check for excessive heating, is developed in [10]. Only the flight is autonomous in [10], while fault detection is done manually. A solution is proposed in [6], where a quadcopter can be used for OHTL inspection, but only theoretical details are provided. The author in [5] has evaluated performance of control algorithms used by quadcopters to survey and inspect OHTL. Image processing for insulator fault detection is discussed in [5], but not implemented. A survey of online robots used to traverse the OHTL and calculate the age of ceramic insulators by measuring their resistance and leakage current, is presented in [15]. As is evident from the above, there is a lack of work on a complete autonomous system for OHTL insulator pollution detection. The novel contribution of this work is to autonomously inspect transmission lines, for detecting polluted insulators, using a quadrotor. The flight of a quadrotor, detection and classification of insulators, and reporting GPS coordinates of polluted insulators, is automated. The GPS location of the transmission line towers need to be provided for the insulator inspection mission, so this work is an initial step in the direction of achieving a complete autonomous system capable of carrying out a full transmission line insulator inspection mission with minimal human involvement. Existing convolutional neural network (CNN) models [9] are trained via well known machine learning platforms [13] and used to carry out image analysis, which is further used to detect insulators and classify insulators as polluted or clean. If an insulator is classified as clean during a quadcopter based inspection flight, no action is taken. If an insulator is classified as polluted, GPS coordinates of this insulator are reported. Such a system can help utility companies to know the exact location of insulators which need cleaning. OHTL insulator inspection missions using helicopters with human inspectors onboard are extremely dangerous, because a small pilot error may cause a helicopter to get too close to a powerline causing accidents. Therefore, the contribution of this work is directed towards enhancing the safety of human operators involved in OHTL inspection by automation of as much as possible of the powerline insulator inspection process. Automating such a process can help inspect OHTL insulators in remote areas, or areas that are hard for people to easily access, and may also reduce inspection costs.

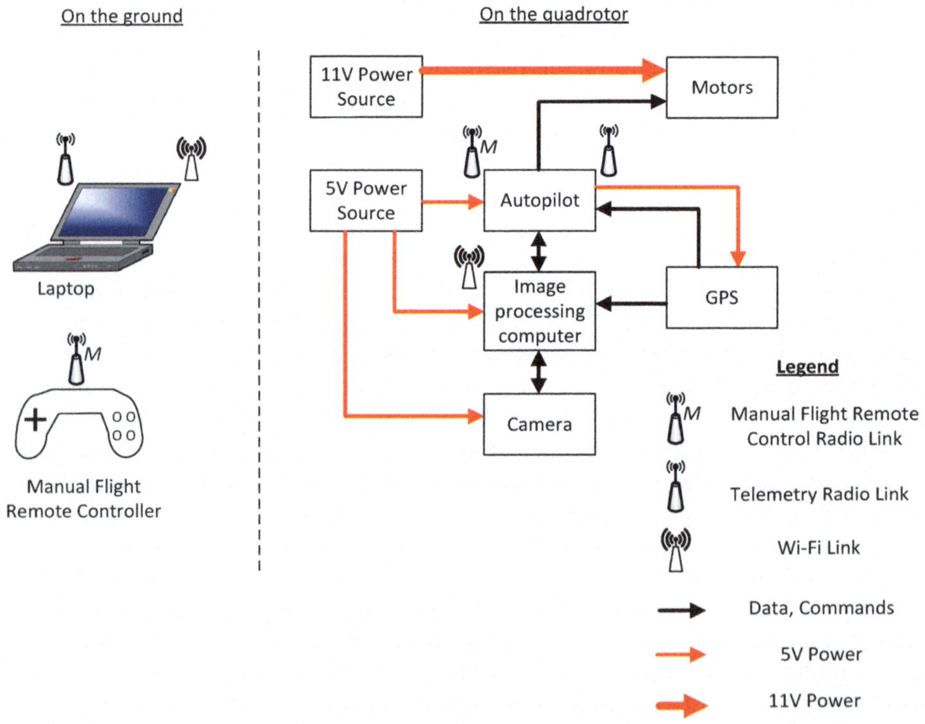

Fig. 2. Overall hardware schematic.

This paper is organized as follows. First the overall design is discussed followed by some details related to the hardware and software systems. Then, the experimental setup is described followed by the presentation of experimental results, and finally the paper is concluded by summarizing the results achieved.

2 Overall System Design

This section describes the hardware and software architecture of the quadcopter based insulator inspection system that is developed in this work.

The overall hardware schematic is shown in Fig. 2, and it contains two parts. The on-shore component represented by a laptop equipped with a 2.4 GHz WiFi link, and a 900 MHz telemetry link, forms the Ground Control Station (GCS). The GCS handles the start and end of quadcopter flight, waypoint assignments, receiving telemetry data from the autopilot, and also the processing of images/videos sent to the GCS from a computer onboard the drone. The electronics on-board the insulator inspection drone consist of two power sources (11.1 V for driving the motors, 5 V for all other electronics), an autopilot, an imaging computer, a camera, a GPS unit, and motors powering the quadcopter flight.

2.1 Hardware

The actually assembled drone, with all the parts labeled is shown in Fig. 3. For the labels related to each part, the readers are directed to Table 1. The drone is equipped with two separate power sources, one for motors, and the other one

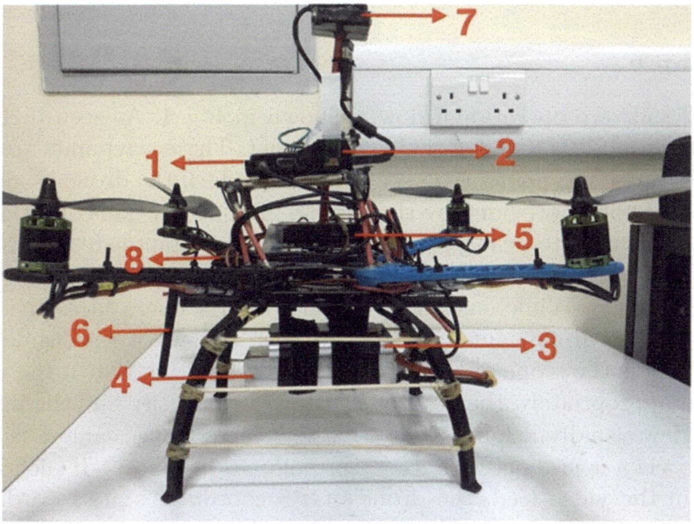

Fig. 3. Insulator inspection drone, with major parts labeled.

Table 1. Components of the prototype insulator inspection drone in Fig. 3.

Part no.	Part name
1	Raspberry Pi 2
2	Raspberry Pi camera module
3	5V battery pack for onboard electronics
4	11.1 V battery pack for drone motors
5	Pixhawk flight controller
6	900 MHz radio telemetry receiver
7	GPS sensor with on-board compass
8	PPM (Pulse position modulated) encoder with 5 GHz radio receiver

for driving electronics. The drone is equipped with a 900 MHz radio telemetry receiver, which allows monitoring of flight parameters from the GCS, and if desired, the user can assume manual control of the drone through the 5 GHz radio link. The flight controller (autopilot) used is a Pixhawk, which readily allows integrating sensors like the GPS. And, when a Pixhawk is used with an appropriate software like Mission Planner, this allows autonomous waypoint following control to be easily achieved by the drone, without having to develop a control algorithm/system from scratch. A PPM encoder is used because the Pixhawk used supports only remote controllers that transmit signals via PPM (pulse position modulation). The particular manual flight remote controller available uses pulse width modulation (PWM), and for compatibility, a PPM encoder is necessitated. The image processing computer is formed by an onboard Raspberry Pi.

2.2 Software

The overall software operational flow is shown in Fig. 4. As seen in Fig. 4, the drone's default mode is to follow a given waypoint. These waypoints are provided by the user. These waypoints represent points at which the drone is supposed to loiter. The waypoints provided by the users represent locations of OHTL towers where insulators need to be inspected. So once a loiter point is reached, the drone hovers at this point for 10 s, during which a video feed from the drone's camera is sent to the GCS. While ensuring that the drone camera faces the tower, or ensuring that the drone camera is able to have a view of the insulators as it loiters, is outside the scope of this work; it is worth noting that if the waypoints are selected appropriately, then a drone's camera can easily find the insulator in its field of view as a drone loiters. The video captured by the camera is processed on the GCS via a neural network to detect polluted insulators. If such insulators are found in the video feed, the drone's GPS coordinates are recorded. These coordinates are sent, or displayed to, the user at the end of an insulator inspection mission, which terminates when the final waypoint is reached. As mentioned in

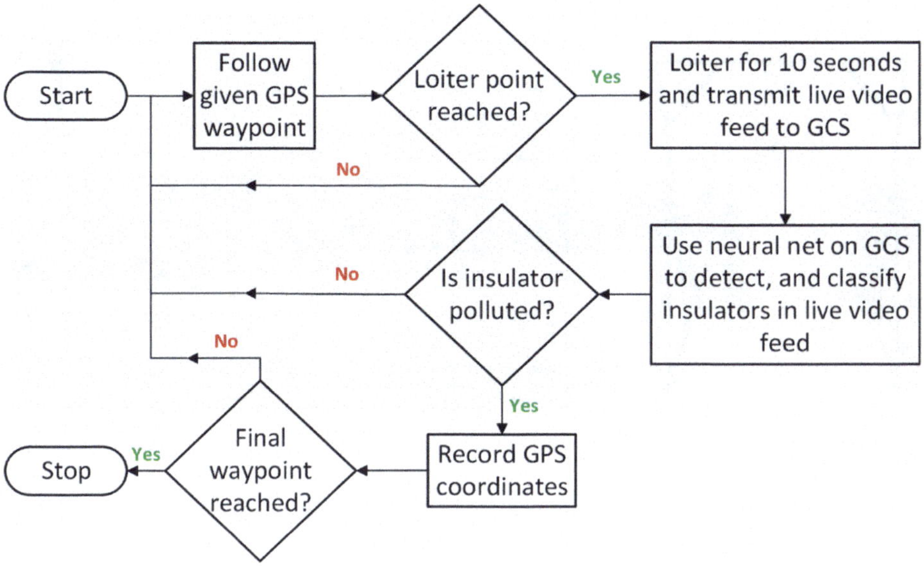

Fig. 4. Overall software operational flow.

the subsection above, the autonomous flight and waypoint following is achieved via a software known as Mission Planner [11], which allows tuning the flight controllers proportional, integral, and derivative (PID) gains to achieve stable flight.

2.3 Neural Networks, Deep Learning, and TensorFlow

Neural networks and related applications are widely used and available nowadays, so theoretical details related to these are not provided. However, readers are directed to [2] for information. Constructing a deep network suitable for object detection, and training such a network, is not necessarily a trivial task. Thus in this work the network is not developed from scratch, but a well known network from the literature, known as the Single Shot MultiBox Detector (SSD) is used. Details related to SSD are available in [9]. The actual variant/version of SSD used in this work is "`ssd_mobilenet_v1_coco_11_06_2017`" where 'coco' refers to the dataset used. This is chosen because it is simple to implement, and its object detection time is faster compared to other neural networks that are available for object detection, as mentioned in [9]. This is a key factor when an object detector is used for real time detection implemented via an autonomously flying drone. The neural network is trained using the TensorFlow platform [13]. For training, the pre-trained weights available are used along with around 1000 images each of clean and polluted insulators. Data augmentation is not used, and only one each of clean and polluted insulators were used for training. Now that the basic hardware and software components used are described, the following section focuses on the experimental setup and results.

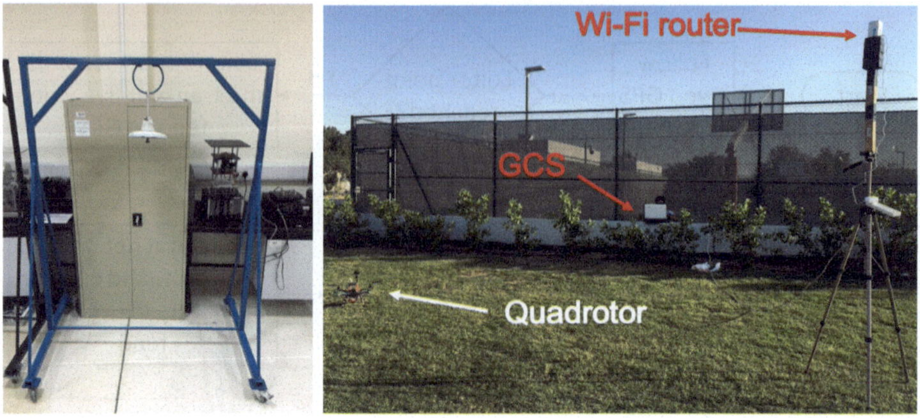

(a) A tower arrangement with an insulator.

(b) The outdoor setup of the GCS.

Fig. 5. Experimental setup.

3 Experiments

3.1 Experimental Setup

The experimental setup is shown in Figs. 5a, b. Figure 5a shows the tower arrangement created, at the top center of which a ceramic insulator is suspended. This tower structure is moved outside to conduct experiments. This arrangement is used because testing drones near live power lines is dangerous, and may need special permits. So at an experimental stage such a setup as shown in Fig. 5a, is more feasible. Figure 5b shows the outdoor ground station setup used. The GCS is simply a laptop. A Wi-fi router is hoisted on a pole to widen the range of availability of the Wi-Fi signal. The drone communicates through this Wi-Fi network with the GCS. In urban settings, such a network can be replaced by the cellular networks which are already in place, so that no separate network setup is required on field where a survey is to be performed. However, in remote locations where cellular service is not available, it is not hard to establish a dedicated Wi-Fi network, as shown here in Fig. 5b.

To be able to perform autonomous flights, and subsequent detection of polluted/clean insulators, the neural network proposed in the earlier section needs to be trained. As shown in [9], the SSD mobilenet neural network used in this work builds on an existing network by adding more layers to it. Thus training requires a sufficiently capable machine. For training purposes, 1042 images of the insulators (polluted and clean), were taken under varying lighting conditions in the parking lot facing the setup shown in Fig. 5b. The images were acquired with arbitrarily varying numbers of different objects, i.e. people, cars, etc. in the background. After acquiring the images, the polluted and clean insulator locations were labeled by drawing a bounding box around them, and adding a textual label representing the appropriate class. For this there are many programs available, the one used in this work is called 'labelimg'. The images were segregated into different sets for training and testing. Of the 1042 images acquired, 55 were used for testing, while the remaining 987 images were used to train the above mentioned neural network on a Core i7 2.6 GHz machine with 16 GB of RAM, and a 2 GB Nvidia GTX 950 graphics card. For details related to using TensorFlow to train the chosen network, readers are directed to [13].

The following additional steps need to be performed to ensure that the labeled images, the neural network, and required classes are all available to TensorFlow. The labeled data from 'labelimg' is output in XML format. This first needs to be converted to TFrecords (TensorFlow Records) format, to be readable by the designed neural network. Once these files are ready, the configuration file provided by TensorFlow is modified to include our chosen neural network model. Also at this stage, the required classes (i.e. polluted and clean insulator), and the batch size to be processed per step by the neural network (i.e. number of pictures fed into the network per step), are to be provided. A batch size of twenty is used for this work, and an additional supplementary '.pbtxt' file is required to be created, which contains the items IDs and labels for each class/object desired. This is done by appending a line of the form "item {id: n name: 'MyClass'}" would create a class with ID 'n', and name 'MyClass', here 'n' is a positive integer. Note that the above steps are not only required for testing, but also for being able to train the chosen neural network. Details related to using TensorFlow for detecting objects in a video feed, are available in [12].

3.2 Experimental Results

After performing the above mentioned steps, the trained neural network is tested by feeding it images taken by the drone's camera as it flies. Most voted classification is used, and a Bayesian filter is not used in combination with the neural network developed. The model output scores of the classes, and the highest percentage associated with a class was chosen as the classification. The neural network takes the video feed as the input, the video feed is at the rate of 25 frames per second, so essentially multiple images are classified per second. Examples are shown in Figs. 6a, and b, which show polluted and clean insulators being detected correctly. The labels on the pictures also show the confidence of the classification, i.e. 99%, and 91% respectively. As seen from Figs. 6a, and b, the drone is quite a distance away from the insulator, and is above the level of the insulator, and also there are other objects in the background, which have similar shapes, or colors compared to polluted insulators, and yet, the neural network correctly detected the insulators. A total of 33 such images containing clean insulators, and 27 images containing polluted insulators, which were never seen by the neural network before, were fed to the neural network to test it. The results are provided in Table 2, where it is seen that out of 33 images containing a clean insulator, 31 were classified correctly and 2 were not classified. Furthermore, out of 27 images containing a polluted insulator, 22 were classified correctly and 5 were not classified. The results are quite substantial as the clean

(a) Polluted insulator detected. (b) Clean insulator detected.

Fig. 6. Real-time insulator detection and classification results.

Table 2. Neural network based insulator classification results.

		Neural network based detection results		
	↓ **Class** →	Unclassified	Clean Insulator	Polluted Insulator
Ground	Clean Insulator	2	31	0
Truth	Polluted insulator	5	0	22

insulator was classified correctly 94% of the time and the polluted insulator was classified correctly 81% of the time. The accuracy can be improved by acquiring more training data and retraining the model.

The next set of results shown in Figs. 7a, and b focus on the flight related aspects along with the reporting of the location of polluted insulators. Figure 7a shows the drone in the top left corner, approaching a GPS waypoint near the tower setup with a polluted insulator, as seen in the bottom-right corner of Fig. 7a. The actual mission related to this is shown in Fig. 7b. The yellow curve is the expected path, the green tear drop shaped markings are the loiter points i.e. GPS waypoints where the drone is supposed to hover for ten seconds and send its video feed to the ground station for insulator detection and classification. The magenta colored curve shows the actual path taken by the drone. The location of the tower with the polluted insulator is shown near the second waypoint, marked with a star. Also as marked in Fig. 7b, the drone maintains a distance of 5.23 m from the tower. This distance is greater than the 3 m clearance required between the tower and the drone, to avoid electrical interference issues, and related risks. Note that for the class of relatively lower voltage distribution lines that this system is aimed towards, a 3 m separation is sufficient. Because the drone maintains a distance more than the required minimum, so this is a good result. The drone in this test identifies the insulator as polluted, the output for which appears similar to the result shown in Fig. 6a. Also the position

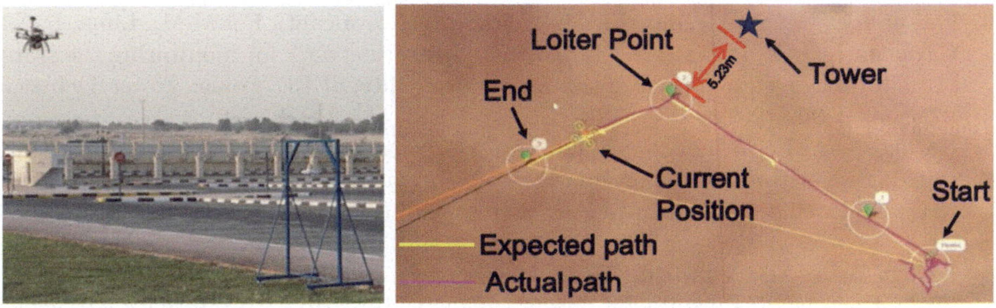

(a) Drone loitering near tower. (b) GPS waypoint following results.

Fig. 7. Drone flight test results.

corresponding to the magenta point near the second loiter point is recorded by the GCS, once the polluted insulator is detected.

4 Conclusion

In summary, this paper reports on the initial stages of the development of a complete system for automatic detection of polluted and clean insulators on overhead transmission lines. The proposed system uses a drone equipped with an autopilot, which lets the drone navigate to, and hover around waypoints provided near overhead transmission line towers, while maintaining the necessary clearance from the transmission lines. As the drone hovers near a waypoint, it transmits a video feed from its camera to a ground control station, where a pre-trained deep neural network identifies and classifies insulators appearing in the video feed, as polluted or clean. Upon locating a polluted insulator, the drone's position is recorded by the ground control station. The system developed has shown a 94% accuracy in detecting clean insulators, and an 81% accuracy in detecting polluted insulators. Both polluted and clean insulators are classified because detecting levels of pollution on the insulators are of interest. The above system was tested on strings of insulators and showed promise although full testing and evaluation is left for future efforts.

References

1. Cavallini, A., Chandrasekar, S., Montanari, G.C., Puletti, F.: Inferring ceramic insulator pollution by an innovative approach resorting to PD detection. IEEE Trans. Dielectr. Electr. Insul. **14**(1), 23–29 (2007)
2. Charniak, E.: Introduction to Deep Learning. The MIT Press, Cambridge (2019)
3. Fangzheng, Z., Wanguo, W., Yabo, Z., Peng, L., Qiaoyun, L., Lingao, J.: Automatic diagnosis system of transmission line abnormalities and defects based on UAV. In: 2016 4th International Conference on Applied Robotics for the Power Industry (CARPI), pp. 1–5, October 2016

4. Fontana, E., Martins-Filho, J.F., Oliveira, S.C., Cavalcanti, F.J.M.M., Lima, R.A., Cavalcanti, G.O., Prata, T.L., Lima, R.B.: Sensor network for monitoring the state of pollution of high-voltage insulators via satellite. IEEE Trans. Power Delivery **27**(2), 953–962 (2012)
5. Karakose, E.: Performance evaluation of electrical transmission line detection and tracking algorithms based on image processing using UAV. In: 2017 International Artificial Intelligence and Data Processing Symposium (IDAP), pp. 1–5, September 2017
6. Khalyasmaa, A.I., Dmitriev, S.A., Romanov, A.M.: Robotic intelligence laboratory for overhead transmission lines assessment. In: 2016 57th International Scientific Conference on Power and Electrical Engineering of Riga Technical University (RTUCON), pp. 1–6 (2016)
7. Li, H., Wang, B., Liu, L., Tian, G., Zheng, T., Zhang, J.: The design and application of SmartCopter: an unmanned helicopter based robot for transmission line inspection. In: 2013 Chinese Automation Congress, pp. 697–702 (2013)
8. Lijun, J., Jianyong, A., Tian, Z., Kai, G., Hua, H.: Pollution state detection of insulators based on multisource imaging and information fusion. In: 2016 IEEE International Conference on Dielectrics (ICD), pp. 544–547, July 2016
9. Liu, W., Anguelov, D., Erhan, D., Szegedy, C., Reed, S., Fu, C.Y., Berg, A.C.: SSD: single shot MultiBox detector. In: Computer Vision – ECCV 2016, pp. 21–37. Springer International Publishing (2016)
10. Lv, L., Li, S., Wang, H., Jin, L.: An approach for fault monitoring of insulators based on image tracking. In: 2017 24th International Conference on Mechatronics and Machine Vision in Practice (M2VIP), pp. 1–6, November 2017
11. Mission Planner. http://ardupilot.org/planner/
12. Streaming Object Detection Video - Tensorflow Object Detection API Tutorial. https://pythonprogramming.net/video-tensorflow-object-detection-api-tutorial/?completed=/introduction-use-tensorflow-object-detection-api-tutorial/
13. TensorFlow. https://www.tensorflow.org/
14. Tiantian, Y., Yang, G., Yu, J.: Feature fusion based insulator detection for aerial inspection. In: 36th Chinese Control Conference (CCC), pp. 10972–10977 (2017)
15. Wang, L., Wang, H.: A survey on insulator inspection robots for power transmission lines. In: 2016 4th International Conference on Applied Robotics for the Power Industry (CARPI), pp. 1–6, October 2016
16. Werneck, M.M., dos Santos, D.M., de Carvalho, C.C., de Nazaré, F.V.B., da Silva Barros Allil, R.C.: Detection and monitoring of leakage currents in power transmission insulators. IEEE Sens. J. **15**(3), 1338–1346 (2015)

Aerodynamic Effects in Multirotors Flying Close to Obstacles: Modelling and Mapping

P. J. Sanchez-Cuevas(✉), Victor Martín, Guillermo Heredia, and Aníbal Ollero

GRVC - Robotics Lab Seville, University of Seville, Seville, Spain
psanchez16@us.es

Abstract. This paper aims to model the aerodynamic effects in the flight of aerial robots close to obstacles in the oil and gas industries. These models are presented in the form of an aerodynamic effects map which represents the changes in the thrust when an aerial vehicle flies very close to different obstacles. Although there are works related to the fly close to different obstacles in the literature, some of the effects needed to develop the aerodynamic map have not been previously studied and tested experimentally in a test stand. The paper also considers the case where the rotor is affected by more than one obstacle.

Keywords: Aerodynamic effects · UAS for inspection

1 Introduction

The application range of unmanned aerial vehicles (UAV) has growing quickly in the last decade [1]. Although in most of these applications the UAV accomplish perceptive tasks such as exploration, monitoring or surveillance among others, there are some of them that directly involve interaction between the UAV and the environment and which are mainly carried out by aerial manipulators [2–4]. These are a new concept of UAV with an integrated robotic manipulator which are used in tasks as contact inspection and sensor installation in inspection and maintenance (I&M) of infrastructures or industrial plants [5, 6].

These I&M tasks usually require that the aerial platform flies very close to different obstacles, for instance, the AEROARMS [7] and HYFLIERS [8] project are focused on aerial manipulation for outdoors applications in I&M in oil and gas plants or the RESIST [9] project which is about the I&M of large civil infrastructures as bridges or tunnels using UAV. Both cases imply that the aerial platforms need to fly in the proximity or even maintain a contact with a building to carry out the I&M operations. Moreover, they need to do it without lacking accuracy or safety conditions during the operation. However, these kind of situations, which involve an UAV flying close to different structures, surfaces, or obstacles in general, changes the flow field surrounding the vehicle leading to changes in the force and torque developed by the rotors which can significantly change the performance of the aerial platform [10] decreasing the accuracy during the inspection operation.

M. F. Silva et al. (Eds.): ROBOT 2019, AISC 1092, pp. 63–74, 2020.
https://doi.org/10.1007/978-3-030-35990-4_6

These aerodynamic effects have been previously studied [11–14] and also by the authors. In [15] a general overview of these kinds of effects was presented and some of them were specifically studied for different applications like the ground effect in [16] or ceiling effect in [6, 17]. These previous studies have shown that it is usually necessary to model this aerodynamic behaviour to guarantee that the final application produces a results which are good enough even flying very close to obstacles.

Thus, this paper is focused on the next step, which is an aerodynamic characterization of the environment and its aerodynamic effect. So, the main contribution is to present and develop a method which allows the generation of an aerodynamic effects map in an environment with multiple obstacles. These maps will be the base for future works in term of studying control solutions or planning methods in complex environments with multiple obstacles.

This paper is structured as follows: Sect. 2 introduces the problem analyzed in the paper and presents a brief compilation of the previous results. Section 3 is focused on the experimental modelling of the aerodynamic effects which has not been previously studied and can arise in the typical scenarios of the aerial manipulation. Section 4 introduces the assumptions that have been taken into consideration as well as the mapping results of a sample scenario. Last, Sect. 5 is about the conclusion and the future works in which the results of this paper can be exploited.

2 Previous Results

This section analyzes the problem of flying close to different obstacles presenting how these effects are included in the dynamic model of a multirotor. Moreover, the previous results related to the aerodynamic effects close to obstacles are presented and novel results in the typical scenarios of I&M applications.

2.1 Dynamic Model with Aerodynamic Effects

The dynamic equations of a multirotor are known in the bibliography as follow:

$$M(\xi)\ddot{\xi} + C\left(\xi, \dot{\xi}\right)\dot{\xi} + G(\xi) = F + F_{ext}$$

Where ξ is the state vector $\xi = [x\, y\, z\, \phi\, \theta\, \psi]'$, M is the generalized inertia matrix, C is the Coriolis and centrifugal terms, G represents the gravity component, F is the generalized vector force developed by the rotors and F_{ext} are the external and unknown forces. However, in case of flying under the influence of an aerodynamic effect, the generalized vector force changes because the forces change due to the relative position between the multirotor and the different obstacles of the environment, so the generalized vector forces is a function of the state of the multirotor, $F = F(\xi)$, such as it has been previously presented in other authors' papers [12]. Thus, the dynamic model is rewritten as:

$$M(\xi)\ddot{\xi} + C\left(\xi, \dot{\xi}\right) + G(\xi) = F(\xi) + F_{ext}$$

This is the reason that justifies the need of model how the aerodynamic effect changes depending on the position of aerial platform in a scenario and specifically with the relative position respect to an obstacle.

2.2 Previous Aerodynamic Effect Results

Figure 1 shows the most common results of the different aerodynamic effects which have been previously studied by the authors and the literature in general. These experimental results show how the thrust of a rotor changes working close to a ground, a ceiling or a wall surfaces. In the figure, T_{IGE}/T_{OGE} is the relation between the thrust "In Ground Effect" and "Out Ground Effect", T_{ICE}/T_{OCE} is the same but in term of ceiling effect and last, T_{IWE}/T_{OWE} aims to model the changes that appears in wall effect. On the other hand, z/R is the distance from the rotor to the obstacle dimensionless with the rotor radius.

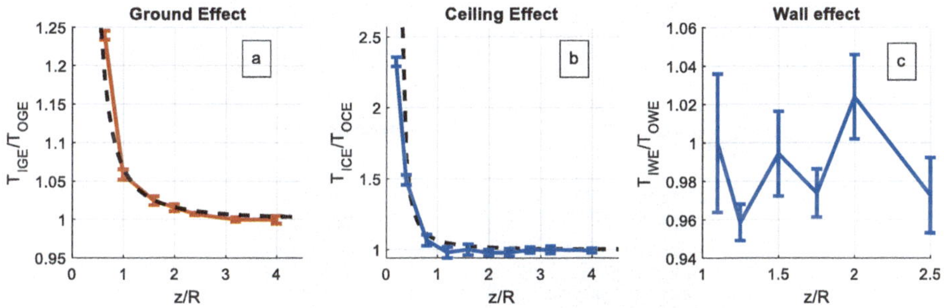

Fig. 1. Previous results in aerodynamic effect in different situations: (a) Ground effect; (b) Ceiling effect; (c) Wall effect.

Ground Effect

The ground effect, which is presented in Fig. 1a, is the most known aerodynamic effect and it is also the widest studied in the literature. This effect arises when an aerial platform flies over a flat surface which acts as a ground. In the aerial robots, the ground effect not only appears in the take-off and landing maneuver, but also when the aerial platform need to fly over a surface, for example, during an inspection or manipulation task.

Ceiling Effect

The ceiling effect, which is shown in Fig. 1b, appears when an aerial platform flies under a surface but very close to it. The results of the Fig. 1b, show that the behaviour of the ceiling effect is very abrupt and unlike that the ground effect which pushes the vehicle away from the obstacle, the ceiling effect pull it toward the obstacle leading to an unsafe flight condition if it is not taken into account.

Wall Effect
Last, the experimental results of the Fig. 1c shows that the wall effect can be considered negligible following the assumption of the helicopters theory which assumes that the flow is perpendicular to the rotor plane.

Paper Contribution
Although these previous results could be a starting point to model the aerodynamic effects flying close to obstacles, this paper goes beyond trying to model not only the aerodynamic effect over or under an obstacle, but also how this effect starts to be significant as a rotor approaches them.

Moreover, due to this paper is focusing on an I&M of an oil and gas industry this paper also models the behaviour of rotors working close to tubular obstacles like pipes.

Next section presents the methodology followed by the experimental results of the different cases of study of this research.

3 Experimental Modelling

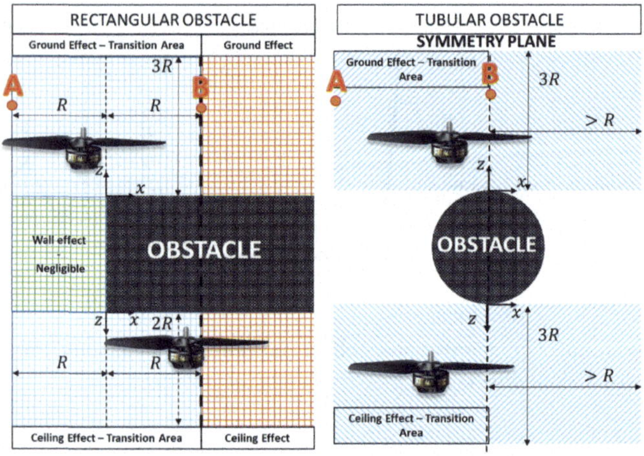

Fig. 2. Test configuration and nomenclature

Due to the typical scenario in an oil and gas inspection application includes two different obstacles, which are flat surfaces like grounds or ceilings and tubular obstacles like pipes, these will be the one taken into consideration on this work. Moreover, in contrast to the classical studies that only model the aerodynamic effect like a rotor working under or over an obstacle, this paper also models the transition part when the rotor approaches the obstacle from the rotor it is out of the effect till it is fully affected by it. These results will be obtained through several experiments in a test stand which is able to measure the thrust of the rotor. In the rectangular obstacle, the area in which the different experiments are carried out are represented in the blue areas in Fig. 2. This area starts when the center of the rotor placed at one radius of distance in the x axis of the Fig. 2 and the propeller is completely out of the obstacle (point A) and finish when the propeller is full placed over the obstacle (point B). The experiments in the tubular obstacles are also accomplished in the blue areas taking into account the symmetry conditions of the problem.

Experimental Procedure
In order to grant that the results can be compared between them, it has been necessary to define an experimental procedure common to all the experiments. In this research, a test bench with a load cell connected to an Arduino Mega 2560 which maintain a serial communication with a computer running MATLAB. In this PC, there is a graphic user interface (GUI) in which it is possible to define the experimental setup. Figure 3 shows the program developed to unify the experimental conditions and the settings used during the experiments.

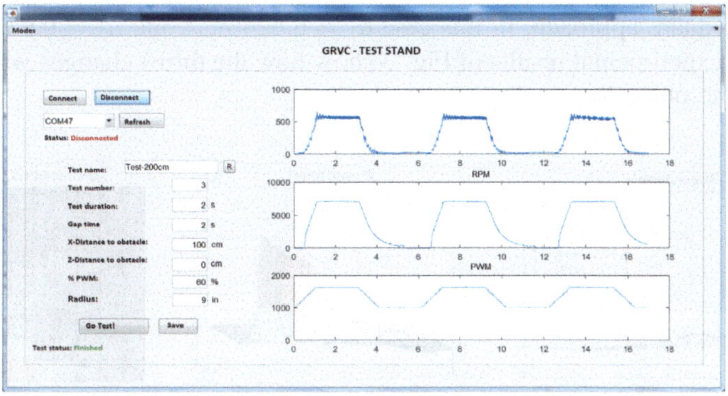

Fig. 3. Graphic user interface developed and used during the experiments.

Figure 3 shows the GUI used to unify the experimental procedure, this system allows to select the most important variables of the experiment and show the results in the graphs placed on the right side. Some of the values which can be settable are the $\%PWM$ of the rotor, the number of iteration, the time in steady state and the gap between the different experiments.

3.1 Aerodynamic Effects Close to Flat Surfaces

Following the experimental procedure presented before, the ground and the ceiling effect has been studied taking into account the behaviour when the rotor approaches the obstacle. Due to the results of these aerodynamic effects have been previously studied in the literature and by the authors when the rotor is completely under the aerodynamic effect, it will be assumed that the effect follows the classical models because they have been validated previously. Thus, the theoretical model the ground effect presented in

[18] and the experimental one of [19] for the ceiling effect will be used to create the aerodynamic effect map. The expression of both models are presented as follows:

$$\text{Ground Effect}: \frac{T_{\{IGE\}}}{T_{\{OGE\}}} = \frac{1}{1 - \frac{1}{16}\left(\frac{R}{z}\right)^2}$$

$$\text{Ceiling Effect}: \frac{T_{\{ICE\}}}{T_{\{OCE\}}} = \frac{1}{1 - \frac{1}{a_1}\left(\frac{R}{a_2 + z}\right)^2}$$

Where the acronyms *ICE*, *OCE*, *IGE* and *OGE* are "in/out ceiling/ground effect" respectively. R is the radius of the rotor and z is the distance between the rotor plane to the obstacle as it is presented in the Fig. 2, in this case, we assume that the rotor is totally over the obstacle. The values of a_1 and a_2 were obtained through an experimental least square approach, in this case $a_1 = 6.924\ \text{m}^{-1}$ and $a_2 = 0.03782$ m.

Last, the experimental results of Fig. 3 show how the thrust changes when the rotor approaches the obstacle.

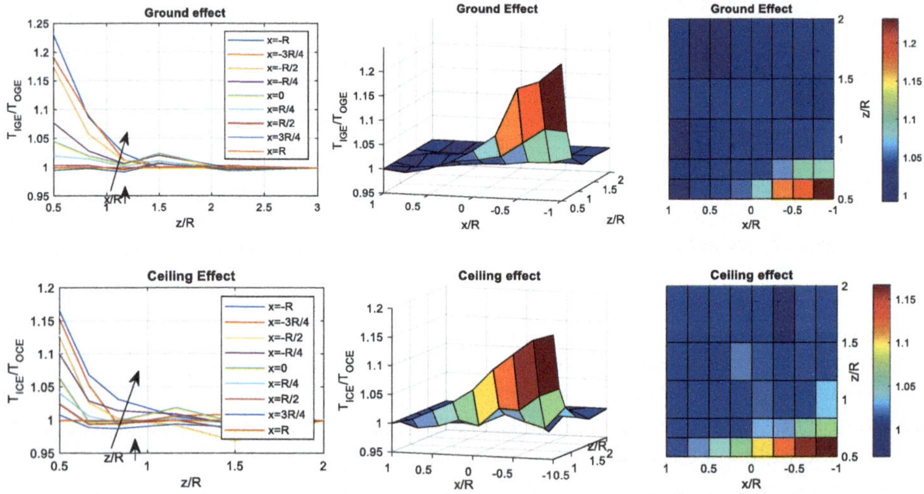

Fig. 4. Ground and ceiling effect with a flat surface - experimental results

The results of Fig. 4 show the differences of the aerodynamic effect when a rotor approaches ground or ceiling obstacles. These results show that for the tested distances the ground effect is stronger than the ceiling which is in line with the previous results of the literature as it is presented in Fig. 1 in the Sect. 2.2. This figure shows the evolution of the ground and the ceiling effect across the longitudinal coordinate but also combines it with the vertical one.

3.2 Aerodynamic Effect Close to Pipes

The results of the aerodynamic effect which arises when a rotor is working very close to tubular objects like pipes are very relevant from the I&M of oil and gas industry point of view. In this case, it is assumed that the results will be symmetric due to the geometry of the problem, thus, the experiments will be tested with a rotor approaching the middle point of the tube as it is presented in Fig. 2 and following the experimental procedure previously established the results are shown in Fig. 5

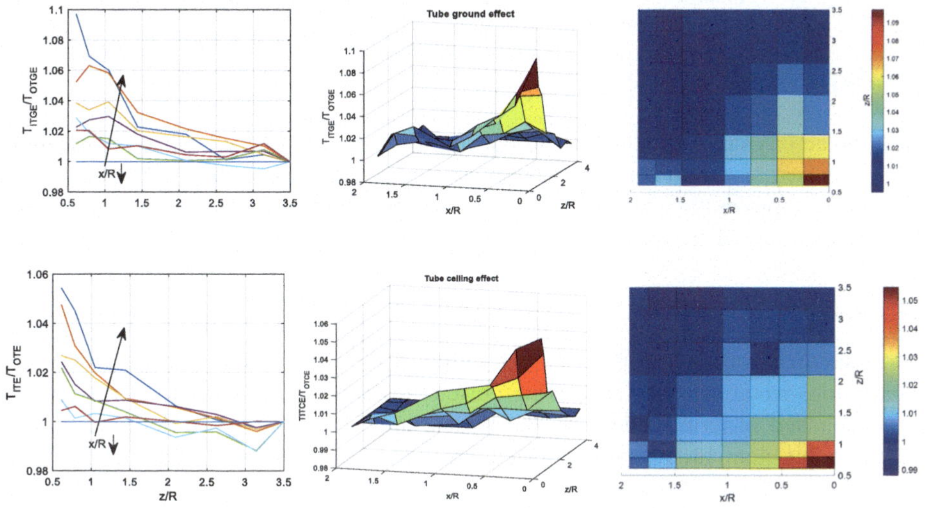

Fig. 5. Tube-ground and tube-ceiling effect - experimental results

Figure 5 shows that the behaviour of the aerodynamic effect is the same respect to the obstacle, however, the magnitude is lower. This is the expected results because the wake of the rotor has more space and the changes in the flow field are lower. Nevertheless, during the experiments it was clear that the aerodynamic effect produced by a pipe not only depends on the relative position of the rotor respect to the pipe, but also depends on the relative size between them. A single test with two different configurations was accomplished and their results are shown in Fig. 6 where it is possible to observe that if the rotor is very small respect to the pipe (blue line) the aerodynamic effect is more similar to the effect close to a flat surface and vice versa.

However, in this paper the aerodynamic map and the most of the test have assumed that the diameter of the rotor is 9 inches and the diameter of the pipe is 6 inches because it is the size of the most of tubes in an oil and gas industrial environment and it also was the recommendation of the industry end-users.

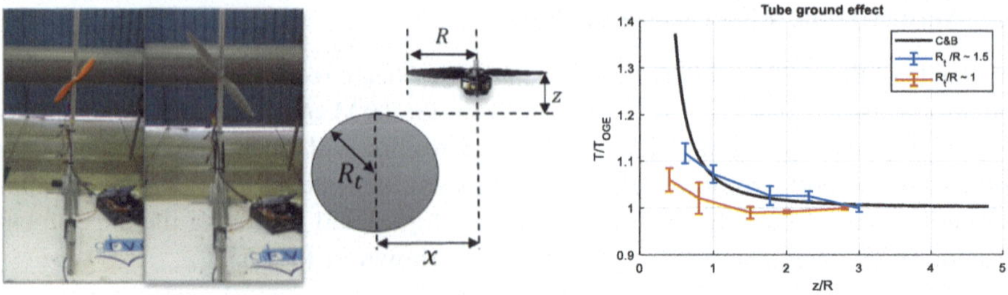

Fig. 6. Comparison with different relation between the pipe and rotor diameter

4 Mapping

Once the aerodynamic effects close to flat surfaces and tubes have been experimentally tested, the next step is to combine them in an aerodynamic map which could be used in a future to improve the control strategies or the planning methods taking into account this aerodynamic effect map.

4.1 Assumptions

This section is focused on defining the limits to the flying area and to establish the assumptions when the rotor is working close to different obstacles. These assumptions are established in term of defining the flyable area during the operation and how to solve the problem in the points which are under the influence of more than one obstacle.

Flyable Area

Figure 7 shows the flyable areas close to a flat/rectangular obstacle and a tubular one. This area establishes the limits of the map and envelope the operation area of the aerial platform.

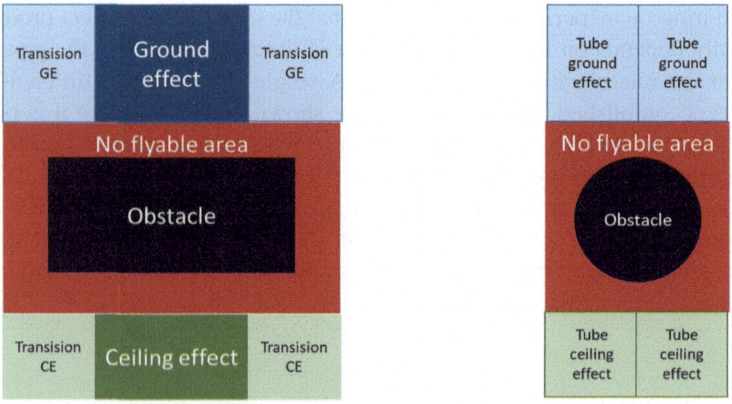

Fig. 7. Flyable area detail and forbidden zones.

The map around these obstacles will be model as it is shown in the Fig. 8.

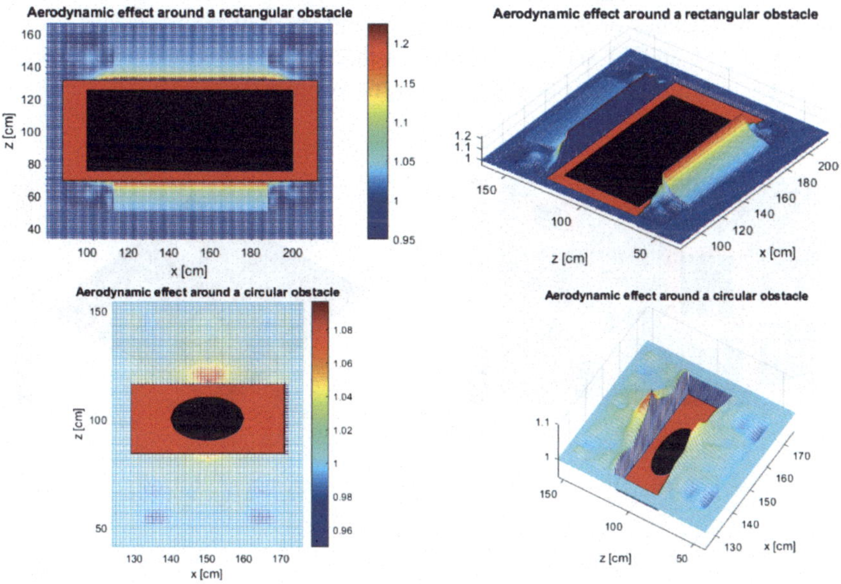

Fig. 8. Sample of aerodynamic effect close to a rectangular obstacle (left) and a pipe (right)

The flat surface where the value of the non-dimensional thrust is 0.95 is out of the flying area but it has been defined with this value to avoid infinite elements in the plot.

Obstacle Overlapping

The assumption to model the aerodynamic effect in the points of the map which are affected by more than one obstacle is that it is possible to apply the principle of superposition, however, it is assumed that the superposition method is no longer valid if one obstacle is in the shadow of another one. This is clearly explained in the Fig. 9, where the red area shows the influence area of the obstacle (a) and the blue area the influence of the obstacle (b). In the section in purple, it is where the principle of superposition is applicable and the grey zone is a shadow area and shows an area in which the aerodynamic effect of the obstacle (a) is considered blocked by the presence of the obstacle (b).

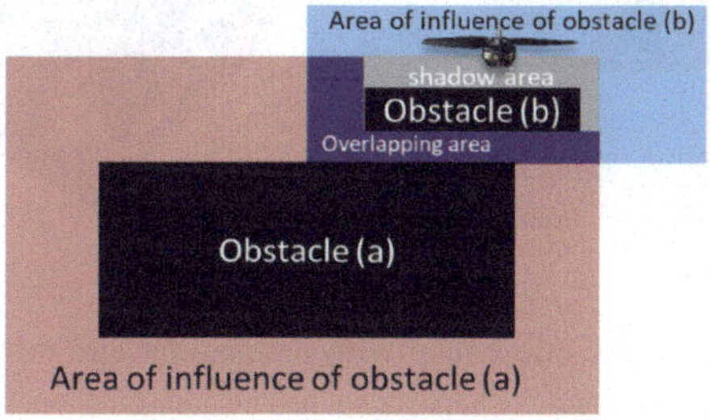

Fig. 9. Obstacle overlapping and shadows conditions

4.2 Results

Last, this section presents the results of different aerodynamic maps in which it is represented the changes of the thrust due to the influence of the obstacles.

Figures 10, 11 and 12 show the results in three different scenarios which can be used later to design different planning or control techniques.

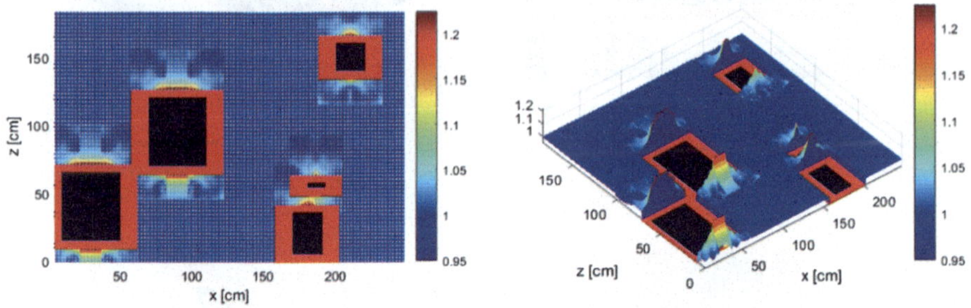

Fig. 10. Aerodynamic effect map with rectangular obstacles

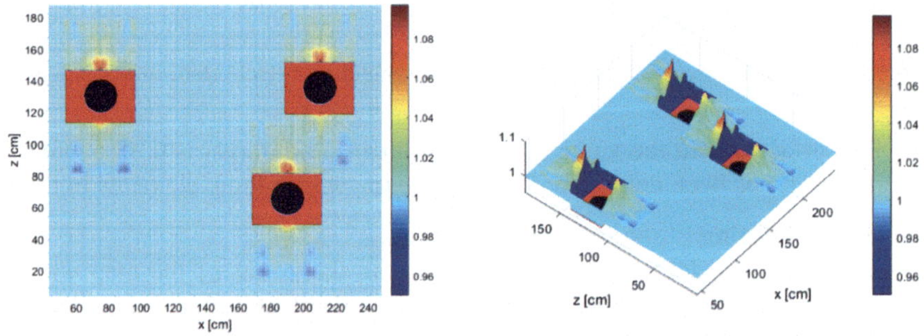

Fig. 11. Aerodynamic effect map with tubular obstacles

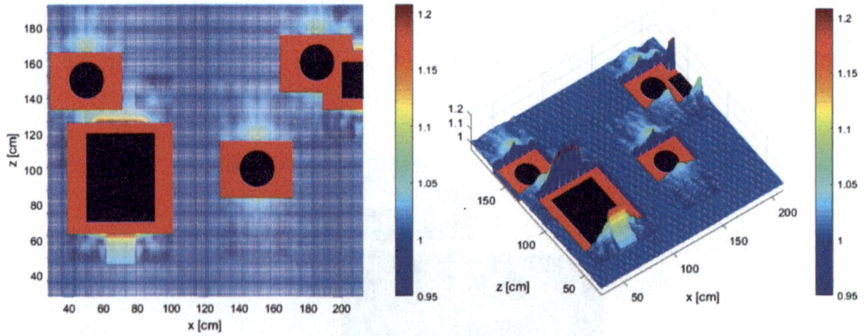

Fig. 12. Aerodynamic effect map with rectangular and tubular obstacles

5 Conclusions and Future Applications

This paper has presented a new approach to the modelling of the aerodynamic effects that can arise during the operation of an UAV flying close to obstacles in oil and gas plants. This approach consists of creating an aerodynamic effects map which links the relative position of the vehicle with the aerodynamic effect that the environment produces in this point.

The different aerodynamic effects have been independently studied and the different assumptions about the areas under the influence of more than one obstacle are also presented.

Future work related to this research will be focused on the application of this map to design different control techniques or planning methods which take into account the aerodynamic effects on the aerial vehicle to improve its behavior or to optimize the use of resources like the power or time consumption during the operation. This kind of map has a direct application in the direction of the actual application of the UAV in the Inspection and Maintenance of oil and gas industries.

Acknowledgments. This work has been supported by the HYFLIERS (H2020-ICT-25-2016-2017) and RESIST (H2020-MG-2017-769066) projects, funded by the European Commission under the H2020 Programme, the ARCTIC (RTI2018-102224-B-I00) project, funded by the Spanish Ministerio de Economia y Competitividad, the ARM-EXTEND project funded by the Spanish RD plan (DPI2017-89790-R) and the FPU Program, funded by the Spanish Ministerio de Educación, Cultura y Deporte. A special thanks to Ricardo Moreno for his support.

References

1. Valavanis, K., Vachtsevanos, G.: Handbook of Unmanned Aerial Vehicles. Springer, Netherlands (2015)
2. Ruggiero, F., Lippiello, V., Ollero, A.: Aerial manipulation: a literature review. IEEE Robot. Autom. Lett. **3**(3), 1957–1964 (2018)
3. Orsag, M., Korpela, C., Oh, P.: Modeling and control of MM-UAV: mobile manipulating unmanned aerial vehicle. J. Intell. Robot. Syst. Theory Appl. **69**(1–4), 227–240 (2013)
4. Fumagalli, M., Naldi, R., Macchelli, A., et al.: Developing an aerial manipulator prototype: physical interaction with the environment. IEEE Robot. Autom. Mag. **21**(3), 41–50 (2014)
5. Trujillo, M.Á., Martínez-de Dios, J.R., Martín, C., Viguria, A., Ollero, A.: Novel aerial manipulator for accurate and robust industrial NDT contact inspection: a new tool for the oil and gas inspection industry. Sensors **19**(6), 1305 (2019)
6. Sanchez-Cuevas, P.J., Ramon-Soria, P., Arrue, B., Ollero, A., Heredia, G.: Robotic system for inspection by contact of bridge beams using UAVs. Sensors. **19**(2), 305 (2019)
7. https://aeroarms-project.eu/. Accessed 03 Oct 2019
8. https://www.oulu.fi/hyfliers/. Accessed 03 Oct 2019
9. http://www.resistproject.eu/. Accessed 03 Oct 2019
10. Powers, C., Mellinger, D., Kushleyev, A., et al.: Influence of aerodynamics and proximity effects in quadrotor flight. In: Experimental Robotics, pp. 289–302. Springer, Heidelberg (2013)
11. Fradenburgh, E.A.: The helicopter and the ground effect machine. J. Am. Helicopter Soc. **5**(4), 24–33 (1960)

12. Curtiss Jr., H.C., Sun, M., Putman, W.F., Hanker Jr., E.J.: Rotor aerodynamics in ground effect at low advance ratios. J. Am. Helicopter Soc. **29**(1), 48–55 (1984)
13. Lee, T.E., Leishman, J.G., Ramasamy, M.: Fluid dynamics of interacting blade tip vortices with a ground plane. J. Am. Helicopter Soc. **55**(2), 22005–2200516 (2010)
14. Hayden, J.S.: Effect of the ground on helicopter hovering power required. In: Proceedings of the AHS 32nd Forum (1976)
15. Sanchez-Cuevas, P.J., Heredia, G., Ollero, A.: Experimental approach to the aerodynamic effects produced in multirotors flying close to obstacles. In: Iberian Robotics Conference, pp. 742–752 (2017)
16. Sanchez-Cuevas, P.J., Heredia, G., Ollero, A.: Characterization of the aerodynamic ground effect and its influence in multirotor control. Int. J. Aerosp. Eng. **2017**, 1–17 (2017)
17. Jimenez-Cano, A.E., Sanchez-Cuevas, P.J., Grau, P., Ollero, A., Heredia, G.: Contact-based bridge inspection multirotors: design, modelling and control considering the ceiling effect. IEEE Robot. Autom. Lett. **4**(4), 3561–3568 (2019)
18. Cheeseman, I., Bennett, W.: The effect of the ground on a helicopter rotor in forward flight. ARC R&M 3021 (1955)
19. Sanchez-Cuevas, P.J., Heredia, G., Ollero, A.: Multirotor UAS for bridge inspection by contact using the ceiling effect. In: International Conference on Unmanned Aircraft Systems (ICUAS), Miami (2017)

Development of a Semi-autonomous Aerial Vehicle for Sewerage Inspection

Angél R. Castaño, Honorio Romero, Jesús Capitán(✉), Jose Luis Andrade, and Aníbal Ollero

Robotics, Vision and Control Group, University of Seville, Seville, Spain
{castano,jcapitan,jandrade,aollero}@us.es, romero_r@hotmail.com

Abstract. This paper describes the design, development and field tests of an aerial vehicle for semi-autonomous inspection of sewer pipes. This vehicle is able to localize and align itself with respect to the longitudinal axis of the pipe, and to keep a predefined height over the water. The human operator only needs to send high-level commands to move the vehicle forward or backward along the pipe, and this navigation is then done safely in an autonomous manner. We tested our system in a realistic mock-up scenario provided by a city water management company. Our field tests show that the platform is operational and fulfill the requirements for sewer inspections.

Keywords: UAV · Aerial inspection · Indoor navigation

1 Introduction

Sewer systems in major cities are becoming of interest lately for robotics applications. These systems consist of large, underground networks of connected pipes to lead the waste away from the city. Periodic maintenance tasks are required in order to monitor multiple possible issues, such as illegal connections, pipes integrity or *fatbergs* (see examples in Fig. 1). Thus, pipes need to be inspected to check for the existence of cracks or fatbergs. A fatberg is a congealed mass made up of wet wipes and cooking fat. When big enough they can become problematic as pipe obstructions. Recently, severe episodes have been reported in cities like London [5] or Sidmouth (England) [2].

In this environment, visual inspection is typically performed by human operators covering the network of pipes throughout long distances. For instance, EMASESA[1], the company in charge of water management at the city of Seville (Spain), inspects more than 20 km of pipes per year. A large part of these are cylindrical or oval pipes with a diameter not lower than 1.8 m. Moreover, pipes can be partially filled with water up to 50 cm, which makes harder to traverse them.

[1] http://www.emasesa.com.

M. F. Silva et al. (Eds.): ROBOT 2019, AISC 1092, pp. 75–86, 2020.
https://doi.org/10.1007/978-3-030-35990-4_7

Fig. 1. Examples of issues in sewer system. Left, illegal connection in the sewerage network. Right, pipe with a crack.

An interesting option is to use robotic platforms to carry out these inspection tasks within sewer systems. There exist some commercial products like Flyability [3], that can be flown manually to inspect sewer pipes. However, it is not straightforward to fly those platforms aligned and centered within the pipes, in order to keep them above the water. There are also works on building autonomous or semi-autonomous prototypes for sewarage inspection. For instance, the EU-funded SIAR project [4] has already developed a semi-autonomous ground robot [6]. An aerial platform is also being developed in EU-funded ARSI project [1].

In general, *Unmanned Aerial Vehicles* (UAVs) have been widely used for visual infrastructure inspection. Some reviews about these technologies can be seen in [8,9]. In [10], for instance, they present a small-scale UAV to perform inspection in an enclosed industrial environment. Semi-autonomous visual inspection of vessels with UAVs has also been explored [7]. In [12], an algorithm for UAV inspection in indoor or GPS-denied environments is presented. Regarding indoor scenarios, there are many works proposing approaches for UAV navigation. For instance, [14] present a method for multi-sensor fusion for UAV indoor navigation. [11] also propose an approach for UAV navigation in GPS-denied environments. Moreover, other works focus more on UAV operation in confined, indoor spaces [13,15], which is the case for sewer pipes.

In this paper, we propose a novel prototype to inspect the sewer system in Seville. The work is done jointly with EMASESA, which is the company in charge of the water management in the city. We present the design and development of an aerial vehicle that is able to navigate through the pipes system in a semi-autonomous manner, at the same time that a human operator performs visual inspection remotely. In particular, the vehicle can localize and align itself with respect to the pipe longitudinal axis thanks to its onboard sensors, and the human can tele-operate it to move forward or backward.

The main contributions of our paper are the following: (i) we propose a novel design for an aerial vehicle to operate in sewer systems; (ii) we develop localization and navigation functionalities to provide the robot with semi-autonomous capabilities; and (iii) we test our system in a realistic mock-up scenario provided by the company in charge of the sewer maintenance.

The remainder of this paper is structured as follows: Sect. 2 presents a description of the inspection problem and the associated requirements; Sect. 3 details the design of the aerial platform and the sensors onboard; Sect. 4 explains the navigation and control system; Sect. 5 presents experimental results; and Sect. 6 concludes and proposes future work.

2 Problem Description and Requirements

This section describes the requirements of the operation for sewer inspection and how the problem is addressed currently. We focus on the sewerage network of Seville as this will be used later to design and test our robot.

2.1 Current Approach

Currently, sewer inspection in Seville is totally accomplished by human operators. They cover the whole system by moving around in groups of 3 people, which entails remarkable security measures and procedures, as well as a considerable amount of time. Moreover, the procedure to access the sewer pipes with people is tedious and needs to be repeated multiple times at different parts of the sewerage network.

Our objective is to provide an aerial platform that is able to perform some of these tasks with a certain degree of autonomy. Thus, human operators would only need to access pipes themselves when an issue is detected by the robot, reducing considerable security procedures and operation time. Accessing with the robot will require shorter time and less restrictive security measures.

2.2 System Requirements

Given the characteristics of the sewer system to be inspected, we describe here specifications for our system. We aim to build an *Unmanned Aerial Vehicle* (UAV) to navigate through the pipes and inspect them. The UAV would be inserted at different points into the sewer system. Then, it would traverse a pipe segment to perform the inspection and would be extracted at the end of that pipe segment. Each segment to inspect is usually around 50 m long, with an expected operation time no longer then 10 min. Pipes are cylindrical with a diameter of 1.8 m and may be filled with up to 0.5 m of water. While navigating, the UAV should record images and transmit them in real time to a control station with human operators that would inspect visually the pipe. Only in the event of any detected issue, human operators would access the pipe for double-checking.

First, the UAV should be as lightweight as possible so that it can be transported to the access point and deployed easily into the pipes through a manhole (circular shape with a diameter of 80 cm). Second, the inspection of the pipes should be semi-autonomous, so that the human operator does not need special skills to navigate the UAV through the pipe. Thus, the UAV will be guided through the pipe by receiving human, high-level commands such as stop, go forward or go backward, and it should execute them autonomously.

More concretely, the UAV should be able to localize itself in altitude and with respect to the longitudinal axis of the pipe. This will allow it to keep a safe distance above the water and with the pipe walls and ceiling. The longitudinal localization of the UAV within the pipe is not required with high precision, as the human operator will be in charge of forward/backward movements. Moreover, the UAV should be able to navigate safely through this confined environment, partially filled with water, with potential obstacles and slight bends. Apart from avoiding these small objects on its path, the UAV must behave safely in case of communication losses and it should float on water in case of a crash or a malfunction. Last, in order to be able to detect anomalies, the images recorded by the UAV should have enough resolution and quality. An artificial lighting system may be required due to the illumination constraints in the pipe. The video transmission system should also be able to reach the control station in real time.

3 Aerial Platform and System Architecture Design

This section explains the design of the system architecture and the aerial platform with the sensors onboard. The system has two main components as shown in Fig. 2: the aerial platform and the control station. The communication between them is established using a wireless router. The control station is a laptop with a gamepad controller to operate the aerial platform. The aerial platform receives commands from the control station and sends back monitoring and status information.

The aerial platform (see Fig. 3) is based on the frame of the well-known quadcopter DJI F450. This frame has been equipped with propeller guards and polyethylene foam noodles to allow the UAV to float (pink parts in Fig. 3). There are two components which are interconnected to carry out data processing and UAV control: an *Up Board* card and a *Pixracer* autopilot[2] placed below. There are also four *TeraRanger One* lidar sensors[3]. One of them is pointing down to measure the height over the water and feed the altitude controller on the autopilot. The other three (front left, front right and rear right) are placed on the same horizontal plane to measure the distance to the pipe walls and the orientation with respect to the longitudinal axis of the pipe. Moreover, the ESCs are placed under the arms and a Wi-Fi USB adapter provides communication between the UAV and the control station.

The details of the onboard electronics are shown in Fig. 4. A magnetometer and a TeraRanger One lidar are connected to the autopilot using an I2C interface. Also a radio receiver and a telemetry channel are connected to the autopilot for backup control and troubleshooting. The autopilot is connected to the Up Board

[2] https://docs.px4.io/en/flight_controller/pixracer.html.

[3] https://www.terabee.com/shop/lidartofrangenfinders/terarangerone.

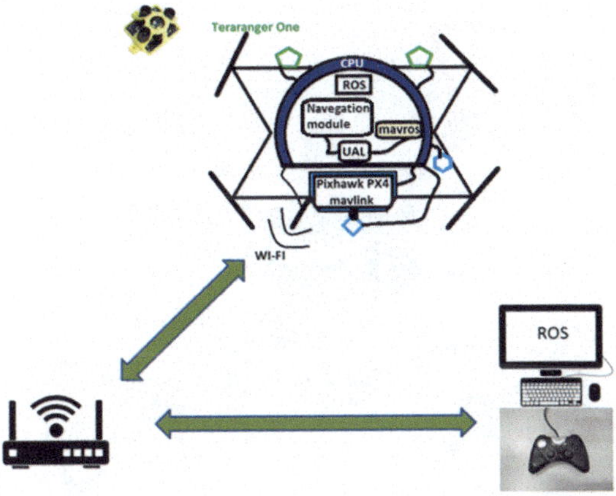

Fig. 2. System architecture for sewarage inspection. The UAV and the control station communicate through a wireless router.

card using a serial connection and the MAVLink protocol. The three remaining lidar sensors are connected to the Up Board through a TeraRanger Hub and a Wi-Fi USB adapter connects the Up Board with the control station. Finally, a system based on 4S LiPo batteries and a 5V BEC converter is used to power the electronics and motors.

The weight of the whole aerial platform is 2.3 kg with a size of 70 cm from tip to tip.

4 Control and Navigation System

This section describes the control and navigation system to operate the UAV in a semi-autonomous mode through the pipes. According to the specifications described in Sect. 2, the control and navigation system has to fulfill the following requirements:

- The aerial platform should be always located as close as possible to the middle of the sewer pipe, aligned with its longitudinal axis and at a predefined height over the water.
- When there is no input from the operator, the aerial platform should keep hovering. The operator can move forward or backward the aerial platform pushing the stick of a gamepad controller through the control station.

The control architecture is based on a standard cascaded position-velocity loop. The inner velocity loop is implemented on the PX4 autopilot[4] while the outer position loop is executed on the Up Board (the position loop of the PX4 autopilot is bypassed) at 50 Hz rate. The parameters and details of each controller are given in the following subsections.

[4] https://dev.px4.io/en/flight_stack/controller_diagrams.html.

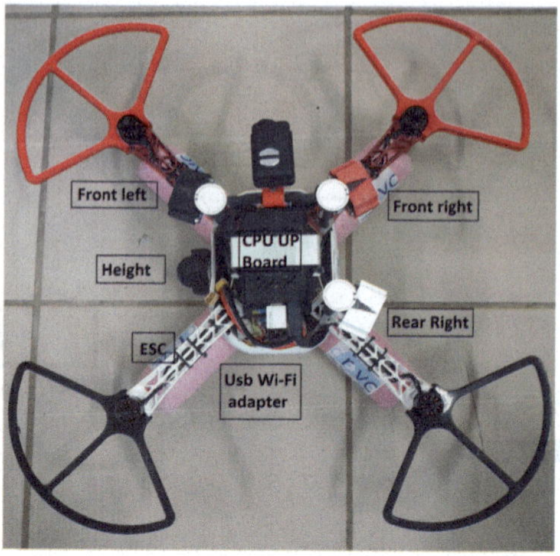

Fig. 3. Our aerial platform based on a DJI F450 frame.

4.1 Forward and Backward Control

If there is no input from the operator, the UAV hovers over the water with zero forward speed. When the operator push the stick in the gamepad, the displacement of the stick is mapped into a desired speed (forward or backward) using a linear function with a saturation at 0.4 m/s. This desired speed is fed as a setpoint to an internal forward velocity PID of the PX4 autopilot.

4.2 Height Control

As shown in Fig. 5, the inner velocity loop of the height controller is implemented in the PX4 autopilot. Its internal EKF (*Extended Kalman Filter*) estimator uses the measurements from the lidar sensor pointing down to estimate the height (h^*). The velocity estimation v^* is also used. Then this estimation and the desired height over the water (h_{sp}) are fed to the *Height PID*. Finally, to avoid sudden altitude changes and guarantee a smooth movement, the velocity setpoint for vertical speed v_{sp} is saturated to ± 0.3 m/s. F_{sp} is the thrust setpoint for the inner UAV attitude controller. There are two additional issues to consider with the height control. First, the measurements of the lidar sensor are affected by the depth of the water and the filth. Several tests have been carried out to tune a model between the measurements and the actual height over the water. Second, the aerial platform will be introduced inside the pipe through the manhole (see Sect. 5 for more details), so the initialization process has to consider that it will start at a variable height over the water.

The parameters used for the Height PID controller are: $P = 0.03$, $I = 0.015$ and $D = 0.015$.

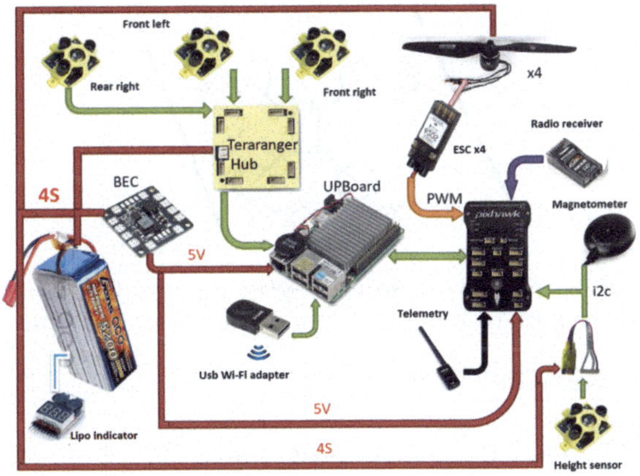

Fig. 4. Electronics and connections onboard the aerial platform.

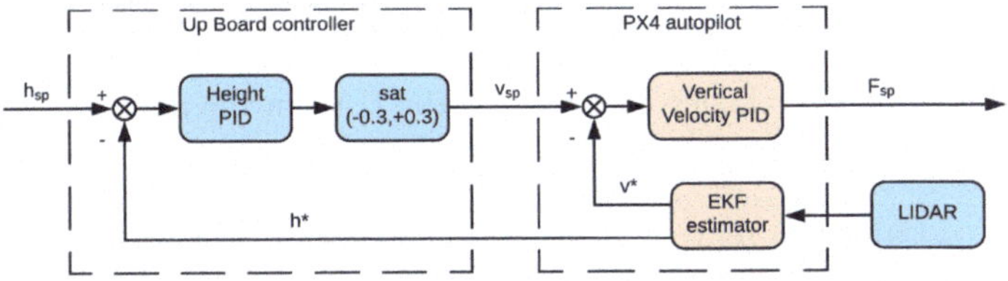

Fig. 5. Cascaded height controller for the UAV.

4.3 Alignment and Centering

The alignment and centering of the UAV inside the pipe is based on the measurements of three lidar sensors (s_{fl}, s_{fr}, s_{rr}) as shown in Fig. 6. Using these sensors the following errors are defined:

$$e_d = s_{fl} - s_{fr} \tag{1}$$

$$e_\theta = s_{fr} - s_{rr} \tag{2}$$

The goal of the control system is to minimize both errors. By minimizing e_d, the UAV tries to locate itself in the middle of the pipe. Minimizing e_θ, the UAV tries to align itself with the longitudinal axis of the pipe. As shown in Figs. 7 and 8 and, the cascaded structure of both controllers are similar. The lateral controller uses the inner velocity PID of the autopilot and an outer PID that tries to minimize e_d. e_d measurements are obtained from the front left (s_{fl}) and front right (s_{fr}) lidar sensors as previously described. Also, the maximum lateral speed setpoint v_{sp} is saturated to ± 0.4 m/s.

The yaw controller (see Fig. 8) tries to minimizes e_θ using an outer *Yaw PID* and the internal *Yaw Rate PID* of the PX4 autopilot. The maximum yaw rate (ω_{sp}) is saturated to ± 0.5 rad/s and the yaw error (e_θ) is obtained from the

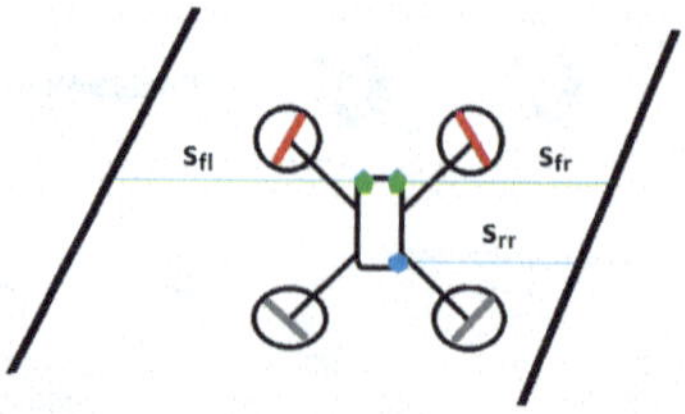

Fig. 6. The lidar sensors measure distances to the pipe walls and are used to estimate the orientation with respect to the longitudinal axis inside the pipe.

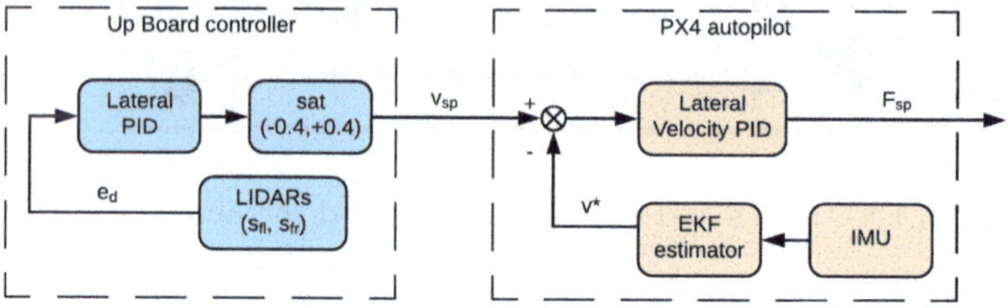

Fig. 7. Cascaded lateral controller to keep the UAV in the middle of the pipe.

front right (s_{fr}) and the rear right (s_{rr}) lidar sensors. The yaw rate estimation (ω^*) is based on the EKF and the measurements from the onboard IMU. The parameters of the Yaw and Lateral PID controllers are the same: $P = 0.00105$, $I = 2.3 \cdot 10^{-7}$ and $D = 0.0013$.

5 Experimental Results

In this section, we describe experimental tests that we run to evaluate the performance of our system. We carried out field experiments in a testbed site provided by EMASESA in Seville (Spain). As it is shown in Fig. 9, this testbed consists of a real sewer pipe placed on the ground. The pipe is 10 m long, 2 m wide and includes a manhole to access inside (like in actual sewer pipes). It can also be filled with water up to a 0.5 m height (see Fig. 9).

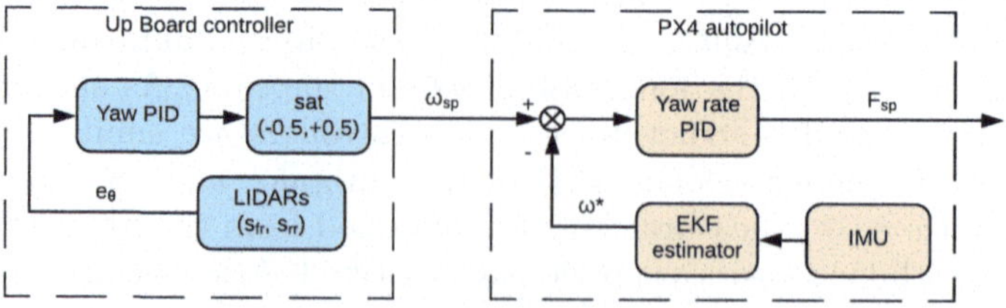

Fig. 8. Cascaded yaw controller to keep the UAV aligned with respect to the longitudinal axis of the pipe.

Fig. 9. Testbed for sewer pipe. Left, pipe with a manhole. Right, our UAV inside the sewer pipe flying above the water.

Several trials have been made to validate the UAV navigation inside the sewer pipe. We obtained similar results in all of them. Figure 10 shows the lateral error e_d and the forward speed during one of our tests, while Fig. 11 shows the control inputs for the yaw rate and lateral speed controllers required to keep the UAV centered and aligned inside the pipe. As it can be seen in Fig. 10, the UAV is not well aligned with the pipe initially, so when the operator abruptly moves forward the UAV (up to 0.4 m/s) the lateral error grows up to 25 cm. Then the controller aligns and centers the UAV to keep it with a maximum lateral error of around 15 cm (from $t = 4$ s to $t = 16$ s). During that time, Fig. 11 depicts that the lateral speed is below 0.2 m/s and the yaw rate is smaller than 0.07 rad/s. From $t = 16$ s to $t = 22$ s the operator moves forward again the UAV. Figures 10 and 11 demonstrate that the maximum error was below 15 cm and the lateral speed below 0.2 m/s. The yaw rate keeps also under 0.07 rad/s. All the results shown in Figs. 10 and 11 are performed in autonomous mode and only the forward/backward command is provided by a human operator.

Additionally, the procedure to insert the UAV through the manhole and an autonomous detachment have been tested. As shown in Fig. 12, a metal pole is introduced through the manhole with the UAV attached at the tip. Once the system is completely deployed, the human operator starts the UAV and the mechanical system allows the UAV to detach itself from the pole autonomously. Figure 13 depicts the height of the UAV during this phase in one of the trials. The UAV is placed 1.5 m above the water and the autonomous detachment process starts at $t = 6$ s. The UAV descends at a maximum vertical speed of 0.3 m/s until a 0.9-m height is reached and then it smoothly stabilises at 0.45 m above the water. It takes 6 seconds (from $t = 6$ s to $t = 12$ s) to complete the complete detachment operation.

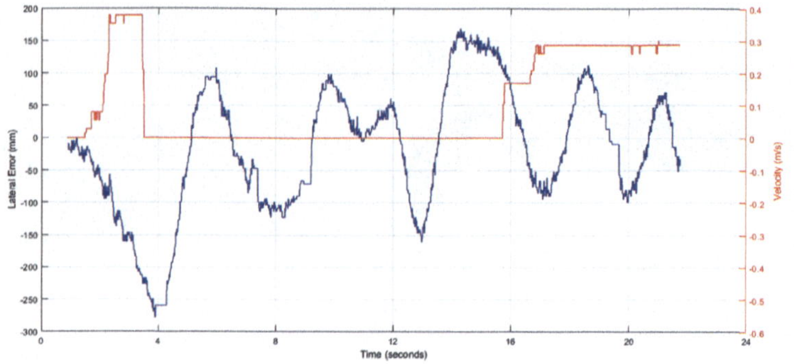

Fig. 10. Lateral error (blue) and forward speed (red) of the UAV navigating inside the sewer pipe.

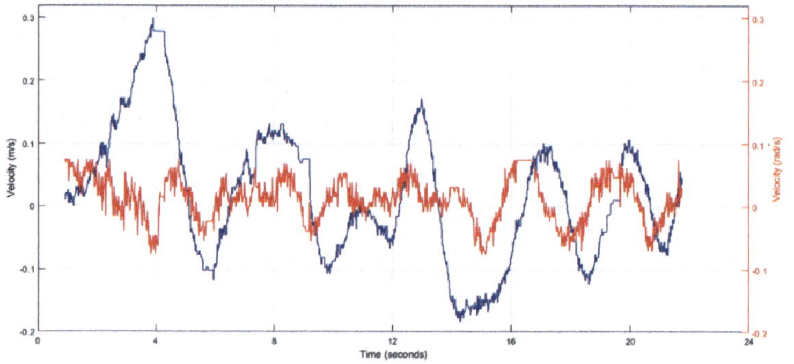

Fig. 11. Control inputs to lateral speed (blue) and yaw rate controller (red) while navigating inside the sewer pipe.

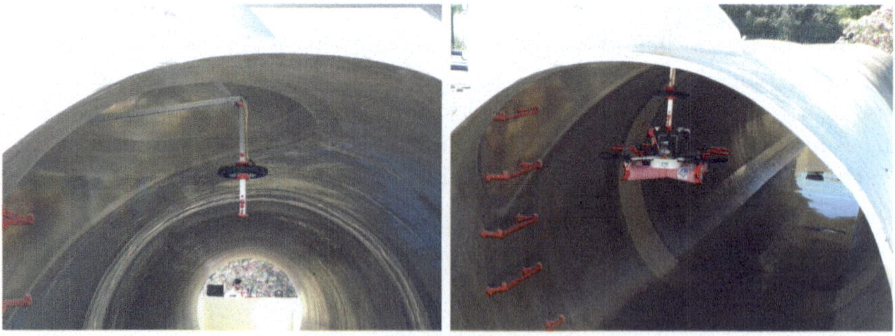

Fig. 12. Insertion operation of the UAV into the pipe. Left, metal pole inserted through the manhole. Right, UAV attached to the metal pole before detaching.

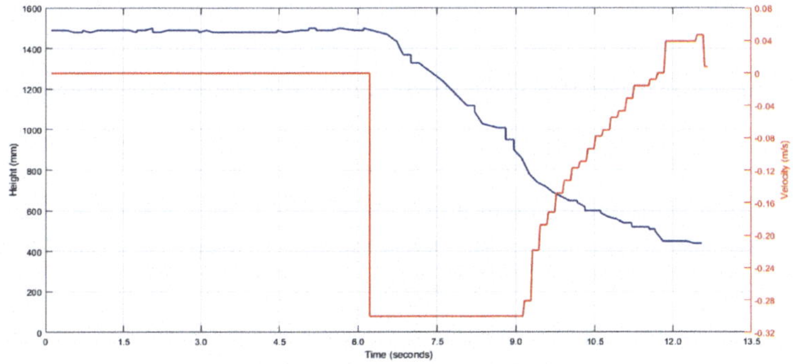

Fig. 13. Height (blue) and vertical speed (red) during an automatic deployment operation through the manhole of the pipe.

6 Conclusions

In this paper, we developed a solution for inspecting sewer systems with a sem-autonomous aerial vehicle. The UAV is able to keep itself aligned with respect to the longitudinal axis of pipes, thus avoiding collision with potential objects coming from the pipe, such as branches, stones or water. Additionally, a human operator can send high-level commands to the UAV with the help of a gamepad in a control station in order to make it move forward or backward through the pipes. If the operator detects a blockage (using the real-time video) she/he could move the UAV backwards to the starting point. In our experiments, we observed that if the forward velocity is abruptly increased at the beginning to its maximum value (0.4 m/s), the separation from the longitudinal axis of the pipe can reach up to 0.25 m, but once our controller centers and aligns the UAV, it can move forward at 0.3 m/s without deviating more than 0.15 m from this longitudinal axis. This allows the UAV to inspect a sewer pipe 50 m long in less than 3 min, keeping distances safe enough with the pipe. Also, we validated the procedure to insert the UAV into the sewer pipe and to start its navigation.

Future work includes the improvement of our UAV prototype, building a smaller platform. Also, we plan further testing of the inspection payload (lighting, cameras, etc.), as well as the design and development of a procedure to recover the UAV from the sewer pipe.

Acknowledgements. This work has been developed in the Spanish SAAIC project and it has received funding from CTA and CDTI. It has also been funded by ARM-EXTEND in the Spanish RD plan (DPI2017-89790-R). We would also like to thank Rafael Salmoral for his support during flying tests.

References

1. ARSI Project. http://echord.eu/essential_grid/arsi-aerial-robot-for-sewer-inspection/index.html
2. BBC News. https://www.bbc.com/news/uk-england-devon-46787461
3. Flyability platform. https://www.flyability.com
4. SIAR Project. https://siar.idmind.pt
5. The Guardian. https://www.theguardian.com/environment/2017/sep/12/total-monster-concrete-fatberg-blocks-london-sewage-system
6. Alejo, D., Mier, G., Marques, C., Caballero, F., Merino, L., Alvito, P.: A ground robot solution for semi-autonomous inspection of visitable sewers. In: Grau, A., Morel, Y., Cecchi, F., Puig-Pey, A. (eds.) ECHORD++: Innovation from LAB to MARKET. Springer, Heidelberg (2019)
7. Bonnin-Pascual, F., Garcia-Fidalgo, E., Ortiz, A.: Semi-autonomous visual inspection of vessels assisted by an unmanned micro aerial vehicle. In: 2012 IEEE/RSJ International Conference on Intelligent Robots and Systems, pp. 3955–3961 (2012)
8. Jordan, S., Moore, J., Box, S.H.J., Perry, J., Kirsche, K., Lewis, D., Tse, Z.T.H.: State-of-the-art technologies for UAV inspections. IET Radar Sonar Navig. **12**, 151–164 (2018)
9. Máthé, K., Buşoniu, L.: Vision and control for UAVs: a survey of general methods and of inexpensive platforms for infrastructure inspection. Sensors **15**(7), 14887–14916 (2015)
10. Nikolic, J., Burri, M., Rehder, J., Leutenegger, S., Huerzeler, C., Siegwart, R.: A UAV system for inspection of industrial facilities. In: 2013 IEEE Aerospace Conference, pp. 1–8, March 2013
11. Perez-Grau, F.J., Ragel, R., Caballero, F., Viguria, A., Ollero, A.: An architecture for robust UAV navigation in GPS-denied areas. J. Field Robot. **35**(1), 121–145 (2018)
12. Sa, I., Hrabar, S., Corke, P.: Inspection of pole-like structures using a vision-controlled VTOL UAV and shared autonomy. In: 2014 IEEE/RSJ International Conference on Intelligent Robots and Systems, pp. 4819–4826 (2014)
13. Shen, S., Michael, N., Kumar, V.: Obtaining liftoff indoors: autonomous navigation in confined indoor environments. IEEE Robot. Autom. Mag. **20**(4), 40–48 (2013)
14. Shen, S., Mulgaonkar, Y., Michael, N., Kumar, V.: Multi-sensor fusion for robust autonomous flight in indoor and outdoor environments with a rotorcraft MAV. In: 2014 IEEE International Conference on Robotics and Automation (ICRA), pp. 4974–4981 (2014)
15. Tripicchio, P., Satler, M., Unetti, M., Avizzano, C.A.: Confined spaces industrial inspection with micro aerial vehicles and laser range finder localization. Int. J. Micro Air Veh. **10**(2), 207–224 (2018)

Proposal of an Augmented Reality Tag UAV Positioning System for Power Line Tower Inspection

Alvaro Rogério Cantieri[1](✉), Marco Aurélio Wehrmeister[2], André Schneider Oliveira[2], José Lima[3], Matheus Ferraz[2], and Guido Szekir[2]

[1] Federal Institute of Paraná, Curitiba, Paraná, Brazil
alvaro.cantieri@ifpr.edu.br
[2] Universidade Tecnológica Federal do Paraná, Curitiba, Paraná, Brazil
{wehrmeister,andreoliveira}@utfpr.edu.br,
matf.ferraz@gmail.com, guidoszekir@gmail.com
[3] CeDRI - Research Centre in Digitalization and Intelligent Robotics, Polytechnic Institute of Bragança and INESC TEC, Bragança, Portugal
jllima@ipb.pt
http://www.ifpr.edu.br, http://www.utfpr.edu.br/,
http://portal3.ipb.pt/index.php/pt/ipb

Abstract. Autonomous inspection Unmanned Aerial Vehicle systems are an essential research area, including power line distribution inspection. Considerable efforts to solve the demanding presented in the autonomous UAV inspection process are present in technical and scientific research. One of these challenges is the precise positioning and fly control of the UAV around the energy structures, which is vital to assure the security of the operation. The most common techniques to achieve precise positioning in UAV fly are Global Positioning Systems with Real-Time Kinematic. This technique demands a proper satellite signal receiving to work appropriately, sometimes hard to achieve. The present work proposes a complementary position data system based on augmented reality tags (AR Tags) to increase the reliability of the UAV fly positioning system. The system application is proposed for energy power tower inspections as an example of use. The adaptation to other inspection tasks is possible whit some small changes. Experimental results have shown that an increase in the position accuracy is accomplished with the use of this schema.

Keywords: Power line inspection UAV · Autonomous UAV inspection · AR Tag UAV position

1 Introduction

The power line inspection process is a mandatory task made periodically by the energy distribution enterprises around the world. It is essential to maintain the

M. F. Silva et al. (Eds.): ROBOT 2019, AISC 1092, pp. 87–98, 2020.
https://doi.org/10.1007/978-3-030-35990-4_8

security of the energy system and avoid energy delivery interruptions. Manual inspection process of power distribution towers is made by human operators that climb the structures to visualize details of the components. This kind of task is demanding and risky to the operator. Manual inspection is being replaced by the use of small-size unmanned aircraft vehicles equipped with image cameras and other sensors, which increases the security and efficacy of the process.

Several enterprises around the world execute inspection services based on UAV. Commonly, the aircraft is remotely controlled by a human pilot, that is responsible for control the navigation around the tower to achieve the necessary visual information of the components. An operation technician that works side by side to the pilot watch the video stream transmitted by the UAV and search for possible fails. The pilot is responsible for precise positioning and collision avoidance during the fly, which brings some restrictions to the flying route and tower proximity, for safety reasons.

An UAV capable of autonomous missions brings security and efficacy to the inspection process. Maintaining the position of the aircraft at the defined points and routes is one of the challenges present on these systems. A possible solution is the use of a Real-Time Kinematic GPS (RTK-GPS) system embedded on the aircraft. RTK-GPS is a technique that calculates the phase difference of two GPS signals received by separate modules and uses it to improve the output data accuracy to centimeter-level. This equipment provides excellent performance, and allows the creation of high secure autonomous position systems for the UAV's. Real-world situations affect the performance of RTK-GPS, like the small number of visible satellites, coverage of the receptor by environment structures, high trees, rain clouds, or other interference elements. When these situations occur, the centimeter-level accuracy is lost, and the navigation may become insecure.

This paper proposes the use of a position measurement methodology using a group of Augmented Reality Tag (AR-TAG), displaced around the tower on the ground. A down-pointing camera embedded on the UAV capture images of the tags, and a ROS package calculate it's position relative to the aircraft. This information to provide a secondary position data to the flight control algorithm.

A set of real-world experiments was executed, to evaluate the accuracy level of the position reading during the fly, measure the influence of environmental conditions, tag dimension, outdoor illumination, camera vibration, tag position and alignment among others.

A simulated experiment was proposed to evaluate the control capability of aircraft using the tags as reference, using the Virtual Robot Experimentation Platform (V-REP) software. The simulated system architecture is based on Robot Operating System (ROS) [11]. An overview of the proposed system is shown in Fig. 1.

Fig. 1. Overview of the system

2 Background and Related Works

UAV power line inspection had grown in number in the last decade. The use of small size remote piloted UAV brings efficacy and security to the process, dispensing the necessity of a human to climb the structures to make the task.

A common way to execute autonomous UAV outdoor fly is to use a Global Navigation Satellite System (GNSS) position system. This technology offers a cheap and straightforward solution to navigate autonomous aircraft, compatible with the most of flight control hardware available on the market. The use of GNSS system to perform precise positioning to an UAV is not recommendable, because of the system accuracy. North America GPS for example, assure a 4-m accuracy performance, as described on [12].

A possible solution to increase the position accuracy of GPS data is the use of RTK -GPS. This technique executes calculations of the phase difference between two GPS signals received in different points, allowing to increase the position accuracy to centimeter-level. This technique demands a communication link between the *base* and the *rover* that receives and calculates the position corrections. The system also needs to receive a minimum number of satellite signals with adequate signal to noise rate to work correctly. The presence of obstacles like buildings, trees, clouds, and communication lost between the modules can cause the decrease of data accuracy level [13].

Some commercial UAV's offers RTK capabilities with autonomous fly. Small RTK modules for drone applications are also available on the global market. This technique is the best choice for precise positioning for UAV, but the problems described justify the development of a new approach for the position problem.

The first UAV autonomous power line inspection systems presented in the literature uses GNSS position system. The proposal's focus is to make the UAV follow a path near to the transmission lines and send images of the structure to a base station. As an example of these systems, the work [10] describes the build and test of two small aircraft, a fix-wing and a tricopter. Both aircraft are controlled trough radio link, and executes flight missions using GPS positioning. These missions consist of autonomous following of a path near to the power line structures, to acquire images of their components.

The work [7] describes a quadrotor helicopter capable of autonomous path following using GPS for power line inspection apply. A regular camera and a thermal infra-red camera capture images of the structure and sends it to the base station. The aircraft performs three flight modes, manual flight, manually GPS assisted attitude fly an autonomous fly.

The propositions described above do not offer high precision positioning to detailed power line inspections. Complementary positioning systems must be added to the aircraft to achieve the position accuracy necessary to perform this task. Position algorithms based on computer vision systems are a possible approach to improve the position accuracy of a UAV autonomous fly. The base of this technique is to acquire and process environment images and identify elements and visual clues, estimating the position of the object relative to the UAV for each new image frame. Artificial intelligence and other techniques are applied to identify the reference points and calculate distances and feed the UAV fly control algorithms.

The work [6] use a stereoscopic image processing to calculate the UAV position related to the tower and power lines. The stereoscopic calculation algorithm uses two consecutive images to evaluate the position of an obstacle. The paper does not presents detailed information about the accuracy of the system to be compared whit other similar propositions.

The work [1] proposes a "Point-Line-Based SLAM" technique to provide a center tower position based on image processing. The system uses two high-performance GPU hardware embedded on the UAV to visual information processing. The results present an average position error of 0.72 m. The disadvantage of the technique is the hardware demanding to process the visual information.

Processing images to calculate objects position is a complex task, which demands a lot of hardware capacity because of the high number of variables to be considered in this calculation. The use of *Artificial Tags* displaced on the environment to position calculation is an excellent approach to decrease the computational demanding of image positioning systems. This technique uses algorithms that process the visual information of specific draws printed, commonly called tags, disposed in known positions. The tags can provide 3D position, orientation, and even additional high-level information like numbers, strings, URL, etc, demanding low computational cost compared with the traditional image processing systems.

Some tags positioning system described in the literature, originally proposed to Augmented Reality applications, has been applied to provide mobile robotic position and orientation data. April Tags [5] and AR Track Alvar [3] are example of these tools. The build of outdoor navigation systems using AR-Tags is a challenging task. Many environment parameters, like sunlight, tag distance, and visual obstacles that shadows the tag, impose difficulties in the correct processing of the tag images. Some recent published works provide background to assure the viability of using the artificial tags to offer outdoor position data to UAV fly.

The work [9] describes an outdoor UAV navigation system where the aircraft follows a moving automobile and executes an autonomous landing, using a group of AR-Tag placed in the automobile roof to get position and orientation data. The proposed system has success in perform correct positioning to the landing even when only using the visual position information provided by the tag.

The paper [4] proposes a UAV vision-based target positioning solution. The work develops a position control algorithm to find a ground target, represented by an and follow it. The authors argue that the solution provides 92% reliability in real-world experiments, showing the viability of using this kind of proposal to UAV outdoor positioning. The paper suggest the application of the solution in autonomous infrastructure inspection system developments, but do not test it in real inspection applications.

In this work, we propose a complementary position solution to apply in UAV autonomous power line detailed inspections, based in AR-Tags placed around the tower to serve as visual reference to the aircraft. The main contribution of this works is to investigate the viability of using AR-Tags in the proposed environmental and technical demands provided by the inspection problem. The second contribution is to provide a first approach solution to the autonomous inspection system, tested through the simulation environment.

3 Architecture Specification

Power line tower inspection is a complex task that demands a carefully visual data collection of several components of the structure and also near area. This project proposes an approach to the single tower inspection problem, a detailed process where the operator needs to visualize components of the tower, like insulators, spacers, dampers, conductors, fixing elements, etc. This visualization demands that the UAV reach some specific positions around the tower, and stay static while the camera is acquiring images of the point of interest.

This work is based on information exchange between the research group and the local energy distribution enterprise. It was essential to define clearly the operational parameters of the UAV inspection system proposition, like secure distance between the tower and the UAV, tower height, and navigation velocity during the inspection. A first approach was defined to validate the concept, using small size energy towers. Some operational parameters were chosen based on this definition, shown on Table 1.

Table 1. Project parameter values defined to the system

Parameter	Value
Ideal distance between UAV tower during inspection	4 m
Minimal security distance between the UAV and the tower	2 m
Horizontal range of Tags visualization during the fly	0 m–7 m
Vertical range of Tags visualization during the fly	10 m–40 m
Maxim height of the inspected tower	30 m
Maxim UAV velocity of displacement during inspection	0.5 m/s
Range of UAV inspection around the tower	360°

3.1 System Components Description

The system is composed of an UAV, a base station and four Augmented Reality tags disposed on the ground around power line tower, as shown in Fig. 1. The UAV camera captures the Tags image and sends it to the base station computer that runs the relative position calculation and transmits this data to the aircraft controller. Simulation and outdoor experiments were executed to evaluate the performance of architecture. The components and tools used to build the solution are described next.

Robot Operating System: System communication is made through Robot Operating System nodes, where the base station computer works as ROS Master [11]. ROS Kinetic runs on a PC that works as a base station, running Ubuntu 16.4 LTS. Computer system is a core i7 processor with 16 GB RAM and Intel® HD Graphics 520 (Skylake GT2) board.

Bebop Drone: A Bebop Drone quadcopter running ROS Bebop Autonomy driver was used to evaluate the position accuracy obtained by the proposed solution. Bebop drone is an excellent choice for this kind of test because it offers stabilized fly performance and embedded Full HD resolution camera gimbal stabilized, that minimizes the image displacement during the data acquisition. ROS Bebop Autonomy driver [8] is another advantage of this model, which makes it easy to exchange data between the drone and the base station during those experiments.

Ar Track Alvar Solution: The solution uses Ar-Track Alvar Augmented Reality [3] to implement the reference AR Tag system on this work. It provides flexible usage and excellent computational performance. The solution also allows multi-marker utilization, an advantage to the proposed architecture presented in this paper. The Ar Track Alvar package developed by Scott Niekum offers ROS compatibility. The Images of the AR-TAG are achieved by Bebop camera and processed by this package, that publishes position and orientation data on *ar_pose_marker* Ros Node.

V-REP Simulation Environment: The simulated environment was created on V-REP software to evaluate the proposed YUAV autonomous navigation system performance. V-REP is a flexible and robust robotic simulation platform that offers a set of programming tools and robot component models, making it easy to test algorithms and robotic systems. An hexacopter model with PID position control script, four AR-Tags images displaced on the ground, a down-pointing camera embedded on the hexacopter frame and a power line tower model composes the environment. A ROS master execute all information exchanges between the components.

3.2 Simulation Description

Some modules compose the system, performing specific tasks on the simulation. All the simulation architecture system runs using ROS interface. The components exchange information between ROS to execute their tasks.

The hexacopter camera captures the AR Tag images and sends them to the package witch processes the calculation of the position of the tag, publishing this information in */visualization_marker* ROS topic.

A C++ code receives this information and filter the position of the nearest tag visualized by the camera. The code performs a reference transformation and publishes the results in */position* and */orientation* ROS topic.

Another C++ code publishes a group of waypoints on a */mission* topic, one at time, to provide the position and orientation points that must be reached by the hexacopter in a mission simulation. Each point is published with a fixed time interval between the last, to assure that the hexacopoter have adequate time to make the displacement between then.

A script written in LUA language runs in V-REP simulator, implementing a PID position controller to the hexacopter. This PID code receives the position calculation and the next waypoint from the C++ codes and performs the velocity calculations to displace the hexacopter to the desired point. This code is simple, and if the position provided is not correct, like when any tag is visible, the control of the aircraft is lost. Figure 2 shows an overview of the ROS topics and nodes running on this system.

4 Experimental Methodology and System Evaluation Tests

The architecture proposal followed methodological steps to guide the system conception. First, outside measurements using a 1-m size tag was run to evaluate the feasibility of the scheme. After that, a simulated environment based on the collected data, to test a control algorithm of the UAV using AR Tag readings. The experiment descriptions and results are shown next.

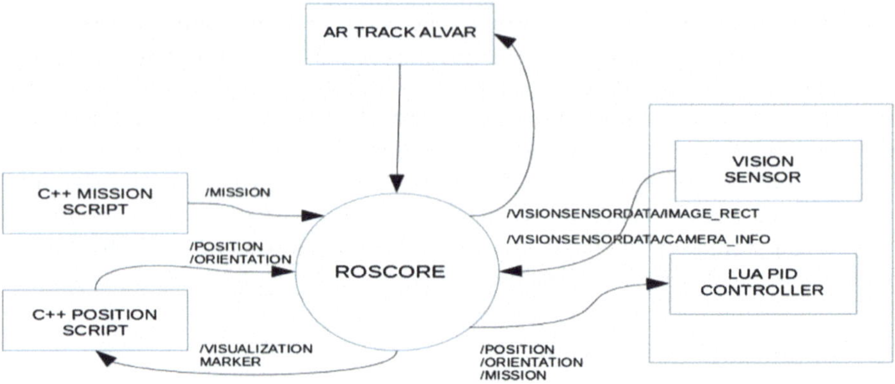

Fig. 2. ROS architecture of the simulation

4.1 Position Reading by the UAV Using AR Tag in an Outdoor Environment

A position accuracy evaluation provided by the AR Tag in an outdoor environment was executed using Bebop drone in a static position. The objective of this experiment is to evaluate if the position readings of the Bebop camera could provide adequate data between the distance range necessary to the system (0 m to 40 m).

The experiment uses a 1.0-m side size printed tag, placed in an outdoor space on a wall, with a 50 m measurement tape reference fixed on the ground. The tag is placed in front of Bebop drone with direct sunlight incidence. Bebop drone was placed on a desk, pointing the camera directly to the tag. The alignment of the drone was made using two wires fixed on the side of the tags and stretched over the entire length of the measuring shaft. The measurements were made in 4.0 m intervals, from 4.0 to 40.0 m distance. Four hundred position measurements samples of *X, Y, and Z* was collected for each interval. Results are presented in the Table 2.

Table 2. Absolute distance error and standard deviation for outdoor long-range tag readings

	DISTANCE									
	4m	8m	12m	16m	20m	24m	28m	32m	36m	40m
X	0.20	0.36	0.50	0.60	0.59	0.69	0.72	0.74	0.79	0.86
Xsd	0.01	0.02	0.04	0.05	0.06	0.08	0.08	0.07	0.08	0.08
Y	0.28	0.38	0.45	0.58	0.68	0.71	0.73	0.77	0.82	0.84
Ysd	0.02	0.03	0.03	0.04	0.04	0.05	0.06	0.07	0.06	0.07
Z	0.15	0.16	0.35	0.39	0.59	0.65	0.72	0.76	0.78	0.88
Zsd	0.01	0.01	0.02	0.04	0.03	0.06	0.06	0.06	0.09	0.1

Calculation of the position standard error for the 40.0-m distance, considered the worst case, is done using the standard deviation of absolute position error, shown on Eq. 1.

$$P.S.E = (Abs\_STD\_Error)/(400samples), \quad (1)$$

The calculation results in a *Position Standard Error (P.S.E)* equal to 0.007 m. Using a confidence interval of 95%, the calculated measurement error results

$$C.I = (0.86) \pm (1.96 \times 0.007) = 0.86 \pm 0.014\,\text{m}, \quad (2)$$

where 0.86 is the mean of absolute position error for the data.

The confidence interval is between 0,86 m and 0,87 m, adequate to estimate the real position of the UAV, considering that a regular GPS has 4.0-m assurance for the same situation.

4.2 Position Accuracy of an UAV Flight Using AR-Tag

The next experiment comprises the evaluation of the percentage error of an entire flight. An Emild Reach RTK-GPS module [2] embedded in Bebop Drone provides a position ground truth. The drone was manually piloted to take off and reach an estimated 30.0-m height. After that, the drone executed a rectangular path, at 1.5 m per second velocity. A 1.0-m size AR-Tag was placed near to the center of the path. The base station computer records The RTK and AR-Tag position data during the flight. The process was repeated for five rounds.

The Fig. 3 shows a graph of one executed fly, comparing the RTK-GPS position readings (in blue) with the AR-Tag position readings (in red).

A calculation of the mean percentage error for Horizontal and Vertical position, using data of all rounds, results in

$$Horiz\_Perc\_Mean\_Error = 7.6\% \quad (3)$$

and,

$$Vert\_Perc\_Mean\_Error = 9,3\%. \quad (4)$$

At 30.0-m distance of the tag, the horizontal error is near to 2.1 m, and the vertical error is near to 2.7 m. The results present a gain of accuracy when compared to a regular GPS error, about 4 m. Besides that, these results must be improved to allow a proper application to the power line inspection problem, because a large position error like that could offer a high risk of collision. The fusion of odometry data and AR Tag data is probably a good approach to improve the accuracy of the solution. Future works proposes the evaluation of these techniques.

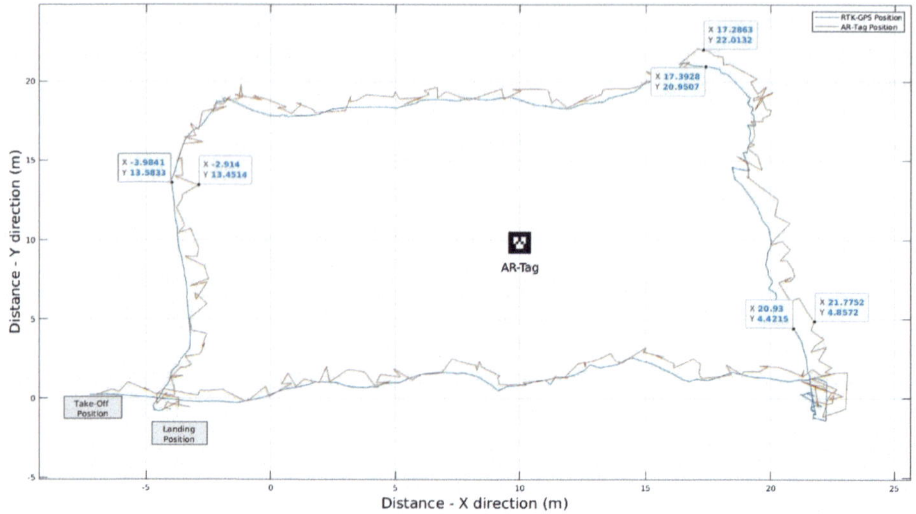

Fig. 3. Path position measured differences between RTK-GPS and AR-Tag

4.3 Simulation Tests

The next experiment describes the simulation of an environment with a power line tower, built to evaluate the autonomous fly of an UAV around the tower using a group of AR-Tag as position reference system. Four 1.0 m size AR-Tags are placed on the ground around the tower, each one presenting a different tag number to allow the differentiation between then. The system calculates the position of each tag concerning the center of the tower. Only one tag data position is used at each time by the system. The system calculates the nearest tag to the center of the UAV frame and decompresses the other tag information.

An UAV model with position and stabilization controller script receives a waypoints list from a C++ program and execute the UAV navigation. Another C++ program identifies the closest tag from the UAV base and calculates the UAV position using the data published on the *ar_pose_marker* node. This data is published on a ROS topic and feeds the LUA PID control algorithm.

Vision sensor aperture angle is 62°, 512×512 pixel resolution. The lowest height of fly defined in this experiment is equal to 10 m. At this height, the image sensor can capture at least one tag image on a 20-m radius distance from the center of the tower. The four tags are placed at a 7-m distance from the tower center on the diagonal direction. This radius is the range of work for the position system.

Four different waypoint missions were executed, in 5 rounds each. The routes were programmed manually, and the maxim distance between two consecutive waypoints is equal to 4 m. Figure 4 shows an image of a mission and the trace 3D tag measured position versus the real position.

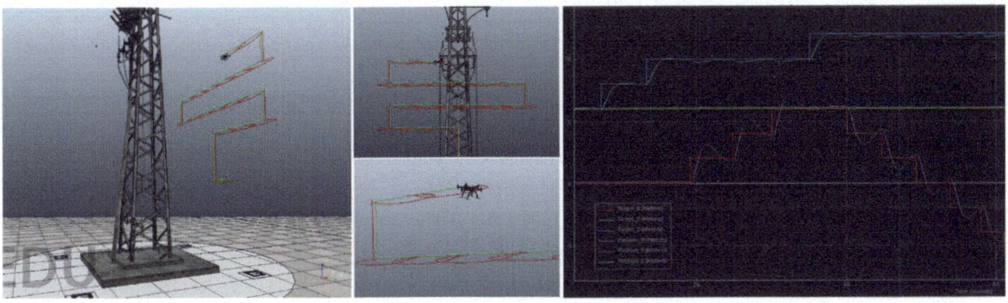

Fig. 4. Graphic visualization of UAV position error

The control algorithm was able to accomplish correctly 17 of 20 missions. In three missions, the UAV lost the position reference, and the control algorithm fails. These situations occurred when the UAV lost partial vision of the reference tag because of position swing when a new displacement begins. This swing occurs because of the simplicity of PID control algorithm, which allows high-velocity commands when the UAV is far from the waypoint.

A possible way to correct this problem is to use a robust control algorithm, as a fuzzy logic-based for example. This kind of algorithm provides soft displacements to the UAV even when the waypoints are far one from another. Application of a Kalman Filter also could bring more robustness to the position inference in this case. Future works search this solutions approach.

5 Results Analysis and Conclusions

This paper presents a complementary UAV position solution based on using Ar Tag visual information. The solution offers a second level position data, to be applied when traditional systems, like GPS, presents lost of accuracy. The solution uses a regular camera and processing hardware to provide the position data, offering a cheap and easy to use solution to the proposed problem.

Results obtained on the experiments and simulations show that the use of the proposed solution in real-world situation is technically viable. Simulation shows that the solution allows the creation of a trustworthy positioning system to develop UAV inspection systems, in particular on the power line inspections problem. Some obtained data presents a measurement error that could be minimized using additional data processing techniques and robust algorithms application. The use of high-resolution image to process the tag's visual information, and artificial lighting are also a proper choice to improve the system accuracy. This research is a work in progress, and next steps include new rounds of experiments to evaluate additional tools and solutions, to improve the results obtained this far.

References

1. Bian, J., Hui, X., Zhao, X., Tan, M.: A point-line-based SLAM framework for UAV close proximity transmission tower inspection. In: 2018 IEEE International Conference on Robotics and Biomimetics, ROBIO 2018, pp. 1016–1021 (2019). https://doi.org/10.1109/ROBIO.2018.8664716
2. Emild: EMILD REACH DOCUMENTATION WEB SITE (2019). https://docs.emlid.com/reach/#discussion
3. V.T.R.C. of Finland: Augmented Reality/3D Tracking (2017). http://virtual.vtt.fi/virtual/proj2/multimedia/
4. Hinas, A., Roberts, J.M., Gonzalez, F.: Vision-based target finding and inspection of a ground target using a multirotor UAV system. Sensors **17**(12) (2017). https://doi.org/10.3390/s17122929
5. April Robotics Laboratory: AprilTags Visual Fiducial System (2018). https://april.eecs.umich.edu/software/apriltag
6. Larrauri, J.I., Sorrosal, G., Gonzalez, M.: Automatic system for overhead power line inspection using an unmanned aerial vehicle RELIFO project. In: 2013 International Conference on Unmanned Aircraft Systems (ICUAS), pp. 244–252, May 2013. https://doi.org/10.1109/ICUAS.2013.6564696. http://ieeexplore.ieee.org/lpdocs/epic03/wrapper.htm?arnumber=6564696
7. Luque-Vega, L.F., Castillo-Toledo, B., Loukianov, A., Gonzalez-Jimenez, L.E.: Power line inspection via an unmanned aerial system based on the quadrotor helicopter. In: Proceedings of the Mediterranean Electrotechnical Conference - MELECON, pp. 393–397 (2014). https://doi.org/10.1109/MELCON.2014.6820566
8. Monajjemi, M.: Bebop Autonomy - ROS Driver for Parrot Bebop Drone (2018). https://bebop-autonomy.readthedocs.io/en/latest/
9. Muskardin, T., Balmer, G., Persson, L., Wlach, S., Laiacker, M., Ollero, A., Kondak, K.: A novel landing system to increase payload capacity and operational availability of high altitude long endurance UAVs. J. Intell. Robot. Syst.: Theory Appl. **88**(2–4), 597–618 (2017). https://doi.org/10.1007/s10846-017-0475-z
10. Rangel, R.K., Kienitz, K.H., Brandão, M.P.: Development of a multi-purpose portable electrical UAV system, fixed & rotative wing. In: 2011 IEEE Aerospace Conference (2011)
11. ROS.org: ROS (2019). http://www.ros.org
12. USA: US Department of Defense: Global Positioning System Standard Positioning Service, pp. 1 – 160, September 2008. http://www.Gps.Gov, http://www.gps.gov/technical/ps/2008-SPS-performance-standard.pdf
13. Zimmermann, F., Eling, C., Klingbeil, L., Kuhlmann, H.: Precise positioning of UAVs - dealing with challenging RTK-GPS measurement conditions during automated UAV flights. ISPRS Ann. Photogramm. Remote Sens. Spatial Inf. Sci. **4**(2W3), 95–102 (2017). https://doi.org/10.5194/isprs-annals-IV-2-W3-95-2017

Evaluation of Lightweight Convolutional Neural Networks for Real-Time Electrical Assets Detection

Joel Barbosa[2(✉)], André Dias[1,2], José Almeida[1,2], and Eduardo Silva[1,2]

[1] INESC TEC - INESC Technology and Science, Porto, Portugal
[2] ISEP - School of Engineering, Polytechnic Institute of Porto, Porto, Portugal
1131197@isep.ipp.pt

Abstract. The big growth of electrical demand by the countries required larger and more complex power systems, which have led to a greater need for monitoring and maintenance of these systems. To overcome this problem, UAVs equipped with appropriated sensors have emerged, allowing the reduction of the costs and risks when compared with traditional methods. The development of UAVs together with the great advance of the deep learning technologies, more precisely in the detection of objects, allowed to increase the level of automation in the process of inspection. This work presents an electrical assets monitoring system for detection of insulators and structures (poles and pylons) from images captured through a UAV. The proposed detection system is based on lightweight Convolutional Neural Networks and it is able to run on a portable device, aiming for a low cost, accurate and modular system, capable of running in real time.

Keywords: Electrical assests · Unmanned aerial vehicle · Convolutional neural networks · Object detection · Real-time

1 Introduction

In recent years, we have witnessed a big growth of the electrical demand by the countries due to its demographic and economic expansion, requiring larger and more complex power systems. This complexity and growth have led to a greater need for inspection and maintenance of these systems in order to reduce their vulnerabilities. The inspection of electrical assets, such as insulators, pylons, dams, etc., is done with specialized human labor, manned helicopter crawling robots and, more recently, UAVs. UAVs are presented as one of the best options taking into account that the cost-benefit ratio is superior to the others, as it presents a lower cost for high accuracy, efficiency, and safety.

These platforms are equipped with the state of the art sensors, such as LIDARs and cameras (video and thermal) which allow gathering real-time footage and data. [1] presented their system for inspection of high voltage power

M. F. Silva et al. (Eds.): ROBOT 2019, AISC 1092, pp. 99–112, 2020.
https://doi.org/10.1007/978-3-030-35990-4_9

line using UAVs, offering services such as image and video data, thermal inspection, corridor mapping and creation of a Digital Terrain Model, troubleshooting reports, acquisition of LIDAR data and report of the major risks to the electrical assets, like vegetation.

The research about automating the task of visually inspecting power transmission systems has also increased in the last years. Some projects, such as [2] and [3], have explored the UAVs to these tasks based on autonomous navigation through the power lines. [2] developed a quadrotor helicopter and achieved autonomous inspection along with GPS waypoints. The data gathered by the system were sent to a ground control station, where it ran a set of vision algorithms, in order to report any anomaly in real time. [3] proposed a multi-platform UAV system, which comprised a fixed wing UAV, a hexarotor UAV and a tethered multirotor UAV, and a multi-model communication system. The UAVs were used to long-distance imaging, short distance imaging, and communication relay. This approach has shown a much higher efficiency than traditional methods. [4] presented a simulated system for tracking transmission lines based on the detection of the wires, using artificial vision, to position the UAV in the electricity transmission corridor. [5] developed a multiple sensors platform using LIDAR, thermal camera, an ultraviolet camera, and cameras to acquire information about power line components and surrounding objects, based on a large unmanned helicopter. They presented the planning method for the flight path and the tasks of the sensors before the inspection and the method used for tracking power lines and insulators during the inspection. Due to the importance of insulators in the power distribution systems, [6] address their studies to the detection of insulators and analysis of their defects, using a ground vehicle. They proposed a new method based on rotation invariant local directional pattern (RI-LDP) to represent the insulator image as a feature vector and used support vector machine (SVM) as a classifier. With the detected insulator, automatic defect detection is applied to partition each cap of an insulator and analyzing each cap for the defect. This method is able to categorize the defect as cracks, contamination, whitening, bullet damage, and alligatoring effects.

With the development of deep learning technology and, consequently, the rise of convolutional neural networks (CNN) resulting on a breakthrough in the field of object detection, some projects have used it to improve the detection of the electrical assets. [7] is one of these projects, where they proposed a real-time electrical equipment detection and defect analysis. Using Darknet's open source framework and object detection system, You Only Look Once (YOLO) version 2, it was possible to differentiate 17 types of the insulator with 98% of accuracy. Additionally, it was developed a defect analyzer, using a rotation normalization and ellipse detection method, capable of detecting gunshot defects in the equipment. [8] also addressed their research to insulator detector and defect analysis using aerial images. They proposed a two-stage cascading network to perform, in the first stage, localization of insulators based on a region proposal network (RPN) and, in the second stage defect detection, also based on RPN. They have created a data augmentation method to prevent the scarcity of defect images, allowing a big increase in precision and recall, as well as, detection of a defect under various conditions. [9] and [10] presented a solution to continuous navigation along one side of overhead transmission lines using deep learning. They developed a system capable to detect and track in real time tower transmission, to provide their localization, based on Faster R-CNN to reliably detect the transmission towers and Kernelized Correlation Filters (KCF) to continuously track their localization in the image. To continually navigate along the power lines, they computed and optimized their vanishing point to provide UAV with a robust heading, using the Line Segment Detector (LSD) to detect the lines. Finally, to measure the distance from transmission lines, a distance estimation process from UAV to the tower, by triangulation, was performed, following a multiple view strategy.

In most of the presented literature, the systems require large computational resources to run the proposed algorithms and, consequently, they present an high cost in real time requirements, power consumption, portability, payload for the UAV's and price. To overcome this problem, this work aims to present a deep learning based perception system using lightweight CNN's oriented to detect object, compatible with the new embedded devices capable of running AI at the edge on different platforms, including ARM-based platforms. An electrical assets monitoring system is proposed, capable of detect insulators and structures (poles and pylons) from images captured through a UAV. The proposed detection system is based on lightweight Convolutional Neural Networks and it is able to run on a portable device, aiming for a low cost, accurate and modular system, capable of running in real time. To train the networks, a dataset which consists of images taken by the UAV and a data augmentation process was created. An evaluation of the different state of the art lightweight CNN's, compatible with this embedded devices, will be presented, whose tests were based on subsets of images effected with different conditions that can occur during an inspection (fog, blur, noise and scale variation).

2 System Overview

In this section, an overview of the used system is presented. Initially, the description of the UAV is made, followed by a presentation of the hardware and software that have been developed for real-time inspection. Finally, a description of the high-level architecture of the system responsible for the perception component is also discussed.

2.1 UAV

Based on the Electrical Asset Inspection project from INESC TEC laboratory, whose main objective is to add a set of cutting-edge methods to improve the current state of the visual inspection of electrical assets, as well as make the inspection process more autonomous. This project consists in developing a UAV,

Fig. 1. UAV STORK I

with the operational capacity to address the requirements of inspection and monitoring of electrical assets, in order to guarantee optimization of the inspection process in electrical assets such as lines, substations and wind turbines. The STORK I, Fig. 1, is the hexacopter UAV developed and used in applications such as search and rescue operations, environmental monitoring, 3D mapping, inspection, and surveillance and patrol. This platform is equipped with a pan&tilt with thermographic and a high-resolution visual camera and also a LiDAR system for 3D reconstruction and obstacle detection and avoidance. The vehicle provides a payload capacity of 3 kg with an endurance of 25 min and a modular approach to allow the integration of new sensors such as a hyperspectral camera.

2.2 Movidius™ Neural Compute Stick and OpenVINO™

The Neural Compute Stick (NCS) is a low-cost and low-power USB device based on Myriad 2 Vision Processing Unit (VPU). This device allows rapid prototyping, validation, and deployment of deep neural network inference applications at the edge.

To take advantage of Movidius™, it has been used the toolkit OpenVINO™. The OpenVINO™ toolkit provides the ability of CNN-based deep learning inference and helps further unlock cost-effective, real-time vision applications. OpenVINO™ supports heterogeneous execution across computer vision accelerators, CPU, GPU, Intel® Movidius™ Neural Compute Stick, an FPGA, using a common API and supports models in popular formats such as Caffe, Tensorflow, MXNet, and ONNX. This toolkit includes two components, namely Model Optimizer and Inference Engine. The Model Optimizer is a cross-platform command-line tool that facilitates the transition between the training and deployment environment, performs static model analysis, and adjusts deep learning models for optimal execution on end-point target devices. Model Optimizer produces an Intermediate Representation (IR) of the network as output. The Inference Engine is a C++ library with a set of C++ classes to infer input data (images) and get a result. This library provides an API to read the Intermediate Representation, set the input and output formats, and execute the model on devices.

2.3 High-Level Architecture

Using the Movidius^{TM} Neural Compute Stick and OpenVINO^{TM} toolkit as benchmarks, a software API was developed using the C++ language and the Robot Operating System (ROS) middleware. A ROS node was developed, with the responsibility for the subscriptions of the images coming from the camera or from a rosbag file, and for the publication of the images with the detected objects (for visualization purposes) and the regions of interest present in the respective image frame. Firstly, using the OpenVINO^{TM} toolkit tools, the optimized model, via its intermediate representation, will be read, loaded and inferred by the inference engine followed by the initialization of the device to be used, in this case, the Movidius NCS or the CPU. Thus, each subscribed image will be inferred by the inference engine and produces the respective output. Then the result will be published as an image with the objects represented and as a region of interest message. In Fig. 2, it is depicted the high-level architecture of the system.

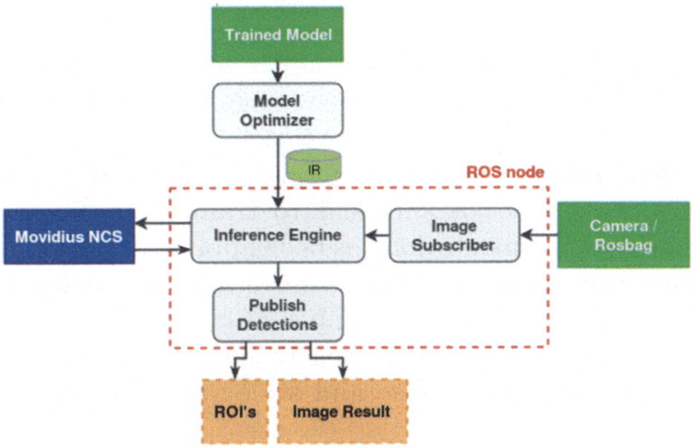

Fig. 2. High-level system architecture.

Fig. 3. Example of images collected by the UAV during the visual inspection.

3 Dataset and Data Augmentation

To train the networks a dataset was created with images gathered from the UAV in different inspection missions performed in manual mode. These images consist

Fig. 4. Example of the transformations applied in the data augmentation process

of samples of a few pylons, unity poles and different types of insulators. In Fig. 3, it is possible to see some examples of these images. The initial dataset contains 585 images with different resolutions and, to each, an annotation file was created containing the location and its class, structure or insulator.

The original data set is not sufficient to train the networks and for this reason a process of data augmentation has been applied through the development of a tool capable of applying different transformations to the original image This process, besides increasing the number of examples per class, allows the reduction of overfitting and improves the immunity to the conditions of the environment, such as fog, blur, noise and scale variation, and consists in the following augmentation techniques:

- **Rescale**: 300×300 and 512×512;
- **Rotation**: 35 degree steps clock-wise;
- **Hue and Saturation**: Change of hue and saturation components;
- **Blur**: Adds two values of a blur;
- **Contrast Normalization**: Normalization of the image contrast;
- **Fog**: Fog simulation;
- **Gaussian Noise**: Adds two intensities of Gaussian Noise;
- **Salt and pepper**: Adds random black and white pixels;
- **Elastic Transformation**: Image quality reduction;

The final dataset, after data augmentation, contained a total of 15795 images. The presented transformations have been applied to each image and its respective annotation file, containing the location of the object in the image, of the original dataset. In Fig. 4, it is depicted an example of the data augmentation process applied to an image of the original dataset. Finally, the dataset was divided into two sets, one for training with 70% of the images and the remaining 30% for test/validation.

4 Experiments

This section aims to describe the set of evaluation experiments on different lightweight networks to detect the proposed objects (insulators and structures) and the respective used material.

4.1 Lightweight Convolutional Neural Networks

Single Shot Detector (SSD) [11] **Based.** The SSD approach is based on a feed-forward convolutional neural network that uses an image classification network as a base network (originally VGG-16) at the early stages. This network works as a feature extractor, whose features are followed by convolutional layers that decrease in size progressively, generating features at different scales. The multi-scale feature maps are divided into cells and for each cell, a fixed amount of default bounding boxes with different sizes and aspect ratios is generated. For each box, the network generates scores for the presence of each object category and produces adjustments to better match the object shape. This step generates a lot of bounding boxes, that's why it is necessary a non-maximum suppression step which removes the low confidence boxes and fuses highly overlapping ones.

In the context of this work, the evaluation of the SSD approach will be to test its performance with different base networks: MobileNet [12], MobileNetV2 [13] and PeleeNet [14]. Each will be trained using the Caffe [15] framework, and later converted to the corresponding Intermediate Representation using the tool OpenVINO.

Yolo Based. The YOLO-based approach consists of a series of methodologies that have been gradually improving its original version, accompanying the development of object-detection technologies, which is currently in the third version [16]. In this version, initially, during the training phase, the network is fed with images to predict 3D tensors corresponding to a certain number of scales (three scales in Yolov3 and two scales in tiny-Yolov3), coming from the backbone network as the feature extractor, which aims to detect objects with different sizes. For each N scale, the image is divided into N×N grid cells and each grid cell corresponds to a voxel that contains the bounding box coordinate, objectness score, and class confidence. If the center of the object's ground truth falls inside a certain grid cell, it is assigned with three prior/anchor boxes of different sizes, choosing, in the training phase, the one that better overlaps with the ground truth bounding box and predicts the corresponding offsets to the prior box. The main differences between the Yolov3 and the tiny-Yolov3 networks are the number of scales and the feature extractor, which are smaller in both cases in the tiny-Yolov3 network.

Here, the evaluation of this approach will be based on tests in the Yolov3 and tiny-Yolov3 networks. These were trained using the Darknet [17] framework, followed by a conversion of the trained weights to Tensorflow and, finally, this one converted to the corresponding Intermediate Representation.

4.2 Experimental Configuration

To train the proposed networks with the previously explained dataset, a workstation powered by an Intel® i7-8700K CPU with 3.70 GHz and a NVIDIA® GeForce GTX 1080 Ti GPU with 3584 NVIDIA® CUDA cores was used.

To perform the deploy tests were used the presented workstation, an onboard UAV computer powered by an Intel® i7-4700MQ, Quad-Core, 2.40 GHz CPU running Ubuntu 16.04 operating system, and an Odroid XU-3. All systems had ROS Kinetic distribution and OpenVino 2019 toolkit.

4.3 Dataset with Different Conditions

To properly evaluate the performance of different networks in different scenarios, different subsets of images were created. Each subset is made up of images that are derived from the data augmentation process presented before. Therefore, the techniques used were Blur, Salt and Pepper, Fog, Scaling, Rotation, and Gaussian Noise. The choice of these techniques was based on the principle of the greater possibility of occurrence when applying this system in real scenarios.

5 Results

As explained before, the performance of the different CNN on different platforms was evaluated. In Table 1 it is possible to see the different precisions resulting from the training process. The Yolo based networks out-performed the remaining networks, using an 0.45 intersection over union (IOU) threshold and a 0.5 confidence threshold. Between the SSD based, the MobileNetv2-SSD was the one with the best precision. In Fig. 5 are represented the precision-recall curves of each network for the two classes, insulators, and structures. The precision-recall curve allows the evaluation of the performance of an object detector by considering if its precision stays high as recall increases, which means that if the confidence threshold changes, the precision and recall will still be high. As we can observe, the curves confirm the results of the obtained precision values (area under the curves). In general, network performance was better at detecting structures. This is due to the fact that the aspect ratio of the structures in relation to the size of the images is larger than the aspect ratio of the insulators. This is accentuated when it is performed the image resize to 300×300, in the case of SSD based, and 416×416, in the case of Yolo based.

Regarding the evaluation process with datasets with different conditions, whose results of precision can be analyzed from the Table 2, greater effectiveness was verified by Yolo based networks. The tiny-Yolov3 in almost every scenario outperformed the remaining networks. From Figs. 6, 7, 8, 9, 10 and 11, it is possible to observe that the quality of detection of the insulators has decreased considerably, whereas for the structures the quality has only been lowered in scenarios with Blur. In the Figs. 12 and 13 it is possible to observe an example which confirms the explained detection results for tiny-YOLOv3 and MobileNet-SSD, respectively.

Table 1. Original dataset precision results

Networks	Precision
MobileNet-SSD	73.2
MobileNetV2-SSD	73.3
PeleeNet-SSD	55.1
tiny-YOLOv3	90
YOLOv3	83.05

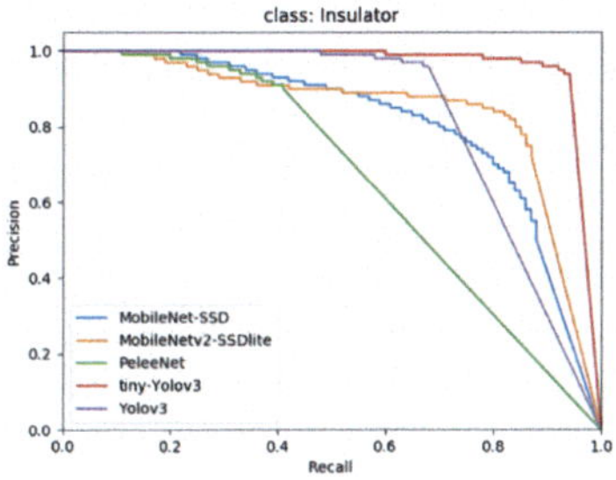

(a) Insulators

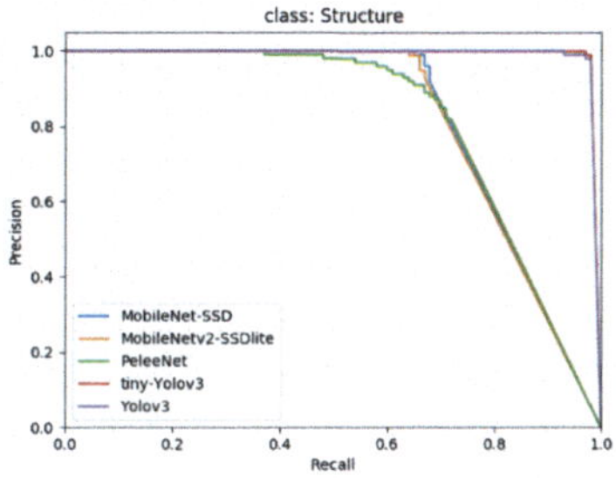

(b) Structures

Fig. 5. Precision-Recall curve for each class

The inference speed in frames per second (fps) is presented in the Table 3 for each network in different platforms, where it is possible to conclude that SSD based networks outperformed the YOLO based ones, in both CPU and NCS. Within YOLO-based, tiny-YOLOv3 is significantly faster than YOLOv3, where when applied to NCS the second does not even reach real-time. Among the SSD-based networks, when applied to CPU, the MobileNet reached the best inference speed on the onboard UAV computer and workstation, but when NCS is used, the MobileNetv2 outperformed the other, reaching the highest mean speed on the onboard UAV computer and on the Odroid. It is also possible to observe that NCS performance slows down on different platforms because it depends on the hardware components of the board.

Table 2. Data augmentation precision results

Networks	Blur	Fog	Scaling	Rotations	Salt and Pepper	Gaussian
MobileNet-SSD	51.6	49.3	53.6	55.7	51.5	50.7
MobileNetV2-SSD	48.3	46.6	49.82	56.2	49.2	47.8
PeleeNet-SSD	46.8	43.8	48.8	41.4	48.3	41.8
tiny-YOLOv3	**59.4**	**58.5**	**61.1**	56.0	**61.5**	**60.3**
YOLOv3	58.3	57.7	58.6	**57.7**	59.2	58.4

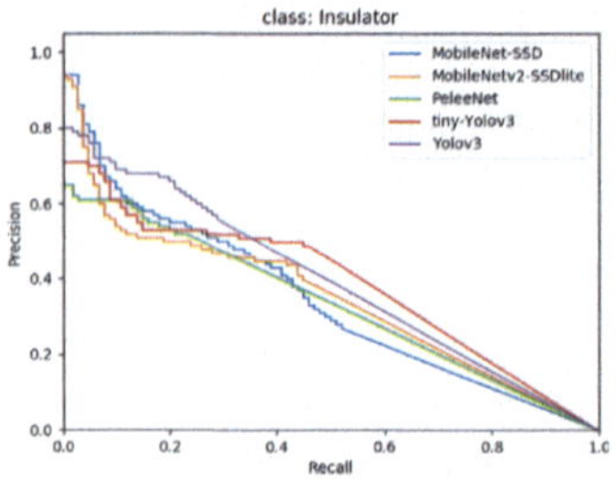

(a) Insulators

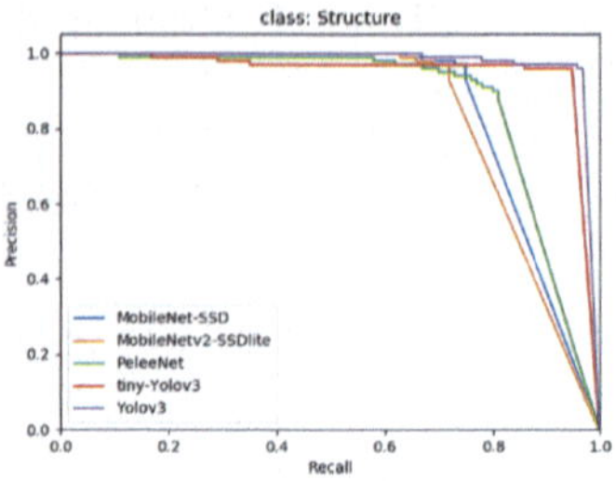

(b) Structures

Fig. 6. Precision-Recall curve for each class with blur occurrence

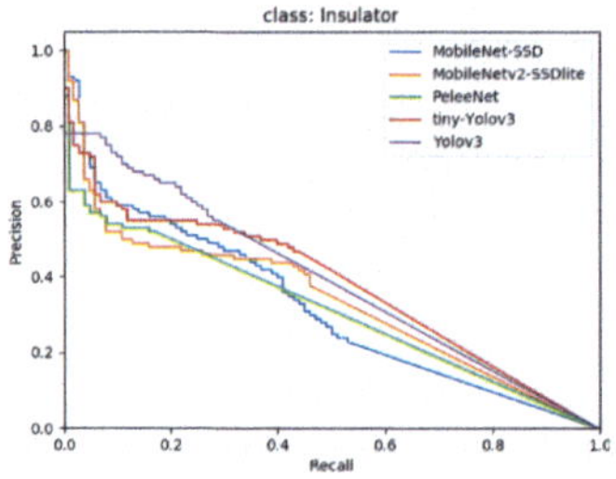

(a) Insulators

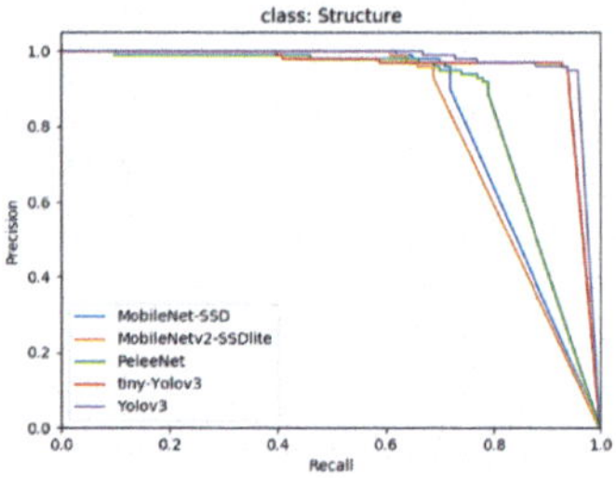

(b) Structures

Fig. 7. Precision-Recall curve for each class with fog conditions

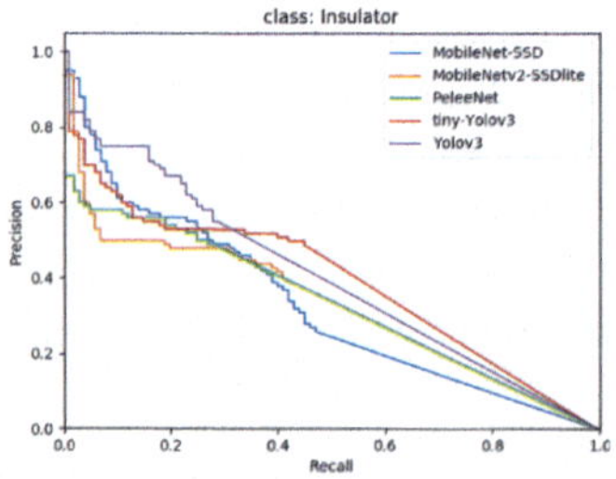

(a) Insulators

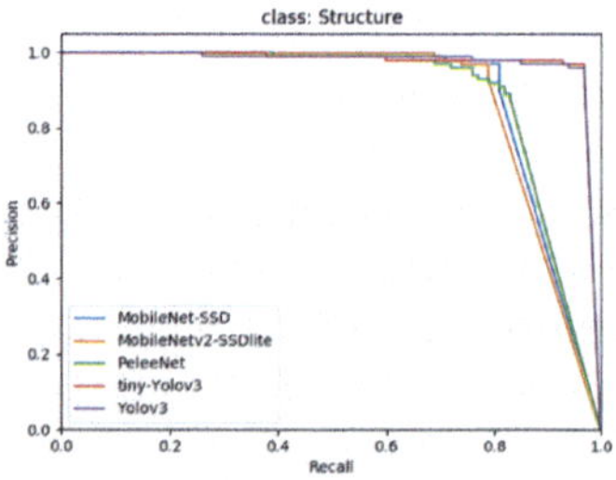

(b) Structures

Fig. 8. Precision-Recall curve for each class with scale variance

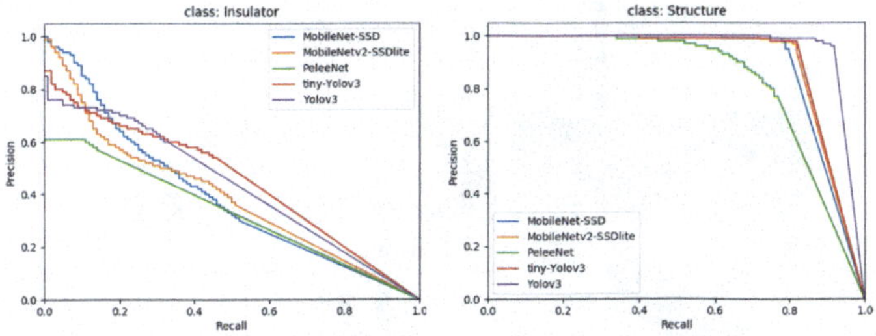

(a) Insulators (b) Structures

Fig. 9. Precision-Recall curve for each class with rotation variance

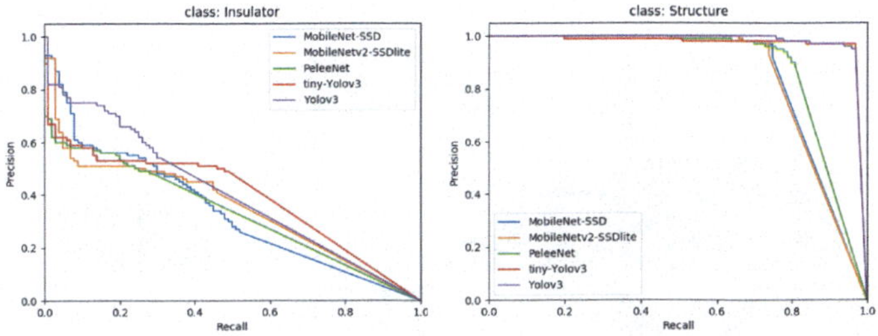

(a) Insulators (b) Structures

Fig. 10. Precision-Recall curve for each class with black and white pixels occurrence

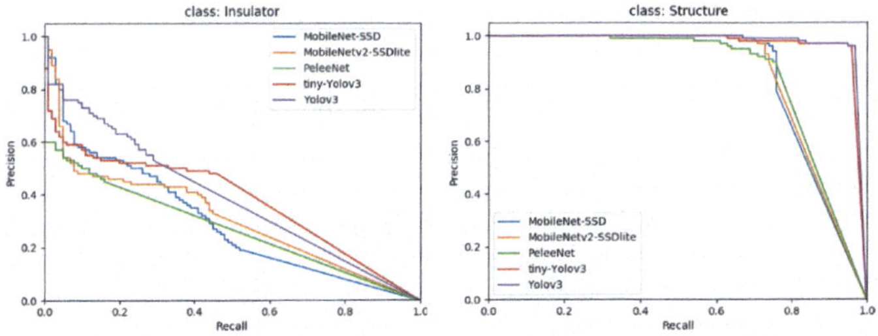

(a) Insulators (b) Structures

Fig. 11. Precision-Recall curve for each class with Gaussian noise conditions

Fig. 12. Detection examples using MobileNet-SSD

Fig. 13. Detection examples using tiny-YOLOv3

Table 3. Inference speed of the networks on different platforms in frames per second

Networks	Onboard UAV PC		Workstation		Odroid XU-3	
	CPU	NCS	CPU	NCS	CPU	NCS
MobileNet	48	12.5	145	NA	NA	10.8
MobileNetv2	26	13.2	60	NA	NA	11.5
Pelee	37	9.2	100	NA	NA	8.7
tiny-Yolov3	22	7.6	79	NA	NA	6.7
Yolov3	2.3	0.74	10	NA	NA	0.73

6 Conclusion

With these experiments and results, it is possible to conclude that the real-time detection of electrical assets is feasible using lightweight Convolutional Neural

Networks on edge devices. These low-cost devices allow great portability and modularity, maintaining the real-time and detection quality requirements on different platforms. The different CNN under testing has shown that they could perform with good precision on different conditions and in the case of tiny-YOLOv3 it was possible to achieve an average precision of 90% on the general dataset at 7 fps.

In the future, this system will be applied in real scenarios being able to be used as the perception system for autonomous inspection of the electrical assets, such as insulators or structures like pylons and poles.

References

1. Malveiro, M., Martins, R., Carvalho, R.: Inspection of high voltage overhead power lines with UAV's. In: Proceedings of the 23rd International Conference on Electricity Distribution (2015)
2. Luque-Vega, L.F., Castillo-Toledo, B., Loukianov, A., Gonzalez-Jimenez, L.E.: Power line inspection via an unmanned aerial system based on the quadrotor helicopter. In: MELECON 2014 - 2014 17th IEEE Mediterranean Electrotechnical Conference (2014)
3. Deng, C., Wang, S., Huang, Z., Tan, Z., Liu, J.: Unmanned aerial vehicles for power line inspection: a cooperative way in platforms and communications. J. Commun. **9**(9), 687–692 (2014)
4. Menendez, O.A., Perez, M., Cheein, F.A.A.: Vision based inspection of transmission lines using unmanned aerial vehicles. In: 2016 IEEE International Conference on Multisensor Fusion and Integration for Intelligent Systems (MFI) (2016)
5. Xie, X., Liu, Z., Xu, C., Zhang, Y.: A multiple sensors platform method for power line inspection based on a large unmanned helicopter. Sensors **17**(6), 1222 (2017)
6. Jabid, T., Ahsan, T.: Insulator detection and defect classification using rotation invariant local directional pattern. Int. J. Adv. Comput. Sci. Appl. **9**(2), 265–272 (2018)
7. Siddiqui, Z., Park, U., Lee, S.W., Jung, N.J., Choi, M., Lim, C., Seo, J.H.: Robust powerline equipment inspection system based on a convolutional neural network. Sensors **18**(11), 3837 (2018)
8. Tao, X., Zhang, D., Wang, Z., Liu, X., Zhang, H., Xu, D.: Detection of power line insulator defects using aerial images analyzed with convolutional neural networks. IEEE Trans. Syst. Man Cybern.: Syst. **PP**(99), 1–13 (2018)
9. Hui, X., Bian, J., Zhao, X., Tan, M.: Vision-based autonomous navigation approach for unmanned aerial vehicle transmission-line inspection. Int. J. Adv. Robot. Syst. **15**(1), 1729881417752821 (2018)
10. Hui, X., Bian, J., Zhao, X., Tan, M.: Deep-learning-based autonomous navigation approach for UAV transmission line inspection. In: 2018 Tenth International Conference on Advanced Computational Intelligence (ICACI) (2018)
11. Liu, W., Anguelov, D., Erhan, D., Szegedy, C., Reed, S.E., Fu, C., Berg, A.C.: SSD: single shot multibox detector. CoRR (2015)
12. Howard, A.G., Zhu, M., Chen, B., Kalenichenko, D., Wang, W., Weyand, T., Andreetto, M., Adam, H.: Mobilenets: efficient convolutional neural networks for mobile vision applications. CoRR (2017)

13. Sandler, M., Howard, A.G., Zhu, M., Zhmoginov, A., Chen, L.: Inverted residuals and linear bottlenecks: mobile networks for classification, detection and segmentation. CoRR (2018)
14. Wang, R.J., Li, X., Ao, S., Ling, C.X.: Pelee: a real-time object detection system on mobile devices. CoRR (2018)
15. Jia, Y., Shelhamer, E., Donahue, J., Karayev, S., Long, J., Girshick, R., Guadarrama, S., Darrell, T.: Caffe: Convolutional architecture for fast feature embedding. arXiv preprint arXiv:1408.5093 (2014)
16. Redmon, J., Farhadi, A.: Yolov3: an incremental improvement. CoRR (2018)
17. Redmon, J.: Darknet: Open source neural networks in c (2013–2016). http://pjreddie.com/darknet/

Agricultural Robotics and Field Automation

Cleaning Robot for Free Stall Dairy Barns: Sequential Control for Cleaning and Littering of Cubicles

Ilja Stasewitsch(✉), Jan Schattenberg, and Ludger Frerichs

Institut für mobile Maschinen und Nutzfahrzeuge,
Technische Universität Braunschweig, Braunschweig, Germany
{i.stasewitsch,j.schattenberg,ludger.frerichs}@tu-braunschweig.de

Abstract. Rising cost pressure and growing farm sizes in dairy farming with constant or even declining working capacity require new robotic solutions in this field. A logical approach is to automate the cleaning due to the high manual workload at this task. Therefore, a cleaning robot is being developed which removes liquid manure from the running surfaces and maintains the cubicles. The cubicles are cleaned of liquid manure with a brush and maintained by the help of a conveyor belt with bedding. The navigation of the robot is presented which is based on a 2.5-D SLAM and localization with two 2-D lidars. The focus is however on the sequential control of this cleaning task which is solved by using semantic maps, work routes, an active collision avoidance strategy and a brush force control. This functioning procedure is shown with experiments in a artificial stall and in a real stall barn.

Keywords: Mobile robotic application · Indoor agricultural robot · Free stall barn · Process robotization

1 Introduction

Manure removal in free stall dairy barns (see Fig. 1) has a major impact on the cleanliness of the running surfaces. In various studies, e.g. in [2,8], a connection between the cleanliness and the occurrence of claw or udder diseases could be proven. Higher manure removal frequencies also help to reduce ammonia emissions which can cause claw diseases. The littering down of cubicles reduces moisture and protect dairy cows from udder diseases. Therefore, an increase in cleanliness has many positive effects on animal health and milk quality. There are, however, also negative effects on working effort and economy when using conventional methods. According to [7] the manual working time is 3 to 7 h per dairy cow per year, with the main part being the cleaning and soiling of the cabins. This work takes between 2 and 5 h per dairy cow per year, depending on the technology used. Currently, there are different robots on the market which are equipped with simple sensors to clean just the running surface. These robots

M. F. Silva et al. (Eds.): ROBOT 2019, AISC 1092, pp. 115–126, 2020.
https://doi.org/10.1007/978-3-030-35990-4_10

Fig. 1. Free stall dairy barn with free and occupied cubicles

Fig. 2. Cleaning robot of the project with the mechanical components

are inflexible and can only follow fix routes, e.g. if the battery is nearly empty the robot can not break the route and drive to the charge station. Therefore, the cleaning and littering down of cubicles and documentation has great potential to reduce workload and improve animal health.

1.1 Project Overview and Challenges

A robot (see Fig. 2) has been developing in cooperation with *Technische Universität Braunschweig* (development of the control), *Bayerische Landesanstalt für Landwirtschaft* (evaluation of the process robotization) and *Peter Prinzing GmbH* (robot design). The project is the sequel of [10]. The robot is equipped with six electric motors. Two of them are used as drives for skid steering. Two other motors are used for folding in and out and rotating the brush, which cleans cubicles from liquid manure. The other two motors are used to transport the bedding from the storage tank and to eject it with a conveyor belt. Several challenges appear from the viewpoint of the process robotization in this project. All of them are related to the special characteristics in this kind of environments:

- Many cows move freely and the way they are in the boxes varies. This makes collision avoidance and use of the brush more difficult.
- The liquid manure causes a large slip, so that a localisation based on pure odometry is not suitable for cleaning and documentation.
- Uneven ground makes it difficult to use 2-D lidar for SLAM, localization and determination of occupied cubicles.

1.2 Paper Structure

The paper treats mainly the sequential control of the cubicle cleaning. The robot navigation is also described due to its necessity for the above mentioned sequential control. The entire system, implemented in ROS (see [9]), is shown in Fig. 3. Section 2 explains the navigation, which consists of "Localization", "Global Path

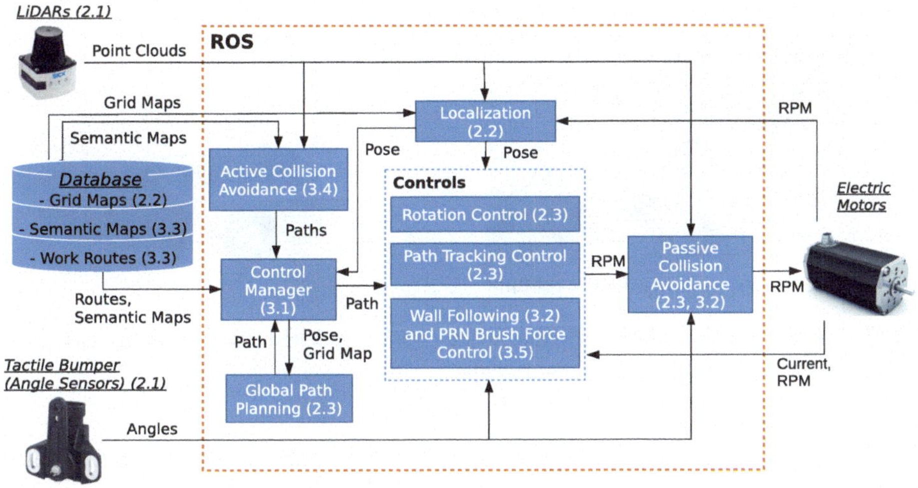

Fig. 3. Outline of the control system with the main components (sections are bracketed)

Planning", "Rotation Control" and "Path Tracking Control". The system has interfaces to the environment through two lidars and a tactile bumper, which are described in Sect. 2.1. Section 3 present the modules "Wall Following", "Brush Force Control", "Active Collision Avoidance", "Passive Collision Avoidance" and the interaction of these components are presented for the sequential control. The collision avoidance prevents injuries to dairy cows. Experimental results of that are shown in Sect. 4. The system has a database of occupancy "Grid Maps", which are generated by a SLAM (see Sect. 2.2), "Semantic Maps" and "Work Routes", which are created by a GUI (see Sect. 3.3).

2 Navigation

2.1 Sensors

The robot is equipped with different sensors (see Fig. 4) for navigation and process robotization. Two 2-D lidars (SICK TiM571) are used for SLAM algorithm, localization as well as for active and passive collision avoidance. The field of view of the lidars are shown in Fig. 5. The idea is, that the upper front lidar is used for collision avoidance and detects features such as walls for localization along the aisles. The lower rear lidar shall determine the lateral position in aisles and map cubicle rows, which is important for the generation of the work routes. In addition to that two lidar can improve the occupancy grid maps of the SLAM algorithm and the localization by sensing more and various features. A tactile bumper detects movements in three different directions (see Fig. 6) using three angle sensors and is used primary for the wall following. The motion is detected on the right and left side as well as at a lateral displacement. The electric motors

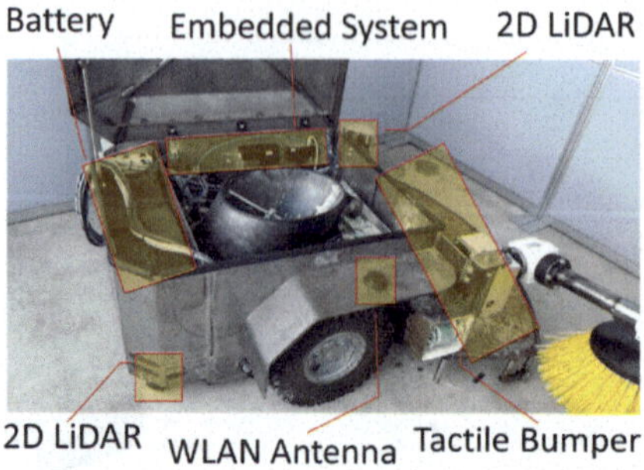

Fig. 4. Position of the lidars and the tactile bumper as well as the electronic components

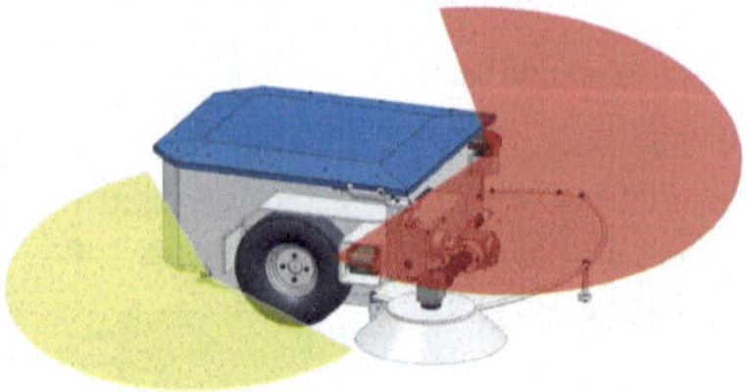

Fig. 5. Field of view of the two 2-D lidar

are equipped with encoders and sensors for electric current which are also used for the localization and process robotization. E.g. the current sensors of the brush electric motor are used to control the brush force during the cleaning. The odometry is calculated with the encoders of the drive wheel motors.

2.2 SLAM and Localization

A map of the barn is required for work route planning, the semantic map of the barn, localisation, active collision avoidance and passive collision avoidance. Due to both lidars a 2.5-D SLAM was developed based on the FastSLAM algorithm of [4], which yields a occupancy grid map. In this algorithm, both lidars are used to build two occupancy grid maps simultaneously. The evaluation of each particle and so the evaluation of the quality of the maps is also done in both maps simultaneous, meaning that the localization is executed by scan matching in both maps. Figure 7 shows two superimposed occupancy grid maps of a real free stall barn. This map shows that the lower rear lidar maps the edges of cubicle rows which are important for the planning the semantic map and the work route. The developed localization is using both occupancy grid maps in one particle filter. The evaluation of each particle is done simultaneously with both lidars and both maps. As already mentioned, the lower rear lidar is used for the determination of the position lateral in the aisle and the top front lidar is used for translational position in the aisle. At the end of the procedure, an iterative closest point algorithm improves the solution of the best particle.

2.3 Global Path Planning, Path Tracking Control and Rotation Control

To drive the robot to the work route a global path planning and path tracking control is necessary. The library SBPL is used for the global path planning, which is described in [1]. Compared to simpler algorithms for shortest paths, this library has the advantage that motion primitives can be defined on which the path is then composed. A motion primitive represents the smallest component. Due to

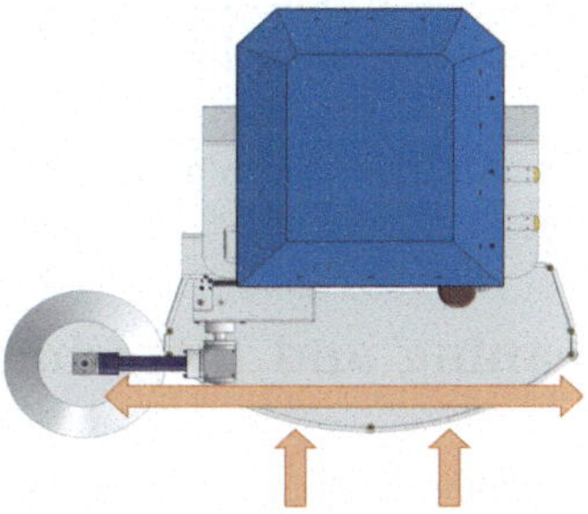

Fig. 6. Three directions of motion detection of the tactile bumper

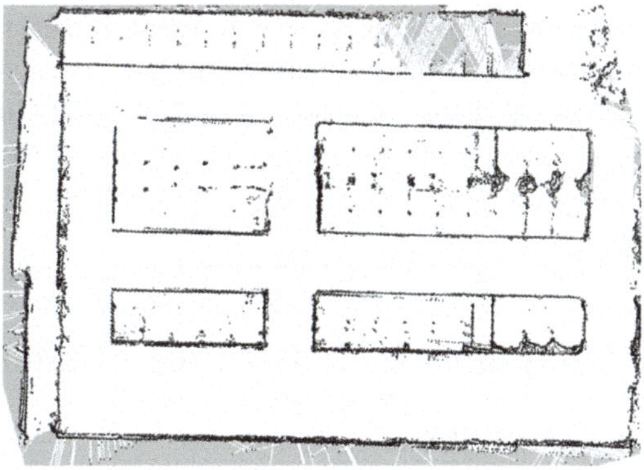

Fig. 7. Superimposed occupancy grid maps of the free stall barn with 60 cubicles at the bavarian state research center for agriculture

the differential steering, only different rotations and straight lines are used for the cleaning robot. This choice has the advantage that a feed-forward control is not necessary and even a simple path tracking control can show a good control behaviour. More important is that the work routes (see Sect. 3.3) are also defined by straight lines and rotations, so that for global planned paths and work routes the path tracking control can be used.

To track the lines of the global planned path, the Lyapunov-based path tracking control of [5] is applied, which is especially developed for skid steered robots. The nonlinear control adjusts the rotational velocity ω as well as the translational velocity v depending on the position and orientation error. For that the error pose

$$\boldsymbol{p}_e = \begin{bmatrix} x_e \\ y_e \\ \theta_e \end{bmatrix} = \begin{bmatrix} \cos\theta & -\sin\theta & 0 \\ \sin\theta & \cos\theta & 0 \\ 0 & 0 & 1 \end{bmatrix} \left(\begin{bmatrix} x_r \\ y_r \\ \theta_r \end{bmatrix} - \begin{bmatrix} x \\ y \\ \theta \end{bmatrix} \right) \tag{1}$$

is calculated from difference of the reference pose x_r, y_r and θ_r and the current pose of the robot x, y and θ and then transformed into the robot frame. From Eq. 1 the lateral error y_e and orientation error θ_e are used to calculate v and ω:

$$v = v_r \cos\theta_e \tag{2}$$

$$\omega = v_r(K_y y_e + K_\theta \sin\theta_e). \tag{3}$$

In Eqs. 2 and 3 $v_r = 0.1\frac{m}{s}$ represents the desired translational velocity and K_y and K_θ are the control variables. For an asymptotic stable control $K_\theta^2 = 4K_y$ has to be applied and $K_y = 9.0$ was determined through experiments till the control showed a good tracking performance. Due to the rotations between the lines in the path a rotation control is implemented as a simple P-control. The error between the θ_r and θ is used to calculate the rotational velocity ω:

$$\omega = p_\theta(\theta_r - \theta). \tag{4}$$

The control parameter p_θ in Eq. 4 is set to $p_\theta = 5.0$. After calculating the velocities with the regarding controls, passive collision avoidance algorithms are applied by using the lidar. For this, polygons around the robot are defined. If enough lidar's points are located in one of the polygons, then the velocities v and ω are reduced. Furthermore, each control has a non-drive strategy, as driving close to obstacles and the angular shape of the robot can bring it to a standstill. These strategies solve this by reversing and rotating.

3 Sequential Control for Cubicle Cleaning and Littering

3.1 Overview

The sequential control to clean cubicles is depicted as a flowchart in Fig. 8. The first step is to fold out the brush. Because of a blocking arm (see Fig. 9) the folding out is executed multiple times by driving forward after a failed trial. Afterwards, the main control loop starts which consists of the brush force control, the wall following and passive collision avoidance. This control loop also drives the electric motors for transporting the bedding from the storage tank and the conveyor belt to eject the bedding onto the cubicles. If there are no relevant sidestep paths from the active collision avoidance due to a occupied cubicles (see Fig. 10), then it will be checked if the segment of the work route (path) is reached. If the end of the path is reached, then the sequential control is done for this segment and the control of the next one can be started, otherwise the main control loop will be executed again. In case the robot drive against a wall (see Fig. 11) before reaching the end of the path, which can occur due to a localization error, then the end of the path is also reached. The path is also interrupted, if the robot hits a wall (see Fig. 11) before reaching the end of the path. This is detected by increased currents in the drive motors, since the wall is a resistance and the control tries to follow the wall. Such an error occurs due to localization errors or poorly planned route. If there is a path for occupied cubicles, the brush will be folded in. If folding in failed the robot drives backwards for a new trial, because the cubicles behind the robot should be unoccupied. The robot rotates subsequently in the direction to the sidestep path, with this rotation the path tracking control has significantly lower overshoot. The sidestep path is tracked and the passive collision avoidance is executed till the end of the sidestep path is reached. If there are no cubicles in front of the robot, the control is done for this work route segment. Otherwise the brush is folded out again and the main control loop is executed again.

3.2 Wall Following with Passive Collision Avoidance

A wall following control is developed for driving along walls and for cleaning the cubicles. The angle for the lateral displacement φ of the tactile bumper is used for this purpose. The translational velocity is set to a constant value $v = 0.1\frac{m}{s}$ and the angular velocity ω is calculated with a P-control

$$\omega = p_{wall}(\varphi_r - \varphi), \tag{5}$$

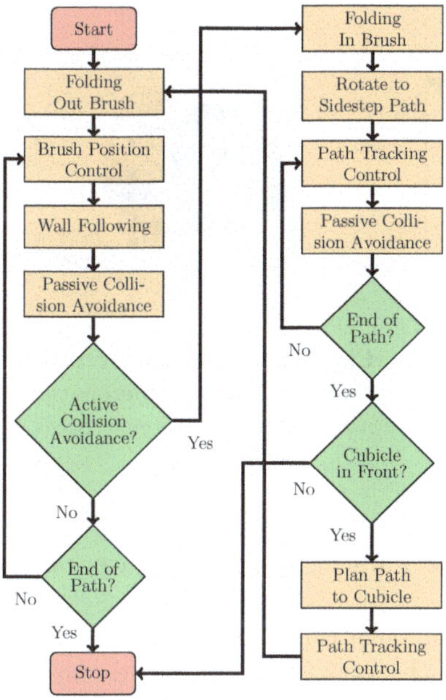

Fig. 8. Flowchart of the sequential control for cubicle cleaning

Fig. 9. Blocked brush at fold in

Fig. 10. Aisle in a barn with occupied cubicles

Fig. 11. Blocked brush by a wall

where the reference angle is chosen to $\varphi_r = 20.2^\circ$ and the control gain is chosen to $p_{wall} = 0.1\frac{rad}{s\,^\circ}$. Experimental results are shown in Fig. 16. However, the control is activated, if the angle reaches the reference angle $\varphi_r = 20.2^\circ$, before that the angular velocity is set to $\omega = -0.05\frac{rad}{s} \approx -2.86\frac{^\circ}{s}$. After determination of the velocities a passive collision avoidance strategy is applied by using the upper front lidar. Static obstacles are filtered from the lidar's point cloud with the help of the occupancy grid maps. If enough lidar points (more than 15 points) are within in the predefined polygons (see Fig. 13), then the velocities are percental reduced. The closer the object the more is the velocity is reduced. This is especially important for the cubicle cleaning. If the next cubicle is occupied, the velocity is reduced, so that the dairy cow does not come under stress and has enough time to escape or to get up. Therefore, the polygons are further in the direction of the right side.

3.3 Semantic Map and Work Route Planning

For the sequential control a GUI was created to define areas over the SLAM-based occupancy grid map and work routes to clean the cubicles, to drive along walls and cleaning running areas. An example for the area definition and route planning is shown in Fig. 14. Three different area types can be defined in the GUI: running area, wall or forbidden area and cubicle area. The quantity of cubicles can be defined for a cubicle area. This information is used in the active

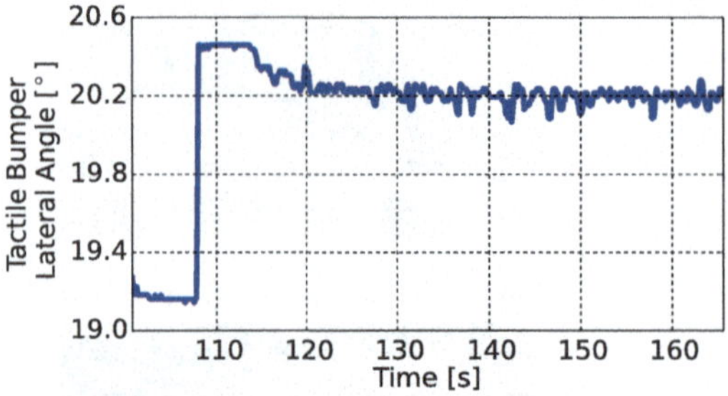

Fig. 12. Control performance of the wall tracking with a desired angle of 20.2°

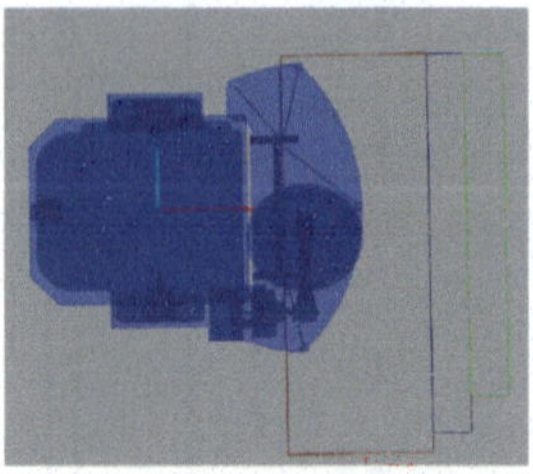

Fig. 13. Polygons for passive collision avoidance - percentual velocity reduction: green = 66%, blue = 33%, red = 10%

collision avoidance to sidestep the occupied cubicle by planning a parallel path to this cubicle.

The work route consists of four different path types: global path, wall cleaning, cubicle cleaning and rotation. The path tracking control and the rotation control from Sect. 2.3 are used for the global paths and for the rotations and the wall following control from Sect. 3.2 is applied for the wall and cubicle cleaning. For each route segment different termination criterion can be defined in the GUI: termination by a global pose, termination by right, left and/or lateral tactile bumper. The work route is planned in the GUI for the front right corner of the tactile bumper because this is the control point for the wall following. The control point for the path tracking control is, however, the midpoint between the driving wheels. That is why the route is translated for this point through transformation the route segments by the displacement from the front right corner of the tactile bumper to the control point ($\Delta x = -0.76$, $\Delta y = 0.85$). If transformed route segments intersect, then the overlapping parts are removed. If a gap exists between route segments, then a global path is added to the work route.

3.4 Active Collision Avoidance

A local planner, such as the popular dynamic window approach [3], is not utilized in the control system. Since there are a lot of dynamic obstacles in a free stall barn and the aisles are narrow for the size of the cleaning robot. These facts are shown in Fig. 10. As a classic local planner was discarded a strategy has to be developed to sidestep an occupied cubicle. As can be seen in Fig. 10, it is not possible to simply fold in the brush and continue the wall following without to come into contact with a dairy cow. The dairy cows are not quite in their cubicle or their bottoms are sticking out of the cubicle. Therefore, the cubicle occupancy is determined using the semantic map from Sect. 3.3 by checking if any points

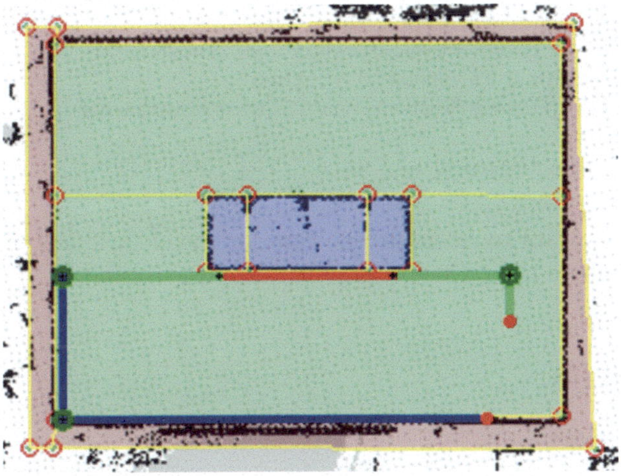

Fig. 14. Exemplary planning of a semantic map (green area: running area, blue area: cubicle area, red area: wall or forbidden area, red circle: area corner) and a work route (green line: global path, red line: cubicle cleaning, blue line: wall cleaning, green dots: rotations, red dots: start and end point)

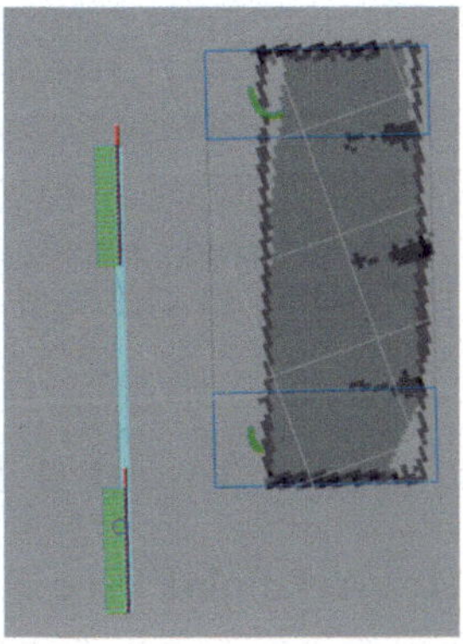

Fig. 15. Active collision avoidance: 1st and 4th, i.e. lower and upper, cubicle (blue rectangle) are occupied (green dots) and two paths parallel to the cubicles are planned as a avoidance path

of the upper front lidar are lying in the cubicle's polygon. For this static points are removed from the lidar's pointcloud with the help of the occupancy grid map which is created through the SLAM from Sect. 2.2. Finally, global paths are planned for occupied cubicles. An example for this is shown in Fig. 15.

3.5 Brush Force Control

It is necessary to adjust the brush position by folding in and folding out the brush for a proper cleaning. The goal is to keep a constant pressure on the cubicle mats. Due to unevenness of the ground and of the cubicle mats and different heights of the cubicles in a barn, it is not possible to drive the brush to a fix position. Therefore, a force control is developed to control the current of the brush motor, since this current represent the pressure on the cubicle mats. The higher is the current, the greater is the pressure, because a higher torque is prerequisite for the higher friction. Algorithm 1 shows the brush force control which consists of two P-controls and which are switched depending on the state.

if Brush-RPM < Low-Velocity-Threshold[20 rpm] **then**
  Fold-RPM = P-Control[-0.01] · (Desired-Brush-RPM[55 rpm] - Brush-RPM)
else if Brush Current < Low-Current-Threshold[5.0 A] **then**
  Fold-RPM = P-Control[1.5] · (Desired-Brush-Current[5.5 A] - Brush-Current)
else
  Fold-RPM = P-Control[0.15] · (Desired-Brush-Current[5.5 A] - Brush-Current)
end if

Algorithm 1: Brush force control to calculate the rpm for the electric motor for folding-in and -out (numbers in brackets represent the chosen values)

4 Experimental Results

4.1 Artificial Free Stall Barn

A simulation based development was pursued in the previous project [10], but building up a more complex robot with the fold-out-in mechanism and tactile bumper in Gazebo [6] is inspected as time-consuming and not target-aimed. Therefore, an artificial free stall barn (see Fig. 17) was build to develop the sequential control. The driven path of an experiment with the different controls and the desired paths and routes are shown in Fig. 12. Here, the robot starts from the right lower corner and tracks the global planned path firstly. Then the planned route from Fig. 14 is executed, but the first cubicle of the four cubicles is occupied, so that a path parallel to the cubicle is planned and tracked. Next, the path to dock to the cubicle is planned and tracked. After that, the robot drives along the edge of the cubicle by using the wall following control (see Sect. 3.2). Simultaneously to the wall following the cubicles are littered additionally by driving the respective electric motors and are cleaned by using the brush force control. The control performance, which is shown in Fig. 18, is sufficient despite of the oscillating current signal. The current is oscillating even the robot is not driving and the brush position is not adapted. Next, the global path to the wall cleaning is tracked and then the robot rotate for starting to drive along the wall.

4.2 Free Stall Dairy Barn at the Bayerische Landesanstalt Für Landtechnik

Experimental tests are executed in this project in a free stall dairy barn at the project partner *Bayerische Landesanstalt für Landtechnik*. In this barn it is able to close gates, so that an empty aisle (see Fig. 19) or an aisle with few dairy cows is available. Experiments in the barn with dairy have shown, that the passive collision avoidance is a suitable approach for preventing contact with and stress on dairy cows. The robot is reducing the velocities smooth and is driving slowly, if dairy cows are located in the driving direction. The dairy cows are really interested in the robot and block it the most of the time. Figure 20 shows the driven path for a test run. In this experiment no dairy cows are located in the aisle, so that no sidestep path was planned and tracked.

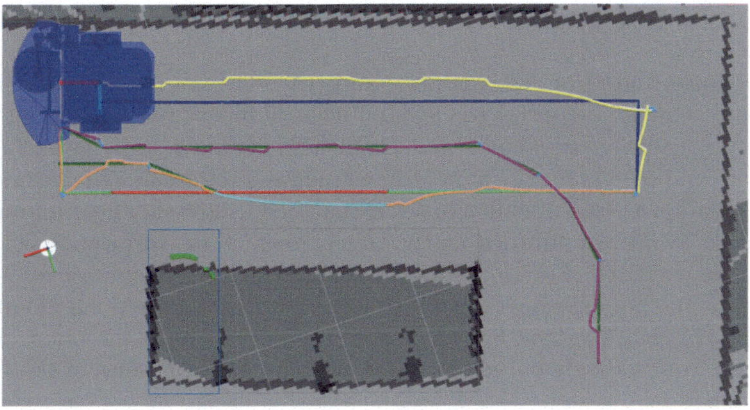

Fig. 16. Experiment in the artificial free stall barn (dark green: planned by SBPL, green: planned path, dark green: sidestep and cubicle docking, red: planned cubicle cleaning, blue: planned wall following, magenta: path tracking control (SBPL path), light blue: rotation, orange: path tracking control (planned route, sidestep and cubicle docking), cyan: cubicle cleaning, yellow: wall following)

Fig. 17. Artificial free stall barn with four cubicles

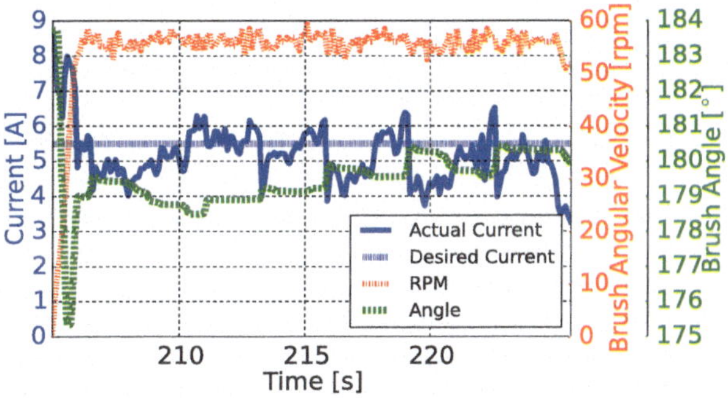

Fig. 18. Performance of the brush force control

Fig. 19. Aisle for testing in a free stall dairy barn

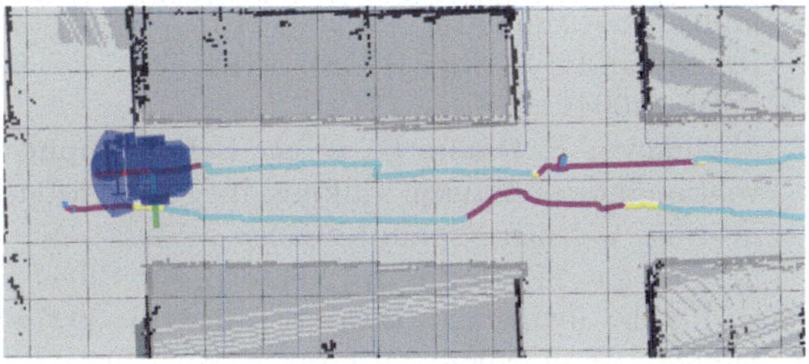

Fig. 20. Experiment in a free stall dairy barn (magenta: path tracking control, light blue: rotation, cyan: cubicle cleaning, yellow: wall following)

5 Conclusion and Outlook

Experimental results from the artificial and the real barn have shown that navigation and sequential control to clean cubicles is functioning regarding to navigate with a collision avoidance and clean cubicles in a stall barn. The most important components for the animal-machine-interaction are passive and active collision avoidance, functionality of which has been showed. These functions are supposed to avoid any injuries at the dairy cows. Now, test series with dairy cows can be executed. For this purpose, the current particle filter for localization must be further developed by filtering the cows out of the point clouds of the 2-D lidar. This will be done by calculating the curvature in the point cloud, so that points from sparsely curved features, such as walls or edges of the cubicles, are remained.

Acknowledgements. The project was supported by funds of the German Government‘s Special Purpose Fund held at Landwirtschaftliche Rentenbank.

References

1. Bhattacharya, S., Likhachev, M., Kumar, V.: Topological constraints in search-based robot path planning. In: Autonomous Robots, vol. 33, pp. 273–290. Springer, Heidelberg (2012)
2. DeVries, T., Aarnoudse, M., Barkema, H., Leslie, K., Von Keyserlingk, M.: Associations of dairy cow behavior, barn hygiene, cow hygiene, and risk of elevated somatic cell count. J. Dairy Sci. **95**, 5730–5739 (2012)
3. Fox, D., Burgard, W., Thrun, S.: The dynamic window approach to collision avoidance. IEEE Robot. Autom. Mag. **4**(1), 23–33 (1997)
4. Grisetti, G., Stachniss, C., Burgard, W.: Improving grid-based slam with rao-blackwellized particle filters by adaptive proposals and selective resampling. In: IEEE International Conference on Robotics and Automation, pp. 2432–2437 (2005)
5. Kanayama, Y., Kimura, Y., Miyazaki, F., Noguchi, T.: A stable tracking control method for an autonomous mobile robot. In: IEEE International Conference on Robotics and Automation, pp. 384–389 (1990)
6. Koenig, N., Howard, A.: Design and use paradigms for gazebo, an open-source multi-robot simulator. In: 2004 IEEE/RSJ International Conference on Intelligent Robots and Systems (IROS) (IEEE Cat. No. 04CH37566), vol. 3, pp. 2149–2154. IEEE (2004)
7. Mačuhová, J., Haidn, B., et al.: Labour input on bavarian dairy farms with conventional or automatic milking. In: International Conference of Agricultural Engineering, Valencia, Spain (2012)
8. Magnusson, M., Herlin, A., Ventorp, M.: Effect of alley floor cleanliness on free-stall and udder hygiene. J. Dairy Sci. **91**(10), 3927–3930 (2008)
9. Quigley, M., Conley, K., Gerkey, B., Faust, J., Foote, T., Leibs, J., Wheeler, R., Ng, A.Y.: Ros: an open-source robot operating system. In: ICRA Workshop on Open Source Software. Kobe, Japan, vol. 3, p. 5 (2009)
10. Robert, M., Lang, T.: Development of simulation based algorithms for livestock robots. In: Landtechnik, vol. 68, pp. 278–280. Kuratorium für Technik und Bauwesen in der Landwirtschaft (KTBL) (2013)

A Version of Libviso2 for Central Dioptric Omnidirectional Cameras with a Laser-Based Scale Calculation

André Aguiar[1(✉)], Filipe Santos[1], Luís Santos[1], and Armando Sousa[2]

[1] INESC TEC - INESC Technology and Science, Porto, Portugal
{andre.s.aguiar,fbsantos,luis.c.santos}@inesctec.pt
[2] Faculty of Engineering of University of Porto, Porto, Portugal
asousa@fe.up.pt

Abstract. Monocular Visual Odometry techniques represent a challenging and appealing research area in robotics navigation field. The use of a single camera to track robot motion is a hardware-cheap solution. In this context, there are few Visual Odometry methods on the literature that estimate robot pose accurately using a single camera without any other source of information. The use of omnidirectional cameras in this field is still not consensual. Many works show that for outdoor environments the use of them does represent an improvement compared with the use of conventional perspective cameras. Besides that, in this work we propose an open-source monocular omnidirectional version of the state-of-the-art method Libviso2 that outperforms the original one even in outdoor scenes. This approach is suitable for central dioptric omnidirectional cameras and takes advantage of their wider field of view to calculate the robot motion with a really positive performance on the context of monocular Visual Odometry. We also propose a novel approach to calculate the scale factor that uses matches between laser measures and 3-D triangulated feature points to do so. The novelty of this work consists in the association of the laser ranges with the features on the *omnidirectional* image. Results were generate using three open-source datasets built in-house showing that our unified system largely outperforms the original monocular version of Libviso2.

1 Introduction

Monocular Visual Odometry (VO) techniques represent a challenging and appealing research area in robotics navigation field. The use of a single camera to track robot motion is a hardware-cheap solution. However, to get high levels of accuracy complex and robust processes are required. Usually, a perspective camera is used and the camera is modeled using the pinhole camera model that projects 3-D points into the image plane. Although this model is suitable for low field of view (FoV) cameras, when this factor is larger than 45° radial distortion impact becomes visible [13]. In fact, for wide FoV cameras this distortion denies the use of the pinhole camera model in VO. To avoid

M. F. Silva et al. (Eds.): ROBOT 2019, AISC 1092, pp. 127–138, 2020.
https://doi.org/10.1007/978-3-030-35990-4_11

that radial distortion degenerates motion estimation a camera model suitable for omnidirectional cameras has to be used. Also, the processes may be adapted to the use of these kind of cameras. If these two prerequisites are respected, the use of wide FoV cameras can be advantageous. It allows to capture more information about the surrounding environment and to track features over a higher number of image frames which can have a high impact. Indoor environments are characterized by scenes where the feature density is sparse and redundant. So, in this kind of spaces capturing a larger scenario is of high importance. Outdoor environments are characterized by more challenging conditions. The variations of illumination, the more recurring presence of moving objects and/or persons and the usual high depth of the scene are conditions that hamper the estimation of the robot motion with accuracy. In this context, the use of omnidirectional cameras in VO in outdoor environments is still not consensual in the literature. Many works compare the performance of conventional perspective cameras with omnidirectional cameras in this and different conclusions are achieved. For example, Alejandro et al. studied the performance of a hyper-catadioptric camera system against a perspective one [10] where the former got superiority in terms of trajectory and orientation in relation with the last. On the contrary, Zichao Zhang et al. showed that the choice between the two types of cameras is not straightforward [17]. Their experiments show that for a fixed image resolution in indoor environments wider FoV cameras should be used. However, for larger scale scenarios, small FoV cameras are preferred.

The use of any kind of standalone camera in VO has an inherent problem that is also still an open issue in the literature: *motion scale*. This phenomenon can be interpreted as the *equal motion produced on an image by a large motion of a distant point and a small motion of a nearby point*. With a single camera it is not possible to have the notion of scene depth which makes impossible the determination of the scale factor without any prior assumption on the motion, robot configuration and/or surrounding environment or without the use of external sources of information such as sensors.

In the context of the mentioned fields of VO, the contribution of this work is twofold. After walking through the state of the literature on Sect. 2 we present an omnidirectional version of Libviso2 suitable for central dioptric omnidirectional cameras on Sect. 3. The implementation is open-source and is available on the official ROS wrapper for Libviso2[1] repository. We also propose a laser-based approach to calculate motion scale on Sect. 4. To test the proposed methodologies we compare the performance of the original Libviso2 approach using a perspective camera and our own using a fisheye camera on Sect. 5. Finally, the conclusions are described in Sect. 6.

[1] https://github.com/srv/viso2.

2 Related Work

Estimating a robot motion using a single camera is a real challenge. In order to optimize VO performance it is convenient to have as many information about the surrounding environment as possible.

In this context, there are nowadays some methods that incorporate omnidirectional cameras in VO systems. For example, Zichao Zhang et al. adapted the state-of-the-art method SVO to work with fisheye or catadioptric cameras [17]. This is done by (1) implementation of a camera model suitable for central omnidirectional cameras, (2) use of reprojection error metrics based on bearing vectors in SVO's BA optimization and (3) sampling the epipolar line based on the unit sphere. An omnidirectional version of DSO is also present in the literature [7]. In this work, they use an unified omnidirectional camera model as a projection function that can both represent perspective and omnidirectional cameras. The camera motion is estimated by the minimization of the photometric error between consecutive frames. In the same way, LSD-SLAM present a version that works with wide FoV cameras [2]. The camera model presented in [7] is used to compute direct image alignment to perform tracking and pixel-wise distance filtering for constructing the map. Jean-Philippe Tardif et al. present a VO system that is executed in urban environments [15]. It uses four cameras with aligned optical centers to simulate an omnidirectional camera. Here, the use of an omnidirectional camera allows to decouple rotation and translation estimation. This method achieves one of the longest paths (2.5 km) reported in VO history showing high accuracy. An approach on Structure for Motion (SfM) using a fisheye camera is presented by Vinay Raju el at. where an epipolar geometry approach is used with 3-D unit vectors to track a camera motion and construct a depth map of the scene [8].

In what concerns to scale estimation, there exist many works that aim to determinate the scale factor using distance sensors such as planar lasers or lidars. For example, Kai Wu et al. present a monocular VO system fused with a laser to improve motion estimation in Astronaut Navigation [16]. They propose a simple calibration between the camera and the laser using the triangulation principle in order to get the image position of every laser measure and calculate the scale drift. Similarly, a work performed by Riccardo Giubilato et al. uses a lidar altimeter to correct scale. Here, laser range data is used as a scale constraint on an optical flow algorithm with a keyframe based tracking and mapping method [4]. Another really interesting approach called LIMO [5] uses a lidar sensor to extract the scene depth and match it with image features estimating motion using bundle adjustment techniques.

In this work we propose an omnidirectional version of Libviso2 [3] that conserves its original matching procedure and computes an epipolar geometry approach that works with central dioptric omnidirectional cameras. We also propose a novel approach to calculate the scale using a planar laser sensor. The novelty of our work consists in the matching procedure of laser measures with features on the omnidirectional image.

3 Visual Odometry Approach

In order to develop a general approach that suits well to central dioptric omnidirectional cameras, a model that describes them accurately is required. After a deep analysis of the literature, the unified camera model [12] proposed by David Scaramuzza et al. was chosen. After having a proper model of the omnidirectional camera, a version of Libviso2 that uses raw monocular omnidirectional images was developed. This approach preserves the original feature matching procedure and presents a new motion tracking algorithm based on epipolar geometry.

3.1 Camera Model

The referenced camera model can be described as follows.

Definition 1. *Let X be a scene point observed by the omnidirectional camera, x'' its projection into the sensor plane, x' in the camera plane and v be the unit vector that emanates from the viewpoint to the scene point. The projection of a point in the camera plane into the unit sphere is given by*

$$\frac{\left[u' \ v' \ f(u',v')\right]^T}{\| \left[u' \ v' \ f(u',v')\right]^T \|}, \ f(u',v') = a_0 + a_1 r' + ... + a_N {r'}^N \tag{1}$$

where $x' = A^{-1}(x'' - t)$ is the affine transformation that converts points in the sensor plane into the camera plane and $f(u',v')$ is the function that estimates depth in function of the euclidean distance $r' = \sqrt{u'^2 + v'^2}$ of the image point to its respective center.

This way, it is possible to convert image points in 3-D unit vectors.

Definition 2. *Let $X = [x \ y \ z]$ be a scene point observed by the omnidirectional camera where z represents its depth and $h(u',v')$ the inverse polynomial of $f(u',v')$. To project this 3-D point into the image the following transformation as to be performed.*

$$x'' = \begin{bmatrix} x\frac{\theta h_1 + \theta^2 h_2 + ... + \theta^N h_N}{\sqrt{x^2+y^2}} \\ y\frac{\theta h_1 + \theta^2 h_2 + ... + \theta^N h_N}{\sqrt{x^2+y^2}} \end{bmatrix} A + \begin{bmatrix} x_c \\ y_c \end{bmatrix}, \ \theta = tan(\frac{z}{\sqrt{x^2+y^2}}) \tag{2}$$

where h_i is the ith coefficient of the inverse polynomial $h(u',v')$, A is the affine matrix of the camera model and $[x_c \ y_c]^T$ is the image center.

To obtain the affine (matrix A) and intrinsic (coefficients of $f(u',v')$ and $h(u',v')$) calibration parameters, we used the calibration toolbox described in [14].

3.2 Method Overview

Libviso2 computes its own features by searching for blobs and corners in the current image. After computing them it tries to find matches in consecutive frames. Given the set of feature matches it estimates motion by calculating the essential matrix that encodes rotation and translation using epipolar geometry.

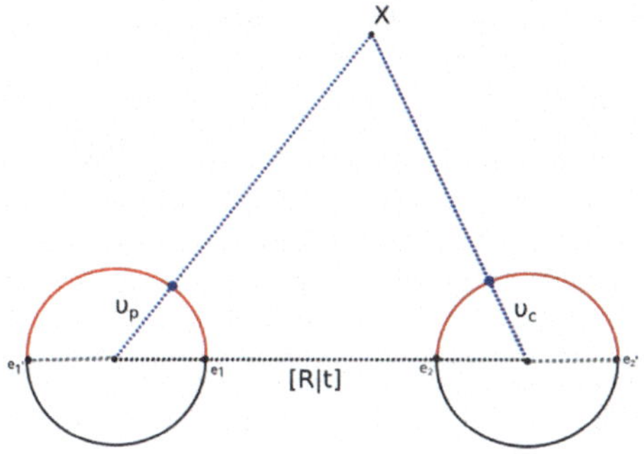

Fig. 1. Epipolar geometry configuration for central dioptric omnidirectional cameras considering the plane that intercepts the unit sphere center and the epipolar curve represented in red.

We propose an epipolar geometry algorithm that works with central dioptric omnidirectional cameras. This being said, given the set of feature matches $\{\mu_p\} \longleftrightarrow \{\mu_c\}$ with $\mu_p = \{x''_{p_1}, x''_{p_2}, ..., x''_{p_n}\}$ and $\mu_c = \{x''_{c_1}, x''_{c_2}, ..., x''_{c_n}\}$ corresponding to distorted points where p subscript denotes previous and c current, we project each match into the unit sphere using Definition 1. This results in a new set of matches that are mapped in the unit sphere which we denote by $\{\eta_p\} \longleftrightarrow \{\eta_c\}$ with $\eta_p = \{\upsilon_{p_1}, \upsilon_{p_2}, ..., \upsilon_{p_n}\}$ and $\eta_c = \{\upsilon_{c_1}, \upsilon_{c_2}, ..., \upsilon_{c_n}\}$. The epipolar geometry configuration can be observed in Fig. 1. It is visible that in this case instead of having epipolar lines as in the traditional planar configuration we have epipolar curves. This means that a point on the unit sphere correspondent to the current image frame that matches a point on the unit sphere correspondent to the last image frame lies on the epipolar curve $e_2e'_2$ [8]. So, after having the image feature matches projected onto the unit sphere, we follow the traditional RANSAC method using eight-point algorithm which is also present in the original monocular version of Libviso2. The key difference of our approach is that we used 3-D points projected in the unit sphere instead of 2-D image points. In this way we can directly calculate the essential matrix using the following definition.

$$\upsilon_{p_i} E \upsilon_{c_i} = 0 \tag{3}$$

Given a match $\{\eta_p\} \longleftrightarrow \{\eta_c\}$ we solve for E in the following way

$$\begin{bmatrix} x_{p_i} \; y_{p_i} \; z_{p_i} \end{bmatrix} E \begin{bmatrix} x_{c_i} \\ y_{c_i} \\ z_{c_i} \end{bmatrix} = 0 \tag{4}$$

which results in

$$\begin{bmatrix} x_{p_1} x_{c_1} \; x_{p_1} y_{c_1} \; x_{p_1} z_{c_1} \; y_{p_1} x_{c_1} \; y_{p_1} y_{c_1} \; y_{p_1} z_{c_1} \; z_{p_1} x_{c_1} \; z_{p_1} y_{c_1} \; z_{p_1} z_{c_1} \end{bmatrix} E' = 0 \tag{5}$$

where $E' = \begin{bmatrix} E_{11} \; E_{12} \; E_{13} \; E_{21} \; E_{22} \; E_{23} \; E_{31} \; E_{32} \; E_{33} \end{bmatrix}^T$. Using single value decomposition (SVD) is possible to extract the solution of E. After that we calculate the inliers using the Sampson Distance. At the end of RANSAC iterations we calculate E using all the inliers computed imposing the *rank 2 constraint* to E. Decomposing E using SVD allows the extraction of four different solutions for rotation R and translation t. As usual, we use linear triangulation to extract the correct solution. The key difference to the traditional approach is, one more time, that we are working in 3-D. So, to triangulate two 3-D points we consider the camera center relatively to the previous frame to be at the origin so we have $P_p = [diag(1, 1, 1)|0]$ and $P_c = [R|t]$. Given this, we have

$$\begin{cases} \alpha v_{p_i} = P_p X_i \\ \alpha v_{p_c} = P_c X_i \end{cases} \Longrightarrow \begin{cases} v_{p_i} \times P_p X_i = 0 \\ v_{c_i} \times P_c X_i = 0 \end{cases} \tag{6}$$

which extended results in

$$\begin{bmatrix} x_{p_i} P_p^3 - z_{p_i} P_p^1 \\ x_{p_i} P_p^2 - y_{p_i} P_p^1 \\ y_{p_i} P_p^3 - z_{p_i} P_p^2 \\ x_{c_i} P_c^3 - z_{c_i} P_c^1 \\ x_{c_i} P_c^2 - y_{c_i} P_c^1 \\ y_{c_i} P_c^3 - z_{c_i} P_c^2 \end{bmatrix} X_i = 0 \tag{7}$$

where $X_i = \begin{bmatrix} x_{t_i} \; y_{t_i} \; z_{t_i} \end{bmatrix}$ and the superscript denotes the *jth* row of the projection matrix. Solving the linear equation for all the matches considering the four possible $[R|t]$ solutions and choosing the one that presents the higher number of 3-D triangulated points with positive depth results in the final solution for the camera motion *up to a scale factor.*

4 Scale Estimation

To scale the motion calculated using the approach previously described, a 2-D planar laser was used. The approach consists in associating the laser measures with 3-D triangulated points to compare both depths and calculate the scale factor.

4.1 Laser and Camera Association

The first phase of this approach is to associate laser measures with 2-D features on the image. This was performed in three main steps: (a) transformation of the laser measures into the camera coordinate system, (b) projection of them into the image and (c) search for associations of the projected measures with feature points.

In the first step, to perform the transformation, the distance from the origin of the laser to the camera center was measured. The transformation $H = [R|t]$ consists in a rotation of laser axis to match with the camera ones direction and a translation relatively to the physical position of the two devices. So, given a distance measure taken by the laser κ_i, it is transformed into a 3-D point in camera's referential in the following way.

$$\psi_i = H \begin{bmatrix} \kappa_i cos(\theta_i) \\ \kappa_i sin(\theta_i) \\ 0 \end{bmatrix} \tag{8}$$

where $\psi_i = [x_{l_i}\ y_{l_i}\ z_{l_i}]^T$ and θ_i is the angle corresponding to the laser measure i. After having the laser measures as 3-D scene points in the camera reference frame, they were converted in 2-D pixel points in the image using Definition 2. So, given the set of 3-D laser points we now have a set of 2-D image points corresponding to the ones that are mapped inside the image, i.e, those who fit in camera's FoV.

The next step consisted in associating the laser measures projected onto the image with 2-D feature points present in the current image frame. As linear searching is computationally expensive, a simplification was performed using the assumption that the standard deviation of y location of laser measures on the image is small. So, while projecting them into the image their average y location was computed. In this way, in a first stage we search for features which present an y distance to the average smaller than 10 pixels. This allowed us to highly reduce the number of features for each we search in the neighborhood of the laser measures. This process consisted in searching in the selected set of features for the ones who present a x pixel location distance smaller than 5 pixels from a laser measure. An example of the resultant projection of laser measures (in black) and associated features (white dots) can be observed in Fig. 2.

4.2 Scale Factor Calculation

From the process previously described we obtained a set of matches between feature points in the current image frame and laser measures projected onto the image. These feature points were already triangulated using their feature matches from the previous image frame. So, we already have a set of matches between the 2-D laser measures and 3-D triangulated feature points and, consequently, the laser measures in the world are also matched with the triangulated points. This being said, given the set of matches $\{\psi_1, ..., \psi_N\} \longleftrightarrow \{X_1, ..., X_N\}$

(a) (b)

Fig. 2. Projection of laser measurements in the fisheye image and 2-D features association with them.

the scale factor is calculated as follows.

$$s = \frac{1}{N}\sum_{i=1}^{N}\frac{||X_i||}{||\psi_i||} \tag{9}$$

In other words, *the scale factor is the average of the relation between the norm of the triangulated matched features and the distances measured by the laser that are matched.* This factor is directly applied to the translation vector extracted from the essential matrix E in the following way.

$$\begin{bmatrix} t_{x_{scaled}} \\ t_{y_{scaled}} \\ t_{z_{scaled}} \end{bmatrix} = \begin{bmatrix} t_x \\ t_y \\ t_z \end{bmatrix} \frac{1}{s} \tag{10}$$

Although in most iterations at least one match between the image features and the laser measures is found, if the feature density is sparse this may not happen. To prevent the motion from not being scaled, in these cases the last computed scale factor is used.

5 Results

5.1 Test Setup

The tests were performed with AgRob V16 [9,11], an agricultural robotic platform for research and development of robotic technologies for Douro Demarcated region (Portugal), UNESCO Heritage place. It is equipped with a set of sensors for navigation and localization such as: IMU, Odometry, GNSS, 2D LIDARs and different types of cameras. The 2D Lidar considered for laser and camera association is a SICK Outdoor LIDAR LMS151-10100 with an angular resolution of 0.25°, an aperture angle of 270° and a scanning range until 50 m. Using this platform we built three open-source datasets[2] which we denote by sequences A, B and C. Each sequence contains fisheye and perspective camera images, laser measures, pose estimation by Hector SLAM [6], wheel odometry and others. For sequences A and C the ground truth used was Hector SLAM. However, for sequence B motion estimation of this method degenerates so we used wheel odometry as ground truth. As this sequence path is mainly rectilinear, odometry does not present much error.

5.2 Motion Estimation

To evaluate the performance of our approach we executed the original version of Libviso2 and our omnidirectional one over the three sequences. Figure 3 present the results for motion estimation of both configurations standalone and fused with a gyroscope resultant for our previous work [1].

Obtaining an accurate and precise robot pose using a monocular configuration is a difficult task. Under the conditions where the three sequences were created this statement gets even stronger. Our main goal is to localize in real time a ground robot in an outdoor environment. The robot motion is characterized by being slow and presenting considerable turbulence. This two factors conjugated represent really harsh conditions to motion estimation. As the robot moves slowly, the effects of turbulence are more visible and can be confused with actual motion. Also, the fact that it is inserted in an outdoor environment makes the conditions even more challenging to obtain a good robot tracking.

In this context, Figs. 3(a), (c) and (e) show the difficulties that the original monocular version of Libviso2 present under the mentioned conditions. Without the KF that fuses the gyroscope, rotations are underestimated and the scale factor is not well calculated. Figures 3(b), (d) and (f) represent the result for motion estimation for the developed omnidirectional version of Libviso2 coupled with the scale calculation approach. Sequence B revealed to be a difficult one. In fact, even the Hector SLAM approach does not provide good results for this sequence. So, for this one our approach does no reveal significant advantages compared with the perspective version. On the other hand, for sequences A and C the improvements are big. Firstly, in terms of rotation the estimation of the omnidirectional version is much smoother and realistic. For example, for sequence A we can see an initial rotation estimation error that propagates over all the sequence but the remaining rotations, even the ones that are almost pure, are well estimated. In fact, the KF fusion shows that if this initial error is eliminated, the motion estimation is very close to the ground truth. Sequence C is more challenging because it presents four almost pure rotations. Although as expected the motion estimation performed by the omnidirectional version presents errors in these cases it largely outperforms the perspective version.

In terms of scale estimation Fig. 3 shows that our approach outperforms the original one. Libviso2 assumes that camera height and pitch are fixed to calculate the ground plane and compute the motion scale. For sequence B this

[2] https://bit.ly/2IzXwow.

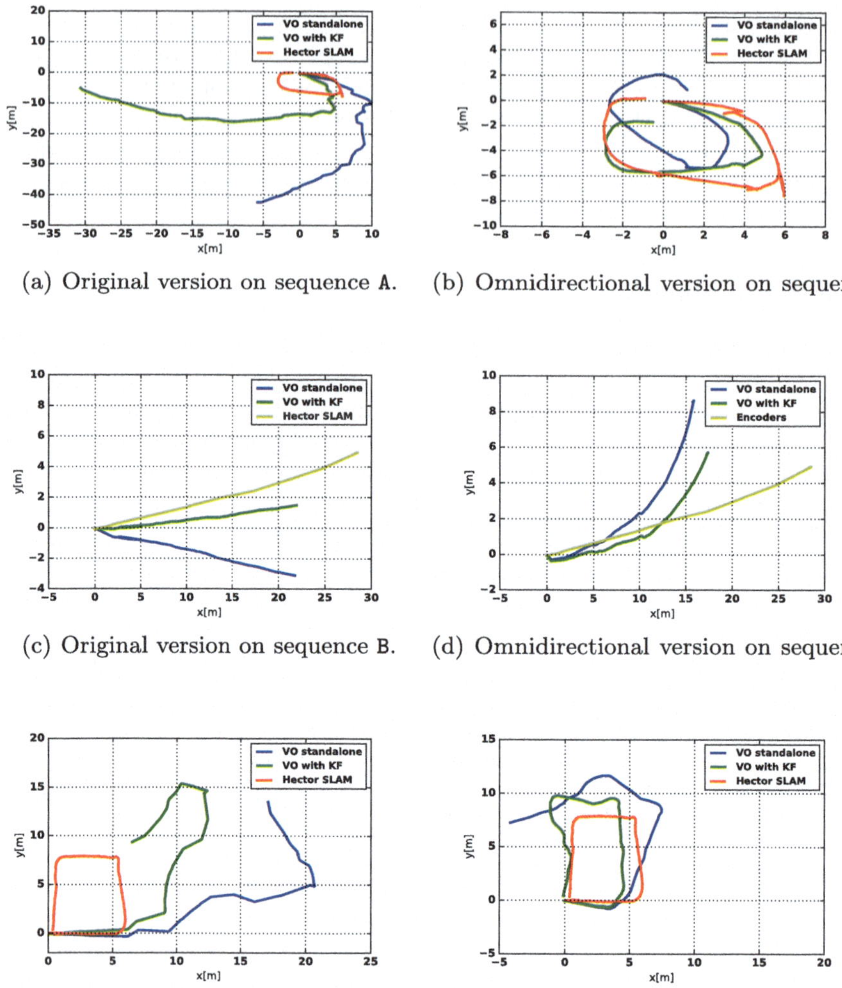

(a) Original version on sequence A. (b) Omnidirectional version on sequence A.

(c) Original version on sequence B. (d) Omnidirectional version on sequence B.

(e) Original version on sequence C. (f) Omnidirectional version on sequence C.

Fig. 3. Performance of the original version of Libviso2 and our omnidirectional version standalone and with the gyroscope fusion on sequences A, B and C.

approach reveals acceptable results. On the contrary, for the other two sequences the estimation of scale largely fails. Our approach, although presenting a visible error, calculates scale with a lot more precision and reveals to be more stable, i.e., the results do not vary from sequence to sequence.

These results although not being perfectly accurate are really positive when analyzed on the context of this work: a monocular visual odometry system in an outdoor environment on a robot with a challenging motion.

6 Conclusion

In this work an open-source omnidirectional version of Libviso2 suitable for central dioptric cameras was developed along side with a laser-based motion scale calculation. This work was accepted as an omnidirectional version of Libviso2 and is now public available on the official ROS wrapper for Libviso2[3]. The approach to calculate the motion scale presents the novelty of matching laser range measures with 2-D features in the *omnidirectional* image. To test the proposed methodologies three available and built in-house datasets were created. The unified developed system outperforms the original one consisting in an improvement of a well known state-of-the-art VO method by the creation of a new version of it.

In a future work we aim to test our system with catadioptric cameras since the camera model used is also suitable for these kind of cameras. Thus, this makes us believe that the omnidirectional VO system will also work with these devices which is really positive since they provide 360° of FoV. Also, we would like to improve our VO approach recurring to a bundle adjustment technique based on the minimization of the reprojection error using the 3-D triangulated feature points.

Acknowledgment. This work is co-financed by the European Regional Development Fund (ERDF) through the Interreg V-A Espanha-Portugal Programme (POCTEP) 2014–2020 within project 0095_BIOTECFOR_1_P. This work also was co-financed by the ERDF European Regional Development Fund through the Operational Programme for Competitiveness and Internationalisation - COMPETE 2020 under the PORTUGAL 2020 Partnership Agreement, and through the Portuguese National Innovation Agency (ANI) as a part of project "ROMOVI: POCI-01-0247-FEDER-017945" The opinions included in this paper shall be the sole responsibility of their authors. The European Commission and the Authorities of the Programme aren't responsible for the use of information contained therein.

References

1. Aguiar, A., Sousa, A., Santos, F., Oliveira, M.: Monocular visual odometry benchmarking and turn performance optimization. In: 19th IEEE International Conference on Autonomous Robot Systems and Competitions, April 2019
2. Caruso, D., Engel, J., Cremers, D.: Large-scale direct SLAM for omnidirectional cameras. In: IEEE/RSJ International Conference on Intelligent Robots and Systems (IROS). IEEE, September 2015
3. Geiger, A., Ziegler, J., Stiller, C.: StereoScan: Dense 3D reconstruction in real-time. In: IEEE Intelligent Vehicles Symposium (IV). IEEE, June 2011
4. Giubilato, R., Chiodini, S., Pertile, M., Debei, S.: Scale correct monocular visual odometry using a lidar altimeter. In: IEEE/RSJ International Conference on Intelligent Robots and Systems (IROS), pp. 3694–3700, October 2018
5. Gräter, J., Wilczynski, A., Lauer, M.: LIMO: lidar-monocular visual odometry. CoRR abs/1807.07524 (2018). http://arxiv.org/abs/1807.07524

[3] https://github.com/srv/viso2.

6. Kohlbrecher, S., von Stryk, O., Meyer, J., Klingauf, U.: A flexible and scalable slam system with full 3D motion estimation. In: IEEE International Symposium on Safety, Security, and Rescue Robotics, pp. 155–160, November 2011
7. Matsuki, H., von Stumberg, L., Usenko, V., Stueckler, J., Cremers, D.: Omnidirectional DSO: Direct sparse odometry with fisheye cameras. In: IEEE Robotics and Automation Letters (RA-L) & International Conference on Intelligent Robots and Systems (IROS) (2018)
8. Raju, V.K.T.P.: Fisheye camera calibration and applications. Master's thesis, Arizona State University (2014)
9. Reis, R., Mendes, J., Neves dos Santos, F., Morais, R., Ferraz, N., Santos, L., Sousa, A.: Redundant robot localization system based in wireless sensor network. In: IEEE International Conference on Autonomous Robot Systems and Competitions (ICARSC), pp. 154–159, April 2018
10. Rituerto, A., Puig, L., Guerrero, J.J.: Comparison of omnidirectional and conventional monocular systems for visual SLAM
11. Santos, L., Ferraz, N., Neves dos Santos, F., Mendes, J., Morais, R., Costa, P., Reis, R.: Path planning aware of soil compaction for steep slope vineyards. In: IEEE International Conference on Autonomous Robot Systems and Competitions (ICARSC), pp. 250–255, April 2018
12. Scaramuzza, D., Martinelli, A., Siegwart, R.: A flexible technique for accurate omnidirectional camera calibration and structure from motion. In: Fourth IEEE International Conference on Computer Vision Systems (ICVS 2006). IEEE (2006)
13. Scaramuzza, D., Fraundorfer, F.: Visual odometry [tutorial]. IEEE Robot. Autom. Mag. **18**(4), 80–92 (2011)
14. Scaramuzza, D., Martinelli, A., Siegwart, R.: A toolbox for easily calibrating omnidirectional cameras. In: IEEE/RSJ International Conference on Intelligent Robots and Systems. IEEE, October 2006
15. Tardif, J.P., Pavlidis, Y., Daniilidis, K.: Monocular visual odometry in urban environments using an omnidirectional camera. In: IEEE/RSJ International Conference on Intelligent Robots and Systems. IEEE, September 2008
16. Wu, K., Di, K., Sun, X., Wan, W., Liu, Z.: Enhanced monocular visual odometry integrated with laser distance meter for astronaut navigation. Sensors **14**, 4981–5003 (2014)
17. Zhang, Z., Rebecq, H., Forster, C., Scaramuzza, D.: Benefit of large field-of-view cameras for visual odometry. In: IEEE International Conference on Robotics and Automation (ICRA). IEEE, May 2016

Deep Learning Applications in Agriculture: A Short Review

Luís Santos[1,2(✉)], Filipe N. Santos[1], Paulo Moura Oliveira[1,2], and Pranjali Shinde[1]

[1] INESC TEC - INESC Technology and Science, Porto, Portugal
{luis.c.santos,fbnsantos,pranjali.shinde}@inesctec.pt
[2] UTAD - University of Trás-os-Montes e Alto Douro, Vila Real, Portugal
oliveira@utad.pt

Abstract. Deep learning (DL) incorporates a modern technique for image processing and big data analysis with large potential. Deep learning is a recent tool in the agricultural domain, being already successfully applied to other domains. This article performs a survey of different deep learning techniques applied to various agricultural problems, such as disease detection/identification, fruit/plants classification and fruit counting among other domains. The paper analyses the specific employed models, the source of the data, the performance of each study, the employed hardware and the possibility of real-time application to study eventual integration with autonomous robotic platforms. The conclusions indicate that deep learning provides high accuracy results, surpassing, with occasional exceptions, alternative traditional image processing techniques in terms of accuracy.

Keywords: Deep learning · Agriculture · Image processing · Survey

1 Introduction

Agriculture is a relevant activity for the global economy. Along the years, this sector suffered several changes to fulfil the world's growing population, which has doubled in the last 50 years [20]. There are several predictions for the continuous growth of the world population, is expected to have 9 billion people on earth in the year 2050, a 60% increase. Besides, the predictions indicate an increase of people living in urban areas and a decrease in the ratio between working people and retired people [31,34]. This means that the world's agriculture productivity has to increase sustainably, and more independent of human work. Technology was introduced to agriculture more than one century ago, being the first tractor presented in 1913. Nowadays, mechanical technology has an incredible evolution with a huge amount of commercial technology available [43]. This increased the productivity resulting in minimal human labour. However, it may not be enough to sustain the world's demand along the future years. To improve the production efficiency, several studies were performed since the 1990s, originating the concept

M. F. Silva et al. (Eds.): ROBOT 2019, AISC 1092, pp. 139–151, 2020.
https://doi.org/10.1007/978-3-030-35990-4_12

of "precision agriculture", a farm management notion based on the observation, measurement and actuation to the variability in the crops, with the goal of optimizing the returns while preserving resources [30]. More recently, existing technologies common in the industry have been applied to agriculture, such as remote sensing [1], Internet of Things (IoT) [33], and robotics platforms [38,39] leading to the concept of "smart farming" [47]. Smart farming is important to face the challenges of agricultural production in terms of productivity, environmental impact, food security and sustainability. To address these challenges it is necessary to analyze and understand the agricultural ecosystems which imply constant monitoring of different variables. This creates huge amounts of data that needs to be stored and processed in real time for some operations [18]. This data can be constituted by images, which can be processed with different image analysis techniques for the identification of plants, diseases, etc., in different agricultural contexts. Some of the image processing techniques are based on machine learning processes such as K-means, support vector machines (SVM) and artificial neural networks (ANN) [41]. Deep learning (DL) [42], is a modern approach that has been successfully employed in various domains. DL is inserted in machine learning techniques, being similar to ANN, with better learning capability, and thus has a higher classification accuracy [17]. There are some approaches with specific hardware like Field-programmable Gate Array (FPGA) [48] or Graphics Processing Unit (GPU) [5] used in order to accelerate the processing time of complex DL models. Several DL techniques have been applied to different agricultural methods with a rise of popularity in recent years. Kamilaris *et al.* [17,19] present a review work of the DL applications in agriculture. Other wider review works also include some DL techniques [25,41]. However, these works do not check parameters like time or hardware restrictions imposed by the complexity of deep learning models. With the exponential growth of this area, it is possible to find in literature a big number of new research works applying DL to agriculture. So, this paper focuses on the review of recent applications of DL techniques to several agricultural domains considering the employed hardware to run the application. To the best of our knowledge, the works mentioned in this paper are not covered by others review documents. Sect. 2 of this article presents the methodology used for this review work. Section 3 presents a brief explanation of the Deep Learning concept. In Sect. 4 all the related work with deep learning applied to agriculture is presented and analyzed. Section 5 presents the conclusions of the work.

2 Methodology

Two main steps are on the basis of this review: search for recent related work with no more than four years; and review and analysis of the collected work. The collection of the related work was performed between April and May 2019 mainly resorting to the scientific search engine Google Scholar. The search was open to any deep learning method and 29 papers were selected from 14 agricultural areas. The analysis of the related work intends to answer the following issues:

(i) problem identification; (ii) type of data used; (iii) deep learning architecture; (iv) deep learning model; (v) overall accuracy; (vi) comparison to alternative machine learning techniques as well as similar works existing in the literature; (vii) employed hardware and real-time application. Most of the works present the Top-1 and Top-5 accuracy, and we present only the Top-1 accuracy. Top-1 is the conventional accuracy: the model answer must be exactly the expected answer. Top-5 accuracy means that any of the model 5 highest probability answers must match the expected answer. Some works use F1 score as a quality metrics, which is adequate for binary classification problems. Usually, the authors give a detailed explanation about their quality metrics, for example, Smith *et al.* [44] present an explanation of the metric F1 score.

3 Deep Learning

DL is part of machine learning methods based on artificial neural networks (ANN). Deep learning has varied applications from natural language to image processing [22]. Different deep learning architectures such as Deep Neural Networks (DNN), Deep Belief Networks (DBN), Recurrent Neural Networks (RNN), recursive neural networks, Fully Convolutional Networks (FCN) and Convolutional Neural Networks (CNN) have been successfully applied to diverse research areas, including agriculture. DL extends the complexity of ANN and represents data hierarchically, through multiple layers of abstraction. For example, in image processing lower levels can identify edges, while higher levels will identify items like objects or faces [6,19]. There are several frameworks, architectures and data-sets publicly available for researchers to build their own model. Frameworks like CAFFE, TensorFlow, Theano, DL Matlab toolbox are some of the most popular tools to experiment with DL and they are used by some of the works reviewed in this paper. AlexNet, VGG, CaffeNet, GoogleNet and ResNet are popular DL models publicly available for research, with the advantage that most of them are already pre-trained with open data-sets, which means that the network is ready for successfully identifying several features. Image-Net[1] or COCO[2] data-set are common among these architectures. DL main disadvantage might be the long training time and the necessity of powerful hardware suitable for parallel programming (GPU, FPGA), while classic methods like Support Vector Machine (SVM) or Scale-invariant feature transform (SIFT) have simpler training processes. However, testing time is faster in DL methods, and generally more accurate [19].

The following sub-sections present a brief introduction to CNN, FCN, and RNN the principal architectures found in this literature review.

3.1 Convolutional Neural Networks

CNN's are a class of deep, feed-forward ANN which has been applied to solve computer vision problems. This is the most common architecture used almost in

[1] Image-net data-set: http://image-net.org/.
[2] COCO data-set: http://cocodataset.org/.

every paper approached in this review. In contrast to ANN, CNN is capable of learning complex problems relatively fast due to the weight sharing and complex models, which allow parallelization. Figure 1 shows an example of a CNN model, where the various convolutions are performed at some layers of the network, creating different representations of the dataset. The convolutional layers act as feature extractors from input images whose dimensionality is reduced by the pooling layers. The fully connected layers act as classifiers, exploiting high-level features to classify input images in its corresponding class.

CNN's increase the probability of correct classifications, provided that adequate large data-set is available. A common technique to increase the data-set and improve CNN accuracy is the data augmentation. As CNN's are invariant to translation, size or illumination, the dataset can be enlarged by applying a series of random transformations to the original images, such as rotation and translation [7,17].

3.2 Fully Convolutional Networks

FCN is a CNN based architecture that uses down-sampling (convolution) followed by up-sampling (deconvolution) to produce semantic mask as output [4]. Typically, the input image is downsized to pass through several convolution layers, and the output is one predicted label for the input image. FCN networks do not downsize the image, thus the output is not a single label, which allow to up-sample the output and predict the pixel-wise of the class [27]. Figure 2 shows an example of a semantic segmentation with FCN.

3.3 Recurrent Neural Networks

RNN is a multi-temporal network model which presents connections between nodes with direct graphs along a temporal sequence. This allows temporal dynamic behaviour. Unlike mono-temporal models (CNN's), RNN has access to information in its own internal memory, which means that it can consult the previous observation for the classification of the current observation. This makes it suitable for temporal sequence tasks as speech recognition [24,37].

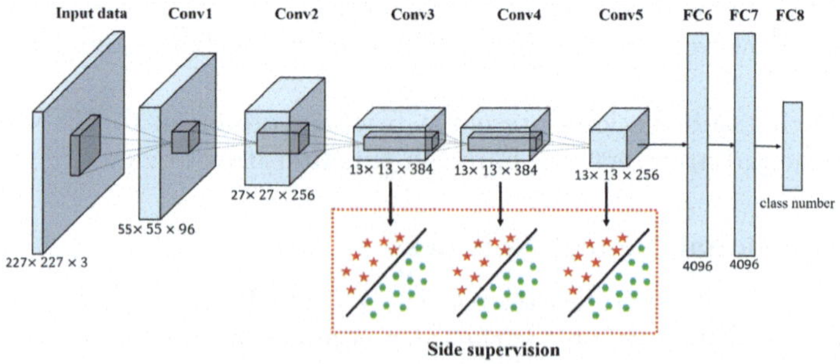

Fig. 1. AlexNet: an example CNN architecture model [14]

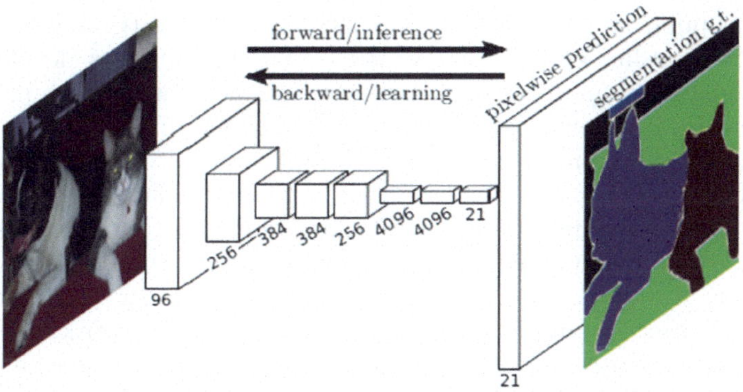

Fig. 2. FCN semantic segmentation example [27]

4 Deep Learning Applications in Agriculture

DL applications in agriculture are spread over several areas, being that 14 areas were identified in a total of 43 recent papers. For this review, only 29 papers got selected, however, the list of the remaining papers is publicly available[3]. The most popular areas are the disease identification (6 papers), plants recognition (4 papers) and land cover and weed identification (3 papers each). CNN's have been the most popular architecture used in 24 papers while the few alternatives are based in FCN or RNN.

The oldest article referred in this review was published in 2016 (one article). There are 3 articles published in 2017, 17 published in 2018 and 8 articles from 2019. Table 1 present a list of the selected papers and short answers for the questions presented in Sect. 2.

The authors from the several reviewed papers used different DL architectures such as CNN, FCN, RNN, Single Shot MultiBox Detector (SSD) and Bidirectional Long Short-Term Memory (BLSTM). However, the current analysis shows that CNN has been clearly the most popular DL architecture in the past years. From the 29 papers approached in this review, only 7 resorted to others architectures, and even in this small group, CNN is just excluded in 2 works [26,37]. Some authors resorted to the combination of CNN/FCN [45] or CNN/RNN [53], while others make use of Region-Based Convolutional Neural Network (R-CNN) and/or Region Based Fully Convolutional Neural Networks (R-FCN). As the name suggests, R-CNN bypasses the problem of selecting a huge number of regions to classify, instead, only some regions are classified, improving the classification time. Fast R-CNN and Faster R-CNN are improved versions of R-CNN built in order to achieve real-time object identification, being that Faster R-CNN is used in 3 papers [4,12,15]. Following a similar logic, R-FCN improves speed classification by reducing the amount of work in the classification process and it is used is 2 papers [12,15]. Most of the authors resorted to known architec-

[3] Review List: https://bit.ly/2ZtS8tA.

tures models publicly available such as VGG, AlexNet, MatConvnet, DenseNet, Darknet (Yolo) using known frameworks (FW). The authors developed their own DL model in about 12 articles, however, it is not always clear in the paper if the author built its own model from scratch, or modified one of the existing models. Most of the authors created their own data-set in various ways such as manually-captured images, images from large public data-sets like ImageNet or COCO [45], Unmanned Ground Vehicles (UGV) [45] or Unmanned Aerial Vehicles (UAV) [36] equipped with sensors to capture the desired data. Data augmentation was a common practice among the majority of the authors to enlarge the dataset and improve the results of the neural network. The majority of the papers compares their DL approach with alternative machine learning methods such as: SVM, SIFT, Random Forest (RF), backpropagation (BP), Particle Swarm Optimization (PSO), Multilayer Perceptron (MLP), AdaBoost [40], Principal Component Analysis (PCA) plus Kernel SVM (kSVM), Random Forests Uncorrelated Plant (RFUP) and Random Forests Correlated Field (RFCF). The overall accuracy of the reviewed papers is generally good and surpasses alternative machine learning methods with occasional exceptions where the traditional methods have the same or higher accuracy [40]. The hardware restriction imposed by the complexity of DL is clear in this review, as only 4 papers worked just with regular Central Processing Units (CPU). Most of the authors resort to high rated GPUs, and Lammie *et al.* [22] used a set of GPU and FPGA to accelerate their CNN model. Even with this power-full hardware, only 11 papers claim to be capable of performing the test operation in real time. The rest of the authors do not provide information about the testing time of their networks.

This analysis leads to the conclusion that DL techniques in agriculture are rising at an incredibly high rate with lots of possible applications and combination with aerial our ground platforms such as UAVs or UGVs, as DL can be used in the process of automatizing these machines, by collecting and processing the data to give an order to the autonomous vehicle. However, in this review, the authors only used unmanned platforms for collecting data. To fully automatize a robotic platform using deep learning might have some restrictions due to the hardware demand. GPUs, the most used hardware platform to process DL models, are subject to large power requirements [22] and this could represent a limitation to a robotic platform that is typically electric and powered by batteries. To run the necessary DL tools in a remote computer, and thus save the battery life of unmanned platforms is not always feasible for remote harsh agricultural terrains. FPGAs could represent a low power alternative but the research with these platforms in agriculture is not yet popular with only one article [22] found for this review. As the review shows, DL surpasses most of the traditional machine learning methods in terms of accuracy, and real-time application, which is crucial for autonomous vehicles is hard to accomplish without DL approaches, as the testing time of other methods is longer. As example, Mendes *et al.* [32] present a work for Autonomous Ground Robot (AGV) localization in a steep slope vineyard. They use an SVM classifier to detect natural

Table 1. Different deep learning applications in agriculture

Agricultural Area	i) Problem Description	ii) Data Used	iii) DL Architecture	iv) DL Model	v) Overall Accuracy (Top1 accuracy)	vi) Comparison to other methods accuracy	vii) Hardware / Real Time?	Ref
Plant Disease Identification	Find suitable DL architecture for real-time tomato diseases and pest recognition	5000 images dataset created by the authors in several farms from Korea	Faster R-CNN, SDD and R-FCN	VGG ResNet ResNeXt	86 % (R-FCN)	N/A	GPU / Yes	[12]
	Recognize cucumber diseases using leaf symptons images	14208 images dataset created by the authors	CNN	Developed by authors	93.4 %	AlexNet: 94% SVM: 81.9% RF: 84.8%	GPU (Nvidia Quadro) / N/A	[29]
	Detection of ESCA disease in Bordeaux Vineyards	6000 images dataset created by the authors in two Bordeaux Vineyards	CNN	MobileNet	90.7%	SIFT Encoded : 87.9%	GPU (Nvidia GTX) / Yes	[35]
	Identification of plant/disease combination in 25 different species with 58 classes	Open database with 87484 photos of healthy and infected leaves	CNN	AlexNet, VGG, GoogleNet and Overfeat	99.48% (VGG)	N/A	GPU (Nvidia GTX) / Yes	[11]
	Identify 10 common rice diseases in leaves	500 images of healthy and diseased rice leaves from public datasets and books	CNN	Developed by authors	95.48%	BP: 92% SVM: 91% PSO: 88%	N/A / N/A	[28]
	Identification of 8 kinds of maize leaves diseases	500 images collected from public sources (Plant Village dataset and Google Websites)	CNN	Modified GoogleNet and Cifar10 (CAFFE FW)	98.9% (GoogleNet)	N/A	GPU / N/A	[52]
Plant Recognition	Identification of 16 plant species	DataSet collected by governmental project over 1200 agro-stations	CNN	Developed by authors	97.47%	SVM : 89.94%	GPU (Nvidia Quadro) / N/A	[50]
	Identify until 184 different species	LeafSnap dataset, FOLIAGE dataset and FLAVIA dataset	CNN	LeafNet (Developed by authors)	LeafSnap: 86.3% FOLIAGE: 95.8% FLAVIA: 97.9%	SVM accuracy: LeafSnap: 79.66% FOLIAGE: 98.75% FLAVIA: 98.69%	GPU (Nvidia Gtx) / N/A	[3]
	Species and trait recognition from herbarium scans	Images of 1000 species on GBIF dataset (subset of 170 species)	CNN	Modified ResNet model (TensorFlow FW)	82.4% (96.3% Top5)	Alternative DL approach: 90.3% (Top5 accuracy)	GPU (Nvidia Titan) / N/A	[51]
	Identification of 1000 species	LifeClef 2015 dataset (90 000 images)	CNN	AlexNet, GoogleNet and VGGNet (CAFFE FW)	80%	N/A	GPU (Nvidia Tesla) / N/A	[13]

Table 2. Different deep learning applications in agriculture

Agricultural Area	i) Problem Description	ii) Data Used	iii) DL Architecture	iv) DL Model	v) Overall Accuracy (Top-1 accuracy)	vi) Comparison to other methods	vii) Hardware / Real Time?	Ref
Land Cover Classification	Identify 13 different crop types	Multi-temporal LandSat enchaced Vegetation Index	RNN and CNN	LSTM and Conv1D	85.54% (Conv1D)	MLP accuracy: 83.81% XGBoost: 84.17% RF: 84.09% SVM: 83.09%	GPU (Nvidia Quadro) / N/A	[54]
	Crop mapping of 14 different species	Multi-temporal Sentinel-1 SAR images	CNN	Keras (TensorFlow FW)	91%	N/A	CPU (Intel Xeon) / N/A	[23]
	Identify 19 crop types	Sentinel-2A observations	RNN	LSTM	84.4%	Other RNN: 83.4% CNN: 76.8% SVM: 40.9%	N/A / N/A	[37]
Weed classification	Weed detection and classification in soybean crops	400 crop images captured by the authors with UAV	CNN	CaffeNet (CAFFE FW)	98%	SVM: 98% AdaBoost: 98.2 % Random Forest: 96 %	GPU (Nvidia TITAN) / N/A	[40]
	Weed detection and classification by spectral band analysis	200 Hyperspectral images with 61 bands	CNN	MatConvnet	94.72 %	HoG: 74.34 %	CPU (intel i7) / N/A	[10]
	Accelerate a DL approach with FPGA for weed classification with 8 classes	18 000 weed images from DeepWeedX dataset	CNN	VGG-16, DenseNet-128-10,	90.08% (DenseNet)	Other work approach: ResNet: 95.7 %	GPU and FPGA (intel DE1-SoC) / Yes	[22]
Seed classification	Estimate number of seeds into soybeans on pods**	Pods photography over ligthbox. Dataset created by authors	CNN	Developed by authors (Theano FW)	82.7%	SVM: 50.4 %	GPU (Nvidia GTX) / N/A	[47]
	Classify and sort 6 kinds of seeds	Pictures of hundreds of thousands of seeds (created by authors)	CNN	ResNet-18	99 %	YOLO9000: 86% SSD: 94 %	4 CPU and 1 GPU /Yes (500 fps)	[16]
Fruit counting	Detect mango fruits in trees canopies and estimate fruit load	1300 images acquired by authors at 5 mango orchards	CNN	Darknet (MangoYolo - YOLO modified by authors)	98.3 %	R-CNN: 95.3% SSD: 98.3 % YOLOv3: 96.7% YOLOv2: 95.9 %	GPU (Nvidia GTX) /Yes (14 fps)	[21]
Fruit classification	Combine DL, tracking and SfM for robust visible fruit counting on orange and apple orchards	N/A	FCN	Developed by authors	Orange count: 99 % Apple count: 97 %	N/A	N/A / N/A	[26]
	Classification of 18 types of fruit	3600 images acquired by authors and downloaded from public websites	CNN	Developed by authors	94.94 %	PCA + kSVM: 89.11% WE+BBO: 89.47% FRFE + BPNN: 88.99%	GPU (Nvidia GTX) / Yes	[53]

Table 3. Different deep learning applications in agriculture

Agricultural Area	i) Problem Description	ii) Data Used	iii) DL Architecture	iv) DL Model	v) Overall Accuracy (Top-1 accuracy)	vi) Comparison to other methods	vii) Hardware / Real Time?	Ref
Soil/Root segmentation	Segment Soil/ /root in X-ray tomography with CNN+SVM	Imagenet Dataset	CNN	N/A	0.57 (Specific quality metric)	N/A	CPU (intel Xeon) and GPU(AMD) / N/A	[8]
	Detect and quantify chicory roots	50 annotated Chicory root images created by authors	CNN	U-Net	99.7 %	FrangiNet: 99.6%	GPU (Nvidia Titan) / N/A	[44]
Cattle detection	Detect and count cattle in UAV images	13520 images captured during UAV flight	CNN	Developed by authors	95.5 %	N/A	CPU (Intel i7) / 3.2s/frame	[36]
Precision Irrigation	Inferring moisture conditions from images	Simulated large datasets of 1200 aerial images of Vineyards	CNN	CNNUP and CNNCF	0.034 (mean error)	SVM: 0.038 RFUP: 0.060 RFCF: 0.094 2 layer NN: 0.086 (mean error)	1 CPU (intel i7) and 3 GPU (Nvidia Titan) / N/A	[46]
Obstacle Detection	Detect humans, obstacles and traversable obstacle agricultural fields for safe machinery operation**	ImageNet and COCO dataset	CNN and FCN	Darknet (improved YOLO), VGG and DeepAnomal (AlexNet)	70.81 % (accuracy of fusion of different approaches)	N/A	N/A / Yes	[45]
Measure Features	Stalk count and stalk width of crops	400 Stereo camera Images collected with robotic platform	Faster RCNN and FCN	StalkNet (created by authors)	10% error for count; 2.76 mm error for measure	N/A	GPU (Nvidia GTX)/ 2 fps	[4]
Crop yield estimation	Yeld prognosis in vineyards by object counting	50 GB of Images from 2 vineyards at different crop stages	Faster R-CNN, R-FCN, SSD	TensorFlow FW	99.6% (F.R-CNN for object detection)	N/A	N/A / Yes	[15]
Automatic labeling of agricultural regulations	Translation of phytosanitary regulations into formal rules	N/A	CNN and BLSTM	Developed by authors	88.3% (F1 score)	N/A	GPU (Nvidia Quadro) / N/A	[9]

features, which requires an accurate and real-time detection, something that this work is not capable to provide. Even with its GPU implementation [2], the real-time detection is not assured. With other DL approaches, as seen on the works covered by this review, it could be possible to perform a fast and more accurate detection, solving this localization problem. To conclude, DL is in a good path to collaborate with diverse agricultural areas, and the focus of research should proceed to low power demanding hardware and integration with autonomous platforms.

5 Conclusion

The current paper presented a review of DL based research efforts applied to agricultural domains. It examined the agricultural area and described the problem they focus on, listed technical details such as DL architecture and model, described the data source, reported the overall accuracy of each work compared to alternative methods, verified the employed hardware and possible real-time application. The findings indigent that DL reached high accuracy in the majority of the reviewed works, scoring higher precision than other traditional technique. The mains advantage to non DL techniques is the relatively low time for the classification process, which allows an easier execution in real time. The hardware restriction might be the main disadvantage, as they require mostly powerful GPUs, which are subject to large power demands. The alternative lower power demand systems (FPGAs) are still with low research work in the agricultural domain. For a proper integration in real time with autonomous robotic platforms (AGVs) the research should focus on decreasing the complexity and hardware demand for the existing effective DL techniques.

Acknowledgements. This work is co-financed by the European Regional Development Fund (ERDF) through the Interreg V-A Espanha-Portugal Programme (POCTEP) 2014-2020 within project 0095_BIOTECFOR_1_P. This work also was co-financed by the ERDF European Regional Development Fund through the Operational Programme for Competitiveness and Internationalisation - COMPETE 2020 under the PORTUGAL 2020 Partnership Agreement, and through the Portuguese National Innovation Agency (ANI) as a part of project "ROMOVI: POCI-01-0247-FEDER-017945" The opinions included in this paper shall be the sole responsibility of their authors. The European Commission and the Authorities of the Programme aren't responsible for the use of information contained therein.

References

1. Atzberger, C.: Advances in remote sensing of agriculture: context description, existing operational monitoring systems and major information needs. Remote Sens. **5**(2), 949–981 (2013)
2. Azevedo, F., Shinde, P., Santos, L., Mendes, J., Santos, F.N., Mendonça, H.: Parallelization of a vine trunk detection algorithm for a real time robot localization system. In: 2019 IEEE International Conference on Autonomous Robot Systems and Competitions (ICARSC), pp. 1–6. IEEE (2019)

3. Barré, P., Stöver, B.C., Müller, K.F., Steinhage, V.: LeafNet: a computer vision system for automatic plant species identification. Ecol. Inform. **40**, 50–56 (2017)
4. Baweja, H.S., Parhar, T., Mirbod, O., Nuske, S.: Stalknet: a deep learning pipeline for high-throughput measurement of plant stalk count and stalk width. In: Field and Service Robotics, pp. 271–284. Springer (2018)
5. Cui, H., Zhang, H., Ganger, G.R., Gibbons, P.B., Xing, E.P.: GeePS: scalable deep learning on distributed GPUs with a GPU-specialized parameter server. In: Proceedings of the Eleventh European Conference on Computer Systems, p. 4. ACM (2016)
6. Deng, L., Yu, D., et al.: Deep learning: methods and applications. Found. Trends® Signal Process. **7**(3–4), 197–387 (2014)
7. Ding, J., Chen, B., Liu, H., Huang, M.: Convolutional neural network with data augmentation for SAR target recognition. IEEE Geosci. Remote Sens. Lett. **13**(3), 364–368 (2016)
8. Douarre, C., Schielein, R., Frindel, C., Gerth, S., Rousseau, D.: Transfer learning from synthetic data applied to soil-root segmentation in X-ray tomography images. J. Imaging **4**(5), 65 (2018)
9. Espejo-Garcia, B., Lopez-Pellicer, F.J., Lacasta, J., Moreno, R.P., Zarazaga-Soria, F.J.: End-to-end sequence labeling via deep learning for automatic extraction of agricultural regulations. Comput. Electron. Agric. **162**, 106–111 (2019)
10. Farooq, A., Hu, J., Jia, X.: Analysis of spectral bands and spatial resolutions for weed classification via deep convolutional neural network. IEEE Geosci. Remote Sens. Lett. **16**(2), 183–187 (2018)
11. Ferentinos, K.P.: Deep learning models for plant disease detection and diagnosis. Comput. Electron. Agric. **145**, 311–318 (2018)
12. Fuentes, A., Yoon, S., Kim, S., Park, D.: A robust deep-learning-based detector for real-time tomato plant diseases and pests recognition. Sensors **17**(9), 2022 (2017)
13. Ghazi, M.M., Yanikoglu, B., Aptoula, E.: Plant identification using deep neural networks via optimization of transfer learning parameters. Neurocomputing **235**, 228–235 (2017)
14. Han, X., Zhong, Y., Cao, L., Zhang, L.: Pre-trained alexnet architecture with pyramid pooling and supervision for high spatial resolution remote sensing image scene classification. Remote Sens. **9**(8), 848 (2017)
15. Heinrich, K., Roth, A., Breithaupt, L., Möller, B., Maresch, J.: Yield prognosis for the agrarian management of vineyards using deep learning for object counting (2019)
16. Heo, Y.J., Kim, S.J., Kim, D., Lee, K., Chung, W.K.: Super-high-purity seed sorter using low-latency image-recognition based on deep learning. IEEE Robot. Autom. Lett. **3**(4), 3035–3042 (2018)
17. Kamilaris, A., Prenafeta-Boldú, F.: A review of the use of convolutional neural networks in agriculture. J. Agric. Sci. **156**(3), 312–322 (2018)
18. Kamilaris, A., Kartakoullis, A., Prenafeta-Boldú, F.X.: A review on the practice of big data analysis in agriculture. Comput. Electron. Agric. **143**, 23–37 (2017)
19. Kamilaris, A., Prenafeta-Boldú, F.X.: Deep learning in agriculture: a survey. Comput. Electron. Agric. **147**, 70–90 (2018)
20. Kitzes, J., Wackernagel, M., Loh, J., Peller, A., Goldfinger, S., Cheng, D., Tea, K.: Shrink and share: humanity's present and future ecological footprint. Philos. Trans. Royal Soc. B Biol. Sci. **363**(1491), 467–475 (2007)
21. Koirala, A., Walsh, K., Wang, Z., McCarthy, C.: Deep learning for real-time fruit detection and orchard fruit load estimation: benchmarking of 'mangoyolo'. Precis. Agric. 1–29 (2019)

22. Lammie, C., Olsen, A., Carrick, T., Azghadi, M.R.: Low-power and high-speed deep FPGA inference engines for weed classification at the edge. IEEE Access **7**, 51171–51184 (2019)
23. Lavreniuk, M., Kussul, N., Novikov, A.: Deep learning crop classification approach based on coding input satellite data into the unified hyperspace. In: 2018 IEEE 38th International Conference on Electronics and Nanotechnology (ELNANO), pp. 239–244. IEEE (2018)
24. Li, X., Wu, X.: Constructing long short-term memory based deep recurrent neural networks for large vocabulary speech recognition. In: 2015 IEEE International Conference on Acoustics, Speech and Signal Processing (ICASSP), pp. 4520–4524. IEEE (2015)
25. Liakos, K., Busato, P., Moshou, D., Pearson, S., Bochtis, D.: Machine learning in agriculture: a review. Sensors **18**(8), 2674 (2018)
26. Liu, X., Chen, S.W., Aditya, S., Sivakumar, N., Dcunha, S., Qu, C., Taylor, C.J., Das, J., Kumar, V.: Robust fruit counting: Combining deep learning, tracking, and structure from motion. In: 2018 IEEE/RSJ International Conference on Intelligent Robots and Systems (IROS), pp. 1045–1052. IEEE (2018)
27. Long, J., Shelhamer, E., Darrell, T.: Fully convolutional networks for semantic segmentation. In: Proceedings of the IEEE Conference on Computer Vision and Pattern Recognition, pp. 3431–3440 (2015)
28. Lu, Y., Yi, S., Zeng, N., Liu, Y., Zhang, Y.: Identification of rice diseases using deep convolutional neural networks. Neurocomputing **267**, 378–384 (2017)
29. Ma, J., Du, K., Zheng, F., Zhang, L., Gong, Z., Sun, Z.: A recognition method for cucumber diseases using leaf symptom images based on deep convolutional neural network. Comput. Electron. Agric. **154**, 18–24 (2018)
30. McBratney, A., Whelan, B., Ancev, T., Bouma, J.: Future directions of precision agriculture. Precis. Agric. **6**(1), 7–23 (2005)
31. McNicoll, G.: World population ageing 1950–2050. Population Dev. Rev. **28**(4), 814–816 (2002)
32. Mendes, J., Dos Santos, F.N., Ferraz, N., Couto, P., Morais, R.: Vine trunk detector for a reliable robot localization system. In: 2016 International Conference on Autonomous Robot Systems and Competitions (ICARSC), pp. 1–6. IEEE (2016)
33. Patil, K., Kale, N.: A model for smart agriculture using IoT. In: 2016 International Conference on Global Trends in Signal Processing, Information Computing and Communication (ICGTSPICC), pp. 543–545. IEEE (2016)
34. Perry, M.: Science and innovation strategic policy plans for the 2020s (EU, AU, UK): will they prepare us for the world in 2050? Appl. Econ. Finance **2**(3), 76–84 (2015)
35. Rançon, F., Bombrun, L., Keresztes, B., Germain, C.: Comparison of sift encoded and deep learning features for the classification and detection of Esca disease in bordeaux vineyards. Remote Sens. **11**(1), 1 (2019)
36. Rivas, A., Chamoso, P., González-Briones, A., Corchado, J.: Detection of cattle using drones and convolutional neural networks. Sensors **18**(7), 2048 (2018)
37. Rußwurm, M., Körner, M.: Multi-temporal land cover classification with long short-term memory neural networks. Int. Arch. Photogram. Remote Sens. Spatial Inf. Sci. **42**, 551 (2017)
38. Santos, L., Santos, F.N., Magalhães, S., Costa, P., Reis, R.: Path planning approach with the extraction of topological maps from occupancy grid maps in steep slope vineyards. In: 2019 IEEE International Conference on Autonomous Robot Systems and Competitions (ICARSC), pp. 1–7. IEEE (2019)

39. Santos, L., Ferraz, N., dos Santos, F.N., Mendes, J., Morais, R., Costa, P., Reis, R.: Path planning aware of soil compaction for steep slope vineyards. In: 2018 IEEE International Conference on Autonomous Robot Systems and Competitions (ICARSC), pp. 250–255. IEEE (2018)
40. dos Santos Ferreira, A., Freitas, D.M., da Silva, G.G., Pistori, H., Folhes, M.T.: Weed detection in soybean crops using convnets. Comput. Electron. Agric. **143**, 314–324 (2017)
41. Saxena, L., Armstrong, L.: A survey of image processing techniques for agriculture (2014)
42. Schmidhuber, J.: Deep learning in neural networks: an overview. Neural Networks **61**, 85–117 (2015)
43. Schmitz, A., Moss, C.B.: Mechanized agriculture: machine adoption, farm size, and labor displacement (2015)
44. Smith, A.G., Petersen, J., Selvan, R., Rasmussen, C.R.: Segmentation of roots in soil with u-net. arXiv preprint arXiv:1902.11050 (2019)
45. Tseng, D., Wang, D., Chen, C., Miller, L., Song, W., Viers, J., Vougioukas, S., Carpin, S., Ojea, J.A., Goldberg, K.: Towards automating precision irrigation: deep learning to infer local soil moisture conditions from synthetic aerial agricultural images. In: 2018 IEEE 14th International Conference on Automation Science and Engineering (CASE), pp. 284–291. IEEE (2018)
46. Uzal, L.C., Grinblat, G.L., Namías, R., Larese, M.G., Bianchi, J., Morandi, E., Granitto, P.M.: Seed-per-pod estimation for plant breeding using deep learning. Comput. Electron. Agric. **150**, 196–204 (2018)
47. Walter, A., Finger, R., Huber, R., Buchmann, N.: Opinion: smart farming is key to developing sustainable agriculture. Proc. Nat. Acad. Sci. **114**(24), 6148–6150 (2017)
48. Wang, C., Gong, L., Yu, Q., Li, X., Xie, Y., Zhou, X.: DLAU: a scalable deep learning accelerator unit on FPGA. IEEE Trans. Comput. Aided Des. Integr. Circuits Syst. **36**(3), 513–517 (2016)
49. Yalcin, H., Razavi, S.: Plant classification using convolutional neural networks. In: 2016 Fifth International Conference on Agro-Geoinformatics (Agro-Geoinformatics), pp. 1–5. IEEE (2016)
50. Younis, S., Weiland, C., Hoehndorf, R., Dressler, S., Hickler, T., Seeger, B., Schmidt, M.: Taxon and trait recognition from digitized herbarium specimens using deep convolutional neural networks. Bot. Lett. **165**(3–4), 377–383 (2018)
51. Zhang, X., Qiao, Y., Meng, F., Fan, C., Zhang, M.: Identification of maize leaf diseases using improved deep convolutional neural networks. IEEE Access **6**, 30370–30377 (2018)
52. Zhang, Y.D., Dong, Z., Chen, X., Jia, W., Du, S., Muhammad, K., Wang, S.H.: Image based fruit category classification by 13-layer deep convolutional neural network and data augmentation. Multimedia Tools Appl. **78**(3), 3613–3632 (2019)
53. Zhong, L., Hu, L., Zhou, H.: Deep learning based multi-temporal crop classification. Remote Sens. Environ. **221**, 430–443 (2019)

Forest Robot and Datasets for Biomass Collection

Ricardo Reis(✉), Filipe Neves dos Santos, and Luís Santos

INESC TEC - INESC Technology and Science, Porto, Portugal
{ricardo.g.reis,fbnsantos,luis.c.santos}@inesctec.pt

Abstract. Portugal has witnessed some of its largest wildfires in the last decade, due to the lack of forestry management and valuation strategies. A cost-effective biomass collection tool/approach can increase the forest valuing, being a tool to reduce fire risk in the forest. However, cost-effective forestry machinery/solutions are needed to harvest this biomass. Most of bigger operations in forests are already highly mechanized, but not the smaller operations. Mobile robotics know-how combined with new virtual reality and remote sensing techniques paved the way for a new robotics perspective regarding work machines in the forest. Navigation is still a challenge in a forest. There is a lot of information, trees consist of obstacles while lower vegetation may hide danger for robot trajectory, and the terrain in our region is mostly steep. The existence of accurate information about the environment is crucial for the navigation process and for biomass inventory. This paper presents a prototype forest robot for biomass collection. Besides, it is provided a dataset of different forest environments, containing data from different sensors such as 3D laser data, thermal camera, inertial units, GNSS, and RGB camera. These datasets are meant to provide information for the study of the forest terrain, allowing further development and research of navigation planning, biomass analysis, task planning, and information that professionals of this field may require.

Keywords: Forestry robotics · Laser odometry · Thermal imagery · Forest dataset · LiDAR · Biomass · Point-cloud

1 Introduction

According to the Portuguese Biomass Working Group report from 2013 [12], about 35% of the territory is covered by forest. From all the forest area, most of it belongs to private owners. With 93% belonging to more than 400 thousand owners and only 7% belonging to the Portuguese state, Portugal is the only country in the European Union where the forest territory is in its majority private property. One of the main raw materials provided by the forest is the biomass, used mainly in energy production. Considering the percentage of the national territory covered by forest, it was identified as forest biomass every

M. F. Silva et al. (Eds.): ROBOT 2019, AISC 1092, pp. 152–163, 2020.
https://doi.org/10.1007/978-3-030-35990-4_13

material obtained from the process of cleaning forests, wood without commercial value coming from areas affected by wildfires. The residual forest biomass is obtained from the excess of forest exploration or sub products of the industry which transforms forest products. Two relevant terms were mentioned: forest cleaning and wildfires. Unfortunately, Portugal has a problem with wildfires. In June of 2017, a single wildfire caused the death of 66 people[1]. While the poor forest management, associated with the fact that it belongs mostly to private owners who, for a variety of reasons don't clean their terrains, may not always be the cause of a wildfire, a forestry terrain that is not managed with proper cleaning will cause the wildfire to spread faster, increasing the danger and damage extension. Considering the current problems, our work in development pretends to create an autonomous machine that is capable of navigating the forest terrain and perform several autonomous tasks related to biomass harvesting activities. In this work, we will describe our current setups used to acquire forest environment data on-board of our robot, and ultimately we will provide access to our public forest datasets.

Section 2 presents the state of the art related to our aim in forestry. In Sect. 3 we present the robotic platform we used for acquiring the data and we describe the configurations and methods used to record the data. In Sect. 4 we provide the datasets description as well as a brief benchmark for some lidar only slam techniques, presenting the results in Sect. 5, where we provide the links to the different datasets. Finally, the conclusions are discussed in Sect. 6.

2 Related Work

Unfortunately, as presented in Biomass Working Group 2013 [12], not all forest biomass is easily exploited because of it's elevated costs and harsh terrains conditions. In the middle of dense vegetation usually placed in hills/forest, GNSS signals are not always available and dead reckoning sensors data like odometry and Inertial Measurement Unit (IMU) are highly affected by the terrain conditions. In agricultural robotics, the bluetooth beacons signal strength (RSSI) was used to help correct odometry based-localization [18]. In contrast, natural features (vineyard trunks and masts) using a visual detector can help localize the robot [14]. In this harsh terrain context, conditions related to the safety of the equipment should also be taken into account when generating trajectories as in [20], where the authors use a Digital Elevation Map, which considering the platform characteristics, will allow to define a safe trajectory.

There are some concepts of agricultural machines (where problems, as the lack of GNSS signal or the irregular terrain applies) of small/medium port robots to work in an autonomous way like the works of [3,7] or [1]. [7] proposes a coverage path planning for autonomous machines on tridimensional terrain and [3] an autonomous robot and its localization system for navigation in steep slope vineyard. In [1] it is proposed a motion planning system for an agricultural machine

[1] https://www.bbc.com/news/world-europe-44438505.

that takes into account some terrain constraints for harvesting tasks. This is a regular agricultural machine with extra sensors for navigation as localization.

As for forestry autonomous machines, most of the autonomous effort are for machines of log cutting and transport. [8] refers to an old project for an autonomous forest machine, the Rofor project, where a forest machine was partially automated to be guided using teleoperated and autonomous functions.

Regarding forest environments, a mobile robot for operating in the Amazon Rain Forest is presented in [4] with different kinematic control strategies to improve certain desired aspects, like wheel traction efficiency. It is addressed for mobile wheel legged robots. Four legged robots are also considered for this environment. For example BigDog is a legged robot prepared to navigate in an unstructured environment like a forest avoiding obstacles such as trees and boulders [23].

It was already mentioned that is not possible to obtain an accurate localization with GNNS systems, so it is necessary to explore other localization methods. It is possible to localize a vehicle using a matching method between a local map built with on board sensors, like LIDAR, and global maps previously built. The prediction step could be carried out by a kalman filter [22]. The next two approaches are based on matching methods: the visualGPS approach is capable of estimating the localization with an accuracy of 0.5 m using GNSS position as a starting point for a combined Kalman Filter and ACML based on LIDAR sensors [19][2]. Another approach also uses GNSS positions for a starting point and gives the localization with an accuracy of 0.3 m, however this accuracy depends in the resolution of the input data, since the method works with a geometric match between a map of observed stem trees obtained by a 3D LIDAR and an aerial image of the trees canopy [9].

[15] presents a real time SLAM approach to the localization and mapping of a forestry, it combines 2D laser localization with mapping and differential GNSS information to form a global tree map. This approach is limited and it has troubles running in real time using a regular computer. In fact, SLAM techniques are not adequate for these environments "because the resulting errors in the map are not bounded and thus are not well-suited for large area operation and for a matching against parcel border" [22].

For building global maps a multi sensor fusion approach helps to construct a map based on aerial and satellite multi-spectral imagery as well on aerial laser scanning using LIDAR. The multi-spectral imagery helps to identify the trees species [10]. The single tree delineation is made according to the aerial images and the 3D Laser Scan information. This allows to generate a global and georeferenced tree.

The work presented in [11] describes a terrain classification using LIDAR data for ground robots. It is capable of identify three classes: "scatter" which represents porous volumes such as, grass and tree canopy, "linear" which identifies objects like wires or tree branches and "surface" that capture the ground surface, rocks or large trunks. This method is to be performed in real time and

[2] http://handbookofrobotics.org/view-chapter/56/videodetails/96.

uses local point distribution statistics to produce salience and capture the different classes of objects. In [5] a visual perception method is used to learn forest trails for mobile robots. Single monocular images are used and a neural network is trained to find the direction of the trail present in the image. It is an interesting approach for the robot navigation between two distant points of the forest, where it can be useful to use an existing trail, but this perception system is not capable of identifying anything else, like trees and rocks. There are other visual terrain classification methods that are capable of identifying navigable zones in outdoors terrains, based on the random forest method [24] or neural networks [2].

Ground Penetrating Radar (GPR) technology can be used to identify hidden stems under different classes of vegetation due to the correlation between the type of vegetation and the electromagnetic wave velocity. The work presented in [17] allies GPR and topography to measure sediment thickness in a periglacial region situated in Greenland. The authors resort to a GPR shielded antenna to perform the measurements and through the mentioned correlation different classes of vegetation are identified such as: "wetland", "Betula", "heath", "grassland", "shore" and "bedrock". Despite of being an interesting approach that may help to identify covered gaps or holes, it is not ready to a practical application in forest environment, as this processes were not made in real time and measurements were made by walking with the GPR system through the terrain.

Electromagnetic Penetrating technologies are important. They can be applied in the detection of hidden obstacles like tree stems, which may be covered by dense vegetation. In [6] the authors work is based on x-rays to detect stems on a tomato field. A portable x-ray source is placed perpendicular to the plant grow and parallel to the soil. It is possible to detect the stem, even when it is covered by foliage, based on the fact that the main stem plant absorbs the x-ray energy [6]. This may not be an adequate approach to be directly applied in forestry since it only detects the stem present in between the x-ray generator and the x-ray detector. In forestry this configuration is not practical due to the increased dimension of the vegetation portion. [13] presents a system with LIDAR and radar to detect obstacles hidden in dense foliage aimed to be used in the autonomous off road ground vehicles, that claims to be able to detect a trunk behind 2.5 m thickness foliage. The US Army Research Laboratory has developed a UWB SAR (synthetic aperture radar) to help the vision of an autonomous navigation system for ground robots, helping to detect minefields, traps and natural obstacles hidden in vegetation [16].

In our approach we selected an adequate machine to harvest forest biomass and added sensors for localization and autonomous navigation, similar do the approaches in [1] and [8].

3 AgRob V18 and Acquisition Setup

In order to acquire the data-sets, a mechanical tower of sensors was assembled. As the motivation behind this paper is the development of an autonomous robot for forestry related tasks, the assembled tower was developed with modular intent, allowing it to be mounted on, and control different robotic platforms.

AgRob V18 is the designated name of the robot iteration observed in Fig. 1a. This platform consists of an adapted Niko Gmbh HRS70 model, an heavy-duty machine for forest and agricultural tasks. The machine is controlled via CANopen protocol. This is a diesel water cooled engine machine with a chassis width of 90 cm and length of 215 cm. As the model name hints, this machine has 70 hp.

Despite the present work being associated to research in robotics for forestry, other agricultural fields are also in our scope of research. In this way the sensors and computer were assembled in a tower structure. This tower only requires to be powered by a 12 V source and it's ready to fully work mounted on any setup, thus becoming modular.

(a) AgRob V18 - forestry robot. (b) Sensor tower

Fig. 1. AgRob V18 and tower

The sensor tower, in Fig. 1b, has an industrial fan-less computer in it's core. This unit receives data from all sensors and outputs the commands for the platform actuators. The main sensor is a Velodyne Puck 16 (VLP16), a Lidar sensor with 16 channels acquiring 360° surrounding data with a vertical field of view of 30° (+/− 15°). According to it's data sheet, the VLP16 has a range above 100 m, distance at which it has an estimated error of +/− 3 cm. When the tower is mounted on the AgRob V18 this sensor height to ground is about 2.2 m. While the VLP16 stays in the rear and tallest point of the platform, we use an Hokuyo UST-10LX, a laser range finder with 10 m range, in the front of the robot. This sensor is used to detect close obstacles and is installed about 1 m height from ground level. After the laser sensors we have 2 vision based sensors. First one is a ZED stereo camera. With this camera we can also extract depth data from the robot trajectory. The other sensor is a FLIR M232, a thermal camera, that will allow the detection of humans and animals, as well as potentially detection of wildfires, making it possible for the platform to raise alerts. The tower has an IMU for robot state estimation and last it has a GPS module GP-808G on the very top op the VLP16 for both helping localization and online upload of the robot position information.

The Robot Operating System (ROS) framework was used for it's set of tools and available libraries. With ROS, the task of setting up the sensors and establish communications allowing the data recording is simplified. For most of the chosen sensors the already existent libraries were ready to use, with some modifications required for our solution and what we wanted to achieve. After acquiring the data most of our time was spent into data processing and analysis, which will be further detailed in Sect. 5.

4 DataSet Description and SLAM Bechmarking

Recalling the focus of the work in achieving autonomous capabilities for the forestry robot Agrob V18, one of the milestones required is the precise navigation. In order to achieve a better informed localization, the more data one can get from it's sensors the merrier, although having robustness, the capability of navigating with the least amount of sensors, is a strength in complex environments such as mountain forests. In this section it will be presented in further detail the contents of the datasets and some SLAM methods already tested.

4.1 DataSet Description

For the tasks of understanding the complex forest environment as well as the robot state, the data-sets provide a variety of data. In Table 1, the most relevant topics are presented and described. This topics are present in all the recorded bags.

Table 1. Description of datasets built.

Topic	Type	Description
/fix	sensor_msgs/NavSatFix	GNSS localization data under WGS 84
/imu_um6/data	sensor_msgs/Imu	Data from the IMU
/imu_um6/mag	geometry_msgs/Vector3Stamped	Magnetometer data from IMU
/imu_um6/rpy	geometry_msgs/Vector3Stamped	Roll, pitch, yaw data from IMU
/imu_um6/temperature	std_msgs/Float32	Temperature data from IMU
/lidar1/hkscan	sensor_msgs/LaserScan	Single scan from Hokuyo
/rtsp2/camera_info	sensor_msgs/CameraInfo	Meta information for FLIR camera
/rtsp2/image_raw/compressed	sensor_msgs/CompressedImage	Contains a compressed image from FLIR
/scan	sensor_msgs/LaserScan	Single scan from VLP16
/tf	tf2_msgs/TFMessage	Contains tf information
/velodyne_points	sensor_msgs/PointCloud2	Point cloud data from VLP16

Although in Sect. 3 it was referred the usage of a ZED camera, included in the sensor tower, only in recent bags ZED data started to be recorded. Right now it is possible to already access this data in our public archive for a vineyard test setup. The temperature and magnetometer data should take into account the fact that the IMU is placed on the inside of the tower. The tower cover is aluminum material, and other electronic devices are also inside, thus possibly affecting the accuracy of this data.

4.2 SLAM Using Lidar Benchmark

The steep mountain forestry terrain is a navigation challenge. The tree canopy and terrain inclination can be responsible for long periods without access to satellite data, thus preventing the use of GNSS data consistently. The fact that the Agrob V18 has no odometry available is another challenge to estimate position. Agrob V18 is a diesel fuelled machine, and even when is not moving, when the motor is on there is a lot of vibration, making it an extremely difficult task to use IMU data. Facing this challenges, the main sensor we could relly on for navigation was Velodyne. In the early stages we found the Lidar Odometry and Mapping in Real-time (LOAM) [25] ROS package and started to test with it. The results were very good, it performed a good point cloud registration in a structured environment such as the campus of a university. The LOAM algorithm classifies the cloud in edge points and planar points to then calculate the transform between two sets of point clouds. Unfortunately however, for non-structured environments such as forests, where there are information overloads with the amount of trees and vegetation, the algorithm failed to perform in real time for our datasets. By reducing the speed of the data to 2% of it's original speed, we managed to obtain the full cloud. An output example of the obtained trajectory at 10% of the speed is shown in Fig. 2a. The described trajectory is rough being a top view. If seen from a side perspective, the trajectory is not co-planar, thus becoming impossible to use in real time. In Fig. 2b and c both described trajectories are correct and in real time. Figure 2b corresponds to the Advanced implementation of LOAM (A-LOAM) method[3]. The authors describe it as being a clean and simple version of LOAM, without complicated mathematical derivation and redundant operations, resulting into it running better and faster. A-LOAM uses Ceres Solver[4], an open source C++ library for modeling and solving large, complicated optimization problems. We tested this method against LeGO-LOAM: Lightweight and Ground-Optimized Lidar Odometry and Mapping on Variable Terrain [21], represented in Fig. 2c. LeGO-LOAM is an improved version of LOAM, focused in processing data from non-structured environments such as a forest that leverages the presence of a ground plane in its segmentation and optimization steps. The authors present in their work a performance comparison between LeGO-LOAM and LOAM, where LeGO-LOAM shows always similar or better results. From our brief testing between A-LOAM

[3] https://github.com/HKUST-Aerial-Robotics/A-LOAM.
[4] http://ceres-solver.org.

and LeGO-LOAM, we tested the results in complex environments up to 10 times the speed, values for which the A-LOAM started to fail in performing the Iterative closest point (ICP) for point cloud registration. From our tests we decided to keep using the LeGO-LOAM and develop our work on top of this method. The tests were performed on a computer with Intel Core i7-8750H CPU 2.20GHzx12 and a GeForce GTX 1070.

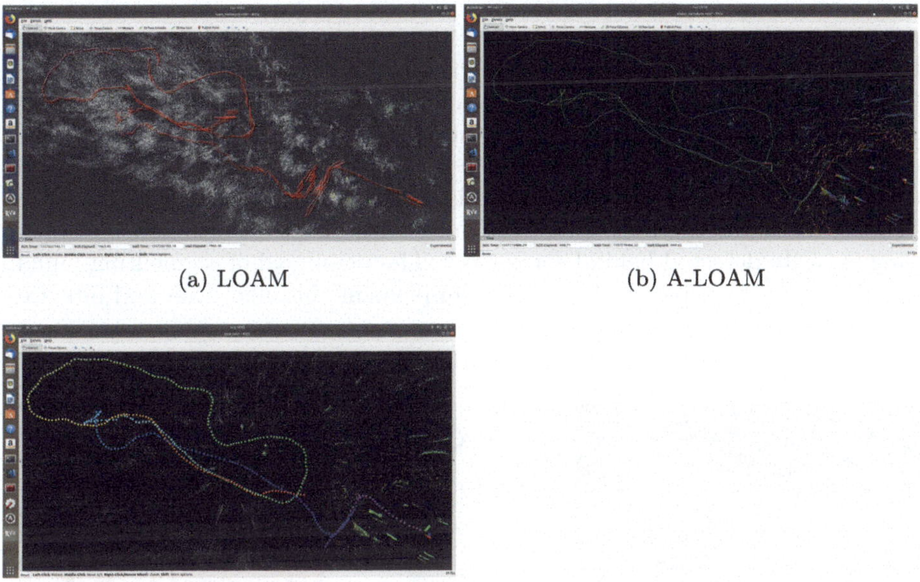

(a) LOAM (b) A-LOAM

(c) LEGO-LOAM

Fig. 2. Laser odometry and mapping results.

5 Results and Discussion

This section presents information about the built and public datasets, acquired in this work. The built data-sets are open-source and available at our public repository[5]. In Table 2, it is provided information to help identify and access the forest datasets. The table shows the files original names and sizes, some of these files were compressed.

The files with reference DS_AG_34 were recorded in Lobão, Santa Maria da Feira, a steep forestry terrain mainly composed by Eucalyptus trees. The other file, DS_AG_35, was recorded in Vila Chã, Vila do Conde, in a planar forestry terrain comprised of both Eucalyptus and Pine trees. Additionally there is another dataset under the reference DS_AG_36. This file has not only the same

[5] http://aladobix.inesctec.pt/projects/repository/wiki.

Table 2. Dataset details.

Ref	Name	Size
DS_AG_34	Agrob_V18_2019-03-08-11-46-02.bag	6.6 Gb
	Agrob_V18_2019-03-08-11-59-18.bag	1.3 Gb
DS_AG_35	Agrob_V18_2019-04-30-16-45-41.bag	6.9 Gb

range of sensors but also the ZED camera, with the only difference being that this bag was recorded on a vineyard environment.

In this subsection we will showcase some of the data and the means used to visualize it. In Fig. 3a is presented the plot of the robot pose and it's described trajectory. The trajectory is shown in red trace while the robot pose is represented by the blue car icon. This data is plotted using a Node-RED interface. Figure 3b shows a thermal image obtained with the FLIR camera. A human being is easily detected behind some trees. The main goal of segmenting human beings seems achievable, we have yet to implement this algorithm and integrate in the robot safety mechanisms.

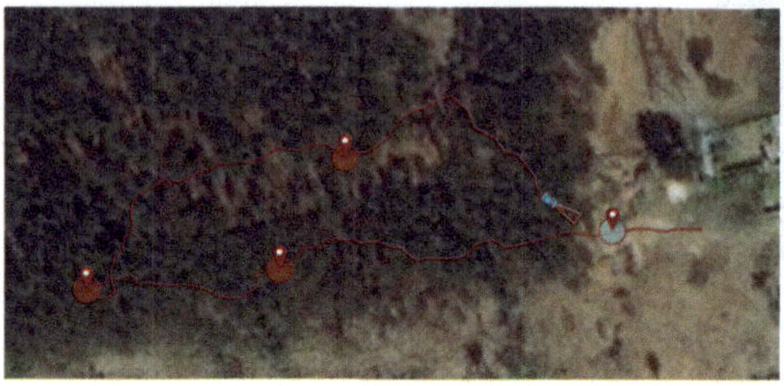

(a) NodeRed view.

(b) FLIR thermal image.

Fig. 3. GPS and thermal image data results.

In order to process and visualize the Velodyne data, we chose to use the LeGO-LOAM [21] method as it performed the best for unstructured environments. We made a few modifications to obtain the desired point cloud output and then visualize it using CloudCompare[6] software as in Fig. 4. In Fig. 4a is shown the full cloud of dataset DS_AG_34. The point cloud is very detailed allowing us to zoom in and observe with some detail the trees contours and features. LeGO-LOAM also allows to obtain light size point clouds with its down sampling routine. The point cloud registration showed great results for the given dataset. For navigation, localization and even forest inventory we are very interested in the tree trunks as the information they provide as natural landmarks and volume information for biomass. Figure 4b shows the segmented point cloud

[6] https://www.danielgm.net/cc/.

for trunks. Even though some outliers are remaining, the trunks are very clear and easy to process and integrate in future developments.

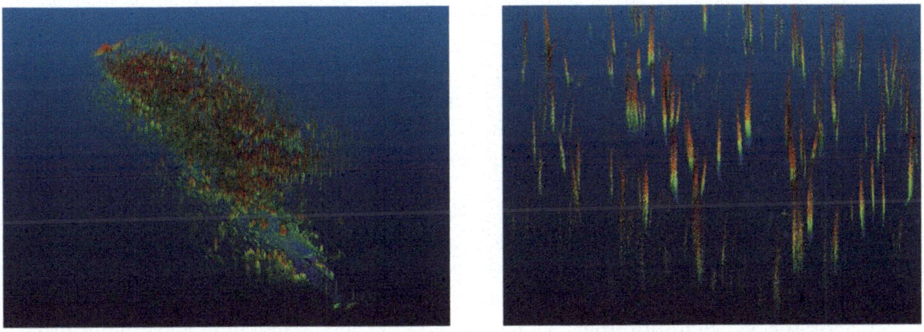

(a) Full field view. (b) Segmented trunks.

Fig. 4. Obtained point clouds.

6 Conclusion

In this work, a data-set of forestry data was assembled and made available to public. The acquired data potential was presented and has potential to aid innovation and research in the forest field. Forest mapping and even forest datasets are not that uncommon nowadays. However, they are usually aerial scans obtained using a drone. These datasets provided in the present work contain detailed data and rich information at ground level, where most of the forest-related tasks are performed, thus possessing value for novel research and planning in this field.

For future work, we are currently developing a precise localization as we aim to achieve full autonomous navigation in forest environment using current resources. The positioning of the IMU is something that requires attention, as moving to a lower and more stable part of the Agrob V18 may reduce the vibration effects, although it affects our desired modularity concept and consistency of data.

With the quality of data we aim to also provide more information about the forest terrain such as the amount and geographical coordinates of the trees, the trunk diameters and estimated volume of wood. This kind of inventory provides useful information to the proprietary about how to better manage and explore the biomass resources.

Acknowledgments. This work is co-financed by the European Regional Development Fund (ERDF) through the Interreg V-A Espanha-Portugal Programme (POCTEP) 2014–2020 within project 0095_BIOTECFOR_1_P. The opinions included in this paper shall be the sole responsibility of their authors. The European Commission and the Authorities of the Programme aren't responsible for the use of information contained therein.

References

1. Auat Cheein, F., Torres-Torriti, M., Hopfenblatt, N.B., Prado, Á.J., Calabi, D.: Agricultural service unit motion planning under harvesting scheduling and terrain constraints. J. Field Robot. **34**(8), 1531–1542 (2017)
2. Chavez-Garcia, R.O., Guzzi, J., Gambardella, L.M., Giusti, A.: Image classification for ground traversability estimation in robotics. In: International Conference on Advanced Concepts for Intelligent Vision Systems, pp. 325–336. Springer (2017)
3. Dos Santos, F.N., Sobreira, H., Campos, D., Morais, R., Moreira, A.P., Contente, O.: Towards a reliable robot for steep slope vineyards monitoring. J. Intell. Robot. Syst. **83**(3–4), 429–444 (2016)
4. Freitas, G., Gleizer, G., Lizarralde, F., Hsu, L., dos Reis, N.R.S.: Kinematic reconfigurability control for an environmental mobile robot operating in the amazon rain forest. J. Field Robot. **27**(2), 197–216 (2010)
5. Giusti, A., Guzzi, J., Cireşan, D.C., He, F.L., Rodríguez, J.P., Fontana, F., Faessler, M., Forster, C., Schmidhuber, J., Di Caro, G., et al.: A machine learning approach to visual perception of forest trails for mobile robots. IEEE Robot. Autom. Lett. **1**(2), 661–667 (2015)
6. Haff, R.P., Slaughter, D.C., Jackson, E.: X-ray based stem detection in an automatic tomato weeding system. Appl. Eng. Agric. **27**(5), 803–810 (2011)
7. Hameed, I.A.: Intelligent coverage path planning for agricultural robots and autonomous machines on three-dimensional terrain. J. Intell. Robot. Syst. **74**(3–4), 965–983 (2014)
8. Hellström, T.: Autonomous navigation for forest machines: a pre-study (2002)
9. Hussein, M., Renner, M., Iagnemma, K.: Global localization of autonomous robots in forest environments. Photogram. Eng. Remote Sens. **81**(11), 839–846 (2015)
10. Krahwinkler, P., Rossmann, J., Sondermann, B.: Support vector machine based decision tree for very high resolution multispectral forest mapping. In: 2011 IEEE International Geoscience and Remote Sensing Symposium, pp. 43–46. IEEE (2011)
11. Lalonde, J.F., Vandapel, N., Huber, D.F., Hebert, M.: Natural terrain classification using three-dimensional ladar data for ground robot mobility. J. Field Robot. **23**(10), 839–861 (2006)
12. Marques, F., Marques, M., Fão, J., Baptista, A., Ramos, J., Fazenda, L., Ferreira, J.: Relatório Grupo de Trabalho da Biomassa. Comissão de Agricultura e Mar. (2013). https://www.parlamento.pt/ArquivoDocumentacao/Documents/coleccoes_relatorio-bio2013-2.pdf
13. Matthies, L., Bergh, C., Castano, A., Macedo, J., Manduchi, R.: Obstacle detection in foliage with ladar and radar. In: Robotics Research. The Eleventh International Symposium, pp. 291–300. Springer (2005)
14. Mendes, J., Dos Santos, F.N., Ferraz, N., Couto, P., Morais, R.: Vine trunk detector for a reliable robot localization system. In: 2016 International Conference on Autonomous Robot Systems and Competitions (ICARSC), pp. 1–6. IEEE (2016)
15. Miettinen, M., Ohman, M., Visala, A., Forsman, P.: Simultaneous localization and mapping for forest harvesters. In: Proceedings 2007 IEEE International Conference on Robotics and Automation, pp. 517–522. IEEE (2007)
16. Nguyen, L., Wong, D., Ressler, M., Koenig, F., Stanton, B., Smith, G., Sichina, J., Kappra, K.: Obstacle avoidance and concealed target detection using the army research lab ultra-wideband synchronous impulse reconstruction (UWB SIRE) forward imaging radar. In: Detection and Remediation Technologies for Mines and Minelike Targets XII, vol. 6553, p. 65530H. International Society for Optics and Photonics (2007)

17. Petrone, J., Sohlenius, G., Johansson, E., Lindborg, T., Näslund, J.O., Strömgren, M., Brydsten, L.: Using ground-penetrating radar, topography and classification of vegetation to model the sediment and active layer thickness in a periglacial lake catchment, western greenland. Earth Syst. Sci. Data **8**(2), 663–677 (2016)
18. Reis, R., Mendes, J., dos Santos, F.N., Morais, R., Ferraz, N., Santos, L., Sousa, A.: Redundant robot localization system based in wireless sensor network. In: 2018 IEEE International Conference on Autonomous Robot Systems and Competitions (ICARSC), pp. 154–159. IEEE (2018)
19. Rossmann, J., Schluse, M., Schlette, C., Buecken, A., Krahwinkler, P., Emde, M.: Realization of a highly accurate mobile robot system for multi purpose precision forestry applications. In: 2009 International Conference on Advanced Robotics, pp. 1–6. IEEE (2009)
20. Santos, L., Ferraz, N., dos Santos, F.N., Mendes, J., Morais, R., Costa, P., Reis, R.: Path planning aware of soil compaction for steep slope vineyards. In: 2018 IEEE International Conference on Autonomous Robot Systems and Competitions (ICARSC), pp. 250–255. IEEE (2018)
21. Shan, T., Englot, B.: Lego-loam: Lightweight and ground-optimized lidar odometry and mapping on variable terrain. In: 2018 IEEE/RSJ International Conference on Intelligent Robots and Systems (IROS), pp. 4758–4765. IEEE (2018)
22. Siciliano, B., Khatib, O.: Springer Handbook of Robotics. Springer, Heidelberg (2016)
23. Wooden, D., Malchano, M., Blankespoor, K., Howardy, A., Rizzi, A.A., Raibert, M.: Autonomous navigation for bigdog. In: 2010 IEEE International Conference on Robotics and Automation, pp. 4736–4741. IEEE (2010)
24. Zhang, H., Dai, X., Sun, F., Yuan, J.: Terrain classification in field environment based on random forest for the mobile robot. In: 2016 35th Chinese Control Conference (CCC), pp. 6074–6079. IEEE (2016)
25. Zhang, J., Singh, S.: LOAM: lidar odometry and mapping in real-time. In: Robotics: Science and Systems, vol. 2, p. 9 (2014)

An Autonomous Guided Field Inspection Vehicle for 3D Woody Crops Monitoring

José M. Bengochea-Guevara(✉), Dionisio Andújar, Karla Cantuña, Celia Garijo-Del-Río, and Angela Ribeiro

Centre for Automation and Robotics, CSIC-UPM, 28500 Arganda del Rey, Madrid, Spain
{jose.bengochea,angela.ribeiro}@csic.es

Abstract. This paper presents a novel approach for crop monitoring and 3D reconstruction. A mobile platform, based on a commercial electric vehicle, was developed and equipped with different on-board sensors for crop monitoring. Acceleration, braking and steering systems of the vehicle were automatized. Fuzzy control systems were implemented to achieve autonomous navigation. A low-cost RGB-D sensor, Microsoft Kinect v2 sensor, and a reflex camera were installed on-board the platform for creation of 3D crop maps. The modelling of the field was fully automatic based on algorithms for 3D reconstructions of large areas, such as a complete row crop. Important information can be estimated from a 3D model of the crop, such as the canopy volume. For that goal, the alpha-shape algorithm was proposed. The on-going developments presented in this paper arise as a promising tool to achieve better crop management increasing crop profitability while reducing agrochemical inputs and environmental impact.

Keywords: Field inspection vehicle · Crop monitoring · 3D reconstruction

1 Introduction

Precise crop monitoring helps farmers to improve crop quality and to reduce operational costs by taking better decisions. Yield estimation is typically based on crop knowledge, historical data, meteorological conditions or crop monitoring through visual or manual sampling. However, these processes are generally time-consuming, labour-intensive, and frequently inaccurate due to a low number of samples to capture the magnitude of variations in a crop. Thus, it is crucial to identify an automated and efficient alternative system to manual processes that can accurately capture the spatial and temporal variations of crops. Vehicles equipped with on-board sensing equipment are a promising choice for acquisition of information. However, navigation of mobile robots in agricultural environments remains a key challenge, due to the variability and nature of the vegetation and terrain. Guidance systems for agricultural vehicles have employed global or local information. Methods that utilize global data guide the vehicle along a previously calculated route using the position of the vehicle relative to an absolute reference. Global navigation satellite systems (GNSS) are usually employed for this approach [1–3]. Guidance methods that utilize local information

M. F. Silva et al. (Eds.): ROBOT 2019, AISC 1092, pp. 164–175, 2020.
https://doi.org/10.1007/978-3-030-35990-4_14

guide the vehicle by the detection of local landmarks, such as crop rows and intervals between them. For that purpose, diverse approaches based on vision [4–6] and laser sensors [7–9] have been proposed for crop row detection. Most studies in autonomous navigation in agricultural vehicles have been focused on tractors or heavy vehicles. Nevertheless, the use of medium-sized platforms for crop scouting is a suitable choice for minimizing soil compaction, enabling the performance of more than one sampling throughout the year due to the minimal impact on the crop.

Moreover, 3D reconstruction of woody crop models using non-destructive methods is a valuable technique to improve the decision-making processes. The use of sensors for crop characterization leads to a better understanding of the processes involved in tree. With information obtained from a 3D reconstruction of the crop, important parameters, such as the growth status, height, shape, biomass, need for nutrients, and health status, can be estimated. These parameters are currently mostly estimated by applying equations that assume the trees to be geometric solids (regular polygons) or by applying empiric models [10], which produce inconsistent results. The use of the information extracted from 3D reconstructions can improve the decisions made related to crop management and contribute to creating new protocols to improve the profitability and health of crops.

This work shows the development performed to achieve the automation and the autonomous navigation of a field inspection vehicle, as well as the 3D modelling of woody crops using the sensors on-board the platform.

2 Field Platform

2.1 Description of the Platform

The field platform (Fig. 1) is based on a Renault Twizy Urban 80 model car. It is equipped with a 13 kW electric motor and it can travel up to a speed of 80 km/h. The vehicle is ultra-compact, with a length of 2.32 m, width of 1.19 m, height of 1.46 m and unladen weight of 450 kg. The completed battery charge allows to travel a distance of over 80 km. The electric motor of the vehicle allows negligible vibration at speeds below 3 km/h [11], which is convenient for high-quality data acquisition.

An aluminum support structure has been integrated for easy placement of sensors on the front of the vehicle. Two devices are installed, adapting their positions to the crop features of each crop. A Microsoft Kinect v2 RGB-D sensor, which up to 30 fps, supplies RGB images with a resolution of 1920 × 1080 pixels together with depth information at a resolution of 512 × 424 pixels. The depth data are obtained via a Time-of-Flight (ToF) system inside the sensor. The depth measurement range of the sensor is 0.5–4.5 m [12]. However, when sampling outdoors, the maximum range decreases. Specifically, studies conducted outdoors under different daytime illumination conditions [13] show that the sensor provides valid depth measurements up to 1.9 m during sunny days, while the distance increases up to 2.8 m under the diffuse illumination of an overcast day. The other on-board sensor is a digital single-lens reflex camera, Canon EOS 7D model, which supplies high-quality RGB images at 2 fps with a resolution of 2592 × 1728 pixels. Both sensors are connected to the on-board computer, equipped with an Intel Core i7-4771@3.5 GHz processor, 16 GB of RAM, and an NVIDIA GeForce GTX 660 graphic card. The platform is also equipped with a RTK-GNSS receiver R220 Hemisphere model, which provides location data at 20 Hz sample rate with an accuracy of 20 mm + 2 ppm (2DRMS, 95%) according to the manufacturer's specifications, and a Vectornav 200 model inertial measurement unit (IMU).

Fig. 1. Field platform and on-board equipment

The inspection plan to be followed by the platform is generated by a path planner [14], which can be formulized as the well-known capacitated vehicle routing problem. The fundamental problem consists of determining the best inspection route that provides complete coverage of the field considering agronomical features (such as field shape, crop row direction, and type of crop), certain characteristics of the platform (such as the turning radius or the number of on-board sensors) as well as the distance travelled.

The mobile platform is prepared to inspect both annual (e.g., maize or cereal) and multi-annual (e.g., orchards or vineyards) crops. In the case of arable crops, the platform can only scout the crop during the early season, when the plants are in the beginning of their growing, which is acceptable since this coincides with the time when treatments for weeds are carried out. The present work is focused on the inspection and 3D reconstruction of woody crops.

2.2 Automation of the Platform

A previously described on-board computer is responsible for decision making processes, sending the commands to the vehicle when working in remote control mode, or autonomously based on RTK-GNSS receiver and IMU information. Thus, the system architecture (Fig. 2) was developed to control the speed and the turning of the vehicle; i.e. to control the acceleration, braking and steering systems of the vehicle. The process of automation was performed focusing to keep the manual driving mode.

Speed variations are commanded by the on-board computer. It sends orders to the microcontroller which generates signals to the internal motor controller of Renault Twizy, simulating that the throttle pedal has been pressed. Regarding steering system, the vehicle does not have power steering system, so it requires acting on the steering axe. A brushless DC motor with an encoder managed by a position controller is used to act on the steering axe through a timing system. When turning the direction of the vehicle is required, the on-board computer sends the necessary commands to the microcontroller, which sends a command to the motor position controller. Then, the motor turns and, therefore, the steering axe, turning the steering wheel to the desired position. A braking actuator system was also installed. The on-board computer commands the microcontroller to generate a signal, which starts a servomotor acting on the brake pedal through a strong wire. Unlike the throttle pedal, braking pedal on Renault Twizy is mechanical, not digital, so it requires an actuator element to press the brake pedal. The control of all aforementioned systems was performed by Robot Operating System (ROS).

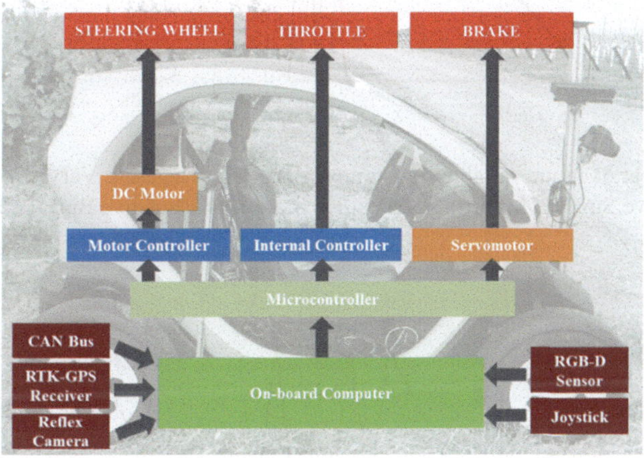

Fig. 2. System architecture

2.3 Autonomous Navigation

When the platform navigates autonomously, the on-board computer takes the decisions to control the direction of the vehicle by fuzzy control techniques. Several previous works [4, 15–17] have showed that fuzzy controllers perfectly mimic human driving behaviour while driving and route tracking. Thus, they help to manage the nonlinear dynamics that characterizes the wheeled mobile robots, usually affected by an important number of disturbances, such as turning and static frictions or variations in the amount of cargo. Human behavior regarding speed and steering control can be approached using artificial intelligence techniques, among which the techniques based on fuzzy logic [18, 19] provide a better approximation to human reasoning, giving a more intuitive control structure. The proposed fuzzy controller enables the platform to follow a path with the correct orientation at constant speed from the data supplied by the RTK-GNSS receiver and the IMU on board the vehicle.

The inputs of the controller are the position and orientation errors respect the trajectory that the vehicle must follow (fuzzy sets are shown in Figs. 3 and 4). The controller produces the angle of the steering wheel as an output. The Takagi-Sugeno implication was used since it is compatible with applications that require on-time responses [19]. The output of the controller is the singleton-type membership functions shown in Fig. 5. 49 rules were used to cover all possible combinations of the inputs variables.

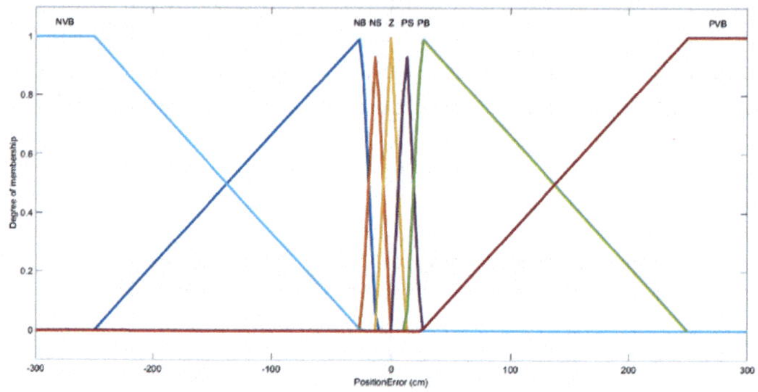

Fig. 3. Fuzzy sets of input variable: position error [NVB (Negative Very Big), NB (Negative Big), NS (Negative Small), Z (Zero), PS (Positive Small), PB (Positive Big), PVB (Positive Very Big)].

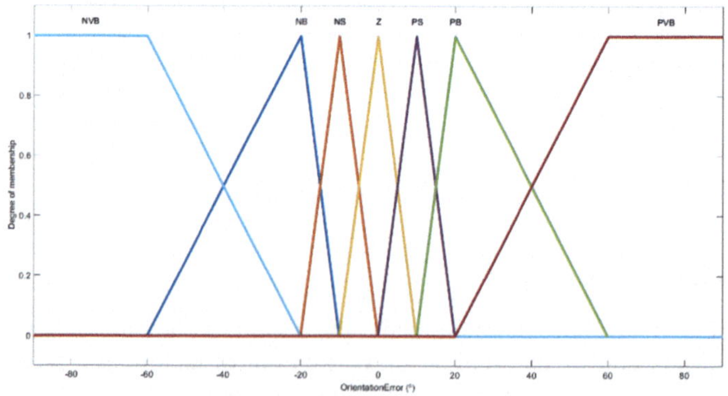

Fig. 4. Fuzzy sets of input variable: orientation error [NVB (Negative Very Big), NB (Negative Big), NS (Negative Small), Z (Zero), PS (Positive Small), PB (Positive Big), PVB (Positive Very Big)].

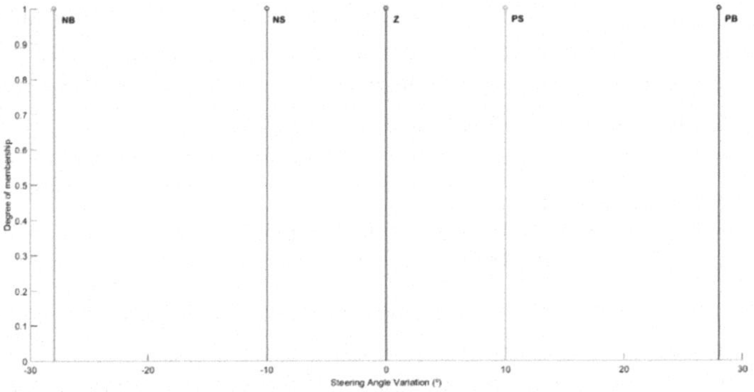

Fig. 5. Fuzzy sets of output variable: steering angle variation [NB (Negative Big), NS (Negative Small), Z (Zero), PS (Positive Small), PB (Positive Big)]

2.4 Three-Dimensional Modelling

Low Cost RGB-D Sensor. The algorithm described in [20] was used for the 3D reconstruction of woody crops, typically formed by several rows. This method provides satisfactory results in large-areas reconstruction from the fusion of different overlapped images supplied by the Kinect sensor.

Once known the camera position, the ray-casting technique [21] is employed to project a ray from the camera focus for each pixel of each input depth image to define the voxels in the 3D world that cross each ray. When the surface of the scene has been obtained using the ray-casting technique, the information is employed to estimate the position and orientation of the camera (6 degrees of freedom) when a new image arrives. This estimation is conducted by a variant of the iterative closest point (ICP) algorithm [22]. In this way, the 3D reconstruction of a woody row is obtained. If any drift appears in the model, it is corrected by applying the novel method developed in [23].

RGB Reflex Camera. Agisoft PhotoScan software solution was used for 3D reconstruction of woody crops from planar RGB images taken by a reflex camera and the associated positions of a RTK-GNSS receiver on board the vehicle. The 3D reconstruction of a woody crop using photogrammetry involves four consecutive stages: (1) orientation of images, (2) aerial triangulation, (3) reconstruction of the dense surface, and (4) generation of the mesh.

The first step defines the exact position of the woody crop row in 3D space from the intrinsic and extrinsic calibration parameters of the camera obtained from EXIF information and WGS84 UTM projected coordinates of each image. The intrinsic calibration parameters include the coordinates of the projection center and the focal length; whereas the extrinsic parameters are the coordinates in the real world (x, y, z) and rotation angles (yaw, pitch, roll) of the camera in each image. (2) The aerial triangulation aligns blocks of overlapping images for each crop row. During this process the image matching forms tie points based on similar characteristics (color, texture, shape, etc.) or interest points invariant to scale, rotation, translation and illumination changes for each pair of images using the SIFT [24] and Bundler [25] algorithms. (3) The reconstruction of dense surfaces includes the generation of dense cloud points of the crop row from correspondence between common points extracted from two or more images to estimate the corresponding 3D coordinate or projective model through the SGM-like stereo algorithm [26]. (4) Following, the generation of the polygonal mesh is carried out from the depth maps using the dense stereo-matching algorithms (Exact, Smooth and Height-field).

Volume Estimation. From the 3D reconstructions of the woody crops, the volume of the canopies can be estimated. For that purpose, the alpha-shape algorithm [27], which has been used in other agriculture researches [28–31], was employed. This method generates an outline that surrounds a set of 3D points; more or less tight to them depending on the value of the alpha parameter. In order to calculate the alpha-shapes of the canopies and the enclosed volumes, the alphashape3d package of R [32] based in the original work of [27] was used.

3 Results

3.1 Automation of the Vehicle

After testing separately the different systems implemented for the automation of the vehicle, different experiments were carried out in a test circuit to verify the correct performance of the entire automation of the vehicle. The vehicle was commanded only by means of a joystick, acting on the direction, acceleration and brake, to complete a route. Figure 6 shows the trajectory of the vehicle during one of the experiments (position data were gathered by the RTK-GNSS receiver of the vehicle).

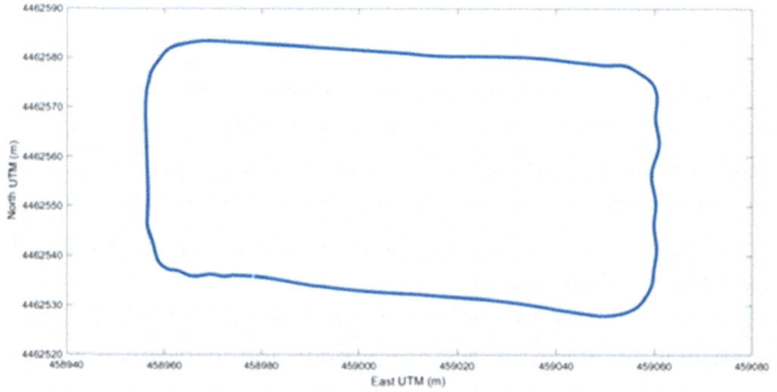

Fig. 6. Trajectory of the vehicle commanded by a joystick

3.2 Autonomous Navigation

In order to verify the correct performance of the developed controller, the vehicle followed a reference trajectory, usually a straight path, similar as the trajectories that it must follow in the field. Figures 7 and 8 shows the position and orientation error of the vehicle in one of the tests. It had a wrong orientation and position at the starting time. However, it corrected its orientation and position following the reference trajectory.

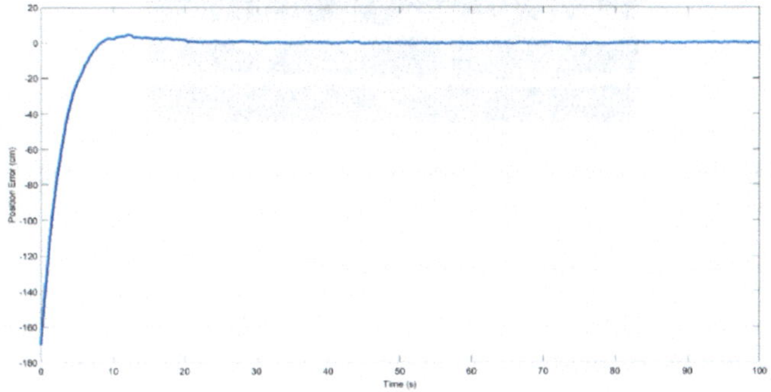

Fig. 7. Position error of the vehicle

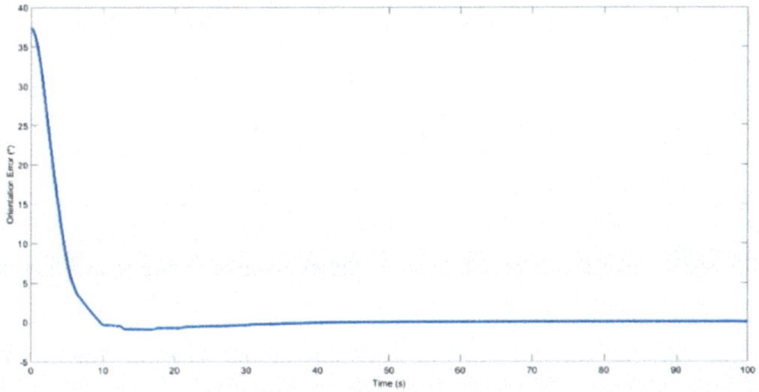

Fig. 8. Orientation error of the vehicle

3.3 3D Reconstruction Using a RGB-D Sensor

The inspection platform was operated in a vineyard field at 3 km/h for data collection. The Kinect v2 sensor was mounted at approximately 1.4 m high with a 10° pitch angle oriented to the crop rows at a distance of approximately 1 m from the crop row. From each inspected row, the starting and ending geographical positions of the row supplied by the RTK-GNSS receiver from the vehicle were stored. 3D reconstructions of the sampled rows of vineyards were performed. Figure 9 shows an example of one of them, while a zoom-in on a section of the 3D reconstruction of this row can be seen in Fig. 10.

Fig. 9. 3D reconstruction of a 105 m length vineyard row using a RGB-D sensor

Fig. 10. Zoom-in on a section of the 3D reconstruction of Fig. 9

3.4 3D Reconstruction Using a Reflex Camera

Similarly to the test of Kinect v2 sensor, the reflex RGB camera was mounted in the aluminum structure of the platform at approximately 1.2 m high with a 45° pitch angle oriented to the crop rows. The camera was configured with the default automatic settings: shutter speed (1/250 s to 1/60 s), depth of field (f/22 to f/36), ISO sensitivity (100 to 400), focal length of 18 mm, shooting speed of two frames per second and flash and autofocus inactive; this configuration remained constant during crop monitoring. Each image taken by the camera had associated coordinates supplied by the RTK-GNSS receiver. The vehicle operated at 3 km/h, allowing a minimum longitudinal overlap of 70% between pairs of consecutive images. Figure 11 shows the stages to generate a 3D model of one of the sampled vineyard rows.

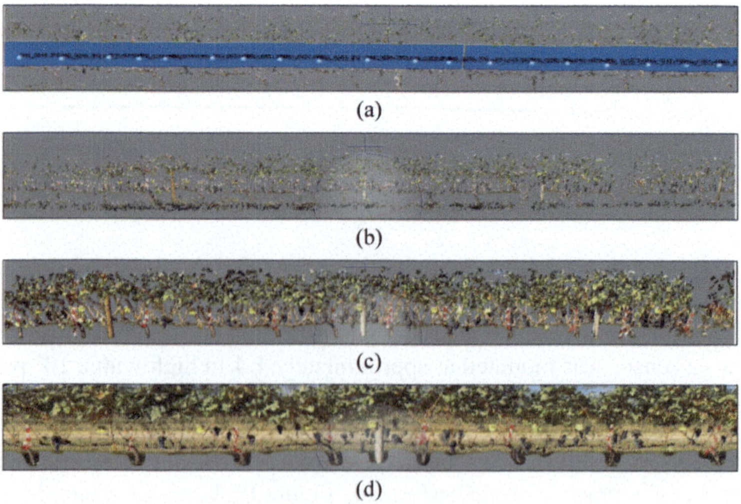

Fig. 11. Stages of construction of the 3D model of the vineyard: (a) orientation of images; (b) tie points; (c) dense cloud of points; (d) polygonal mesh

3.5 Volume Estimation

To test the alpha-shape algorithm, four trees of different known shapes (sphere, cylinder, cone) were used (Fig. 12). First, their 3D reconstructions were obtained using the same procedure described in Sect. 3.3. Then, the alpha-shape method was applied to generate the outlines that surround the canopies of each tree. After trying with different alpha values, 0.5 was selected as the most suitable one for these forms (Fig. 13).

Fig. 12. Four trees used to test the alpha-shape algorithm

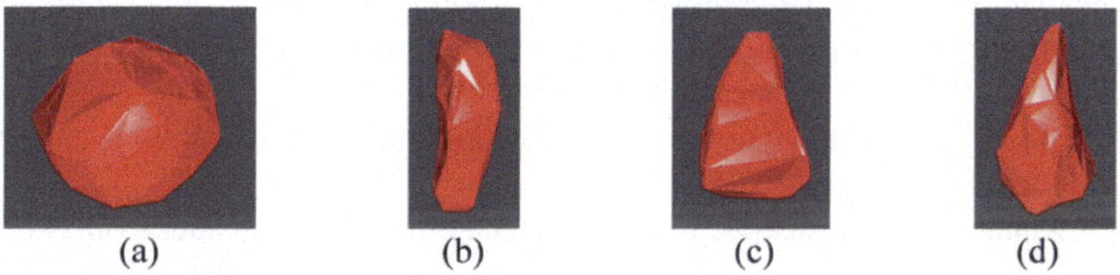

Fig. 13. Alpha-shapes generated with alpha = 0.5 for (a) the lower sphere of the tree with two round-shape canopies of Fig. 12, (b) the cylinder-shape tree of Fig. 12, (c) the big cone-shape tree of Fig. 12, and (d) the small cone-shape tree of Fig. 12

In view of the fact that the generated alpha-shapes had a similar shape to those of the trees, the volume enclosed for them can provide good estimations of the real canopy volumes of the trees.

4 Conclusions

This paper describes a crop inspection system. A commercial electric vehicle is equipped with different on-board sensors to scan crops. Automation of the platform and autonomous navigation were developed and tested. Different methods to generate 3D reconstructions of woody crops from the information supplied by a low-cost RGB-D sensor and a reflex camera were studied, developed and tested in real fields under uncontrolled lighting conditions. The good performance of the alpha-shape algorithm to estimate the canopy volume was checked. The promising results will allow to achieve an autonomous field platform able to generate 3D models of the crop for a better crop management.

As future work, the comparison of the two different methodologies to generate 3D models must be performed in order to discover which one is the most suitable to reconstruct woody crops and to extract information as accurate as possible.

Acknowledgments. This work was financed by the Spanish Ministerio de Economía y Competitividad (AGL2014-52465-C4-3-R) and the Spanish Agencia Estatal de Investigación (AEI) and Fondo Europeo de Desarrollo Regional (FEDER) (AGL2017-83325-C4-1-R and AGL2017-83325-C4-3-R). Karla Cantuña thanks the service commission for the remuneration given by the Cotopaxi Technical University. The authors also wish to acknowledge the ongoing technical support of Damián Rodríguez.

References

1. Rovira-Más, F., Chatterjee, I., Sáiz-Rubio, V.: The role of GNSS in the navigation strategies of cost-effective agricultural robots. Comput. Electron. Agric. **112**, 172–183 (2015)
2. Case New Holland. http://assets.cnhindustrial.com/caseih/NAFTA/NAFTAASSETS/Products/Advanced-Farming-Systems/Brochures/AFS_Brochure.pdf. Accessed 03 Sept 2019
3. Bakker, T., van Asselt, K., Bontsema, J., Müller, J., van Straten, G.: Autonomous navigation using a robot platform in a sugar beet field. Biosyst. Eng. **109**(4), 357–368 (2011)
4. Bengochea-Guevara, J.M., Conesa-Muñoz, J., Andújar, D., Ribeiro, A.: Merge fuzzy visual servoing and GPS-based planning to obtain a proper navigation behavior for a small crop-inspection robot. Sensors **16**(3), 276 (2016)
5. Naio Technologies. http://www.naio-technologies.com/machines-agricoles/robot-de-desherbage-oz/. Accessed 03 Sept 2019
6. Wang, Q., Zhang, Q., Rovira-Más, F., Tian, L.: Stereovision-based lateral offset measurement for vehicle navigation in cultivated stubble fields. Biosyst. Eng. **109**(4), 258–265 (2011)
7. Hansen, S., Bayramoglu, E., Andersen, J.C., Ravn, O., Andersen, N., Poulsen, N.K.: Orchard navigation using derivative free Kalman filtering. In: 2011 International Conference on American Control Conference (ACC), IEEE, pp. 4679–4684 (2011)
8. Libby, J., Kantor, G.: Deployment of a point and line feature localization system for an outdoor agriculture vehicle. In: 2011 International Conference on Robotics and Automation (ICRA), IEEE, pp. 1565–1570 (2011)
9. Weiss, U., Biber, P.: Plant detection and mapping for agricultural robots using a 3D LIDAR sensor. Robot. Auton. Syst. **59**(5), 265–273 (2011)
10. West, P.W.: Tree and Forest Measurement, vol. 20. Springer, Heidelberg (2009)
11. Anderson, C.D.J.: Electric and Hybrid Cars: A History. McFarland, North Carolina (2010)
12. Pagliari, D., Pinto, L.: Calibration of Kinect for Xbox One and comparison between the two generations of microsoft sensors. Sensors **11**, 27569–27589 (2015)
13. Fankhauser, P., Bloesch, M., Rodriguez, D., Kaestner, R., Hutter, M., Siegwart, R.: Kinect v2 for mobile robot navigation: evaluation and modelling. In: 2015 International Conference on Advanced Robotics (ICAR), IEEE, pp. 388–394 (2015)
14. Conesa-Munoz, J., Bengochea-Guevara, J.M., Andujar, D., Ribeiro, A.: Efficient distribution of a fleet of heterogeneous vehicles in agriculture: a practical approach to multi-path planning. In: IEEE International Conference on Autonomous Robot Systems and Competitions, IEEE, pp. 56–61 (2015)
15. Fraichard, T., Garnier, P.: Fuzzy control to drive car-like vehicles. Robot. Autom. Syst. **34**, 1–22 (2001)

16. Naranjo, J.E., Sotelo, M., Gonzalez, C., Garcia, R., Sotelo, M.A.: Using fuzzy logic in automated vehicle control. IEEE Intell. Syst. **22**, 36–45 (2007)
17. Kodagoda, K.R.S., Wijesoma, W.S., Teoh, E.K.: Fuzzy speed and steering control of an AGV. IEEE Trans. Control Syst. Technol. **10**, 112–120 (2002)
18. Zadeh, L.A.: Fuzzy sets. Inf. Control **8**, 338–353 (1965)
19. Sugeno, M.: On stability of fuzzy systems expressed by fuzzy rules with singleton consequents. IEEE Trans. Fuzzy Syst. **7**, 201–224 (1999)
20. Niessner, M., Zollhöfer, M., Izadi, S., Stamminger, M.: Real-time 3D reconstruction at scale using voxel hashing. ACM Trans. Graph. **32**(6), 169 (2013)
21. Roth, S.D.: Ray casting for modeling solids. Comput. Graph. Image Process. **18**(2), 109–144 (1982)
22. Chen, Y., Medioni, G.: Object modelling by registration of multiple range images. Image Vis. Comput. **10**(3), 145–155 (1992)
23. Bengochea-Guevara, J.M., Andújar, D., Sanchez-Sardana, F.L., Cantuña, K., Ribeiro, A.: A low-cost approach to automatically obtain accurate 3D models of woody crops. Sensors **18**(1), 30 (2017)
24. Lowe, G.: Object recognition from local scale-invariant features. In: Proceedings of the Seventh IEEE International Conference on Computer Vision, vol. 2, pp. 1150–1157 (1999)
25. Bundler: Structure from motion (SFM) for unordered image collections. http://phototour.cs.washington.edu. Accessed 03 Sept 2019
26. Remondino, F., Spera, M. G., Nocerino, E., Menna, F., Nex, F., Gonizzi-Barsanti, S.: Dense image matching: comparisons and analyses. In: 2013 Digital Heritage International Congress, DigitalHeritage, vol. 2, pp. 740–741 (1987)
27. Edelsbrunner, H., Mücke, E.P.: Three-dimensional alpha shapes. ACM Trans. Graph. **1994**(13), 43–72 (1994)
28. Bengochea-Guevara, J.M., Andújar, D., Sanchez-Sardana, F.L., Cantuña, K., Ribeiro, A.: 3D monitoring of woody crops using a medium-sized field inspection vehicle. Adv. Intell. Syst. Comput. **694**, 239–250 (2017)
29. Colaço, A.F., Trevisan, R.G., Molin, J.P., Rosell-Polo, J.R., Escolà, A.: a method to obtain orange crop geometry information using a mobile terrestrial laser scanner and 3D modeling. Remote Sensing **9**(8), 763 (2017)
30. Martinez-Guanter, J., Ribeiro, A., Peteinatos, G.G., Pérez-Ruiz, M., Gerhards, R., Bengochea-Guevara, J.M., Machleb, J., Andújar, D.: Low-cost three-dimensional modeling of crop plants. Sensors **19**(3), 2883 (2019)
31. Rueda-Ayala, V.P., Peña, J.M., Höglind, M., Bengochea-Guevara, J.M., Andújar, D.: Comparing UAV-based technologies and RGB-D reconstruction methods for plant height and biomass monitoring on grass ley. Sensors **19**(3), 535 (2019)
32. Lafarge, T., Pateiro-Lopez, B., Possolo, A., Dunkers, J.P.: R implementation of a polyhedral approximation to a 3D set of points using the alpha-shape. J. Stat. Softw. **56**, 1–19 (2014)

Autonomous Driving and Driver Assistance Systems

NOSeqSLAM: Not only Sequential SLAM

Jurica Maltar[1](✉), Ivan Marković[2], and Ivan Petrović[2]

[1] Department of Mathematics, University of Osijek, Osijek, Croatia
jmaltar@mathos.hr

[2] Faculty of Electrical Engineering and Computing, University of Zagreb, Zagreb, Croatia
{ivan.markovic,ivan.petrovic}@fer.hr

Abstract. The essential property that every autonomous system should have is the ability to localize itself, i.e., to reason about its location relative to measured landmarks and leverage this information to consistently estimate vehicle location through time. One approach to solving the localization problem is *visual place recognition*. Using only camera images, this approach has the following goal: during the second traversal through the environment, using only current images, find the match in the database that was created during a previously driven traversal of the same route. Besides the image representation method – in this paper we use feature maps extracted from the OverFeat architecture – for visual place recognition it is also paramount to perform the scene matching in a proper way. For autonomous vehicles and robots traversing through an environment, images are acquired sequentially, thus visual place recognition localization approaches use the structure of sequentiality to locally match image sequences to the database for higher accuracy. In this paper we propose a not only sequential approach to localization; specifically, instead of linearly searching for sequences, we construct a directed acyclic graph and search for any kind of sequences. We evaluated the proposed approach on a dataset consisting of varying environmental conditions and demonstrated that it outperforms the SeqSLAM approach.

Keywords: Visual place recognition · Localization · SeqSLAM · Deep convolutional neural networks

1 Introduction

Localization, i.e., reasoning about own's location given a set of measurements from one or mulitple sensors, is a prerequisite capability for any autonomous system. It can be approached in different ways depending on the used sensor setup, and by relying on onboard sensors many approaches have been developed for mobile agents based on laser range sensors, cameras (mono, stereo or depth), ultrasonic sensor and even radars. Localization is often performed in a known environment; however, it can be challenging when the environment is highly dynamic both in the sense of having dynamic objects and changing its

M. F. Silva et al. (Eds.): ROBOT 2019, AISC 1092, pp. 179–190, 2020.
https://doi.org/10.1007/978-3-030-35990-4_15

own appearance through time. Simply during a single day, visual appearance of the same location can change drastically due to day and night conditions. Nevertheless, although challenging, since images contain a rich set of information, they can be leveraged to recognize places even during strong appearance changes such as day/night or seasonal changes.

Visual place recognition, as the name suggests, tackles the problem of recognizing a previously seen location given a single or multiple images captured by a camera, thus it can be seen as a specific approach to solving the localization problem. More formally, given a *query* image $I_{q_i} \in \mathcal{Q}$, taken as the vehicle traverses its route *on the fly*, we are trying to find the most plausible match from the labeled database of *reference* images $\mathcal{D}$ [11]. Such a match, often denoted with $I^*_{d_j} \in \mathcal{D}$, represents the current position hypothesis. Furthermore, it is an instance of a well-known computer vision problem – *visual instance retrieval* [18] where given a query image, we are trying to find all the possible matches that correspond to the category of this instance. However, subtle differences exist, the most prominent one being that both $\mathcal{Q}$ and $\mathcal{D}$ in visual place recognition are sequentially ordered. This insight into data sequentiality can help us build more robust systems [12,16] and, as it will be seen, the proposed method strongly relies on it. Given that, in order to make visual place recognition a robust localization method, it should be developed to be *view-point invariant*, i.e., to be able to recognize the same location from different viewpoints, and *condition invariant*, i.e., to be able to recognize the same location irrespective of the time-of-the-day or season. For example, $\mathcal{D}$ could have been captured on a stormy winter evening, while $\mathcal{Q}$ is captured on a sunny autumn noon. Consequently, two main design aspects regarding visual place recognition are *image representation* and *image matching* - as the appropriate image representation is obtained, the goal is to find the appropriate match.

To represent an image of a place, we can employ classical computer vision approaches consisting of image feature extractors and descriptors - e.g., SURF [2] was used by [5], while ORB [3,19] was used by [13]. *Histogram of oriented gradients* is often used as a global image descriptor [14–16,25]. However, given the development of deep convolutional neural networks, research has been directed into utilizing feature maps obtained by passing an image through the network and using them as a global description of an image. Sünderhauf et al. [22] have concluded that those feature maps extracted from middle layers of the AlexNet architecture [10] behave better for *condition variations* while feature maps from higher layers are more suitable when view-point variance occurs. The same authors propose a system [23] that uses *an object proposal* method in order to achieve even stronger *view-point invariance*. Another notable application of DCNNs is NetVLAD [1] by Arandjelović et al. who have modified the original VLAD [9] by replacing the indicator function with softmax. Inspired by NetVLAD, Garg et al. [7] propose another descriptor called LoST which aggregates residuals of semantic categories. Fetching the appropriate representation of an image from DCNN, Hausler et al. [8] filter out "bad" slices from feature maps extracted from some k-th convolution layer. Chen et al. [4] obtain the representation by *multi-scale pooling* where feature map is divided into $S \times S$ subregions and the maximum activation is pooled resulting in a more compact representation for each feature map.

Regarding image matching, often used approach is SeqSLAM [12], that searches for the locally optimal sequence match - a sequence that bears the information about the vehicle traversing in a local scope. Siam and Zhang [21] upgraded SeqSLAM such that N approximate nearest neighbors (ANN) of the query image I_T were taken. Yin et al. [27] incorporate particle filter within SeqSLAM in order to reduce computational complexity. SMART [17] improves SeqSLAM by incorporating the odometry information from *wheel encoders*. Improved variants of SeqSLAM search methods (*cone-based* and *hybrid* method) can be found in [24]. In [6] for each query image N most similar matches from reference database are fetched. Thereafter, by using another system that approximates the depth from an image, authors find the reference image with the most plausible neighborhood. Nasser et al. [16] addressed the traversal matching by using *data association graph*. Each node within this graph represents route match between $\mathcal{Q}$ traversal and $\mathcal{R}$ traversal, while both traversals consist of sequence of images. When data association graph is constructed, an appropriate traversal A is obtained by solving *min-cost flow* problem. The work of Vysotska and Stachniss [26] can be considered as follow-up to [16] where the improvement is manifested in the fashion of search. System proposed in [16] operates in *offline fashion* meaning that both $\mathcal{D}$ and $\mathcal{Q}$ are first obtained and thereafter appropriate associations are found while this system operates in *online fashion* which means that right after the last query image is obtained, the appropriate match is found. As emphasized in [16] solving for min-cost flow problem is equivalent to the *shortest path* problem in *directed acyclic graph*. Nodes of the shortest path in their formulation represent match hypotheses.

In this paper we propose a not only sequential approach to localization; specifically, instead of linearly searching for sequences, as SeqSLAM does, for image matching we construct a directed acyclic graph and search for any kind of sequences. Thus instead of using the shortest path as route hypothesis, we use shortest paths to measure the association of the matches between $\mathcal{Q}$ and $\mathcal{D}$. For image representation, we use deep learning feature maps extracted from the OverFeat architecture [20], since it was constructed for localization and detection tasks. We evaluated the proposed approach on the *Bonn dataset* [26] consisting of varying environmental conditions and compared it to SeqSLAM. To evaluate quantitatively, we constructed precision-recall curves and computed the area under the curve. The results show that on the tested database the proposed approach outperforms SeqSLAM. Source code of our approach is available online[1].

2 Proposed Visual Place Recognition

The general scheme for visual place recognition, given in Fig. 1, is as follows:

[1] https://bitbucket.org/unizg-fer-lamor/noseqslam/.

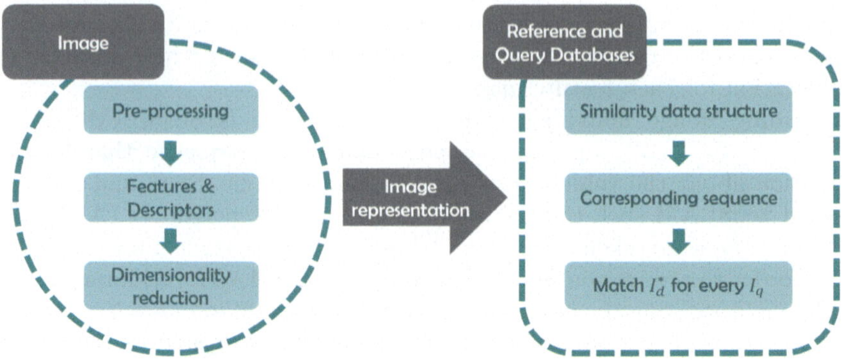

Fig. 1. Visual place recognition scheme.

1. Image representation for each image is obtained by pre-processing, feature extraction, description and dimensionality reduction.
2. Similarity s_{I_q,I_d} is calculated and by sequence matching, the best match $I_d^* \in \mathcal{D}$ for query image $I_q \in \mathcal{Q}$ is taken.

As stated above, more robust matching between $I_{q_i} \in \mathcal{Q}$ and $I_{d_j} \in \mathcal{D}$ can be found by incorporating data sequentiality, thus by observing an ordered local neighborhood around this match in contrast to the naive approach

$$I_{d_j}^* = \operatorname*{argmax}_{I_{d_j} \in \mathcal{D}} s_{I_{q_i},I_{d_j}} \tag{1}$$

where

$$s_{I_{q_i},I_{d_j}} = \cos(\theta) = \frac{I_{q_i} I_{d_j}}{||I_{q_i}|| ||I_{d_j}||}. \tag{2}$$

Because images are represented as vectors we can measure the similarity by taking their cosine similarity according to (2). However, as noted by [15], "matching images just according to the best similarity score produces considerable false positives [...]".

2.1 SeqSLAM

Let T denote the current time and d_s denote the number of matches some sequence is consisted of and let M denote the difference matrix where the j-th column $M[:,j] = \hat{D}^j$, $j \in \{T - \lfloor \frac{d_s}{2} \rfloor, \ldots, T + \lfloor \frac{d_s}{2} \rfloor\}$ is the vector of differences between j-th query image I_{q_j} and the reference database. Then, given a certain velocity V we can calculate the sequence weight

$$S_{j,T,V} = \sum_{t=T-\lfloor \frac{d_s}{2} \rfloor}^{T+\lfloor \frac{d_s}{2} \rfloor} \hat{D}_k^t \tag{3}$$

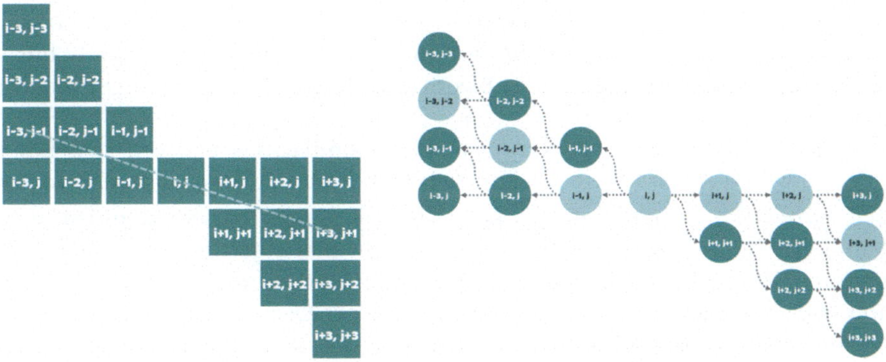

(a) SeqSLAM **(b)** NOSeqSLAM

Fig. 2. (a) SeqSLAM searches for the optimal linear sequence passing through (I_{q_i}, I_{d_j}) while (b) NOSeqSLAM searches for the optimal single-source shortest path from root (I_{q_i}, I_{d_j}) to the left subgraph and from root to the right subgraph.

where $k = j + V(t - T)$. The most appropriate sequence for I_T and I_{d_j} is the one that minimize the weight of sequence by velocity, therefore

$$S_{j,T} = \min_{V} S_{j,T,V}. \tag{4}$$

The procedure operates in the same manner if we are taking similarity between images, but then we maximize by V because *minimizing the difference* is equivalent to *maximizing the similarity* [11]. Another relevant fact is that SeqSLAM is agnostic regarding image representation. In the original work no features from images were extracted, thus original *human-understandable* down-sampled image is used while the difference between two images is measured via *sum of absolute differences.*

2.2 NOSeqSLAM

Our method differs from SeqSLAM in such a way that the appropriate association between I_{q_i}[2] and I_{d_j} is not found by measuring the weight of optimal linear sequence, but generally by measuring the weight of any kind of sequence passing through the match of I_{q_i} and I_{d_j}. Henceforth, we denote this match with (I_{q_i}, I_{d_j}). For illustration of similarities and differences between the two methods confer Fig. 2.

By *linear sequence* we mean pure linear correlation between indices in difference matrix M just as it is illustrated in Fig. 2(a). From a physical point of view this means a vehicle should traverse same subroute in both $\mathcal{Q}$ and $\mathcal{D}$ with linear correlation in acceleration/deceleration. One special case of this condition is to traverse same subroute in $\mathcal{Q}$ and $\mathcal{D}$ without acceleration at all. This is a limiting

[2] We substitute T with q_i for the sake of clarity.

factor, as a vehicle is akin to accelerate/decelerate all the time. We therefore model our system so that it searches for any kind of sequences - not only linear, and for that reason we name it NOSeqSLAM, where NO is acronym for *"not only"*.

Similar to difference matrix M, we place similarities between matches in matrix A, where $A[j,i] = s_{I_{q_i},I_{d_j}}$. For each match (I_{q_i}, I_{d_j}) we construct *directed acyclic graph* $G_{(i,j)}$ rooted at (I_{q_i}, I_{d_j}). This root is thereafter expanded in the left and the right direction with respect to i-th row resulting in left and right subgraphs. We build graph iteratively until the depth of $\lfloor \frac{d_s}{2} \rfloor$ is reached. As the graph expands, each node in the graph will be predecessor for η_{exp} nodes. This procedure can be parallelized on two threads, one for the left and one for the right subgraph of a DAG. Our system thereby depends upon two parameters: d_s (sequence length) and η_{exp} (expansion rate). The example for $d_s = 7$ and $\eta_{exp} = 2$ is shown in Fig. 2(b).

The weight of an edge from (I_{q_i}, I_{d_j}) to (I_{q_k}, I_{d_l}) is defined as

$$w\left[(I_{q_i}, I_{d_j}), (I_{q_k}, I_{d_l})\right] = 1 - A[l,k]. \tag{5}$$

Naseer et al. [16] use $1/s_{I_{q_k},I_{d_l}}$ in this situation; however, (5) is also reasonable – as association approaches one, weight approaches zero, and vice-versa. Therefore, it is appropriate to reach some node reciprocal to its similarity measure.

As the graph for (I_{q_i}, I_{d_j}) with associated weight is constructed, we still have to measure association for (I_{q_i}, I_{d_j}), thus, how good I_{d_j} fits for I_{q_i}. In NOSeqSLAM this measure is defined as the sum of similarities of those nodes that lie on the minimal of the shortest paths in $U_{(i,j)}$ that connect the leaf from left subgraph to the leaf in the right subgraph passing through (I_{q_i}, I_{d_j}), where $U_{(i,j)}$ is undirected version of $G_{(i,j)}$. Although this was our initial thought about the formulation, a simpler way to achieve this is to construct $G_{(i,j)}$, find the minimal of the shortest paths in left subgraph $l^*_{(i,j)}$ and the minimal of the shortest paths in right subgraph $r^*_{(i,j)}$, to sum similarity measures through this minimal shortest paths together with similarity between I_{q_i} and I_{d_j} what yields the following association measure:

$$S_{j,i} = A[j,i] + \sum_{(k,l)\in l^*_{(i,j)}} A[l,k] + \sum_{(k,l)\in r^*_{(i,j)}} A[l,k]. \tag{6}$$

Pseudocodes for both the proposed method and SeqSLAM are shown in Algorithms 1 and 2.

3 Experimental Results

The first step to experimental evauluation was to build up the association matrix A which truly reflects the relationship between $\mathcal{Q}$ and $\mathcal{D}$. As mentioned, the key idea is to build this matrix up *row-by-row* whenever a new $I_{q_i} \in \mathcal{Q}$ is captured. This I_{q_i} is then compared with every $I_{d_j} \in \mathcal{D}$. The result is depicted in Fig. 3.

Algorithm 1. NOSeqSLAM

for each $(I_{q_i}, I_{d_j}) \in \mathcal{Q} \times \mathcal{D}$ **do**
  $G_{(i,j)} = DAG(I_{q_i}, I_{d_j}, d_s, \eta_{exp})$
  $l^*_{(i,j)} = \min_m SP(G_{(i,j)}, (I_{q_i}, I_{d_j}), (I_{q_{i-\lfloor \frac{d_s}{2} \rfloor}}, I_{d_m}))$
  $r^*_{(i,j)} = \min_m SP(G_{(i,j)}, (I_{q_i}, I_{d_j}), (I_{q_{i+\lfloor \frac{d_s}{2} \rfloor}}, I_{d_m}))$
  $S_{j,i} = A[j,i] + \sum_{(k,l)\in l^*_{(i,j)}} A[l,k] + \sum_{(k,l)\in r^*_{(i,j)}} A[l,k]$
end for

Algorithm 2. SeqSLAM

$V_{steps} = linspace(V_{min}, V_{max}, V_{step})$
for each $(I_T, I_{d_j}) \in \mathcal{Q} \times \mathcal{D}$ **do**
  $S_{j,T} = \infty$
  for each $V \in V_{steps}$ **do**
    $s = S_{j,T,V}$
    if $s < S_{j,T}$ **then**
      $S_{j,T} = s$
    end if
  end for
end for

3.1 Dataset

For evaluation purposes we used the *Bonn dataset* contained with the publicly available implementation of [26]. The route is driven in an urban area and is captured in two different environmental conditions. When driven for the first time (and therefore saved as $\mathcal{D}$), route is captured in the evening, while for the second time (saved as $\mathcal{Q}$) it is captured on a gloomy day. Although there is no record about the distance traveled throughout this route, by observing sequences alongside the route and from the cardinality of both datasets ($|\mathcal{D}| = 488$, $|\mathcal{Q}| = 544$) we can assert that the route is 1–2 km long.

View-point variance is not that accentuated in this dataset because, the vehicle stays in the same track for both traversals. However, variations in condition are severe as can be seen in Fig. 4. Not only illumination is different, but also various moving objects appear throughout both traversals.

3.2 Image Representation

Besides raw images, the dataset comes with the feature maps extracted from the OverFeat architecture [20] for both $\mathcal{D}$ and $\mathcal{Q}$. This architecture is nearly the same as the AlexNet [10], but besides classification, it was constructed also for other tasks such as *localization* and detection.

In order to choose the best image representation algorithm, we conducted an experimental comparison. First, we constructed the HOG representation, but for which it can be seen from Fig. 3(a), that it does not quite yield discriminative associations for $\mathcal{Q}$ with respect to $\mathcal{D}$. Generally, the trend of replacing *handcrafted* features and descriptors with representations extracted from DCNN architectures is also present in image representation. Second, beside the OverFeat map that was readily available, we have also extracted feature maps from AlexNet conv3 layer as it was reported that this architecture is suitable for image representation too [22]. Moreover, therein it is also asserted that when the network is trained on *a scene-centric* training set (in contrast to *an object-centric* training set), even more accurate results can be obtained. For that purpose, we employ another AlexNet network trained on a scene-centric dataset Places365 [28].

Given that, we tested the suitability of the aforementioned representations. Even though the contrast of the association matrix plot, as in Fig. 3(a), can act as a qualitative indicator whether a representation is fine, a quantitative measure is needed. For that purpose we calculated precision and recall for different representations using $d_s = 7$ in combination with $\eta_{exp} = 3$ and accumulated the results in order to obtain the *area under the curve* (AUC) measure. No significant

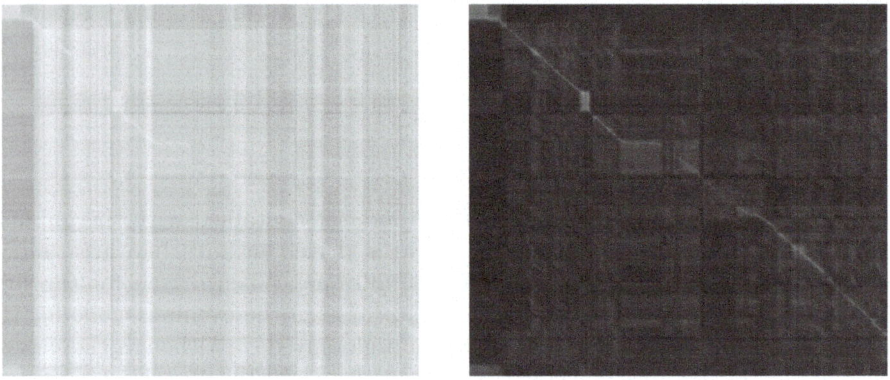

(a) HOG (b) OverFeat

Fig. 3. Plotting the association matrix A when using different image representations reflects their quality. The more accentuated is the contrast, the better.

(a) $\mathcal{D}$ (b) $\mathcal{Q}$

Fig. 4. Different environmental conditions and occlusions at each traversal. Images taken from [26].

improvement when using AlexNet trained on Places365 ($AUC = 0.89193$) with respect to the one trained on ImageNet ($AUC = 0.88786$) was noticed. The HOG representation, as expected, had the lowest score ($AUC = 0.77529$). Finally, similar results were obtained with OverFeat ($AUC = 0.91425$) and AlexNet representations, which was expected since OverFeat is almost identical in its convolutional layers. Finally, given the results, we decided to use OverFeat feature maps in the ensuing experiments both for NOSeqSLAM and SeqSLAM.

3.3 Comparison of NOSeqSLAM and SeqSLAM

In the next experiment we compared the proposed NOSeqSLAM and SeqSLAM for visual place recognition. By focusing on Fig. 5, we can see that the association matrix reflects the designs of each approach, i.e., SeqSLAM leaves traces in the form of linear sequences distributed through the path, while at the same time, NOSeqSLAM traces are softer.

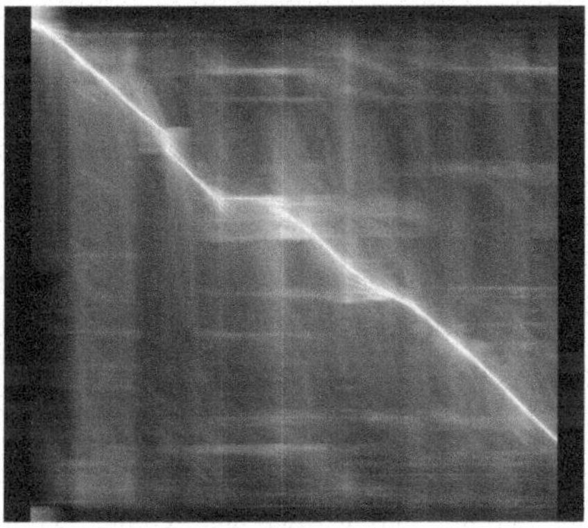

(a) SeqSLAM

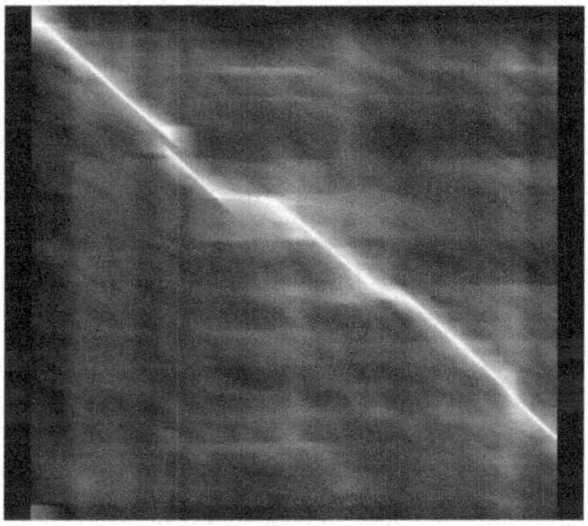

(b) NOSeqSLAM

Fig. 5. Association matrices of $\mathcal{Q}$ and $\mathcal{D}$ for SeqSLAM and NOSqeSLAM. Differences are visible as the two algorithms assign different associations.

Figure 5 shows that the sequence (no matter if linear or non-linear) of length d_s does not fit into first $\lfloor \frac{d_s}{2} \rfloor$ indices and last $\lfloor \frac{d_s}{2} \rfloor$ indices of $\mathcal{Q}$ (notice lateral black areas). This implicates that no viable match can be found for first $\lfloor \frac{d_s}{2} \rfloor$ and last $\lfloor \frac{d_s}{2} \rfloor$ query images and matches for those $2 \lfloor \frac{d_s}{2} \rfloor$ images will be declared as false negatives. From *precision-recall* curves in Fig. 6(a) we can see that in general NOSeqSLAM performs better regardless of chosen d_s and that the maximal recall is roughly the same depending on d_s no matter what method is used. For fair comparison we group the results according to d_s and show the results in Table 1, from which we can see that in terms of the AUC measure, NOSeqSLAM outperforms SeqSLAM. In visual place recognition we are striving to achieve the AUC measure as large as possible - ideally equal to 1. This, amongst others, means that no false positives have been encountered at all, i.e., each match is consistent with the ground truth.

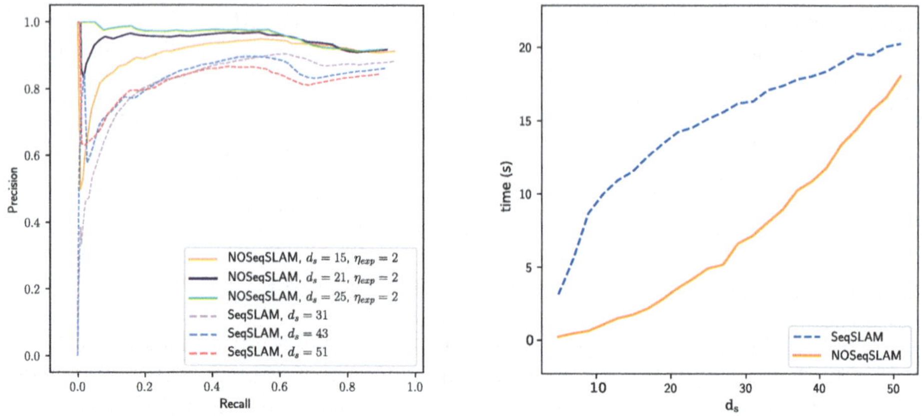

(a) **(b)**

Fig. 6. (a) Precision-recall plots with $d_s \in \{31, 43, 51\}$. (b) Running time as a function of the sequence length $d_s \in \{5, 7, \ldots, 51\}$.

Given the association matrix A, NOSeqSLAM takes $\Theta\left(|\mathcal{Q}||\mathcal{D}|d_s^2\eta_{exp}^2\right)$ asymptotic running time while SeqSLAM operates in $\Theta\left(|\mathcal{Q}||\mathcal{D}|d_s V_{steps}\right)$. If we diminish the role of factors $d_s^2\eta_{exp}^2$ and $d_s V_{steps}$, thus $\Theta\left(d_s^2\eta_{exp}^2\right) = \Theta\left(d_s V_{steps}\right) = \Theta\left(1\right)$, we can say that both algorithms operate in $\Theta\left(|\mathcal{Q}||\mathcal{D}|\right)$ asymptotic time. In terms of *real-time* performance measured on a `i7@2.8 GHz` laptop processor, NOSeqSLAM evaluation takes 0.29 s and SeqSLAM evaluation takes 3.16s for $d_s = 5$. For $d_s = 51$, NOSeqSLAM evaluation takes 18.1 s, while SeqSLAM evaluation takes 20.29 s. This is approximately 33 ms per query image which is readily employable for autonomous vehicles. In general, for $d_s \in \{5, 7, \ldots, 51\}$ NOSeqSLAM operates faster than SeqSLAM as can be seen in Fig. 6(b). Although NOSeqSLAM running time will exceed SeqSLAM running time once when d_s is sufficiently large, we think that $d_s = 51$ is more than enough to describe a local neighborhood so there is no need for larger d_s.

Table 1. SeqSLAM (denoted with S) and NOSeqSLAM (denoted with N) AUC results.

d_s	alg.	η_{exp}	AUC	d_s	alg.	η_{exp}	AUC	d_s	alg.	η_{exp}	AUC	d_s	alg.	η_{exp}	AUC
	S	-	0.73113		S	-	0.75650		S	-	0.77137		S	-	0.84974
51	N	3	0.84503	43	N	3	0.85559	31	N	3	0.83986	19	N	3	0.87886
	N	2	0.86518		N	2	0.86630		N	2	0.84798		N	2	0.88981
d_s	alg.	η_{exp}	AUC	d_s	alg.	η_{exp}	AUC	d_s	alg.	η_{exp}	AUC	d_s	alg.	η_{exp}	AUC
	S	-	0.87877		S	-	0.90616		S	-	0.90482		S	-	0.91714
15	N	3	0.89566	11	N	3	0.90744	7	N	2	0.91319	5	N	2	0.92258
	N	2	0.90672		N	2	0.91058		N	3	0.91425		N	3	0.92288

4 Conclusion

In this paper we have presented a not only sequential approach to visual place recognition. This design objective has been achieved with the powerful tool of graph theory - *shortest path* on a directed acyclic graph in such a way that the accumulated similarity throughout the shortest path represents the plausibility of the match. By using state-of-the-art image representations extracted from DCNNs, we made our system view-point and condition invariant.

From experiments conducted on Bonn dataset, we have shown that our system can operate in a rather demanding urban area with strong appearance changes. Not only that the proposed approach is capable of achieving this objective, but it has outperformed SeqSLAM in terms of both precision-recall and execution time on the tested dataset. Given the results we have observed, this system may be used for other purposes such as simultaneous localization and mapping *loop closing detection* and *relocalization*.

References

1. Arandjelović, R., Gronat, P., Torii, A., Pajdla, T., Sivic, J.: NetVLAD: CNN architecture for weakly supervised place recognition **70**(5), 641–648 (2015)
2. Bay, H., Ess, A., Tuytelaars, T., Van Gool, L.: Speeded-up robust features (SURF). Comput. Vis. Image Underst. **110**(3), 346–359 (2008)
3. Calonder, M., Lepetit, V., Strecha, C., Fua, P.: BRIEF: binary robust independent elementary features. In: Lecture Notes in Computer Science (Including Subseries Lecture Notes in Artificial Intelligence and Lecture Notes in Bioinformatics) (2010)
4. Chen, Z., Jacobson, A., Sünderhauf, N., Upcroft, B., Liu, L., Shen, C., Reid, I., Milford, M.: Deep learning features at scale for visual place recognition. In: Proceedings - IEEE International Conference on Robotics and Automation, vol. 1, pp. 3223–3230 (2017)
5. Cummins, M., Newman, P.: FAB-MAP: probabilistic localization and mapping in the space of appearance. Int. J. Robot. Res. **27**(6), 647–665 (2008)
6. Garg, S., Babu, M., Dharmasiri, T., Hausler, S., Sünderhauf, N., Kumar, S., Drummond, T., Milford, M.: Look no deeper: recognizing places from opposing viewpoints under varying scene appearance using single-view depth estimation (2019)
7. Garg, S., Sünderhauf, N., Milford, M.: LoST? Appearance-invariant place recognition for opposite viewpoints using visual semantics (2018)
8. Hausler, S., Jacobson, A., Milford, M.: Feature map filtering: improving visual place recognition with convolutional calibration (2018)
9. Jegou, H., Douze, M., Schmid, C., Perez, P.: Aggregating local descriptors into a compact image representation. In: 2010 IEEE Computer Society Conference on Computer Vision and Pattern Recognition, pp. 3304–3311. IEEE (2010)
10. Krizhevsky, A., Sutskever, I., Hinton, G.E.: ImageNet classification with deep convolutional neural networks. In: NIPS (2012)
11. Lowry, S., Milford, M.J.: Supervised and unsupervised linear learning techniques for visual place recognition in changing environments. IEEE Trans. Robot. **32**(3), 600–613 (2016)
12. Milford, M.J., Wyeth, G.F.: SeqSLAM: visual route-based navigation for sunny summer days and stormy winter nights. In: Proceedings - IEEE International Conference on Robotics and Automation, pp. 1643–1649. IEEE (2012)

13. Mur-Artal, R., Montiel, J.M.M., Tardós, J.D.: ORB-SLAM: a versatile and accurate monocular SLAM system. CoRR abs/1502.0 (2015)
14. Naseer, T., Burgard, W., Stachniss, C.: Robust visual localization across seasons. IEEE Trans. Robot. **34**(2), 289–302 (2018)
15. Naseer, T., Ruhnke, M., Stachniss, C., Spinello, L., Burgard, W.: Robust visual SLAM across seasons. In: IEEE International Conference on Intelligent Robots and Systems, December 2015, pp. 2529–2535 (2015)
16. Naseer, T., Spinello, L., Burgard, W., Stachniss, C.: Robust visual robot localization across seasons using network flows. In: Proceedings of the AAAI Conference on Artificial Intelligence, pp. 2564–2570 (2014)
17. Pepperell, E., Corke, P.I., Milford, M.J.: All-environment visual place recognition with smart. In: 2014 IEEE International Conference on Robotics and Automation (ICRA), pp. 1612–1618 (2014)
18. Razavian, A.S., Sullivan, J., Carlsson, S., Maki, A.: Visual Instance Retrieval with Deep Convolutional Networks (June 2017) (2014)
19. Rosten, E., Porter, R., Drummond, T.: Faster and better: a machine learning approach to corner detection. IEEE Trans. Pattern Anal. Mach. Intell. **32**(1), 105–119 (2010)
20. Sermanet, P., Eigen, D., Zhang, X., Mathieu, M., Fergus, R., Lecun, Y.: OverFeat: integrated recognition, localization and detection using convolutional networks. In: International Conference on Learning Representations (ICLR2014), CBLS, April 2014 (2014)
21. Siam, S.M., Zhang, H.: Fast-SeqSLAM: a fast appearance based place recognition algorithm. In: 2017 IEEE International Conference on Robotics and Automation (ICRA), pp. 5702–5708 (2017)
22. Sünderhauf, N., Shirazi, S., Dayoub, F., Upcroft, B., Milford, M.: On the performance of ConvNet features for place recognition. In: IEEE International Conference on Intelligent Robots and Systems, December 2015, pp. 4297–4304 (2015)
23. Sünderhauf, N., Shirazi, S., Jacobson, A., Dayoub, F., Pepperell, E., Upcroft, B., Milford, M.: Place recognition with ConvNet landmarks: viewpoint-robust, condition-robust, training-free. In: Robotics: Science and Systems XI, Robotics: Science and Systems Foundation (2015)
24. Talbot, B., Garg, S., Milford, M.: OpenSeqSLAM2.0: an open source toolbox for visual place recognition under changing conditions. IEEE Robot. Autom. Lett. **1**(1), 213–220 (2018)
25. Vysotska, O., Naseer, T., Spinello, L., Burgard, W., Stachniss, C.: Efficient and effective matching of image sequences under substantial appearance changes exploiting GPS priors. In: 2015 IEEE International Conference on Robotics and Automation (ICRA), pp. 2774–2779 (2015)
26. Vysotska, O., Stachniss, C.: Lazy data association for image sequences matching under substantial appearance changes. IEEE Robot. Autom. Lett. **1**(1), 213–220 (2016)
27. Yin, P., Srivatsan, R.A., Chen, Y., Li, X., Zhang, H., Xu, L., Li, L., Jia, Z., Ji, J., He, Y.: MRS-VPR: a multi-resolution sampling based global visual place recognition method (2019)
28. Zhou, B., Lapedriza, A., Khosla, A., Oliva, A., Torralba, A.: Places: a 10 million image database for scene recognition. IEEE Trans. Pattern Anal. Mach. Intell. **40**, 1452–1464 (2017)

Web Client for Visualization of ADAS/AD Annotated Data-Sets

Duarte Barbosa[1,2(✉)], Miguel Leitão[2], and João Silva[1]

[1] Altran Portugal, R. de Serpa Pinto 44, 4400-012 Vila Nova de Gaia, Portugal
duarteasbarbosa@gmail.com

[2] Instituto Superior de Engenharia do Porto, R Dr. Antonio Bernardino de Almeida 431, 4200-072 Porto, Portugal
https://www.altran.com
https://www.isep.ipp.pt

Abstract. This project aims to develop a web platform that is capable of showing data from an Advanced Driving Assist Systems (ADAS) and an Autonomous Driving (AD) system. This data can have multiple sources including cameras, LiDARs, GNSS, all of which must be visualized simultaneously and easily controlled by the platform's interface. Typically, companies would have to develop their unique visualization platform, or use standards such as Robot Operative System (ROS) to support the visualization of data logs. The problem with approaches such as ROS is that, although many development teams in the area are using it as base for their projects, the contribution of analysts outside the development team is hard to achieve since using ROS would require an initial setup that, not only can be time-consuming, but also could be difficult for these analysts teams to perform. The premise of this project is to change this kind of mindset, providing a generic visualization platform, that can load logged data from different sources in an easily configurable format, without the need for initial setup. The fact that this application is web-based allows for various analysts teams spread across the world to analyze data from these autonomous systems. Although the visualization is not ROS based, we used ROS as the framework for data processing and transformation, before deploying it in the server.

Keywords: Automotive · Data logs · Point cloud · Robot operating system · WebGL

1 Introduction

Many road accidents occur due to human error, the need to reduce such accidents led to the creation of an automated system to assist the driver called ADAS. An ADAS is a driving support system which provides information about the surrounding environment, aiming to increase not only car safety, but also road safety. These support systems provide a variety of features such as automate lighting, adaptive cruise control and collision avoidance, Pedestrian crash

M. F. Silva et al. (Eds.): ROBOT 2019, AISC 1092, pp. 191–202, 2020.
https://doi.org/10.1007/978-3-030-35990-4_16

avoidance mitigation (PCAM), incorporate satnav/traffic warnings, connection to smartphones, driver alert to other cars or dangers, lane departure warning system, automatic lane centering, or simply show what is in blind spots.

One of the most important key aspects of working between teams is communication. When developing these autonomous systems, it is hard to analyze the car given information without the proper tools and it could be even more challenging to trade ideas and information between clients and co-workers. Web technology is a great option that is capable of filling this gap. With the help of many existing frameworks, API's and libraries, it allows a fast analysis between teams spread across the world without the need for a complex setup.

The initiative for this project started at Altran Portugal aiming to develop a web platform that is capable of showing data from ADAS. Synchronized and processed visualization of data produced by an ADAS is important for debugging and developing algorithms for these systems and, since web technologies allow for a decentralized work, due to its global nature, it makes perfect sense to adapt to a web environment. The goal of this project is to provide a service that, using a server capable of storing annotated data sets, allows for any device capable of using a web browser, in any part of the world, to visualize these data sets in a 3D environment.

2 State of the Art

Autonomous driving is a topic that has increased in interest over the years. Companies are investing in technologies that can assist the driver, avoiding obstacles, alerting the presence of other cars or simply automatically turn on car lights. Although there seems to be a great investment in those features and tools referenced above, there also seems to be a less focus on generic platforms, for debugging purposes, that can aid developers to create these kind of autonomous systems. Usually, companies develop their own simulation software for debugging purposes. For instance, Tartan Racing has "Tartan Racing Operator Control Station (TROCS)", which is a graphical interface based on QT for developers to monitor telemetry from Boss (Tartan Racing's self-driving car) while it is driving and replay data offline for algorithm analysis [7]. Another example is Waymos with their carcraft, which is a simulator that not only allows to predict the behavior of the car in a simulation environment, but also allows for the analysis of the real data gathered from the car [3]. Uber visualization team also created a technology that can be used in this area called Deck.gl, which his a WebGL based frame-work that has a layer approach, meaning that it can render different types of data, such as point clouds, maps and polygons on top of each other [5]. They further created a framework called streetscape.gl, which is a visualization toolkit for autonomy and robotics data encoded in the XVIZ protocol [6] that can be used to debug data from autonomous driving vehicles. Nonetheless, all these technologies are frameworks for building visualization platforms and not a full build platform. Throughout this section, we can see that there are not many generic visualizations tools for ADAS and AD visualization on the market, especially web-based. This project aims to change this premise implementing a Deck.gl based, generic web platform for ADAS and AD visualization.

3 Data Sources

To develop a web platform that is capable of showing data from an ADAS, we need to get a source of data. There are a couple of open-source data-sets available online that contain data from various kinds of sensors, such as inertial and GPS navigation systems, LIDARs and cameras. In our project, we used the KITTI data-set [2] provided by Karlsruhe Institute of Technology which is composed of various sequences captured by driving around the mid-size city of Karlsruhe, in rural areas and on highways. Up to 15 cars and 30 pedestrians are visible per image. This was accomplished by equipping a standard station wagon with two high-resolution color and grayscale video cameras. Accurate ground truth is provided by a Velodyne laser scanner and a GPS localization system.

4 Solution Architecture

This project will be based on the architecture represented by Fig. 1, it is a web platform capable of showing data from an ADAS. The user interacts with the application using any device capable of using a web browser, the app then acts as a client and sends a request to the server to get the data needed for visualization. The server is responsible for answering the request by making queries to the data lake and retrieving the correct data, which will then be passed to the client.

This project is divided in three main components, the visualization client, server side software and data lake.

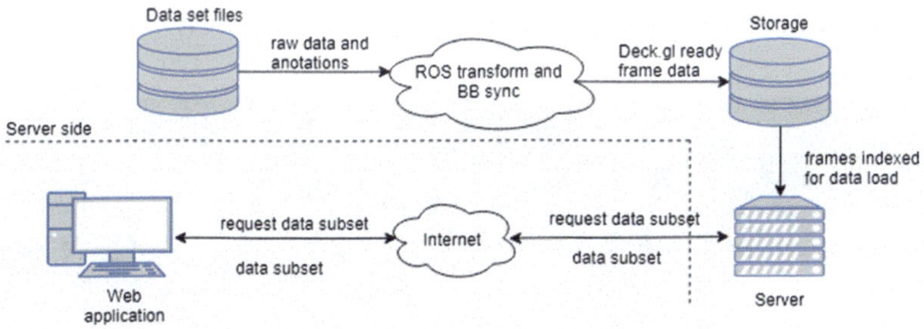

Fig. 1. Project architecture

5 Visualization Client

This component of the application is what represents everything related to rendering images and objects, plotting graphics and developing a user interface.

The goal was to develop an application capable of rendering data in different layers, so it could be manipulated separately. This was one of the reasons we chose deck.gl as our main development tool. The other main reason is that deck.gl is built on top of WebGL, but has a significantly lower learning curve than the later. This means that it uses the GPU to easily render polygons and images, while fulfilling the previous referred requirement of a layered approach, meaning that it can render different types of data, such as point clouds, maps and polygons on top of each other. It is worth mention that deck.gl uses JSON format to represent its data structures so that the different types of layers can later be rendered, which means our data had to be converted into this format beforehand.

To maximize customization, allowing users to configure the environment for highlighting the important aspects of their analysis, five different layers were created. The base layer that we use in deck.gl is a map over which all the other layers will be drawn. deck.gl uses the Mapbox GL JS JavaScript library to render this base map [1]. The remaining layers are the point cloud layer, which uses a specific deck.gl layer object whose sole purpose is to render point clouds, the ego layer for drawing the origin vehicle on the map, the bounding boxes layers for rendering all the other cars and obstacles on the road (these last two were both drawn using the polygon layer object) and the trajectory layer for drawing the ego vehicle trajectory line using the path layer object.

5.1 Graphical User Interface

The layout of the application is composed of a navigation bar on the left side that can be hidden and has tree different tabs and an animation control bar, which controls the playback of all data. The user is also able to create floating windows for plotting data or showing various camera feeds. All these features are visible on Fig. 2.

Fig. 2. Layout overview

By default, the sidebar is visible and the load tab is selected. With the load tab (Fig. 3a), the user can chose one of many pre-captured sequences, to load them in a browser's memory so that they can be animated. To manipulate visual information from the selected sequence, we created a separated tab called the sequence tab (Fig. 3b). Here the user can create new floating windows containing the camera feeds, or change layer properties such as visibility, color, and opacity using a tree-structured menu where the parent is a layer of the sequence and the children are visual options associated with that layer.

The third and last tab is the plot tab (Fig. 3c), as the name states this is where the user can create charts with data coming from the origin vehicle, it can be plotted a group of pre-selected variables such as velocities and accelerations. There are also custom options which allow the user to select the variables that he would like to plot, as well as their colors and labels. When the user is done choosing the elements to plot, he can press the "Create Plot" button to generate a floating re-sizeable window with the chosen information.

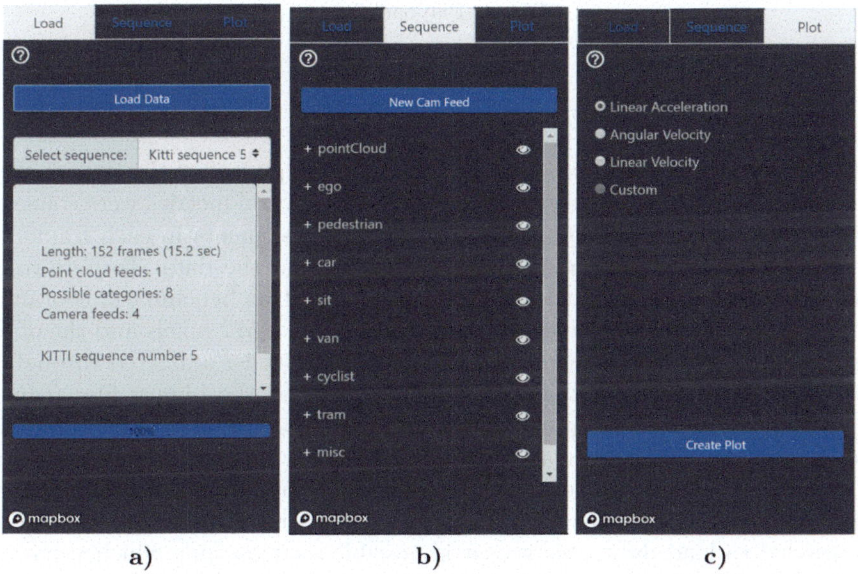

Fig. 3. Side bar tabs: **(a)** load tab; **(b)** sequence tab; **(c)** plot tab

To fully control the playback of all data, we created a playback bar very similar to the ones available in media players. This bar is visible in Fig. 4 and can start and stop the frame animation, run the animation step by step, jump back or forward using the timeline of the animation and choose from three different camera modes.

When it comes to the three-camera modes available, we developed them with different use cases in mind:

- The "free move" allows for a more detailed analysis, that's because in this mode the map doesn't move with the animation, this means that regardless the current frame in the animation, the map stays static until the user moves it;
- The "follow locked" is thought for a temporal overview of the sequence, with a common fixed "third person view" approach;
- the follow free move allows a temporal analysis from a specific point of view, useful for keeping focus on some specific perspective over the ego vehicle (origin vehicle);

New camera modes can be added to accommodate new specific needs along the project.

Fig. 4. Playback bar

5.2 Performance Analysis

In this kind of application, it is very important to keep track of the time that the different processes take to execute. For this reason several metrics were collected, such as the time required to animate all the visual elements in each frame, the time to decode the point cloud and the time to load the data. To perform all these tests, we used a sequence from KITTI data-set as benchmark, containing 153 frames, 4 camera feeds with a resolution of 1392 $\times$ 512 pixels and the object information of 9 Cars, 3 Vans, 2 Pedestrians and 1 Cyclist scattered throughout the sequence. All this data takes about 77.6 MB for the binary files (camera images and compressed point clouds) and 57.7 MB for the JSON files (bounding boxes and origin vehicle) for a total of 135.3 MB, which means that since this sequence has 153 frames, these values translate to roughly 880 KB per frame.

In order to enhance user experience it is important to keep track of the time needed to load data, this way it is possible to determine which values are acceptable and optimize the process. In Table 1 we can see fifteen iterations of data being loaded locally from the JSON files and the binary files (point clouds and images). The highest value in this measurements was about 8.1 s, while the smallest was about 3.7 s. For now all the data is being loaded in from the local machine, so it is worth mentioning that this values will very likely grow when transitioning for a remote server.

The time to animate each frame and the time to decode the point cloud are related, since the sum of these two can not exceed the time between frames otherwise the animation wouldn't run smoothly. The number of frames per second (fps) of the animation will dictate the time between frames. For instance, if the animation is running at 10 fps, then each frame will be animated every 100 ms.

Table 1. Time taken for multiple loading runs of the full data described previously

Load time (s)	8.03 7.42 4.10	7.53	7.58 5.92 3.68	3.81 5.38 7.97	8.13	7.77 7.30 5.77	3.82
Average (s)	6.28						

Having this in mind, we measured the times involved in our cycle. One of the measurements taken was the time to update the visual elements, which involves updating the values and rendering them. This process is performed in an average of 35.6 ms, as illustrated in Fig. 5. With this average value we have a decent margin to decode the point cloud.

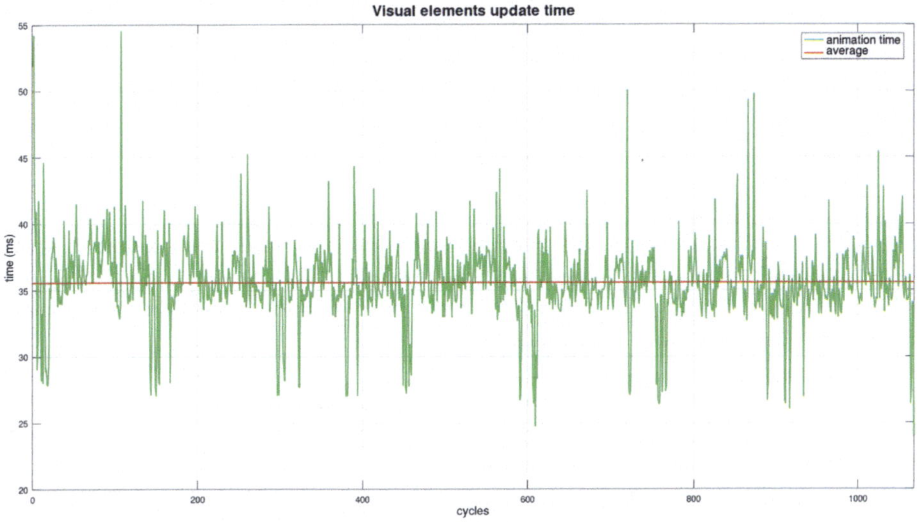

Fig. 5. Time to update visual elements

The process of decoding the point cloud was divided in two steps, a pre-decode that occurs when the data is being loaded and the final decode that occurs when each frame is being animated. The process of decoding will later be addressed in Sect. 7. Figure 6 shows the time that both these steps take to execute. For the animation cycle sake, the top plot is the most relevant, since it represents the time to perform the final decode of the point cloud during the animation. We obtained an average of 14.6 ms which, adding to the visual elements update time, yields an average cycle time of 50.2 ms.

6 Server and Data Lake

Since the beginning of the project, the idea was to create a remote server that could provide information from a centralized repository that would allow us to

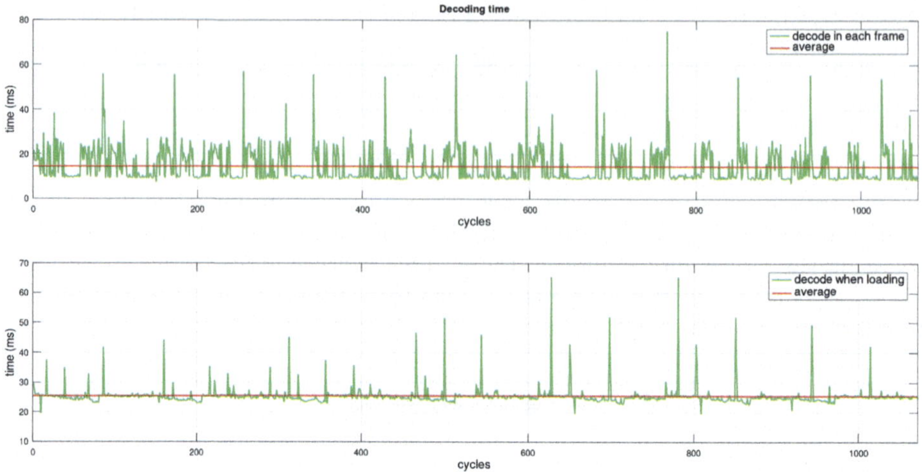

Fig. 6. Decoding times: **(top)** decode time in each frame; **(bottom)** decode time when loading data

store all our structured and unstructured data at any scale. We needed to store large amounts of data largely due to the size of the point clouds, which can have millions of points, it was considered that the best solution would be to configure a Hadoop cluster to store all this data.

For managing reading and writing of our large data-sets, we choose Apache Hive, which is a hadoop service that has a good amount o community support and provides an SQL dialect called Hive Query Language (abbreviated HiveQL or just HQL), making it more accessible.

For creating the script that can query the Hive table inside the cluster and send that data to the visualization client, we choose python. It is a programming language with immense online support and is widely used for creating servers. With all this in mind, the next step was to choose a python client library that could connect to the cluster. The first library that was tested is called *PyHive*, however this option was abandoned shortly after due to compatibility issues with some python versions. The next idea was to use *pyspark*, which is a python interpreter for spark. This method has various drawbacks: first, the python script has to be inside the cluster, since spark is a Hadoop service; second, since the script has to be inside the cluster it would have been necessary to create a communication channel, using sockets, for instance, to send the data over another client that would receive and parse this data outside the cluster; lastly, the time for reading from the table was not satisfactory since the first read request, from the client to the server, would often take from thirty to thirty six seconds.

With further research we found another python library called *pyHS2* and, despite having good reading times, it is no longer being maintained, which could lead to unsolvable issues in a later stage of the project.

Currently *impyla* is being used, a Python client for HiveServer2 implementations (e.g., Impala, Hive) for distributed query engines. This is the best option so far, it has python interface to query to the Hive tables (using HiveServer2), it can be used outside the cluster and has satisfactory reading times. Table 2 acts as a summary of what has been stated about these libraries.

Table 2. Hive reading times

Library	Average time (s)	Availability
PyHive	–	Was not compatible with targeted python versions of the project
pyspark	31.672	Has slow reading times when making the first request, needs to be inside the cluster
pyHS2	1.2186	Good reading times, but it was deprecated
impyla	1.1165	Good reading times, actively supported

As referenced in Table 2, the average reading time for pyspark is 31.672 s, for pyHS2 is 1.2186 and for impyla is 1.1165. Figure 7 presents some tests of the loading times for the three test libraries, where pyHive is not represented, since it never reached implementation phase. Furthermore, these values were plotted into two separated graphics because pyspark has much higher reading times than the other two python libraries. If these values were plotted in the same graphic, than we wouldn't be able to visually analyse the differences between the three, because the difference in scale would be extremely high. This reinforces the idea that pyspark is not a great option for this specific implementation.

7 Point Cloud Compression

One of the biggest challenges of storing data into memory was the size of the point cloud. The first time that the point cloud was loaded into the browser's memory, the application stopped working due to memory overflow. The first approach that we followed, and since we considered that the ground points are usually not very important for most of analyses, was to perform a plane extraction operation on the point cloud using PCL's sample consensus with a plane model [4]. This worked to some extent, but even so the memory was still near the limit. Also this was not an ideal solution, since it implied loosing data that might be useful to some specific analyses.

It was obvious that we needed to somehow compress the point cloud data. To be able to perform this compression we studied two compression tools, they are Draco and Corto. Table 3 illustrates the differences between Draco and Corto in various scenarios, such as encoding with different precisions (qp), with or without position and intensity (pos and int). The first row of the table represents the original point cloud for comparison. The second column illustrates the time needed for the encoding and the last three are related to the size of the point cloud. The third one illustrates the average size of the point cloud in bytes, whereas the forth represents it as a percentage. Lastly, the final column represents the compression Ratio.

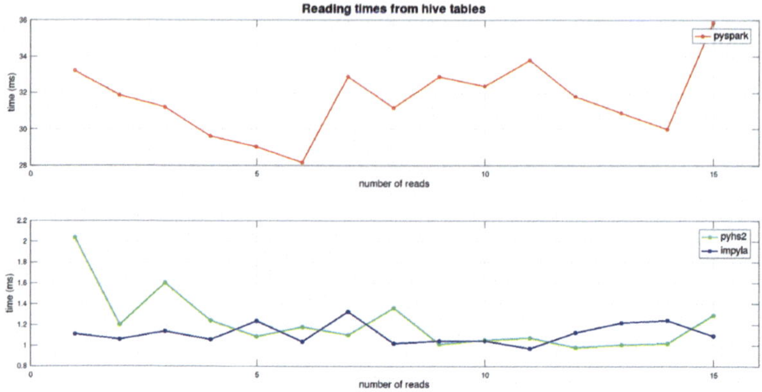

Fig. 7. Reading times from hive tables

Table 3. Encoding tests for Draco and Corto

	Time to encode (ms)	Size		
		Average (bytes)	Percentage	Compression ratio
draco pos+int (original)	NA	1,937,907	100%	1.00
corto pos (qp = auto)	36	433,172	22.35%	4.47
draco pos+int (qp = 14, qg = 8)	52	289,293	14.93%	6.70
corto pos (qp = 14)	36	240,280	12.40%	8.07
draco pos (qp = 14)	41	186,326	9.61%	10.40
draco pos+int (qp = 11, qg = 8)	52	158,932	8.20%	12.19
corto pos (qp = 11)	30	84,706	4.37%	22.8

Both Corto and Draco have very similar results, although Draco seems to have the upper head in the same circumstances, not only that but, at the time these tests were done, Draco seemed to be more actively supported.

After being compressed, the point cloud needs to be decoded with the JavaScript Draco decoder to use its data on the visualization client. This tool can decode meshes and point clouds previously encoded with the Draco encoder. There are two decoders available, one that uses WebAssembly and other that does not. The first step when decoding the point cloud is to detect whether or not the browser has WebAssembly support, because when it does it generally has a superior decoding performance. Then, the necessary files are loaded accordingly and the decoder module is created. When frames are being loaded to memory, they are also being pre-decoded. This process consists of using the decoder module to create a buffer array from the encoded data, get its geometry type (point cloud or mesh) and decode it accordingly. The output is then stored as a DracoFloat32Array in memory in a partially decoded state, that is finally decoded point by point when the Deck.gl instance calls for each frame.

Figure 8 illustrates a comparison, in terms of browser's allocated memory, between the point cloud without compression and all its points (Fig. 8a), the point cloud without compression and with no points on the ground (Fig. 8b) and the point cloud with compression and with all its points (Fig. 8c). As we can see, the point cloud with all its points takes about seventeen times less space when compressed with Draco encoder.

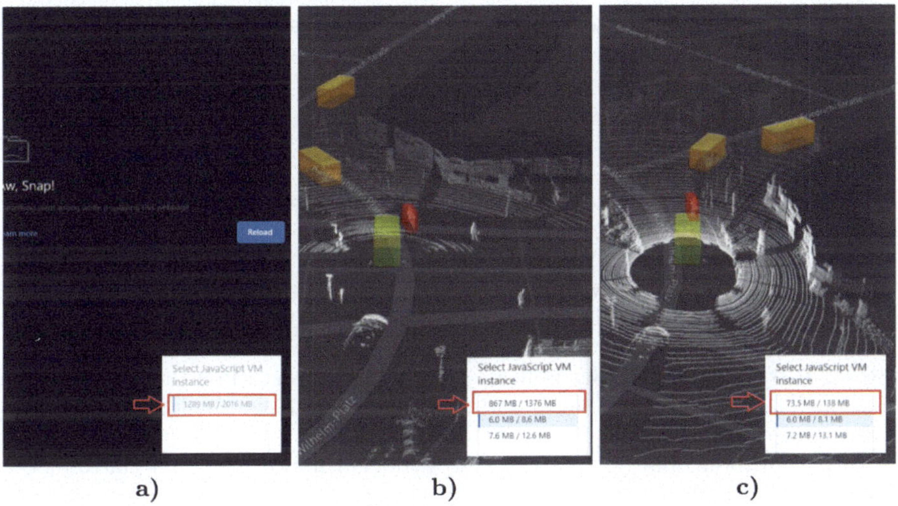

Fig. 8. Point cloud memory usage: **(a)** No compression with all points; **(b)** No compression without ground points; **(c)** Compression with all points

8 Conclusions and Future Work

Companies have to develop their unique visualization platform to support the development and visualization of data logs from autonomous systems. The premise of this project is to change this kind of mindset, providing a generic visualization platform, that can load logged data from different sources in an easily configurable format. The fact that this application is web-based allows for various teams spread across the world to analyze data from these autonomous systems, without the need for a complex setup.

The main goal of providing a visualization platform for representing annotated data sets from ADAS and AD is completed, by providing the client application. The problem of point cloud compression was successfully addressed and allowed for a decent browser memory management on the client side.

The used approaches were also allowed to successfully fulfill the processing times requirements. All the operations performed on the client side, point cloud decompression and element drawing, takes an average of 50.2 ms, which is well within the requirements of most of the default data set's 100 ms.

In terms of server software and data lake, we considered that Hive is the best choice for now, due to its resemblance to traditional databases and its SQL interpreter. impyla, the python module chosen for writing the client-side script, was also satisfactory, in our tests it gave us an average reading time of 1.12 s.

As future work, we intend to include visual object analysis by clicking, such as the distance between two objects. In terms of graphical plotting, there is going to be added the option to export the graphics that are already being drawn as well as being able to plot data from the surrounding objects (cars, cyclists, pedestrians, etc.) if available. Lastly, we will import all the binary data such as point clouds and images to the cluster.

References

1. Mapbox GL JS. https://docs.mapbox.com/mapbox-gl-js/api/. Accessed Sept 2019
2. Geiger, A., Lenz, P., Stiller, C., Urtasun, R.: Vision meets robotics: the kitti dataset. Int. J. Robot. Res. (IJRR) **32**, 1229–1235 (2013)
3. Madrigal, A.C.: Inside waymo's secret world for training self-driving cars (2017). https://www.theatlantic.com/technology/archive/2017/08/inside-waymos-secret-testing-and-simulation-facilities/537648/. Accessed Jun 2019
4. Rusu, R.B., Cousins, S.: 3D is here: point cloud library (PCL). In: IEEE International Conference on Robotics and Automation (ICRA), Shanghai, China, 9–13 May 2011
5. UBER: Introduction. https://deck.gl/#/documentation/overview/introduction. Accessed Jun 2019
6. UBER: streetscape.gl. https://avs.auto/#/streetscape.gl/overview/introduction. Accessed Jun 2019
7. Urmson, C., et al.: Autonomous driving in urban environments: boss and the urban challenge. J. Field Robot. Part I **25**(8), 425–466 (2008). Special Issue on the 2007 DARPA Urban Challenge

A General Approach to the Extrinsic Calibration of Intelligent Vehicles Using ROS

Miguel Oliveira[1,2], Afonso Castro[1(✉)], Tiago Madeira[1], Paulo Dias[1,2], and Vitor Santos[1,2]

[1] University of Aveiro, Aveiro, Portugal
{mriem,afonsocastro,tiagomadeira,paulo.dias,vitor}@ua.pt
[2] Institute of Electronics and Telematics Engineering of Aveiro, Aveiro, Portugal

Abstract. Intelligent vehicles are complex systems which often accommodate several sensors of different modalities. This paper proposes a general approach to the problem of extrinsic calibration of multiple sensors of varied modalities. Our approach is seamlessly integrated with the *Robot Operating System* (ROS) framework, and allows for the interactive positioning of sensors and labelling of data, facilitating the calibration process. The calibration problem is formulated as a simultaneous optimization for all sensors, in which the objective function accounts for the various sensor modalities. Results show that the proposed procedure produces accurate calibrations.

Keywords: Extrinsic calibration · ROS · Optimization · Bundle adjustment · Intelligent vehicles

1 Introduction

State of the art intelligent vehicles require a large number of on-board sensors, often of multiple modalities in order to operate consistently. The combination of the data collected by these sensors requires a transformation or projection of data from one sensor coordinate frame to another. The process of estimating these transformations between sensor coordinate systems is called extrinsic calibration.

An extrinsic calibration between two sensors requires an association of some of the data from one sensor to the data of another. By knowing these data associations, an optimization procedure can be formulated to estimate the parameters of the transformation between those sensors that minimizes the distance between associations. Since the accuracy of the associations is critical to the estimation procedure, most calibration approaches make use of calibration patterns, i.e., objects that are robustly and accurately detected.

Although there have been many works published in the literature on the topic of calibration, there still is no available straightforward software package for the calibration of intelligent vehicles, or robots with multiple sensors in general.

M. F. Silva et al. (Eds.): ROBOT 2019, AISC 1092, pp. 203–215, 2020.
https://doi.org/10.1007/978-3-030-35990-4_17

There are multiple factors that contribute to this, addressed in the following lines.

The large majority of works on calibration focus on sensor to sensor pairwise calibrations [2,7,16–18,24,26,27]. When considering pairwise combinations of sensors, there are several possibilities, according to the modality of each of the sensors in the pair. Most of them have been addressed in the literature: RGB to RGB camera calibration [7,16,17,24,27], RGB to depth camera (RGB-D cameras) calibration [3,4,12,13,20,28], camera to 2D *Light Detection And Ranging* (LiDAR) [5,11,21,26], 2D LiDAR to 3D LiDAR calibration [2], camera to 3D LiDAR [9,14], camera to radar [6], etc.

It seems as though most possible sensor combinations have been tackled.

Nonetheless, all these approaches have the obvious shortcoming of operating only with a single pair of sensors, which is not directly applicable to the case of intelligent vehicles, or more complex robotic systems in general. To be applicable in those cases, pairwise approaches must be arranged in a graph-like sequential procedure, in which one sensor calibrates with another, that then relates to a third sensor, and so forth. For instance, in [18], a methodology for calibrating the ATLASCAR2 autonomous vehicle [25] is proposed, wherein all sensors are paired with a reference sensor. In that case, the graph of transformations between sensors results in a one level pyramid, which contains the reference sensor on top and all other sensors at the base. Sequential pairwise approaches have three major shortcomings: (*i*) transformations are estimated using only data provided from the selected sensor tandem, despite the fact that data from additional sensors could be available and prove relevant to the overall accuracy of the calibration procedure; (*ii*) since it computes transformations after the estimations of the sensor pairs in a sequence, this approach is sensitive to cumulative error; (*iii*) sequential calibration procedures output a transformation graph derived from the calibration methodology itself, rather than from the preference of the programmer. Figure 1 shows a conceptual example in which these problems are visible.

There are a few works which address the problem of calibration from a multi-sensor, simultaneous optimization, perspective. In [15], a joint objective function is proposed to simultaneously calibrate three RGB cameras with respect to an RGB-D camera. Authors report a significant improvement in the accuracy of the calibration. In [23], an approach for joint estimation of both temporal offsets and spatial transformations between sensors is presented. This approach is one of few that is not designed for a particular set of sensors, since its methodology does not rely on unique properties of specific sensors. In [19], a joint calibration of the joint offsets and the sensors locations for a PR2 robot is proposed. This method takes sensor uncertainty into account and is modelled in a similar way to the bundle adjustment problem.

Our approach is similar to [19], in the sense that we also employ a bundle adjustment-like optimization procedure. However, our approach is not focused on a single robotic platform, rather it is a general approach that is applicable to any robotic system, which also relates it with [23].

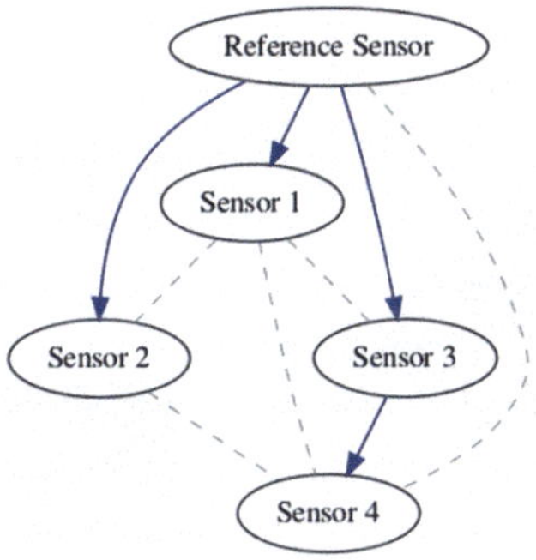

Fig. 1. A conceptual example a of calibration setup: the estimated transforms using a pairwise approach use the arrangements show in solid blue arrows, but other possible arrangements (dashed gray lines) and the data that supports them are not considered by the calibration.

Robot Operating System ROS [22] based architectures are the standard when developing robots. There are several ROS based calibration packages available (e.g., see[1],[2] or[3]), but none that provides a solution for the calibration of intelligent vehicles. Thus, the seamless integration with ROS become a core component of the proposed approach. To that end, the proposed calibration procedure is self-configured using the standard ROS robot description files, and provide several tools for sensor positioning and data labelling based on RVIZ interactive markers.

The remainder of the paper is organized as follows: Sect. 2 describes the proposed approach; Sect. 3 details the results of the proposed calibration procedure; Sect. 4 provides conclusions and future work.

2 Proposed Approach

A schematic of the proposed calibration procedure is displayed in Fig. 2. It consists of five components: configuration; interactive positioning of sensors; interactive labelling of data; data collection; and optimization procedure. Each will be described in detail in the following lines.

Configuration of the Calibration Procedure: Robotic platforms are described in ROS using a *xml* file called *xacro*. We propose to extend the *xacro* description file of a robot in order to provide information necessary for configuring how the calibration should be carried out. It was created a novel *xacro*

[1] http://wiki.ros.org/camera_calibration/Tutorials/StereoCalibration.
[2] http://wiki.ros.org/openni_launch/Tutorials/IntrinsicCalibration.
[3] https://github.com/code-iai/iai_kinect2.

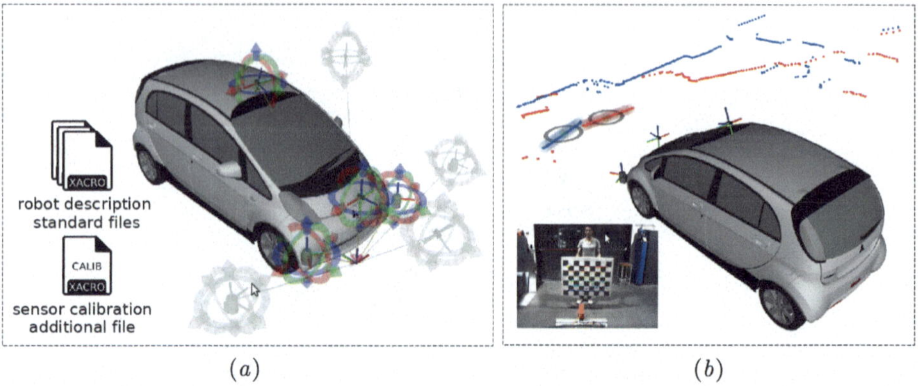

Fig. 2. The proposed calibration procedure: (*a*) initialization from *xacro* files and interactive first guess; (*b*) data labelling and collecting.

element called *calibration* specifically for this purpose, and extended the functionalities of this parser[4] accordingly. Each *calibration element* describes a sensor to be calibrated. The element contains information about the calibration parent and child links, which define the partial transformation that is optimized. An example of a calibration *xacro* can be found here[5].

Interactive Positioning of Sensors: Optimization procedures have the known problem of local minima. These problems tend to occur when the initial solution is far from the optimal parameter configuration. Thus, it is expected that by ensuring an accurate first guess for the sensor poses, there is less likelihood of running into local minima. We propose to solve this problem in an interactive fashion: the system parses the configuration files and creates an rviz interactive marker associated with each sensor. As we can see in[6], it is then possible to move and rotate the interactive marker and the corresponding sensor/data. This provides a simple interactive method to manually calibrate the system or to easily generate plausible first guesses. Real time visual feedback is provided by the observation of the bodies of the robot model (e.g. where a LiDAR is placed w.r.t. the vehicle), and also by the data measured by the sensors (e.g. how well the measurements from two LiDARs match together).

Interactive Data Labelling: Since the goal is to propose a calibration procedure which operates on multi-modal data, a calibration pattern adequate to all available sensor modalities must be selected. A chessboard pattern is one of the best options for this purpose, since it is a common calibration pattern in particular for RGB and RGB-D cameras. To label image data it is used of one of

[4] http://wiki.ros.org/urdfdom_py.
[5] https://codebeautify.org/xmlviewer/cb29a66c.
[6] https://youtu.be/zyQF7Goclro.

the many available image-based chessboards detectors. In the case of 2D LiDAR data it is not possible to robustly detect the chessboard, since there are often multiple planes in the scene derived from other structures such as walls, doors, etc. To solve this, we propose an interactive approach which requires minimal user intervention: rviz interactive markers are positioned along the LiDAR measurement planes and the user drags the marker to indicate where in the data the chessboard is observed. This is done by clustering the LiDAR data, and selecting the cluster which is closer to the marker. This interactive procedure is done only once, since it is then possible to track the chessboard robustly. We can see here[7] how this interactive data labelling procedure is done.

Collecting Data: The data streaming from the sensors is collected at different frequencies. However, to compute the associations between the data of multiple sensors, temporal syncronization is required. For now, this is solved trivially by collecting data (and the corresponding labels) at user defined moments, in which it is assumed that all the data is synchronized. We refer to these as data collections. This information is stored in a json file that afterwards will be read by the optimization procedure. The json file contains abstract information about the sensors, such as the sensor transformation chain among others, and specific information about each collection, i.e., sensor data, partial transformations and data labels. One example of a json file can be found here[8]. It is important to note that the set of collections should contain as much different poses as possible. So, collections should preferably have different distances and orientations w.r.t the chessboard so that the calibration becomes more reliable. This concern is transversal to the majority of calibration procedures.

Sensor Poses from Partial Transformations: The calibration of a complex, multi-sensor system requires the creation of a transformation graph. For this purpose, ROS uses a directed acyclic graph referred to as *tf tree* [8]. One critical factor for any calibration procedure is that it should not change the structure of that existing *tf* tree. The reason is that the *tf* tree, derived from the *xacro* files by the robot state publisher[9], also supports additional functionalities such as robot visualization or collision detection. If the *tf* tree changes due to the calibration, those functionalities are compromised or have to be redesigned. To accomplish this, we propose to compute the pose of any particular sensor (i.e., the transformation from the Reference Link, also known as world, to that Sensor) as a composite transformation where an aggregate transformation **A** is obtained after the chain of transformations for that particular sensor, extracted from the topology of the *tf* tree:

[7] https://youtu.be/9pGXShLIEHw.

[8] https://jsoneditoronline.org/?id=bb8ccea635a84b2f965e7c471cbbe8e0.

[9] http://wiki.ros.org/robot_state_publisher.

$$\mathbf{A} = \prod_{i=\text{world}}^{\text{sensor}} {}^{i}\mathbf{T}_{i+1} = \overbrace{\prod_{i=\text{world}}^{\text{parent}} {}^{i}\mathbf{T}_{i+1}}^{\text{prior links}} \cdot \overbrace{{}^{\text{parent}}\mathbf{T}_{\text{child}}}^{\text{calibrated}} \cdot \overbrace{\prod_{i=\text{child}+1}^{\text{sensor}} {}^{i}\mathbf{T}_{i+1}}^{\text{later links}} \tag{1}$$

where ${}^{i}\mathbf{T}_{i+1}$ represents the partial transformation from the i-th to the i-th $+1$ link, and *parent* and *child* are the indexes of the parent and child links in the sensor chain, respectively. Our approach preserves the predefined structure of the *tf* tree, since, during optimization, only one partial transformation contained in the chain is altered (in blue in Eq. 1). This computation is performed within the optimization's cost function and, therefore, a change in one partial transform affects the global sensor pose and thus the error to minimize. The optimization may target multiple links of each chain, and is agnostic to whether the remaining links are static or dynamic, since all existing partial transformations are stored for each data collection.

To the best of our knowledge, our approach is one of few which maintains the structure of the transformation graph before and after optimization. Note also that our approach can deal with complex cases. For example, Sensor 2, contains an aggregate transformation that includes the partial transformation optimized w.r.t. *Sensor 1*. Since these partial transformations may change over time, they are stored for each corresponding set of labelled sensor data.

Optimization Procedure: The goal of the optimization is to estimate the pose of each sensor. As such, the set of parameters to optimize, defined as $\mathbf{\Phi}$, must contain parameters that translate the pose of each sensor. As discussed in the beginning of this section, we propose to maintain the initial structure of the transformation graph, and thus only optimize one partial transformation per sensor. In the example of Fig. 3, these partial transformations are denoted by solid arrows. Since the usage of camera sensors is considered, it is also possible to introduce the intrinsic parameters of each camera in the set $\mathbf{\Phi}$. Our goal is to define an objective function that is able to characterize sensors of different modalities.

Pairwise methodology for devising the cost function results in complex graphs of exhaustive definition of relationships. For every existing pair of sensors, these relationships must be established according to the modality of each of the sensors, and, although most cases have been addressed in literature, as discussed in Sect. 1, a problem of scalability remains inherent to such solution.

To address this issue, we propose to structure the cost function in a sensor to calibration pattern paradigm, similar to what is done in bundle adjustment. That is, the positions of 3D points in the scene are jointly refined with the poses of the sensors. These points correspond to the corners of the calibration chessboard. What is optimized is actually the transformation that takes these corners from the frame of reference of the chessboard to the world, for every collection. The first guess for each chessboard is obtained by computing the pose of a chessboard detection in one of the cameras available. The output is a transformation from the chessboard reference frame to the camera's reference frame. Since we already have the first guess for the poses of each sensor, calculated as an aggregate transformation A (see Eq. 1), to obtain the transformation from the chessboard reference frame to the world, we only need to apply the following calculation:

$$^{\text{chess}}\mathbf{T}_{\text{world}} = \overbrace{^{\text{camera}}\mathbf{A}_{\text{world}}}^{\textit{Equation}\ 1} \cdot \overbrace{^{\text{chess}}\mathbf{T}_{\text{camera}}}^{\text{chess detection}}, \tag{2}$$

where chess and camera refer to chessboard and camera coordinate frames, respectively. Thus, the set of parameters optimized $\mathbf{\Phi}$, contains the transformation represented in Eq. 2, for each collection, along with the poses of each sensor:

$$\begin{aligned}\mathbf{\Phi} = \Big[& \overbrace{\mathbf{x}_{m=1}, \mathbf{r}_{m=1}, \mathbf{i}_{m=1}, \mathbf{d}_{m=1}, \ldots, \mathbf{x}_{m=M}, \mathbf{r}_{m=M}, \mathbf{i}_{m=M}, \mathbf{d}_{m=M}}^{\text{Cameras}}, \\ & \overbrace{\mathbf{x}_{n=1}, \mathbf{r}_{n=1}, \ldots, \mathbf{x}_{n=N}, \mathbf{r}_{n=N}}^{\text{LiDARs}}, \quad \overbrace{\ldots}^{\text{Other modalities}}, \\ & \overbrace{\mathbf{x}_{k=1}, \mathbf{r}_{k=1}, \ldots, \mathbf{x}_{k=K}, \mathbf{r}_{k=K}}^{\text{Calibration object}} \Big]\end{aligned} \tag{3}$$

where m refers to the m-th camera, of the set of M cameras, n refers to the n-th LiDAR, of the set of N LiDARs, k refers to the chessboard detection of the k-th collection, contained in the set of K collections, $\mathbf{x}$ is a translation vector $[t_x, t_y, t_z]$, $\mathbf{r}$ is a rotation represented through the axis/angle parameterization $[r_1, r_2, r_3]$, $\mathbf{i}$ is a vector of a camera's intrinsic parameters $[f_x, f_y, c_x, c_y]$, and $\mathbf{d}$ is a vector of camera's distortion coefficients $[d_0, d_1, d_2, d_3, d_4]$. The axis/angle parameterization was chosen because it has 3 components and 3 degrees of freedom, making it a fair parameterization, since it does not introduce more numerical sensitivity than the one inherent to the problem itself [10].

The cost function for this optimization can be thought of as a piecewise function, where, for every modality of sensor added to the calibration, a new sub-function is defined accordingly, that allows for the minimization of the error associated with the pose of sensors of that modality. Thus, the optimization procedure can be defined as:

$$\min_{\mathbf{\Phi}} \sum_{l=1}^{L} f_l(\mathbf{\Phi})$$
$$f_l = \begin{cases} f_l^{\text{camera}}, & \text{if } l \text{ is camera} \\ f_l^{\text{lidar}}, & \text{if } l \text{ is lidar} \\ \ldots, & \text{other modalities} \end{cases} \tag{4}$$

where l refers to the l-th sensor, of the set of L sensors, f_l is the cost function applied to the l-th sensor, f_l^{camera} is the sub-function for cameras, applied to the l-th sensor, f_l^{laser} is the sub-function for lasers, applied to the l-th sensor.

When the sensors are cameras, their calibration is performed as a Bundle Adjustment [1], and as such, the sub-function created is based on the average geometric error corresponding to the image distance between a projected point and a measured one. The 3D points corresponding to the corners of the calibration chessboard are captured by one or more cameras in each collection. Each camera being defined by its pose relative to a reference link and intrinsic parameters. After the desired acquisitions are completed, the 3D points are projected from the world into the images and the 2D coordinates are compared to the ones obtained by detection of the calibration pattern in the corresponding images. The positions of the 3D points in the world are obtained by applying the transformation described in Eq. 2 to the chessboard corner points defined in the chessboard detection's reference frame. The goal of this cost sub-function is to adjust the initial estimation of the camera parameters and the position of the points, in order to minimize the average reprojection error f_l^{camera}, given by:

$$f_l^{\text{camera}} = \sum_{j=1}^{J} \ell^2(\mathbf{x}_j, \hat{\mathbf{x}}_j) \tag{5}$$

where ℓ^2 is the Euclidean norm, j denotes the index of the chessboard corners, $\mathbf{x}_j$ denotes the pixels coordinates of the measured points (given by chessboard detection), and $\hat{\mathbf{x}}_j$ are the projected points, given by the relationship between a 3D point in the world and its projection on an image plane:

$$\hat{\mathbf{x}}_j = \mathbf{K}^{\text{camera}} \cdot {}^{\text{world}}\mathbf{T}_{\text{camera}} \cdot {}^{\text{chess}}\mathbf{T}_{\text{world}} \cdot \mathbf{p}_j \tag{6}$$

where $\mathbf{p}_j$ refers to the x, y, z coordinates of a chessboard corner, defined in the local chessboard coordinate frame, and $\mathbf{K}$ is the intrinsic matrix of the corresponding camera. Note that the parameters to be optimized define the chess to world transform, and that the world to camera transform is computed from an aggregate of several partial transformations, one of which is defined by other parameters being optimized, and also that the intrinsic matrix is also dependent on parameters which are accounted for in the optimization.

Finally, for LiDARs, the cost sub-function is based on distance between the M points measured by the range sensor that correspond to the calibration pattern object (extracted in the labelling stage), and the XoY plane of the chessboard coordinate frame, which is co-planar with the chessboard plane. Thus, the LiDAR cost sub-function $f_l{}^{\text{lidar}}$ is defined as:

$$f_l{}^{\text{lidar}} = \sum_{m=1}^{M} |z^{\text{chess}}|, \tag{7}$$

where z^{chess} are the range measurement points transformed to the chessboard's coordinate frame, as follows:

$$\begin{bmatrix} x \\ y \\ z \end{bmatrix}^{\text{chess}} = {}^{\text{world}}\mathbf{T}_{\text{chess}} \cdot {}^{\text{lidar}}\mathbf{T}_{\text{world}} \cdot \begin{bmatrix} x \\ y \\ z \end{bmatrix}^{\text{lidar}} . \tag{8}$$

This cost function, f_l, is minimized by the least-squares method of the optimize scipy library. Least Squares finds a local minimum of a scalar cost function, with bounds on the variables, by having an m-dimensional real residual function of n real variables[10]. As such, we choose this minimization tool as it is the best fit for our problem.

3 Results

The proposed approach was tested on the ATLASCAR2 intelligent vehicle [25]. In this work, were used four sensors: two RGB cameras, named *Top Right Camera* and *Frontal Camera*, and two 2D LiDARs, designated *Left Laser* and *Right Laser*. The *Top Right Cam Optical* and the *Frontal Cam Optical* frames are the ROS conceptual optical frames associated with each camera. These links have the same world translation as *Top Right Camera* and *Frontal Camera* links, but they have z-axis along with the camera lenses. The transforms of the car center to the optical frames were not calibrated, but the transformations between each Camera- Optical Camera were toked into account to compute the cost function. The *tf* tree of this systems in displayed in Fig. 3.

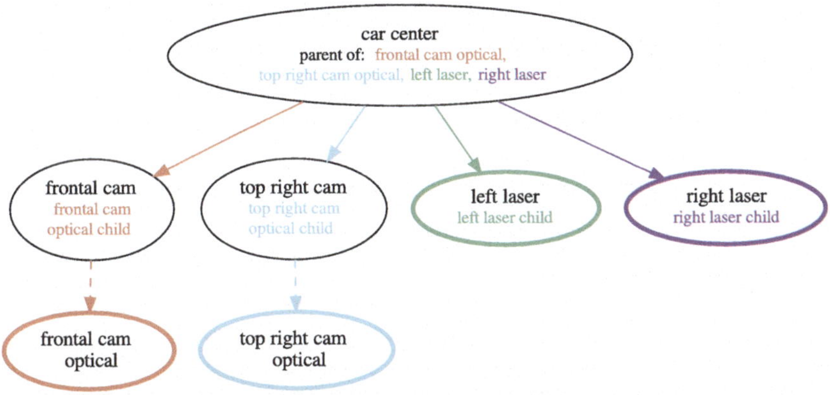

Fig. 3. Transformations graph for the ATLASCAR2 intelligent vehicle, along with information about the xacro driven configuration of the calibration: solid arrows denote partial transformations which will be calibrated; dashed arrow transformations are taken into account during optimization, but not altered.

In order to assess the influence of each sensor in the accuracy of the calibration procedure, we have calibrated different combinations between these four sensor which are referred to as setups. In addition, the results obtained by executing an optimization which starts from a first guess obtained using the interactive

[10] https://docs.scipy.org/doc/scipy/reference/generated/scipy.optimize.least_squares.html.

mode, denoted by the super-index fg, were discriminate. The following setups were tested: 2 cameras only, without and with first guess, respectively #1 and #1^{fg}, 2 LiDARs only, #2 and #2^{fg}, 1 LiDAR and one camera, #3 and #3^{fg}, and finally the experiment containing all four sensors, #4 and #4^{fg}. Note that creating all these experiments is straightforward due to the generality of the proposed approach. Table 1 shows the initial and final error for several setups.

Table 1. Average errors in pixels (px) or meters (m) per sensor before and after the optimization. Several experiments are presented for different combinations of sensors. Experiments annotated ^{fg} included an interactive first guess procedure to provide a first guess for the optimization.

Setup	Sensor	Error	
		Initial	Final
#1	Top Right Camera	0.51 px	72.24 px
	Frontal Camera	758.53 px	180.15 px
#1^{fg}	Top Right Camera	0.30 px	18.44 px
	Frontal Camera	292.10 px	32.96 px
#2	Left Laser	1.947 m	0.029 m
	Right Laser	1.691 m	0.032 m
#2^{fg}	Left Laser	0.257 m	0.008 m
	Right Laser	0.121 m	0.007 m
#3	Left Laser	1.947 m	1.233 m
	Top Right Camera	0.51 px	0.40 px
#3^{fg}	Left Laser	0.257 m	0.256 m
	Top Right Camera	0.30 px	0.29 px
#4	Left Laser	1.947 m	0.927 m
	Right Laser	1.691 m	0.771 m
	Top Right Camera	0.51 px	63.56 px
	Frontal Camera	758.53 px	176.81 px
#4^{fg}	Left Laser	0.256 m	0.091 m
	Right Laser	0.121 m	0.088 m
	Top Right Camera	0.30 px	20.97 px
	Frontal Camera	292.10 px	34.53 px

From the analysis of Table 1, we can see that in the vast majority of cases the errors associated with all sensor are reduced. The exception is the error associated with the *Top Right Camera*, since it increases in most experiments. However, in experiments #1, #1^{fg}, #4 and #4^{fg}, the error associated with the *Frontal Camera* decreases significantly more, which reduces the overall error. Experiment #2 obtained good results, which prove that the proposed objective function designed to handle LiDARs is working. Here we can see that the interactive first guess lead to the optimization reaching even better solutions. Experiments using the interactive first guess procedure has consistently enhanced the optimization or, in the worst case scenario, had no effect on the output of the optimization. Experiment #4 shows that the system is able to calibrate a set of sensors of different modalities, which was the initial objective.

4 Conclusions and Future Work

This paper proposes an extrinsic calibration methodology which is general in the sense that the number of sensors and their modalities are not restricted. The approach is compliant with the ROS framework, and has the advantage of not altering the *tf* tree. To accomplish this we formalize the problem as an optimization procedure of a set of partial transformations which account for specific links in the transformation chains of the sensors. Additionally, we also contribute with interactive tools for positioning the sensor and labelling data, which significantly ease the calibration proceedings.

Results show a decrease in errors in the order of some centimeters for metric measurements, and of hundreds of pixels for image based errors. This shows that the proposed approach is adequate for the calibration of complex robotic systems as are most intelligent vehicles.

Future work will focus on the extension to additional sensor modalities, e.g., 3D LiDARs, RGB-D cameras, *Radio Detection And Ranging* (RaDAR), etc. Given the scalability of the proposed framework, it is expected this to be more or less straightforward. Finally, the ultimate goal is to produce a multi-sensor, multi-modal calibration package that may be released to the community.

Acknowledgement. This Research Unit is funded by National Funds through the FCT - Foundation for Science and Technology, in the context of the project UID/CEC/00127/2019.

References

1. Agarwal, S., Snavely, N., M. Seitz, S., Szeliski, R.: Bundle adjustment in the large, pp. 29–42, November 2010
2. Almeida, M., Dias, P., Oliveira, M., Santos, V.: 3d-2d laser range finder calibration using a conic based geometry shape. In: Image Analysis and Recognition, pp. 312–319, June 2012
3. Basso, F., Menegatti, E., Pretto, A.: Robust intrinsic and extrinsic calibration of RGB-D cameras. IEEE Trans. Robot. **34**(5), 1315–1332 (2018)
4. Chen, G., Cui, G., Jin, Z., Wu, F., Chen, X.: Accurate intrinsic and extrinsic calibration of RGB-D cameras with GP-based depth correction. IEEE Sens. J. **19**(7), 2685–2694 (2019)
5. Chen, Z., Yang, X., Zhang, C., Jiang, S.: Extrinsic calibration of a laser range finder and a camera based on the automatic detection of line feature. In: 2016 9th International Congress on Image and Signal Processing, BioMedical Engineering and Informatics (CISP-BMEI), pp. 448–453, October 2016

6. Gao, D., Duan, J., Yang, X., Zheng, B.: A method of spatial calibration for camera and radar. In: 2010 8th World Congress on Intelligent Control and Automation, pp. 6211–6215, July 2010
7. Dinh, V.Q., Nguyen, T.P., Jeon, J.W.: Rectification using different types of cameras attached to a vehicle. IEEE Trans. Image Process. **28**(2), 815–826 (2019)
8. Foote, T.: TF: the transform library. In: 2013 IEEE Conference on Technologies for Practical Robot Applications (TePRA), pp. 1–6, April 2013
9. Guindel, C., Beltrán, J., Martín, D., García, F.: Automatic extrinsic calibration for lidar-stereo vehicle sensor setups. In: 2017 IEEE 20th International Conference on Intelligent Transportation Systems (ITSC), pp. 1–6 (2017)
10. Hornegger, J., Tomasi, C.: Representation issues in the ML estimation of camera motion, vol. 1, pp. 640–647, February 1999
11. Häselich, M., Bing, R., Paulus, D.: Calibration of multiple cameras to a 3D laser range finder. In: 2012 IEEE International Conference on Emerging Signal Processing Applications, pp. 25–28, January 2012
12. Khan, A., Aragon-Camarasa, G., Sun, L., Siebert, J.P.: On the calibration of active binocular and RGBD vision systems for dual-arm robots. In: 2016 IEEE International Conference on Robotics and Biomimetics (ROBIO), pp. 1960–1965, December 2016
13. Kwon, Y.C., Jang, J.W., Choi, O.: Automatic sphere detection for extrinsic calibration of multiple RGBD cameras. In: 2018 18th International Conference on Control, Automation and Systems (ICCAS), pp. 1451–1454, October 2018
14. Lee, G., Lee, J., Park, S.: Calibration of VLP-16 lidar and multi-view cameras using a ball for 360 degree 3D color map acquisition. In: 2017 IEEE International Conference on Multisensor Fusion and Integration for Intelligent Systems (MFI), pp. 64–69, November 2017
15. Liao, Y., Li, G., Ju, Z., Liu, H., Jiang, D.: Joint kinect and multiple external cameras simultaneous calibration. In: 2017 2nd International Conference on Advanced Robotics and Mechatronics (ICARM), pp. 305–310, August 2017
16. Ling, Y., Shen, S.: High-precision online markerless stereo extrinsic calibration. In: 2016 IEEE/RSJ International Conference on Intelligent Robots and Systems (IROS), pp. 1771–1778, October 2016
17. Mueller, G.R., Wuensche, H.: Continuous stereo camera calibration in urban scenarios. In: 2017 IEEE 20th International Conference on Intelligent Transportation Systems (ITSC), pp. 1–6, October 2017
18. Pereira, M., Silva, D., Santos, V., Dias, P.: Self calibration of multiple lidars and cameras on autonomous vehicles. Robot. Auton. Syst. **83**, 326–337 (2016)
19. Pradeep, V., Konolige, K., Berger, E.: Calibrating a Multi-arm Multi-sensor Robot: A Bundle Adjustment Approach, pp. 211–225. Springer, Heidelberg (2014)
20. Qiao, Y., Tang, B., Wang, Y., Peng, L.: A new approach to self-calibration of hand-eye vision systems. In: 2013 International Conference on Computational Problem-Solving (ICCP), pp. 253–256, October 2013
21. Zhang, Q., Pless, R.: Extrinsic calibration of a camera and laser range finder (improves camera calibration). In: 2004 IEEE/RSJ International Conference on Intelligent Robots and Systems (IROS) (IEEE Cat. No. 04CH37566), vol. 3, pp. 2301–2306, September 2004
22. Quigley, M., Conley, K., Gerkey, B.P., Faust, J., Foote, T., Leibs, J., Wheeler, R., Ng, A.Y.: Ros: an open-source robot operating system. In: ICRA Workshop on Open Source Software (2009)
23. Rehder, J., Siegwart, R., Furgale, P.: A general approach to spatiotemporal calibration in multisensor systems. IEEE Trans. Robot. **32**(2), 383–398 (2016)

24. Su, R., Zhong, J., Li, Q., Qi, S., Zhang, H., Wang, T.: An automatic calibration system for binocular stereo imaging. In: 2016 IEEE Advanced Information Management, Communicates, Electronic and Automation Control Conference (IMCEC), pp. 896–900, October 2016
25. Santos, V., Almeida, J., Ávila, E., Gameiro, D., Oliveira, M., Pascoal, R., Sabino, R., Stein, P.: Atlascar - technologies for a computer assisted driving system, on board a common automobile. In: 13th International IEEE Conference on Intelligent Transportation Systems, pp. 1421–1427, September 2010
26. Vasconcelos, F., Barreto, J.P., Nunes, U.: A minimal solution for the extrinsic calibration of a camera and a laser-rangefinder. IEEE Trans. Pattern Anal. Mach. Intell. **34**(11), 2097–2107 (2012)
27. Wu, L., Zhu, B.: Binocular stereovision camera calibration. In: 2015 IEEE International Conference on Mechatronics and Automation (ICMA), pp. 2638–2642, August 2015
28. Zhang, C., Zhang, Z.: Calibration between depth and color sensors for commodity depth cameras. In: 2011 IEEE International Conference on Multimedia and Expo, pp. 1–6, July 2011

Self-awareness in Intelligent Vehicles: Experience Based Abnormality Detection

Divya Kanapram[1,3(✉)], Pablo Marin-Plaza[2], Lucio Marcenaro[1], David Martin[2], Arturo de la Escalera[2], and Carlo Regazzoni[1]

[1] University of Genova, Genoa, Italy
divya.kanapram@ginevra.dibe.unige.it
{lucio.marcenaro,carlo.regazzoni}@unige.it
[2] Universidad Carlos III, Leganes, Spain
{pamarinp,dmgomez,escalera}@ing.uc3m.es
[3] Queen Mary University of London, London, UK

Abstract. The evolution of Intelligent Transportation System in recent times necessitates the development of self-driving agents: the self-awareness consciousness. This paper aims to introduce a novel method to detect abnormalities based on internal cross-correlation parameters of the vehicle. Before the implementation of Machine Learning, the detection of abnormalities were manually programmed by checking every variable and creating huge nested conditions that are very difficult to track. Nowadays, it is possible to train a Dynamic Bayesian Network (DBN) model to automatically evaluate and detect when the vehicle is potentially misbehaving. In this paper, different scenarios have been set in order to train and test a switching DBN for Perimeter Monitoring Task using a semantic segmentation for the DBN model and Hellinger Distance metric for abnormality measurements.

Keywords: Autonomous vehicles · Intelligent Transportation System (ITS) · Dynamic Bayesian Network (DBN) · Hellinger distance · Abnormality detection

1 Introduction

The self-awareness field is vast in terms of detecting abnormalities in the field of Intelligent Vehicles [22]. It is possible to classify critical, medium, or minor abnormalities by defining the line between normal and abnormal behaviour with the help of top design architectures. The problem of self-awareness systems is to measure every sensor, data acquired, and behavior of the system at every moment, comparing each measurement with the nominal range. Due to the huge amount of data, these tasks are not easy and becomes typically dead-end in big projects where the re-usability is not possible. Furthermore, it is possible to be unaware of situations where the vehicle is not working in the normal ranges for a very short period of time. Self-awareness management could be

M. F. Silva et al. (Eds.): ROBOT 2019, AISC 1092, pp. 216–228, 2020.
https://doi.org/10.1007/978-3-030-35990-4_18

divided into three main categories which are hardware, software, and behavior. The first category is based on the detection of malfunctions on electronic devices, actuators, sensors, CPUs, communication, etc. The second category focuses on software requirements where the most important measurements for message delivery are time, load, bottlenecks, delays, heartbeat, among others. Finally, the last self-awareness field analyzes the behavior of the vehicle which is related to the performance of the task assigned at each moment such as keep in lane, lane change, intersection management, roundabout management, overtaking, stop, etc. Accordingly, the management of self-awareness is a cross-layer problem where every manager should be built subsequently to the other layers to create a coherent self-awareness system [20].

To reduce the amount of process effort in intelligent self-awareness system, the emergent techniques in Machine Learning allow the creation of models using Dynamic Bayesian Networks (DBN) to automatize this process [14]. The novelty of this paper is the use of DBN models to generate a cross-correlation between a pair of internal features of the vehicle using a Hellinger distance metric for abnormality detection. Finally, compared the performance of different DBN models in order to select the best model for abnormality detection.

The remainder of this paper is organized as follows. Section 2 presents a survey of the related work. In Sect. 3, described the proposed method, defining principles exploited in the training phase and the steps involved in the test phase for detecting the abnormality. Section 4 summarizes the experimental setup in addition to the description of the research platform used. Section 5 gathered the results (i.e., abnormality measurements) from pair based DBNs for each vehicle and compared the results, and finally, Sect. 6 concludes the paper.

2 State of the Art

This section describes some of the related work regarding the development of self-awareness in *agents*. In [4], the authors propose an approach to develop a multilevel self-awareness model learned from multisensory data of an agent. Such a learned model allows the agent to detect abnormal situations present in the surrounding environment. In another work [19], the learning of the self-awareness model for autonomous vehicles based on the data collected from human driving. On the other hand, in [13], the authors propose a new architecture for mobile robots with a model for risk assessment and decision making when detecting difficulties in the autonomous robot design. In [21], the authors proposed a model of driving behavior awareness (DBA) that can infer driving behaviours such as lane change. In all the above works either used the data from one entity or the objective was limited; for example in [21] the objective was to detect lane change either on left or right side of the considered vehicles. In this work, we have considered the data from the real vehicles and developed pair based switching DBN models for each vehicle and finally made the performance comparison among different DBNs learned.

3 Proposed Method

This section discusses how to develop "intelligence" and "awareness" into vehicles to generate "Self-aware intelligent vehicles." The first step is to perform synchronization operation over the acquired multisensory data to make them synchronized in time in a way to match their time stamps. The data sets collected for training and testing are heterogeneous, and two vehicles are involved in the considered scenarios. The observed multisensory data from the vehicles are partitioned into different groups to learn a pair-based switching DBN model for each pair-based vehicle feature. Then compare the performances to qualify the best pair-based feature for automatic detection of abnormality. Switching DBNs are probabilistic models to integrate observations received from multiple sensors in order to understand the environment where they operate and take appropriate actions in different situations. The proposed method is divided into two phases: offline training and online testing. In the offline training phase, learn DBNs from the experiences of the vehicle in their normal behaviour. In the next phase, online testing, we have used a dynamic data series that are collected from the vehicles while they pass through different experiences than in the training phase. Accordingly, a filter called Markov Jump Particle Filter (MJPF) applied to the learned DBN models to estimate the future states of the vehicles and finally detects the abnormality situations present in the environment.

3.1 Offline Training Phase

In this phase, learn switching DBNs from the datasets collected from the experiences of the vehicle in their normal behaviour. The various steps involved in learning a DBN model are described below.

Generalized States. The intelligent vehicles used in this work are equipped with one lidar of 16 layers and 360° of Field of view (FOV), a stereo camera, and encoder devices to monitor different tasks being performed. In this work, it is assumed that each vehicle is aware of the other vehicle by its communication scheme and cooperation skills. By considering the vehicles endowed with a certain amount of sensors that monitors its activity, it is possible to define Z_k^c as any combination (indexed by c) of the available measurements in a time instant k. Let X_k^c be the states related to measurements Z_k^c, such that: $Z_k^c = X_k^c + \omega_k$; where ω_k represents the sensor noise. The generalized states of a sensory data combination c can be defined as:

$$\boldsymbol{X}_k^c = [X_k^c \ \dot{X}_k^c \ \ddot{X}_k^c \ \cdots \ X_k^{c,(L)}]^\intercal, \tag{1}$$

where (L) indicates the L-th time derivative of the state.

Vocabulary Generation and State Transition Matrix Calculation. In order to learn the desecrate level of the DBN (i.e., the orange outlined box in

Fig. 1), it is required to map the generalized states into a set of nodes. We have used a clustering algorithm called Growing Neural Gas (GNG) to group these generalized states and to obtain nodes. In GNG, the learning is continuous, and the addition or removal of nodes is automated [8]. These are the main reasons to choose GNG algorithm over other clustering algorithms such as K-means [9] or self-organizing map (SOM) [11]. The output of each GNG consists of a set of nodes that encode constant behaviours for specific sensory data combinations time derivative order. In other words, at a time instant, each GNG takes the data related to a single time derivative $X_k^{c,(i)} \in \boldsymbol{X}_k^c$ and cluster it with the same time derivative data acquired in previous time instances. The nodes associated with each GNG can be called as a set of *letters* containing the main behaviours of generalized states. The collection of nodes encoding i-th order derivatives in an observed data combination c is defined as follows:

$$A_i^c = \{\bar{X}_1^{(i),c}, \bar{X}_2^{(i),c}, \cdots \bar{X}_{P_i^c}^{(i),c}\} \tag{2}$$

where P_i^c is the number of nodes in the GNG associated to the i-th order derivative of data in data combination c. Each node in a GNG represents the centroids of associated data point clusters. By taken into consideration all the possible combinations of centroids obtained from GNGs, we can get a set of *words*, that define generalized states in an entirely semantic way. The obtained *words* can form a *dictionary* and can be defined as:

$$D^c = \{\beta, \dot{\beta}, \cdots \beta^{(L)}\} \tag{3}$$

where $\beta_c^{(i)} \in A_i^c$. D^c contains all possible combinations of discrete generalized states. D^c is a high-level hierarchy variable that explains the system states from a semantic viewpoint. In this work, we have only considered states, and it's first-order derivatives.

Furthermore, estimated the state transition matrices based on the timely evolution of such *letters* and *words*. The state transition matrix is a matrix that contains the transition probability between the discrete level nodes of the switching DBN shown in Fig. 1. When the first data emission occurs, the state transition matrix provides the probability of the next discrete level, i.e., the probability of activation of a node from the GNG associated with the first-order derivatives of the states. The vocabulary (i.e., *letters*, *words* and *dictionary* and the transition matrices constitute the discrete level of the DBN model.

DBN Model Learning. All the previous steps are the step by step learning process of the switching DBNs by each entity taken into consideration. The number of DBNs learned by each entity in the network can be written as:

$$\boldsymbol{DBN}m = \{DBN_1, \cdots, DBN_n\}. \tag{4}$$

where m represents the mth vehicle in the network and n is the total number DBN learned by the mth vehicle. The same DBN architecture is considered for making inferences with different sensory data combinations belong to different vehicles. The learned DBN can be represented as shown in the Fig. 1. The DBN has mainly three levels such as measurements, continuous and discrete levels.

3.2 Online Test Phase

In this phase, we have proposed to apply a dynamic switching model called Markov Jump Particle Filter (MJPF) [3] to make inferences on the learned DBN models. MJPF is a mixed approach with particles inside each Kalman filter. The MJPF is able to predict and estimate continuous and discrete future states and to detect deviations from the normal model. In MJPF, we use Kalman Filter (KF) in state-space and Particle Filter (PF) in super state space in Fig. 1.

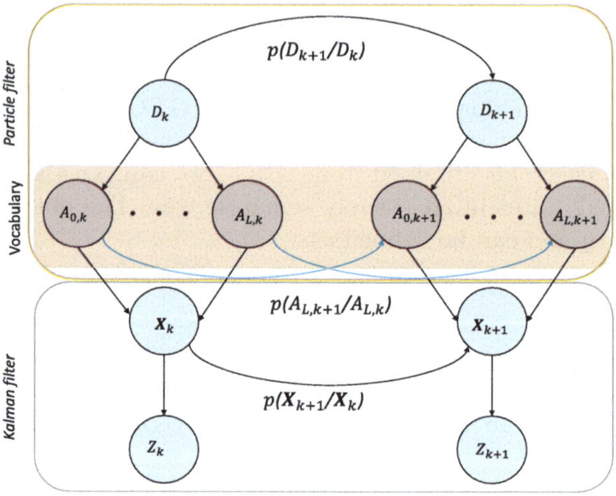

Fig. 1. Proposed DBN

Abnormality Detection and Complementarity Check. In probability theory, a statistical distance quantifies the distance between two statistical objects, which can be two random variables, or two probability distributions, etc. Some important statistical distances include: Bhattacharya distance [6], Hellinger Distance [17], Jensen–Shannon divergence [7], Kullback–Leibler (KL) divergence [10] etc. and they are the distances generally used between two distributions. Although, the HD is defined between vectors having only positive or zero elements [1]. The datasets in this work are normalized, so the values vary between zero and one; there aren't any negative values. Moreover, HD is symmetric compared to KL divergence. By these reasons, HD is more appropriate than using other distance metrics as abnormality measure. Moreover, the works in [15] and [2] used HD as an abnormality measurement.

Abnormality measurement can be defined as the distance between predicted state values and the observed evidence. Accordingly, let $p(\boldsymbol{X}_k^c|\boldsymbol{X}_{k-1}^c)$ be the predicted generalized states and $p(Z_k|\boldsymbol{X}_k^m)$ be the observation evidence. The HD can be written as:

$$\theta_k^c = \sqrt{1 - \lambda_k^c}, \tag{5}$$

where λ_k^c is defined as the Bhattacharyya coefficient [5], such that:

$$\lambda_k^c = \int \sqrt{p(\boldsymbol{X}_k^c|\boldsymbol{X}_{k-1}^c)p(Z_k^c|\boldsymbol{X}_k^c)}\, \mathrm{d}\boldsymbol{X}_k^c. \tag{6}$$

When a given experience in evaluation an abnormality measurement obtains at each time instant, and can be seen in the Eq. (5). The variable $\theta_k^c \in [0, 1]$, where values close to 0 indicate that measurements match with predictions; whereas values close to 1 reveal the presence of an abnormality. After calculating the abnormality measures by HD, it is possible to check the complementarity among different DBN models learned.

4 Experimental Setup

In order to validate the proposed method, it has been used two intelligent research platform called iCab (**I**ntelligent **C**ampus **A**utomo**B**ile) [16] (see Fig. 2a) with autonomous capabilities. To process and navigate through the environment, the vehicles count with two powerful computers along with the screen for debugging and Human-Machine Interaction (HMI) purposes. The software prototyping tool used is ROS [18].

(a) The autonomous vehicles (iCab) (b) The environment

Fig. 2. The agents and the environment used for the experiments.

The data sets collected with the two *iCab* vehicles are synchronized in order to observe the vehicles as different entities in a heterogeneous way to match their time stamps. The intercommunication scheme is proposed in [12] where both vehicles share all its information over the network by a Virtual Private Network (VPN). For this experiment, as long as the synchronization level reaches the nanoseconds, the recorded dataset in both vehicles has been merged and ordered using the timestamp generated by the clock on each vehicle which has been previously configured with a Network Time Protocol (NTP) tool called *Chrony*. Both vehicles perform a PMT task which consists of the autonomous movement of platooning around a square building (see Fig. 2b). The data generated from the lidar odometry such as the ego-motion of the vehicle and the different combinations of the control variables such as steering angle, rotor velocity and power of the rotor are considered the main metrics to learn and test the models. Moreover, it aims to detect the unseen dynamics of the vehicles with the proposed method.

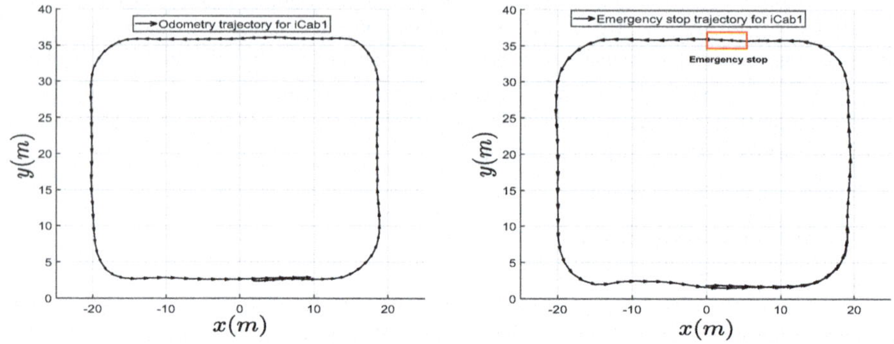

(a) Perimeter monitoring (b) Emergency stop

Fig. 3. Odometry data for iCab1

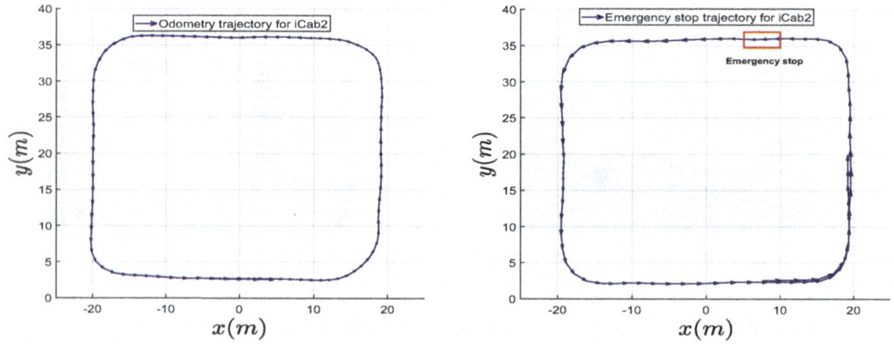

(a) Perimeter monitoring (b) Emergency stop

Fig. 4. Odometry data for iCab2

4.1 Perimeter Monitoring Task (PMT)

In order to generate the required data for learning and detect abnormalities, both vehicles perform a rectangular trajectory in a platooning mode. The leader just follows the rectangular path and the follower receives the path and keep the desired distance with the leader. This task has been divided into two different scenarios.

- **Scenario I:** Both vehicles perform the platooning operation by following a rectangular trajectory in a closed environment, as shown in Fig. 2b, four laps in total by recording the ego-motion, stereo camera images, lidar Point Cloud Data, encoders, and self-state. Notice that the GPS has troubles to acquire good signal because of the urban canyon. Figures 3a and 4b show the plots of odometry data for the perimeter monitoring task for iCab1 and iCab2 respectively. Moreover, Fig. 5 shows the steering angle w.r.t the iCab1's position (Fig. 5a) and rotor velocity w.r.t the iCab1's position (Fig. 5b). The rotor power data plotted w.r.t iCab1's position is shown in Fig. 6.
- **Scenario II:** Both vehicles perform the same experiment, but now a pedestrian crosses in front of the leader vehicle (i.e., iCab1). When the leader vehicle detects the pedestrian, automatically executes a stop and wait until the pedestrian fully crosses and move out from the danger zone. Meanwhile, the follower (i.e., iCab2) detects the stop of the leader and stops at a certain distance. When the leader continues the PMT, the follower mimics the action of the leader. Figures 3a and 4b show the plots of odometry data for the emergency stop criteria for iCab1 and iCab2 respectively.

5 Results

As explained in the previous section, there are two different scenarios taken into consideration with two vehicles. Moreover, the data combinations from odometry and control of vehicles have been treated independently and finally compared the abnormality measures to understand the correlation between them. We set the abnormality threshold to 0.4, considering the average Hellinger distance value of 0.2 when vehicles operate in normal conditions. The DBN models are trained over the scenario I based on PMT where no pedestrians are crossing in front of the vehicles. As said at the beginning of this paper, one of the objectives is the automatic extraction of abnormalities by learning from experiences. Hence, for PMT, the DBN models have been trained to extract the HD by pairing two different variables: Steering Angle-Power (SP), Velocity-Power (VP) and Steering Angle-Velocity (SV).

Testing Phase. The switching DBN model in this work is designed for the control part of the vehicles. However, we have considered odometry data and tested the performance of the learned DBN. Figure 7 shows the plots of abnormality measures by considering odometry data for the vehicle leader (iCab1) and the

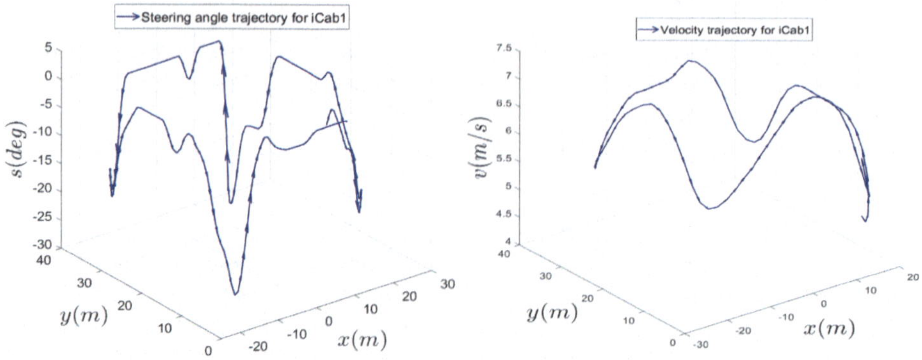

(a) Steering angle(s) w.r.t position (b) Velocity(v) w.r.t position

Fig. 5. Control data for iCab1 for perimeter monitoring task

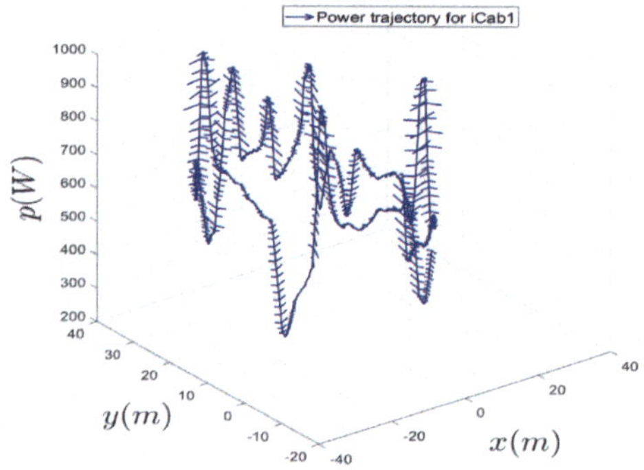

Fig. 6. Rotor power (p) w.r.t position for iCab1 for perimeter monitoring task

vehicle follower (iCab2), respectively. During the interval (cyan shaded area) while pedestrian passes and vehicle stops, there isn't any significant difference in HD value for iCab1 (leader) as well as iCab2 (follower). This behaviour is due to the fact that, during that interval the vehicles always inside the normal trajectory range. However, there are specific intervals when the vehicle deviates from the normal trajectory range, and the HD measures provided a high value of about 0.2 during those intervals. So the learned DBN model was able to predict if any trajectory deviation occurred.

- **Steering Angle-Velocity (SV):** It is necessary to check that the HD is working in both directions when the metrics involved reflect abnormal behaviour such as power and velocity and in cases when the metric used does not notice when there is an abnormal behavior such as steering angle. For this reason, the pair (S-V) shown in Fig. 8 displays that this pair is not detecting

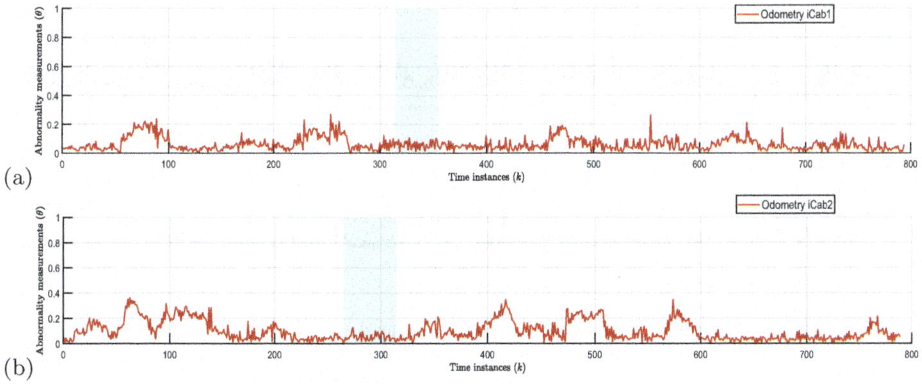

Fig. 7. Abnormality measurements for odometry: (a) *iCab*1, (b) *iCab*2

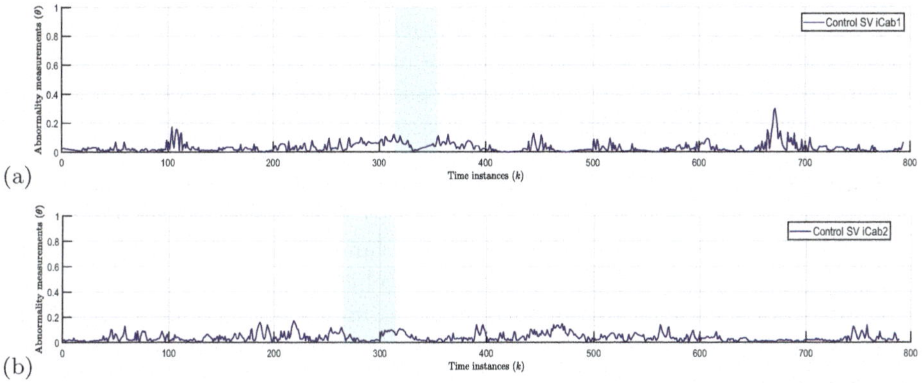

Fig. 8. Abnormality measurements for control (SV): (a) *iCab*1, (b) *iCab*2

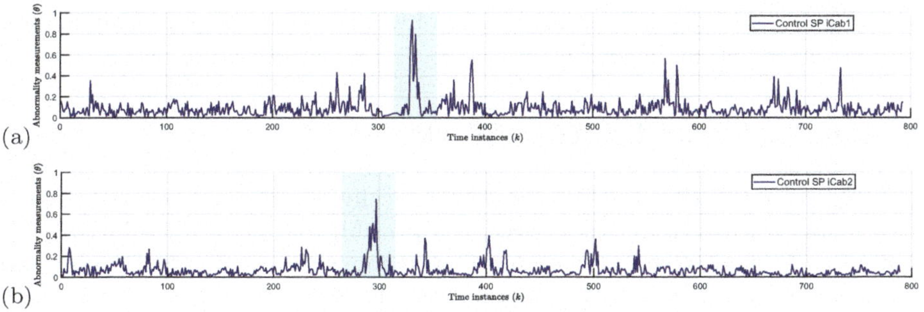

Fig. 9. Abnormality measurements for control (SP): (a) *iCab*1, (b) *iCab*2

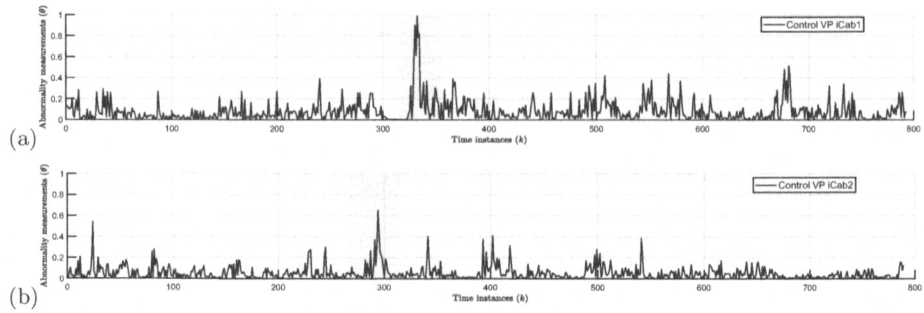

Fig. 10. Abnormality measurements for control (VP): (a) $iCab1$, (b) $iCab2$

abnormalities when a pedestrian crosses in front of the vehicle (cyan shaded), which is expected.

- **Steering Angle-Power (SP):** Figure 9 shows that the HD is high when a pedestrian is crossing in front of the leader vehicle (iCab1). This high value is considered as an abnormality in the behaviour of the leader, and as it is expected, the follower (iCab2) also gives an abnormal behaviour for the platooning task. However, the HD measures for the follower is not as significant as the leader, because the follower was not doing emergence break rather reducing its speed until reaching the minimum distance with the leader.
- **Velocity-Power (VP):** The last pair tested is velocity and the power consumption which are highly related. In Fig. 10, it is shown the moment when a pedestrian cross in front of the leader in cyan colour, which match with the highest value of the HD. For the follower vehicle, the abnormality measurement is very significant due to the new performance of the vehicle against the emergency brake of the leader. The consecutive peaks in the HD are caused by the high acceleration when the leader vehicle starts moving, but the current distance between them is still lower than the desired. Next peak in HD is caused by the acceleration of the leader and the emergency stop in the follower due to the pedestrian. To summarize, the switching DBN learned from SP and VP data were able to predict the unusual situations present; however, odometry and SV combination of control was not showing good combination to detect abnormal behavior.

6 Conclusion and Future Work

It has been proved that the HD for automatically detecting abnormalities in a DBN model learned from experiences, is a possible and plausible solution. The main idea of the proposed method has been demonstrated with the training and testing phase, and the results support that the methodology applied is more useful instead of checking and delimiting each metric of the vehicle depending on the event and defining each upper and lower limits in which an abnormal behaviour is considered.

The future work of this new approach could be extended by establishing communication between the objects involved in the tasks and develop collective awareness models. Such models can make the mutual prediction of the future states of the objects involved in the task and enrich the contextual awareness where they operate. Another direction could be the development of an optimized model from the different feature combinations that can be used for the future state predictions of the considered entities. Additionally, the classification of detected abnormality by considering different test scenarios and comparing the performance of abnormality detection by using different metric as abnormality measure could be considered.

Acknowledgement. Supported by SEGVAUTO 4.0 P2018/EMT-4362) and CICYT projects (TRA2015-63708-R and TRA2016-78886-C3-1-R).

References

1. Abdi, H., Salkind, N.J.: Encyclopedia of Measurement and Statistics. Sage, Thousand Oaks (2007). Agresti, A.: Categorical Data Analysis. Wiley, New York (1990), Agresti, A.: A survey of exact inference for contingency tables. Statist. Sci. **7**, 131–153 (1992)
2. Baydoun, M., Campo, D., Kanapram, D., Marcenaro, L., Regazzoni, C.S.: Prediction of multi-target dynamics using discrete descriptors: an interactive approach. In: ICASSP 2019-2019 IEEE International Conference on Acoustics, Speech and Signal Processing (ICASSP), pp. 3342–3346. IEEE (2019)
3. Baydoun, M., Campo, D., Sanguineti, V., Marcenaro, L., Cavallaro, A., Regazzoni, C.: Learning switching models for abnormality detection for autonomous driving. In: 2018 21st International Conference on Information Fusion (FUSION), pp. 2606–2613. IEEE (2018)
4. Baydoun, M., Ravanbakhsh, M., Campo, D., Marin, P., Martin, D., Marcenaro, L., Cavallaro, A., Regazzoni, C.S.: A multi-perspective approach to anomaly detection for self-aware embodied agents. arXiv preprint arXiv:1803.06579 (2018)
5. Bhattacharyya, A.: On a measure of divergence between two statistical populations defined by their probability distributions. Bull. Calcutta Math. Soc. **35**, 99–109 (1943)
6. Bhattacharyya, A.: On a measure of divergence between two statistical populations defined by their probability distributions. Bull. Calcutta Math. Soc. **35**, 99–109 (1943)
7. Endres, D.M., Schindelin, J.E.: A new metric for probability distributions. IEEE Trans. Inf. Theory **49**, 1858–1860 (2003)
8. Fritzke, B.: A growing neural gas network learns topologies. In: Advances in Neural Information Processing Systems, pp. 625–632 (1995)
9. Hartigan, J.A., Wong, M.A.: Algorithm as 136: a K-means clustering algorithm. J. Roy. Stat. Soc. Ser. C (Appl. Stat.) **28**(1), 100–108 (1979)
10. Hershey, J.R., Olsen, P.A.: Approximating the kullback leibler divergence between Gaussian mixture models. In: 2007 IEEE International Conference on Acoustics, Speech and Signal Processing-ICASSP 2007, vol. 4, pp. IV–317. IEEE (2007)
11. Kohonen, T.: The self-organizing map. Proc. IEEE **78**(9), 1464–1480 (1990)

12. Kokuti, A., Hussein, A., Marín-Plaza, P., de la Escalera, A., García, F.: V2x communications architecture for off-road autonomous vehicles. In: 2017 IEEE International Conference on Vehicular Electronics and Safety (ICVES), pp. 69–74. IEEE (2017)
13. Leite, A., Pinto, A., Matos, A.: A safety monitoring model for a faulty mobile robot. Robotics **7**(3), 32 (2018)
14. Lewis, P.R., Platzner, M., Rinner, B., Tørresen, J., Yao, X.: Self-Aware Computing Systems. Springer (2016)
15. Lourenzutti, R., Krohling, R.A.: The hellinger distance in multicriteria decision making: an illustration to the topsis and todim methods. Expert Syst. Appl. **41**(9), 4414–4421 (2014). https://doi.org/10.1016/j.eswa.2014.01.015
16. Marin-Plaza, P., Hussein, A., Martin, D., de la Escalera, A.: Global and local path planning study in a ROS-based research platform for autonomous vehicles. J. Adv. Transp. **2018**, 1–10 (2018)
17. Pardo, L.: Statistical Inference Based on Divergence Measures. Chapman and Hall/CRC, New York (2005)
18. Quigley, M., Conley, K., Gerkey, B., Faust, J., Foote, T., Leibs, J., Wheeler, R., Ng, A.Y.: Ros: an open-source robot operating system. In: ICRA Workshop on Open Source Software, vol. 3, p. 5. Kobe, Japan (2009)
19. Ravanbakhsh, M., Baydoun, M., Campo, D., Marin, P., Martin, D., Marcenaro, L., Regazzoni, C.S.: Learning multi-modal self-awareness models for autonomous vehicles from human driving. arXiv preprint arXiv:1806.02609 (2018)
20. Schlatow, J., Möstl, M., Ernst, R., Nolte, M., Jatzkowski, I., Maurer, M., Herber, C., Herkersdorf, A.: Self-awareness in autonomous automotive systems. In: Proceedings of the Conference on Design, Automation & Test in Europe, pp. 1050–1055. European Design and Automation Association (2017)
21. Xie, G., Gao, H., Huang, B., Qian, L., Wang, J.: A driving behavior awareness model based on a dynamic Bayesian network and distributed genetic algorithm. Int. J. Comput. Intell. Syst. **11**(1), 469–482 (2018)
22. Xiong, G., Zhou, P., Zhou, S., Zhao, X., Zhang, H., Gong, J., Chen, H.: Autonomous driving of intelligent vehicle bit in 2009 future challenge of china. In: Intelligent Vehicles Symposium (IV), pp. 1049–1053. IEEE (2010)

Joint Instance Segmentation of Obstacles and Lanes Using Convolutional Neural Networks

Leonardo Cabrera Lo Bianco[1], Abdulla Al-Kaff[2(✉)], Jorge Beltrán[2], Fernando García Fernández[2], and Gerardo Fernández López[1]

[1] Mechatronics Research Group, Simón Bolívar University, Caracas 13-10194, Venezuela
gfernandez@usb.ve
[2] Intelligent Systems Lab, Universidad Carlos III de Madrid, Avenida de la Universidad, 30, 28911 Leganés, Madrid, Spain
{akaff,jbeltran,fgarcia}ing.uc3m.es
http://www.lsi-uc3m.es

Abstract. Autonomous vehicles aim at higher levels of intelligence to recognize all the elements in the surrounding environment; in order to be able to make decisions efficiently and in real time. For this reason, a convolutional neural networks capable of perform semantic segmentation of these elements have been implemented. In this work it is proposed to use the ERFNet architecture to segment the main obstacles and lanes in a road environment. One of the requirements for training this type of networks is to have a complete and large dataset with these two types of labels. In order to avoid manual labeling, an automatic way of carrying out this process is proposed, using convolutional neural networks and different dataset already labeled. The generated dataset contains 19000 images tagged with obstacles and lanes, to be used to train a network of ERFnet architecture. From the experiment, the obtained results show the performance of the proposed approach providing accuracy of 74.42%.

Keywords: Semantic segmentation · Deep learning · Lane segmentation · Autonomous vehicle

1 Introduction

Autonomous vehicles have been encountered a great development in many areas of research. Many systems have been implemented in these vehicles, such as speed control or emergency braking mechanism. However, one of the main challenges to reach a higher level of automation is the ability to identify all the elements in the surrounding environment, similar to or better than driver. This makes the vehicle to be able to understand the environment, and to make the right decisions in the shortest time possible, being this one of the most important things to ensure the safety of people both inside and outside the vehicle. Aiming to solve the problems

M. F. Silva et al. (Eds.): ROBOT 2019, AISC 1092, pp. 229–241, 2020.
https://doi.org/10.1007/978-3-030-35990-4_19

of detection and classification of all those elements, numerous advances have been made in the field of perception. Starting by using image processing techniques to detect contours, filter by color, among other techniques. Over the years, more tools were added until the use of neural networks trained to perform the tasks of detection and classification, for example, the detection and classification of traffic signs [17], which allowed what actions the vehicle should take when it encounters speed limit or stop signs.

One of the actions that must be performed in the vehicles on the streets, is to stay in or changing lanes, without performing any maneuvers that cause danger to other elements in the road. For this reason, numerous advances have been made, from controllers to keep the vehicle within the lane to neural networks capable of identifying the lines of the roads [2]. This last one is an important issue to deal with, because, thanks to the advances in the techniques based on deep learning, it is possible to implement specific convolutional networks that are able to obtain characteristics of the image, and thus in an autonomous manner to recognize patterns that later can be classified in different elements.

However, when using this type of networks, several disadvantages appear. The first problem is to reduce the computational cost as much as possible, but without losing accuracy. This is important to place this type of architecture in autonomous vehicles. Another problem is the need for a sufficiently large dataset, which has labeled all classes of interest and whose images are taken in different climatic and lighting conditions. The latter is the main drawback when training a neural network for segmentation that identifies obstacles, such as vehicles and pedestrians, and also identify the lanes. In this paper, several datasets with one or other of the classes of interest is shown; however, from our knowledge, there is no dataset that contains all these classes in the same group of images, this means that in many cases new dataset has to be generated and must be labeled manually, which requires more work and time to obtain a dataset of large numbers of images.

From this point, in this paper, a semantic segmentation approach of obstacles and lanes using convolutional neural networks is presented. In addition, this work proposed an algorithm to automatically generate a new dataset containing both groups of labels (obstacles and lanes).

The remainder of this paper is organized as follows: in Sect. 2, a review of related works are introduced, followed by the proposed algorithms in Sect. 3. Then Sect. 4 discusses the experimental results. Finally, in Sect. 5 conclusions and future works are summarized.

2 Related Work

One of the first models of Convolutional Neural Networks (CNN) that focused on performing semantic segmentation is the Fully Convolutional Networks (FCN) [9], where the Fully Connected (FC) layers of traditional network models are replaced by equivalent convolutional capabilities, which allowed to obtain a map of characteristics instead of probabilities. These completely convolutional networks allow creating pixel-level predictions by classifying each pixel into a particular class.

Architectures such as U-net [15] proposed different approach where it divided the network into two stages: the encoder, where the extraction of the characteristics is carried out. The decoder, which consists of symmetrical expansion layers that allow the location of these characteristics. This type of encoder-decoder architecture is one of the most used, and throughout the investigations have been designed new capable of downsampling, upsampling, batch normalization and dropout, which has been improved the performance of these architectures.

Other techniques are used to improve the results of the training of this type of networks, works such as those carried out in [18], which consists of training a ResNEt-101 network, using data augmentation techniques, such as illumination change, color, texture, orientation, and scaling. This type of tools allows to expand the datasets and contributes to have a more robust network.

In many cases, there is the necessity of combining elements of another type of networks, to obtain better results at the moment of segmenting. This happens in architectures like Mask R-CNN [6], where the detection is mixed using bounding boxes, and generate parallel segmenting of the elements within that area.

In the field of autonomous vehicles, there are more complex networks that have combined various tasks to obtain more information on the environment; in which the vehicles are developed. In [11], the authors proposed an architecture that generates the regions corresponding to each of the classes and assigns a global label to the image which allows classifying the environment in different categories, this architecture is used to determine the type of road where the vehicle is situated. In addition, works such as the one implemented in [8], proposed more complex network of the type *end-to-end*, capable of multitasking such as lane and road marking detection and recognition, as well as being able to predict the vanishing point of the image. There are also jobs focused on identifying lanes by detecting the lines that delimit it, for example, in [20] a robust architecture is proposed that allows the identification of the boundaries of each lane, making use of a hybrid architecture that combines a convolutional neuronal network with a recurrent neuronal network.

Recently, the use of semantic segmentation for autonomous vehicles has increased, to recognize and classify elements in the road, this has resulted in the creation of various datasets available for training. Several datasets focused on the road segmentation, many of them oriented on labeling the main elements that are found in the road, such as Cityscape [4] which has 5000 images with more than 30 classes, considering many elements present in urban areas, KITTI Segmentation Benchmark [1] contains 400 images using the same classes as Cityscape, KITTI Road Benchmark [5] with approximately 500 images, where the road area and ego-line are labeled, CULane Dataset [12] formed by 133235 images labeled traffic lanes are, and BDD100K [19] is a very extensive dataset that has 10000 images for semantic segmentation, using the main classes of Cityscape and also another 100000 images labeling the lanes and lane markings, these images are taken with different weather conditions and at different times.

3 Proposed Approach

There is a wide variety of datasets to use; however, to perform training that identifies the lanes and the road elements, there is no dataset that includes labeled images with the two types of classes (obstacles and lanes). Therefore, having a dataset that contains these classes will allow to train networks that are used for applications, such as determining when the lane change maneuver is possible. In order to achieve this, it is necessary to identify the principal obstacles located in the road, such as pedestrians and vehicles, as well as to determine the lane where the vehicle is located and those lanes parallel to it; taking into account this application it is necessary to generate a new dataset. However, labeling the classes manually takes a lot of time, especially for datasets of a large number of images. For this reason, it is required to implement a way to generate this dataset faster. There are different investigations that try to solve this problem, one of them consists of an architecture designed with several stages [10], which allows simultaneous learning using different datasets. Each stage adds the corresponding segmentation, obtaining as a result a segmented image for each stage and merged into a single one, but these architectures need more computational resources in order to be executed in real time.

The proposed solution is to take images from the BDD100K dataset, to generate a new dataset that merges segmentation classes of obstacles and lanes. This new dataset is called LSIOLD[1].

The BDD100K dataset, has two completely different groups of images, the first corresponding to 10000 images labeled for the segmentation of elements present in the road, and the second is 100000 images different from the previous ones, labeled for the segmentation of lanes.

In order to solve the problem that the images present in these two groups are different, the use of a convolutional neural network, which functions as a tool, is proposed. This method is divided into two stages, the first consists of training the network to segment the obstacles, using the 10000 images from the BDD100K dataset, and the second is to use the trained model to generate weak labels on the other group of images that have the labels of the lanes.

3.1 Inference Network

The architecture chosen for this training was ERFNet [14] based on the Enet neural network [13], which implements a new residual block called *Non-Bottleneck-1D* which allows to get information from previous layers and decreasing the computational cost without losing the accuracy and learning qualities of the network, making real-time segmentation possible. In addition, incorporating blocks of *Downsampler*, *batch normalization*, *dropouts* at the end of each block *Non-Bottleneck-1D* and Deconvolution (upsampling), which acts on maps of characteristics, instead of individual elements. Table 1 shows the order of each of the

[1] This dataset is published in Github: https://github.com/lsi-uc3m/lsiold.

network layers. Finally, a system of weights from the Enet network has been used, considering the distribution of each class in the dataset according to Eq. 1.

$$Wc = \frac{1}{ln(c + p)} \quad (1)$$

where c is an adjustable parameter and p corresponds to the probability of a class appearing within the dataset, that is, the relationship between the number of pixels in a class and the totals.

Table 1. Layer disposal of ERFNet architecture

Layer	Type
1–2	2 × Downsampler block
3–7	5 × Non-Bottleneck-1D
8	Downsampler block
9	Non-Bottleneck-1D (dilated 2)
10	Non-Bottleneck-1D (dilated 4)
11	Non-Bottleneck-1D (dilated 8)
12	Non-Bottleneck-1D (dilated 16)
13	Non-Bottleneck-1D (dilated 2)
14	Non-Bottleneck-1D (dilated 4)
15	Non-Bottleneck-1D (dilated 8)
16	Non-Bottleneck-1D (dilated 16)
17	Deconvolution (upsampling)
18–19	2 × Non-Bottleneck-1D
20	Deconvolution (upsampling)
21–22	2 × Non-Bottleneck-1D
23	Deconvolution (upsampling)

This architecture uses a cost function called *cross-entropy loss*; however, when it is applied to several classes it is generalized as *softmax regression* as it is shown in Eq. 2.

$$J = \frac{1}{m}[\sum_{i=1}^{m}\sum_{j=1}^{k} H_{ln,j} 1\{y^i = j\} logP(y^{i} = j|x^{i}); \theta)] \quad (2)$$

where c is an adjustable parameter and p corresponds to the probability of a class appearing within the dataset, that is, the relationship between the number of pixels in a class and the totals.

3.2 Obstacle Segmentation

All stages of the network are implemented using the Torch7 framework [3] with CuDNN backends and CUDA. The model is trained using Adam's optimization [7] of stochastic gradient descent, and the training parameters used are as follows: weight decay of $1e^{-4}$, momentum of 0.9, a batch size of 4, a learning rate of $5e^{-4}$ which is divided in half each time, training error becomes stagnant which occurred approximately every 50 epochs, the adjustable parameter c to calculate the weight of each class is fixed at 1.1, and trained during 200 epochs. Finally, taking into account the characteristic of learning transfer, of this architecture, starting the encoder weights using the pre-trained with the dataset of ImageNet [16].

The main objective of this network is to segment the obstacles that in the roads, this is why the segmented images from the dataset BDD100K is used, divided into 7000 images for training and 1000 images for validation. The labels are modified keeping the types of interest and discarding the others types; these modifications consist in iterate in each one of the images and replace the discard classes, with the label *unlabeled*, and the classes corresponding to the obstacles are ordered of ascending way, giving as result a new image with the labels of the obstacles. In order to compare results, two tests were carried out, which are explained in the following:

First Training. Seven classes were specified for this training, the images resulting from this training are shown in Fig. 1, and all results are reported using the Intersection-over-Union (IoU) metric according to the Eq. 3.

$$IoU = \frac{TP}{TP + FP + FN} \tag{3}$$

where TP corresponds to true positives, FP to false positives and FN as false negatives at pixel level. For this training, the average IoU is 52.63% and Table 2 shows the accuracy percentages for each class.

(a) Segmentation (day)

Fig. 1. First training results.

Table 2. Results first training, segmentation of obstacles

Class	IoU%
Pedestrians	67.24
Riders	31.72
Bicycles	29.37
Buses	70.72
Cars	92.01
Motorcycles	29.01
Trucks	48.32

Second Training. In the first training there are high class unbalances, many of the classes are very underrepresented. One of the main factors causing this problem is that there are few images for some of the classes. This is why the labels of the datasets are modified again, joining the classes with few images into one. For this, the IDs of each class are substituted, obtaining four new classes: pedestrians, small vehicles (bicycles and motorcycles), medium vehicles (cars) and large vehicles (trucks and buses). After performing the classification, the network was trained, obtaining as a result a new segmentation as shown in Fig. 2. Also, an average IoU value of 66.91%, and the precision percentages for each class are shown in Table 3.

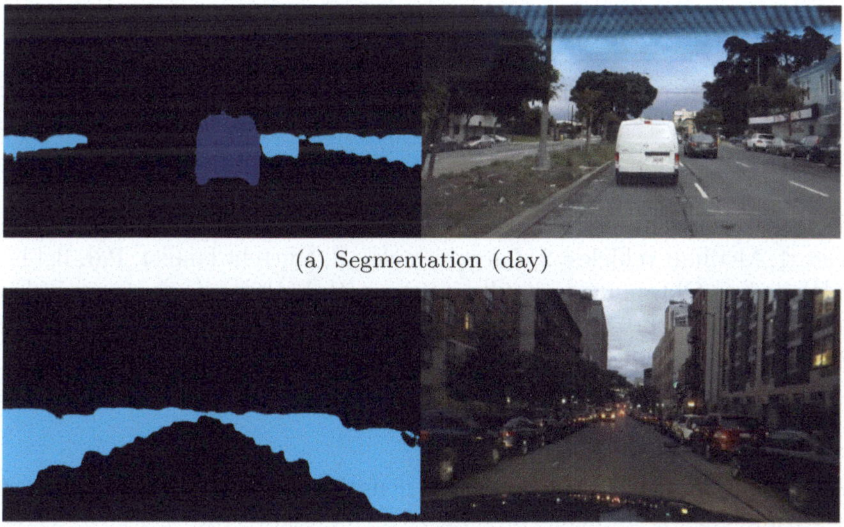

(a) Segmentation (day)

(b) Segmentation (afternoon)

Fig. 2. Second training results.

Table 3. Results second training, segmentation of obstacles

Class	Pedestrians	Small vehicles	Medium vehicles	Medium vehicles
IoU%	68.75	43.18	92.57	63.14

3.3 Dataset Generation

Once the first stage is finished, the second training step is chosen, considering that it has better results and provide efficient identification of obstacles. Then, 20000 images are randomly selected from the dataset containing the lanes, to generate obstacle weak labels using the trained network as a tool. Finally, as shown in Fig. 3, some images are incorrectly segmented, due to very low illumination, bad contrast or even blurred images, these images are discarded, resulting in a new group of 19000 images that are used as obstacle labels.

Fig. 3. Example of a discarded image.

At this point, 19000 images with two groups of labels have been classified, one for the obstacles, and another with the lanes. The next step is to overlap these two labels to obtain a single group of images with both groups of labels. This is achieved by adding each pixels corresponding to the lanes, to the labels of the obstacles; then modify the IDs of each class to be: Pedestrians: 0, Small vehicles: 1, Medium vehicles: 2, Large vehicles: 3, Current lane: 4, Parallel lanes: 5, Unlabeled: 6. Obtaining as results images with both labels as shown in Fig. 4. Finally, the images are divided into two groups: a training group with 15000 images and a validation group with 4000 images.

3.4 Scene Segmentation

With this new dataset, a convolutional neural network capable of segmenting obstacles and lanes is trained. The same ERFNet architecture is used and trained for 200 epochs starting the encoder weights using the pre-trained with the ImageNet dataset, using the same training parameters explained in Sect. 3.2.

Fig. 4. Class merger results.

Table 4. Segmentation of obstacles and lanes

Class	IoU%
Pedestrians	76.77
Small vehicles	33.44
Medium vehicles	93.80
Large vehicles	77.03
Current lane	89.01
Parallel lanes	76.44

4 Experimental Results

In order to validate the proposed method, several rosbag files are processed for different trajectories, all these trajectories were recorded using the vehicle (Mitsubishi i-MiEV) on urban roads in the city of Leganés, in Madrid. In addition, all the results shown below were applied to these rosbags.

4.1 Quantitative Evaluation

Once the network training is finished, the group of 4000 randomly selected images are taken to perform the tests and calculate the IoU values, in the same way as in the previous sections. An average IoU value of 74.42% was obtained, in addition to the precision values of each of the classes shown in Table 4. Figure 4 shows different results for images taken under different conditions. It is shown in Fig. 5(a) the parallel lanes are segmented. Finally, Figs. 5(b) and (c) show how the network works for images at night, with low illumination and different weather conditions. In addition, a network with the same training parameters was trained, but using the BDD100K dataset, with the images of the lanes, in order to be able to compare if the precision of these classes is altered when making the merger. Observing the Table 5 and comparing it with the previous results it is concluded that the difference between the IOU is very small what means that with the new dataset, it can be recognized in a similar way to the original dataset, without losing precision.

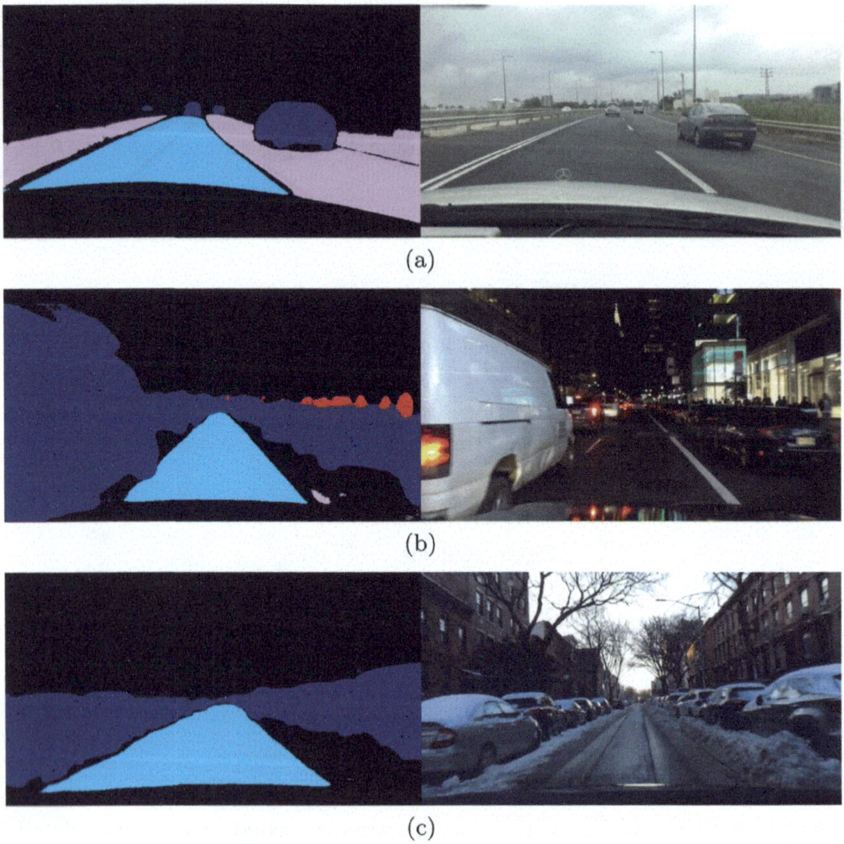

Fig. 5. Segmentation of obstacles and lanes.

Table 5. Segmentation of lanes

Class	Current lane	Parallel lanes
IoU%	81.72	70.62

4.2 Qualitative Evaluation

For network testing, an autonomous electric vehicle (Mitsubishi i-MiEV) is used for real experiments, and it is equipped with a 32 layers LiDAR (Robosense RS-LiDAR-32), located on the top of the vehicle, a front camera with a 3.2 MP CMOS sensor, mounting a 3.1 mm optic with FOV of 90°, and an RTK GPS. The processing unit is equipped with a TITAN XP GPU, and the second processing unit is an Intel NUC 6I7KYK. A router and a switch are mounted for communication purposes. In addition to three 12 V/70mAh connected to an inverter that generates 230 V AC and a 1500 VA UPS is used in order to manage the connections and disconnections when charging the batteries.

These experiments were implemented on Robot Operating System (ROS) node, with the architecture and weights of the trained network. This node works in the following way: when an image from the vehicle's camera is published, it passes through the trained network to obtain the segmented image as a result. In addition, the amount of frame per second is calculated, allowing to obtain results on the processing time.

Figure 6 shows the results of the tests performed with the images captured by the vehicle, these results behave in a similar way to those obtained with the validation images, regardless of the fact that the characteristics of the vehicle's camera are different from those used in the training, which shows that the model trained is robust and generates good results. Finally, it is important to point out that the qualitative tests were carried out, using a graphics card NVIDIA GEFORCE GTX 1080 of 8 GB, calculating the amount of frame per second (fps), obtaining as result 11.02 fps with a resolution 640×480 for each image.

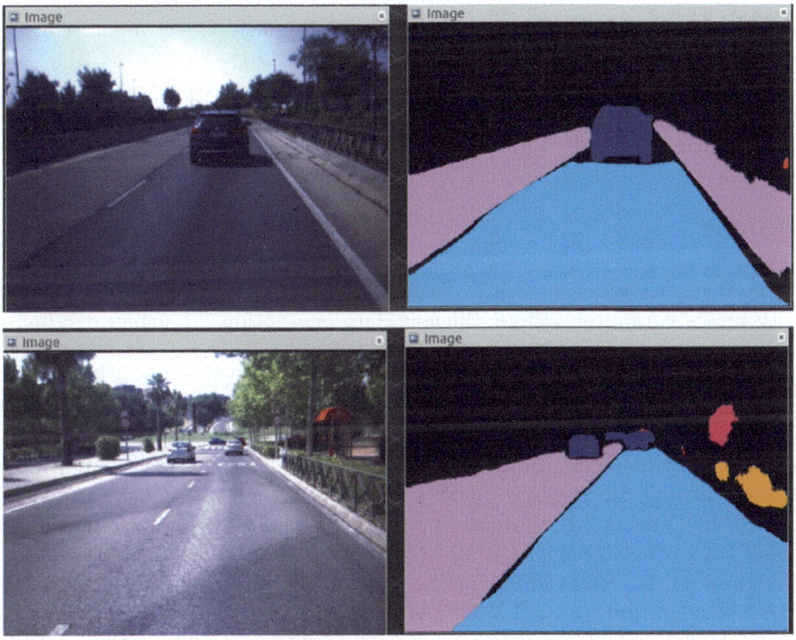

Fig. 6. Results from autonomous vehicle

5 Conclusions

This paper proposed an approach of generating a new dataset from existing ones, using convolutional neural networks to combine different datasets, in order to obtain a complete one with combined classes, thus avoiding having to carry out the labeling manually. Using this new dataset allows training different neural networks capable of identifying the different elements with which autonomous vehicles interact, which is of vital importance when making decisions to perform a maneuver.

In the case of the network used to segment obstacles and lanes, the obtained results show the efficiency and the robustness of the proposed algorithm to identify the elements of interest in different climatic and illumination conditions. Additionally, the implementation of a ROS node with the architecture and weights of the network trained, allowed to test the operation in real environments as well as calculate the processing time.

Finally, the future works are focused on the use different architectures, data augmentation techniques or different images, to improve the accuracy of the generated dataset, as well as, with the help of the developed node, be able to implement an algorithm to determine if it is possible to perform different maneuvers such as a lane change, starting from the segmented image.

Acknowledgment. Research supported by the Spanish Government through the CICYT projects (TRA2016-78886-C3-1-R and RTI2018-096036-B-C21), and the Comunidad de Madrid through SEGVAUTO-4.0-CM (P2018/EMT-4362). Also, we gratefully acknowledge the support of NVIDIA Corporation with the donation of the GPUs used for this research.

References

1. Alhaija, H.A., Mustikovela, S.K., Mescheder, L., Geiger, A., Rother, C.: Augmented reality meets computer vision: efficient data generation for urban driving scenes. Int. J. Comput. Vision **126**(9), 961–972 (2018)
2. Behrendt, K., Witt, J.: Deep learning lane marker segmentation from automatically generated labels. In: 2017 IEEE/RSJ International Conference on Intelligent Robots and Systems (IROS), pp. 777–782. IEEE (2017)
3. Collobert, R., Kavukcuoglu, K., Farabet, C.: Torch7: a matlab-like environment for machine learning. In: BigLearn, NIPS Workshop. No. EPFL-CONF-192376 (2011)
4. Cordts, M., Omran, M., Ramos, S., Rehfeld, T., Enzweiler, M., Benenson, R., Franke, U., Roth, S., Schiele, B.: The cityscapes dataset for semantic urban scene understanding. In: Proceedings of the IEEE Conference on Computer Vision and Pattern Recognition, pp. 3213–3223 (2016)
5. Fritsch, J., Kuehnl, T., Geiger, A.: A new performance measure and evaluation benchmark for road detection algorithms. In: 16th International IEEE Conference on Intelligent Transportation Systems (ITSC 2013), pp. 1693–1700. IEEE (2013)
6. He, K., Gkioxari, G., Dollár, P., Girshick, R.: Mask R-CNN. In: 2017 IEEE International Conference on Computer Vision (ICCV), pp. 2980–2988. IEEE (2017)
7. Kingma, D.P., Ba, J.: Adam: A method for stochastic optimization. arXiv preprint arXiv:1412.6980 (2014)
8. Lee, S., Kim, J., Shin Yoon, J., Shin, S., Bailo, O., Kim, N., Lee, T.H., Seok Hong, H., Han, S.H., So Kweon, I.: VPGNet: Vanishing point guided network for lane and road marking detection and recognition. In: Proceedings of the IEEE International Conference on Computer Vision, pp. 1947–1955 (2017)

9. Long, J., Shelhamer, E., Darrell, T.: Fully convolutional networks for semantic segmentation. In: Proceedings of the IEEE Conference on Computer Vision and Pattern Recognition, pp. 3431–3440 (2015)
10. Meletis, P., Dubbelman, G.: Training of convolutional networks on multiple heterogeneous datasets for street scene semantic segmentation. In: 2018 IEEE Intelligent Vehicles Symposium (IV), pp. 1045–1050. IEEE (2018)
11. Oeljeklaus, M., Hoffmann, F., Bertram, T.: A combined recognition and segmentation model for urban traffic scene understanding. In: 2017 IEEE 20th International Conference on Intelligent Transportation Systems (ITSC), pp. 1–6. IEEE (2017)
12. Pan, X., Shi, J., Luo, P., Wang, X., Tang, X.: Spatial as deep: spatial CNN for traffic scene understanding. In: Thirty-Second AAAI Conference on Artificial Intelligence (2018)
13. Paszke, A., Chaurasia, A., Kim, S., Culurciello, E.: ENet: A deep neural network architecture for real-time semantic segmentation. arXiv preprint arXiv:1606.02147 (2016)
14. Romera, E., Alvarez, J.M., Bergasa, L.M., Arroyo, R.: ERFNet: efficient residual factorized convnet for real-time semantic segmentation. IEEE Trans. Intell. Transp. Syst. **19**(1), 263–272 (2018)
15. Ronneberger, O., Fischer, P., Brox, T.: U-Net: convolutional networks for biomedical image segmentation. In: International Conference on Medical Image Computing and Computer-Assisted Intervention, pp. 234–241. Springer (2015)
16. Russakovsky, O., Deng, J., Su, H., Krause, J., Satheesh, S., Ma, S., Huang, Z., Karpathy, A., Khosla, A., Bernstein, M., et al.: Imagenet large scale visual recognition challenge. Int. J. Comput. Vision **115**(3), 211–252 (2015)
17. Stallkamp, J., Schlipsing, M., Salmen, J., Igel, C.: The German traffic sign recognition benchmark: a multi-class classification competition. In: The 2011 International Joint Conference on Neural Networks (IJCNN), pp. 1453–1460. IEEE (2011)
18. Wang, P., Chen, P., Yuan, Y., Liu, D., Huang, Z., Hou, X., Cottrell, G.: Understanding convolution for semantic segmentation. In: 2018 IEEE Winter Conference on Applications of Computer Vision (WACV), pp. 1451–1460. IEEE (2018)
19. Yu, F., Xian, W., Chen, Y., Liu, F., Liao, M., Madhavan, V., Darrell, T.: BDD100K: a diverse driving video database with scalable annotation tooling. arXiv preprint arXiv:1805.04687 (2018)
20. Zou, Q., Jiang, H., Dai, Q., Yue, Y., Chen, L., Wang, Q.: Robust lane detection from continuous driving scenes using deep neural networks. arXiv preprint arXiv:1903.02193 (2019)

Scalable ROS-Based Architecture to Merge Multi-source Lane Detection Algorithms

Tiago Almeida[1](✉), Vitor Santos[2], and Bernardo Lourenço[1]

[1] DEM, University of Aveiro, Aveiro, Portugal
tm.almeida@ua.pt

[2] DEM, IEETA, University of Aveiro, Aveiro, Portugal

Abstract. Road detection is a crucial concern in Autonomous Navigation and Driving Assistance. Despite the multiple existing algorithms to detect the road, the literature does not offer a single effective algorithm for all situations. A global more robust set-up would count on multiple distinct algorithms running in parallel, or even from multiple cameras. Then, all these algorithms' outputs should be merged or combined to produce a more robust and informed detection of the road and lane, so that it works in more situations than each algorithm by itself. This paper proposes a ROS-based architecture to manage and combine multiple sources of lane detection algorithms ranging from the classic lane detectors up to deep-learning-based detectors. The architecture is fully scalable and has proved to be a valuable tool to test and parametrize individual algorithms. The combination of the algorithms' results used in this paper uses a confidence based merging of individual detections, but other alternative fusion or merging techniques can be used.

Keywords: Visual perception · Data combination · ROS · Deep learning · Computer vision · Road detection · Driving assistance

1 Introduction

The fields of Autonomous Driving (AD) and Advanced Driver Assistance Systems (ADAS) have carried out a wide range of studies that can bring new possibilities to the drivers. One of the most relevant is the detection of the road boundaries or lane detection. For that, the car is equipped with relevant sensors to extract information from the real scene.

Different approaches have been implemented over the years to detect those lane and road boundaries, and they can be divided into two main types: methods that use classic computer vision techniques, and more recent approaches that use learning-based techniques, namely using AI (Artificial Intelligence)[1].

[1] https://emerj.com/ai-sector-overviews/machine-vision-for-self-driving-cars-current-applications.

M. F. Silva et al. (Eds.): ROBOT 2019, AISC 1092, pp. 242–254, 2020.
https://doi.org/10.1007/978-3-030-35990-4_20

Despite this abundance of techniques, there is not a single algorithm that performs accurately in all situations, but it is expectable that a combined result of multiple algorithms can increase the overall performance and robustness when compared to individual algorithms.

This work describes a road boundary detection method that includes an ensemble of different types of road detection algorithms, thus providing a more robust detection than any individual algorithm, and being suitable for both unstructured and structured roads. The former are roads with no lane lines on it, commonly presented on rural locations and the latter are normal roads that have clear lane marks and road boundaries. This is a problem since it is necessary to perceive two different types of features. In addition to this, three main problems have to be considered: lighting changes, shadows and vehicle occlusions [15]. Also, this approach can optionally merge the detection from multiple cameras.

2 Related Work

There are several algorithms to detect the road lanes through classical computer vision techniques. The most used pipeline is composed of different phases: image pre-processing, feature extraction and model fitting [1,2,9]. Some of these algorithms ally the road detection with the road lane lines tracking [3,7,11]. Recently, methods based on Deep Learning (DL) have been developed, providing more robust results, but still at a higher computational cost [5,8,10]. The problem presented by many of the algorithms based on classical computer vision techniques is the inability to deal with different types of road scenarios. Algorithms based on modern DL techniques are now appearing but expected to have intensive applications in the future.

The combination of different types of algorithms for road and line detection is not very common in the literature. However, it has been applied successfully in other fields, such in machine learning, where, for example, there is a technique that combines the predictions from multiple trained models to reduce variance and improve prediction performance [13,16]. The objective is similar the one presented in this work: to combine different results in order to obtain a more accurate one.

3 Proposed Approach

The proposed approach in this paper consists of a special architecture, which is responsible to combine the detection of multiple algorithms from multiple cameras, in order to produce a more confident detection. The technique can be divided in three parts: the definition and parametrization of detection algorithms, the combination of multiple algorithms from a camera, and finally, the combination of several cameras, each with one or more algorithms.

3.1 Detection Algorithms

One of the challenges of this infrastructure is to work with multiple types of road lane detectors. Therefore, it is crucial to define a common interface for the inputs and outputs of these algorithms. For the inputs, it was observed that every algorithm expects an image from the camera. Furthermore, two types of outputs were found: either a polygon of the lane, i.e. the lines delimiting the lane, or a binary image of the region of the road lane. Either way, it is important that all algorithms output the same data type; hence, the binary image was chosen because it is the most flexible representation. This image of the road was named the *confidence map*.

To transform the polygons and lane lines into *confidence maps*, an extra step was added to these algorithms to handle the conversion, and unify the road representation. This can be seen in Fig. 1.

Fig. 1. Example of the conversion between polylines to the confidence map. In this example, the lane lines detected by an algorithm (left) are converted to the corresponding *confidence map* (right).

3.2 Combination of Multiple Algorithms

This part of the procedure receives the multiple confidence maps for the multiple algorithms, and combines them into a single confidence map, that is expected to represent a more confident detection. The steps of this procedure are described next:

1. The computational mechanism tries to synchronise the multiple incoming detections from the various algorithms. This is, however, not always possible, because the latency is different for each algorithm. A threshold for the latency was defined in order to ignore messages that have a rate smaller than the camera frame rate; in the case of this implementation, a threshold for the latency of 0.06 s was used, which corresponds to a frame rate of 15 Hz (due to a hardware limitation). In other words, if a message is older than the threshold, it is discarded. This is possible because the detection algorithms keep the acquisition time stamp of the image.

2. The maps are combined through the logical **AND** operation, to obtain the *intersection* zone.
3. Next, a larger region of confidence is generated after a convolution with a square kernel of size `kernel_size`. This operation smooths the intersection zone of the confidence map, creating a smooth transitions on the boundaries of the detection. An example of this operation being applied is shown in Fig. 2.
4. The maps are also combined through the logical **XOR** operation to obtain the *non-intersection* zone.
5. The region of less confidence is generated by multiplying the pixel values by a constant value L_C given by

$$L_C = \frac{\text{ceil}(kernel\_side/2)}{kernel\_size} - \alpha, \tag{1}$$

where α is a parameter defined by the user.
6. Finally, the two maps are combined with a **SUM** operation to obtain the final *confidence map*.

Fig. 2. Example of the effects of the smoothing operation in the confidence map.

The overall procedure is illustrated in Fig. 3.

3.3 Combination of Multiple Cameras

The final step in the proposed architecture is the combination of the detection of the multiple cameras. Note that this step can sometimes be not useful because the camera's fields of view may not intersect. Thus, it can be disabled, then the algorithm produces confidence maps equal to the number of cameras in the system. To merge the confidence maps from multiple cameras, a perspective transformation is necessary to place the cameras into a common reference frame. This can be achieved through the warp transformation using the intrinsic and extrinsic calibrations of each camera.

In the field of autonomous driving, the most common choice for the reference frame is the top view, or *birds-eye view*. This technique is quite commonly named

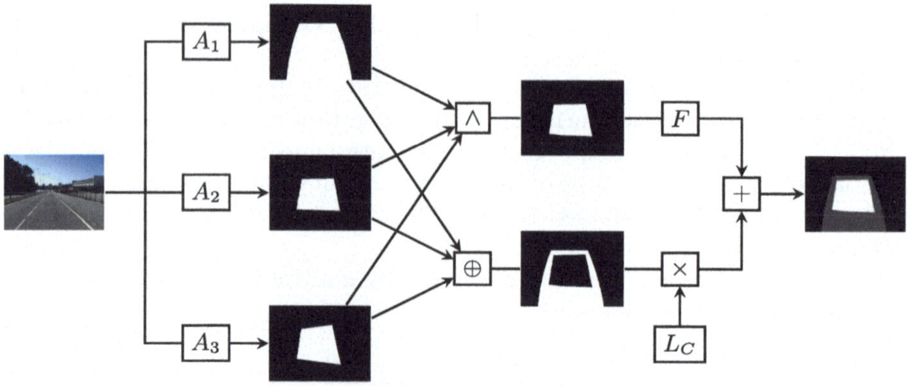

Fig. 3. Example of the technique to combine multiple algorithms. First, the image is processed by the A_1, A_2, A_3 algorithms, where each produces a map. The maps are then combined using the AND ($\wedge$) and the XOR ($\oplus$), and the corresponding smoothing filter F and the multiplication by L_C are applied, producing the maps with the most and less confidence. Finally, the maps are summed (+), producing the final confidence map.

Inverse-Perspective-Mapping, or *IPM*. An example of this technique can be seen in [12].

After warping all the confidence maps into this new perspective, the combination is done by averaging the multiple maps. An example of this can be seen in Fig. 4.

Fig. 4. Example of the warping transformation. The left and center image show the detection map in the camera point of view. The image on the right shows the warped image resulting from the *IPM* technique.

4 Experimental Infrastructure

This work was developed under the ATLASCAR2 project[2], which is an Mitsubishi i-MiEV equipped with cameras, LiDAR and other sensors. The sensors

[2] http://atlas.web.ua.pt/.

used for this work were two PointGrey Flea3 cameras, installed on the roof-top of the car, as can be seen on Fig. 5. The software architecture that handles the connection of the sensors and the algorithms developed is ROS.

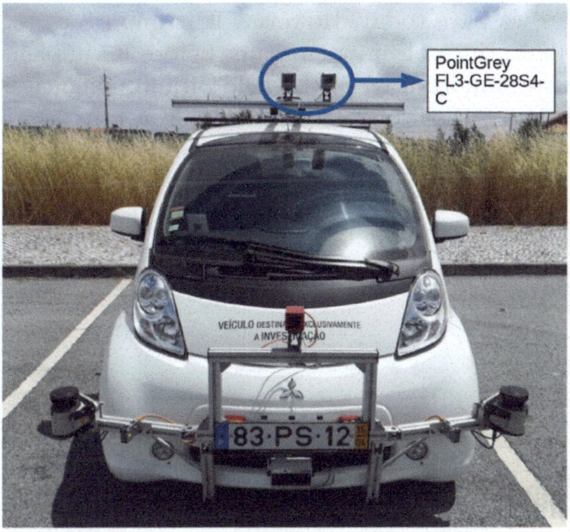

Fig. 5. The AtlasCar2 vehicle. The two cameras used for this work are placed on car roof.

Next, the two methods used for this work are described: the classical method, based on standard computer vision algorithms, and a Deep Learning-based approach.

4.1 Detector Based on Classical Methods

The detector that represents the classical method used in this work can be found in a *Udacity* course[3]. This algorithm is composed of the following steps:

1. Image rectification and warping: this step undistorts the camera image according to the intrinsic parameters. Then, an inverse perspective transform is applied, changing the perspective of the image to a top-down view.
2. Road lanes segmentation: this step is composed of two methods: a segmentation of the red channel of the image and a sobel edge detector. These two partial segmentations are then merged into a final binary segmentation.
3. Curve fitting: this method is applied to the binary segmented image to extract the road lines according to a set of parameters.

An example of the application of this algorithm can be seen in Fig. 6.

[3] https://medium.com/deepvision/udacity-advance-lane-detection-of-the-road-in-autonomous-driving-5faa44ded487.

4.2 Detector Based on Deep Learning Method

Deep Learning (DL) is a thriving field that is revolutionising and constantly achieving state-of-art results in computer vision challenges. It is undeniable that the implementation of these new algorithms could greatly expand the capabilities of the technologies we have today. However, it is important to understand the capabilities and limitations of any algorithm before its implementation in any task, and even more importantly, in the field of autonomous driving. This can be achieved in this work, through a side-by-side comparison of the classical algorithms, described earlier, and the DL approach.

Fig. 6. Example of the lane detection using classical computer vision techniques.

The challenge of road segmentation belongs, in the domain of DL, to the field of pixel-wise semantic segmentation. Or in other words, each pixel in the input image is labelled with the corresponding class (for example, road, car, sidewalk). For our purpose, only the pixels that correspond to the road are important, but there is enough flexibility to interpret more classes, which could be valuable in other ways (for example, traffic sign identification).

The work we developed in this area is mostly the same as the one described in [14], with slight modifications: we used the pretrained resnet50 [6] as the backbone for the UNet network; and we used a modified version of CamVid dataset [4] with just the most important 11 classes, in order to reduce the computational cost inherent to the training of the neural network. The Unet, as most fully convolutional networks, is characterised by an encoder-decoder network. The encoder part produces features that are produced by successive convolutions, where the raw pixel input is transformed into high level features. The decoder network interprets these high level features and associates the corresponding class to the image. Therefore, the encoder successively reduces the spatial dimension of the image, while increasing the depth, and the decoder decreases the spatial size of the images, while decreasing the depth. The encoder is usually transferred from a pre-trained classification network, and the decoder weights are learned through training.

An example of a detection using this algorithm is shown in Fig. 7.

5 Experiments and Results

To demonstrate the usefulness, scalability and reliability of the proposed architecture, several experiments were carried out. They range from the simple "one camera–one algorithm" up to "multiple cameras–multiple algorithms". The experiments do not assess the architecture directly, but show how the architecture can be used to test the performance of the algorithms and their combination. Therefore, performance indices, related to the confidence maps, were created to allow the evaluation and tuning of the algorithms depending on some variable parameter.

Fig. 7. Example of the DL algorithm successfully detecting the road map. The input image (left) is segmented into a pixel-wise semantic segmentation (center), where, for example, the cars correspond to the orange color and the road in purple. Because only the road is relevant, a map of the road is the result of this algorithm (right).

The proposed indices are based on the areas of the confidence maps described in Sect. 3.3, and the variable parameter is the size of the smoothing filter used. The first index (I_1) is defined by (2):

$$I_1 = \frac{W_{CA}}{A_T} \tag{2}$$

where W_{CA} is the "weighted confidence area" of the confidence map I, and A_T is the total area of the confidence map, or, formally:

$$W_{CA} = \sum_r \sum_c I(r, c) \tag{3}$$

being I the normalised matrix of the confidence map image, $I(r, c) \in [0; 1]$, and

$$A_T = \sum_r \sum_c \lceil I(r, c) \rceil = \sum_r \sum_c \text{ceil}\left(I(r, c)\right). \tag{4}$$

The second performance index (I_2) is given by (5):

$$I_2 = \frac{A_C}{A_T} \tag{5}$$

where A_C is the common area, that is, the area of the confidence map where the confidence is 100%. Formally:

$$A_C = \sum_r \sum_c \lfloor I(r,c) \rfloor = \sum_r \sum_c \text{floor}\,(I(r,c)) \tag{6}$$

As mentioned, the variable parameter used was the size of the convolution filter (`kernel_size` variable), and sizes from 3×3 up to 51×51 were tested. The results presented next were obtained on the same real road section (Fig. 8) which cover two situations from the several tested: one camera and two algorithms, and two cameras and two algorithms.

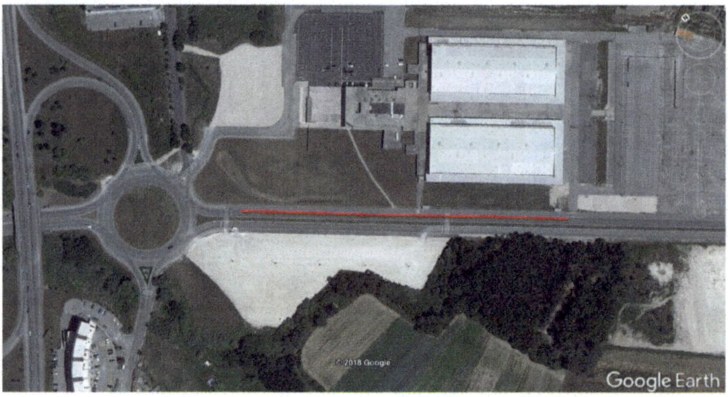

Fig. 8. The road section where the experiments were made.

5.1 Results from One Camera and Two Algorithms

The first experiment consists of combining the outputs of two distinct processor ROS nodes using images captured by one camera. The first result presented is the variation of each indicator with the filter size. Then, for each value of the filter size, an average of the two indices was calculated for the full set of 360 frames. Those are the values plotted for $I_{1,avg}$ and $I_{2,avg}$ shown in Fig. 9a.

Since black pixels will turn non-black with the increase of the filter size, the A_T parameter increases. The opposite happens for the common area (A_C) because white pixels will be transformed into darker pixels as the filter size increases. Consequently, the weighted confidence area (W_{CA}) remains almost invariable.

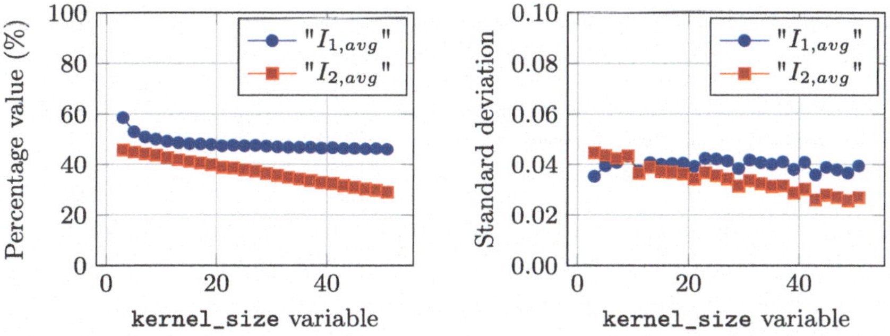

(a) Variation of the indices (%). (b) Standard deviation values.

Fig. 9. Results of indices $I_{1,avg}$ and $I_{2,avg}$ for the experiment with one camera and two algorithms.

The other analysis (Fig. 9b) concerns the standard deviation of each indicator for each filter size. The low values of the standard deviations values reflect a large uniformity in the samples analysed. These results, and potentially others, demonstrate the usefulness of the architecture to study and tune parameters, either from each algorithm (not the case here) or from the merging technique it self (as was the case here).

Fig. 10. Results using two detectors (classic and DL-based on top). The bottom left image represents the road zone given by DL algorithm. As can be seen, the imperfections in the deep learning detector are mitigated by the combination with the classical detector. The final result (bottom right) shows an enhanced representation of the road.

An example of the usefulness of the architecture is shown in Fig. 10, where a classic computer vision and deep learning-based techniques were combined.

5.2 Results from Two Cameras and Two Algorithms

As explained in Sect. 3.3, there are two options to launch the architecture in multi-camera mode: with or without a combination of the maps coming from each camera. If the two maps are not combined, the results would be the same as those presented in the previous sub-section. Therefore, the results presented in this sub-section are based on the combination of the maps coming from each camera. Since a warp transformation is applied to the polygons to be combined, the confidence map representation is expected to be more straight (road top-view). This implies that the blur caused by the application of the filter is smoother along the road lane border compared to the previous study.

In general, the results obtained (Fig. 11) are similar to those presented in the previous experiment. The only clear difference is the rate of change of both indices, which in this case is lower. This is due to the fact that, as mentioned earlier, the confidence map here is represented by a top-view perspective.

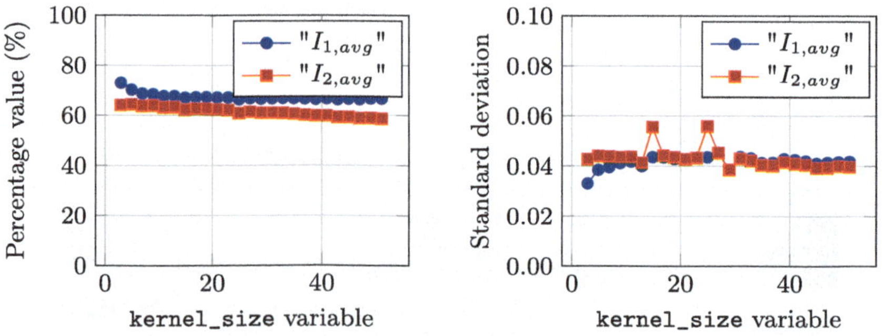

(a) Variation of the indices (%). (b) Standard deviation values.

Fig. 11. Results obtained for the experiment related to the use of two cameras and the two algorithms.

6 Conclusion

This paper proposes a scalable architecture to merge multiple road detection algorithms, creating the conditions to obtain more robust detected road maps than by using the algorithms individually. The two types of output representations (polylines or regions of pixels) are converted into a unique representation to allow the merging procedures. This architecture has shown to be valuable also to combine traditional computer vision techniques and DL based classifiers to detect the road in more robust. Even though Deep Learning usually outperforms the traditional techniques it can fail by overestimation and traditional techniques can be used to limit that effect, although this issue still needs further studies and analysis. One planned next step is to migrate into a unified representation, probably based in occupancy grids, to merge data obtained also from LIDAR-based perception. In conclusion, although there is still space for improvements, the architecture proved to be a valid instrument to combine multiple source road detection algorithms with the possibility to tune them interactively, or, hopefully in the future, by an automatic procedure to optimise road detection algorithms and their parameters towards a more generic detector.

Acknowledgments. This work was partially supported by project UID/CEC/00127/2019.

References

1. Aly, M.: Real time detection of lane markers in urban streets. In: 2008 IEEE Intelligent Vehicles Symposium, pp. 7–12. IEEE, November 2008
2. Assidiq, A.A., Khalifa, O.O., Islam, M.R., Khan, S.: Real time lane detection for autonomous vehicles. In: 2008 International Conference on Computer and Communication Engineering, pp. 82–88, May 2008
3. Bounini, F., Gingras, D., Lapointe, V., Pollart, H.: Autonomous vehicle and real time road lanes detection and tracking, pp. 1–6, October 2015
4. Brostow, G.J., Fauqueur, J., Cipolla, R.: Semantic object classes in video: a high-definition ground truth database. Pattern Recogn. Lett. **30**, 88–97 (2009)
5. David Jenkins, M., Carr, T.A., Iglesias, M.I., Buggy, T., Morison, G.: A deep convolutional neural network for semantic pixel-wise segmentation of road and pavement surface cracks. In: 2018 26th European Signal Processing Conference on (EUSIPCO), pp. 2120–2124, September 2018
6. He, K., Zhang, X., Ren, S., Sun, J.: Deep residual learning for image recognition. CoRR abs/1512.03385 (2015)
7. Hou, C., Hou, J., Yu, C.: An efficient lane markings detection and tracking method based on vanishing point constraints. In: 2016 35th Chinese Control Conference (CCC), pp. 6999–7004, July 2016
8. John, V., Kidono, K., Guo, C., Tehrani, H., Mita, S., Ishimaru, K.: Fast road scene segmentation using deep learning and scene-based models. In: 2016 23rd International Conference on Pattern Recognition (ICPR), pp. 3763–3768, December 2016
9. Kluge, K., Lakshmanan, S.: A deformable-template approach to lane detection. IEEE pp. 54–59, September 1995
10. Li, L., Zheng, W., Kong, L., Ozguner, U., Hou, W., Lian, J.: Real-time traffic scene segmentation based on multi-feature map and deep learning. In: 2018 IEEE Intelligent Vehicles Symposium (IV), pp. 7–12, June 2018
11. Liu, S., Lu, L., Zhong, X., Zeng, J.: Effective road lane detection and tracking method using line segment detector. In: 2018 37th Chinese Control Conference (CCC), pp. 5222–5227, July 2018
12. Oliveira, M., Santos, V., Sappa, A.D.: Multimodal inverse perspective mapping. Inf. Fus. **24**, 108–121 (2015)
13. Pouyanfar, S., Chen, S.: Semantic event detection using ensemble deep learning. In: 2016 IEEE International Symposium on Multimedia (ISM), pp. 203–208, December 2016

14. Ronneberger, O., Fischer, P., Brox, T.: U-net: Convolutional networks for biomedical image segmentation. ArXiv abs/1505.04597 (2015)
15. Su, C.Y., Fan, G.H.: An effective and fast lane detection algorithm. In: Bebis, G., Boyle, R., Parvin, B., Koracin, D., Remagnino, P., Porikli, F., Peters, J., Klosowski, J., Arns, L., Chun, Y.K., Rhyne, T.M., Monroe, L. (eds.) Advances in Visual Computing, pp. 942–948. Springer, Heidelberg (2008)
16. TaSci, E., Ugur, A.: Image classification using ensemble algorithms with deep learning and hand-crafted features. In: 2018 26th Signal Processing and Communications Applications Conference (SIU), pp. 1–4, May 2018

Improving Localization by Learning Pole-Like Landmarks Using a Semi-supervised Approach

Tiago Barros[1(✉)], Luís Garrote[1], Ricardo Pereira[1], Cristiano Premebida[1,2], and Urbano J. Nunes[1]

[1] Institute of Systems and Robotics, Department of Electrical and Computer Engineering, Polo II, University of Coimbra, Coimbra, Portugal
{tiagobarros,garrote,ricardo.pereira,urbano}@isr.uc.pt
[2] Department of Aeronautical and Automotive Engineering, Loughborough University, Loughborough, UK
C.Premebida@lboro.ac.uk

Abstract. The aim of this paper is to contribute with an object-based learning and selection methods to improve localization and mapping techniques. The methods use 3D-LiDAR data which is suitable for autonomous driving systems operating in urban environments. The objects of interest to be served as landmarks are pole-like objects which are naturally present in the environment. To detect and recognize pole-like objects in 3D-LiDAR data, a semi-supervised iterative label propagation method has been developed. Additionally, a selection method is proposed for selection the best poles to be used in the localization loop. The LiDAR localization and mapping system is validated using data from the KITTI database. Reported results show that by considering the occurrence of pole-like objects over time leads to an improvement on both the learning model and the localization.

Keywords: LiDAR odometry · Semi-supervised learning · Incremental label propagation · SLAM · Pole-based localization

1 Introduction

Localization and mapping methods for mobile robots and autonomous vehicles operating outdoors have additional requirements in terms of reliability and robustness compared to indoor environments. The highly dynamic nature and the fact that the appearance of outdoors environments may change over time, makes the localization process an even more difficult task to be solved.

Most real-world localization methods for outdoor applications combine data from multiple technologies (such as GPS, LiDAR, and cameras) which allow to overcome individual sensor limitations and increase the localization accuracy. Even though, and to some extent, the current localization methods have proven to be accurate and reliable in confined outdoor environments [2], on the other

M. F. Silva et al. (Eds.): ROBOT 2019, AISC 1092, pp. 255–266, 2020.
https://doi.org/10.1007/978-3-030-35990-4_21

(a) (b)

Fig. 1. Selected urban-like scenario from the KITTI Benchmark database (a) point cloud from LiDAR; (b) image from stereo camera. All the poles with similar appearance and shape, as per identified by a yellow bounding box (BB), were labeled as 'poles', whilst the remaining pole-like objects such as those identified by a green BB were not considered for this paper and therefore not labeled.

hand, in long-term operation the capability of dealing with highly dynamic contexts and changing environments without affecting the localization performance remains a challenge.

The Global Navigation satellite system (GNSS) is a global localization system that uses satellite constellations (e.g., GPS, GALILEO, GLONASS) having accuracy ranging from meter level (standard version) to centimeter level when more sophisticated technologies such as Differential Global Position (DGPS) or Real-time Kinematics (RTK) are used. DGPS and RTK solutions, in contrast to the standard GPS version, are suitable for autonomous driving since the required accuracy is at least 20 cm [20]. The biggest drawback, however, of GNSS-based systems is the lack of reliability in cluttered environments, where they suffer from multipath effect, signal blockage, lost of differential signal, and other effects as described in [6]. As consequence, GNSS-based localization methods have a reasonable reliability and accuracy in open spaces however, in environments with high-rise buildings or in forests where the signal can be obstructed, the localization can be compromised [14,18].

In contrast to GNSS systems, map-aided localization systems based on LiDAR and/or vision (e.g., stereo, monocular, RGB-D), rely on a dependable representation over time of the environment (i.e., map) with the fundamental assumption that it is static. Therefore, the detection of unique landmarks (e.g., poles as shown in Fig. 1) that are static and invariant is of utmost importance for a reliable pose estimation in a highly dynamic or appearance changing environments. To this end, using LiDAR sensors, which are immune to day-night lighting change, is a suitable choice for long-term operations.

In urban contexts, there are a multitude of natural and artificial landmarks that can be helpful to improve the localization accuracy and reliability. In particular, pole-like objects (e.g., poles, traffic signs, trees' trunks) are commonly found in urban areas. Also, due to the relatively small cost of pole-like objects, additional pole-like objects could be placed to reduce the influence of circumstances which affect negatively the pose estimation problem. This is particularly beneficial in challenging areas e.g., cluttered and dynamic scenarios. When considering a dynamic urban use-case, the aims of a long-term localization system are: to keep an accurate localization, to maintain a suitable environment representation (if one has been given *a-priori*), and to adapt/learn over time using landmarks which are more suitable to improve the localization performance. In this regard, semi-supervised techniques (e.g., Iterative Label Propagation (ILP) [3]) can provide a promising contribution to solve this problem by updating and improving, over-time, the pole-like models.

This work addresses the LiDAR-based landmark detection for the localization problem by proposing a novel learning pipeline which is based on a semi-supervised approach for pole-like object classification and landmark selection. The semi-supervised approach is based on an iterative label propagation [3,19] which is able to update the 'pole-model' by incorporating the poles that are more likely to improve the localization performance. Concurrently, the pole-like objects are recursively selected based on a quantitative metric.

In terms of structure, Sect. 2 briefly covers the related work concerning pole-like detection, classification and LiDAR-based localization and mapping. Section 3 presents an overview of the label propagation approach. Section 4 details the proposed approach, including the semi-supervised learning and localization and mapping techniques. The following section presents evaluation results using the KITTI database. Finally, conclusions are drawn in Sect. 6.

2 Related Work

Virtual and natural landmarks can be used in field robotics and autonomous vehicles to increase the accuracy of localization and SLAM approaches. In particular, the detection of pole-like objects has been a recent focus of multiple research teams [8,11,15,17]. In [15], a framework for vehicle self-localization was proposed based on an Adaptive Monte-Carlo Localization (AMCL) approach that uses semantic cues from distinctive landmarks (e.g., trees and street lamps) and depends on an a prior map. The approach uses as inputs the measurements from LiDAR and a stereo camera. Alternatively, the ACML-based approach described in [9] uses the semantic detection and observation of objects to detect pole-like objects. In [5] a particle filter and a reinforcement learning (RL) approach are used for localization and map updating, respectively. A particle filter is also used in [13] for long-term pole-based localization in urban environments. Poles are recognized by extracting features from 3D-LiDAR voxel representation.

The detection and recognition of pole-like features extracted from 3D point clouds is still a demanding task. In [11], pole-like objects are detected using a modified region growing method and two features are extracted from the data (vertical continuity with isolation and roughness). In [12], a pipeline is proposed to detect and classify pole-like objects from 3D point clouds. A *voxelization* strategy is employed to reduce the complexity of the 3D point cloud and a heuristic segmentation algorithm is then used for detection. Two supervised algorithms (Linear Discriminant Analysis and Support Vector Machines) are then used to classify the different types of poles.

One of the main problems in feature-based localization is the data association step. Incorrect data associations e.g., a non pole-like object associated to a pole-like object, can lead to an increase in the localization drift/error. Also, in ever changing environments, classic learning approaches may not be able to generalize which means, in the context of this work, they may not correctly identify previously unseen pole-like objects which leads to increasing errors. In [19], an iterative label propagation (LP) approach is proposed to label unseen data from a dataset with a set of prior labeled data. This approach uses areas with high density of features to propagate the labels from the known labeled data to the near unlabeled samples. In [16], a mathematical framework is proposed for a semi-supervised learning approach designated by group induction. In [3], an extension to the LP approach is proposed for problems where the dataset is not completely known *a-priori*, e.g. sequential observations from a LiDAR, using a nearest-neighbor graph that is updated at each iteration (when new samples are available). On the other hand, a SLAM approach using semi-supervised learning is proposed in [8]. This approach leverages prior knowledge of a mine environment to increase map accuracy, requiring minimal user information using a single camera setup.

3 Label Propagation

The iterative label propagation (ILP) approach, as proposed in [3], is a semi-supervised learning method that comprises an offline and online stages. Both stages are based on the label propagation (LP) algorithm proposed in [19] however, while the offline stage follows a fully connected graph approach, the online stage relies on a partially connected graph which updates (only) the nodes of interest thus, saving computational effort. For the purpose of this work, and to support the changes made in the offline stage of the ILP algorithm [3], the LP [19] is reviewed.

Let denote a partially labeled training set $\mathcal{D} = \{\mathcal{L} \cup \mathcal{U}\}$, with $\mathcal{D} \in \mathbb{R}^{n\times 12}$ comprising a labeled subset $\mathcal{L} = \{(f_1, g_1), ..., (f_l, g_l)\}$ where $f_i \in \mathbb{R}^{1\times 12}$ is the ith feature vector and $g_i \in \{-1, 1\}$ is the label of the ith observation. Label "1" represents the positives (i.e., pole class) and "-1" the negatives (everything else). The unlabeled subset is designated as $\mathcal{U} = \{f_{l+1}, ..., f_{l+u}\}$, where $f_i \in \mathbb{R}^{1\times 12}$ is the feature vector of the unlabeled observation.

The transition matrix P, which represents the learning model, is given by $P = D^{-1}W$, where $W \in \mathbb{R}^{n\times n}$ is an affinity matrix $W = \{w_{ij}\}$ with $n = l + u$. D is a diagonal matrix $D = \text{diag}(d_1, ...d_n)$, where $d_i = \sum_{j=1}^{n} w_{ij}$. Each edge weight w_{ij} is computed by

$$w_{ij} = \exp\left(-\frac{\sum_{m=1}^{M}(f_i^m - f_j^m)^2}{\sigma_m^2}\right) \tag{1}$$

The hyperparameters σ_m in (1), with $m = [1, 2, ..., 13]$, are used to control the edge weights by giving more relevance to the features that enhance the classification performance. The hyperparameters are computed using gradient descent optimization as follows:

$$\sigma_m[t + 1] = \sigma_m[t] - \gamma\nabla H(\sigma_m[t]) \tag{2}$$

where $H(\sigma_m[t])$ is the average label entropy ($H(G) = \frac{1}{u}\sum_{i=l+1}^{l+u} H_i(G(i))$) and the label entropy is given by

$$H_i(G(i)) = -G(i)\log G(i) - (1 - G(i))\log(1 - G(i)). \tag{3}$$

The label matrix is $G \in \mathbb{R}^{n\times 2}$, where the ith row represents the label probability of the ith observation. Following the convention that the labeled nodes appear first and the unlabeled ones afterwords, the label matrix $G = [G_L\, G_U]^T$, where $G_L \in \mathbb{R}^{l\times 2}$ denotes the first l known labels of $\mathcal{D}$ and $G_U \in \mathbb{R}^{u\times 2}$ is initialized as zero matrix. The transition matrix (P) is partitioned in four blocks, defined by

$$P = \begin{bmatrix} P_{LL} & P_{LU} \\ P_{UL} & P_{UU} \end{bmatrix} \tag{4}$$

where subscripts of the blocks indicate transitions between labeled (LL), unlabeled (UU) and mixed point pairs (UL and LU). Finally, labels are computed by

$$\begin{aligned} G_U[t+1] &= P_{UL}G_U[t] + P_{UL}G_L[t] \\ G_L[t+1] &= G_L[0]\,\forall t \end{aligned} \tag{5}$$

where $G_U[t+1]$ and $G_U[t]$ denote the next and current label estimation of the unlabeled data. $G_L[t+1]$ and $G_L[0]$ denote the next and the initial true labels of the labeled data respectively. Since the true labels of the labeled data are already known their value is never updated.

4 Proposed Approach

In this work, a novel pole-like learning and selection method is proposed and has been tested in a LiDAR-based localization and mapping framework. The framework, or pipeline, is comprised of the following main blocks as shown in Fig. 2: Semi-supervised Learning; Local Pole Map Mapping; Map Matching; and Pole Selection.

4.1 Learning

In order to detect existing poles in the LiDAR data, a pre-processing step is necessary to transform point clouds into a feature representation. To be accomplished, three modules were developed: Ground Extraction, Clustering, and Feature Extraction. The Ground Extraction module processes the raw point cloud (denoted by $PC[t]$) to identify and then to remove, using a random sample consensus (RANSAC) algorithm [10], the points belonging to the ground. The remaining points $E[t]$ are fed to the clustering module which uses a density-based spatial clustering of applications with noise (DBSCAN) algorithm [4] to segment the points into clusters $C[t]$. The Feature Extraction module receives $C[t]$ and calculates the following features: height (h); length (l); width (w); eigen vectors

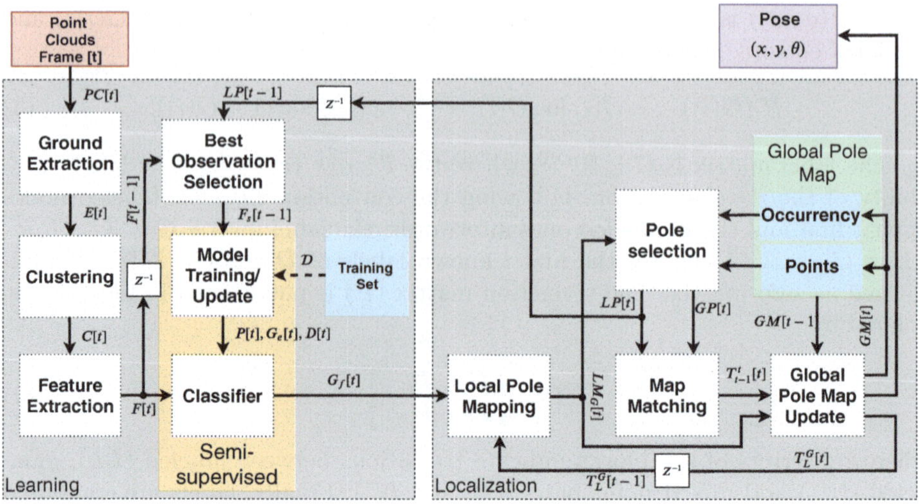

Fig. 2. Proposed learning and localization pipeline.

of the covariance matrix in the x-axis (V_x), the y-axis (V_y), and the z-axis (V_z); Euclidean distance (D) from the cluster's centroid to the sensor's frame origin. The output is a feature vector $F[t] = [h \ l \ w \ Vx_{[1\times3]} \ Vy_{[1\times3]} \ Vz_{[1\times3]} \ D]_{n\times13}$ which is 13 feature wide (columns) and n clusters long (rows).

The ILP algorithm, described in Sect. 3, was used for the classification of the pole-like objects and the hyper-parameters in Table 1 were learned with the training set (20% of the entire dataset) using (2). The Model Training/Update stage computes a first estimation of the transition matrix (P) and the labels (G_e) by using the training data $\mathcal{D}$ at $t = 0$. For $t > 0$ it updates P by incorporating the selected poles returned by the Pole Selection module. The classification module receives P, G_e and $\mathcal{D}$ as inputs and uses the online ILP algorithm to estimate the labels ($G_f[t]$) of the detected clusters.

4.2 Localization and Mapping Framework

The localization and mapping algorithms developed in this work comprise four key modules/sub-systems: (i) Local Pole Mapping module, which selects the detected poles from $C[t]$; (ii) Map Matching module, responsible for computing a translation and rotation between local pole map and global pole map; (iii) Pole Selection module, which selects the poles that minimize the matching error and refines the learning module.

The local pole map ($LM_G[t]$) and the global pole map ($GM[t]$) are metric representations in the global frame $\{G\}$, which corresponds to the point cloud origin of $t = 0$, of the detected poles in the environment. While $LM_G[t]$ represents all the poles detected in the point cloud at instant t, $GM[t]$ represents all the detected poles in the environment over time. Additionally, the global map tracks the occurred overlaps from those poles.

Global Map Update Module. The Global Map Update module uses T_{t-1}^{t} to transform the poles in $LM_G[t]$ to $GM[t]$. Poles in $LM_G[t]$ are merged with the corresponding poles in the $GM[t]$ and the respective overlap occurrence is updated. The remaining poles are added to $GM[t]$ and the occurrence is set to one. Two poles are considered to be overlapping when the L2-norm between their centroids is less then 3σ, where σ is the standard deviation of the point distribution of the respective pole.

Local Pole Mapping Module. The Local Pole Mapping module receives as input $G_f[t]$, selects all the clusters in $C[t]$ that were labeled with the class "1" (pole) and transforms (using $T_L^G[t-1]$), all the selected clusters to the global frame of the previous iteration. The selected poles w.r.t. the global frame define, thus, the local map ($LM_G[t]$).

Map Matching Module. The Map Matching module uses the iterative closest point (ICP) algorithm [1] to compute the transformation $T_{t-1}^{t}[t]$ that relates the frame $t-1$ to t. As inputs, the ICP uses the poles that exist in both local map (LP) and global map (GP) which are returned by the Pole Selection module. The transformation $T_L^G[t]$ tracks the current pose w.r.t. the global reference, which is updated using $T_L^G[t] = T_L^G[t-1] \times T_{t-1}^{t}$.

Pole Selection. The purpose of the proposed Pole Selection module is to detect pole-like objects in the local pole map, that are potentially static over time and filter those that are "dynamic" (i.e., non-stationary) or that have a tendency to be static for short periods of time. Only poles in the local map that have correspondence in the global map and occurred more than α times are selected. The selected poles in the local pole map are represented by $LP[t]$ and those in the global pole map are represented by $GP[t]$.

5 Experimental Validation

The proposed approach has been validated on a urban-like sequence ("2011_09_28_*drive*_0038_*sync*"), which contains 110 LiDAR scans from the KITTI database (see Fig. 1) [7]. Because the KITTI dataset only provides labels for the following classes "Car", "Van", "Truck", "Pedestrian", "Person (sitting)", "Cyclist", "Tram" and "Misc", it was necessary to label all existing poles - similar to those shown in Fig. 1b and marked with yellow bounding boxes.

Table 1. Estimated σ_m for increasing number of features.

σ_1	σ_2	σ_3	σ_4	σ_5	σ_6	σ_7	σ_8	σ_9	σ_{10}	σ_{11}	σ_{12}	σ_{13}
0.135	0.137	0.136	0.129	0.135	0.122	0.138	0.133	0.129	0.128	0.132	0.106	0.199

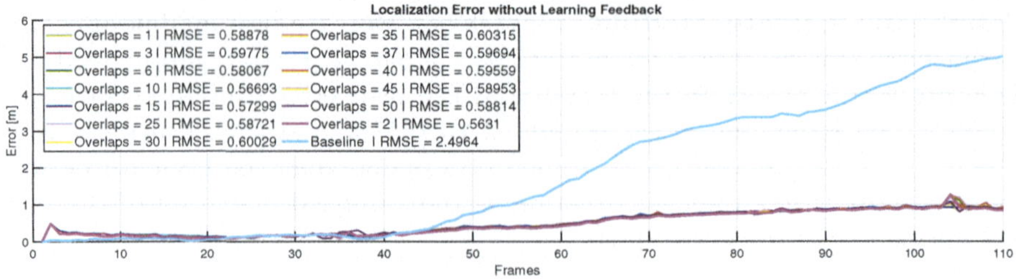

(a)

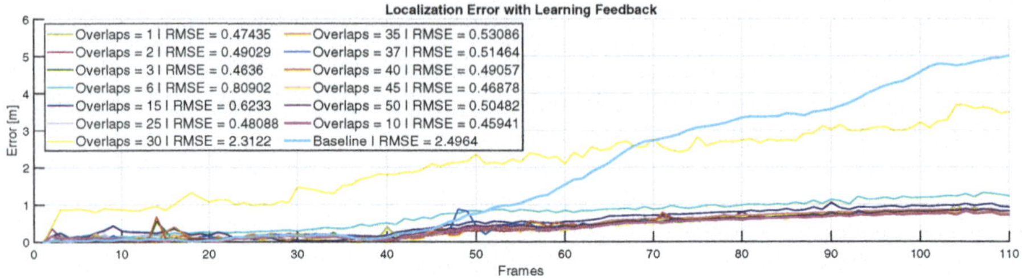

(b)

Fig. 3. Path error for the experiments: (a) without learning feedback and (b) with learning feedback.

The influence of the proposed selection and learning methods was evaluated through two experiments: (1) without learning feedback (i.e., the selected poles are not used to improve the model) and (2) with learning feedback (selected poles are used in the learning model). Both experiments were conducted in the same conditions i.e., a path was estimated for each α (thirteen in total) and the initial model P was trained with 20% of randomly selected ground-truth poles. For comparison purposes, a baseline path was computed using only the ICP directly on the raw point cloud and a global map (which was also built from raw data). Figure 3a and b show the path errors from the aforementioned experiments. The results demonstrate that poles are reliable landmarks for localization in comparison to the baseline. Also, when the selected poles are incorporated in the learning model the results improve even more.

Considering the experiment where the learning feedback was not activated, for $\alpha = 2$ the path error (RMSE) is 0.5631 m i.e., all poles that have two or more overlaps were used for localization. The second lowest path error (with 3 mm difference) was achieved when $\alpha = 10$ which, on the other hand, generated the smallest path error in the experiment with learning feedback (RMSE = 0.459 m) as well. Figure 4 depicts the path with the smallest RMSE for both experiments.

Figure 6 presents the detected poles in the global pole-map and the impact of the Selection Pole module for different α values (1, 2, 6 and 37). The green dots represent the ground-truth poles; the blue represent detected poles (true positives and false positives); and the red and black dashed-lines represent the

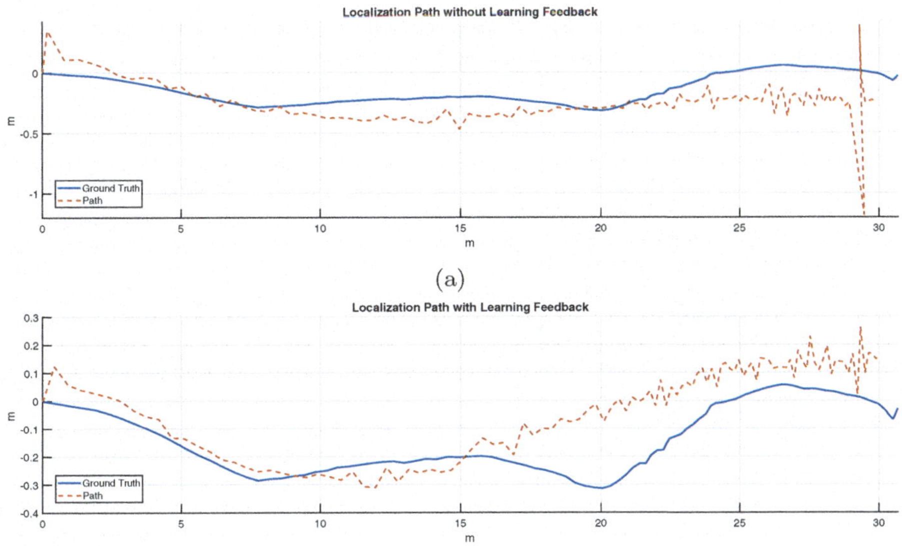

(a)

(b)

Fig. 4. Localization path, with the lowest error, from (a) without and (b) with learning feedback. The presented baseline represents, in both plots, the same path.

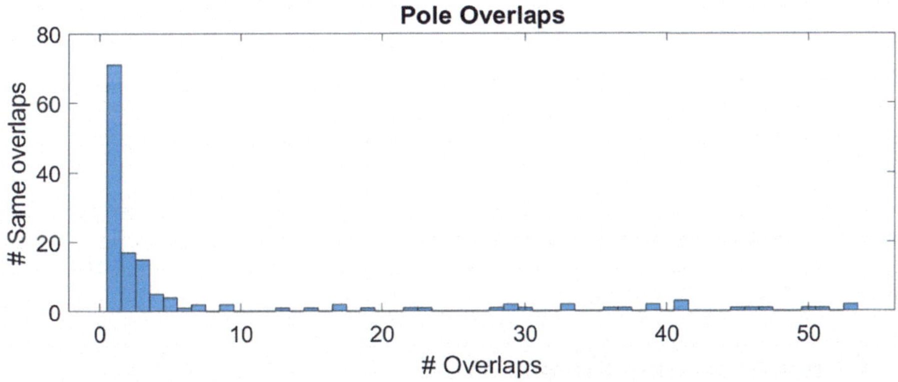

Fig. 5. Histogram of the same number of pole overlaps in the global map.

ground-truth and the estimated path, respectively. Considering that the overlaps depend directly on the classification performance, most overlaps occur between one and three times. From Fig. 6d it is evident that true poles tend to overlap more often than the remaining false positives, therefore, making it easier to distinguish between the two.

By comparing the histogram given in Fig. 5 with the maps illustrated in Fig. 6, a common pattern about the objects' nature emerges. It is noticeable, from Fig. 6a and b, that almost all false positives have a low number of overlaps (left-hand side of the histogram). On the other hand, actual poles tend to overlap more often. This characteristic was used to remove the false positives and to refine the learning model.

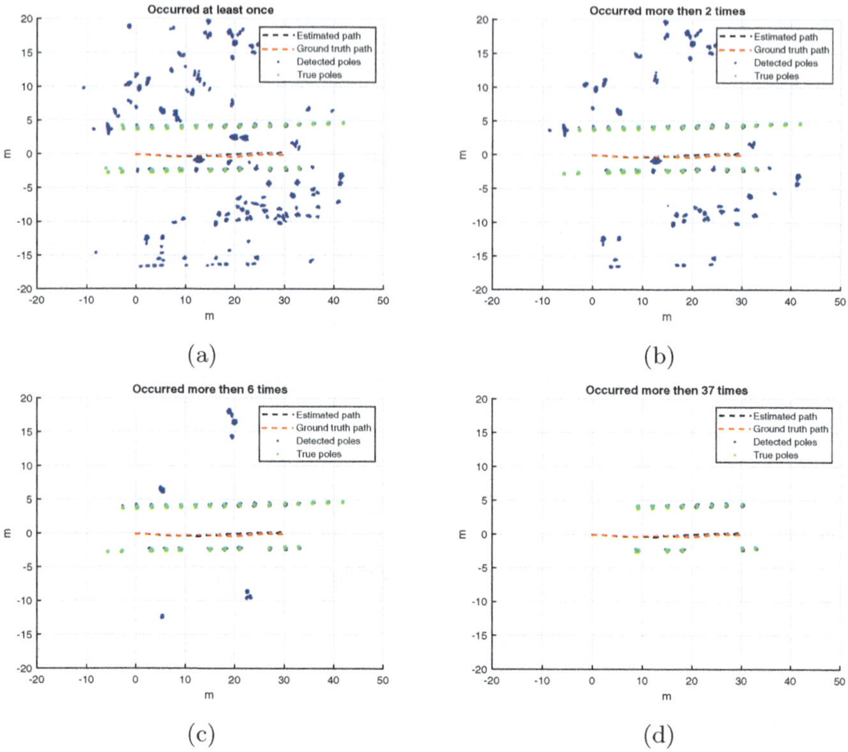

Fig. 6. Global map showing the estimated path, the ground-truth, and the detected poles that occurred at: (a) least once; (b) least twice; (c) least 6 times; (d) least 37 times.

6 Conclusion and Future Work

In this work, a novel learning pipeline based on a semi-supervised approach for pole-like object classification and landmark selection was proposed with the aim of leveraging pole-like objects in urban environments to improve LiDAR-based localization. By using the localization feedback to update the semi-supervised learning model, the reported results show a clear improvement when compared to the baseline. The results demonstrate that the localization performance is strongly affected by the selection module which, when properly tuned, can help the Map Matching module.

For future work we plan to incorporate more stationary objects (e.g., trees, traffic signs) in order to increase the generalization capability of the proposed approach.

Acknowledgements. This work was supported partially by the project MATIS (CENTRO-01-0145-FEDER-000014) co-financed by the European Regional Development Fund (FEDER) through of the Centro Regional Operacional Program (CENTRO2020), Portugal. It was also partially supported by the University of Coimbra, Institute of Systems and Robotics (ISR-UC) and FCT (Portuguese Science Foundation) through grant UID/EEA/00048/2019.

References

1. Bergström, P., Edlund, O.: Robust registration of point sets using iteratively reweighted least squares. Comput. Optim. Appl. **58**(3), 543–561 (2014)
2. Bresson, G., Alsayed, Z., Yu, L., Glaser, S.: Simultaneous localization and mapping: a survey of current trends in autonomous driving. IEEE Trans. Intell. Veh. **2**(3), 194–220 (2017)
3. Chiotellis, I., Zimmermann, F., Cremers, D., Triebel, R.: Incremental semi-supervised learning from streams for object classification. In: 2018 IEEE/RSJ International Conference on Intelligent Robots and Systems (IROS), pp. 5743–5749 (2018)
4. Ester, M., Kriegel, H.P., Sander, J., Xu, X., et al.: A density-based algorithm for discovering clusters in large spatial databases with noise. In: KDD, pp. 226–231 (1996)
5. Garrote, L., Torres, M., Barros, T., Perdiz, J., Premebida, C., Nunes, U.J.: Mobile robot localization with reinforcement learning map update decision aided by an absolute indoor positioning system. In: 2019 IEEE/RSJ International Conference on Intelligent Robots and Systems (IROS) (2019)
6. Gaylor, D.E., Lightsey, E.G., Key, K.W.: Effects of multipath and signal blockage on GPS navigation in the vicinity of the international space station (ISS). Navigation **52**(2), 61–70 (2005)
7. Geiger, A., Lenz, P., Stiller, C., Urtasun, R.: Vision meets robotics: the KITTI dataset. Int. J. Robot. Res. (IJRR) **14**(3), 195–210 (2013)
8. Jacobson, A., Zeng, F., Smith, D., Boswell, N., Peynot, T., Milford, M.: Semi-supervised SLAM: leveraging low-cost sensors on underground autonomous vehicles for position tracking. In: 2018 IEEE/RSJ International Conference on Intelligent Robots and Systems (IROS), pp. 3970–3977 (2018)
9. Kampker, A., Hatzenbuehler, J., Klein, L., Sefati, M., Kreiskoether, K.D., Gert, D.: Concept study for vehicle self-localization using neural networks for detection of pole-like landmarks. In: International Conference on Intelligent Autonomous Systems, pp. 689–705. Springer (2018)
10. Lam, J., Kusevic, K., Mrstik, P., Harrap, R., Greenspan, M.: Urban scene extraction from mobile ground based LIDAR data. In: Proceedings of 3DPVT, pp. 1–8 (2010)
11. Li, Y., Wang, W., Tang, S., Li, D., Wang, Y., Yuan, Z., Guo, R., Li, X., Xiu, W.: Localization and extraction of road poles in urban areas from mobile laser scanning data. Remote Sens. **11**(4), 401 (2019)
12. Ordóñez, C., Cabo, C., Sanz-Ablanedo, E.: Automatic detection and classification of pole-like objects for urban cartography using mobile laser scanning data. Sensors **17**(7), 1465 (2017)
13. Schaefer, A., Büscher, D., Vertens, J., Luft, L., Burgard, W.: Long-term urban vehicle localization using pole landmarks extracted from 3-D lidar scans (2019)

14. Schipperijn, J., Kerr, J., Duncan, S., Madsen, T., Klinker, C.D., Troelsen, J.: Dynamic accuracy of GPS receivers for use in health research: a novel method to assess GPS accuracy in real-world settings. Front. Public Health **2**, 21 (2014)
15. Sefati, M., Daum, M., Sondermann, B., Kreisköther, K.D., Kampker, A.: Improving vehicle localization using semantic and pole-like landmarks. In: 2017 IEEE Intelligent Vehicles Symposium (IV), pp. 13–19 (2017)
16. Teichman, A., Thrun, S.: Group induction. In: 2013 IEEE/RSJ International Conference on Intelligent Robots and Systems (IROS), pp. 2757–2763 (2013)
17. Weng, L., Yang, M., Guo, L., Wang, B., Wang, C.: Pole-based real-time localization for autonomous driving in congested urban scenarios. In: 2018 IEEE International Conference on Real-time Computing and Robotics (RCAR), pp. 96–101 (2018)
18. Williams, T., Alves, P., Lachapelle, G., Basnayake, C.: Evaluation of GPS-based methods of relative positioning for automotive safety applications. Transp. Res. Part C Emerg. Technol. **23**, 98–108 (2012)
19. Zhu, X., Ghahramani, Z.: Learning from labeled and unlabeled data with label propagation. Tech. rep. CMU-CALD-02-107, Carnegie Mellon University (2002)
20. Ziegler, J., Bender, P., Schreiber, M., Lategahn, H., Strauss, T., Stiller, C., Dang, T., Franke, U., Appenrodt, N., Keller, C.G., et al.: Making bertha drive–an autonomous journey on a historic route. IEEE Intell. Transp. Syst. Mag. **6**(2), 8–20 (2014)

Detection of Road Limits Using Gradients of the Accumulated Point Cloud Density

Daniela Rato(✉) and Vitor Santos

DEM, IEETA, University of Aveiro, Aveiro, Portugal
{danielarato,vitor}@ua.pt

Abstract. Detection of road curbs and berms is a critical concern for autonomous vehicles and driving assistance systems. The approach proposed in this paper to detect them uses a 4-layer LIDAR placed near to the ground to capture measurements from the road ahead of the car. This arrangement provides a particular point of view that allows the accumulation of points on vertical surfaces on the road as the car moves. Consequently, the point density increases in vertical surfaces and stays limited in horizontal surfaces. A first analysis of the point density allows to distinguish curbs from flat roads, and a second solution based on the gradient of point density not only detects curbs as well but also detects berms due to the transitions of the gradient density. To ease and improve the processing speed, point clouds are flattened to 2D and traditional computer vision gradient and edge detection techniques are used to extract the road limits for a wide range of car velocities. The results were obtained on the ATLASCAR real system, and they show good performance when compared to a manually obtained ground truth.

Keywords: LIDAR · Road curbs · Point clouds · Occupancy grid

1 Introduction

Autonomous Driving (AD) and driving assistance systems face very complex challenges both at control and perception levels. The detection of road limits is certainly one of the most important, and research in the field is very active but still incomplete. One specific problem concerns the detection of curbs and road berms using onboard sensors.

The approach in this paper proposes a solution to these problems using a 4 layers 3D Light Detection And Ranging (LIDAR) mounted on the front of a car, close to the ground. LIDAR sensors measure the distance to the closest object allowing the creation of a point cloud with three-dimensional points corresponding to the intersection of laser beams with the objects ahead.

This activity appears in the context of the Atlas project that started at the University of Aveiro in 2003. The main focus of this project is the development of advanced sensing and active systems designed for implementation in automobiles and similar platforms [1].

M. F. Silva et al. (Eds.): ROBOT 2019, AISC 1092, pp. 267–279, 2020.
https://doi.org/10.1007/978-3-030-35990-4_22

The full scale project evolved into the *AtlasCar* vehicle, constituting a prototype for research on Advanced Driver's Assistance Systems (ADAS) [1]. The vehicle was able to perceive its surroundings [2] and react accordingly using multiple actuators to move its mechanical components. After severe modifications and years of wear, in 2015 the original car was replaced by the *AtlasCar2*. This new prototype is based on a *Mitsubishi i-MiEV* and brought a new world of possibilities to the project, namely due to being electric and doesn't need gear changing, making the autonomous control potentially easier and also allowing to obtain power directly from the batteries for the sensors and computers.

Figure 1 shows the multiple sensors mounted on *AtlasCar2* to assist the study on AD. The most relevant sensor for the work of this paper is the SICK LD-MRS4000001, a 4-layer 3D LIDAR prepared for outdoor usage. In the current setup, the sensor is defined to output data at 50 Hz with an angular resolution of 0.5° and an aperture of 85°. Novatel SPAN-IGM-A1 and Novatel GPS-702-GG Dual-Frequency Antenna, not visible in the figure, mounted on top of the car, are also used to localize the car and its frames in relation to the world, allowing the road reconstruction. The Inertial Measurement Unit (IMU) and Global Positioning System (GPS) combo constitute a powerful and precise global positioning solution. To calculate the inclinometry of the car, the SICK DT20 Hi sensors are used with a refresh rate of up to 400 Hz, measuring distances between 200.0 ± 0.3 mm and 700.0 ± 0.3 mm.

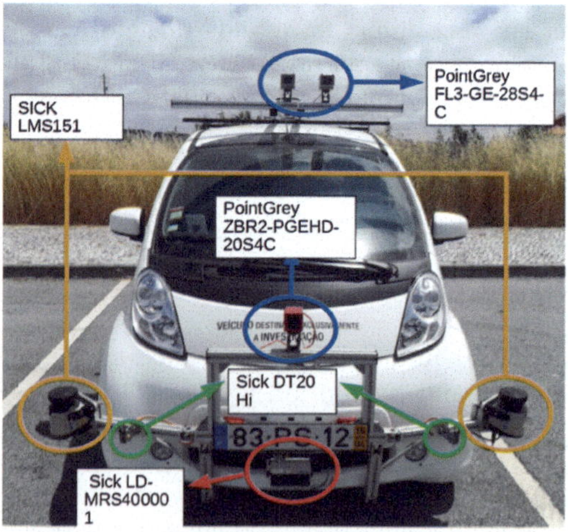

Fig. 1. Placement of *AtlasCar2* sensors.

In the previous *AtlasCar*, the accumulation of laser readings with the car movement produced interesting results in real-time road limits detection/reconstruction ahead of the car Laser readings can be converted into point clouds to which algorithms are applied, resulting in the identification of road curbs or similar obstacles at road level.

The advantage of placing the sensor in front of the car, unlike most solutions, is to focus more in objects at road level and to have a detailed perspective near the ground and not a general vision of the road with bigger surroundings that most times don't affect driving. In this way, road curbs, holes, and similar obstacles are identified with more precision.

The general idea proposed is, not only to take advantage of point cloud density but also of the density gradient using an edge detection tool and other edge detection filters like *Sobel* and *Canny* to identify density changes that usually define obstacles. This technique is a step ahead of density analysis taking into account that it considers both positive and negative variations of density, allowing to also identify negative obstacles.

2 Related Work

When it comes to road detection and identifying the navigable road limits, there are two different approaches: camera-based methods and LIDAR-based methods. The first is more affected by external factors as weather conditions, etc.; on the other hand, the second technique represents a more robust system yet more difficult to work with. The ideal solution would be to incorporate both techniques and compare the results to take advantage of each strength.

When it comes to road limits or curb detection using LIDAR, there are numerous different methodologies. Contrarily to the proposed solution in this paper, most methodologies take advantage of 360° LIDAR on top of the car, like [5,7,9,10], allowing to detect higher objects like trees and other cars, however, showing a lack of precision in ground level objects.

Many methods also take advantage of elevation profiles, by filtering objects according to their elevation difference and grouping the ones with similar values. For example, in [4], which uses a roof-mounted 3D LIDAR, a prediction method is used to find the height difference between two points and create an elevation map with the predicted measures.

Also with a 360° 3D-LIDAR, Zhang et al. in [10] use a sliding beam method for road segmentation and a search based method applied to detect the curbs in each frame. The sliding beam method contains two sections: one bottom layer beam model that allows to find the beam angles, that represent the road direction in the current location of the vehicle in a beam zone inside the region of interest, and a top layer beam model, that allows to build segments of road ahead of the vehicle. After this, spatial features of curbs are defined and extracted; the features' thresholds are defined and the search-based method is applied to detect curbs in each segment of the road.

In the end, the algorithm checks if the parameters are within the thresholds and makes the decision if a point belongs or not to a road curb. This method proposed by Zhang et al. proves to have more efficient results when compared with similar approaches.

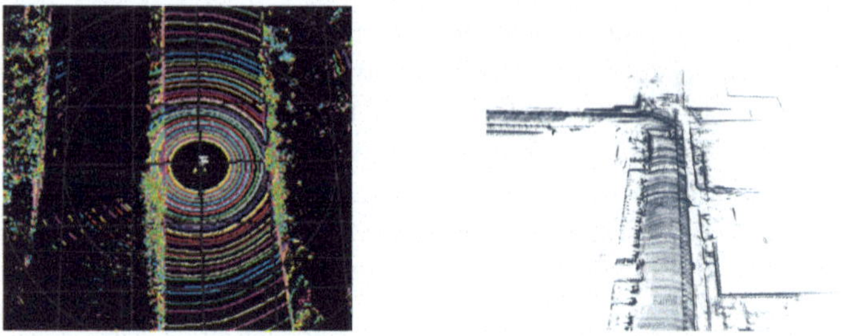

(a) Point cloud map obtained with a *Velodyne* [4].

(b) Point cloud accumulated with a SICK-LDMRS.

Fig. 2. The difference of perspective between a *Velodyne* and the used LIDAR mounted on the front of the car.

Xu et al. in [8] use a similar approach to the one studied, based on a density gradient of point clouds. The methodology consists of calculating the difference of density in adjacent `voxels` in 2D and then adding the 3rd dimension as the difference of elevation between `voxels`. The road classification is then based on three principles: `voxel` within the surface (one large gradient), *voxel* intersecting two surfaces (two large gradient and `voxel` in the intersection of three mutually non-parallel surface (three large gradient).

Concerning this work, Xu et al.'s approach is the one that comes the closest to the proposed work, by also using point cloud density and gradients. However, Xu's methodology diverges by adding the third component of gradient as elevation differences and follows a completely different path by doing road classification and identifying features as opposed to focusing on road limits.

Also, all the described methodologies use a 3D 360° LIDAR, offering a completely different perspective, not so close to the car and focusing on a wider view with bigger obstacles like tree or people (Fig. 2(a)). The LIDAR used in the proposed method only has a field of view 85° and is placed to cover objects close to the car and at ground level, like holes, depressions, road curbs, etc. (Fig. 2(b)).

3 Proposed Approach

The main goal of this study is the establishment of navigable road limits. Logically, numerous different approaches can lead to road limits detection, but through the years, in the Atlas Project, the accumulation of laser readings (i.e. point clouds) with the car movement proved to be an interesting and less studied approach This accumulation, further studied by Marques [6], allows instant decision making, identifying road obstacles on time.

To develop a new and effective methodology for road limits detection the idea was to take advantage of point cloud density to detect not only positive obstacles, defined as obstacles above the road surface like curbs, but also negative obstacles, defined as obstacles below the road surface, as depressions or inverted curbs.

Although other studies have also applied gradient filters, this method takes it a step further by applying edge detection techniques to flattened point clouds.

3.1 Initial Idea

To reconstruct the environment around the car as it navigates, a processing Robot Operating System (ROS) `node` was developed to gather the laser scan readings from the SICK LD-MRS sensor and accumulate them to greatly increase the 3D information, forming a point cloud representative of the perceived environment.

The information of a point cloud of a single reading proves to be insufficient to consistently extract features about the road and the environment. To counteract this effect the number of data points used to represent the road is increased, by taking advantage of the movement of the vehicle. Since the sensor is accompanying the car movement during the measurement process, it registers different road sections each time it moves along the vehicle's trajectory. For example, with a scanning frequency of 50 Hz, if the car is moving at 50 km h^{-1} the sensor performs a measurement, on all four scanning planes, every 0.29 m. By accumulating these successive measurements it is possible to form a point cloud that represents the road and its features with good accuracy and density to allow to extract the road boundaries.

To accumulate the readings gathered from the LIDAR, the `laser_assembler` ROS package was used. This package provides a service which takes successive data from a single laser scan and assembles it, relative to an input frame, in a rolling buffer of a predefined size (Fig. 3), and allowing to control the size of the road reconstruction point cloud. When a new cloud is available, the node poses a request to the service that returns the point cloud buffer for the respective laser scan. This is done for each of four laser scans, providing the `laser_assembler` service with the correct frame for each scan (from *ldmrs[0-3]*).

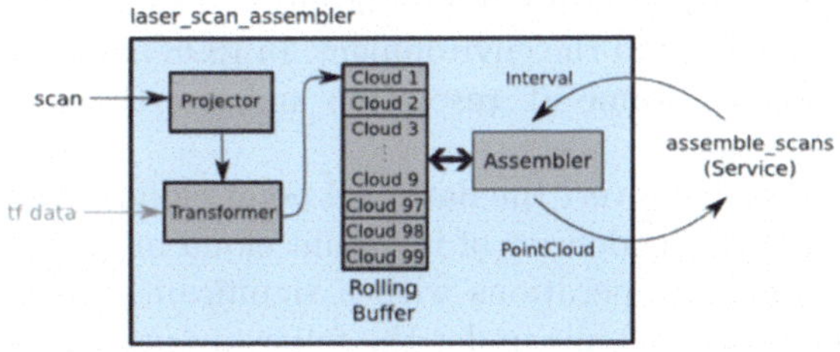

Fig. 3. Diagram of the working principle for the *laser_ assembler* service.

This accumulation method is very efficient and there is almost no delay between the acquisition of the data and its assembly to the point cloud, making it a good solution even for a fast-moving vehicle.

The accumulation node is also responsible for taking the assembled point clouds corresponding to each 4 of the scan planes and merging them into a single point cloud ready to be processed. Figure 4 illustrates the process of road scanning where the LIDAR beams appear as the colored lines leaving the car.

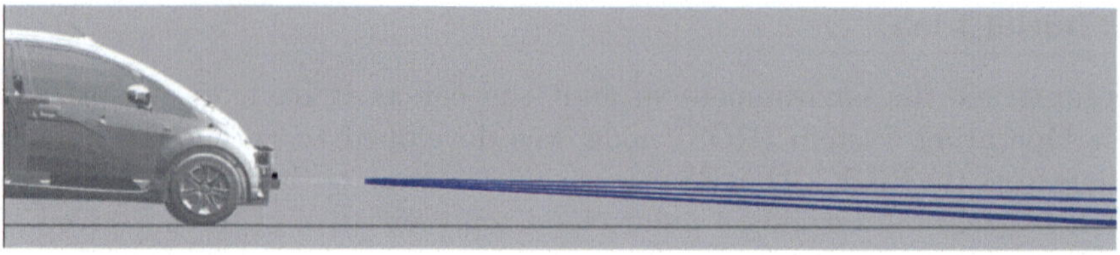

Fig. 4. Simulation of the 4 layers LIDAR mounted on the car.

The accumulated point cloud proves to be a good representation of the environment, with rich information about the road and curb profiles and allowing to reconstruct the road, due to position and orientation of the laser beams (Fig. 4) with the current placement of the LIDAR. This reconstruction also presents some interesting properties regarding the representation of the road boundaries, which can easily be identified in the image due to the higher concentration of points characteristic of these regions.

In a first approach, an algorithm is applied to the reconstructed cloud that scans each point and removes it if there are not enough neighbor points in a predefined radius. What this means is that a point is only considered a road limit if it is in a place with high point density. However, this methodology did not perform well enough in many situations, especially in the presence of negative obstacles [6].

3.2 Density Grids

The basic idea of an occupancy grid is to represent a map of the environment as an evenly spaced grid with each cell representing the likelihood of the presence of an obstacle at that location in the environment. In ROS environment, occupancy grids have an associated frame id, resolution and size that can be customized according to the situation.

The main idea was to convert the flattened point cloud (Fig. 5(b)) into a density grid (Fig. 5(c)) with the density of the point cloud in each grid cell, allowing to perform more complex operations with a significantly lower computational effort. The density grid was built under the following principles:

1. The density inside each cell is equal to the number of points within the coordinates of that cell.
2. The altitude component is not relevant due to the positioning of the sensor, thus only x and y components were considered.
3. The grid base frame is a frame placed on the center of the LIDAR and consequentially the point cloud must be transformed from the `world` frame to the target frame.
4. Since there is no advantage in considering the information behind the car front, the grid was defined with 40 m ahead of the car and 20 m to each side of the car, making a grid with a final size of 40×40 m.

Although the influence of grid resolution in the detection of road limits is discussed further on, the base resolution is considered 0.4 m/cell. Also, Fig. 5(a) shows a camera image from the camera installed in the car of the same moment where the point cloud and density grid of Fig. 5(b) and (c). In this image, it is possible to visualize the real environment of the road and the road curbs detected on the right side.

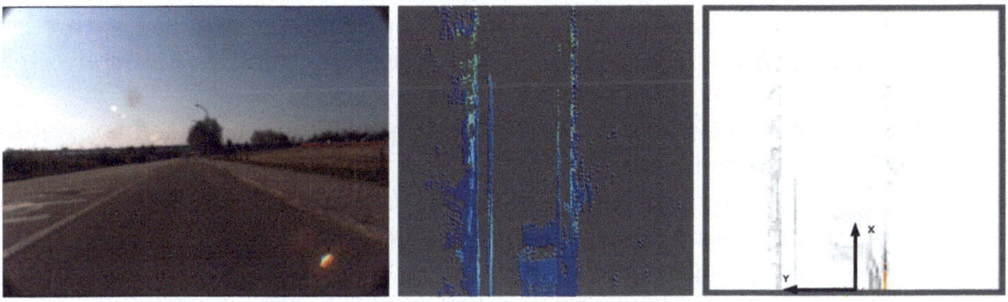

(a) Camera view (b) Point cloud of the road. (c) Density grid with frame.

Fig. 5. Camera image, point cloud and density grid with a reference frame.

To develop the occupancy grid, a ROS package was created with its corresponding `node`. In this `node`, the accumulated point cloud `topic`, created by a cloud accumulator, is subscribed and with `pcl_ros` it is transformed to the `moving_axis` frame. Then, for each point in the point cloud, the cell coordinates (line and column) to which it belongs are calculated. A `nav_msgs::OccupancyGrid` according to the described proceedings is then published with the density information in the correspondent topics. With ROS visualization tools, `rviz` or `mapviz`, the grid can be visualized along with the car movement. To standardize the values of the density in each frame, and minimize the impact of the car's velocity and variations in the RPY values, the values of the density occupancy grid are normalized to the maximum density value in each frame, returning values from 0 to 100.

3.3 Gradient and Edge Detection

With the density occupancy grid defined, the next step was to create new grids with the horizontal and vertical gradient components, and the overall gradient magnitude. The grids were created separately to evaluate the most efficient in this particular case.

To calculate the horizontal (G_y) and vertical (G_x) gradients, the simple linear filters used were, respectively, $\begin{bmatrix}-1 & 1 & 0\end{bmatrix}$ and $\begin{bmatrix}-1 & 1 & 0\end{bmatrix}^{\mathsf{T}}$. As usual, the magnitude of the gradient was used in all further calculations: $G = |G_x| + |G_y|$.

Simple gradients, however, turn out a bit limited and less sensitive, so more complex filters like *Prewitt* and *Kirsch* filters were exploited.

Logically, the more complex the filter, the more difficult the implementation, and tools like *Canny* and *Laplacian* operators were inconvenient to use due to the computational complexity of implementation applied to point clouds. The solution was to use the `GridMapRosConverter` package to easily convert the density grid in an Open Source Computer Vision Library (OpenCV) image and open a new set of possibilities for edge detection tools. This conversion has two steps: convert the occupancy grid into a `GridMap`, and convert the grid map to a `cv::Mat`, an OpenCV image, with a `CV_8UC1` encoding (8-bit unsigned type with one channel). To this image, the referred operators are applied with edge detection algorithms in the OpenCV image processing libraries and converted back into new occupancy grids through the same process.

When using edge operators, in general three parameters need to be given as input: the kernel size, odd number that defines the size of the filter, set to 3, the scale, if necessary to scale the output, and *Delta*, an optional value that is added to the results prior to storing them in a matrix, both as 0.

After applying the edge detection operator, the image must be converted to absolute values as both positive and negative gradients are relevant.

A big advantage in this process is the possibility to apply a custom threshold when converting the image back to a grid and eliminate low-interest regions in the edge detection process corresponding to road noise, etc.

3.4 Ground Truth Definition

To facilitate the process of marking the ground truth, the car coordinates are saved in each frame in a Keyhole Markup Language (KML) file, allowing to visualize the car path in the Google Earth application. For this, the GPS topics are subscribed to obtain the car latitude and longitude and XML file is created at the beginning of the program where latitude and longitude coordinates are added in each frame, with zero altitude. At the end of the program, the file handler is closed and becomes readable by Google Earth.

The process of reading the road limits in the form of ground truth is more complex, involving manually drawing the data corresponding to the right and left side of the road overlapped on satellite images. The program opens the correspondent files and identifies and saves the coordinates within the `<coordinates>` `</coordinates>` tab. Each coordinate is separated and the latitude and longitude are placed in `double` type vectors.

The coordinates placed in the World Geodetic System (WGS 84) frame need to be transformed to the correct car frame, `moving_axis` (in the front of the car). This requires the following steps:

1. Convert the car latitude and longitude to the Universal Transverse Mercator (UTM) frame.
2. Convert every road limit point to the UTM frame.
3. With both points in a metric scale, calculate the difference between each point coordinates and the car coordinates.

4. Rotate the obtained coordinate to the correct frame orientation (see Fig. 5(c)) by performing a z rotation correspondent to the yaw (azimuth in the GPS message - clockwise rotation) (Eq. 1).
5. Add 2.925 m to the x coordinate of each point, correspondent to the translation from the GPS frame to the LIDAR frame.

$$\begin{bmatrix} x_{correct} \\ y_{correct} \\ 1 \end{bmatrix} = \begin{bmatrix} \cos(yaw) & \sin(yaw) & 0 \\ -\sin(yaw) & \cos(yaw) & 0 \\ 0 & 0 & 1 \end{bmatrix} \begin{bmatrix} x_{utm\_point} - x_{utm\_car} \\ y_{utm\_point} - y_{utm\_car} \\ 1 \end{bmatrix} \tag{1}$$

To create a continuous line, an interpolation function was implemented to create new points within the lines, assuming a straight line between each one. These limits are then filled to comprehend the navigable space of the road.

4 Tests and Results

4.1 Qualitative Evaluation

Images in Fig. 6 show a representation of the navigable space obtained from each edge detector at the same moment.

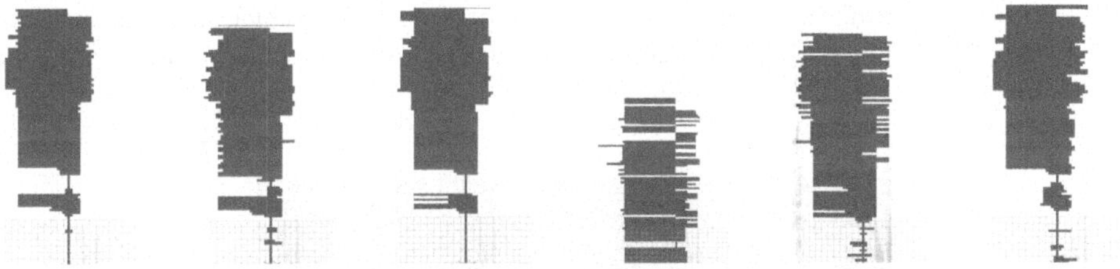

(a) *Laplace* (b) *Prewitt* (c) *Sobel* (d) *Canny* (e) *Grad.* (f) *Kirsch*

Fig. 6. Results of several edge detection grids in the identification of navigable space in a straight road situation.

By analysing the images, it can be concluded that:

- Although being the only one that results relatively well close to the car, the *Canny* edge detection produces poorer results further from the car, with many gaps in the detection.
- All the edge detectors, apart from *Canny* produce poor results in the approximately 10 m following the car.
- The Gradient filter is the one with fewer gaps in the detection and more clear road apart from the initial meters.
- The *Laplace* and *Prewitt* algorithms produce similar results with some gaps in the middle of the road.
- *Sobel* and *Kirsh* filters have defined road but further away from the car than the rest of the algorithms.

4.2 Quantitative Evaluation

Statistical indicators like Precision $= \frac{TP}{TP+FP}$, Sensitivity $= \frac{TP}{TP+FN}$, $NPV = \frac{TN}{TN+FN}$, F-measure $= (1+\beta^2)\frac{\text{Precision}\times\text{Sensitivity}}{\beta^2\times\text{Precision}+\text{Sensitivity}}$, Accuracy $= \frac{TP+TN}{TP+FP+TN+FN}$ and Specificity $= \frac{TN}{TN+FP}$ are commonly used to quantify performance of automotive applications [3]. These indicators are based on a binary classification of trues and positives. A cell is considered positive if it inside the navigable space, and negative if not, thus a true positive (TP) is a cell that is correctly identify as road, a false positive (FP) is a cell misidentified as road, a true negative (TN) is a cell correctly identified out-of-limits and a false negative (FN) is a misidentified out-of-limits cell. Also F-measure, β measures the importance ratio between Precision and Sensitivity, and for this application it was considered $\beta = 1$.

Concerning this work, the most relevant parameters are the F-measure and Accuracy, as these include the global performance of the algorithm in terms of detection of both positives and negatives. The remaining parameters are less relevant but are also taken into consideration, especially if they present particularly low values that may indicate an unreliable algorithm.

With the quantitative evaluation method, the influence of the `Occupancy Grid` resolution and the algorithm's threshold were also tested. The threshold can be applied when converting *OpenCV* images to Occupancy Grids and means removing the filtered values below that number, which is useful to remove low values corresponding to road-noise, etc. To perform these evaluations, several paths were used accordingly to their characteristics. An example of a path is shown in Fig. 7.

Fig. 7. Satellite view of a path used to evaluate the performance of algorithms. In red, the GPS path of the car. In green, the ground truth road limits.

Concerning the algorithm's threshold, the *Simple Gradient* and *Canny* work better with 0 threshold, the *Prewitt* and *Sobel* with 25 and the *Laplace* and *Kirsch* with 50. This study is relevant to eliminate lower values that correspond to road or filtering noise. The threshold value varies from filter to filter as some are more susceptible to noise, like *Kirsch*, than others, like *Gradient*. For this reason, the study was performed individually for each algorithm to optimize individual performance.

By the analysis of Fig. 8, showing a graphic of the influence of cell resolution in the performance of the *Simple Gradient* edge detector, it can be concluded that a resolution of 0.4 m/cell is the one that produces the best results. Given that a pedestrian can be fitted in a square of about 0.5 × 0.5 m, a resolution of 0.4 m is considered enough to even identify a person in the road. Logically, if there is a thin object in the road, like a road pin, this resolution can dissolve the points in that cell and lead to the non-identification of the obstacle, but there has to be a trade-off between a very thin resolution that results in worse outcome and a size that can still detect small objects.

One of the major concerns with this work was its robustness to the different car speeds. To evaluate this, the same road was tested at different speeds, evaluating the system in about 200 m. Logically, the number of frames evaluated varies according to the car velocity since lower speed results in more time to travel the same distance and, consequently, more frames. Figure 8 shows the results of the algorithm analysis by varying car speed from 20 to 60 km h^{-1}.

Analysing the graphic, it can be concluded that the performance remains more or less constant until 50 km h^{-1}. From this value up the performance of the algorithm starts decreasing but still above 50% of performance. Note that tests in the same road could not be performed at higher velocities due to the speed limit in that zone.

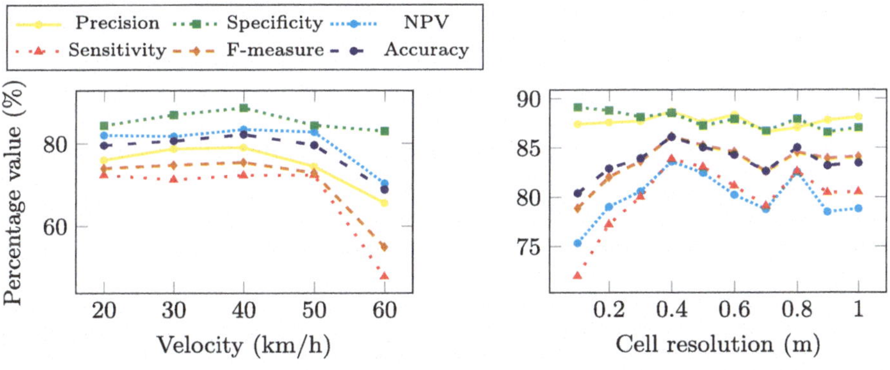

Fig. 8. Influence of car velocity and cell resolutions, in the statistical indicators for Simple Gradient algorithm with 0.4 m/cell and no threshold (10 to 30 m).

Table 1 shows the results of the best parameters' combination for each algorithms with the chosen thresholds, car speed, and cell resolution values within the tested range. Simple Gradient and the *Laplace*, followed by *Kirsch* filter, have the best results. *Canny* edge detector, for the reasons already mentioned, is not suitable for this application and the *Sobel* and *Prewitt* filters, very similar to each other, although having good results, fall short of the best algorithms.

Table 1. Algorithms performance with the chosen parameters (values in %).

Filter	Precision	Specificity	NPV	Sensitivity	F-measure	Accuracy
Laplace	88	88	84	85	87	86
Gradient	89	89	83	83	86	86
Sobel	87	87	80	79	83	83
Prewitt	87	88	77	74	80	81
Kirsh	86	86	84	85	85	85
Canny	87	91	67	55	66	73

5 Conclusion

This paper proposes an effective way to correctly identify road limits and road navigable space. The argument was that the density of the accumulated cloud, obtained from car movement with an 85° range 3D LIDAR, is a new methodology to perform road perception. Unlike most of the work in this area, using 360° Velodyne LIDARs, seeing the road at ground level offers a completely different perspective, possibly allowing to make real-time driving decisions given the identified road limits.

To solve the problem, the used approach was to transform accumulated point clouds in a density occupancy grid, calculate a two-dimensional gradient and subsequently apply edge detection algorithms to find density patterns that define obstacles. One of the biggest innovations, when compared with previous work done in the *AtlasCar2*, is the ability to detect both positive and negative obstacles due to considering density increases and decreases. Regarding the edge detection, the Simple Gradient showed the best performance, although all algorithms except *Canny* have good results in general. The developed work also showed decent performance at different speed, only decreasing in performance at higher speeds.

Although this system alone may not be enough to provide data for a navigation system, when integrated with the other sensors in the *AtlasCar2*, it provides essential navigational information of hard limits, leaving open a wide range of possibilities. On the one hand, road detection for navigational purposes can be largely improved by combining it with lane detection using cameras (soft limits) and creating a multi-sensorial algorithm with the possibly to create a common navigation `Occupancy Grid` with different levels of probability according to the detected features (lane less serious, hard limits more serious, etc.). Fusing the results of several edge detection algorithms may also be interesting to obtain more complete and robust information.

Acknowledgements. This work was partially supported by project UID/CEC/00127/2019.

References

1. de Aveiro, U.: ATLAS project. http://atlas.web.ua.pt/index.html
2. Azevedo, R.: Sensor fusion of laser and vision in active pedestrian detection. http://hdl.handle.net/10773/14414 (2014)
3. Fritsch, J., Kuhnl, T., Geiger, A.: A new performance measure and evaluation benchmark for road detection algorithms. In: 16th International IEEE Conference on Intelligent Transportation Systems, ITSC 2013. pp. 1693–1700. IEEE (2013)
4. Huang, R., Chen, J., Liu, J., Liu, L., Yu, B., Wu, Y.: A practical point cloud based road curb detection method for autonomous vehicle. Information **8**, 93 (2017)
5. Jung, J., Bae, S.H.: Real-time road lane detection in urban areas using LiDAR data. Electronics **7**(11), 276 (2018)
6. Marques, T.: Detection of road navigability for ATLASCAR2 using LIDAR and inclinometer data (2017). http://lars.mec.ua.pt/public/LAR%20Projects/Perception/2018_TiagoMarques/TMarques_dissertation.pdf
7. Peterson, K., Ziglar, J., Rybski, P.E.: Fast feature detection and stochastic parameter estimation of road shape using multiple LIDAR. In: 2008 IEEE/RSJ International Conference on Intelligent Robots and Systems, pp. 612–619 (2008)
8. Xu, S., Wang, R., Zheng, H.: Road curb extraction from mobile LiDAR point clouds. IEEE Trans. Geosci. Remote Sens. **55**(2), 996–1009 (2017)
9. Zai, D., Li, J., Guo, Y., Cheng, M., Lin, Y., Luo, H., Wang, C.: 3-D road boundary extraction from mobile laser scanning data via supervoxels and graph cuts. IEEE Trans. Intell. Transp. Syst. **19**, 802–813 (2018)
10. Zhang, Y., Wang, J., Wang, X., Dolan, J.M.: Road-segmentation-based curb detection method for self-driving via a 3d-LiDAR sensor. IEEE Trans. Intell. Transp. Syst. **19**(12), 3981–3991 (2018)

Autonomous Sailboats and Support Technologies

Force Balances for Monitoring Autonomous Rigid-Wing Sailboats

Matias Waller[1(✉)], Ulysse Dhomé[2], Jakob Kuttenkeuler[2], and Andy Ruina[3]

[1] Åland University of Applied Sciences, Mariehamn, Åland Islands
matias.waller@ha.ax
[2] KTH Royal Institute of Technology, Stockholm, Sweden
[3] Cornell University, Ithaca, USA

Abstract. Real-time monitoring and evaluation of practical trials with autonomous sailboats is often challenging: varying winds and waves influence a visual evaluation directly connected to a subjective verification of expected mechanical behavior while the inherent nature of sailing with different sensors operating in different coordinate frames complicates an objective data-based evaluation. In this paper, we illustrate the use of force balances as a tool for monitoring performance, for estimating and evaluating measurements and detecting unreliable sensors.

Keywords: Autonomous sailboat · Wingsail · Model-based data evaluation · Fault detection · Force balance

1 Introduction

Projects with autonomous rigid-wing sailboats often involve a range of practical challenges including design and construction of wingsails and electronic systems for sensors and control. Consequently, practical trials are used for, e.g., verifying mechanical behavior, control robustness and performance, sensor reliability, etc. Furthermore, projects rely on engaged students and can form part of exam requirements with a related emphasis on different aspects of the operation. As is often the case with projects of similar nature, deadline for meeting exam requirements approaches during practical experiments and, accordingly, time for more extensive quantitative reflection and possibilities for repeated trials might be limited. For the long-term goal of releasing a fleet of small autonomous sailboats for collecting data over long time-periods in waters with significant currents, uncertainties in expected sailing performance are important to consider for successful planning of missions. In this paper, we propose an approach based on basic sailing theory and apply it to data from two different rigid-wing sailing robots. Possibilities for evaluating sailing performance and sensor reliability as well as automatic fault detection and diagnosis are discussed.

The study is limited to rigid wingsails for two reasons. Firstly, we have access to data from two different sailing robots with rigid wings. Secondly, the aerodynamics for rigid wings are hopefully more well-defined than for soft sails. Accordingly, experiments should be more repeatable for rigid-wing sailboats.

M. F. Silva et al. (Eds.): ROBOT 2019, AISC 1092, pp. 283–294, 2020.
https://doi.org/10.1007/978-3-030-35990-4_23

The basis for our analysis is the balance of aerodynamic and hydrodynamic forces at equilibrium, i.e., when speed and direction of boat, wind and any possible currents are approximately constant. Thus, data at equilibrium is used for comparing measurements with behavior expected based on the models. For the approach, mechanical and geometrical information about the hull and the wingsail are needed in addition to the following six "standard" variables:

1. Heading, θ (in degrees): the direction of the longitudinal axis of the boat in earth frame. Ideally, a measure of heading will be provided by the compass (or the gyro).
2. Course over water, ϑ (in degrees): the direction of the movement of the boat relative to water in earth frame. In the absence of currents, this will be the same as course over ground provided by the (D)GPS. The difference between heading and course over water, $\lambda = \theta - \vartheta$, is known as leeway.
3. Boat speed over water, v_b (in m/s): the speed of the boat relative to the water. In the absence of currents, this will be the same as speed over ground provided by the (D)GPS. Note that this includes speed in the direction of the heading (sometimes called surge) as well as side speed (sway).
4. Apparent wind angle, β_{AWA} (in degrees): The angle of the wind relative to the heading of the boat as it is perceived on the boat. By convention, 0 to 180° indicates starboard and 0 to −180° port. Typically, this is measured by a wind vane physically installed relative to the heading of the boat.
5. Apparent wind speed, v_w (m/s): The speed of the wind relative to the boat, i.e., the speed as it is perceived on the boat. Typically measured by an anemometer.
6. Angle of attack for the wingsail, α (in degrees).

As expected, and accordingly confirmed by our analysis, absolute values for heading and course over water are not relevant for the models, i.e., only the leeway. Primarily, leeway defines the angle of attack for the keel. Following the same practise as for apparent wind angle, a negative leeway will indicate a course over water to port. For illustrating the presented approach, four sets of data collected with the two autonomous sailing robots, ASPire, described in, e.g., [3] and Maribot Vane, described in, e.g., [1] are used. ASPire (upper left) and Maribot Vane (upper right) along with locations for collecting sailing data are depicted in Fig. 1. Results are presented after the theoretical basis has been introduced.

2 Theoretical Basis for Evaluating Measurements

The basis for the presented approach is to compare measurements with the behavior that would be expected based on static models of the sailboats, in part based on geometrical properties. The ASPire and Maribot Vane have similar hulls based on the International 2.4mR sailing class. The profile for the keel was chosen as NACA0012 based on a 3-D scan of the hull of Maribot Vane. For Maribot Vane, the wingsail is based on the NACA0018 profile while ASPire

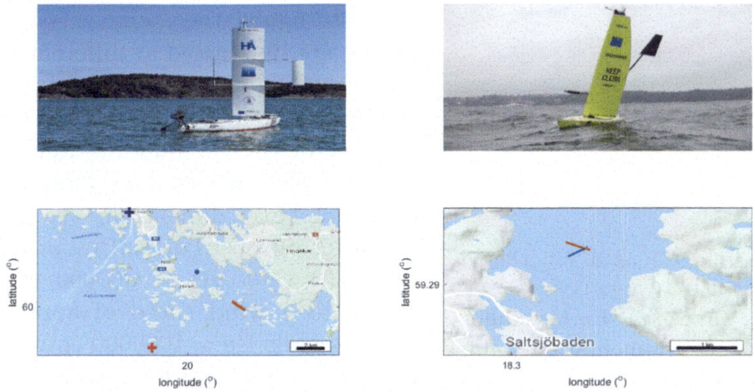

Fig. 1. ASPire (upper left, © Anna Friebe) and Maribot Vane (upper right, © Ulysse Dhomé) and maps of sailing (lower panels). Lower left: ASPire dataset 1 (blue dot), dataset 2 (red line). Västra Hamnen (blue cross) and Nyhamn (red cross) are two Finnish Meteorological Institute weather stations. Lower Right: Maribot Vane dataset 1 (blue) and dataset 2 (red)).

uses two sides of the asymmetrical NACA63$_2$-618 to make it symmetrical. For simplicity, NACA0018 is used for all calculations for both wingsails. Apart from the size of the wingsails, which is smaller on ASPire, the models for the two sailing robots are thus identical. For clarity of presentation, the forces acting on the sailboat are divided into aerodynamic and hydrodynamic forces.

2.1 Aerodynamic Force

The aerodynamic force, $\vec{F}_a$, is the result of a (symmetric) wingsail set at an angle of attack towards the apparent wind. It is decomposed into lift, $F_{l,a}$, perpendicular to the apparent wind, and drag, $F_{d,a}$, in the direction of the apparent wind. It is assumed that these can be calculated by standard expressions for lift and drag for airfoils, i.e.,

$$F_{l,a} = \frac{1}{2}\rho_a C_l A_s v_w^2 \tag{1}$$

and

$$F_{d,a} = \frac{1}{2}\rho_a C_d A_s v_w^2 \tag{2}$$

where ρ_a is the density of air and A_s is the wingsail area. The coefficients for lift and drag, C_l and C_d, will primarily depend on the profile of the wing and the angle of attack and, to some extent, on v_w. For the NACA0012 and the NACA0018 profiles, experimental results for C_l and C_d as a function of angle of attack are illustrated in Fig. 2.

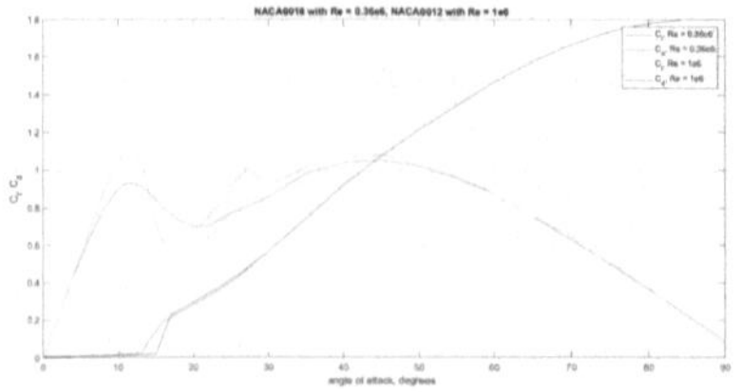

Fig. 2. C_l and C_d as a function of angle of attack ($C_l(-x) = -C_l(x)$ and $C_l(-x) = C_d(x)$) for two different profiles. Experimental data presented in [4].

In the figure, the coefficients for the two profiles are illustrated at different Reynolds numbers. The Reynolds number is defined by $\text{Re} = vL/\mu$, where μ is the dynamic viscosity of the surrounding fluid, v is the speed of the surrounding fluid, and L is a characteristic dimension, i.e., the chord length for airfoils. For ASPire and Maribot Vane, the value $\text{Re} = 360000$ thus corresponds to a wind speed of approximately 6–7 m/s while $\text{Re} = 1000000$ corresponds to a water speed of approximately 1.5 m/s. In order to calculate lift and drag for the wingsail, the only measurements needed are thus apparent wind speed and the angle of attack for the sail. For the keel, the corresponding measurements needed are boat speed and leeway.

Wingsails often operate at quite small angles of attack in order to avoid the stalling region. Typically, leeway is also often small. In such cases, a common approach is to use a linear model for C_l and possibly quadratic for C_d. In the current paper, however, we are interested in what a model will predict if, for example, a large angle of attack or a large leeway is measured. Therefore, realistic values for C_l and C_d at large angles of attack are also needed and the simplified models are not sufficient.

2.2 Hydrodynamic Forces

Similarly to the aerodynamic force, the hydrodynamic force, $\vec{F}_h$, can also be decomposed into lift and drag, $F_{l,h}$ and $F_{d,h}$. These are assumed to be given by

$$F_{l,h} = \frac{1}{2}\rho_w C_{l,k} A_k v_b^2 \tag{3}$$

perpendicular to the movement through water and

$$F_{d,h} = F_{\text{wave}} + \frac{1}{2}\rho_w (C_R A_w + C_{d,k} A_k) v_b^2 \tag{4}$$

in the direction of the movement through water. Note that due to leeway, this direction does not correspond to the heading of the boat. The density of water is ρ_w while $C_{l,k}$ and $C_{d,k}$ are the coefficients for lift and drag for the keel, A_k is the keel surface area, A_w is the wetted area of the hull, C_R is a skin friction coefficient and F_{wave} is wave resistance due to the waves generated by the boat when moving through water and not external waves. C_R and F_{wave} are estimated based on v_b and geometric details for the hull according to the presentation in [2]. The variables needed to calculate $F_{l,h}$ and $F_{d,h}$ are thus boat speed and leeway. In the case of no currents, i.e., surrounding water is stationary relative to earth, measurements are typically provided by the (D)GPS for speed over ground (taken as boat speed v_b) and the difference between heading (provided by the compass and taken as θ) and course over ground provided by (D)GPS (taken as course over water ϑ). This gives leeway, $\lambda = \theta - \vartheta$, i.e., the underwater angle of attack for the keel central for estimating $C_{l,k}$ and $C_{d,k}$.

2.3 Simplifications

In this study, the model is limited to two dimensions, i.e., any effect of heeling is neglected. This is also partly motivated by the moderate wind speeds during data collection. In addition, it is assumed that no moment on the boat is generated by either the sail or the keel, i.e., no active rudder is needed and consequently the rudder has no effect on lift and drag. Apart from the forces on the wingsail, it is also assumed that there are no other aerodynamic forces, e.g., effect of hull over water is neglected. Also, the effect of external waves is neglected.

3 Data Assessment

3.1 Consistency Check

Since $\vec{F}_a = \vec{F}_h$ at equilibrium, an initial screening of data is achieved by calculating

$$R_F = \sqrt{\frac{F_{l,h}^2 + F_{d,h}^2}{F_{l,a}^2 + F_{d,a}^2}} \tag{5}$$

The use of the square root, although computationally more demanding, is motivated by intended use of models for real-time monitoring of forces. Ideally, R_F should be equal to one and significant deviations indicate an inconsistency between data and model. Since C_l and C_d vary with the angle of attack, and projections of $\vec{F}_a$ and $\vec{F}_h$ depend on apparent wind angle and leeway, it is computationally demanding to calculate, e.g., a wind speed that could explain a deviation of R_F from one. Improved explanations for an observed inconsistency require more detailed analysis. This is presented next, and discussed in connection to the specific data sets analyzed.

3.2 Side and Forward Forces

For an improved understanding of the data, the aerodynamic and hydrodynamic lift and drag can be considered in the same frame of reference. Given an apparent wind angle, heading and course according to previous definitions, the aerodynamic and hydrodynamic lift and drag can be projected in a forward force, F_f, in the direction of the heading of the boat, and a side force, F_s, perpendicular to the forward force. These are given by

$$\begin{aligned} F_{f,a} &= \mathrm{sgn}(\beta_{\mathrm{AWA}})\sin(\beta_{\mathrm{AWA}})F_{l,a} - \cos(\beta_{\mathrm{AWA}})F_{d,a} \\ F_{s,a} &= -\left(\mathrm{sgn}(\beta_{\mathrm{AWA}})\cos(\beta_{\mathrm{AWA}})F_{l,a} + \sin(\beta_{\mathrm{AWA}})F_{d,a}\right) \end{aligned} \tag{6}$$

for the aerodynamic force and, similarly, for the hydrodynamic force by

$$\begin{aligned} F_{f,h} &= \sin(\lambda)F_{l,h} - \cos(\lambda)F_{d,h} \\ F_{s,h} &= \cos(\lambda)F_{l,h} + \sin(\lambda)F_{d,h} \end{aligned} \tag{7}$$

In this projection, a negative sideways force should be interpreted as directed towards port while a positive sideways force is directed towards starboard. It follows that

$$\begin{aligned} F_{f,a} + F_{f,h} &= 0 \\ F_{s,a} + F_{s,h} &= 0 \end{aligned} \tag{8}$$

at equilibrium. For a numerical evaluation, it can be noted that identical aerodynamic forward and sideways forces can be obtained for different apparent wind angles, e.g., it might be that the effect of apparent wind angle of approximately 50° cannot be distinguished from that of approximately −150°. Bearing this in mind, the use of random initial values in relevant intervals is consistently explored in numerical routines.

Given the two nonlinear equations of Eq. 8, several possibilities for exploring their dependence on the listed six variables could be conceived. Given that we are interested in exploring the mismatch between measured data and our model, we could consider a general configuration of using four measurements to estimate the remaining two. This gives fifteen estimates for each variable. In addition, uncertain parameters in the model, such as coefficients for lift and drag, could also be explored. As an initial exploration, we therefore propose to use the (D)GPS measurements of course over ground and speed over ground as coupled, i.e., they are either both reliable or, possibly, both wrong. For the two sets of data discussed in detail in Sect. 3.3 along with results, this gives the estimates presented in Table 1. It can also be noted that independent meteorological data might be available for (approximately) verifying apparent wind speed and angle.

Even though the measurements of course and boat speed provided by the (D)GPS often are reliable, they will not, in the presence of currents, provide accurate estimates of leeway and speed over water needed to determine the hydrodynamic forces. Therefore, the alternative to consider them unreliable coincides with estimating a speed, v_s, and an angle, φ, of a current. These estimates are provided on the last line in Table 1.

As an alternative, it can be questioned whether it is desirable to exactly satisfy Eq. (8), since these include simplifications that render the equations approximate at best. For example, even under ideal conditions the coefficients C_l and C_d are not exactly known and may vary with time and use, the wind around the sail is not necessarily uniform and constant, mechanical inaccuracies can include twist and asymmetries, measurements and assumption of equilibrium are only approximate, etc. Therefore, it could be valuable to estimate only one of the six variables and assess the possible improvement in consistency between measurements and model thus obtained. In the general case, it will not be possible to solve Eq. (8) with only one variable. Instead, we suggest to minimize

$$V(x) = c(F_{f,a} + F_{f,h})^2 + (1-c)(F_{s,a} + F_{s,h})^2 \tag{9}$$

Table 1. Measurements (in black) and estimates (italics in blue). Dataset 2 in parentheses. First line presents measurements only with calculated leeway included for comparison in the last column. The following lines presents the two estimates obtained by solving Eq. (8) while using measurements for the other variables. The abbreviation NS implies that no numerical solution was found. The last line presents estimates of angle φ (relative to the measured course over ground) and speed v_s (relative to the measured speed over ground) for a current obtained by assuming all measurements are correct. Combining the estimated current with estimated boat speed and course over water gives the measurements of speed over ground and course over ground.

$\theta(°)$	$\beta_{\mathrm{AWA}}(°)$	v_w (m/s)	$\alpha(°)$	$\vartheta(°)$	v_b (m/s)	$\lambda(°)$
116 (115)	−112 (−96)	3.6 (5.5)	15 (15)	133 (136)	0.99 (1.3)	17 (21)
133 (136)	−112 (−96)	*1.9 (2.3)*	15 (15)	133 (136)	0.99 (1.3)	*0.03 (0.00)*
133 (136)	*26 (19)*	3.6 (5.5)	15 (15)	133 (136)	0.99 (1.3)	*0.40 (0.55)*
116 (115)	*15 (13)*	*14 (20)*	15 (15)	133 (136)	0.99 (1.3)	17 (21)
133 (136)	−112 (−96)	3.6 (5.5)	*2.3 (1.3)*	133 (136)	0.99 (1.3)	*0.04 (0.00)*
116 (115)	*NS (NS)*	3.6 (5.5)	*NS (NS)*	133 (136)	0.99 (1.3)	17 (21)
116 (115)	−112 (−96)	*NS (NS)*	*NS (NS)*	133 (136)	0.99 (1.3)	17 (21)
116 (115)	−112 (−96)	3.6 (5.5)	15 (15)	*116 (115)*	*1.9 (2.2)*	*0.02 (-0.01)*
					$\varphi(°)$	v_s (m/s)
					−35 (−48)	*0.98 (1.1)*

where x is, in turn, one of the six variables listed earlier and $0 \leq c \leq 1$ is a parameter that can be used to emphasize either forward or sideways balances. For comparing the results, $V_0(x)$ is used to denote the value obtained with no minimization, i.e., measurements are used for all variables. Results with $c = 0.5$, i.e., no emphasis on any term, are provided in Table 2. Equation (9) might also have several local minima and despite the use of random initial values for the numerical routines, no guarantee for finding the global minimum can be given.

Table 2. Estimates (italics in blue) determined by minimizing Eq. (9) for one variable while using measurements for the other variables. Underneath, the corresponding values for $V(x)/V_0(x)$ of Eq. (9) are provided. Values for dataset 2 in parenthesis.

	$\theta(°)$	$\beta_{\mathrm{AWA}}(°)$	v_w (m/s)	$\alpha(°)$	$\vartheta(°)$	v_b (m/s)
	133 (136)	*15 (-180)*	*7.4 (0.00)*	*11.00 (11)*	*116 (115)*	*0.14 (0.06)*
$V(x)/V_0(x)$	0.003 (0.004)	0.9 (0.9)	0.9 (1.0)	1.0 (1.0)	0.003 (0.004)	0.005 (0.006)

3.3 Data from ASPire

Two different datasets collected with ASPire on June 12th, 2018 are used in the study. The first set stretches over 30 s, marked with a blue dot in the lower left panel of Fig. 1, and the second set, collected about an hour later, stretches over 1000 s and is marked with a red line. The sailing waters are just south of mainland Åland in the middle of the Baltic Sea. The locations of the closest two official weather stations of the Finnish Meteorological Institute are indicated in large red cross (Nyhamn, Lemland) and large blue cross (Västra Hamnen, Mariehamn). Measurements of course, heading and apparent wind angle are illustrated as functions of time in the left panels of Fig. 3, upper for dataset 1 and lower for dataset 2. The right panels illustrate corresponding wind and boat speed. The angle of attack for the wingsail is based on mechanical inspection, and estimated to 15°.

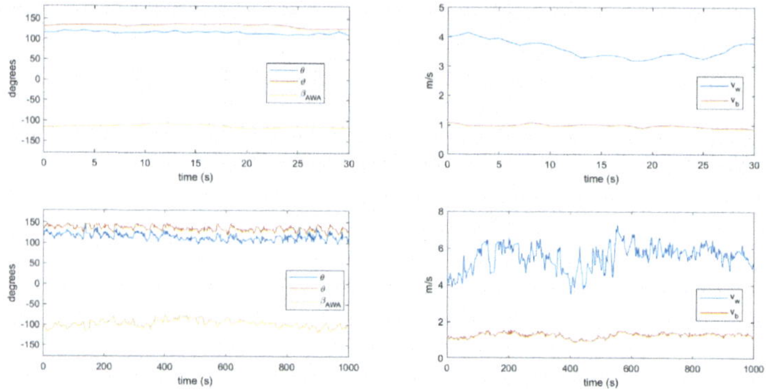

Fig. 3. Left panels: heading (blue) with a mean value of 120° (upper) and 110° (lower), course (red) with a mean value of 130° (upper) and 140° (lower), apparent wind angle (yellow) with a mean value of −110° (upper) and −100° (lower). Right panels: wind speed (blue) with a mean value of 3.6 m/s (upper) and 5.5 m/s (lower) and boat speed (red) with a mean value of 1.0 m/s (upper) 1.3 m/s (lower).

Using data from the two weather stations indicated in Fig. 1 for the same time, three ten-minute averages with similar values indicate a true wind speed between 4.2 m/s (Nyhamn) and 5.5 m/s (Västra Hamnen) and a true wind direction of 34° (Nyhamn) and 337° (Västra Hamnen) for dataset 1. The measured apparent wind speed of 3.6 m/s and apparent wind direction between 4° (using measured heading) and 23° (using measured course) thus seem reasonable. On the other hand, the measure of consistency, Eq. (5), gives $R_F = 14$ with mean values for all measurements. Using four ten-minute averages, the official weather stations indicate a true wind speed between 5.4 m/s (Nyhamn) and 4.9 m/s (Västra Hamnen) and a true wind direction of 326° (Nyhamn) and 342° (Västra Hamnen). The measured apparent wind speed of 5.5 m/s seems reasonable while the apparent wind direction between 19° (based on measured heading) and 40° (based on measured course) deviate slightly more, but can still be explained by local variations. On the other hand, $R_F = 13$ is, again, a clear indication of mismatch between model and measurements. While both sets are reasonable approximations of equilibrium, the second set is much longer and correspondingly contain larger variations. On average, the wind and boat speed are also significantly higher. The values for R_F clearly indicates mismatch between model and measurements and further exploration of the inconsistencies provides the estimates in Table 1.

How should the information in Table 1 be interpreted? Firstly, it can be noted that no solutions, denoted NS, are obtained by varying the angle of attack for the wingsail and either the apparent wind angle or apparent wind speed. This implies that the model cannot be satisfied without varying other variables. One possible solution is obtained with strong apparent wind speed (14 m/s and 20 m/s for each dataset respectively) and small apparent wind angle (15° and 13°). This solution can, in this case, be discarded for two reasons. (1) A strong apparent wind at a very small angle with a corresponding large leeway is theoretically possible, but would probably only correspond to transient, possibly unstable, behavior and thus not readily explained by our static model. (2) Based on personal observations during the sailing. This leaves four possible estimates. As can be noted, all of these correspond to a much smaller leeway, thus providing support for three possible explanations:

1. Heading measurements are unreliable.
2. Measurements of course over ground are either unreliable or do not correspond to course over water.

The presence of a current could explain why course over ground would not equal course over water. An estimate of the corresponding current is provided on the last line in the tables. In the present case, a current of such magnitude (≈1 m/s) is highly unlikely. It is also contrary to observations during the experiments. With consistent speed and connected to a sufficient number of satellites, GPS measurements of speed and course over ground are usually very reliable. Therefore, a reasonable conclusion from the numerical analysis is that measurements of heading are unreliable. In addition, the model possibly seems to overestimate performance, i.e., in practice the boat speed is lower than the model predicts. This might be due to inaccuracies or simplifications included in the model, e.g., no heeling, no hull over water, no active rudder. Future work will also explore whether a similar numerical analysis can be used to quantify model assumptions and the impact of uncertain parameters on the model.

That measurements of heading seem unreliable and that the model possibly overestimates performance can also be indicated by studying the minimization of Eq. (9) with respect to, in turn, one of the six variables. For the minimization, measurements are used for all other variables. Interpreting the results presented in Table 2, it is, again, clear that the large leeway measured does not agree with the model. Furthermore, the angle of attack for the wingsail might be slightly lower than assumed and a more detailed visual inspection seems warranted. Interestingly, the low value for $V(x)/V_0(x)$ when minimizing with respect to boat speed indicates that a very low speed could explain the deviations between model and measurements. This, however, can be a natural consequence of the high measured leeway—at a low speed over water the keel will not generate much lift and thus drift, i.e., leeway, can be large.

In summary, both sets collected with ASPire yield similar conclusions: the analysis of data consistently implies unreliable measurements of heading. This conclusion is also consistent with independent weather data. Indeed, the presence of electromagnetic disturbances was later noted and the compass-unit subsequently moved. The approach thus seems promising for automating (some) fault detection and diagnosis. Had the approach of the present study been applied at an early stage, the unreliability of the compass-unit had also been noted and addressed. This, in turn, could have provided reliable data for more detailed comparisons of model versus measurements.

3.4 Data from Maribot Vane

Maribot Vane has a different design for the wingsail and apparent wind angle is measured relative to the wingsail. This gives a measurement of the angle of attack instead of the estimate used for ASPire. In order to obtain the apparent wind angle (relative to the heading of the boat), the angle of the wingsail is also measured. For Maribot Vane, two sets of data collected on November 15th, 2018, are studied. Both sets last 4 min each, and measurements of course, heading and apparent wind angle are illustrated as functions of time in the left panels of Fig. 4, upper panel for the first dataset and lower for the second. Apparent wind speed and boat speed are depicted in the right panels. Observations recorded at the Stockholm station of the Swedish Meteorological and Hydrological Institute approximately 15 km to the west of the sailing waters provide a mean true wind direction of 230° and a true wind speed of 4 m/s in agreement with collected data. The sailing routes are depicted in the lower right panel of Fig. 1.

Interestingly, and in contrast to the data collected with ASPire, two different tacks, different headings and also different (measured) angles of attack for the wingsails can be noted in the two sets. The data can thus also have the potential to either confirm or reject the existence of a current. As can be seen in the upper right panel of Fig. 4 (dataset 1), the wind speed seems to increase over time followed by an increase in boat speed. In dataset 2, there are also significant variations in apparent wind speed (lower right panel). Strictly taken, these observed changes correspond to a deviation from the assumption of equilibrium but are assumed to not significantly effect the results.

For the first set of data, $R_F = 10$ indicates a better match between model and measurements but still a mismatch. Detailed results are provided in Tables 3 and 4. For the second set, the value $R_F = 22$ indicates, again, a significant mismatch between measurements and model.

Despite some quite significant differences between the two sets of data, interpreting Tables 3 and 4 yields similar conclusions: The measured leeway is significantly higher than the model would predict. The very small boat speeds and zero angle of attack for set one obtained by minimization of Eq. (9), can also be the result of a numerical attempt to explain the large leeway. Indeed, the most significant improvement in fitting data to the model is consistently obtained by

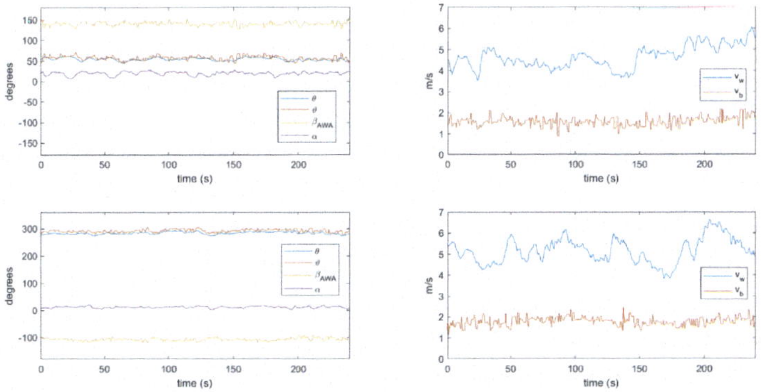

Fig. 4. Left panels: heading (blue) with a mean value of 54° (upper, first dataset) and 280° (lower, second dataset), course over ground (red) with a mean value of 58° (upper) and 290° (lower), apparent wind angle (yellow) with a mean value of 130° (upper) and −110° (lower), mean wingsail angle of attack of 20° (upper) and 12° (lower). Right panels: apparent wind speed (blue) with a mean value of 4.7 m/s (upper) and 5.2 m/s (lower), boat speed (red) with a mean value of 1.6 m/s (upper) and 1.8 m/s (lower).

Table 3. Same as in Table 1 for Maribot Vane.

$\theta(°)$	$\beta_{\text{AWA}}(°)$	v_w (m/s)	$\alpha(°)$	$\vartheta(°)$	v_b (m/s)	$\lambda(°)$
54 (284)	141 (−106)	4.7 (5.2)	20 (12)	58 (292)	1.6 (1.8)	3.3 (8.6)
58 (292)	141 (−106)	*3.3 (3.5)*	20 (12)	58 (292)	1.6 (1.8)	*−0.08 (0.04)*
57 (292)	*47 (27)*	4.7 (5.2)	20 (12)	58 (292)	1.6 (1.8)	*0.30 (0.35)*
54 (284)	*21 (NS)*	*15 (NS)*	20 (12)	58 (292)	1.6 (1.8)	3.3 (8.6)
58 (292)	141 (-106)	4.7 (5.2)	*4.9 (3.8)*	58 (292)	1.6 (1.8)	*−0.17 (0.04)*
54 (284)	*NS (NS)*	4.7 (5.2)	*NS (NS)*	58 (292)	1.6 (1.8)	3.3 (8.6)
54 (284)	141 (−106)	*NS (NS)*	*NS (NS)*	58 (292)	1.6 (1.8)	3.3 (8.6)
54 (284)	141 (−106)	4.7 (5.2)	20 (12)	*54 (284)*	*2.0 (2.3)*	*−0.11 (0.06)*
					$\varphi(°)$	v_s (m/s)
					−18 (−38)	*0.37 (0.55)*

Table 4. Same as in Table 2 but for Maribot Vane.

	$\theta(°)$	$\beta_{\text{AWA}}(°)$	v_w (m/s)	$\alpha(°)$	$\vartheta(°)$	v_b (m/s)
	58 (293)	*−180 (−180)*	*0.00 (8.5)*	*0.00 (11)*	*54 (284)*	*0.00 (0.13)*
$V(x)/V_0(x)$	0.003 (0.001)	0.8 (0.9)	0.9 (1.0)	0.9 (1.0)	0.003 (0.001)	0.009 (0.002)

estimating heading or course over water to have similar values, i.e., changing the measured leeway. Also, surprising results like apparent wind angles at 180° and very low boat speeds can also be seen attempts to explain a large leeway. Accordingly, it might be motivated to study the sensitivity of our model for errors in leeway. For Maribot Vane, earlier trials revealed an unreliable compass which was replaced before collecting the data studied in the present paper. Although it cannot be excluded that the new compass is also unreliable, an alternative can be considered. The alternative, if all measurements are reliable, is that the model does not adequately capture the leeway. In this case, the approach of the current paper clearly indicates that the main deficit of the present model is that it does not explain the leeway experienced in practical experiments.

On the other hand, it can be noted that similar currents, with apparent current angles, were estimated for both datasets. However, courses over ground differ by approximately 130° between the two sets that were collected within the same ten minutes and it can safely be concluded that there can be no significant current. Instead, these estimates serve to explain the measured leeway, and might possibly also indicate a bias in the model for overestimating boat speed for the apparent wind speeds measured.

4 Conclusions

It has been seen that the approach can be useful for detecting unreliable measurements. Furthermore, it seems that the approach can provide some guidelines regarding uncertainties in expected sailing performance. Possibly, the model might underestimate leeway and over-estimate sailing performance. This can be an indication of omitted terms and might warrant a reevaluation of the included simplifications and assumptions. In the long run, a focus on estimating different forces acting on the sailboat can also be beneficial for improving the robustness of the design, control and path-planning algorithms, e.g., choose routes and strategies in real time that will put less strain on the vessel.

References

1. Dhomé, U., Tretow, C., Kuttenkeuler, J., Wängelin, F., Fraize, J., Fürth, M., Razola, M.: Development and initial results of an autonomous sailing drone for oceanic research. In: Marine Design XIII – Proceedings of the 13th International Marine Design Conference, vol. 1, June 2018
2. Fossati, F.: Aero-Hydrodynamics and the Performance of Sailing Yachts. Adlard Coles Nautical, London (2009)
3. Friebe, A., Olsson, M., Gallic, M.L., Springett, J.L., Dahl, K., Waller, M.: A marine research ASV utilizing wind and solar power. In: OCEANS 2017 – Aberdeen, pp. 1–7, June 2017
4. Sheldahl, R., Klimas, P.: Aerodynamic characteristics of seven symmetrical airfoil sections through 180-degree angle of attack for use in aerodynamic analysis of vertical axis wind turbines. Technical report, Sandia National Labs, Albaquerque (1981)

Acoustic Detection of Tagged Angelsharks from an Autonomous Sailboat

Jorge Cabrera-Gámez[1,2], Antonio C. Domínguez-Brito[1,2(✉)], F. Santana-Jorge[2], Diego Gamo[2], David Jiménez[3], A. Guerra[3], and José Juan Castro[3]

[1] Instituto Universitario SIANI and Departamento de Informática y Sistemas, Universidad de Las Palmas de Gran Canaria, Las Palmas, Spain
antonio.dominguez@ulpgc.es
[2] Servicio Integral de Tecnología Marina, Universidad de Las Palmas de Gran Canaria, Las Palmas, Spain
[3] Instituto Universitario ECOAQUA, Universidad de Las Palmas de Gran Canaria, Las Palmas, Spain
http://www.siani.es, http://www.dis.ulpgc.es, http://www.ulpgc.es, http://www.sitma.ulpgc.es, http://www.ecoaqua.ulpgc.es

Abstract. Autonomous sailboats are silent surface vehicles which are well suited for acoustic monitoring. The integration of an acoustic receiver in an unmanned surface vehicle has a large potential for population monitoring as it permits to report geo-referenced detections in real time, so that researchers can adapt monitoring strategies as data arrive. In this paper we present preliminary work, done on the framework of ACUSQUAT project, to explore the usage of an acoustic receiver onboard a small (2 m length-over-all) autonomous sailboat in order to detect the presence of tagged adult exemplars of angelshark (*Squatina squatina*), the target species in ACUSQUAT, in certain areas which have demonstrated that this approach is feasible. Results obtained in simulation and during field trials are presented.

Keywords: Autonomous sailboat · Autonomous navigation · Sensor integration · Acoustic receiver · Acoustic tag detection · Angelshark (*Squatina squatina*)

1 Introduction

Angelsharks (*Squatina squatina*, a specimen shown in Fig. 1a) were common and abundant elasmobranchs along the European Atlantic coast and also in the Mediterranean Sea. This species, alike some other Squatinidae, have suffered from intensive fishing, both because they were captured intentionally or they were a common by-catch in trawling fishing practices, with the consequence of very strong reductions in population or even its disappearance in many areas. All this has driven *Squatina squatina* to be included in the Red List of Threatened Species by the International Union for Conservation of Nature (IUCN) as

M. F. Silva et al. (Eds.): ROBOT 2019, AISC 1092, pp. 295–304, 2020.
https://doi.org/10.1007/978-3-030-35990-4_24

(a) (b)

Fig. 1. (a) Tagged angelshark over a sandy bottom. Tag is appreciable on the main dorsal fins. Picture courtesy of Mike J. Sea and the Angel Shark Project. (b) Vemco's VR2C mini acoustic receiver.

in critical danger of extinction [7]. Moreover, very recently, June 5th 2019, the Spanish Government has officially included *Squatina squatina* and two other Squatinedae spp., *Squatina oculata* and *Squatina aculeata*, in the Spanish Catalog of Threatened Species as in danger of extinction.

The Canary Islands archipelago is one of the few locations in the EU (European Union) where the population of *Squatina squatina* is still well preserved [5], but its conservation is not exempt of problems derived from touristic activities, e.g. maritime traffic, sportive fishing or intense utilization of beaches, in some areas.

Conservation efforts addressing the recovery of marine endangered species require a careful knowledge of species-specific daily and seasonal patterns of habitat occupation, population structure and which environmental factors influence or determine the preference of this species for certain areas. Acoustic tracking has shown to be an effective methodology to research these aspects with other species and it is currently being implemented for *Squatina squatina* in Canary Islands.

The utilization of autonomous vehicles for acoustic registration started more than ten years ago with the works of J. Dobarro and collaborators [3], exploring the feasibility of using an autonomous underwater vehicle (AUV) to acoustically track tagged Atlantic sturgeons in the Hudson River. Lin and collaborators [4] have discussed the utilization of a stereo hydrophone system for tracking a tagged leopard shark. A recent review on the utilization of unmanned vehicles for the detection and monitoring of marine fauna can be found on [9].

Unmanned sailboats are slowly revealing their potential as interesting cost-effective platforms for oceanographic research [2,10]. In the specific realm of acoustic tracking, these acoustically silent vessels have already demonstrated their utility, e.g. [6], and how they can be used to complement a moored network of detectors in very interesting ways. It is normally unfeasible to cover all the area under study with the number of available receivers located at a few, strategically selected, locations leaving the rest of the area uncovered. Additionally, a network of acoustic receivers do not usually provide data in real-time, but when receivers are recovered. Using one or several vehicles, acting as mobile receivers, it is possible to extend the detection coverage using different strategies.

Recently, this approach has been extended to a fleet of vehicles in order to achieve wider coverage and/or improve localization. In [11] it is described, as a proof of concept, the utilization of heterogeneous fleet, including unmanned and manned surface vehicles and AUVs, to acoustically track tagged fishes.

This paper first describes the main characteristics of the elements required for setting up an acoustic tracking network and its integration in the A-Tirma autonomous sailboat. Final sections are devoted to illustrate results obtained in simulation and during the first field trials.

2 Materials and Methods

An acoustic tags receiver (Fig. 1b) have been integrated in A-Tirma G2 (A-Tirma, for short), a two-meter length autonomous sailboat provided with two wing sails (a more detailed description can be found in [1]). The boat is shown in Fig. 2a, and b displays an underwater view of A-Tirma towing the receiver.

The acoustic receiver integrated into A-Tirma was a Vemco's VR2C mini receiver. This receiver's dimensions are: 317 mm length, 54 mm diameter, 0.7 kg in-air weight. It operates at 69 kHz and has a maximum depth of 25 m. It lacks an internal battery, so it must be powered through the towing cable, but power demands are modest and flexible (line voltage 10–32 VDC, @12 V, record mode < 1 mA, 3–15 mA during communications). The VR2C mini has an internal memory (16 Mbytes) capable of storing nearly 1.6 million detections. A bidirectional serial interface, either RS-232 or RS-485, runs also over the cable. Using this interface, the receiver can be reconfigured at anytime from the host system. If properly configured, the receiver can report detections in real-time to the sailboat controller where, using the GNSS receiver available on board, they can be geo-localized and reported.

Adult angelsharks are being tagged with Vemco's V9-2L 69 kHz. coded tags (length: 29 mm.; diameter: 9 mm.; weight in water: 2.9 g and power output: 146

(a) (b)

Fig. 2. (a) A-Tirma G2. (b) Underwater view of the receiver being towed by the sailboat. The sailboat's keel and one of the skegs are visible at the right of the picture.

dB re 1 μPa @1 m). These tags' weight and dimensions are perfectly compatible with the tagging ethics standards. Programmed with a nominal period of 180 s, the expected battery lifetime is 912 days, permitting to extend the study over several seasonal cycles.

A-Tirma is equipped with a communications infrastructure consisting of three types of communication devices. Namely, one for short range communications: a XBee radio link. And, two for long range communications: a 3G/GPRS link for areas where mobile telephony communication services are available, and a satellite Iridium SBD transceiver for high seas when no other way of long-distance communication is possible. Through those three links the sailboat can be supervised and monitored remotely during a real navigation using a software graphical interface. In addition, a public web tracker of the boat is available, where the trajectory and behavior of the boat can be openly published during on-field navigation. A detailed description of A-Tirma's communication infrastructure can be found in [8]. In relation to the operation of the acoustic receiver during a navigation session, it is possible to activate/deactivate the receiver remotely using the graphical interface. Once the receiver is active, its operational status can be monitored in real-time through the graphical interface. Also, when active, if any animal tag is detected, the detection is registered on secondary memory on the boat, and on the graphical interface, the position of the boat where the detection was verified is also shown. Figure 3 shows a snap-shot of A-Tirma's graphical interface, where we can see how the acoustic sensor information has been integrated graphically. This figure also shows how detections are displayed on the interface along with operational status information. From a hardware point of view, Fig. 4 depicts, in a deployment diagram, how Vemco's VR2C acoustic receiver has been integrated in the boat. As we can note in the figure

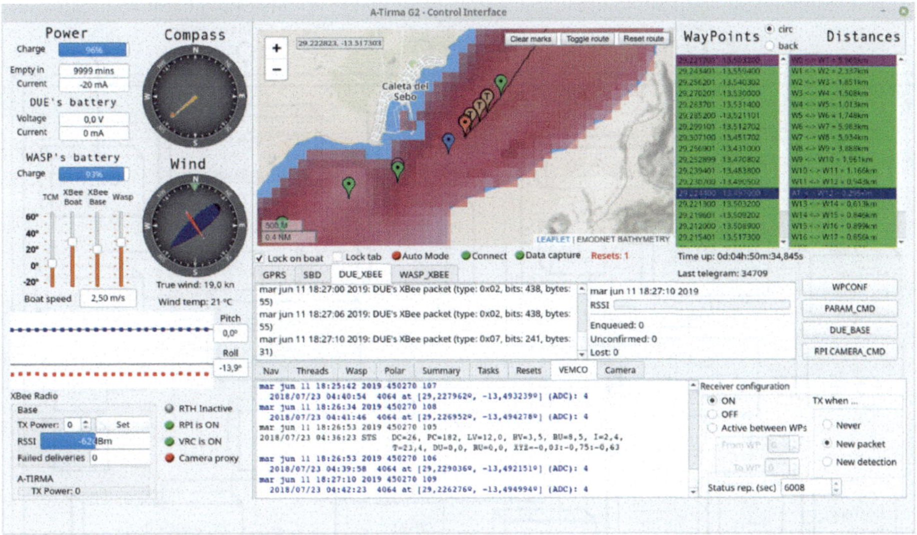

Fig. 3. A-Tirma G2's graphical interface. On the map area: the blue icon is the current waypoint the boat is navigating to, the red icon is the boat itself, green icons are the rest of waypoints which conform the route followed by the boat. On the right side of the map all waypoints coordinates are listed, and the current one is also highlighted in blue. Finally, light brown icons signed with a "T" are acoustic tag detections performed by the boat using the Vemco's VR2C acoustic receiver. In addition, there is a tab in the inferior part of the interface, tab "VEMCO", which presents, when the receiver is active, operational status information (text in black), and data (text in blue) about the tags as they get detected in real-time during the navigation.

the receiver is connected via an asynchronous serial link (RS-232) to A-Tirma's main system controller on board.

The receiver is towed using a reinforced 10 meter-long cable. To avoid cable torsion, a two-part stabilizer (visible in yellow in Fig. 2b) has been designed and manufactured by 3D printing. This part also offers a secure cable locking point so that the towing tension is not directly supported by the locking sleeve of the receiver connector.

3 Simulation Results and Field Experiments

Extensive simulation tests have permitted to verify the correct integration of the receiver in the control and communication architecture of the boat. These tests have been carried out first with the real receiver in the hardware loop and later with a simulated device written as an extra simulator's software module. This procedure allowed to validate the simulated receiver. Additionally, the simulator module permits to select the motion patterns that the simulated tagged animals will exhibit during a simulation, and also to constraint their displacements to

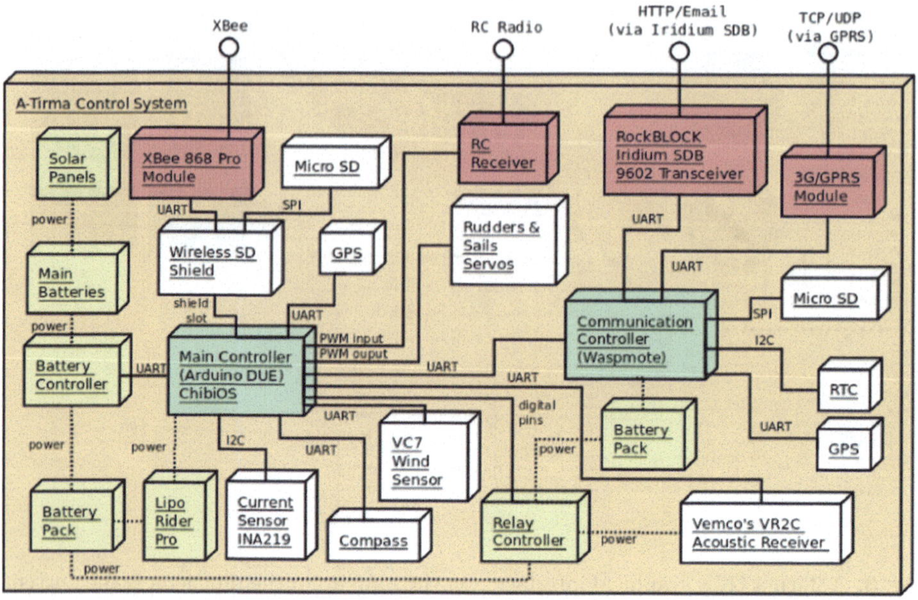

Fig. 4. A-Tirma G2's deployment diagram. Vemco's VR2C acoustic receiver (bottom-right corner) has been integrated on the main controller through an asynchronous serial (RS-232) connection.

certain geographical zones. In particular, for a given animal, its pattern of movement is restricted to a circular area centered on a given geographical location, which is the starting position of the animal when the simulation is started. In addition, some other parameters define the animal's movement pattern, namely: maximum reachable depth, maximum speed, and maximum heading change (at each simulation step, a positive or negative heading change is calculated randomly limited by this value). Moreover for each animal, some other values define how the animal's simulated tag behaves, such as, its emission period, whether it is a high transmission power tag or not, and if it is equipped with a pressure sensor or not. In this simulation the sound has been modelled in a very simple way, considering a constant sound velocity. As already shown, Fig. 3 displays the reception of status and detection packets as they are shown on the control graphical interface during a navigation session, whether real or simulated.

Some field trials have been carried out to test the feasibility of using a receiver onboard the A-Tirma sailboat. This study has been developed in parallel to animal tagging, which is still underway. These experiments have focused on determining the best placement of the receiver in the A-Tirma in order to maximize tag detection probability with a minimal impact on sailboat's maneuverability.

The receiver has an omnidirectional reception sensibility pattern centered around its longitudinal axis. Thus, the best detection results are expected with the receiver close to a vertical position with the hydrophone facing downwards, specially if the receiver is being towed near the surface and the target species is benthonic, as in our case.

The placement explored so far with the A-Tirma has consisted in towing the receiver at the extreme of a 10-m long cable secured at the stern of the sailboat. The receiver has been ballasted carefully to achieve a towing depth of approximately 2 m and a characteristic pitch around 30°, as illustrated in Fig. 2b. This disposition has not affected in any significant manner ship's maneuverability under the testing conditions, which have been characterized by low wind speed. However, increased drag is likely to be expected at higher wind speeds.

The field tests performed in order to study the reception and detection of tags towing the receiver from a boat were carried out in the area of Puerto Rico in the island of Gran Canaria, Spain (illustrated in Figs. 5 and 6). Four acoustic tags were moored in a line, situated each at different depths. The four tags correspond with tag identifiers 4064, 20255, 20254 and 4063, situated respectively at 1, 3.5, 6 and 8.5 m from sea bottom. The first three tags emit their ID in low power mode, while the 4063 emits in high power mode. Low power mode reduces the utilization of energy, with a positive effect on the longevity of the tag, at the cost of a possible reduction in range detection, i.e. pinging is done at a reduced intensity. Additionally, tags 4063 and 4064 also provide pressure information, from which we can derivate depth. In Fig. 6 we can observe how many times and where each one of them was detected during the experiment. As expected, tag 4063 was detected more times since it is a high power tag. During the field tests, a fifth tag (tag no. 25639) was detected three times (Fig. 6). This was a tag previously deployed in the area during animal tagging. This configuration of four-tags were situated in two locations, namely, locations L1 and L2 displayed in Fig. 5, the depth in both locations was of about 26.5 and 29.5 m respectively. Despite the high maritime traffic of the area, these experiments were carried out in this zone as this is the focal area for the ACUSQUAT project where some angelsharks exemplars are being tagged.

A relevant qualitative conclusion drawn from these experiments has been that the orientation of the receiver relative to the tag position has an important influence in the probability of detecting a tag emission. This is evident from inspection of Fig. 5. More precisely, when the boat is moving away from the tags and the receiver gets oriented towards them (the moving away green arrow from location L2 in the figure) the probability of detection is high and many tags emissions are detected. On the contrary, when the boat goes toward the tags (indicated with the green arrow going towards L2 in the figure), there are no detections until it gets quite close to tags in location L2. In this second case, the orientation of the receiver, as it gets towed, has the hydrophone looking in the opposite direction to where the tags are situated. This asymmetry in detection vanished when towing was stopped and the receiver recovered a vertical orientation or, when the receiver was not towed directly towards the position of the tags, as was the case when tags were moored at L1. As conclusion, keeping the receiver vertically orientated seems more adequate to maximize tag detection probability.

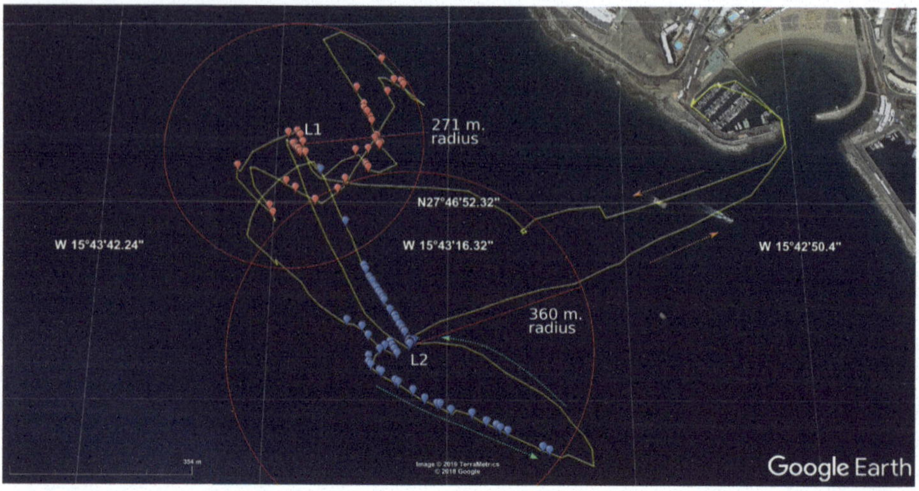

Fig. 5. Field tests for checking the reception and detection of acoustic tags performed in the area of Puerto Rico, Gran Canaria, Spain. The trajectory of the boat towing the receiver is shown in yellow. Red icons show tag detections when the four-tag pattern was situated in location L1. Blue icons correspond to detections with the four-tag pattern in location L2.

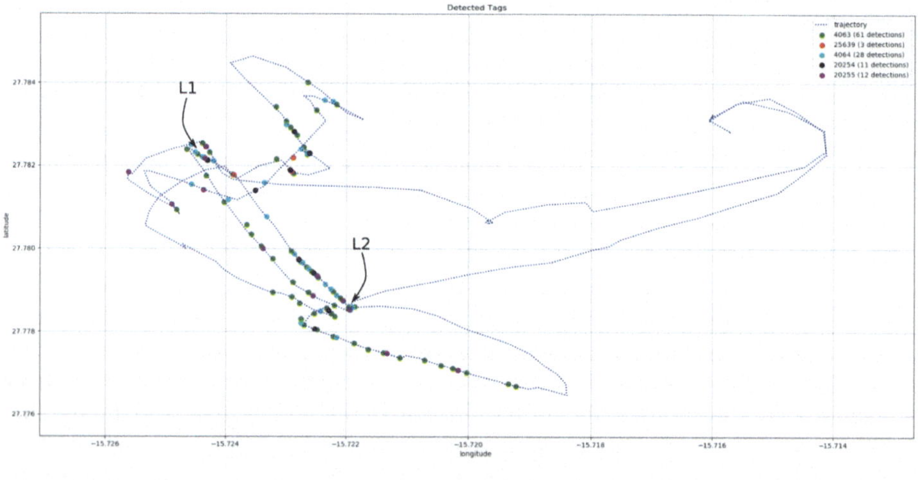

Fig. 6. Detected tags corresponding to the field tests shown in Fig. 5.

4 Conclusions

In a first stage, we have integrated a Vemco VR2C mini receiver into the control and communication layers of A-Tirma autonomous sailboat. Extensive tests have been run in simulation to verify the stability and responsiveness of the control and communication system software. Also, first field trials have been developed to characterize the detection probability with range. These experiments have shown that the sailboat is capable of towing the receiver even with faint winds, even though the absolute orientation of the receiver during towing needs to be optimized. More in-depth testing is planned to study the influence of the environment (bathymetry, distance to shore, ship traffic, etc) in detection probability with range, and specially for evaluating how the quality of tag detections get influenced by tag's orientation, receiver's distance-from-tag, orientation and depth, and also by boat's direction and speed.

Future work will address exploring alternative ways of installing the receiver. One attractive possibility is removing the towing fixation point from the stern to a point located underwater in the bulb at the deeper extreme of the keel. This will allow to reduce the cable length, for the same towing depth, with a positive impact in the maneuverability of the boat. Another possibility we aim to explore is fixating the receiver, close to a vertical position, as suggested above, at the deepest available placement on the keel. This option would eliminate also the drag induced by the towing cable. However, under the conditions imposed by A-Tirma dimensions, it is foreseen that this placement will produce a significant reduction in range detection due to the noise induced by wind, waves and hull motion.

Acknowledgements. This research has been partially funded by Loro Parque Fundation through project BioACU and Fundación Biodiversidad (Spanish Ministry for Ecological Transition) through project: "ACUSQUAT: Seguimiento acústico del comportamiento del angelote (*Squatina squatina*) en áreas críticas de conservación" (Acoustic tracking of Angelshark (*Squatina squatina*) behavior in critical areas of preservation). Reference CA_BT_BM_2017.

References

1. Domínguez-Brito, A.C., Valle-Fernández, B., Cabrera-Gámez, J., Ramos-de Miguel, A., García, J.C.: A-TIRMA G2: an oceanic autonomous sailboat. In: Friebe, A., Haug, F. (eds.) Robotic Sailing 2015, pp. 3–13. Springer International Publishing (2016). https://doi.org/10.1007/978-3-319-23335-2_1
2. Ghani, M.H., Hole, L.R., Fer, I., Kourafalou, V.H., Wienders, N., Kang, H., Drushka, K., Peddie, D.: The sailbuoy remotely-controlled unmanned vessel: measurements of near surface temperature, salinity and oxygen concentration in the northern gulf of mexico. Methods Oceanogr. **10**, 104–121 (2014). Special Issue: Autonomous Marine Vehicles. https://doi.org/10.1016/j.mio.2014.08.001. http://www.sciencedirect.com/science/article/pii/S2211122014000395
3. Grothues, T.M., Dobarro, J., Ladd, J., Higgs, A., Niezgoda, G., Miller, D.: Use of a multi-sensored auv to telemeter tagged atlantic sturgeon and map their spawning habitat in the Hudson river, USA. In: 2008 IEEE/OES Autonomous Underwater Vehicles, pp. 1–7 (2008). https://doi.org/10.1109/AUV.2008.5347597
4. Lin, Y., Hsiung, J., Piersall, R., White, C., Lowe, C.G., Clark, C.M.: A multi-autonomous underwater vehicle system for autonomous tracking of marine life. J. Field Rob. **34**(4), 757–774 (2017). https://doi.org/10.1002/rob.21668

5. Meyers, E.K.M., Tuya, F., Barker, J., Alvarado, D.J., Castro-Hernández, J.J., Haroun, R., Rödder, D.: Population structure, distribution and habitat use of the critically endangered angelshark, squatina squatina, in the canary islands. Aquatic Conserv. Marine Freshw. Ecosyst. **27**(6), 1133–1144 (2017). https://doi.org/10.1002/aqc.2769
6. Mordy, C.W., Cokelet, E.D., De Robertis, A., Jenkins, R., Kuhn, C.E., Lawrence-Slavas, N., Berchok, C.L., Crance, J.L., Sterling, J.T., Cross, J.N., Stabeno, P.J., Meinig, C., Tabisola, H.M., Burgess, W., Wangen, I.: Advances in ecosystem research: saildrone surveys of oceanography, fish, and marine mammals in the bering sea. Oceanography **30** (2017). https://doi.org/10.5670/oceanog.2017.230
7. Morey, G., Barker, J., Hood, A., Gordon, C., Bartolí, A., Meyers, E., Ellis, J., Sharp, R., Jimenez-Alvarado, D., Pollom, R.: Squatina squatina. The IUCN red list of threatened species 2019 **e.T39332A117498371** (2019). https://doi.org/10.2305/IUCN.UK.2019-1.RLTS.T39332A117498371.en
8. Santana-Jorge, F.J., Domínguez-Brito, A.C., Cabrera-Gámez, J.: A component-based C++ communication middleware for an autonomous robotic sailboat. In: Øvergård, K.I. (ed.) Robotic Sailing 2017, pp. 39–54. Springer International Publishing (2018). https://doi.org/10.1007/978-3-319-72739-4_4
9. Verfuss, U.K., Aniceto, A.S., Harris, D.V., Gillespie, D., Fielding, S., Jiménez, G., Johnston, P., Sinclair, R.R., Sivertsen, A., Solbø, S.A., Storvold, R., Biuw, M., Wyatt, R.: A review of unmanned vehicles for the detection and monitoring of marine fauna. Marine Pollut. Bull. **140**, 17–29 (2019). https://doi.org/10.1016/j.marpolbul.2019.01.009. http://www.sciencedirect.com/science/article/pii/S0025326X19300098
10. Voosen, P.: Saildrone fleet could help replace aging buoys. Science **359**(6380), 1082–1083 (2018). https://doi.org/10.1126/science.359.6380.1082. https://science.sciencemag.org/content/359/6380/1082
11. Zolich, A., Johansen, T.A., Alfredsen, J.A., Kuttenkeuler, J., Erstorp, E.: A formation of unmanned vehicles for tracking of an acoustic fish-tag. In: OCEANS 2017 - Anchorage, pp. 1–6 (2017). https://ieeexplore.ieee.org/document/8232099

Airfoil Selection and Wingsail Design for an Autonomous Sailboat

Manuel F. Silva[1,2(✉)], Benedita Malheiro[1,2], Pedro Guedes[1], and Paulo Ferreira[1]

[1] ISEP/PPorto, School of Engineering, Polytechnic of Porto, Porto, Portugal
{mss,mbm,pbg,pdf}@isep.ipp.pt
[2] INESC TEC, Porto, Portugal

Abstract. Ocean exploration and monitoring with autonomous platforms can provide researchers and decision makers with valuable data, trends and insights into the largest ecosystem on Earth. Regardless of the recognition of the importance of such platforms in this scenario, their design and development remains an open challenge. In particular, energy efficiency, control and robustness are major concerns with implications in terms of autonomy and sustainability. Wingsails allow autonomous boats to navigate with increased autonomy, due to lower power consumption, and greater robustness, due to simpler control. Within the scope of a project that addresses the design, development and deployment of a rigid wing autonomous sailboat to perform long term missions in the ocean, this paper summarises the general principles for airfoil selection and wingsail design in robotic sailing, and are given some insights on how these aspects influence the autonomous sailboat being developed by the authors.

Keywords: Rigid wingsail · Autonomous sailboat · Wingsail design

1 Introduction

Significant research is being conducted in autonomous systems (whether they are land, marine or aerial), since these platforms are useful in a variety of tasks, due to their ability to remove humans from dangerous environments, relieve them of tedious tasks, or simply go to locations otherwise inaccessible or inhospitable [16]. Diverse applications have been envisaged for these vehicles, from exploration of remote places to warfare [26].

Concerning marine robots, most of the research has been directed to electrically or fuel powered surface and underwater vessels. This type of vehicles presents severe limitations in range and endurance, depending on battery capacity or onboard fuel for propulsion, making them unsuitable for long term operation in inaccessible areas, that would otherwise be well suited for unmanned operation. These limitations make wind propelled vessels, used by mankind since ancient times [9], into an attractive possibility. Not only do they harvest the

M. F. Silva et al. (Eds.): ROBOT 2019, AISC 1092, pp. 305–316, 2020.
https://doi.org/10.1007/978-3-030-35990-4_25

energy for propulsion from the environment, instead of having to transport it, but also require reduced electrical power for the on board control systems and, thus, the sail. In such case, photovoltaic cells and/or wind generators are capable of producing enough power to run the complete electrical system, making these vehicles sustainable and autonomous in terms of energy and, therefore, capable of operating continuously, for extended periods of time, in the middle of the ocean [25]. Nonetheless, sail powered vessels also present a limitation, when compared with other forms of propulsion: if there is no wind, there is no propulsion.

Research on autonomous, or robotic, sailboats has been ongoing for about 20 to 25 years [19]. The last years have witnessed an increasing interest in the development of autonomous water surface vehicles (ASV). By robotic sailing it is meant that the whole process of sailing boat navigation is performed by an autonomously acting system of technical devices [27]. The key characteristics of a robotic autonomous sailing boat are the following [27]: (*i*) wind is the only source of propulsion; (*ii*) it is not remote controlled – the entire control system is on board and, therefore, has to perform the planning and manoeuvres of sailing automatically and without human assistance; and (*iii*) it is completely energy self-sufficient – this is not a must in the sense of definition of a robotic sailing boat, but it opens a wider range of applications.

The use of autonomous sailboats has been proposed for different purposes, including long term oceanographic research, such as monitoring marine mammals (the absence of self-generated noise during navigation is an unique advantage of robotic sailboats for underwater acoustics applications) and automated data acquisition, surveillance of harbours, borders and other areas of interest, and as intelligent sensor buoys [25]. There are also several autonomous sailing boat prototypes developed with educational and research objectives, including the participation in robotic sailing competitions for benchmarking [8,12,17].

In order to be truly autonomous, sailboats need to incorporate several distinct technologies: (*i*) propulsion; (*ii*) sensing; (*iii*) actuation; (*iv*) communication; and (*v*) control. So far, different technologies and techniques have been proposed for each of these subsystems [25]. This paper addresses specifically the propulsion system, based on a wingsail, for a new autonomous sailing vehicle.

After this introduction, Sect. 2 details the main characteristics of wingsails and its operation. Next, Sect. 3 briefly introduces the hull that has been chosen for the sailboat and Sect. 4 describes how the airfoil for the wingsail was selected. Based on this information, Sect. 5 presents the design and construction of the wingsail. The paper ends with concluding remarks on Sect. 6.

2 Wingsail Characteristics and Operation

This section introduces wingsails, the basic physical aspects of their operation principle, the possible options that can be considered to control its angle of attack in relation to the wind, and the operation of a tail controlled wingsail.

2.1 Definition

Apart from the recent sailing history, sailing boats used conventional fabric sails. This type of sails is advantageous when controlled by a human sailor [27]. By contrast, a wingsail is a rigid structure presenting an airfoil cross-section (like an airplane wing), which improves the lift-to-drag ratio, and being also more robust and reliable (a very important aspect for autonomous vehicles), when compared with conventional sails [5]. Wingsails should not be confused with solid square sails or rigid sails. Although this concept may seem a novelty, the first rigid lift-generating devices for use as auxiliary ship propulsion were proposed and developed by Anton Flettner in 1922 [2], and, since then, several boats have been equipped with wingsails [2,5,25].

Although wingsails present some disadvantages in relation to traditional sails [25], after extensive testing with different wingsails, Neal *et al.* maintain that the potential gains in reliability and efficiency outweigh the problems [15].

2.2 Operating Principle

The primary function of the wingsail is to propel the boat through the generated lift force. The propulsion will be generated by the resulting force when the airflow interacts with the wing. The force depends on the cross-sectional shape of the wing, *i.e.*, the airfoil.

The lift on an airfoil is primarily the result of its angle of attack (α), defined as the angle between the mean chord of the wing (a line drawn between the leading edge and the trailing edge of the wing) and the direction of the relative wind (W_a). When oriented at a suitable angle, the airfoil deflects the oncoming air, resulting in a force on the airfoil in the direction opposite to the deflection. This force is known as aerodynamic force and can be resolved into two components: the component perpendicular to the direction of the apparent wind is called lift (L); the component parallel to the direction of the apparent wind is called drag (D). Most airfoil shapes require a positive angle of attack to generate lift, but cambered (asymmetrical) airfoils can generate lift at zero angle of attack [1]. The lift and the drag produced by an airfoil are given by Eqs. 1 and 2, respectively, where ρ represents the air density, V the velocity of the wingsail, S the surface of the wingsail, and C_L and C_D the lift and drag coefficient of the airfoil.

$$L = 1/2\rho V^2 S C_L \quad (1)$$

$$D = 1/2\rho V^2 S C_D \quad (2)$$

Figure 1 depicts the main forces that are involved in a wingsail ASV when sailing upwind and downwind. F is the aerodynamic force that decomposes in H, a heeling component, and T, the thrust that effectively propels the vessel [6].

Due to the need to operate with the wind blowing from either side of the sailboat, wingsails tend to have a symmetrical airfoil section, although asymmetrical airfoils typically present a maximum higher lift coefficient (C_L). To improve the characteristics of symmetrical airfoils, some wingsail arrangements include a flap or tail section to create additional lift and to help the main wing to reach its maximum lift capacity [2,5].

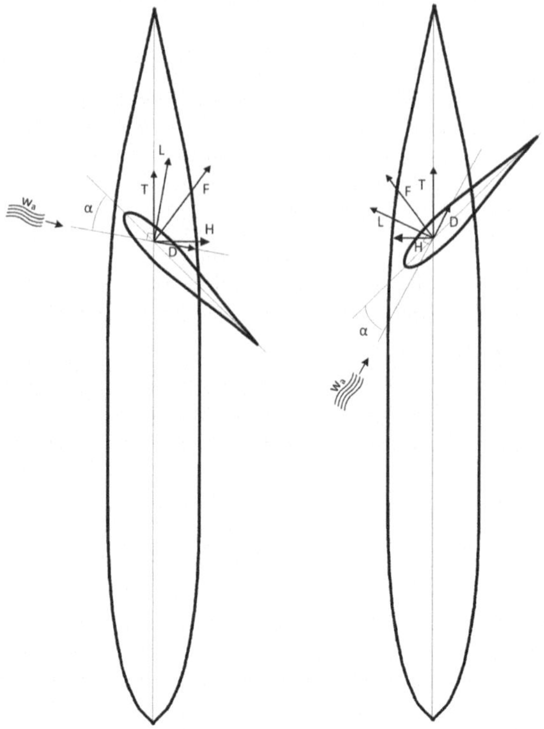

Fig. 1. Wingsail configurations for sailing upwind (left) and downwind (right) [4].

2.3 Control Architectures

Sail control strategies in autonomous sailboats are mainly focused on controlling the wingsail angle of attack (its angle relative to the wind).

One possibility is to have the wingsail attached directly to the mast (somewhat similar to a windsurfer sail). The sail is manipulated through the rotation of the entire mast, typically using an electric motor. This design eliminates common points of failure found in traditional sails, presenting advantages for use in an autonomous sailing vessel.

There is another possibility that presents improved reliability, easier control and reduced power consumption. If the wingsail rotates freely around the mast, it will work like a wind vane, meaning it will turn to the wind without inducing large heeling moments. If the wingsail is further equipped with a smaller control surface (canard or tail wing), this arrangement (known as a tandem wing) allows the wingsail to automatically attain the optimum angle to the wind. In this situation, the wingsails are said to be self-trimming. This configuration also allows to adjust the angle of attack of the wingsail in relation to the wind, according to the wind speed and, thereby, vary the resulting force. This avoids other aspects of traditional sail control, such as mast raking, reefing, control of luffing and sail shape adjustment, and allows a reduction in power consumption [20].

The self-trimming capability is achieved by aligning the centre of mass of the wing arrangement on the axis of rotation (the mast). If this point is also aligned with the aerodynamic centre of the main wing, small forces from the control surface are sufficient to achieve a proper angle of attack. The aerodynamic centre on a symmetric low speed airfoil is located approximately one quarter of the chord length from the leading edge of the airfoil and characterises the position where the magnitude of the aerodynamic moment remains nearly constant for all angles of attack.

The most common tandem wing arrangement is a controllable tail behind a fully rotational wing. This transforms the fully rotational into a self-trimmed wingsail and considerably simplifies the control system design. However, this tailed wing arrangement is tail heavy, which must be compensated for with ballast positioned forward of the main wing, placing the centre of mass of the wing arrangement in the desired position. This ballast in the wing causes, on its turn, the rise of the boat's centre of gravity, making the boat more prone to capsizing, and increases the rotational radius of the wing, potentially causing damage to itself or other boats in close vicinity; however, this is not an issue out at sea where these boats usually spend most of their time apart.

A canard arrangement is similar to a tailed wing with the exception that the "tail" is in front of the main wing rather than behind. This wing arrangement has a more balanced weight placement compared to the tailed wing, meaning that a canard wing requires less ballast to position the wing's mass centre at the desired position. Also the turning radius is smaller for a canard arrangement. However, in varying wind conditions, the canard arrangement, when freely rotational, is more unstable than the tailed wing [2,5].

2.4 Working Principle of a Tail Controlled Wingsail

The working principle of a tail controlled wingsail is relatively simple. Typically, a wind vane is used to measure the wind direction relative to the angle of the wingsail, being this essentially a measure of the wingsail's angle of attack. While the tail has no angle of attack to the wind, the wingsail acts as a weather-vane, pointing into the relative wind, as can be seen in Fig. 2 (top).

To make the sailboat move, the tail needs to be rotated to give it an angle of attack (Fig. 2 (middle)), creating Lift (L_T). This, in turn, will create a torque around the mast of the winsail (T_W), causing the entire wingsail/tail system to rotate in the opposite direction of the tail until a torque balance is achieved between the wingsail and the tail, as depicted in Fig. 2 (bottom). This moment balance will cause the wingsail to be at a positive angle of attack in relation to the apparent wind, generating lift. The component of lift pointing along the hull is the thrust, which will accelerate the sailboat until the drag on the hull equals the thrust from the wing. Once this equilibrium is achieved, the boat will continue to move at a constant velocity, while the wing and tail are in moment equilibrium [5].

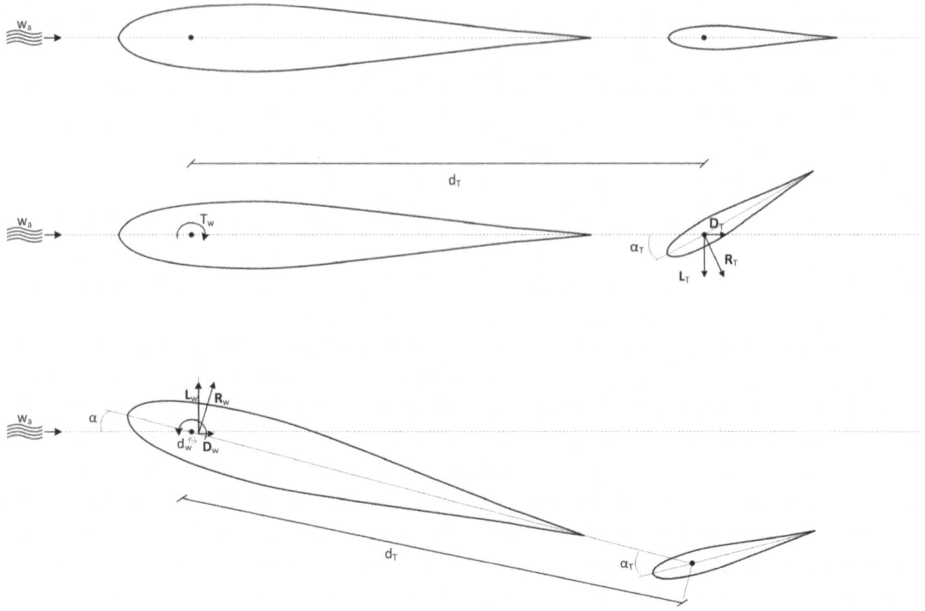

Fig. 2. Forces and torques involved in the control of the wingsail – tail set: tail not actuated (top), actuation of the tail (centre), and wingsail and tail in equilibrium (bottom).

When the wind changes direction, the wingsail and tail will rotate to a new position, that is identical relative to the wind. Since the lift can only be generated perpendicular to the relative wind, the lift vector will also rotate to the same position relative to the wind. Thus, the entire wingsail, tail, and lift vector rotate together with the wind as a rigid unit [5].

The moment balance between the wing and the tail keeps the wingsail at a constant angle of attack relative to the wind. As long as the wind does not cross the centre-line of the boat, then the wing continues to provide thrust in the correct direction for forward motion through passive stability of the wingsail system. Should the wind cross through the centre-line of the boat, then the position of the flap and tails must be reversed, which corresponds to tacking or jibing, depending on whether the wind crosses the centre-line facing aft or forward, respectively [5].

3 Sailboat Hull Selection

Since the characteristics of the hull influence the design choices of the wingsail, this section briefly presents the options considered and the selected vessel hull.

Several possible hull designs exist for autonomous sailing boats. Miller *et al.* present a series of performance trade-off studies concerning hull design features and the corresponding performance effects [13]. An ideal hull would be cheap to manufacture, able to self-right in the event of capsize [7], small enough to allow for easy transportation and to prevent damage to another vessel in the event of a collision, but large enough to be able to sail effectively in heavy seas. Additionally any such hull needs to be fully enclosed to prevent water entering the hull, eliminating the need for costly pumps to remove excess water [18]. Other authors, considering the fact that this solution is very expensive to produce and not always very reliable, chose to make a completely unsinkable sailboat building it from blocks of closed cell foam [10].

Additionally, Miller *et al.* briefly compare monohulls with multihulls and state that catamarans and trimarans have demonstrated significantly higher performance than monohulls in many applications that do not include large changes in displacement. They argue that while the multihulls showed promise, two issues raised concerns: (*i*) the first was weight, and (*ii*) the second relates to capsizing, since as multihulls are typically not self-righting, a multihull sailing boat presents a higher risk of not being able to recover from a capsize [14].

Regarding monohulls, there are three main possibilities for the hull Length Overall (LOA): (*i*) the first one is to use a hull intended for radio controlled model boats (under 2 m) [21]; (*ii*) a second possibility is that of a small dinghy hull (3 m to 5 m); and, (*iii*) the final option is to modify a yacht sized hull (larger than 5 m) [18].

The advantages of shorter vessels – option (*i*) – are that they are cheap, lightweight, easy to handle and easy to build. Test runs can easily be arranged without the need for any special infrastructure (slip ramp, crane) or a chasing boat. On the other hand, a boat of this size is extremely sensitive even to small waves and wind gusts. This makes it difficult to reproduce experimental results and to evaluate the implications of minor changes in the control system. Furthermore, with its restricted space for additional equipment and a relatively short operating time, it is not a serious platform for maritime applications. Boats with hulls according to option (*iii*), mainly present the opposite characteristics. For this reason, in the last years option (*ii*) seems to be gaining momentum with several autonomous sailboats developed based on such hulls. This is also observed when the commercial examples of rigid wingsail boats are analysed [25].

Given the above considerations, it was decided to opt for a "medium" sized monohull. After a market search, it was chosen the DCmini hull [3], depicted in Fig. 3, which can be seen as an inspiration from the 2.4mR class. This boat, on its "traditional sail" version, presents a total sail area of 5.7 m^2 distributed by the main sail with 3.4 m^2 and the genoa with 2.3 m^2.

4 Airfoil Selection

After having chosen the wingsail characteristics, it is needed to select an adequate airfoil profile to design it.

The Reynolds Number (*Re*), given by Eq. 3, represents the ratio of kinematic or inertial forces to the viscous forces in a fluid, being ρ the density of

Fig. 3. DCmini hull [3].

the medium, V the velocity of the flow, L the characteristic length, and μ the viscosity of the medium.

$$Re = \rho V L / \mu \tag{3}$$

According to Elkaim, the proper Reynolds number range for sail operation is 200 000 to 1.2 million [5]. Therefore, to choose the airfoil for the wingsail, the main characteristics of several symmetrical airfoils for values of $200\,000 < Re < 500\,000$ have been analysed. The objective was to find an airfoil that presented higher values for C_L/C_d and C_{Lmax} and a relatively high maximum thickness, to improve the rigidity of the wingsail structure.

Given these selection criteria, the airfoils subjected to further consideration were the J5012, NACA 0009, NACA 64A010 and SD 8020, described in [22–24], CG Ultimate, DH4009, E472, Trainer 60 and Ultra-Sport 1000 [11], and several other symmetrical airfoils whose characteristics are presented in [28]. The final decision fell on the Eppler E169 low Reynolds number airfoil [29], whose main aerodynamic characteristics at $Re = 200\,000$ are depicted in the charts presented in Fig. 4. From the analysis of these charts it can be concluded that this airfoil presents values of $C_L/C_d \approx 50$ for attack angles in the interval $5° < \alpha < 10°$, while having a relatively low value of $C_d < 0.02$ in this interval of α. For $\alpha \approx 10°$ the value of $C_L \approx 1.1$. Therefore, the wingsail should be operated with an angle of attack $\alpha \approx 10°$ to obtain the optimal conjugation of aerodynamic parameters. This value of the angle of attack ($\alpha \approx 10°$) will be used in the sequel for computational purposes only and, during sailing, it will have to be adjusted to the particular course being sailed.

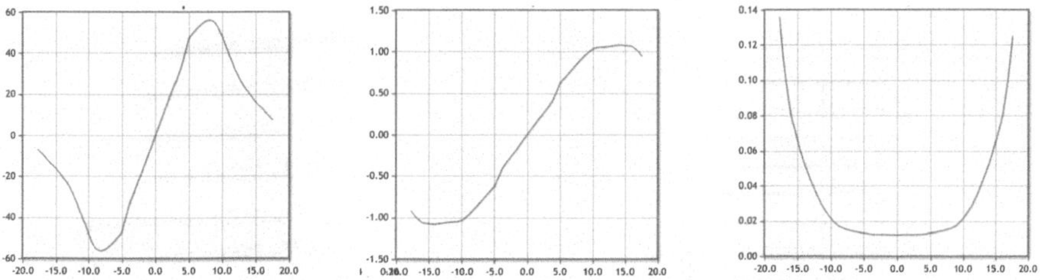

(a) C_L/C_d *vs.* α. (b) C_L *vs.* α. (c) C_d *vs.* α.

Fig. 4. Main aerodynamic characteristics of the Eppler E169 airfoil [29].

5 Wingsail Design and Construction

According to Elkaim, a sloop rig sail can achieve a maximum $C_L \approx 0.8$ if the jib and sail are perfectly trimmed. Realistically, with an operating maximum $C_L \approx 0.6$, the C_L/C_d of the conventional sail is in the 3 to 5 range [5]. For the profile chosen, the ratio between the C_L at an angle of attack $\alpha \approx 10°$ and the C_L for the traditional sail was computed using Eq. 4:

$$C_{Lratio} = C_{LE169}/C_{Lclothsail} \approx 1.83 \tag{4}$$

This value indicates that the wingsail area should be 1.83 times lower than that of the cloth sail, which would correspond to $A_{wingsail} \approx 3.1\,\text{m}^2$. However, since the C_L/C_d of the chosen profile around $\alpha \approx 10°$ is much better that the one for the traditional sail cloth, it was decided that the wingsail should have a maximum area of $A_{wingsail} = 1.5\,\text{m}^2$.

For this wingsail area, two geometrical shapes have been considered for the wing: (*i*) a rectangular shape wingsail with an height of 1500 mm and a width of 1000 mm, and (*ii*) a trapezoidal wingsail with an height of 1500 mm and a width varying between 800 mm to 1200 mm. However, the current state of development of the project does not yet allows to present the polar diagrams for these two shapes of wingsails and its comparison with traditional sails.

Although the trapezoidal wingsail has the advantage of having a lower centre of aerodynamic pressure for the same value of generated lift (meaning that the force that heels the boat is applied lower, lowering also the heeling torque), the decision was to build a rectangular wingsail due to its construction simplicity compared to the trapezoidal counterpart. The designed wingsail, with a wing span of 1530 mm and a wing chord of 1000 mm (Fig. 5a), has a rectangular structure (Fig. 5b).

The wingsail plywood ribs were cut using a computer numerical control machine. The ribs of the main sail were inserted in the mast, separated using plywood spacers and glue (Fig. 5c) and covered using transparent ultraviolet resistant polyvinyl chloride (Fig. 5d). The assembly of the tail around a light plastic axis followed the same process. Then, the wingsail was mounted and secured on top of the frame, using two pins. The initial tests showed that tail

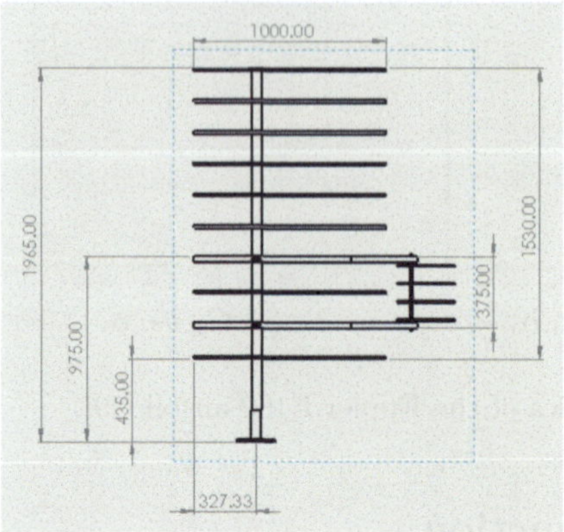

(a) Dimensions.

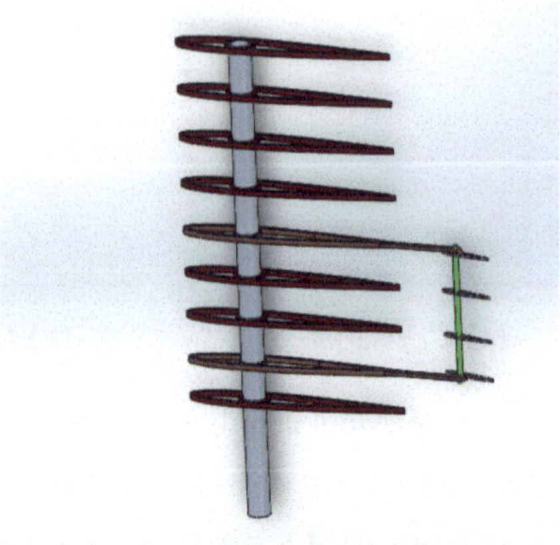

(b) Structure.

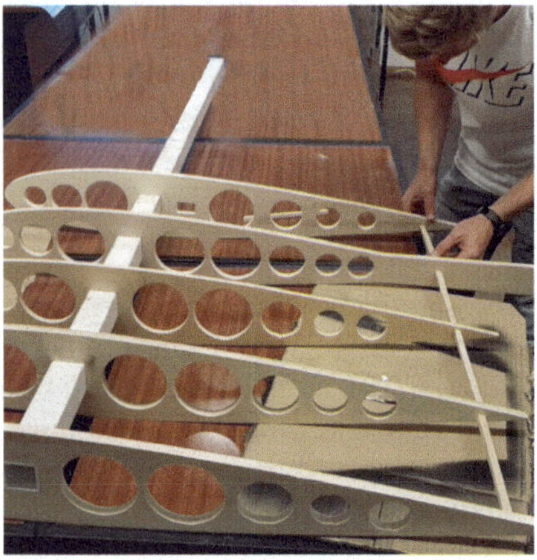

(c) Rib assembly.

(d) Completed wingsail.

Fig. 5. Design and assembly of a rectangular tail controlled wingsail.

rotates freely around its plastic axis and the main mast rotates around its base on the frame, moving the whole wingsail.

6 Conclusions

Autonomous sailing robots, and in particular the ones propelled by wingsails, present several advantages in comparison to the more common electrical or fuel propelled ASV. For this reason, the authors are developing an autonomous wingsail boat, intended for long term ocean navigation. In the scope of this project, this paper presented the background research performed on wingails operating principle and control followed by the decisions that led to the hull choice and the wingsail design option, including the wingsail airfoil and its dimensions. The wingsail will be submitted to performance tests in the next months.

Concerning future developments, the authors plan to address in detail the possible advantages that can arise by using a wingsail with a trapezoidal shape and build a wingsail with this shape for performance comparison purposes.

Funding. This work was partially financed by National Funds through the Portuguese funding agency, Fundação para a Ciência e a Tecnologia (FCT), within project UID/EEA/50014/2019.

References

1. Anderson, D.F., Eberhardt, S.: Understanding Flight. McGraw-Hill, New York (2010)
2. Atkins, D.W.: The CFD assisted design and experimental testing of a wingsail with high lift devices. Ph.D. thesis, University of Salford (1996)
3. DelMar Conde: DCmini (2019). https://www.delmarconde.pt/?page_id=19. Accessed 31 Jan 2019
4. Domínguez-Brito, A.C., Valle-Fernández, B., Cabrera-Gámez, J., de Miguel, A.R., García, J.C.: A-TIRMA G2: an oceanic autonomous sailboat. In: Robotic Sailing 2015 - Proceedings of the 8th International Robotic Sailing Conference, pp. 3–13, September 2015. https://doi.org/10.1007/978-3-319-23335-2_1
5. Elkaim, G.H.: System identification for precision control of a WingSailed GPS-guided catamaran. Ph.D. thesis, Stanford University (2001)
6. Elkaim, G.H., Boyce, C.L.: Experimental aerodynamic performance of a self-trimming wing-sail for autonomous surface vehicles. IFAC Proc. Vol. **40**(17), 271–276 (2007). 7th IFAC Conference on Control Applications in Marine Systems. https://doi.org/10.3182/20070919-3-HR-3904.00048
7. Holzgrafe, J.: Transverse stability problems of small autonomous sailing vessels. In: Robotic Sailing 2013 - Proceedings of the 6th International Robotic Sailing Conference, pp. 111–123, September 2013. https://doi.org/10.1007/978-3-319-02276-5_9
8. INNOC - Österreichische Gesellschaft für innovative Computerwissenschaften: Robotic sailing (2018). https://www.roboticsailing.org/. Accessed 16 Nov 2018
9. Kimball, J.: Physics of Sailing. CRC Press - Taylor & Francis Group, Boca Raton (2010)
10. Leloup, R., Pivert, F.L., Thomas, S., Bouvart, G., Douale, N., Malet, H.D., Vienney, L., Gallou, Y., Roncin, K.: Breizh spirit, a reliable boat for crossing the atlantic ocean. In: Robotic Sailing - Proceedings of the 4th International Robotic Sailing Conference, pp. 55–69, August 2011. https://doi.org/10.1007/978-3-642-22836-0_4
11. Lyon, C.A., Broeren, A.P., Giguère, P., Gopalarathnam, A., Selig, M.S.: Summary of Low-Speed Airfoil Data - Volume 3. SoarTech Publications, Virginia Beach (1997). https://m-selig.ae.illinois.edu/uiuc_lsat/Low-Speed-Airfoil-Data-V3.pdf
12. Microtransat: The microtransat challenge. https://www.microtransat.org/index.php. Accessed 9 Nov 2018
13. Miller, P., Beeler, A., Cayaban, B., Dalton, M., Fach, C., Link, C., MacArthur, J., Urmenita, J., Medina, R.Y.: An easy-to-build, low-cost, high-performance sailbot. In: Robotic Sailing 2014 - Proceedings of the 7th International Robotic Sailing Conference, pp. 3–16, September 2014. https://doi.org/10.1007/978-3-319-10076-0_1

14. Miller, P.H., Hamlet, M., Rossman, J.: Continuous improvements to USNA SailBots for inshore racing and offshore voyaging. In: Robotic Sailing 2012 - Proceedings of the 5th International Robotic Sailing Conference, September 2012. https://doi.org/10.1007/978-3-642-33084-1_5
15. Neal, M., Sauzé, C., Thomas, B., Alves, J.C.: Technologies for autonomous sailing: wings and wind sensors. In: Proceedings of the 2nd International Robotic Sailing Conference, pp. 23–30, July 2009
16. Olson, S. (ed.): Autonomy on Land and Sea and in the Air and Space: Proceedings of a Forum. The National Academies Press, Washington, DC (2018). https://doi.org/10.17226/25168
17. SailBot: Sailbot—international robotic sailing regatta (2018). https://www.sailbot.org/. Accessed 16 Nov 2018
18. Sauzé, C., Neal, M.: An autonomous sailing robot for ocean observation. In: Proceedings of the 7th Towards Autonomous Robotic Systems (TAROS) Conference, pp. 190–197, September 2006
19. Sauzé, C., Neal, M.: Design considerations for sailing robots performing long term autonomous oceanography. In: Proceedings of the International Robotic Sailing Conference, pp. 21–29, May 2008
20. Sauzé, C., Neal, M.: MOOP: a miniature sailing robot platform. In: Robotic Sailing - Proceedings of the 4th International Robotic Sailing Conference, pp. 39–53, August 2011. https://doi.org/10.1007/978-3-642-22836-0_3
21. Schlaefer, A., Beckmann, D., Heinig, M., Bruder, R.: A new class for robotic sailing: the robotic racing micro magic. In: Robotic Sailing - Proceedings of the 4th International Robotic Sailing Conference, pp. 71–84, August 2011. https://doi.org/10.1007/978-3-642-22836-0_5
22. Selig, M.S., Donovan, J.F., Fraser, D.B.: Airfoils at Low Speeds. H.A. Stokely, Virginia Beach (1989). https://m-selig.ae.illinois.edu/uiuc_lsat/Airfoils-at-Low-Speeds.pdf
23. Selig, M.S., Guglielmo, J.J., Broeren, A.P., Giguère, P.: Summary of Low-Speed Airfoil Data - Volume 1. SoarTech Publications, Virginia Beach (1995). https://m-selig.ae.illinois.edu/uiuc_lsat/Low-Speed-Airfoil-Data-V1.pdf
24. Selig, M.S., Lyon, C.A., Giguère, P., Ninham, C.P., Guglielmo, J.J.: Summary of Low-Speed Airfoil Data - Volume 2. SoarTech Publications, Virginia Beach (1996). https://m-selig.ae.illinois.edu/uiuc_lsat/Low-Speed-Airfoil-Data-V2.pdf
25. Silva, M.F., Friebe, A., Malheiro, B., Guedes, P., Ferreira, P., Waller, M.: Rigid wing sailboats: a state of the art survey. Ocean Eng. **187**, 106–150 (2019). http://www.sciencedirect.com/science/article/pii/S0029801819303294
26. Springer, P.J.: Outsourcing War to Machines - The Military Robotics Revolution. Praeger Security International, Santa Barbara (2018)
27. Stelzer, R.: Autonomous sailboat navigation - novel algorithms and experimental demonstration. Ph.D. thesis, De Montfort University (2012)
28. Tools, A.: Airfoil tools (2018). http://airfoiltools.com/. Accessed 4 Feb 2019
29. Tools, A.: Airfoil tools (2018). http://airfoiltools.com/airfoil/details?airfoil=e169-il#polars. Accessed 4 Feb 2019

Collaborative Robots for Industry Applications

Augmented Reality System for Multi-robot Experimentation in Warehouse Logistics

Marcelo Limeira[1], Luis Piardi[1,2(✉)], Vivian Cremer Kalempa[1,3], André Schneider[1], and Paulo Leitão[2]

[1] Universidade Tecnológica Federal do Paraná (UTFPR), Av. Sete de Setembro, Curitiba, PR 80230-901, Brazil
limeira@alunos.utfpr.edu.br,andreoliveira@utfpr.edu.br
[2] Research Center in Digitalization and Intelligent Robotics (CeDRI), Instituto Politécnico de Bragança, Campus de Santa Apolónia, 5300-253 Bragança, Portugal
{piardi,pleitao}@ipb.pt
[3] Santa Catarina State University, Luiz Fernando Hastreiter St., 180, São Bento do Sul, SC, Brazil
vivian.kalempa@udesc.br

Abstract. The application of tools as augmented reality has been developing innovative solutions for the industrial scenario. In this context, this work presents an industrial plant of a warehouse, where augmented reality is used to represent virtual loads to be transported by multiple small mobile robots. The results promote an application developed in ROS, with virtual and real objects sharing the same environment, producing an excellent scenario to development and experimentation to new approaches for automation in warehouses or smart factories.

Keywords: Augmented reality · Multi-robots · Warehouse · Smart factories

1 Introduction

New paradigms of industry demand the optimization of time and service with high-quality requirements [7]. The evolution of autonomous robotics in industrial environments is promising, and industrial activity has an immense interest in the application of robotics to reduce costs, increase productivity, customize and raise the quality of its services. Cooperation between Multi-Robot System (MRS) has a crucial role in this context, such as Amazon's automatic logistics warehouse with Kiva system [6]. However, to obtain these desired impacts, several validations are necessary to ensure the reliability and robustness of MRS operating in a complex and real environment. In this context, one of the biggest problems is faced for the technological update with autonomous agents in industrial environments.

M. F. Silva et al. (Eds.): ROBOT 2019, AISC 1092, pp. 319–330, 2020.
https://doi.org/10.1007/978-3-030-35990-4_26

In addition to the challenge of technology comprehension and the concepts behind MRS operations, it is not rationally feasible to stop or slow down the current processes of a classic warehouse to perform experimental cycle, as debugging, validation, tuning. In this context, the development of new technologies to the automation of smart warehouses is complex, due to the impossibility of real experimentation. New assumptions and solutions to optimize the autonomy of the robots in loads transportation to transform a classical warehouse in an autonomous one are not rigorously tested and prove, without compromise the global operation.

Several academic papers have been addressing the use of augmented reality for multi-robot systems [3] or robot swarms [1,8,10]. In [8], an augmented reality tool is presented to analyze and debug data and behaviors of the robot swarms and its information (such as diagnosing robot control code bugs, identifying sensor hardware faults or calibration issues, and verifying actuator hardware operation) for developers to evaluate their systems. It is necessary that the robots have a Wifi or Bluetooth interface to communicate through the JSON package. In this line of research, two papers [1,10] present an augmented reality model for the well-known kilobot robots [16]. The work described in [1] the authors propose an electronic device with a glass surface (kilogrid), where the kilobots are located to move. This device has the ability to communicate with robots through an infrared interface. Thus, the Kilogrid allows robots to operate in augmented reality and can be used to collect experimental data and create virtual sensors. One point that hinders the spread of this system is its high cost, compared to systems that use overhead cameras to locate and create a virtual environment interacting with a real environment like [5,11].

Already in work [10], it is presented a system for the augmented reality of the kilobots, where the location of the robot is performed by four overhead cameras. The application of this tool is presented with two demonstrations of location and assignment of identifiers for each robot autonomously and, a demonstration presents a foraging scenario where fifty robots collect virtual load from a source location and deposit it at a destination.

Real industrial environments are susceptible to many problems and situations that are difficult to predict, requiring laboratory tests and simulations to locate and prevent potential problems [2]. In this context, Augmented Reality (AR) provides a methodology for testing techniques and tools that can be reproduced in reality, much more flexible than a simulation environment, as it allows developers to design a variety of scenarios by introducing any kind of virtual objects into real-world experiments [4,9]. In this way, AR becomes a valuable tool for the industries that intend to upgrade or optimize their systems, where this avoid complex simulation models, expensive hardware setup and a highly controlled environment, in the various stages of a multi-robot system development [4].

The main contribution of this work and what differentiates it from the other augmented reality works discussed above is that it presents a real industrial application, following in scales the dimensions and sections of the warehouse of a company. Besides, it is not limited to only displaying, tracking, and identification of robots, but also represents the virtual process and loads to be carried by the robot. In this way, a study is being conducted for the automation of the warehouse of a company, using autonomous robots to transport products inside a real warehouse scenario.

An outline of this paper is as follows. In Sect. 2, the description of the industrial plant used as the basis for the AR system is carried out. Section 3 describes the architecture of the system, namely the arena and the robot. Section 4 describes the tools used to develop the AR system. Experimental results are presented in Sect. 5, and Sect. 6 concludes the paper.

2 Warehouse Logistics

This work is inspired by a real Brazilian warehouse logistics scenario that requires a full automation process. In this way, all sectors such as maintenance, reloading of robots, loading and unloading of products, warehouses, sorting, among others, were faithfully reproduced in the model. Forklift trucks can perform different actions inside the represented warehouse, for example, loading and unloading trucks, going to charging station or maintenance area and storing and removing loads from shelves. Figure 1 illustrates the floor design of the warehouse; then, each region will be explained in detail below.

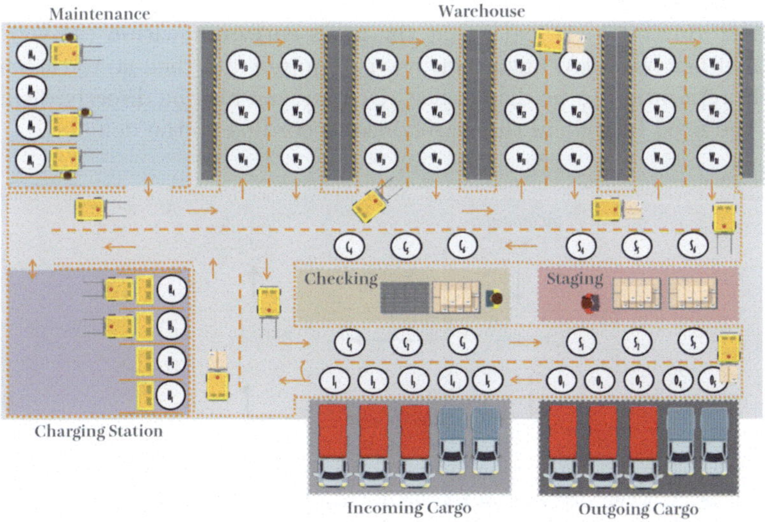

Fig. 1. Logistic representation and floor plan of the reproduced warehouse.

The area called Incoming Cargo will receive the trucks with the loads that must be entered in the warehouse to be stored. As soon as it parks in the indicated location, forklift trucks will begin the process of unloading these trucks, transporting the loads to the place called checking. At this stage, the system must be loaded with all the information about the load it has just received making the proper distribution according to the destination of each one.

All movement that the system controls is performed through tasks, for each motion a given load; a new job is created. These tasks are then distributed to the available forklifts. When no forklift is available, it will be inserted into a list so that, once a forklift is available, it is then executed. This dynamic is repeated throughout the operations inside the warehouse. When a particular load is able to be immediately sent, i.e., the truck that will carry this load is at the exit site, the system will define a task to address this load from the checking area to the staging area, otherwise the system will assign a warehouse position for the storage of this cargo to be transported later.

After a final verification, the object will be released to be sent to the cargo area called Outgoing Cargo; this task will be performed by another forklift trucks chosen by the system. All these loading, storing, and unloading tasks will be delivered simultaneously. Therefore, it is the function of the order to achieve the management of the forklifts so that there is no collision or congestion and that all tasks are executed in the most efficient way possible.

The structure of the company also has areas of maintenance and loading of the batteries of the forklifts. As soon as load levels are detected below the stipulated, the forklift is sent to the recharging area. Likewise, when any maintenance need is recognized, the robot is also forwarded to the area responsible for maintenance.

Figure 2 illustrates the path that the loads carry out within the warehouse. They can either be stored until their use is required and then go to the checkout area and preparation to be dispatched. Or they can also be directly dispatched, without the need to wait in the warehouse, according to the demand.

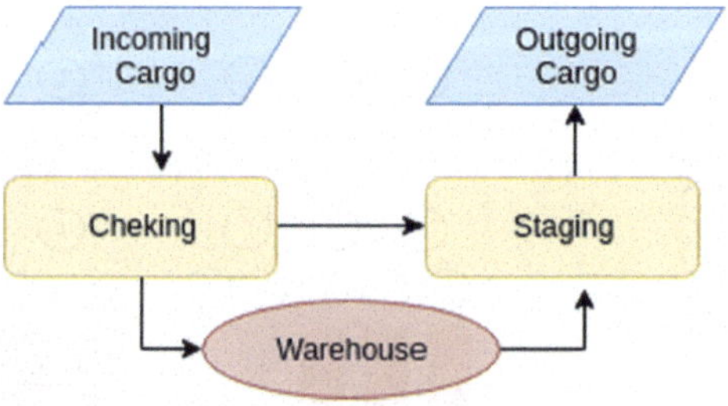

Fig. 2. Flow diagram of the loads inside the warehouse logistics.

3 System Architecture

The entire physical structure of the arena is composed of several elements that perform different functions. These elements can be classified as static and dynamic. The static elements compose the whole fixed structure of the arena as the table that supports all the structure, models, cameras, reflectors. Robots make up the dynamic elements. They can move about the structure of the arena and interact with the other features present. The structures of these elements will be better detailed in the following topics.

3.1 Arena

The arena is composed of all fixed elements present in the structure, and it has a table with a dimension of approximately 2×1.4 m. Its surface was covered by a vinyl tarp that is plotted with the industry plant used as a model, in this case, a warehouse. In this way, it is possible to visualize the different areas of the model even when there is no virtual projection on the structure. As the cost of making and plotting a new cloth is considerably low, we can represent various logistics and industry models very quickly and cheaply. Another important factor concerns the material used to make this canvas. Its smooth surface provides low friction between it and the robot's sliding surface, while the contact between it and the robot's wheels shows excellent adhesion. This fact considerably facilitated its movement because it reduces the sliding presented by the robot.

On the table were also deposited various objects. These objects represent the physical structures of the model. In the case in question were added shelves where the boxes moved by the robots will be stored. These objects are directly related to the model being simulated. For each model of logistics or industry, different objects can be added. There is also the possibility of not inserting any physical object into the model. However, this is not recommended because the main idea of the whole structure is the interaction of real and virtual objects that can represent the most reliable possible situations and scenarios in the industry.

Two bars, on the sides of the table, support a third bar where the cameras and reflectors are fixed. These bars have a device that allows easy adjustment of height. In this way, we can regulate the height at which the cameras will be arranged. This adjustment enables models on a smaller scale to be executed. In this case, the cameras would be positioned at a shorter distance, thus allowing the visualization of more details.

The arena has two cameras, one is used to read the QR-CODE tags, and the other provides the image used in the generation of augmented reality. It was decided to use two cameras in this project because the configuration of colors, brightness, and contrast that obtained better readings of the tags present a lousy condition for visualization. In this way, we can apply the best configuration for each camera, taking into account its specific purpose.

3.2 Robot

Figure 3 shows a real representation of the robot used in this project to execute forklift truck actions. The main components that form part of its structure will be detailed below.

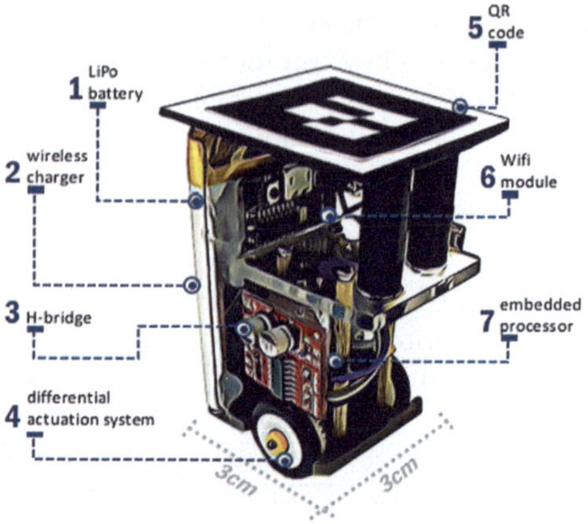

Fig. 3. Micro robot diagram.

The robot architecture was designed to be small in size and easy to assemble, using low-cost components but providing all the communication, mobility, and scalability requirements for a multi-robot or swarm system. This care has been taken to enable its production on a large scale.

The microcontroller used is an Esp12-E V3 nodeMCU, this device has an ESP82266 processor with RISC architecture and operates in 32 bits with a frequency of operation that can vary between 80 and 160 MHz. Its memory is of the flash type of 4 MB besides, having a built-in wireless network adapter. Each robot has a QR-code tag on its top that represents the ID of the respective robot. This tag is printed on plain paper and is the only physical element needed, in addition to the camera, for the exact location of the robot in the arena.

Despite the small size, the robot has all its structure of communication and control via ROS, through messages of type cmd_vel, which is the type of standard message used to control mobile robots. This standardization allows that once the system responds satisfactorily in the simulated environment, it can be easily applied in operations with robots in real sizes without the need of significant changes, being necessary only its connection to the server. The robot also has an induction charging device, so no connection is required to recharge the battery.

4 Software Architecture

All the software used in this project uses the ROS communication infrastructure. The Robot Operating System (ROS) is a flexible framework for writing robot software. It is a collection of tools, libraries, and conventions that aim to simplify the task of creating complex and robust robot behavior across a wide variety of robotic platforms [13].

The collaborative nature of ROS provides the development and improvement of several tools used in robotics, that not be achieved if developed by isolate way.

The ROS community brings together all those tools that are continually being updated by diverse people around the world. A significant advantage of ROS is the standardization of protocols. In this way, a specific tool produced with the ROS structure can be connected to any device that has a ROS communication interface. Another great advantage is the modularization of the system where each module is called a node. Each node executed by the system can communicate with the others, and there may be a node running on different devices, such as robot control systems where each robot performs its own software. Some of the packages used in this project have been made available in the ROS community, as shown in the following list:

- **Usb_Cam.** The node is responsible for the publication of the information captured by webcams on ROS topics. From this node, two topics are created, one for each camera [12].
- **Ar_Track_Alvar.** The package that performs the image analysis identifies and returns the spatial coordinates of all tags identified. Due to the robust calibration process where the exact measures of the tags are informed, it can identify, even using a 2d image, the coordinates in 3d of the tags identified [15].
- **RVIZ (Ros visualization).** Is a three-dimensional visualizer used to visualize robots, the environments they work in, and sensor data. It is a highly configurable tool, with many different types of visualizations and plugins [14]. With this tool, we can create virtual objects and make them interact with any real object. It is the tool used to produce augmented reality.

The other nodes were developed exclusively for this work and not yet published on the ROS platform:

- **Swarm_Control.** This node is responsible for the entire movement of each robot. Therefore Swarm_Control realizes the planning of the trajectory and executes it. While performing this task, this packet sends the system information about the robot's current state and the task being executed.
- **Robot_N.** The robot's microcontroller runs this software, it is responsible for executing all commands sent to the robot. There will be an execution of this node for each robot connected to the system, where the letter N represents the robot ID. This node is also responsible for publishing the state of the robot's battery.
- **Scheduler.** The node was responsible for generating all static virtual elements.

Once the system launches, the camera_arena and camera_tag nodes access their cameras and publish the image in the image_raw topic and publish the parameters of the image as resolution and encoding in the topic camera_info. The node responsible for identifying the tags is ar_track_alvar, this node subscribes to the topics image_raw and, camera_info analyzes the image and publishes it in the topic visualization_marker the data of the tags identified in the picture.

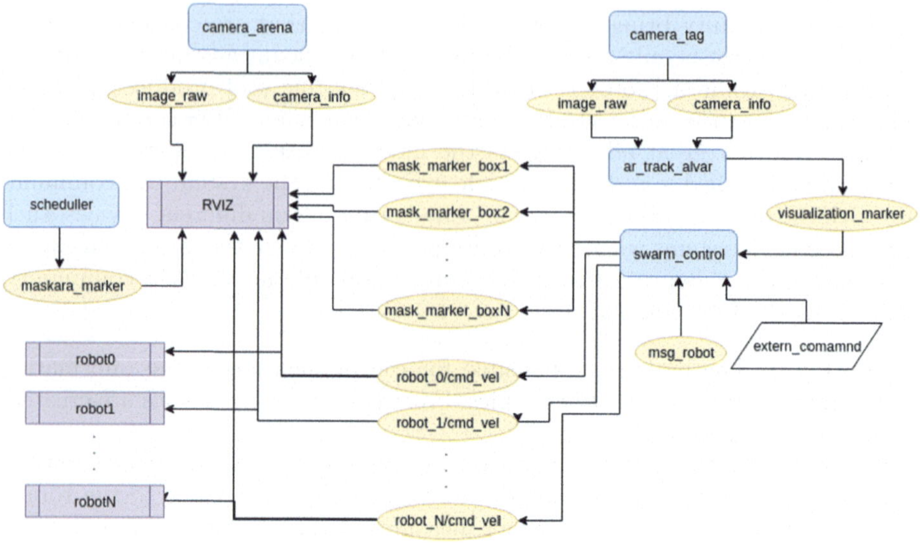

Fig. 4. Software architecture data flow.

Another node that runs at system startup is the scheduler. It publishes all fixed virtual map tags on the topic maskara_marker. These markings are alphanumeric codes within a white circle that identify positions on the map. Another signing performed by this node is the marking of the different sectors of the map. A color represents each sector (Fig. 4).

The nodes robot0, robot1 to robotN, are the nodes that control each robot. When connected to the system, each robot subscribes to its control node (called as robot_N/cmd_vel) and waits for a control command to be sent.

The last module that the system initializes is RVIZ, this module is responsible for the generation of all graphics components of the system, from static objects to dynamic objects such as boxes loaded by robots.

The Swarm_Control node is the primary node, and it performs the trajectory planning for a given task, the control of the robot during the execution of this task besides all the management and coordination of all the robots in operation. Also checks the need for recharging or repair. All this process is done centrally as follows: When initialized, this node waits for the publication in the topic msg_robot, this topic receives three parameters: robot ID, target position and a boolean variable that determines if the robot should carry a box. The current location of the robot is automatically identified by the algorithm from the coordinates provided by the system. After collecting these coordinates, a calculation is performed to determine the closest marked point of the robot, and this point is used as a starting point for the robot's trajectory planning. This planning is done considering not only the possible routes but also the movement of the other robots that may be performing other tasks in parallel. Once the trajectory planning has been carried out, identifying all the points between the origin and the target, the control algorithm is initialized, and the robot starts executing the requested task.

4.1 Augmented Reality

There are two types of virtual objects present in this project, static and dynamic. Static objects are map position markers and sector markers. Dynamic objects are the position marker of the robot, the boxes that will be moved throughout the warehouse, and a dashboard with information about the tasks that are being performed. It is worth mentioning that any virtual object present, such as position markers, virtual boxes, and information panel, can easily have their visualization enabled or disabled through RVIZ.

Static Objects. Static objects have different characteristics concerning dynamic objects. They are loaded at system startup and have their positions already defined in the system. The only case where adjustments of these objects are necessary is when there are some changes in the locations of the cameras or some adjustments in their configurations. In these cases, there are offset parameters that can correct any difference of positions in the x or y-axis. If there is a change in camera height, there is a parameter that adjusts the resolution of the projected image. If there were no such parameters, it would be necessary to remap all points in the arena.

Dynamic Objects. These objects are all those that present changes according to the execution of the tasks, for example, the status panel in the upper corner of the preview screen. This panel will inform the identification codes of the robots that are performing a particular task, the positions x and y of each robot and finally the status in which it is. In the case in question, there are four possibilities: carrying, waiting, recharging, and defective. Various other information could be added in this panel as the color of each robot, and the refresh rate coordinates the speed of each robot among many others. The determining factor for choosing information from this panel is the type of test performed.

The main activity of the company selected as the model for the project is the transportation of products, and these products are represented by virtual boxes generated by the system. When the system receives the tasks of transport a particular load from one point to another, a rectangular object, which symbolizes the box, is generated at a certain distance from the center of mass of the robot. This object accompanies all linear and angular movement of the robot during the entire path from the start point to the end point when the robot reaches the final position the load is moved from the robot to the selected area. This process is repeated throughout the storage area from the cargo entry, sorting, storage, and exit area.

The color of the box will vary according to the state in which it is. The following list outlines the possible colors presented and what each represents:

- **Green:** Carrying.
- **Red:** Incoming Cargo.
- **Yellow:** Warehouse.
- **Blue:** Outgoing Cargo.

5 Results

The final configuration of the arena can be seen in Fig. 5. It is possible to note that due to the elements inserted in the table and the plot of the plant on the canvas, and is possible to identify and locate the structure of the transport company, representing the real plant of the industry.

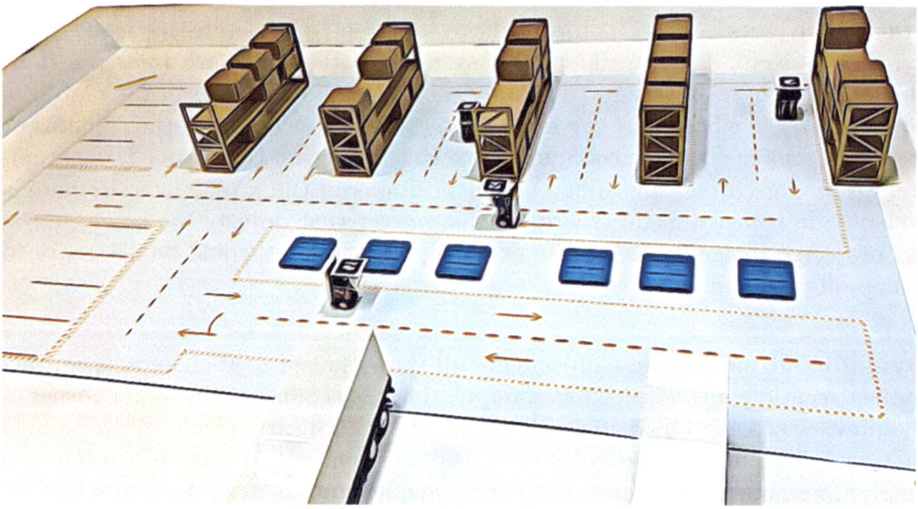

Fig. 5. Arena - scale representation of the warehouse logistics.

Figure 6 show the final result with the application of all virtual elements in augmented reality. Is possible to observe the movement of the virtual boxes in a real environment, represented in green color, performed by each robot. When the box is deposited in its respective area, its color is altered, thus identifying its state. The projection of the way-points plotted in the image helps to identify the path through which the robot should travel. Therefore, it is possible to identify the trajectory planning failures.

The information panel located in the upper left corner of the image displays a variety of real-time information about the tasks being performed. It is possible to observe which robots are performing a particular task and which robots are idle. It is worth remembering that the information shown in this panel can be customized according to the tests according to the occasion. We can enter various other details such as battery charge, tracking system refresh rate, robot speed, among many others.

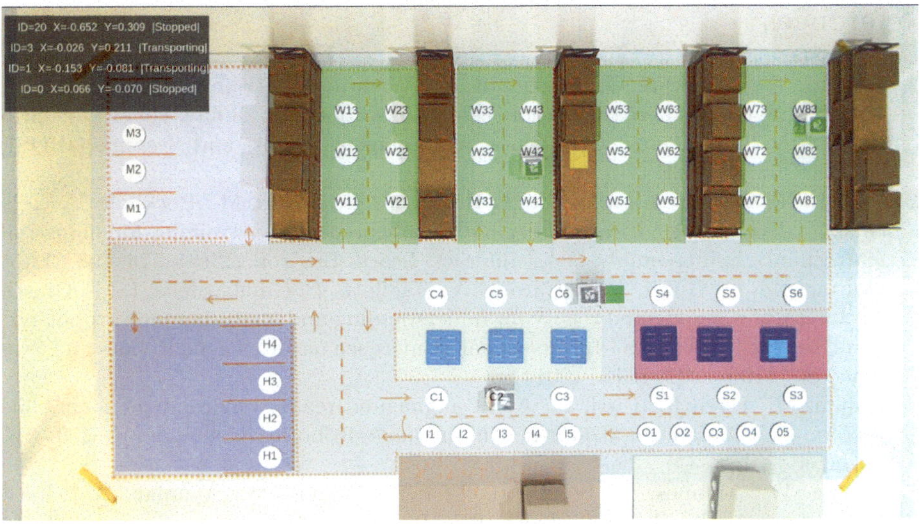

Fig. 6. Virtual arena - real and virtual elements sharing the same environment.

The construction of this structure enabled the development of the robot control algorithm. Tests were performed, in the most varied situations, to verify the ability of the robot to avoid obstacles, to achieve the planning of the trajectory, and to remain in it. Thus, several faults were corrected until the algorithm reached an adequate level of response. The use of a real robot with characteristics very similar to those used in the industry contributed to the improvement of this algorithm.

6 Conclusion and Future Works

This work developed a small scale industrial environment using the ROS framework and Augmented Reality. The proposed system allows analyzing information and state of the robot in real-time. It also enables the interaction of the real robot with virtual objects that represent loads of a warehouse, which are transported by the robots according to its desired destination. With this real-virtual interaction, it is possible to use this tool to develop systems of automation of the real environment, eliminating potential errors and failures that would be observed only in an actual order, thus avoiding wasted resources and time. After the validation of the proposed system, the intention is to provide a robust method to be implemented in the real scenario.

As future work, we intend to deepen in the development of the warehouse automation system. In this way, we want to develop virtual sensors (such as sonar and laser scan) to raise the world perception of the robot, to establish a system of preemption and allocation of tasks and to apply techniques of fault tolerance for the jobs performed by robots.

References

1. Antoun, A., Valentini, G., Hocquard, E., Wiandt, B., Trianni, V., Dorigo, M.: Kilogrid: a modular virtualization environment for the kilobot robot. In: 2016 IEEE/RSJ International Conference on Intelligent Robots and Systems (IROS), pp. 3809–3814. IEEE (2016)
2. Gavish, N., Gutiérrez, T., Webel, S., Rodríguez, J., Peveri, M., Bockholt, U., Tecchia, F.: Evaluating virtual reality and augmented reality training for industrial maintenance and assembly tasks. Interact. Learn. Environ. **23**(6), 778–798 (2015)
3. Ghiringhelli, F., Guzzi, J., Di Caro, G.A., Caglioti, V., Gambardella, L.M., Giusti, A.: Interactive augmented reality for understanding and analyzing multi-robot systems. In: 2014 IEEE/RSJ International Conference on Intelligent Robots and Systems, pp. 1195–1201. IEEE (2014)
4. Gianni, M., Ferri, F., Pirri, F.: ARE: augmented reality environment for mobile robots. In: Conference Towards Autonomous Robotic Systems, pp. 470–483. Springer (2013)
5. Hoenig, W., Milanes, C., Scaria, L., Phan, T., Bolas, M., Ayanian, N.: Mixed reality for robotics. In: 2015 IEEE/RSJ International Conference on Intelligent Robots and Systems (IROS), pp. 5382–5387. IEEE (2015)
6. Li, J.T., Liu, H.J.: Design optimization of Amazon robotics. Autom. Control Intell. Syst. **4**(2), 48–52 (2016)
7. Lu, Y.: Industry 4.0: a survey on technologies, applications and open research issues. J. Ind. Inf. Integr. **6**, 1–10 (2017)
8. Millard, A.G., Redpath, R., Jewers, A., Arndt, C., Joyce, R., Hilder, J.A., McDaid, L.J., Halliday, D.M.: ARDebug: an augmented reality tool for analysing and debugging swarm robotic systems. Frontiers Robotics AI (2018)
9. Nee, A.Y., Ong, S., Chryssolouris, G., Mourtzis, D.: Augmented reality applications in design and manufacturing. CIRP Ann. **61**(2), 657–679 (2012)
10. Reina, A., Cope, A.J., Nikolaidis, E., Marshall, J.A., Sabo, C.: Ark: augmented reality for kilobots. IEEE Robot. Autom. Lett. **2**(3), 1755–1761 (2017)
11. Reina, A., Salvaro, M., Francesca, G., Garattoni, L., Pinciroli, C., Dorigo, M., Birattari, M.: Augmented reality for robots: virtual sensing technology applied to a swarm of e-pucks. In: 2015 NASA/ESA Conference on Adaptive Hardware and Systems (AHS), pp. 1–6. IEEE (2015)
12. ROS: ar_track_alvar ros wiki. http://wiki.ros.org/ar_track_alvar (2019). Accessed 9 May 2019
13. ROS: Ros.org. https://www.ros.org/about-ros (2019). Accessed 8 May 2019
14. ROS: rviz ros wiki. http://wiki.ros.org/rviz (2019). Accessed 9 May 2019
15. ROS: Usb_cam ros wiki. http://wiki.ros.org/usb_cam (2019). Accessed 9 May 2019
16. Rubenstein, M., Ahler, C., Nagpal, R.: Kilobot: a low cost scalable robot system for collective behaviors. In: 2012 IEEE International Conference on Robotics and Automation, pp. 3293–3298. IEEE (2012)

Collision Avoidance System with Obstacles and Humans to Collaborative Robots Arms Based on RGB-D Data

Thaddeu Brito[1(✉)], José Lima[1,2], Pedro Costa[2,3], Vicente Matellán[4], and João Braun[5]

[1] Research Centre in Digitalization and Intelligent Robotics and IPB, Bragança, Portugal
{brito,jllima}@ipb.pt
[2] INESC TEC - INESC Technology and Science, Porto, Portugal
[3] Faculty of Engineering of University of Porto, Porto, Portugal
pedrogc@fe.up.pt
[4] Research Institute on Applied Sciences in Cybersecurity, León, Spain
vicente.matellan@unileon.es
[5] Federal University of Technology - Paraná, Toledo, Brazil
jneto.1996@alunos.utfpr.edu.br

Abstract. The collaboration between humans and machines, where humans can share the same work environment without safety equipment due to the collision avoidance characteristic is one of the research topics for the Industry 4.0. This work proposes a system that acquires the space of the environment through an RGB-Depth sensor, verifies the free spaces in the created Point Cloud and executes the trajectory of the collaborative manipulator avoiding collisions. It is demonstrated a simulated environment before the system in real situations, in which the movements of pick-and-place tasks are defined, diverting from virtual obstacles with the RGB-Depth sensor. It is possible to apply this system in real situations with obstacles and humans, due to the results obtained in the simulation. The basic structure of the system is supported by the ROS software, in particular, the MoveIt! and Rviz. These tools serve both for simulations and for real applications. The obtained results allow to validate the system using the algorithms PRM and RRT, chosen for being commonly used in the field of robot path planning.

Keywords: Collaborative robots · Manipulator path planning · Collision avoidance · RGB-D

1 Introduction

The ability to predict and avoid collisions for a robotic manipulator is just one of many other perspectives of Industry 4.0. Another view of Industry 4.0 is the collaboration between robots and humans, in which humans can share the same

M. F. Silva et al. (Eds.): ROBOT 2019, AISC 1092, pp. 331–342, 2020.
https://doi.org/10.1007/978-3-030-35990-4_27

work environment without additional safety equipment. Therefore, the ability of collaborative robots to acquire the working environment and to plan (or re-plan) their own movements avoiding obstacles or humans around, is of great importance to the industrial sector.

To promote the ability to acquire the working environment, Red Green Blue Depth (RGB-D) sensors can be used [1]. These sensors help in the acquisition and perception of the environment so that the system can make the path planning with restrictions of occupied spaces. This type of device has high popularity, not only by this feature, but also by the low cost. An example of this is the Kinect sensor.

Based on the pick-and-place movements of a collaborative robotic manipulator, it is intended to develop a security system capable of avoiding the collision of manipulators with objects and humans. That is, giving anthropomorphic manipulators the ability to avoid colliding with obstacles in the work environment while performing their movements. Through the union of some tools, as shown in Fig. 1, the collaborative robot can then perform different movements to reach the target point, different from what is applied today.

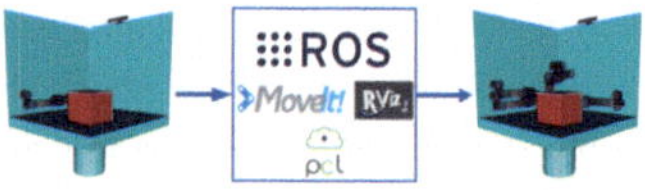

Fig. 1. Tool diagram for manipulators to avoid collisions.

This paper is organized as follows. After an introduction in this section, the related work is presented in Sect. 2. Section 3 presents the system architecture whereas Sect. 4 addressees the experimental verification and system deployment. Section 5 shows the results and the paper is concluded in Sect. 6 that also presents some future work.

2 Related Work

In elaborate and delicate operations, robotic arms must be able to interact and collaborate with human beings. In this sense, [2] points out the basic resources of collaborative manipulators that perform tasks together with humans. In this way, presents the importance of the acquire of the environment of the integration of human beings with robots. However, for the definition of the robot movements, path planning should be as fast as possible [3]. Demonstrating the reduction in search time when using Genetic Algorithm (GA) for the path planning, in relation to works done by earlier heuristic calculations. Based on the master link concept, [4] demonstrates the development of software that can plan trajectories in redundant manipulators. Applying potential field algorithms to smooth path movements of manipulators.

Through voice commands and gestures, a control platform in robotic arms performs the obstacle avoidance process with a hierarchy of consistent behaviors. The collaboration process is started or finalized also by gestures and commands by sound [5]. Another system combined with the manipulator controller is developed by [6], which applies the guided vision system to determine risk situations for operators. By analyzing the triangular mesh of an object, it is possible to prevent a collision in real time by applying the kinetostatic safety field process. This method depends on the position, velocity of the body and also the shape to determine the planned spaces [7]. Through a UR5 configured with 6 Degrees Of Freedom (DOF), [8] develops a platform that compares some algorithms in situations of pick-and-place. With the platform, it is demonstrated the differences between the algorithms: Rapidly-exploring Random Tree (RRT), RRTConnect, Kinematic Planning by Interior-Exterior Cell Exploration (K-PIECE), Probabilistic Roadmaps (PRM), PRM* and Extensive Space Trees (EST). Applying the same obstacle avoidance conditions, the comparison informs the advantages and disadvantages of each algorithm as to search time. In real obstacle avoidance situations, it is demonstrated in [9] the training of a neural network system to perform movements in robotic manipulators avoiding obstacles and people in the workspace. With data from the three manipulator joints and the environment, the neural network can detect collisions within a few milliseconds.

3 System Architecture

Many applications and tools need to be combined to develop the system that acquires the environment through an RGB-D sensor. And with this union, it is possible to process the data and define the free spaces for the motion planning for robotic manipulators (virtual or real). Accordingly, Fig. 2 illustrates a simplified block diagram of the system. Each block owns the tasks that are detailed on the following topics:

- **User:** It is possible to visualize the path found, before the robotic arm executes the movements;
- **Environment:** Work environment acquired by the RGB-D sensor;
- **Command:** The user can get system data in text format, or can also perform actions manually in any process involved;
- **Rviz/MoveIt!:** With these two ROS tools it is possible to verify the performance of the system, both the movements of the robot and the acquisition of the environment. It is also possible to perform graphical movements in real and virtual situations;
- **Kinect sensor:** RGB-D sensor used to acquire the working environment;
- `move_group`: Central part of the system, this element connects all the processes involved. Its main function is to order all processes to work together;
- **Planning scene:** It receives data in the already Point Cloud format and transforms it into Octree data. With this, the system can define the free spaces to plan a path;

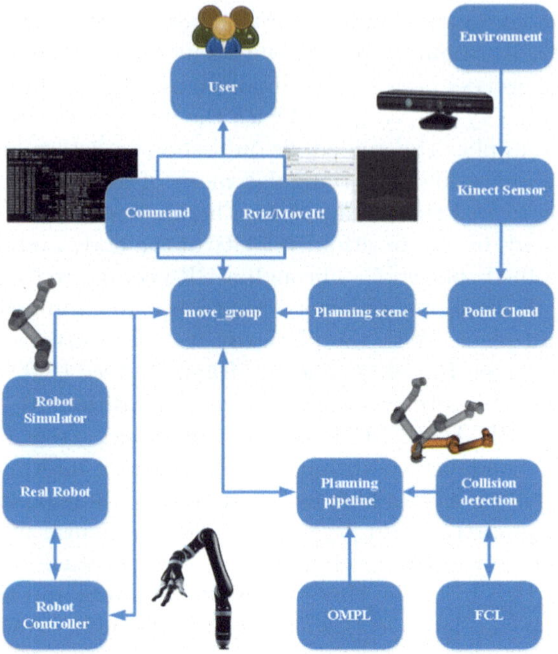

Fig. 2. Diagram of the system architecture [10].

- **Point Cloud:** Process that performs the data structure of multidimensional points provided by the sensor;
- **Robot Simulator:** Trajectory simulation process, in this way it is possible to perform the movements found in the simulated robot and then apply to the real robot. It avoids possible failures or accidents;
- **Real robot:** Execution of the movements in real manipulators;
- **Planning pipeline:** Collect information from The Open Motion Planning Library (OMPL) and Flexible Collision Library (FCL), transmitting to the `move_group` the calculated trajectory and whether or not collisions occur;
- **Collision detection:** Process with the ability to verify that the manipulator is colliding with the parts of its own body and with obstacles fixed at the tip of the tool;
- **Robot Controller:** Controllers of the robot communicate with the central element of the system, in this way they can inform the current position of the robot or receive the execution of movements;
- **OMPL:** Library with algorithms that perform the calculations of trajectory in the free spaces;
- **FCL:** This library contains algorithms for calculating collisions between the manipulators and their parts, in other words, calculate whether the robot will collide or not with its own body.

As MoveIt! finds the routes that the robotic arm must make to reach the final point, it is necessary to configure an algorithm to perform the calculation. Two algorithms were chosen for this configuration, the PRM and the RRT. Both were selected because they are commonly used by the community in path planning, so these algorithms can validate the system. Still in MoveIt!, two parameters have changed: Planning time and Planning Attempt. The first parameter has been set to 10 s, this is the time the system will have to find a possible route. The second parameter is set to the value of 15, which determines the maximum number of paths that the system must find. Therefore, if the system does not find the path in these conditions, the robotic arm will not be given the command to move. All other parameters within MoveIt! are set as default.

4 Experimental Verification and System Deployment

To verify if the developed system is able to avoid collisions, a simulated environment with the UR5 manipulator and a Kinect sensor is developed. Thus, the virtual scenario can indicate if the proposed system works as expected. The virtual manipulator is expected to perform pick-and-place movements in the free spaces, that is, without colliding with obstacles in the environment. In this Section, a brief demonstration of the entire simulation check is demonstrated. For the complete demonstration see [10,11].

The simulated environment for the system verification is elaborated with a box as obstacle, UR5 and Kinect sensor. The Fig. 3 shows the sequence of steps for this analysis. Then, the UR5 is adjusted to the original position of the scene, with coordinates (0, 0, 0) and the Kinect is centralized with UR5 but at 150 cm height, with coordinates (0, 0, 150). The Fig. 3a shows the first positioning of the elements. For now the Kinect sensor is not acquiring data from the environment.

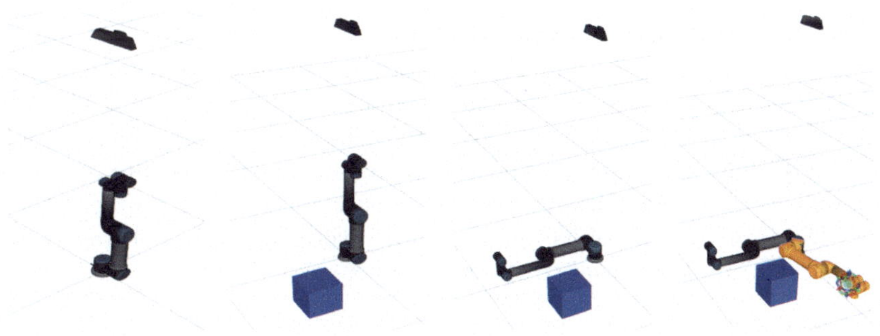

(a) First step. (b) Second step. (c) Third step. (d) Fourth step.

Fig. 3. Sequence of images that demonstrate the steps of the simulation environment to move UR5 without collisions [10].

Then a simple box is inserted into the scenario. As Fig. 3b shows, the positioning of the box is (50, 50, 12.5). This is due to the dimensions of the box, as it is configured to be 25 cm length, 25 cm width and 25 cm height. The positioning on the Z axis is half the size of the box because it is relative to its own center of mass. Then the initial and final position is defined to simulate the pick-and-place movements in UR5, as illustrated by the Fig. 3c for the initial point and Fig. 3d for the final point. The sensor is activated to acquire the data from the environment and then the proposed system is also activated to determine the path for the UR5 to reach the final point without colliding with the box. The sequence of images in the Fig. 4 shows the trajectory found.

(a) Isometric. (b) Front. (c) Side.

Fig. 4. Sequence of images that demonstrate from different views the simulation of the system that avoids collisions [10].

5 Results in Real Situations

After the validation of the proposed system in a simulation environment, it is possible to carry out the process of avoiding obstacles and humans in real situations, that is, to the real manipulator and his work environment. The tests in real situations of the developed system were realized with the robot manipulator Jaco2 6 DOF, with the same perspective to realize the movements of pick-and-place. Similar to the simulation steps, the created scenario is acquired through an RGB-D sensor. A foam object and a human were inserted as obstacles during the trajectory from the initial to the final point.

5.1 System Avoiding Real Obstacles with Real Manipulator

Differently from the simulation, before setting up the scenario to test the obstacle avoidance system, it is necessary to calibrate the RGB-D sensor because in simulation the sensor is used with ideal characteristics, and for the real case there is the problem of desynchronization between the RGB image and the depth image [12]. Figure 5 shows the sensor calibration steps, where it is possible to detect the desynchronization of the images in Fig. 5a.

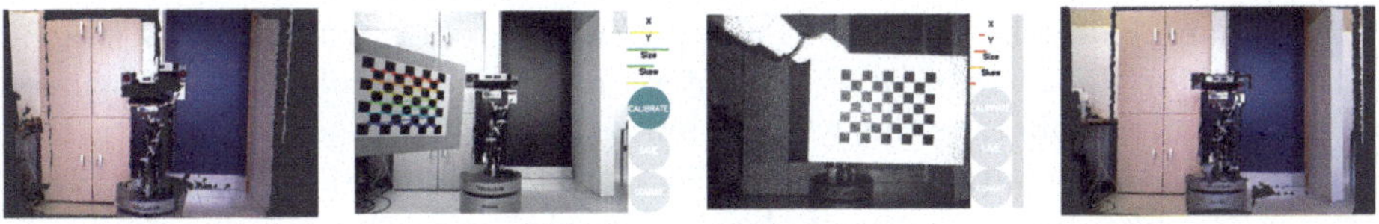

(a) Identification of desynchronization. (b) RGB image calibration. (c) Depth image calibration. (d) Result after the calibration process.

Fig. 5. Images that demonstrate the calibration process of the Kinect sensor.

The reason for the desynchronization is due to the fact that the images are generated by different sensors, calibration methods have been extensively investigated since the first version of the Kinect sensor [12]. Depth sensors can instantly and completely represent the 3D structure of the current surrounding environment. The 3D data structure of the environment is displayed in Colored Point Cloud, and with this data, robots can perceive and interact with other agents within the workspace.

Another important factor for RGB-D sensor calibration is the synchronization of the environment viewed in Rviz and MoveIt! with real workspace, that is, the Point Cloud needs to be synchronized with real robotic arm movements. Otherwise, the proposed system may determine as obstacles the parts of the robot in the Point Cloud that are not synchronized. Therefore, sensor calibration is also a fundamental part of the ground truth of the system.

With the complete calibration of the RGB-D sensor, the working environment of the robot that is the test scenario is created. The elements for this scenario can be seen in the Fig. 6. The foam object is shown in the Fig. 6a, the sensor fixed in Fig. 6b and the Jaco2 6 DOF in Fig. 6c.

(a) Foam object. (b) Kinect fixation. (c) Front view of scenario.

Fig. 6. Images that demonstrate the real scenario.

The manipulator is attached to another mobile robot, which is only serving as the base for the Jaco2 6 DOF. Therefore, no further details will be discussed about the mobile robot, as this will not influence the test. Another factor is that the model of the manipulator is different from that applied in the simulation, however the idea is to identify if the developed system can also be applied in different models of robotic arms. Figure 7 illustrates the schematic of the scenario with the placement and dimensions of all elements inserted. A rectangular shaped foam is inserted as an obstacle into the Jaco2 6 DOF workspace, it is positioned 55 cm away from the arm, or 140 cm from the sensor.

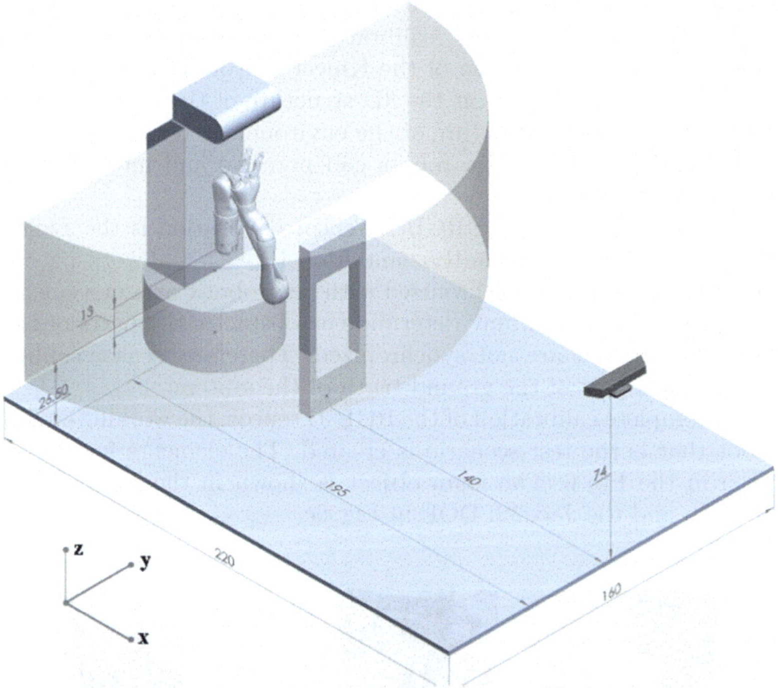

Fig. 7. Geometric arrangement of the elements with the distances, positions and workspace of the robotic arm.

After acquiring the Point Cloud, the next step is to turn these data into Octree format. In this way, the system can check the free spaces in the environment to calculate a path for the robot. The Fig. 8 shows the acquisition of the environment and the transformation to Octree at different angles. It is also possible to note that the simulation is completely synchronized with the perception from the environment.

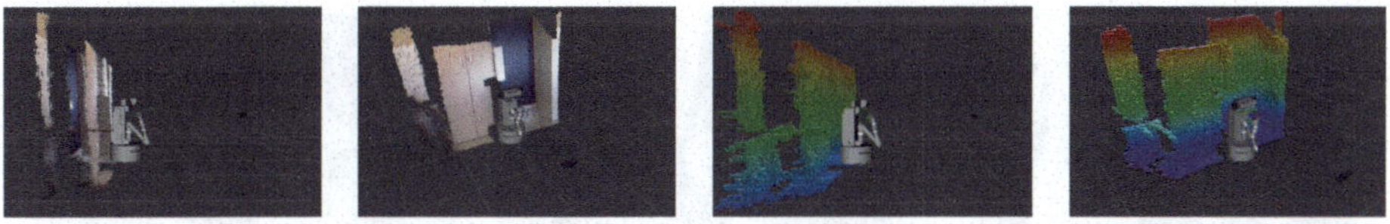

(a) Point Cloud, side view. (b) Point Cloud, isometric view. (c) Octree data, side view. (d) Octree data, isometric view.

Fig. 8. Images that demonstrate the acquired Point Cloud and the Octree transformation.

With the scenario in Octree format, is possible to set the initial and final positions to move Jaco2 6 DOF. The Fig. 9a indicates the starting point, and in Fig. 9b the end point is indicated. These points have been chosen for the pick-and-place movements that the Jaco2 6 DOF can carry through the obstacles or avoid them. For example, with these points the path planning could collide with the obstacle if the developed system fails, in other word, this case the path planning would be perpendicular to the obstacle.

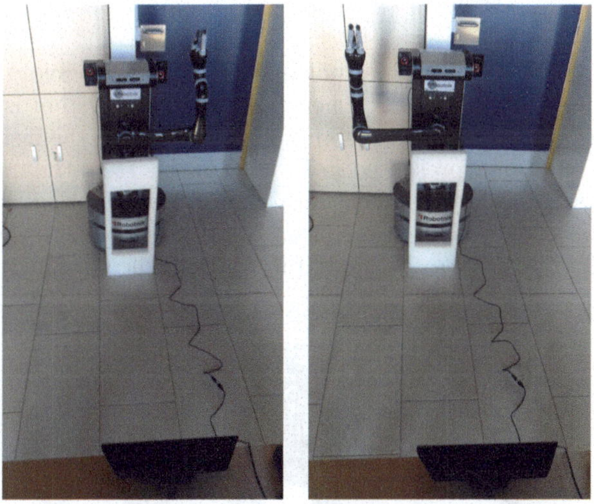

(a) Initial point. (b) Final point.

Fig. 9. Setting the initial point and the final point.

Figure 10 demonstrates this example, it is noted that with the disable sensor the arm moves in a trajectory that collides with the obstacle. There are still other possible trajectories that the robot controller can do to reach the final point, however with the aid of the sensor the motion planning can deviate from the obstacle.

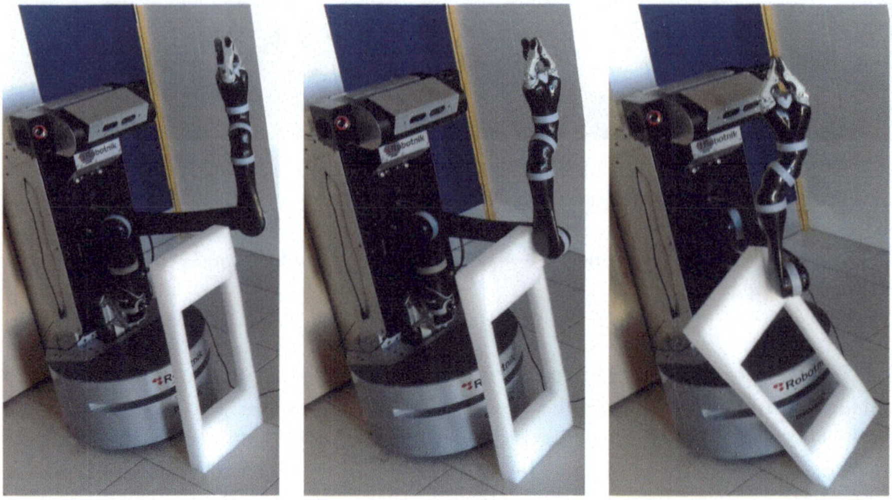

(a) First state of motion. (b) Second state of motion. (c) Third state of motion.

Fig. 10. Images that show the movement with collision in the obstacle.

Therefore, the robotic arm does not have the ability to avoid collision without activating the RGB-D sensor. Then the developed system is activated and requested that the arm return to the initial position, since the last positioning of the robot is the end point presented in Fig. 10. In this step, the manipulator will not move if the system does not find path planning without collision. But if the system finds a trajectory in the free space, the arm will move to the desired point (initial point). The result can be viewed in the sequence of images in Fig. 11.

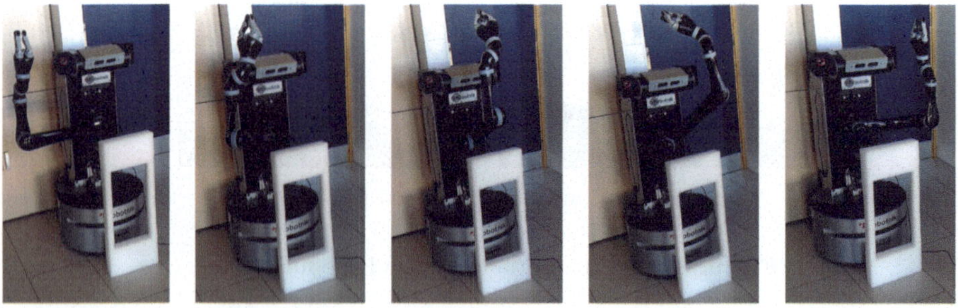

(a) First. (b) Second. (c) Third. (d) Fourth. (e) Fifth.

Fig. 11. Sequence of images demonstrating the movement without collision with the obstacle.

5.2 System Avoiding Human with Real Manipulator

To verify if the developed system can avoid collisions with humans, the same scenario from the previous test is used. Hence, in this test there is the replacement of the foam object by a human. In this sense, the purpose of this test is to create a real scenario with a human in the workspace of the robotic arm and perform pick-and-place movements with the Jaco2 6 DOF (the same points of the previous test) through acquisition by the sensor RGB-D. The Fig. 12 demonstrates in an image sequence the movement of the arm avoiding collision with the human. Similarly as in the previous test, the robot would not move if the system finds no trajectory or checks the collision.

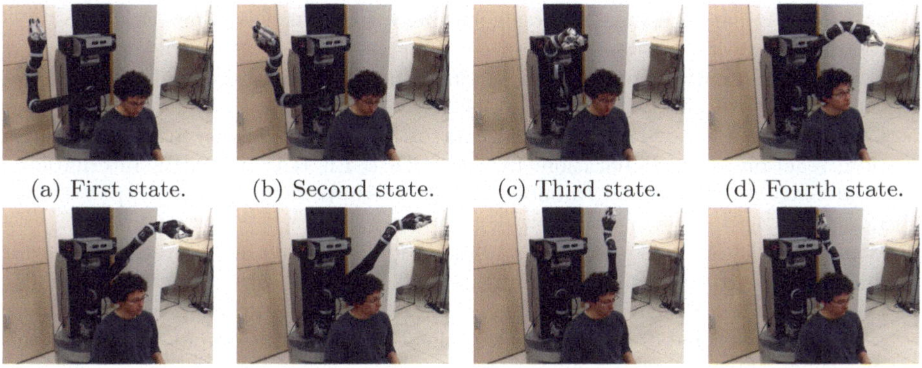

(a) First state. (b) Second state. (c) Third state. (d) Fourth state.

(e) Fifth state. (f) Sixth state. (g) Seventh state. (h) Eighth state.

Fig. 12. Sequence of images demonstrating the movement without collision with the human.

6 Conclusion

In the course of this work, a system was developed capable of path planning in robotic manipulators that avoid collision with objects and humans presented in the work environment. Collaborative (real and virtual) manipulators acted as test models and two path planning algorithms used to determine the route. The system allows to re-plan the robotic movement avoiding collisions, guaranteeing the execution of the operations, with the use of ROS as the platform, RGB-D sensor (Kinect) to acquire the environment and, a foam and human as obstacle. Therefore, the system provides the proposal for robotic applications that collaborate with humans, according to the new paradigms of Industry 4.0.

Based on the results obtained in this work, future work can be done for the developed system. In such sense, optimizing the Point Cloud acquisition can lead to less delay in robot movements, an implementation of a second RGB-D sensor to achieve a more uniform working environment and finally, the determination of an optimal resolution for the generation of Octree based on computational costs are pointed out.

Acknowledgments. This work has been partially funded by Junta de Castilla y León and FEDER funds, under Research Grant No. LE028P17 and by "Ministerio de Ciencia, Innovación y Universidades" of the Kingdom of Spain through grant RTI2018-100683-B-I00.

References

1. Mutto, C.D., Zanuttigh, P., Cortelazzo, G.M.: Time-of-Flight Cameras and Microsoft Kinect (TM). Springer, Boston (2012). ISBN: 14614380639781461438069
2. Khatib, O., Yokoi, K., Brock, O., Chang, K., Casal, A.: Robots in human environments: basic autonomous capabilities. Int. J. Robot. Res. **18**(7), 684–696 (1999)
3. Ali, M.A.D., Babu, N.R., Varghese, K.: Collision free path planning of cooperative crane manipulators using genetic algorithm. J. Comput. Civ. Eng. **19**(2), 182–193 (2005)
4. Conkur, E.S.: Path planning using potential fields for highly redundant manipulators. Robot. Auton. Syst. **52**(2–3), 209–228 (2005)
5. De Luca, A., Flacco, F.: Integrated control for pHRI: collision avoidance, detection, reaction and collaboration. In: 2012 4th IEEE RAS & EMBS International Conference on Biomedical Robotics and Biomechatronics (BioRob), pp. 288–295. IEEE (2012)
6. Wang, L., Schmidt, B., Nee, A.Y.: Vision-guided active collision avoidance for human-robot collaborations. Manuf. Lett. **1**(1), 5–8 (2013)
7. Polverini, M.P., Zanchettin, A.M., Rocco, P.: A computationally efficient safety assessment for collaborative robotics applications. Robot. Comput. Integr. Manuf. **46**, 25–37 (2017)
8. Frutuoso, I.P.: Smart collision avoidance system for a dual-arm manipulator. Master thesis (2018). https://hdl.handle.net/10216/114146
9. Sharkawy, A.N., Koustoumpardis, P.N., Aspragathos, N.: Human-robot collisions detection for safe human-robot interaction using one multi-input-output neural network. Soft Comput. 1–33 (2019)
10. Brito, T.: Intelligent collision avoidance system for industrial manipulators. Master thesis (2019). http://hdl.handle.net/10198/19319
11. Brito, T., Lima, J., Costa, P., Piardi, L.: Dynamic collision avoidance system for a manipulator based on RGB-D data. In: Iberian Robotics conference. pp. 643–654. Springer, Cham (2017)
12. Darwish, W., Tang, S., Li, W., Chen, W.: A new calibration method for commercial RGB-D sensors. Sensors **17**, 1204 (2017). https://doi.org/10.3390/s17061204

Human Manipulation Segmentation and Characterization Based on Instantaneous Work

Anthony Remazeilles[1](✉), Irati Rasines[1], Asier Fernandez[1], and Joseph McIntyre[2]

[1] TECNALIA, San Sebastian, Spain
{anthony.remazeilles,irati.rasines,asier.fernandez}@tecnalia.com
[2] Ikerbasque Research Foundation, Bilbao, Spain
joseph.mcIntyre@tecnalia.com
http://tecnalia.com

Abstract. This paper is related to the observation of human operator manipulating objects for teaching a robot to reproduce the action. Assuming the robotic system is equipped with basic manipulation skills, we focus here on the automatic segmentation of the observed manipulation, for extracting the relevant key frames in which the manipulation is best described. The segmentation method proposed is based on the instantaneous work, and presents the advantage of not depending on the force and pose sensing locations. The proposed approach is experimented with two different manipulation skills, sliding and folding, under different settings.

Keywords: Teaching by demonstration · Manipulation segmentation

1 Introduction

One of the pillar of the upcoming Industry 4.0 is the rise of collaborative robots designed to be inherently safe. In a close future, the traditional security fences will be removed from the production cells, and a closer collaboration with human operators will be possible [13]. A great expectation is placed on the capability of such robotic solution to adapt their actions to a less static environment, as well as to provide simple programming tools to the human operator. The ease of programming is indeed a key factor for bringing the collaborative robots to small and medium companies, which handle smaller production lots [7]. Instead of reprogramming completely the robot actions every-time a new product has to be produced, an approach proposed is to provide the robot with basic motion skills, as atomic and generic actions [9]. The human instructor does not need to know how are implemented these skills, and can focus on defining higher level recipes, by combining these skills.

Different strategies are possible for defining these skills. Some propose to let the robot learn them automatically, during a learning by demonstration phase

M. F. Silva et al. (Eds.): ROBOT 2019, AISC 1092, pp. 343–354, 2020.
https://doi.org/10.1007/978-3-030-35990-4_28

where a human instructor shows to the robot the action to be reproduced [2]. Others consider that experts should program such basics skills, selecting the best sensing and control strategy for conducting the task [4]. In that context, a good definition of the control task relies on the good selection of the task frame, *i.e* the frame in which the task definition is straightforward.

The work we present is dealing with the segmentation of observations of a human manipulating objects. Assuming that the robot is equipped with a database of skills, we are studying how a robot could learn from the human observation where to segment the demonstration and how to extract the manipulation characteristics for adjusting the generic robotic skills to the current manipulation. In particular we will show that a simple measurement, based on the instantaneous work can be used to segment the demonstration with very little a priori information on the manipulation taking place. Then we will detail how two skills, sliding and folding, can be characterized based on the observed kinematics.

Next section gathers related work from the literature, and Sect. 3 introduces the segmentation process using the instantaneous work. Then Sect. 4 presents several experimentations of manipulation segmentation, and Sect. 5 illustrates the extraction of the relevant task frames for the skills considered. Finally Sect. 6 concludes the paper and presents future work.

2 Related Work

The transmission of skills to a robot is frequently addressed as a *learning by demonstration* process [3]: a human operator shows the required motion to the robot (through kinesthetic teaching, teleoperation or human motion analysis), and the robot maps a model, such as a Gaussian Mixture model or an Hidden Markov Model, to the perceived behavior. Then, to generate a motion, a regression process can be applied, or the Dynamic Movement Primitive framework can be used. The observed motion can be effectively reproduced but specific motion constraints, like sliding on a plane, cannot be ensured explicitly.

The variability observed during the demonstration can be used to deduce the motion constraints. In [5], the variance of the mixture of Gaussian is used to adjust the stiffness of the controller on the different dimensions, depending on the advancement of the manipulation. In [12], several control variables are studied across trials to detect significant and systematic variations. The manipulation is then segmented when the significant variables change, and appropriate controllers are used according to the predominance of position of force variables. The execution of this approach requires nevertheless generating demonstrations from various starting and ending positions in order to let the system distinguish along time significant and non significant variables.

Given the limited success of the above methodologies in real conditions, another class of approaches considers that the definition of the basic skills should be done by robotic experts. If it requires a skilled engineer to implement these skills, it also simplifies the task demonstration process since it is then only needed to recognize the involved skills and/or to deduce the appropriate tunning parameters. The design of the basic skills relies on the definition of relevant frames in between which the specification of the motion is simplified. In [10] different key frames of manipulator poses for manipulating objects are stored in a database, and a demonstrated motions is then learned by recognizing stored reference frames, and ad-hoc motion constraints, as basic skills, are compared with the motion observed. In [11] key frames are detected through 3D vision by analyzing the temporal and spatial relations between involved objects, and are used to delimit the successive phases of a manipulation. In [4] the notion of *task frame* is extended to the notion of feature frame to not reduce the constraint-based task definition to the euclidean frame dimensions, and to better represent a complex task as a combination of constraints placed on feature coordinates. But still, a proper definition of the relevant frames poses is needed to use that framework.

The use of specific control schemes, specially designed and optimized for a given manipulation skill is also considered, like in [1]. There, skills for controlling sliding and folding manipulations are proposed, and a Kalman filter involving kinetic and kinematic information enables to cope with a certain perception uncertainty. Nevertheless it requires knowing the location of the main task frames for mapping the perception information into these control frames. In conclusion, if the skill-based manipulation permits using optimal control models (since implemented by experts), it still requires identifying the pose of the relevant task frames. Such information can be learned from observation, which is what we are proposing in this paper as well.

3 Manipulation Segmentation

A complex manipulation is composed of n successive basic manipulation skills, like sliding, inserting or folding. In order to characterize each of these skills, a first operation consists in segmenting an observed manipulation into the different skills involved. We assume that we know which skills are involved, so that we do not consider the skill recognition problem.

3.1 Human Behavior Observation

We start with the description of our observation setup, since several frame definitions provided are used in the segmentation process definition.

Our work is based on data collected during object manipulations performed by human subjects[1]. All experiments conducted have been performed using the setup illustrated in Fig. 1. The objects poses are obtained using the CODA tracking system[2]. Objects were augmented with a set of active markers tracked at 200 Hz by two 3D scanners, providing the object's pose at each instant with respect to a reference frame located on the operation table. The wrench applied onto each object is measured by 6D force sensors from Optoforce[3], at a frequency of 1 kHz. This equipment provides us both the pose of the two manipulated objects, as well as the wrenches applied onto them. We assume that such type of information could be obtained by a sensorized robotic system.

[1] The experiments were approved by a local Ethics Evaluation Committee.
[2] http://codamotion.com/.
[3] https://optoforce.com/.

The *insertion* experiment involves two objects (see Fig. 1). One is held by the human and the other is static and fixed to the table, above the force sensor. It contains a sliding manipulation, followed by a folding procedure. Figure 2 illustrates the main frames involved. $\mathcal{F}_s$ stands for the reference frame of the static receptacle, $\mathcal{F}_b$ is the body frame of the moving object being successively translated and then rotated to be inserted into the receptacle, and $\mathcal{F}_c$ is a frame attached to a contact point, which orientation is aligned with the one of $\mathcal{F}_b$. We assume we know from the CAD model of the mobile object a reference point C in contact with the static receptacle during the whole manipulation.

By aligning the reference frames of the objects with the key directions of the manipulation, the sliding is done along the $\overrightarrow{\mathbf{x}}$ axis of $\mathcal{F}_s$, and the folding is a pure rotation around an axis collinear with the $\overrightarrow{y}$ axis of $\mathcal{F}_s$. Under such condition, the end of the sliding manipulation is characterized by an abrupt reduction of the contact point velocity in the direction of the sliding, i.e. $\overrightarrow{\mathbf{x}}$, and by a significant increase of the force on the same axis. Similarly, the end of a folding manipulation should be characterized by a strong reduction of the angular velocity around the $\overrightarrow{\mathbf{y}}$ axis and by a strong increase of the torque around the same axis. Nevertheless, these characteristics are only valid if the manipulation (sliding, folding) is effectively done along the main directions of the sensor frames, which is a strong assumption. Next section shows that our approach based on the instantaneous work does not rely on such hypothesis, and is valid independently of the measurement and manipulation reference frames locations.

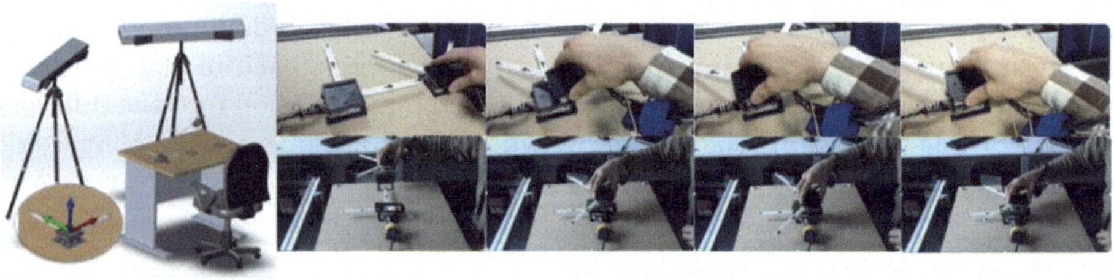

Fig. 1. Left: experimental setup, including a CODA tracking system, and force sensors mounted onto the objects. Right: insertion experiment, and phone motherboard folding. Videos available at https://aremazeilles.gitlab.io/publication/2019-remazeilles-b/

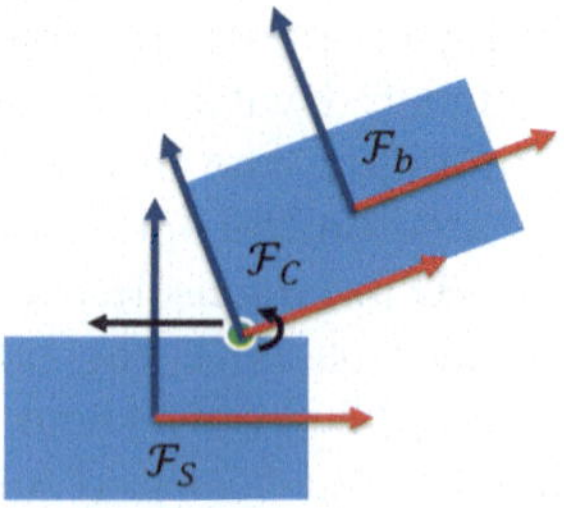

Fig. 2. *Insertion* experiment: frames involved, with the $\overrightarrow{x}$ axis in red and $\overrightarrow{z}$ in blue.

3.2 Manipulation Segmentation with the Instantaneous Work

The instantaneous (or infinitesimal) work is defined as the combination of a wrench and a twist [8]. Considering an inertial frame $\mathcal{F}_A$ and frame $\mathcal{F}_B$ attached to the rigid body, we note $V_{ab}^b \in \mathcal{R}^6$ the instantaneous body velocity and $\mathbf{F}_b$ the wrench measured at $\mathcal{F}_B$. Their inner product gives the instantaneous work:

$$\delta W = V_{ab}^b . \mathbf{F}_b \tag{1}$$

It is interesting noting that two wrenches are said to be equivalent if they generate the same work. This is easily demonstrated using the two following relations:

$$V_{ab}^b = {}^bA_{d_c} V_{ac}^b \tag{2}$$

$$\mathbf{F}_b = {}^bA_{d_c} \mathbf{F}_c \tag{3}$$

where ${}^bA_{d_c}$ is the adjoint transformation, which is defined as:

$${}^bA_{d_c} = \begin{bmatrix} {}^bR_c & \left[{}^bt_c\right]_\times {}^bR_c \\ 0 & {}^bR_c \end{bmatrix} \tag{4}$$

Going back to the instantaneous work definition, we obtain:

$$\delta W = V_{ab}^b . \mathbf{F}_b = \left({}^bA_{d_c} V_{ac}^b\right)^\top \mathbf{F}_b = V_{ac}^b \left({}^bA_{d_c}^\top \mathbf{F}_b\right) = V_{ac}^b \left({}^bA_{d_c}^\top {}^bA_{d_c}^\top \mathbf{F}_c\right) \tag{5}$$

Since ${}^bA_{d_c}^\top = {}^bA_{d_c} = \mathbf{I}_{6\times 6}$, we deduce that $\delta W = V_{ab}^b \mathbf{F}_b = V_{ac}^b \mathbf{F}_c$. So the instantaneous work of a rigid body is independent of the body frame used for the measurements (of both velocity and wrench). In our context that means that the instantaneous work computed at the contact point (position of $\mathcal{F}_c$ in Fig. 2) is equivalent to the one computed directly at the reference frame of the moving object (i.e. frame $\mathcal{F}_B$ in Fig. 2), wherever it is placed on the object.

In our experimental setup, the static object is mounted onto a force sensor which pose $\mathcal{F}_f$ in the world is known. Through perception means, the pose of the moving object $\mathcal{F}_M(t)$ is also known along time. We can thus deduce the instantaneous work of the moving object:

$$\delta W = V_{wm}^b . \mathcal{F}_m = V_{wm}^b . {}^mA_{d_f} \mathcal{F}_f \tag{6}$$

with V_{wm}^b being the body twist of the moving object expressed in its reference frame and ${}^mA_{d_f}$ the adjoint transformation between the force sensor frame and the moving object reference frame. This is the relation we are using for segmenting the manipulation. Next section presents several experimentations done on the manipulation segmentation from such instantaneous work metrics.

4 Segmentation Experimentations

4.1 Insertion Experimentation

Figure 3 presents the data collected of the human demonstration during the *insertion* manipulation (refer to Figs. 1 and 2), as well as the segmentation output. Left and middle plot present respectively the pose of the mobile object with respect to the static one, and the wrench measured by the force sensor located below the receptacle. The key instants, output of the segmentation and automatically detected, are displayed with vertical blue lines. The manipulation start and end are deduced from the wrench signal, detecting when the difference between the current wrench and the estimated mean wrench is higher than the standard deviation.

The sliding stop and folding stop are detected looking at the instantaneous work, as proposed in Sect. 3.2. The sliding stop is visually noticeable by a peak in the y orientation, and mainly peaks in all the wrench dimensions. The folding stop is noticeable by the sudden stabilization of the y orientation, and by the wrench variations. Right plot of Fig. 3 presents the computed instantaneous work. The key instants corresponding to the end of the sliding and the end of the folding are highlighted by deriving the instantaneous work.

4.2 Inclined *Insertion* Experiment

In the following sequence, an additional piece is placed in between the static receptacle and the measurement components, so that the sliding plane and the folding axis are not aligned anymore with the measurement frames (see Fig. 4). With respect to the previous experiment in which the orientation error between the two pieces was almost null once inserted, we see here in the recorded pose some rotation remaining at the end of the manipulation (around 14° of roll and 45° of pitch). This is due to the additional piece we intentionally added.

In the wrench plot of Fig. 5 we can notice some signal quantification, in particular on the force measured along z and on all torque dimensions. In that experiment, a newer version of the force sensor (Optoforce HEX) is used. It provides higher amplitude of measurement, but with a lower resolution. The system is thus switching in between different quantified values. We did not apply any filtering to compensate for this, and used this raw signal for our computations.

Right plot of Fig. 5 presents the computed instantaneous work. First we can see that a significant peak is present around iteration 300. It is provoked by a

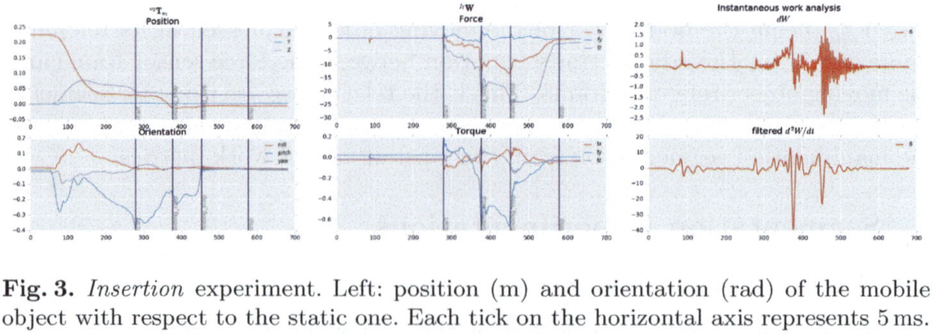

Fig. 3. *Insertion* experiment. Left: position (m) and orientation (rad) of the mobile object with respect to the static one. Each tick on the horizontal axis represents 5 ms. Middle: force (N) and torque (Nm) measured by the static sensor. Right: instantaneous work computed, and its filtered derivative used for detecting absolute peaks.

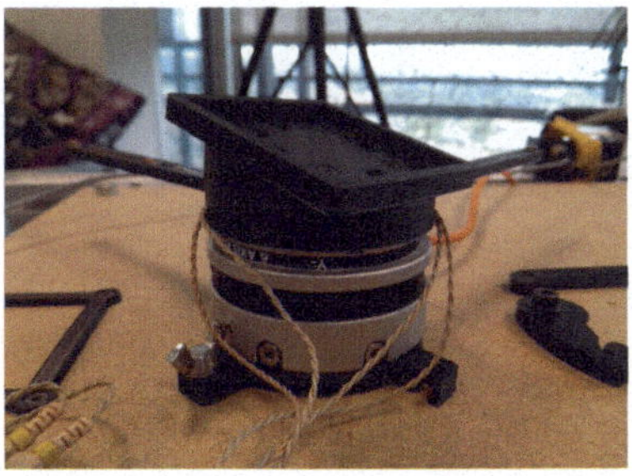

Fig. 4. Inclined insertion experiment: the receptacle main plane is not anymore aligned with the sensing frames.

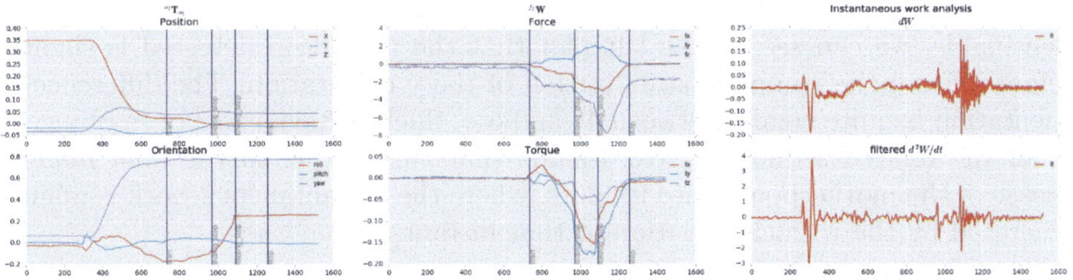

Fig. 5. *Inclined insertion* left: position (m) and orientation (rad) of the mobile object. Each horizontal tick is 5 ms. Middle: force (N) and torque (Nm) measured by the static sensor. Right: instantaneous work computed.

quick variation of the mobile orientation when the user grasps the object. The noise present in the wrench signal enables that variation to be present in the resulting work. Nevertheless, it can be easily filtered by considering that the instantaneous work should be monitored only after the manipulation starts (i.e. when the two objects are in contact), which is detected and displayed with a vertical line around iteration 730. The two next highest peaks are related to the two transitions we are interested in, the end of the sliding and the end of the folding, and they are displayed with vertical bars in left and middle plots of Fig. 5. This demonstrates that the instantaneous work is an appropriate metric for detecting manipulation transitions, in particular because it can make such detection wherever the measurement frames are located (as long as we know the relative transform between the pose and force measurement frames).

4.3 Phone Experiment

In the phone experiment a real cell-phone motherboard is installed through a folding procedure into its receptacle. Similarly, left and middle plots of Fig. 6 present the pose of the mobile object with respect to the static one and the wrench measured by the fixed sensor. The motion involved is mainly a rotation around $\overrightarrow{y}$ axis. Even though the peak in the instantaneous work is clearly

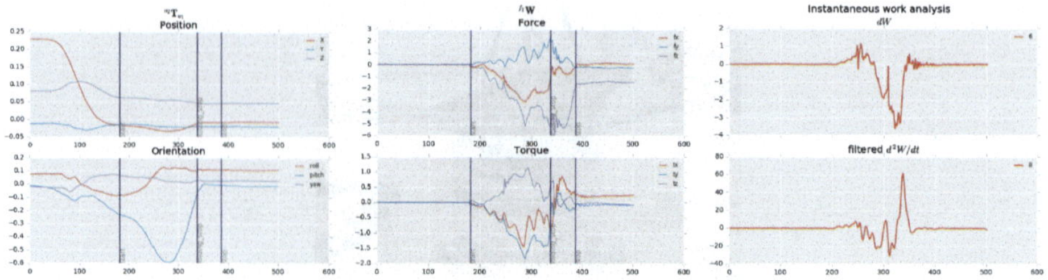

Fig. 6. *Phone* experiment. Left: position (m) and orientation (rad) of the mobile object. Each horizontal tick is 5 ms. Middle: force (N) and torque (Nm) measured by the static sensor. Right: instantaneous work computed.

identifiable, we can see on the left plot that the transition detected is slightly before (55 ms) the complete stabilization of the y orientation. The difference in orientation is quite small (between 4° and 5°). Such behavior has been observed in all the related sequences. We assume that this is due to the final manual release of the motherboard into its slots, where the instantaneous work is mainly generated by the wrench variation at that instant.

4.4 Bi-manual Experiment

The following experiment is a bi-manual version of the phone motherboard folding into the phone case: the phone case is also held by the human, and not static anymore. Figure 7 illustrates the experimental results. The peak detected in the derivative of the instantaneous work is located at iteration 480. Again, we can see on the left plot that the relative orientation is not at its minimum at this instant, but the error is, as in the previous experiment, less than 5°. This experiment shows that the instantaneous work measure can also be used for bi-manual experiments. Nevertheless, more computational effort is needed for detecting the beginning and end of the manipulation, since the wrench recorded by the receptacle is not constant before and after the manipulation as it was the case when the receptacle was static.

5 Manipulation Characterization

Once segmented, each manipulation phase needs to be characterized and the relevant task frames need to be deduced. The location of such task frame is usually considered as an a priori knowledge, like in [1,4]. Nevertheless, we cannot assume that the object and force sensing frames are always nicely aligned with the interaction sites. It is therefore necessary deducing them directly from the observation. The two next subsections present how such characterization can be done for the sliding and folding skills.

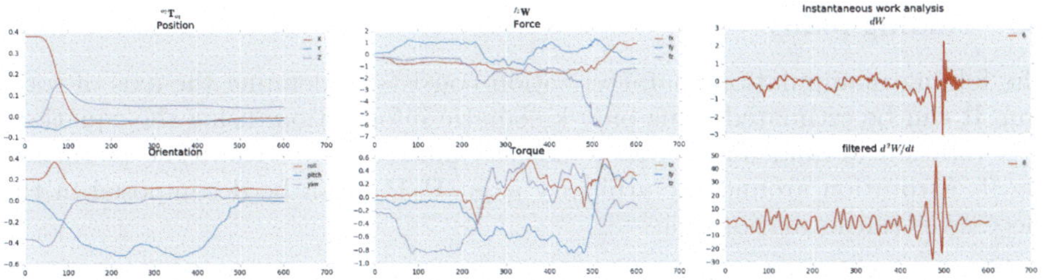

Fig. 7. *Bi-manual phone* experiment. Left: pose of the mobile object with respect to the mobile receptacle. Each horizontal tick is 5 ms. Middle: wrench measured by the receptacle force sensor. Right: instantaneous work computed.

5.1 Sliding Skill

The characterization of the sliding skill consists in defining the plane of sliding, and the direction of the sliding. This is sufficient to place the reference frames required by the robotic controller. If we assume that we know the segment of the moving object in contact with the receptacle, we get from all the k timestamps of the sliding phase a set of points belonging to the sliding plane. They are noted ${}^{2}\mathbf{P}_k = (X_k, Y_k, Z_k)^\top$ and they are expressed in the frame attached to the static object given the pose of the mobile object with respect to the static one.

The estimation of the sliding plane ${}^{2}\pi$ is obtained by minimizing with a least square regression the following equation:

$$\sum_k \delta_k^2 = \frac{1}{A^2 + B^2 + C^2} \sum_k (AX_k + BY_k + CZ_k + 1)^2 , \tag{7}$$

with the plane equation defined as: $AX + BY + CZ + 1 = 0$. Once the plane equation obtained, a frame is attached to it, with the $\overrightarrow{z}$ direction aligned with the plane normal, and $\overrightarrow{x}$ pointing in the direction of the main motion. Figure 8 illustrates the outcome of the plane estimation with such technique.

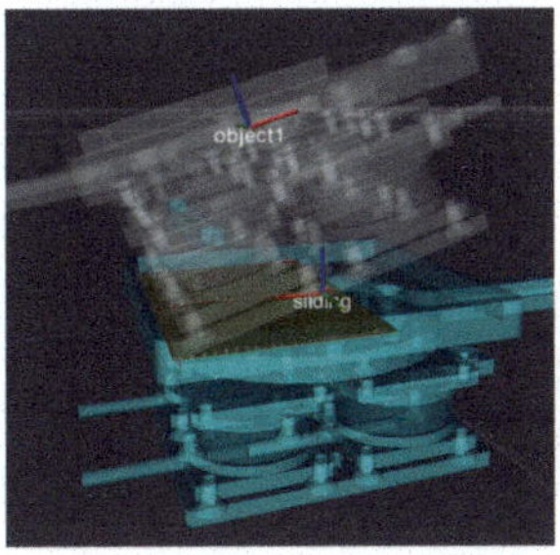

Fig. 8. *Insertion* experiment: sliding plane estimation. Dots are contact points used for estimating the yellow plane, and then deducing the reference frame named "sliding".

5.2 Folding Skill

The folding manipulation characterization consists in defining the axis of rotation. It can be estimated using only kinematic information, using the equations of a rotation around an arbitrary axis, as illustrated on the right picture of Fig. 9. A rotation around the arbitrary axis $(\mathbf{P}_1\mathbf{P}_2)$ can be decomposed in the successive transformations [6]:

- Translate $\mathbf{P}_1$ to the origin, (pure translation transformation $\mathbf{T}$)
- Rotate $\mathbf{P}_1\mathbf{P}_2$ towards the $\overrightarrow{z}$ axis (pure rotation transformation $\mathbf{M}$)
- Rotate about the $\overrightarrow{z}$ axis (pure rotation transformation $\mathbf{R}$)
- Rotate back to the original orientation (transformation $\mathbf{M}^{-1}$)
- Translate to the original position (transformation $\mathbf{T}^{-1}$)

So that a point $\mathbf{P}$ is transformed to point $\mathbf{P}'$ after the rotation of angle θ about the axis $(\mathbf{P}_1\mathbf{P}_2)$, by following the relation:

$$\mathbf{P}' = \mathbf{TMRM}^{-1}\mathbf{T}^{-1}\mathbf{P} \tag{8}$$

Now we consider the pose of the moving object at two distinct timestamps, ${}^{w}\mathbf{M}_{t_1}$ and ${}^{w}\mathbf{M}_{t_2}$ expressed with respect to any static reference frame w. The motion between the two frames is a pure rotation about an arbitrary axis. The axis of rotation $\mathbf{u}\theta$ extracted from ${}^{t_2}\mathbf{R}_{t_1}$ is collinear to the arbitrary rotation axis. The only difference is that the rotation ${}^{t_2}\mathbf{R}_{t_1}$ is applied with $\mathbf{u}$ anchored to the reference frame of the object at $t = t_1$, while in the rotation around an arbitrary axis, the rotation axis is anchored at $\mathbf{P}_1$. Since the change of anchor point is carried by the transformation $\mathbf{T}$, we can deduce the following relation:

$$^{2}\mathbf{P} = \mathbf{T}^{t_2}\mathbf{R}_{t_1}\mathbf{T}^{-1 1}\mathbf{P} \tag{9}$$

By noting $\mathbf{t}$ the translation carried by transformation $\mathbf{T}$, we can deduce:

$$^{2}\mathbf{P} = \mathbf{t} + {}^{t_2}\mathbf{R}_{t_1}\left({}^{1}\mathbf{P} - \mathbf{t}\right) \tag{10}$$

$$\mathbf{t} - {}^{t_2}\mathbf{R}_{t_1}\mathbf{t} = {}^{2}\mathbf{P} - {}^{t_2}\mathbf{R}_{t_1}{}^{1}\mathbf{P} \tag{11}$$

$$\left(\mathbf{I} - {}^{t_2}\mathbf{R}_{t_1}\right)\mathbf{t} = {}^{2}\mathbf{P} - {}^{t_2}\mathbf{R}_{t_1}{}^{1}\mathbf{P} \tag{12}$$

$$\mathbf{t} = \left(\mathbf{I} - {}^{t_2}\mathbf{R}_{t_1}\right)^{-1} {}^{2}\mathbf{P} - {}^{t_2}\mathbf{R}_{t_1}{}^{1}\mathbf{P} \tag{13}$$

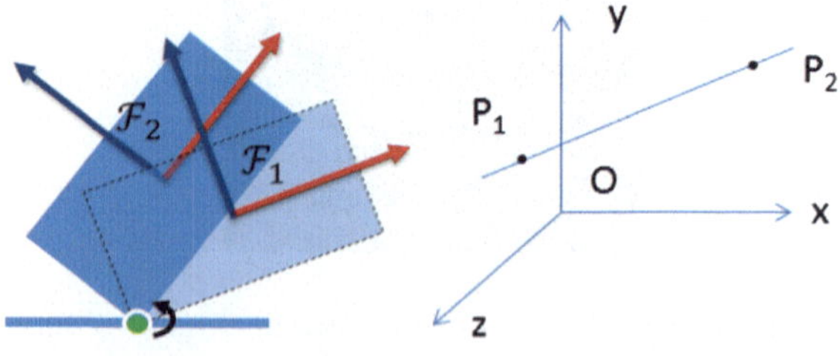

Fig. 9. Rotation around an arbitrary axis. Left: Object motion during folding manipulation. Right: axis of rotation defined by two points belonging to it, $\mathbf{P}_1$ and $\mathbf{P}_2$.

Considering that ${}^{2}\mathbf{P} - {}^{t_2}\mathbf{R}_{t_1}\,{}^{1}\mathbf{P} = {}^{t_2}\mathbf{t}_{t_1}$, we can deduce that $\mathbf{t}$ is:

$$\mathbf{t} = \left(\mathbf{I} - {}^{t_2}\mathbf{R}_{t_1}\right)^{-1}\,{}^{t_2}\mathbf{t}_{t_1} \tag{14}$$

$\mathbf{t}$ is a point belonging to the axis of rotation, expressed in the frame of the object at $t = t_1$. It can be estimated directly from the transformation ${}^{t_2}\mathbf{M}_{t_1}$, as long as this transformation is related to a pure rotation about an arbitrary axis. Thus a reference frame attached to the folding axis can be deduced from the relative motion between two different poses of the object during the folding, using Eq. (14) to deduce an anchor point for the rotation axis $\mathbf{u}$ extracted from the relative frame rotation between these two instants, ${}^{t_2}\mathbf{R}_{t_1}$.

In practice, we do the folding axis estimation by considering the pose of the mobile part at the beginning ($t = t_1$), and at the end of the folding detected with the instantaneous work ($t = t_2$). Figure 10 illustrates the folding axis estimated for the insertion and folding experiment. The relative motion in between these two frames is thus sufficient to deduce the rotation axis, as previously explained.

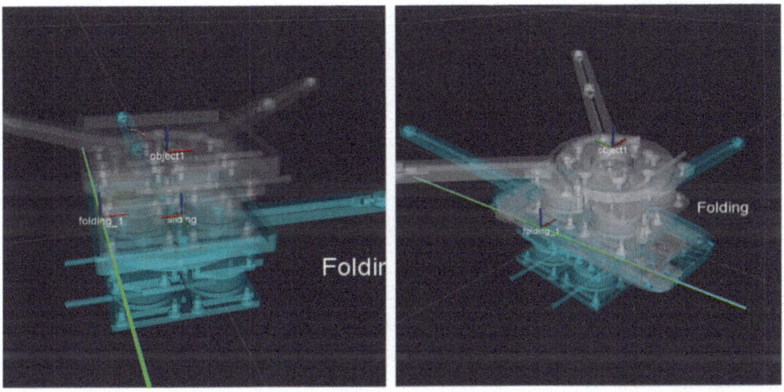

Fig. 10. Folding characterization; (left: insertion experiment, right: phone experiment) the estimated folding line is presented in green

6 Conclusion

In this article we have demonstrated that the instantaneous work can be a relevant features for segmenting manipulation observations composed of kinetic and kinematic information. The experimentations presented demonstrate that the segmentation is feasible even if the main interaction frames are not known. We have also shown how the skills like sliding and folding can be characterized by estimating the task frame from the kinematic information observed.

Similarly to [1] we would like to embed the wrench measurement in the skill characterization to improve the estimation performed. We would like also to generalize the sliding characterization to avoid any CAD a priori information in the computation, and complete the approach to represent other skills. Finally, we would like to use the characterization done to instantiate robot control skills, and reproduce the observed manipulation with a robotic system.

Acknowledgements. Supported by the Elkartek MALGUROB and the SARAFun project under the European Union's Horizon 2020 research & innovation programme, grant agreement No. 644938. The authors would like to thank Dr. Pierre Barralon for the fruitfull discussions that led to the segmentation approach presented here.

References

1. Almeida, D., Karayiannidis, Y.: Folding assembly by means of dual-arm robotic manipulation. In: IEEE ICRA, pp. 3987–3993, May 2016
2. Billard, A., Calinon, S., Dillmann, R., Schaal, S.: Robot Programming by Demonstration, pp. 1371–1394. Springer, Berlin (2008)
3. Calinon, S.: Robot Programming by Demonstration. A Probabilistic Approach. EPFL Press, Lausanne (2009)
4. De Schutter, J., De Laet, T., Rutgeerts, J., Decré, W., Smits, R., Aertbeliën, E., Claes, K., Bruyninckx, H.: Constraint-based task specification and estimation for sensor-based robot systems in the presence of geometric uncertainty. Int. J. Robot. Res. **26**, 433–455 (2007)
5. Kormushev, P., Calinon, S., Caldwell, D.G.: Imitation learning of positional and force skills demonstrated via kinesthetic teaching and haptic input. Adv. Robot. **25**(5), 581–603 (2011)
6. Kovács, E.: Rotation about an arbitrary axis and reflection through an arbitrary plane. In: Annales Mathematicae et Informaticae, vol. 40, pp. 175–186, January 2012
7. Kragic, D., Gustafson, J., Karaoguz, H., Jensfelt, P., Krug, R.: Interactive, collaborative robots: challenges and opportunities. In: IJCAI-18, pp. 18–25 (2018)
8. Murray, R.M., Sastry, S.S., Zexiang, L.: A Mathematical Introduction to Robotic Manipulation, 1st edn. CRC Press, Boca Raton (1994)
9. Pedersen, M., Nalpantidis, L., Andersen, R., Schou, C., Bogh, S., Kruger, V., Madsen, O.: Robot skills for manufacturing: from concept to industrial deployment. Robot. Comput.-Integr. Manuf. **37**, 282–291 (2016)
10. Perez-D'Arpino, C., Shah, J.A.: C-learn: learning geometric constraints from demonstrations for multi-step manipulation in shared autonomy. In: IEEE ICRA, pp. 4058–4065, May 2017
11. Piperagkas, G.S., Mariolis, I., Ioannidis, D., Tzovaras, D.: Key-frame extraction with semantic graphs in assembly processes. IEEE Robot. Autom. Lett. **2**(3), 1264–1271 (2017)
12. Ureche, A.L.P., Umezawa, K., Nakamura, Y., Billard, A.: Task parameterization using continuous constraints extracted from human demonstrations. IEEE Trans. Rob. **31**(6), 1458–1471 (2015)
13. Villani, V., Pini, F., Leali, F., Secchi, C.: Survey on human-robot collaboration in industrial settings: safety, intuitive interfaces and applications. Mechatronics **55**, 248–266 (2018)

Spherical Fully Covered UAV with Autonomous Indoor Localization

Agustin Ramos(✉), Pedro Jesus Sanchez-Cuevas, Guillermo Heredia, and Anibal Ollero

GRVC Robotics Laboratory, University of Seville, Seville, Spain
aramos11@us.es

Abstract. This paper presents a UAV (Unmanned Aerial Vehicle) with intrinsic safety which can interact with people and obstacles while flying in an indoor environment in an autonomous way. A system description including mechanical features, the design of the external protective case, electrical connections and the communication using the Robot Operating System (ROS) between the different devices is presented. Then, the dynamic model of the aerial system taking into account the protective case, the local positioning algorithm (Hector SLAM) and the control models implemented are also described. Different experimental results, which include simulation in Gazebo and real flights are shown to verify the positioning system developed. Two additional experiments have also been tested to validate two emergency safety systems in case of a failure in the position estimation is detected.

Keywords: UAS applications · Onboard localization · Intrinsic safety

1 Introduction

In the last years, the range of applications of UAVs has grown exponentially [1]. Nowadays, UAVs are not only used in observation applications such as mapping, exploration, surveillance or localization, but also in applications in which the aerial platform becomes an aerial robot that is able to physically interact with the environment [2].

The applications of these vehicles to deliver parcels, transport cargo or others which involve working with people are daily growing. For instance, [3] presents a scenario in which a UAV interacts with people to improve the productivity and efficiency of a company and [4] shows deep learning techniques for UAV interaction and collision avoidance. However, in general, most of aerial robots have not been designed to cooperatively work with people.

Several designs for UAVs that fly in proximity or interact with people have been proposed in the literature, and most of them include a protective case to increase safety during operation. For example, [5] compares the behavior against collisions of UAVs with spherical covers, either fixed to the frame or gimballed. A teleoperated spherical UAV commercialized by Flyability [6] follows this last design and has been used for industrial inspection. [7] presents a UAV with an origami-inspired protective case which allows interaction with humans and to reduce its size. [8] shows a hybrid

M. F. Silva et al. (Eds.): ROBOT 2019, AISC 1092, pp. 355–367, 2020.
https://doi.org/10.1007/978-3-030-35990-4_29

quadrotor with a cage that allows to fly and roll. To sum up, most of these systems are teleoperated and only mount a small inspection camera to reduce size as much as possible.

This paper proposes an aerial robot that is fully surrounded by a spherical cover so that it can safely interact and co-work with people, and it also has enough payload to implement a local positioning system onboard to fly indoors using a laser ranger.

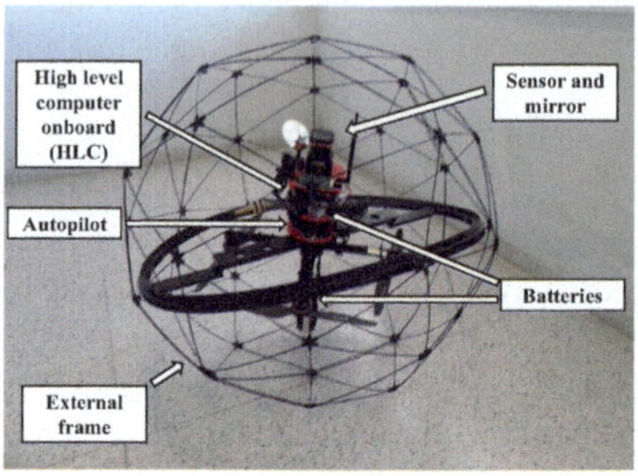

Fig. 1. The aerial platform and the main devices.

Thus, the main contribution of this paper is focused on the use of a system with intrinsic safety, which is suitable to co-work with people in the same environment (see Fig. 1). The proposed system consists of an aerial vehicle with a protective case which acts as a passive system to absorb small impacts. The aerial robot also implements a positioning system to fly autonomously in an indoor environment without depending on external measurements.

The organization of this paper is structured as follows: Sect. 2 presents the hardware and software architecture of the aerial vehicle. Section 3 explains the mathematical model of the system, the algorithm developed to position it and the control model implemented. Section 4 shows the tests performed to evaluate the correct localization. Finally, Sect. 5 includes the conclusions and future works and applications of the solution proposed.

2 System Description

In this section, the mechanical description of the aerial system, the avionics as well as the architecture of communications between the different devices are presented.

In Fig. 2 it can be observed the components of the aerial frame: the autopilot, the high-level computer and the sensor, which communicate with the PC Ground Station. Furthermore, the frame includes the batteries and the motors of the UAV.

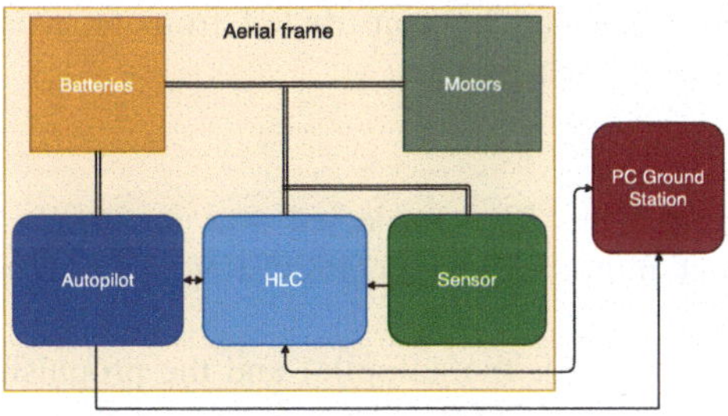

Fig. 2. System architecture. The double link lines represent the electrical connections. The directional and bidirectional connectors represent the communication between devices.

2.1 Aerial System

The aerial vehicle has been designed using an external carbon fiber structure with 3 rotational degrees of freedom, as the frame of a gyroscope. This allows rotating the external side of the platform in case of touching gently with obstacles or living beings, absorbing the impact and transforming it into rotational kinetic energy. Inside the external case, a cross layout quadrotor is in charge of maintaining the attitude of the vehicle and the payload.

Fig. 3. CATIA model for the simulation experiments.

For the simulation tests, a simplified CATIA model has been designed with the same size as the real aerial platform, which is shown in Fig. 3. This model has been imported in Gazebo, the open-source 3D robotics simulator [9].

A. Case Design

The spherical case consists of thin carbon fiber rods with a thickness of 2 mm, being the diameter of the sphere of 870 mm and the weight of 1.7 kg. The mechanical specifications of the aerial vehicle have been determined to have a payload of at least 200 g in addition of the laser sensor.

The rods of the external case form a truncated icosahedron structure. This polyhedron is composed of 12 regular pentagonal faces and 20 regular hexagonal faces. It is an Archimedean solid which can be constructed from an icosahedron with the 12 vertices cut off one third of each edge.

B. Aerial Platform

As was above mentioned, the frame selected was a custom quadrotor with cross configuration using the T-motor MN4006-23 380 KV and T-motor propellers 14x4.8L as power plant.

The power supply system for the autopilot and the propulsive system is a 6 cells LiPo battery with a capacity of 7000 mAh. On the other hand, for the positioning sensor and the high level computer onboard (HLC), another independent circuit is connected to other 3s LiPo battery with 2700 mAh of capacity. The electrical schemes are represented in Fig. 4. The purpose of the Battery Eliminator Circuit (BEC) modules is to transform the voltage of the batteries (22,2 V or 11,1 V) to 5 V (the supply voltage of the autopilot and the HLC).

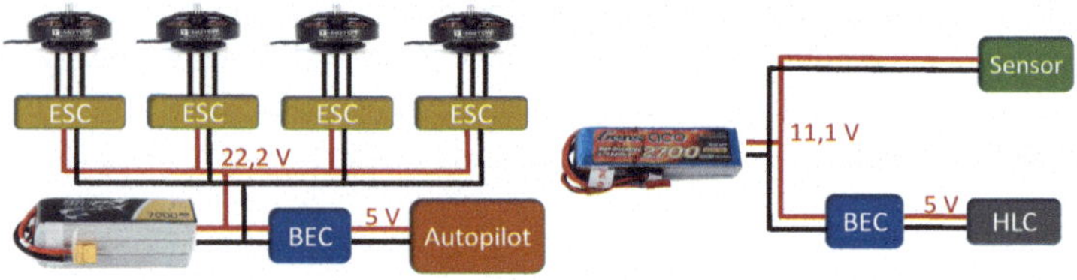

Fig. 4. On the left, the electrical scheme of the autopilot and the motors. On the right, the electrical scheme of the sensor and the HLC.

2.2 Avionics

The avionics architecture is represented in Fig. 5 in which it is also detailed the different communication protocols between the different devices.

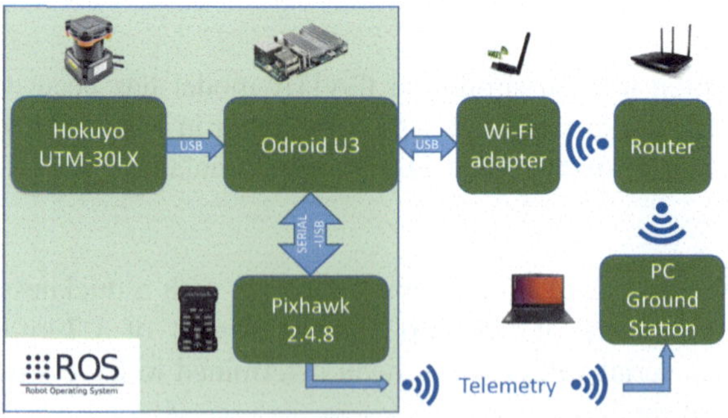

Fig. 5. Scheme of the communication architecture between devices.

A. Hardware description

The Pixhawk 2.4.8 is the autopilot board of the UAV [10]. This has embedded the inertial measurement unit, the magnetometer and runs the flight stack which is a customized version of the PX4 code [11]. The autopilot is physically connected to the Electronic Speed Controller (ESCs) of each of the four rotors and to the HLC through the TELEM2 port using the serial protocol. Furthermore, the port TELEM1 of the Pixhawk is used to connect the air side radio-modem and send information directly to the PC Ground Station using MavLink [12]. The Pixhawk has also connected other sensors and devices (GPS & Compass, receiver, buzzer and switch) as well as the BEC module (the voltage regulator) to turn on the autopilot and power the ESCs.

On the other hand, the HLC is an Odroid U3 [13] which is the 'core' of the avionics. This is a single-board computer and it communicates using its 3 USB ports with the Hokuyo UTM-30LX [14], the Pixhawk and the PC Ground Station, respectively. Thus, the Hokuyo is a scanning laser rangefinder sensor which sends the scan data to the Odroid, as it is detailed in Sect. 3.2, the Pixhawk communicates with it as it is described above and the third port is connected to a Wi-Fi adapter to communicate the Odroid with the PC Ground Station via SSH. This SSH protocol communication is used to log in remotely to the Odroid from the PC Ground Station and in that way to be able to launch several commands in this device. The PC Ground Station receives and monitors the state of the vehicle through the ground-side radio-modem.

To implement the communication between the Hokuyo sensor, the Pixhawk autopilot, and the HLC Odroid U3 has been required to use ROS, as explained in more detail in the next subsection.

B. Software description

Due to the aim of this UAV is to fly indoors autonomously, a local positioning system based on the Hokuyo UTM-30LX laser has been implemented to obtain a position estimation of the vehicle suitable to be included in the position control loop. This sensor provides scan data information sweeping an area of 270° and with a maximum range of 30 m.

The software layer has been implemented under a ROS Kinetic [15] framework running in an Ubuntu 16.04 OS, which are open-source. The main advantage of using ROS is that this is a publisher/subscriber system that easily interconnects different nodes which can be implemented in different programming languages. The software of the autopilot is PX4, which is running in the Pixhawk board [11, 16]. The communication between ROS and PX4 is accomplished through serial communication using the MavLink protocol. In addition, it has been used the open-source Mavros node running as an interpreter between ROS and MavLink [17]. Finally, this research uses a UAV abstraction layer (UAL) [18, 19] that allows a high level custom communication between the UAV and the users (the Ground Control System).

3 Modelling, Localization and Control

3.1 Modelling

The dynamic model of a classical multirotor is usually presented as:

$$M(\xi)\ddot{\xi} + C\left(\xi, \dot{\xi}\right) + G(\xi) = F + F_{ext} \tag{1}$$

Where M is the generalized inertia matrix, C is the Coriolis and centrifugal terms, G represents the gravity component, F is the generalized vector force developed by the rotors and F_{ext} are the external and unknown forces.

Usually, if an aerial platform interacts with the environment the vector of F_{ext} will be composed by the three forces and the three momentums as follows: $F_{ext} = \left[F_{ext_x} F_{ext_y} F_{ext_z} \tau_{ext_x} \tau_{ext_y} \tau_{ext_z}\right]$. However, the main advantages of the presented design is that the spherical case has an articulated link with the multirotor core and the propulsive system, so assuming that the bearing friction is null, the vector of external forces is: $F_{ext} = \left[F_{ext_x} F_{ext_y} F_{ext_z} 0\, 0\, 0\right]$. Therefore, the spherical case not only provide us an external protection to interact with the multirotor, but also acts in the dynamic model reducing the disturbance on the multirotor when it interacts with the environment. This is due to the kinematic energy that can be assumed by the external case.

3.2 Localization

The position and orientation of the UAV is obtained using the Hector SLAM algorithm [20, 21]. The reason for using this 2D algorithm is because it works with low computational capacity compared to other SLAM algorithms. With it, it has been possible to get the horizontal position estimation (x, y) and the orientation of the vehicle (ψ) while obtaining a map of the flight scenario.

The procedure for obtaining the required information is as follows (see Fig. 7 to observe the list of topics):

First, a ROS package called hokuyo_node [22] is used. This is a driver for Hokuyo laser range-finders which gives several ROS topics, services and sensor parameters. The package shows in the ROS topic "/scan" scans data provided by the sensor. The scan topic is a LaserScan message [23] that contains, among other things, the range data in meters which are stored in a float vector called ranges.

Second, another ROS package named hector_slam serves to get the mapping and the positioning of the vehicle. Using this set of packages, a script is launched to call the Hector SLAM algorithm implemented, which transforms the previous scan data in the topic /slam_out_pose. This is a PoseStamped message [24] that contains the position x, y in meters and the orientation yaw (ψ) in quaternion form of the aerial vehicle.

Third, a script to relay the data of /slam_out_pose to a topic named/uav_1/mavros/vision_pose/pose has been implemented to add the z estimation and two safety systems explained below. Moreover, in order to add the estimation of the z position to the last topic, a system with a mirror has been developed beside the Hokuyo sensor. The mirror is placed opposite the sensor with a rotation of 45° with respect to the xy plane, as it can see in Fig. 6. Thereby, the samples of the laser scan which reflect on the mirror go vertically to the roof of the room.

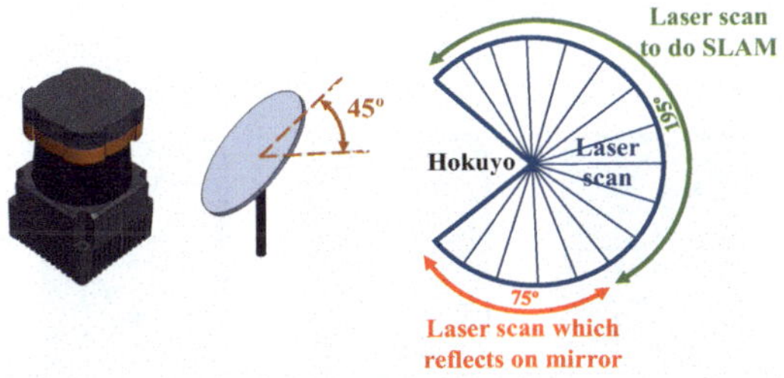

Fig. 6. On the left, model of the Hokuyo sensor and the mirror rotated 45°. On the right, distribution of the samples of the laser sensor.

Due to the Hector SLAM package uses all the scan data provided by the 1080 samples or steps which sweep the 270° detection angle of the sensor [14], it has been necessary to modify part of the code of the Hector SLAM package to avoid to do SLAM in the side where the mirror is located. In this way, part of the mapping script belonging to the Hector SLAM package has been changed, doing that the algorithm only effects between the step number 300 and 1080 (195°), because the mirror is placed between the steps 1 and 300 (75°) of the sensor (Fig. 6).

This way to estimate the z position is measuring the distance between the UAV and the roof. Therefore, to estimate the height of the robot is necessary to change the roof reference to the ground, being $h(t)$ the height in the instant t, $range(t = 0)$ the distance measured in the initial instant and $range(t)$ the distance measured in the instant t.

$$h(t) = range(t = 0) - range(t) \tag{2}$$

To improve the z estimation, it is analyzed a ray beam of several steps or samples which reflect on the mirror, calculating the average of the measured distances and removing the measures lower than 20 cm to be considered the structure of the sphere itself. However, the laser scan to do SLAM does not need to remove any sample considered wrong because of the structure, given that the higher amount of samples to process the algorithm and the very thin carbon fiber rods allow to create the map of the scenario and estimate the position without interferences.

This method to obtain the z estimation has been developed to fly at constant height, i.e., in environments with roofs where there is not a large variation in height.

Once x, y, z and yaw estimation is obtained and published in the topic /uav_1/mavros/vision_pose/pose, this information is interpreted and fusion with the rest of the internal sensors in the Pixhawk by the PX4 EKF to get the position and the orientation of the aerial vehicle. This data is required to fly autonomously or in position hold mode, as will be shown in the control Sect. 3.3.

To send information between both topics, two PX4 parameters must be changed to enable the external position and orientation estimation. These are the EKF2_AID_MASK which select the source of the fusion data in the Extended Kalman Filter and the parameter EKF2_HGT_MODE to enable vision fusion an external altitude estimator (z axis) [25].

Figure 7 shows the scheme about the sending data between topics.

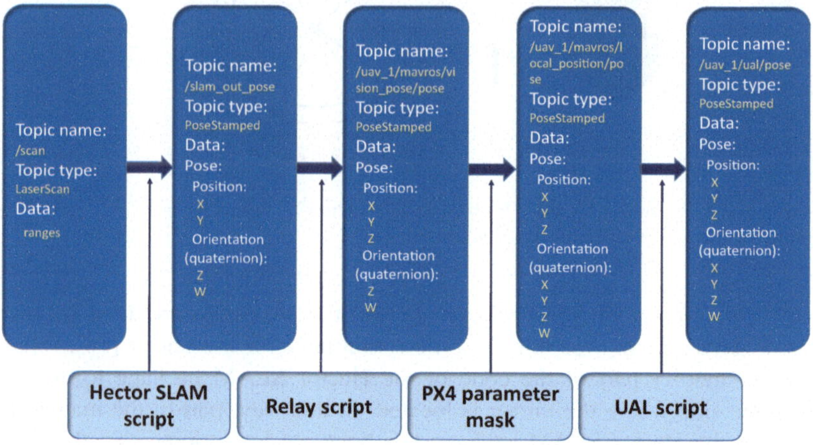

Fig. 7. Scheme of the ROS topics involved in the localization process.

The authors have also created two safety systems in case an emergency caused the vehicle to leave the autonomous flight. The first, if there is a loss of roof reference, the UAV will land automatically.

The second safety system is activated in case the robot reaches a high linear velocity in the axis x or y. That is due to the Odroid U3 can process the localization and the mapping of the algorithm if the UAV flies at not very high velocities. Because of that, the system switches to altitude hold mode when the aerial vehicle is in an autonomous flight and the velocity in the xy plane is high. In Sect. 4 some graphics which show the results of these safety systems are explained.

3.3 Control

The control algorithm of the multirotor enveloped in the spherical case is the standard control scheme with a cascaded PID linear controller as the one presented in Fig. 8.

Although the advantage of the external case could be exploited with a dedicated controller, this paper it was mainly focused on providing the autonomous capabilities of flying in an indoor scenario, leaving the development of applied control algorithms which improve the performance during the operation taking into account the presence of the external case to a future work.

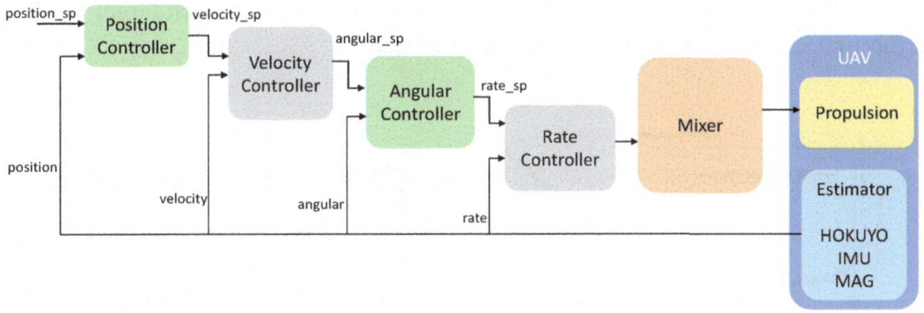

Fig. 8. Control scheme of the UAV.

4 Simulation and Experimental Results

Several simulated and experimental tests have been carried out to evaluate the quality of the localization and the mapping methods in an indoor scenario and also the safety systems mentioned in the previous section. The results are shown along this section as well as some simulation videos which can be found in [26] in which it is possible to see the performance of the algorithm during an autonomous mission. Finally, a real flight video has also been included to be able to observe how to obtain a map in a real environment [26].

In Fig. 9 it can be observed the x and y positions of the UAV compared with the set points commanded respectively. The reference of the position controller is shown in red and the state of the aircraft in blue. During this test, it has been commanded two different waypoints at 2 and 4 m for the x axis, and -2 and -4.5 m for the y axis. As it can see in blue, the position control of the vehicle reaches appropriately the target and it is well tuned in steady state. Although it seems that the controller is very low, this is the effect of limiting the linear velocities of the vehicle which was a mandatory action to grant the stability of the position estimation as it was explained in Sect. 3.2.

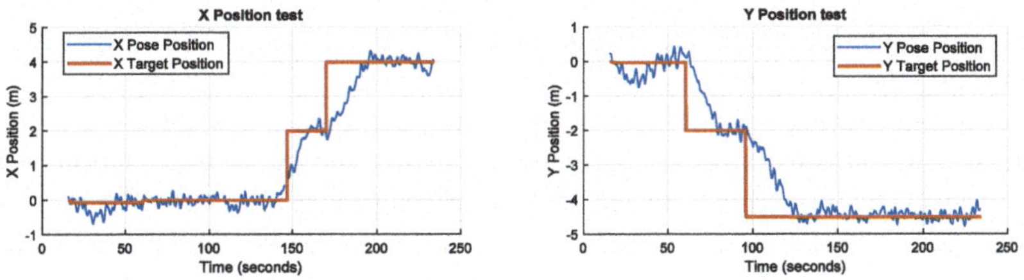

Fig. 9. x and y position of the aerial vehicle. WP tracking.

One of the solutions proposed to accelerate the convergence time of the position controller maintaining the safety conditions, consisted on establishing a limit in the maximum distance between the different waypoints removing the limits on the velocity controller. By this way, in this simulation the pose gets to reach each target faster than with the first experiment. Figure 10 shows the test.

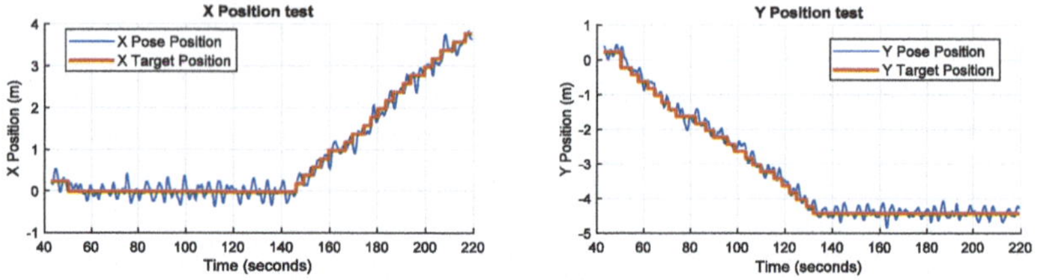

Fig. 10. x and y position of the aerial vehicle. Path tracking.

On the other hand, the results of the estimation and the controller in the z coordinate and yaw angle are shown in Fig. 11. These results clearly show that the tracking of the desired altitude is good enough and validate the solution proposed to estimate of the height and the orientation of the aircraft presented in Sect. 3.2.

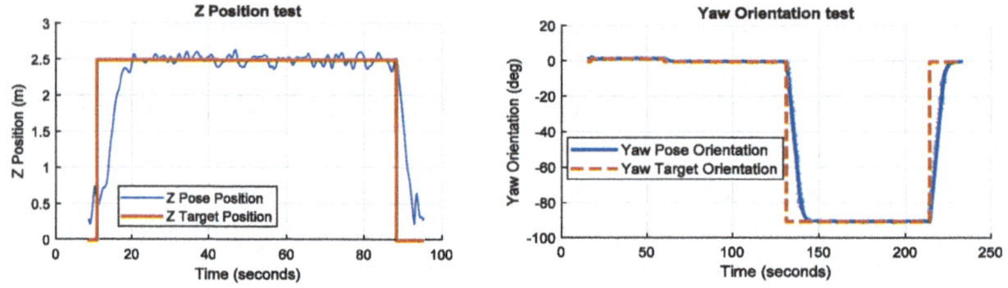

Fig. 11. z position and Yaw orientation of the aerial vehicle.

The yaw angle starts at zero degrees due to the algorithm fixes its own local reference frame when it is initialized. The controller and the estimation results have also been evaluated through the results obtained in Fig. 11. In this experiment, the maximum angular speed in yaw was established in 10 deg/s not only to increase the safety conditions during the autonomous operation avoiding a possible saturation in the motor mixer, but also to improve the mapping results and avoid localization mistakes.

In this test, an autonomous mission has been programmed for this UAV doing first a takeoff, later a translation in x axis 2 and 4 m, a rotation of 90° in yaw angle after that a translation in y axis −2 and −4.5 m and finally a landing. To do this, UAL ROS Services have been used [18, 19].

Regarding the safety systems mentioned in Sect. 3.2, different tests have been carried out (see Figs. 12 and 13). These tests can be also found in [26]. First, in the z emergency test, it can see several flight modes which have been represented by different colors in Fig. 12. The modes 1 and 5 (blue) are LANDED_ARMED, this means that the vehicle is not flying but ready to do it due to the motors are turns on and armed. Mode 2 (green) TAKING_OFF means that the UAV is flying but in the process to reach the takeoff height commanded. Mode 3 (yellow) FLYING_AUTO is when the aerial vehicle is flying in an autonomous mission sending waypoints different to a takeoff and a land order. Finally, mode 4 (red) is LANDING, the process by which the UAV stops flying. This result shows that while the positioning system detects a roof to reference the z position of the vehicle it allows to perform the take-off maneuver and to fly autonomously. However, as it can be observed, at the moment in which the roof disappears, the z emergency system is activated and the vehicle lands automatically.

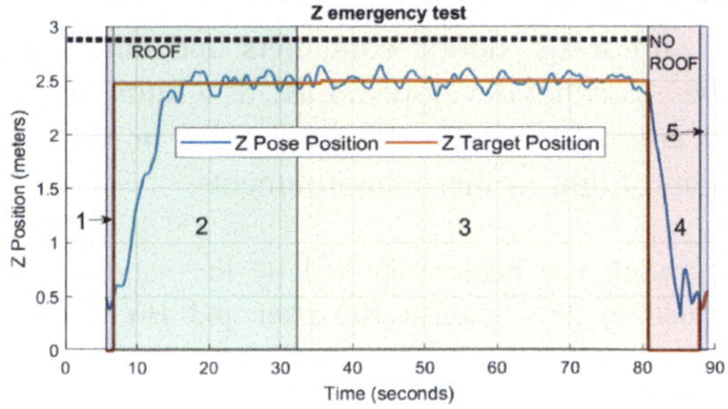

Fig. 12. z emergency test of the aerial vehicle.

The second safety system is the velocity test. It is about changing the flight mode in altitude hold in case of the xy linear velocity exceeds a limit. For this experiment, a waypoint in the x direction has been commanded limiting the xy velocity in autonomous mode in 1.0 m/s. As can be seen in Fig. 13, the flight modes have also been represented by colors being the mode 1 (blue) LANDED_ARMED, the mode 2 (green) TAKING_OFF, the mode 3 (yellow) FLYING_AUTO and the mode 4 (cyan) FLYING_MANUAL. In this last mode the UAV is manually controlled by a safety pilot, except in the height of the vehicle. Therefore, the UAV begins in landed state. Then, the vehicle takes off and after that starts to move in the x direction. In case the xy velocity exceeds the threshold of 0.3 m/s five times checking with a frequency of 10 Hz, the UAV switches to altitude hold mode if the vehicle was flying auto.

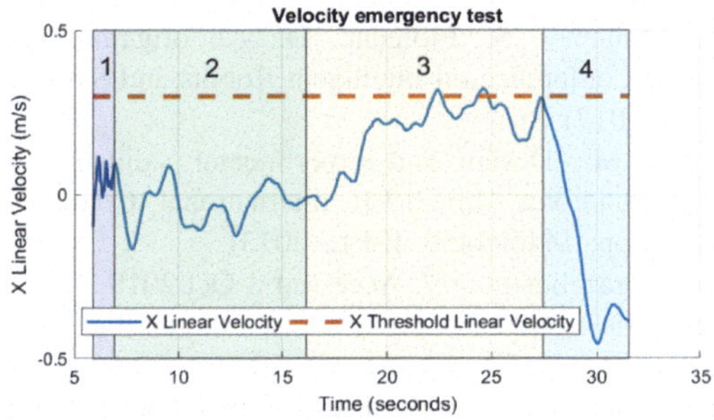

Fig. 13. Velocity emergency test of the aerial vehicle.

5 Conclusions

This paper has presented the design of an aerial vehicle in which a local positioning system has been developed to fly autonomously indoors with intrinsic safety. The mechanical parts of the external protective case of the quadrotor, the electrical configuration and the communication of the electronic devices have been explained. A mathematical model of the UAV, as well as, the integration of the localization system in the platform and a control model have been exposed.

The localization and mapping systems have been tested in simulation and in several indoor environments with walls, doors, containers and other obstacles which allow creating a map with the algorithm developed. Last, as a future work there would be still to test it in industrial areas, like factories, warehouses or similar to check a correct positioning, mapping and flight in these environments.

Acknowledgment. This work has been supported by the national project ARM-EXTEND (DPI2017-89790-R) funded by the Spanish RD plan and HYFLIERS H2020-ICT-2017-1-779411 projects.

References

1. Valavanis, K.P., Vachtsevanos, G.J.: Handbook of Unmanned Aerial Vehicles. Springer, Dordrecht (2015)
2. Sanchez-Cuevas, P.J., Heredia, G., Ollero, A.: Multirotor UAS for bridge inspection by contact using the ceiling effect. In: International Conference on Unmanned Aircraft Systems (ICUAS), Miami, pp. 767–774. IEEE (2017)
3. Nikolic, J., Burri, M., Rehder, J., Leutenegger, S., Huerzeler, C., Siegwart, R.: A UAV system for inspection of industrial facilities. In: IEEE Aerospace Conference, Montana. IEEE (2013)
4. Gandhi, D., Pinto, L., Gupta, A.: Learning to fly by crashing. In: IEEE/RSJ International Conference on Intelligent Robots and Systems (IROS), Vancouver, pp. 3948–3955. IEEE (2017)
5. Briod, A., Kornatowski, P., Zufferey, J., Floreano, D.: A collision-resilient flying robot. J. Field Robot. **31**, 496–509 (2014)
6. Flyability webpage. https://www.flyability.com/. Accessed 3 Oct 2019
7. Kornatowski, P.M., Mintchev, S., Floreano, D.: An origami-inspired cargo drone. In: IEEE/RSJ International Conference on Intelligent Robots and Systems (IROS), Vancouver, pp. 6855–6862. IEEE (2017)
8. Kalantari, A., Spenko, M.: Design and experimental validation of HyTAQ, a hybrid terrestrial and aerial quadrotor. In: IEEE International Conference on Robotics and Automation, Karlsruhe, pp. 4445–4450. IEEE (2013)
9. Gazebo webpage. http://gazebosim.org/. Accessed 3 Oct 2019
10. Pixhawk documentation page. https://docs.px4.io/en/flight_controller/pixhawk.html. Accessed 3 Oct 2019
11. Meier, L., Honegger, D., Pollefeys, M.: PX4: a node-based multithreaded open source robotics framework for deeply embedded platforms. In: IEEE International Conference on Robotics and Automation (ICRA), Seattle, pp. 6235–6240. IEEE (2015)

12. Atoev, S., Kwon, K.R., Lee, S.H., Moon, K.S.: Data analysis of the MAVLink communication protocol. In: International Conference on Information Science and Communications Technologies (ICISCT), Tashkent, pp. 1–3. IEEE (2017)
13. Odroid U3 documentation page. https://www.hardkernel.com/shop/odroid-u3/. Accessed 3 Oct 2019
14. Hokuyo documentation page. https://www.hokuyo-aut.jp/search/single.php?serial=169. Accessed 3 Oct 2019
15. ROS Kinetic page. http://wiki.ros.org/kinetic. Accessed 3 Oct 2019
16. PX4 documentation webpage. https://px4.io/documentation/. Accessed 3 Oct 2019
17. ROS Mavros webpage. http://wiki.ros.org/mavros. Accessed 3 Oct 2019
18. Real, F., Torres-González, A., Ramón-Soria, P., Capitán, J., Ollero, A.: UAL: an abstraction layer for unmanned aerial vehicles. In: 2nd International Symposium on Aerial Robotics (ISAR), Philadelphia. Springer, Cham (2018)
19. UAL documentation page. https://github.com/grvcTeam/grvc-ual/wiki. Accessed 3 Oct 2019
20. Hector SLAM documentation. http://wiki.ros.org/hector_slam. Accessed 3 Oct 2019
21. Kohlbrecher, S., Von Stryk, O., Meyer, J., Klingauf, U.: A flexible and scalable slam system with full 3d motion estimation. In: IEEE International Symposium on Safety, Security, and Rescue Robotics, Kyoto, pp. 155–160. IEEE (2011)
22. Hokuyo node documentation. http://wiki.ros.org/hokuyo_node. Accessed 3 Oct 2019
23. LaserScan message documentation. http://docs.ros.org/melodic/api/sensor_msgs/html/msg/LaserScan.html. Accessed 3 Oct 2019
24. PoseStamped message documentation. http://docs.ros.org/melodic/api/geometry_msgs/html/msg/PoseStamped.html. Accessed 3 Oct 2019
25. PX4 parameter reference guide. https://dev.px4.io/en/advanced/parameter_reference.html. Accessed 3 Oct 2019
26. Link to the video of the experiments. https://www.dropbox.com/sh/4evf4xgn4hslycz/AADmZHtEL8xDlwrCMvzvflCia?dl=0

Towards Endowing Collaborative Robots with Fast Learning for Minimizing Tutors' Demonstrations: What and When to Do?

Ana Cunha[1,3], Flora Ferreira[2], Wolfram Erlhagen[2], Emanuel Sousa[1], Luís Louro[3], Paulo Vicente[3], Sérgio Monteiro[3], and Estela Bicho[3](✉)

[1] Center for Computer Graphics, University of Minho, 4800-058 Guimaraes, Portugal
[2] Department of Mathematics and Applications, Center of Mathematics, University of Minho, 4800-058 Guimaraes, Portugal
[3] Department Industrial Electronics, University of Minho, 4800-058 Guimaraes, Portugal
estela.bicho@dei.uminho.pt

Abstract. Programming by demonstration allows non-experts in robot programming to train the robots in an intuitive manner. However, this learning paradigm requires multiple demonstrations of the same task, which can be time-consuming and annoying for the human tutor. To overcome this limitation, we propose a fast learning system – based on neural dynamics – that permits collaborative robots to memorize sequential information from single task demonstrations by a human-tutor. Important, the learning system allows not only to memorize long sequences of sub-goals in a task but also the time interval between them. We implement this learning system in Sawyer (a collaborative robot from Rethink Robotics) and test it in a construction task, where the robot observes several human-tutors with different preferences on the sequential order to perform the task and different behavioral time scales. After learning, memory recall (of what and when to do a sub-task) allows the robot to instruct inexperienced human workers, in a particular human-centered task scenario.

Keywords: Industrial robotics · Assembly tasks · Learning from demonstration · Sequence order and timing · Rapid learning · Dynamic Neural Fields

1 Introduction

One of the current challenges of the Industry 4.0 Era is the implantation of collaborative robots able to work side by side with (different) human-operators

This work was carried out within the scope of the project "PRODUTECH SIF – Soluções para a Indústria do Futuro" reference POCI-01-0247-FEDER-024541, co-funded by Fundo Europeu de Desenvolvimento Regional (FEDER), through "Programa Operacional Competitividade e Internacionalização (POCI)."

M. F. Silva et al. (Eds.): ROBOT 2019, AISC 1092, pp. 368–378, 2020.
https://doi.org/10.1007/978-3-030-35990-4_30

in a large variety of tasks [11]. This implies that this new generation of robots (CoBots) have to master a wide variety of tasks and interaction scenarios that cannot be completely designed in advance by experts as in traditional industrial applications. For symbiotic and efficient human-robot collaboration in (sequential or supportive) tasks, it is fundamental to endow the robotic co-workers with cognitive and learning capabilities [3,14]. In this paper, we focus on the implementation of learning mechanisms that allow the robotic co-worker to acquire high-level knowledge about the sequential structure of multi-part assembly/disassembly tasks (which may have time constraints) without being explicitly programmed. We adopt the learning paradigm known as programming by demonstration/observation since it allows non-specialists in robot programming to train the robot in an intuitive and open-ended manner [9]. However, often this learning paradigm requires multiple demonstrations of the same task which can be time-consuming and annoying for the human tutor [10]. Thus, for user acceptance, it is crucial to make possible to the robot the acquisition of generalized task knowledge in very few demonstrations. With this in mind, we implement and test on the collaborative robot Sawyer (from Rethink Robotics) a computational model that integrates fast activation-based learning to robustly represent sequential information from single task demonstrations by a human-tutor. Important, this learning system allows not only to memorize long sequences of sub-goals in a task, but also the time interval between them. The integration of these two features – ordinal and temporal information – allows the robot to memorize in one shot 'what to do' and 'when to do' in a certain task scenario, which builds a fast memory mechanism that significantly reduces the number of demonstrations needed from a tutor. After learning, the autonomous reactivation of this memory can be used by the robot to instruct inexperienced workers, or to make decisions when it plays the role of an active assist/co-worker. To build this fast learning system, we apply the theoretical framework of dynamic neural fields (DNF), that has been proven to provide key processing mechanisms for applications in cognitive robotics [5,13] including robot learning [4,16]. As a specific task example we consider a thrusters/pipes assembly task. One or more human tutors show the robot Sawyer the assembly work consisting of a series of assembly steps necessary to construct the structure from its parts/thrusters (Fig. 1). Different tutors may have different preferences in the sequential order of assembly steps, and may act with different time scales. The remainder of the paper is structured as follows: Sect. 2 describes the construction task and the robotic platform Sawyer; Sect. 3 contains the description of the DNF based learning model; Sect. 4 presents experimental results; and the paper ends with a discussion and future work in Sect. 5.

2 Experimental Setup

For the experiments we used the robot Sawyer (displayed in Fig. 2d) designed by the company "Rethink Robotics" to execute collaborative tasks [12]. Sawyer features a 7 degrees of freedom robot arm with 1.26 m reach and its "head" is

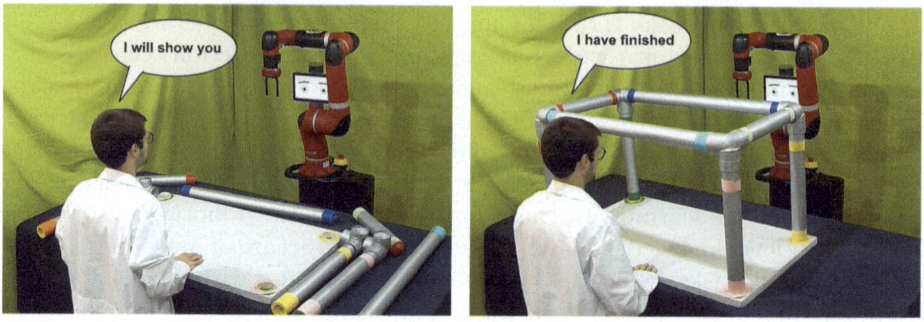

(a) Beginning of the task (b) End of the task

Fig. 1. Illustration of the construction task scenario: a tutor collaborates with the robotic co-worker in assembling a structure composed of eight colored thrusters

the LCD display that sits on top. Sawyer displays different eyes movements in a familiar way which contribute to its human-friendly design. The robot is also equipped by two cameras, one located in the head and another in the arm. The information about object color/type is provided by the head camera system. As a test scenario, we used a task of building a structure of eight thrusters/pipes, where the insertion of each one of the thrusters corresponds to one sub-task. We considered three different Layouts that imply three different assembly sequences and two tutors with different behavioral time scales. The disposal of the thrusters in each scenario can be seen in Fig. 2. A tutor demonstrates how to assemble the thrusters while the robotic co-worker observes and memorizes the serial order and timing of each assembly step. Later, the robot acts as a tutor and recalls the memorized task to a different operator, respecting the order and time interval of each step. Depending on how the thrusters are disposed on the workplace, different sequences can be used to assemble all the parts, which will require the robotic platform to learn multiple possible sequences to build the structure. Moreover, different tutors will assemble the sequence with different time intervals: for example, an older tutor may take longer to reach and insert all the thrusters/parts than a younger one. The speech synthesizer allows the robot to communicate the result of its decision process to the human co-worker.

3 Model Description

The model presented in this paper is based on previous research on natural human-robot interaction [2,16,17] based on Dynamic Neural Fields (DNFs). DNFs provide a rigorous theoretical framework to implement neural computations that endow a robot with crucial cognitive functions such as working memory, prediction and decision making [15]. DNFs are formalized by nonlinear integro-differential equations in which activity of neurons is summarized into the activity function $u(x, t)$, which can be used to reduced computational complexity and can be mathematically analyzed. The concept behind dynamic field models

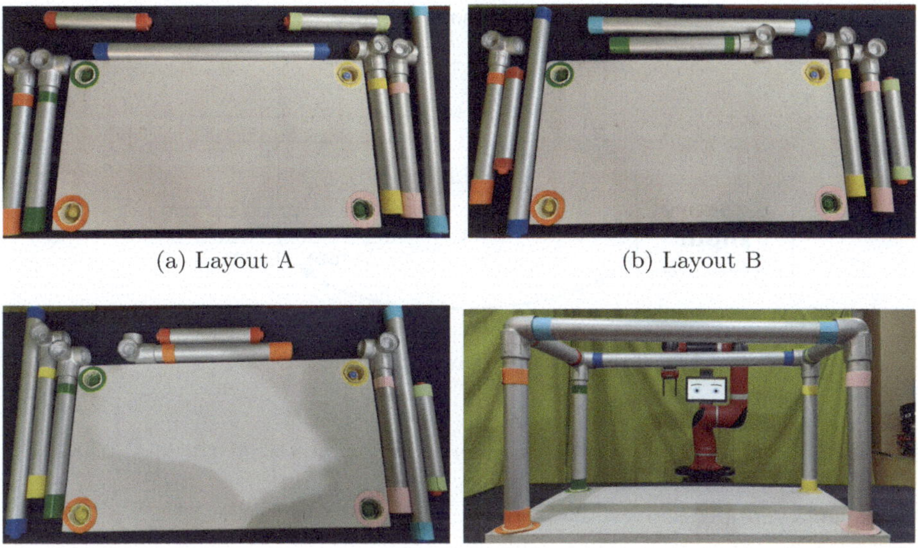

(a) Layout A (b) Layout B

(c) Layout C (d) Construction Task

Fig. 2. Construction task and Layout scenarios used during the experiments

is that task-relevant information is expressed by supra-threshold bumps of neural populations where each bump represents a specific action or sub-task. Input from external sources, such as vision, causes activation in the correspondent populations that remain active with no further external input due to recurrent excitatory and inhibitory interactions within the populations. Those interactions are able to hold an auto-sustained multi-bump pattern which can be turned into a memory mechanism for order and time interval of sequential processes [6,8,17].

Figure 3 presents the learning by demonstration of several sequence memory fields. Each Sequence Memory field (u_{SM}) stores a sequence of stimulus events as a multi-bump pattern. A bump represents an event triggered through excitatory sensory input. The strength of each memory representation reflects the time elapsed since stimulus presentation, resulting in an activation gradient from the first to the last event [6]. According to the context (e.g. position of the objects in the workplace, characteristics of the co-worker), several sequences with different order and timing can be memorized.

Figure 4 depicts an overview of the sequence recall process. Taking into account a specific context, a Sequence Memory field (u_{SM}) is chosen from the set of stored fields. The Sequence Recall field (u_{SR}) receives the multi-bump pattern of u_{SM} as subthreshold input. During sequence recall, the continuous increase of the baseline activity in u_{SR} brings all subpopulations closer to the threshold for the evolution of self-stabilized bumps. When the currently most active population reaches this threshold, the corresponding sensory output is triggered. At the same time, the excitatory-inhibitory connections between associated popula-

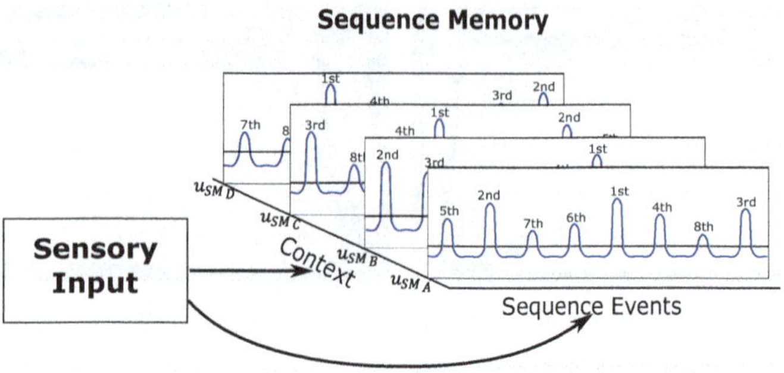

Fig. 3. Sketch of sequence learning process.

tions in u_{SR} and the Past Events field (u_{PE}) guarantee that the suprathreshold activity representing the latest sequence event becomes first stored in u_{PE} and subsequently suppressed.

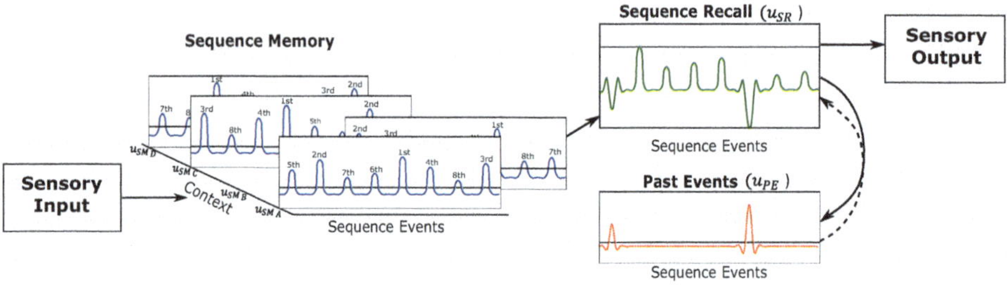

Fig. 4. Sketch of sequence recall process. Dashed lines indicate inhibitory connections, solid lines excitatory connections.

The dynamics of each Sequence Memory field u_{SM}, the Sequence Recall field u_{SR} and the Past Events field u_{PE} are governed by the following equations, respectively [1,6]:

$$\tau_{SM}\frac{\partial u_{SM}(x,t)}{\partial t} = -u_{SM}(x,t) + \int w(x-y)f\left(u_{SM}(y,t)\right)\mathrm{d}y + s(x,t) + h_{SM}(x,t) \tag{1}$$

$$\tau_{SR}\frac{\partial u_{SR}(x,t)}{\partial t} = -u_{SR}(x,t) + \int w(x-y)f\left(u_{SR}(y,t)\right)\mathrm{d}y - \int w(x-y)f\left(u_{PE}(y,t)\right)\mathrm{d}y + u_{SM}(x) + h_{SR}(t) \tag{2}$$

$$\tau_{PE}\frac{\partial u_{PE}(x,t)}{\partial t} = -u_{PE}(x,t) + \int w(x-y)f\left(u_{SM}(y,t)\right)\mathrm{d}y + u_{SR}(x,t)f(u_{SR}(x,t)) + h_{PE} \quad (3)$$

where $u_{SM}(x,t)$, $u_{SR}(x,t)$ and $u_{PE}(x,t)$ represents the activity at time t of a neuron tuned to the feature value x. The parameters $\tau_{SM}, \tau_{SR}, \tau_{PE} > 0$ define the time scale of each field. The connection function $w(x)$ determines the coupling between neurons within the field and to enable multi-bump solutions is use the following function [7]:

$$w(x) = Ae^{-b|x|}\left(b\sin|\alpha x| + \cos(\alpha x)\right), \quad (4)$$

where $b > 0$ determines the rate at which the oscillations in w decay with distance and $A > 0$ and $0 < \alpha \leq 1$ control the amplitude and the spatial phase of w, respectively. The function $s(x,t)$ represents the time dependent localized input at site x from the sensory input ($s(x,t) > 0$ when the encoded variable has an excitatory input, and $s(x,t) = 0$ otherwise). The strength of individual memory representations in u_{SM} is controlled by the baseline dynamics $h_{SM}(x,t)$:

$$\frac{\partial h_{SM}(x,t)}{\partial t} = \left(1 - f(u_{SM}(x,t))\right)\left(-h_{SM}(x,t) + h_{SM_0}\right) + \frac{1}{\tau_{h_{SM}}}f(u_{SM}(x,t)) \quad (5)$$

where $h_{SM_0} < 0$ defines the level to which h_{SM} converges without suprathreshold activity at position x and $\tau_{h_{SM}}$ measures the growth rate when it is present. The baseline activity $h_{SR}(t)$ evolves continuously in time described by the equation:

$$\frac{\partial h_{SR}(t)}{\partial t} = \frac{1}{\tau_{h_{SR}}}, h_{SR}(t_0) = h_{SR_0} < 0 \quad (6)$$

where $\tau_{h_{SR}}$ controls the growth rate of h_{SR}. The baseline activity $h_{PE} < 0$ is constant. $f(x)$ is the output function of the neuron and is taken as the Heaviside step function with threshold 0.

4 Experimental Results

In this section, we describe the experimental results of learning the sequential order and timing of assembling the structure displayed in Fig. 2d, using the implemented model described in Sect. 3. Two tutors with different behavioral time scales, who already had experience with the construction task, were asked to build the structure starting from three different Layouts (A, B and C, Fig. 2). At the same time, the robotic co-worker Sawyer pays attention to the tutor and stores the sequential order and time interval of each step, for all demonstrations. Afterward, the robotic platform acts as a tutor and teaches two other inexperienced workers with no knowledge of the construction task. From the memorized sequences, the system selects the most suitable one according to the characteristics of the new worker and the distribution of the pieces in the table. The robot recalls the selected sequence, verbalizing the color of the piece that should be inserted, according to the order and time stored in the Sequence Memory field during the demonstration trials.

4.1 Learning the Sequence Order and Timing of the Assembly Task

During the demonstration period, when the system detects that one of the colored thrusters is going to be inserted, an input stimulus is generated in the location of the population of neurons encoding the respective colored thruster, which leads to activation in the u_{SM} field, forming a bell-shaped bump that grows gradually as a function of time. Figure 5 pictures the demonstration experiment performed, where both tutors show to the robotic co-worker how to build the structure. In the scenario illustrated in Fig. 5a and b, the colored thrusters are distributed in the work table according to Layout A. A video of this trial can be found in the following link: https://youtu.be/YTZTDJzzGYw. Next, Fig. 5c and d are snapshots of the Sequence Memory fields stored during the demonstration, where the amplitude of the bumps encodes the serial order of the inserted thrusters, with the highest peak (orange) being the color of the first inserted piece and the shortest one (light blue) being the last. Although the serial order used by both tutors is the same, the intervals of time between each assembly step in both trials are considerably different, which will be demonstrated in Sect. 4.2.

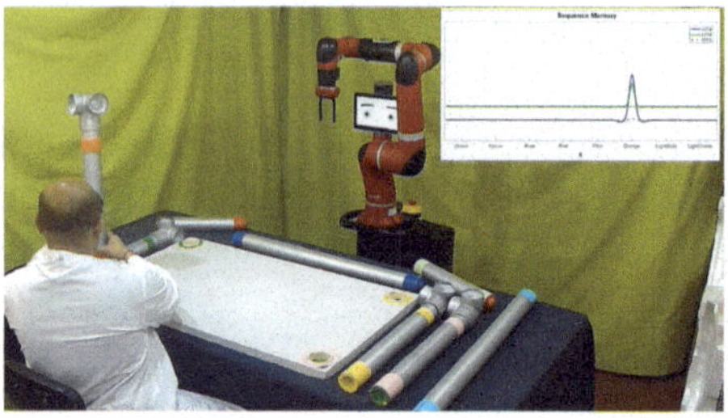

(a) **Older tutor** inserting the Orange thruster (1st piece)

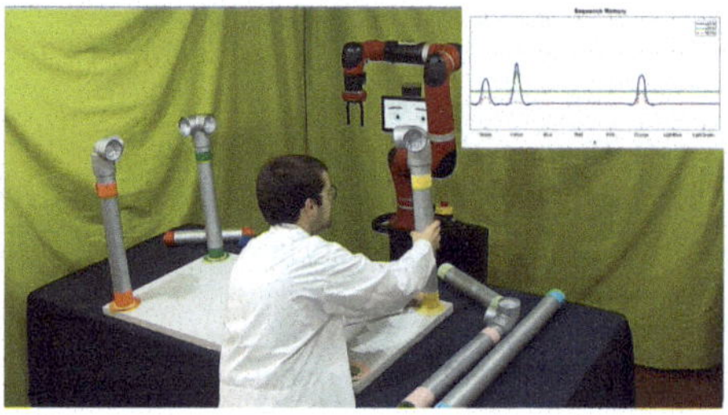

(b) **Younger tutor** inserting the Yellow thruster (3rd piece)

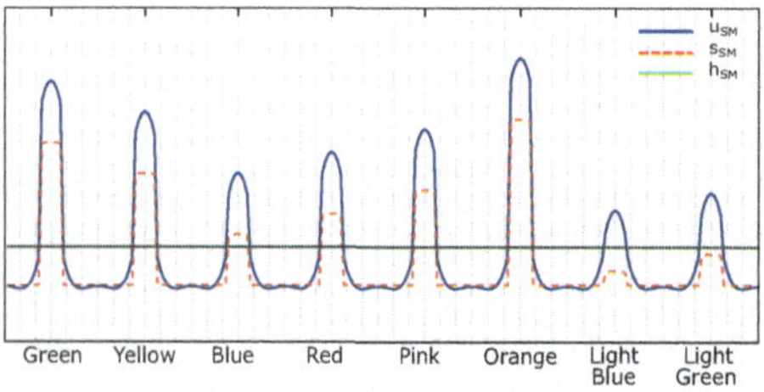

(c) Sequence Memory Field A demonstrated by the **Older tutor**

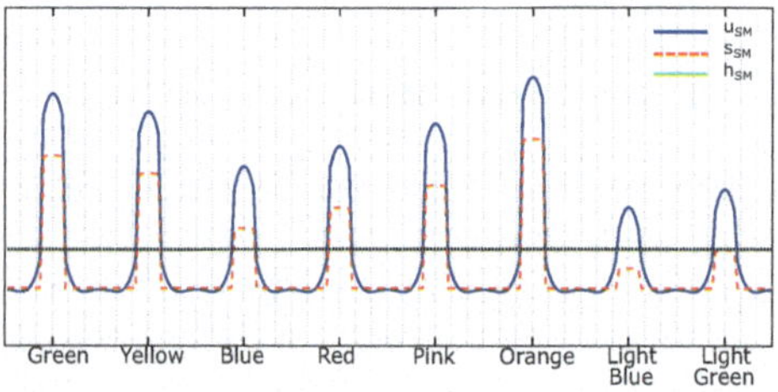

(d) Sequence Memory Field A demonstrated by the **Younger tutor**

Fig. 5. Two different tutors assembling **Sequence A**: Orange → Green → Yellow → Pink → Red → Blue → Light Green → Light Blue

Subsequently, Fig. 6 portraits the demonstration of two other sequences, this time by the younger tutor. In Fig. 6a, the parts are organized in the scenario according to Layout B, so the tutor starts the construction by inserting the pink thruster, opting for a different assembly sequence. A complete video of this trial can be found at https://youtu.be/0F8T_d_y2xs. Similarly, in the scenario displayed in Fig. 6b, the tutor was asked to assemble the structure starting from Layout C, which resulted in a new sequence. After the demonstration of each trial, the information acquired in both fields - Sequence Memory B (Fig. 6c) and Sequence Memory C (Fig. 6d) - is stored in the model, so later the robotic platform can use the memorized information to instruct inexperienced workers, taking into account the Layout of the parts that constitute the sequence.

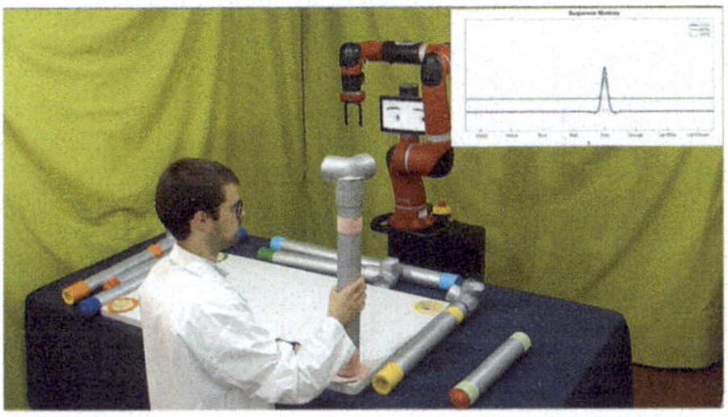

(a) Younger tutor assembling **Sequence B**, starting from the Pink thruster (1st piece).

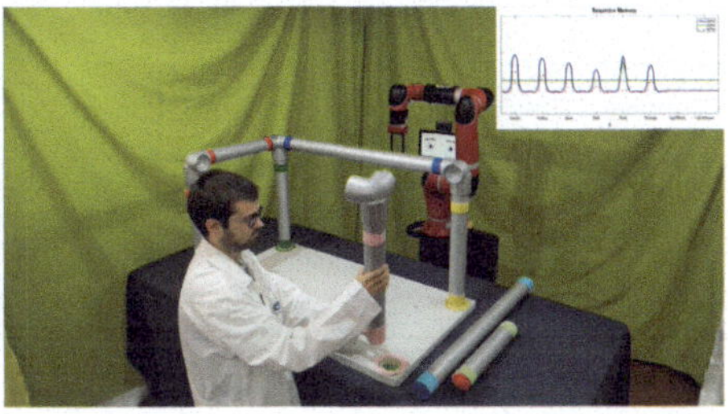

(b) Younger tutor assembling **Sequence C**, inserting the Pink thruster (6th piece)

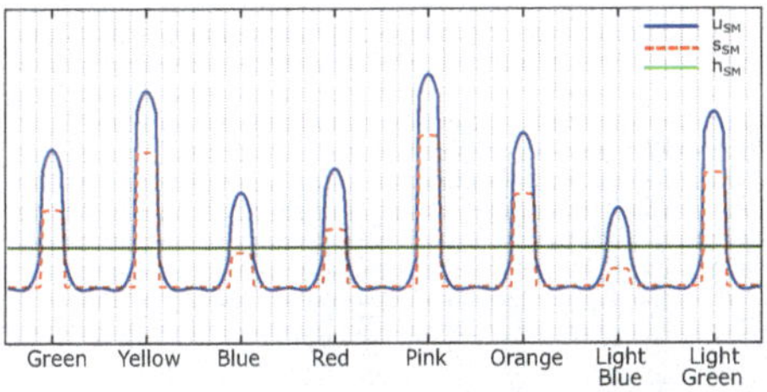

(c) **Sequence Memory Field B:** Pink → Yellow → Light Green → Orange → Green → Red → Blue → Light Blue

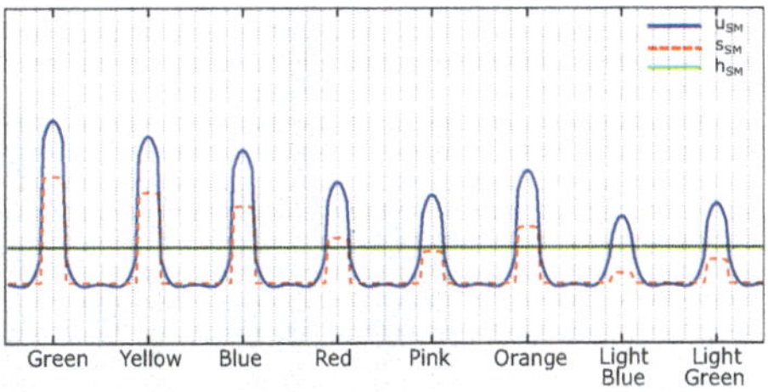

(d) **Sequence Memory Field C**: Green → Yellow → Blue → Orange → Red → Pink → Light Green → Light Blue

Fig. 6. Younger tutor assembling **Sequence B and C**

4.2 Recalling the Memorized Sequences

In order to verify if the robotic co-worker was able to memorize not only the sequence but also the time interval between each construction step, two workers with no previous knowledge of the task were asked to collaborate with Sawyer and follow its instructions to learn the assembly steps of the sequence A, which was previously demonstrated by the two previous tutors, with different time intervals. The system was programmed to verbalize each construction step, taking into consideration the sequence and timing between each insertion.

(a) **Older worker** following Sawyer's instruction to assemble Sequence A

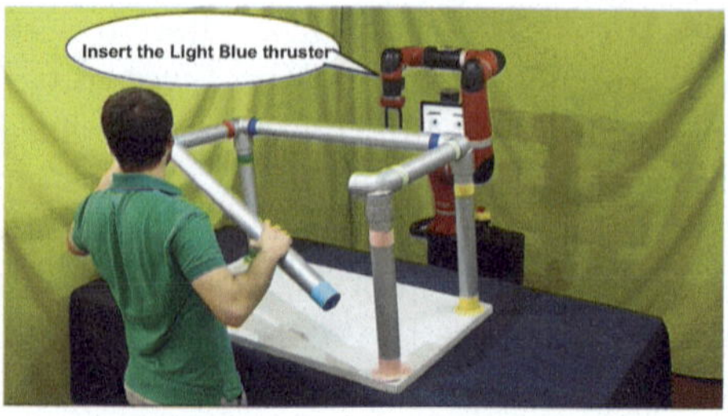

(b) **Younger worker** following Sawyer's instruction to assemble Sequence A

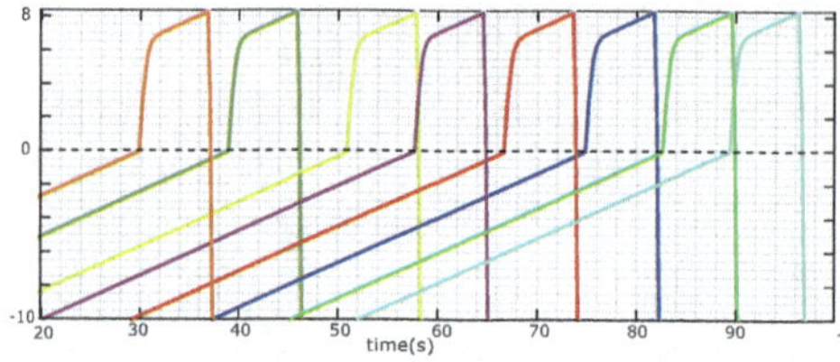

(c) Older worker: Total time of 67s **(slower)**

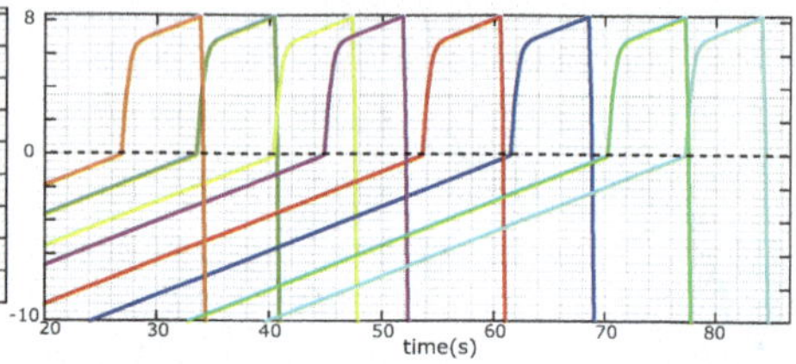

(d) Younger worker: Total time of 58s **(faster)**

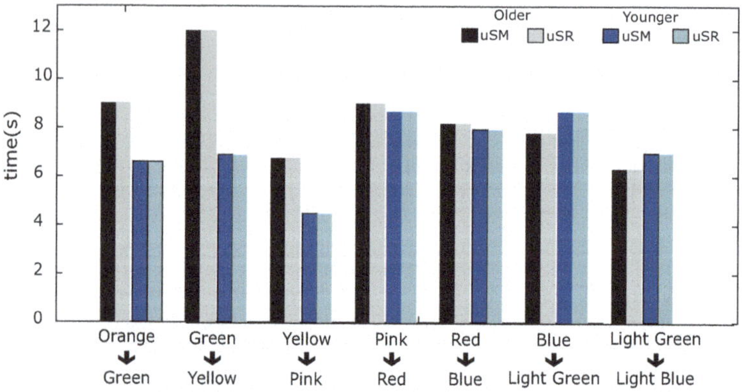

(e) Contrast between the time intervals of consecutive assembly steps during the construction of the sequence A, performed by an older (slower) and a younger (faster) worker

Fig. 7. Co-worker Sawyer recalls the memorized sequence, respecting the time interval of each demonstration

Figure 7a and b picture the robotic system as a tutor, guiding two different workers through the construction sub-tasks of the structure, while Fig. 7c and d illustrate the time course of the maximal activation of each sub-neuronal population when the sequence was assembled by each worker, according to the instructions given by the robot. A video example can be found at https://youtu.be/Vn0_1raKq4I.

Each instruction is verbalized when each sub-neuronal population encoding the corresponding colored thruster in the Sequence Recall field (u_{SR}) reaches the threshold level, as stated in Sect. 3. As can be observed in both figures, all sub-neuronal populations seem to have a pre-activation strength that respects the temporal order of the sequential task, learned during the demonstration trials. The first thrusters (orange) are inserted at t = 30 s (older tutor trial) and t = 27 s (younger tutor trial), while the last ones (light blue) are placed at t = 97 s (older tutor trial) and t = 85 s (younger tutor trial). By comparing both trials (Fig. 7e), we can observe that the older worker was slower than the younger in the majority of the steps and the younger worker took less time to perform the complete task, as it should be expected since the workers are following the order and time memorized in the previous demonstrations (Fig. 5).

5 Discussion

In this paper, we have proposed and tested, on a collaborative robot, a rapid learning system that allows the robot to memorize knowledge about ordinal and temporal aspects of sequential tasks in a learning by demonstration paradigm. One benefit of this learning system is that a single demonstration is sufficient, thus minimizing the efforts of the human tutor to train the robot. We have shown that after learning, the recall of the memorized information can be used by the robot to instruct inexperienced human operators, in the same context. This instructional process was, however, performed in open-loop. This fast learning system offers other benefits that will be explored in future work. For example, the recall of the stored information can be used as input to a long-term learning mechanism that allows the robot also to learn the connections between the several sub-tasks [16], thus extrapolating task knowledge, i.e. that a task can be performed in many different ways. As future work, the collaborative robot Sawyer can also contribute by using its robotic arm to interact with its co-worker and jointly build the construction task, reducing the workload of the human partner thus increasing the efficiency of the process.

Endowing a collaborative robot with the capacity to predict not only the ordinal sequence structure but also the time interval between successive events is central for efficient coordination of actions and decisions in space and time, in human-robot joint action tasks. It allows the robot to anticipate what the human operator will need, or will do, and when it should start an adequate complementary behavior in the service of the joint task.

References

1. Amari, S.: Dynamics of pattern formation in lateral-inhibition type neural fields. Biol. Cybern. **27**(2), 77–87 (1977). https://doi.org/10.1007/BF00337259
2. Bicho, E., Erlhagen, W., Louro, L., e Silva, E.C.: Neuro-cognitive mechanisms of decision making in joint action: a human-robot interaction study. Hum. Mov. Sci. **30**(5), 846–868 (2011). https://doi.org/10.1016/j.humov.2010.08.012

3. El Zaatari, S., Marei, M., Li, W., Usman, Z.: Cobot programming for collaborative industrial tasks: an overview. Robot. Auton. Syst. (2019). https://doi.org/10.1016/j.robot.2019.03.003
4. Erlhagen, W., Mukovskiy, A., Bicho, E., Panin, G., Kiss, C., Knoll, A., Van Schie, H., Bekkering, H.: Goal-directed imitation for robots: a bio-inspired approach to action understanding and skill learning. Robot. Auton. Syst. (2006). https://doi.org/10.1016/j.robot.2006.01.004
5. Erlhagen, W., Bicho, E.: The dynamic neural field approach to cognitive robotics. J. Neural Eng. **3**(3), R36–R54 (2006). https://doi.org/10.1088/1741-2560/3/3/R02
6. Ferreira, F., Erlhagen, W., Bicho, E.: A dynamic field model of ordinal and timing properties of sequential events. In: Lecture Notes in Computer Science (including subseries Lecture Notes in Artificial Intelligence and Lecture Notes in Bioinformatics) (2011). https://doi.org/10.1007/978-3-642-21738-8_42
7. Ferreira, F., Erlhagen, W., Bicho, E.: Multi-bump solutions in a neural field model with external inputs. Phys. D: Nonlinear Phenom. **326**, 32–51 (2016). https://doi.org/10.1016/j.physd.2016.01.009
8. Ferreira, F., Erlhagen, W., Sousa, E., Louro, L., Bicho, E.: Learning a musical sequence by observation: a robotics implementation of a dynamic neural field model. In: IEEE ICDL-EPIROB 2014 - 4th Joint IEEE International Conference on Development and Learning and on Epigenetic Robotics, pp. 157–162 (2014). https://doi.org/10.1109/DEVLRN.2014.6982973
9. Kyrarini, M., Haseeb, M.A., Ristić-Durrant, D., Gräser, A.: Robot learning of industrial assembly task via human demonstrations. Auton. Robots **43**(1), 239–257 (2019). https://doi.org/10.1007/s10514-018-9725-6
10. Orendt, E.M., Fichtner, M., Henrich, D.: Robot programming by non-experts: intuitiveness and robustness of one-shot robot programming. In: 2016 25th IEEE International Symposium on Robot and Human Interactive Communication (RO-MAN), pp. 192–199. IEEE (2016). https://doi.org/10.1109/ROMAN.2016.7745110
11. Papanastasiou, S., Kousi, N., Karagiannis, P., Gkournelos, C., Papavasileiou, A., Dimoulas, K., Baris, K., Koukas, S., Michalos, G., Makris, S.: Towards seamless human robot collaboration: integrating multimodal interaction. Int. J. Adv. Manuf. Technol. 1–17 (2019). https://doi.org/10.1007/s00170-019-03790-3
12. Robotics, R.: Sawyer collaborative robot (2018). http://www.rethinkrobotics.com/sawyer/
13. Sandamirskaya, Y., Zibner, S.K.U., Schneegans, S., Schöner, G.: Using dynamic field theory to extend the embodiment stance toward higher cognition. New Ideas Psychol. **31**(3), 322–339 (2013). https://doi.org/10.1016/j.newideapsych.2013.01.002
14. Schaal, S.: The new robotics towards human-centered machines. HFSP J. **1**(2), 115–126 (2007). https://doi.org/10.2976/1.2748612
15. Schöner, G.: Dynamical systems approaches to cognition (January) (2012). https://doi.org/10.1017/cbo9780511816772.007
16. Sousa, E., Erlhagen, W., Ferreira, F., Bicho, E.: Off-line simulation inspires insight: a neurodynamics approach to efficient robot task learning. Neural Netw. **72**, 123–139 (2015). https://doi.org/10.1016/j.neunet.2015.09.002
17. Wojtak, W., Ferreira, F., Louro, L., Bicho, E., Erlhagen, W.: Towards temporal cognition for robots: a neurodynamics approach. In: 7th Joint IEEE International Conference on Development and Learning and on Epigenetic Robotics, ICDL-EpiRob 2017, pp. 407–412 (2018). https://doi.org/10.1109/DEVLRN.2017.8329836

Core Concepts for an Ontology for Autonomous Robotics

An Ontology for Failure Interpretation in Automated Planning and Execution

Mohammed Diab[1(✉)], Mihai Pomarlan[2], Daniel Beßler[2], Aliakbar Akbari[1], Jan Rosell[1], John Bateman[2], and Michael Beetz[2]

[1] Institute of Industrial and Control Engineering, Universitat Politècnica de Catalunya, Barcelona, Spain
mohammed.diab@upc.edu

[2] Institute for Artificial Intelligence, Universität Bremen, Bremen, Germany

Abstract. Autonomous indoor robots are supposed to accomplish tasks, like serve a cup, which involve manipulation actions, where task and motion planning levels are coupled. In both planning levels and execution phase, several source of failures can occur. In this paper, an interpretation ontology covering several sources of failures in automated planning and also during the execution phases is introduced with the purpose of working the planning more informed and the execution prepared for recovery. The proposed failure interpretation ontological module covers: (1) geometric failures, that may appear when e.g. the robot can not reach to grasp/place an object, there is no free-collision path or there is no feasible Inverse Kinematic (IK) solution. (2) hardware related failures that may appear when e.g. the robot in a real environment requires to be re-calibrated (gripper or arm), or it is sent to a non-reachable configuration. (3) software agent related failures, that may appear when e.g. the robot has software components that fail like when an algorithm is not able to extract the proper features. The paper describes the concepts and the implementation of failure interpretation ontology in several foundations like DUL and SUMO, and presents an example showing different situations in planning demonstrating the range of information the framework can provide for autonomous robots.

1 Introduction

Challenging robotic problems, e.g. assembly tasks in cluttered environments, require planning at task and motion levels. For both levels, the use of knowledge may enhance planning and robot capabilities, giving more autonomy to the robots [1,2]. The enhanced robot capabilities include the capture of rich semantic descriptions of the scene, knowledge about the physical behavior of objects, and reasoning about potential manipulation actions.

Different tasks may have different grades of complexity, both at symbolic and geometric levels, as well as regarding the dependence between them. Logic states and

This work was partially funded by Deutsche Forschungsgemeinschaft (DFG) through the Collaborative Research Center 1320, EASE, and by the Spanish Government through the project DPI2016-80077-R. M. Diab is supported by the Spanish Government through the grants FPI 2017.

M. F. Silva et al. (Eds.): ROBOT 2019, AISC 1092, pp. 381–390, 2020.
https://doi.org/10.1007/978-3-030-35990-4_31

actions have to be mapped to geometric instances, and a state transition can only occur if the action is geometrically feasible. A smart combination of task and motion planning capabilities that produces fewer failures in both levels is required to make the process efficient – e.g., the symbolic planner should not ask too many impossible queries from the motion planner. For robotic systems that need an interplay between symbolic and geometric planning, interpreting failures, diagnosing their causes, and figuring out the proper solutions for failure recovery is essential for both automated planning and execution phases. This is true for robotic systems based on heuristic search classical planning approaches [3] or for ontology-guided assembly planning [4].

Various studies have investigated the use of knowledge in the form of ontologies for the detailed description of failures in several robotics fields such as electronic industrial applications [5] or in manufacture kitting applications [6]. Several sources of failures types have been analyzed and categorized. Geometric failures related to reachability and action feasibility are described in their relationship to their causes (missing IK solution or detected collision) [7]. Failures related to motion planners not finding collision free trajectories have also been ontologically described [8,9]. The Unified Foundational Ontology (UFO) [10] was proposed as a reference conceptual model (domain ontology) of software defects, errors and failures, which takes into account an ecosystem of software artifacts. Although the aforementioned works describe failure types, they are very task specific, do not use any sort of interpretation mechanism to diagnose the reason of failure, and don't reuse such a mechanism in a generic way. Also, there are few works considering failures of planning and run-time phase for the automated manipulation domain and representing such failure knowledge in a generalizable, shareable ontology format.

2 Problem Statement and Approach Overview

We are mainly interested in assembly manipulation tasks for bi-manual robots, which often encounter complexity or failures in the planning and execution phases. Planning phase failures typically refer to failures of the planner itself, but we will use planning phase failures to also refer to situations where the planner reasons that some action would be infeasible, e.g. because objects block access to what the robot should reach. A correct selection of grasps and placements must be produced in such an eventuality. Depending on the type of problem, goal order must be carefully handled especially in the assembly domain; very large search spaces are possible, requiring objects to be moved more than once for achieving the goals. Execution phase failures refer to hardware failures related to the system devices– e.g. robot or camera needs to be re-calibrated–, or software failures related to the capabilities offered by specific software components, or failures in action performance such as an unexpected occluding object, or slippage.

To accurately capture, share, and reuse knowledge about failures, we propose an ontological model of failure interpretation under several foundations like SUMO [11] and DUL [12]. SUMO provides a conceptual structure that can be used and integrated with other specific ontologies developed for the robotics and automation domain, meanwhile DUL organizes concepts in a descriptive way, attempting to catch cognitive categories. It brings along modeling constraints which guide the development of domain ontologies in such a way that they may support complex reasoning tasks. The ontological formulation provides a common understanding for the robot to interpret the causes of the failures in automated planning and execution and find solutions. This paper contributes with an ontology for failures in automated planning and execution phases. The contributions are:

1. *Ontology formulation:* Introduction of an ontology to describe different types of failures in the automated manipulation planning domain.
2. *Modeling in different foundations:*
   - *For robotics domain:* Modeling the absolute abstract concepts under robotics upper-level ontologies such as CORA, which uses the SUMO ontology as an upper level.
   - *For engineering domain:* Modeling the absolute abstract concepts under very generic foundational ontologies such as DUL.
3. *Use of the failure ontology:* Description of how to use the failure ontology in a task and motion planning (TAMP) process, such as heuristic search classical approaches or the knowledge-enabled approaches, and in a static code analysis.

For the case study, we use the concept of workflow, that represent the structure of a task by subtasks linked via conditional transitions between them. In this framework, knowledge modeling for robot assembly execution includes a model of conditional workflows, software services, and robot tasks that can be automatically executed by the workflows, and a way of interfacing existing components of a robot control system.

3 Ontology Driven Failure Interpretation

The higher level concepts in the proposed ontological module are presented. The ontological module consists of two ontologies: failure ontology and geometric one as a sub-module. The combination is required when we report the geometric failure in the manipulation domain. The concepts are absolute term that cope with the modeling formalism. Moreover, some basic concepts, without being formally defined, like Agent, Plan, Task, Goal or event are used (as defined in [11, 12]). The two ontologies concepts are described below using description logic (DL).

3.1 Concepts Describing Failures

Task Failure. A situation which interprets a series of Events as the failed plan or execution of some task(s). In other words, there needs to be an Event, with an *Agent* as an effective cause; further, the *Agent* should be pursuing some *Goal*, which it does by following a *Plan* to complete a *Task*. This situation is described by a *Failure narrative*. For example, a robot has the task to serve a cup from a tray to a table. While in transit, the robot drops the cup and it shatters. The collection of events, together with the knowledge of the robot task and goal, constitutes a failure situation.

Failure Narrative. A communicable description of a *Task failure* situation. It defines several roles, to be filled by an *Agent*, *Task*, and a *Goal*. It uses a *Failure symptom* to classify the *Action* that the *Agent* performed, and may provide an *Explanation* for the failure in a *Failure diagnostic*.

Failure symptom and Failure diagnostic is a simple classification label which can be applied to events or event series in order to interpret them as a failure of some sort. Failure symptoms include software error codes and signal or exception types. For example, when a robot controller raises a "hardware error signal", we would say the robot classified an event as being a hardware failure. Error status codes in query returns are another subtype of failure symptom; e.g., the status code that an IK solver would use to indicate it found no solution. Sometimes software signals or exceptions come with more data attached to them in order to identify the failure cause; for example the hardware error signal might also include which motor is thought defective. This extra information is the *Failure diagnostic*. The concepts above are defined via their relations to foundational ontologies (DUL or SUMO; see Sect. 4). Also, a taxonomy of failure symptoms is defined, as follows. One dimension of classifying failures is the "when" of happening: **Inception failures** (prevent an action from starting; e.g. *CapabilityFailure*), **Performance failures** (prevent an action from completion; e.g. *ResourceDepletionFailure*), **End state failures** (the outcome of a completed action is not conformant to the goal; e.g. *ConfigurationNotReached*). Failures are also classified by the nature of participants. Currently there are three top-level classes for this dimension: **Cognition failure**, **Communication failure**, and **Physical failure**. *CognitionFailures* classify events involving only an *Agent* and whatever internal representations it uses. *CommunicationFailures* classify an Event that involves some *Agents* exchanging information. Finally, Events with only *Physical object* or *Agent* participants can be classified as *PhysicalFailure*; e.g., the robot not being able to manipulate an object because of it being too far (*ReachabilityFailure*) or because the gripper is broken (*EndEffectorFailure*).

For failures where it makes sense to identify a physical location, there is a classification along the "where" of occurrence. So far, this is only seen in the taxonomy for embodied *Physical agents*, and failures may be classified as **Body part failure** (example, *TorsoFailure*). There is a dimension of classifying failures along the "what" of the relevant interaction. So far, this is done only for *PhysicalFailures*, which can be **Mechanical failures** or **Electrical failures**.

3.2 Concepts Describing Geometric Queries

Since one of the main focus on the proposed failure interpretation ontology is the geometric failures, an ontological module is defined to cover geometric notions used in robotics such as collision, placement feasibility, etc. The concepts in this ontology also describe geometric querying, query status, and status diagnosis. A more specific geometric ontology module contains terms for particular queries such as IK (inverse kinematics) for a specific robot. To describe geometric failure, the following terms in the ontology are included.

Geometric Querying. An *Event* in which some (software) spatial reasoner– e.g. a collision checker– participates, and which is classified by/executes a **GeometricReasoningTask**. It is defined as *GeometricQuerying* — $\sqsubseteq$ ($\exists$*isClassifiedBy.GeometricReasoningTask*) $\sqcap$ ($\exists$*hasParticipant.SpatialReasoner*) $\sqcap$ ($=$1*hasStatus.*

QueryStatus). We also say that *GeometricReasoningTask* $\sqsubseteq$ $\forall$ *isExecutedBy. GeometricQuerying*.

Spatial Reasoner. A software component used to answer geometric queries. For example, for checking reachability, an IK solver can be used as a *SpatialReasoner*. The spatial reasoner is represented as a *SpatialReasoner*$\sqsubseteq$*ComputationalAgent*. Several types of *SpatialReasoner: CollisionChecker, IKSolver, MotionPlanner* are proposed in this ontology.

Query Status. The outcome of a *GeometricQuerying*. If failure, it means the query was not answered at all, and the reason is given. Such reasons include the geometric component responsible for the query not answering in time or being unavailable. If success, it means the query has been answered, and the answer can further be interpreted to ascertain what it implies for an action's feasibility. Note that "query failure" is used for situations where no answer at all is given; "geometric failure" for situations where an answer is given, but it is not satisfactory for some constraint. E.g., an IK solver returning it failed to find a solution is not a query failure (the IK solver works well) but it is a geometric failure (a *ReachabilityFailure*).

Status Diagnosis. Information to support the analysis of geometric query answers.

4 Modeling of Failure Ontology Under Different Upper Level Foundations

Aiming at making the failure ontology sharable and widely used, it has been formalized under SUMO [11] and DUL [12] foundations, as shown in Table 1. The upper-level

Table 1. Modeling the failure ontology under the DUL and SUMO foundations.

Concept	DUL	DL description	SUMO	DL description
Failure symptom	Event type: A Concept that classifies an Event. An event type describes how an Event should be interpreted, executed, expected, seen, etc., according to the Description that the EventType isDefinedIn (or used in)	EventType $\sqsubseteq$ Concept EventType $\sqsubseteq$ ($\forall$classifies.Event) FailureSymptom $\sqsubseteq$ EventType FailureSymptom $\sqsubseteq$ $\forall$classifies. ($\exists$hasParticipant.Agent)	Class: similar to Sets, but not assumed to be extensional, i.e. distinct classes may have the same members. Membership decided by some condition.	FailureSymptom $\sqsubseteq$ Class FailureSymptom $\sqsubseteq$ $\forall$ instance^{-1}.AgentPatientProcess
Failure Narrative Role	Role: A Concept that classifies an Object	Role $\sqsubseteq$ Concept Role $\sqsubseteq$ $\forall$classifies.Object FailureNarrativeRole $\sqsubseteq$Role	Class: similar to Sets, but not assumed to be extensional, i.e. distinct classes may have the same members. Membership decided by some condition.	FailureNarrativeRole $\sqsubseteq$ Class FailureNarrativeRole $\sqsubseteq$ $\forall$ instance^{-1}.Object
Failure Narrative	Narrative: A descriptive context of situations	Narrative$\sqsubseteq$Description FailureNarrative$\sqsubseteq$ Narrative	Proposition: An abstract entities that express a complete thought or a set of such thoughts.	FailureNarrative $\sqsubseteq$ Proposition
Task Failure	Situation: A view, consistent with ('satisfying') a Description, on a set of entities.	Situation $\sqsubseteq$ $\exists$satisfies.Description TaskFailure $\sqsubseteq$ Situation TaskFailure $\sqsubseteq$ $\exists$satisfies.FailureNarrative	Propositional attitude: An IntentionalRelation in which an agent is aware of a proposition.	TaskFailure $\sqsubseteq$ PropositionalAttitude TaskFailure $\sqsubseteq$ $\exists$satisfies.FailureNarrative
Failure diagnosis	Diagnosis: A Description of the Situation of a system, usually applied in order to control a normal behavior, or to explain a notable behavior (e.g. a functional breakdown).	Diagnosis $\sqsubseteq$ Description FailureDiagnosis $\sqsubseteq$ Diagnosis	Proposition: An abstract entities that express a complete thought or a set of such thoughts.	FailureDiagnosis $\sqsubseteq$ Proposition

ontologies SUMO and DUL foundations are used because of their abilities to cover several concepts related to robotics but also more general domains. Several ontology frameworks are defined under those upper-levels, such as [1], to facilitate the incorporation/importing of different ontologies with the same common vocabularies while avoiding semantic conflicts. This way of modeling becomes essential in large-scale research and development projects.

5 Case Study: Use of Failure Ontology in Robotic Assembly Domain

5.1 In Planning Phase

To illustrate our proposal some simulation examples are performed using Rviz [13] for visualization and The Kautham Project for geometric planning [14], which has the ability to report the potential geometric failures. Ontologies are encoded using the Web Ontology Language (OWL) [15]. The ontologies are designed using the Protégé[1] editor.

A classical task planning approach and a knowledge-enabled approach have been used. With the former a sequence of actions is computed using a combined heuristic task and motion planning [3] (actions like transit, to move to grasping configurations, and transfer, to pick and place objects have been used). For latter a workflow is proposed that describes task and actions required in task execution. Some geometric situations are used for testing. The sequence to plan the assembly operations is:

1. call IK module to check reachability for grasping the objects;
2. call a collision checker to validate a path to grasp an object.

Some failures can happen through calling these modules. The failures get reported to the planner, where the failures are interpreted and a decision on the next action is made. Some situations in manipulation domain may happen often, such as the case where an object is blocking the chosen configuration to grap/place an object. This situation requires the selection of alternative feasible (or reachable) grasping poses and/or placements.

For example, as shown in Fig. 1, if the object *BottomWing* has four grasping poses from each side g1-g4, the g1 and g4 are occluded by the holders of *PropellerHolder* and *TopWingHolder* respectively, that means g1 and g4 are not feasible and the robot is not able to reach object *BottomWing* through them. Meanwhile, the robot may not be able to grasp the object *BottomWing* through g2 and g3 because of the infeasibility of IK configurations. By querying over the proposed ontology, the robot will be able to analyze the cause and report to the planner (infer a prober solution will be included in our upcoming work).

The failure symptom produced when planning to use g2 and g3 for grasping is a *ReachabilityFailure*, which is a *CapabilityFailure*, whereas a failure produced when planning to grasp using g1 or g4 is an *OcclusionFailure*, which is an *AffordanceFailure*.

[1] http://protege.stanford.edu/.

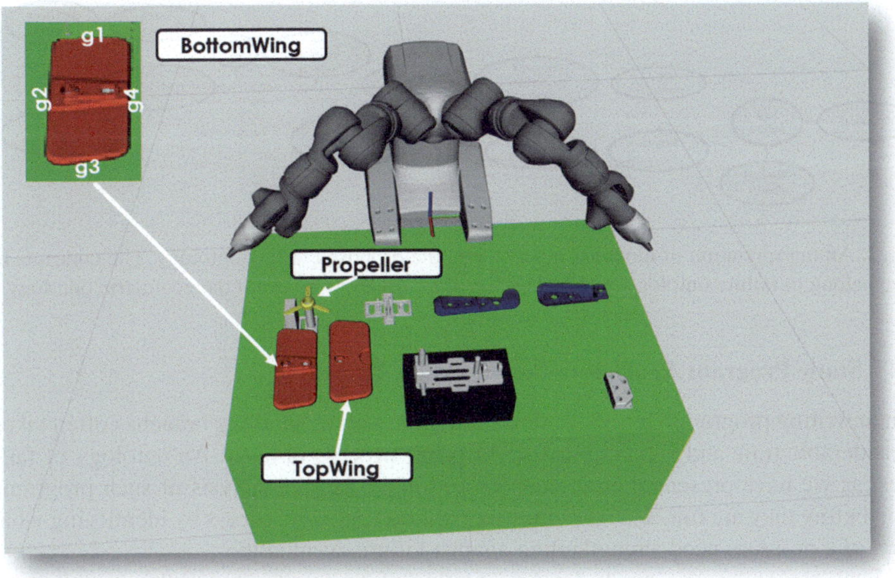

Fig. 1. Motivating example in assembly domain showing cases that need the planner to use the failure ontology to interpret query and action results. Description of reachability and collision problems. The sequence of assemble the Battat toy can be found in https://sir.upc.es/projects/ontologies/

Both of these are *InceptionFailures* which prevent the planned task from even being undertaken. *CapabilityFailure* can be addressed by generating new capabilities, e.g. selecting new grasping poses that are reachable. Affordance failures, meanwhile, can be addressed by manipulating the environment to better expose its affordances, so by generating intermediary goals of moving the occluders out of the way, the robot can eventually grab the *BottomWing*.

To interpret the causes of those situations presented in Fig. 1, the geometric ontology is integrated with the failure ontology as described in Fig. 2. This ontology describes the failure symptoms, as well as the *FailureNarratives* which make use of these symptoms to classify failures. The narratives may include other information to enhance the diagnosis process, such as the initial goal and participating objects in the agent's task, and a diagnostic to indicate which component failed. Failure symptoms are ontologically characterized also in terms of what failure diagnostics they are compatible with; for example, a *ReachabilityFailure* can only be used by a failure narrative where the explanation role is played by a failure diagnostic that names some IK component as the failure cause. These modules (i.e, IK and collision check) are low-level modules that the symbolic level considers to be spatial reasoners.

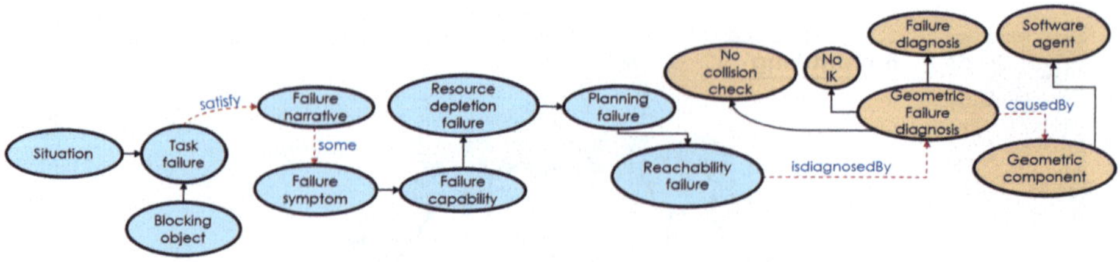

Fig. 2. An interpretation of blocking object using the proposed failure ontology. The concepts in blue belong to failure ontology, meanwhile the ones in yellow are from the geometric ontology.

5.2 Static Program Analysis in the Workflow Execution

When writing programs for robots, developing the failure handling branches often takes considerable time and is a complex, error prone process in itself. An ontology of failures, as we have presented here, enables reasoning for the analysis of such programs even before they are run, and may improve the development process by identifying what parts of a program need strengthening against failure eventualities.

Such reasoning also needs a representation of programs or workflows, and upper ontology axioms. Our technique for static analysis works if the tasks in a workflow are characterized properly, i.e., each task belongs to a task concept that has sufficient axioms to define it. Definitions of such task concepts, as well as object concepts restricting what may participate in a task and filling which roles, are not part of the failure ontology module but should be part of the knowledge model of the system as a whole. The reasoning process can only be done within some models of the world. That is, there is a classification of potential failures (this is the failure ontology), a classification of possible tasks, their roles, and possible fillers for them. Within this model, we can ask whether certain combinations of tasks and failures make sense, and this is what our static code analysis answers. From previous work, we take a workflow to be a transition system where the nodes correspond to tasks, linked by transitions that are conditional on, among other things, task outcomes including failure signals, if any. Tasks can be "atomic", but they can also be described by workflows. When executing a workflow and encountering a non-atomic task, the execution process will begin following a path through the workflow describing the non-atomic task, in a pattern similar to invoking a piece of a program called a subroutine via a shorthand name for it.

The kinds of reasoning questions we will focus on here are the following:

1. Are there possible failures not included among the outcomes of the workflow describing how to perform a particular task?,
2. Are the failure outcomes of a workflow actually possible outcomes of the task described by the workflow?,
3. Are there enough branches in a workflow to account for the possible failure outcomes of a subtask of the workflow?

For all of these queries, we assume there exists some ontological characterization of a task, and a workflow that describes how to perform this task. An ontological characterization of a task means indicating what kind of task it is, and what restrictions there are on the participants in an event classifiable by this task. The ontological characterization of failure symptoms is from our ontology. In effect, the above queries involve answering one question: for a particular Task, what are the possible failures that can happen? This can be analyzed using the Distributed Ontology, Modeling and Specification Language (DOL) [16] pattern below. Here, *ExampleTask* and *ExampleFailure* are some named subconcepts of *Task* and *FailureSymptom* respectively, while *Ot*, *Of* are an ontology of tasks and our ontology of failures, respectively.

```
ontology PossibleFailure[Class: ExampleTask SubClassOf: Task]
                        [Class: ExampleFailure SubClassOf: FailureSymptom]
                        [Class: ExampleEvent SubClassOf: Event]
 given Of, Ot =
 ObjectProperty: isClassifiedBy InverseOf: classifies
 Class: ExampleEvent
  SubClassOf: (isClassifiedBy some ExampleTask) and
              (isClassifiedBy some ExampleFailure)
end
```

A DL reasoner would be presented with the ontology resulting from applying the pattern above, which combines an ontology of tasks and other relevant concepts such as robot parts and objects, with our ontology of failures and the axioms above. The reasoner would be required to find a model for *ExampleEvent*. If no such model exists, then the failure symptom is impossible for the task. This query would be run for every pair of named task/failure symptom concepts in an ontology.

6 Conclusion

This paper proposes the formalization and implementation of the standardized failure ontology to extend the capabilities of autonomous robots related to manipulation tasks that require task and motion planning, including actions description, along with execution. This combination require the integration of geometric ontology that also proposed here. In the modeling level, the absolute concepts of both ontologies are modeled under DUL and SUMO foundations to facilitate the usability for roboticists community. A case study is introduced to illustrate the use of failure ontology in automated planning and workflow execution phases by proposing the common situations that could be encountered by such planner. Moreover, some static code analysis is proposed to analyze the possible failures while executing the tasks.

References

1. Diab, M., Akbari, A., Ud Din, M., Rosell, J.: PMK - a knowledge processing framework for autonomous robotics perception and manipulation. Sensors **19**(5), 1166 (2019)
2. Tenorth, M., Beetz, M.: Representations for robot knowledge in the KnowRob framework. Artif. Intell. **247**, 151–169 (2017). Special Issue on AI and Robotics
3. Akbari, A., Lagriffoul, F., Rosell, J.: Combined heuristic task and motion planning for bi-manual robots. Auton. Robots, 1–16 (2018)
4. Beßler, D., Pomarlan, M., Akbari, A., Muhayyuddin, Diab, M., Rosell, J., Bateman, J., Beetz, M.: Assembly planning in cluttered environments through heterogeneous reasoning. In: Trollmann, F., Turhan, A.Y. (eds.) KI 2018: Advances in Artificial Intelligence, pp. 201–214. Springer, Cham (2018)

5. Zhou, X., Ren, Y.: Failure ontology of board-level electronic product for reliability design. In: The Proceedings of 2011 9th International Conference on Reliability, Maintainability and Safety, pp. 1086–1091, June 2011
6. Kootbally, Z., Schlenoff, C., Antonishek, B., Proctor, F., Kramer, T., Harrison, W., Downs, A., Gupta, S.: Enabling robot agility in manufacturing kitting applications. Integr. Comput. Aided Eng. **25**, 1–20 (2018)
7. Srivastava, S., Riano, L., Russell, S., Abbeel, P.: Using classical planners for tasks with continuous operators in robotics. In: Workshops at the Twenty-Seventh AAAI Conference on Artificial Intelligence (2013)
8. Caldiran, O., Haspalamutgil, K., Ok, A., Palaz, C., Erdem, E., Patoglu, V.: Bridging the gap between high-level reasoning and low-level control. In: International Conference on Logic Programming and Nonmonotonic Reasoning, pp. 342–354. Springer, Berlin (2009)
9. Caldiran, O., Haspalamutgil, K., Ok, A., Palaz, C., Erdem, E., Patoglu, V.: From discrete task plans to continuous trajectories. In: Proceedings of BTAMP (2009)
10. Duarte, B.B., Falbo, R.A., Guizzardi, G., Guizzardi, R.S., Souza, V.E.: Towards an ontology of software defects, errors and failures. In: International Conference on Conceptual Modeling, pp. 349–362. Springer, Cham (2018)
11. Niles, I., Pease, A.: Towards a standard upper ontology. In: Proceedings of the International Conference on Formal Ontology in Information Systems, vol. 2001, pp. 2–9. ACM (2001)
12. Masolo, C., Borgo, S., Gangemi, A., Guarino, N., Oltramari, A.: WonderWeb deliverable D18 ontology library. Technical report, IST Project 2001-33052 WonderWeb: Ontology Infrastructure for the Semantic Web, August 2003
13. Kam, H.R., Lee, S.H., Park, T., Kim, C.H.: RViz: a toolkit for real domain data visualization. Telecommun. Syst. **60**(2), 337–345 (2015)
14. Rosell, J., Pérez, A., Aliakbar, A., Muhayyuddin, Palomo, L., García, N.: The Kautham project: a teaching and research tool for robot motion planning. In: Proceedings of the IEEE Emerging Technology and Factory Automation (ETFA), pp. 1–8, September 2014
15. Antoniou, G., van Harmelen, F.: Web Ontology Language: OWL, pp. 67–92. Springer, Berlin (2004)
16. Mossakowski, T., Codescu, M., Neuhaus, F., Kutz, O.: The distributed ontology, modeling and specification language – DOL. In: The Road to Universal Logic, pp. 489–520. Springer, Cham (2015)

Deducing Qualitative Capabilities with Generic Ontology Design Patterns

Bernd Krieg-Brückner[1,2(✉)] and Mihai Codescu[1]

[1] Collaborative Research Center EASE, University of Bremen, Bremen, Germany
bkb@uni-bremen.de
[2] Bremen Ambient Assisted Living Lab,
German Research Center for Artificial Intelligence, Bremen, Germany

Abstract. Generic Ontology Design Patterns, GODPs, encapsulate design details, hiding the complexity of modeling. As a separation of concerns, they are to be developed by ontology experts, while their safe (re-)use is entrusted to domain experts, who may focus on appropriate instantiations; these effectively document the design decisions. The deployment to the domain of robotics is demonstrated by simple GODPs for adding new objects, relating them consistently to a given ontology. Advanced GODPs show how semantic device properties such as qualitative capabilities can be deduced from quantitative data in a general way.

Keywords: Generic Ontology Design Patterns · Safe reuse · Graded relations · Qualitative capabilities · Device properties

1 Introduction

Since ontology experts are scarce, ontologies are largely authored and maintained by domain experts, whose domain knowledge is dearly needed and appreciated, but who often merely receive a cursory training in ontology semantics and design. This may lead to poor design choices and avoidable errors. Because of their lack of experience, domain experts may not be able to identify opportunities to reuse existing best practices and ontologies. It is no wonder that mistakes occur, and ontologies become incomplete or, worse, inconsistent; the problem is aggravated in the maintenance of large and, often regrettably so, incomprehensible ontologies due to suboptimal or lacking structure.

Ontology Design Patterns (ODPs) [9] have been introduced as a means to establish best practices for ontology design, and as a way to provide a set of carefully-designed building blocks for ontologies that may be reused in different contexts. Several languages for representing ODPs, their instantiations and

---

This work has been partially supported by the German Research Foundation, DFG, as part of the Collaborative Research Center 1320 "EASE - Everyday Activity Science and Engineering" (http://www.ease-crc.org), and the EU project CrowdHEALTH (https://www.crowdhealth.eu).

M. F. Silva et al. (Eds.): ROBOT 2019, AISC 1092, pp. 391–403, 2020.
https://doi.org/10.1007/978-3-030-35990-4_32

the relationships between them have been proposed. The OPLa language [10] makes use of OWL annotation properties to mark patterns and their relationships. OntoUML [8] has been extended with a pattern language based on graph transformations in [21]. OTTR [20] is a language for representing patterns as parameterized ontology templates, instantiated via macro expansion.

Generic Ontology Design Patterns (GODPs) were introduced in [13] to support domain experts, encapsulating the design decisions of "classic" ODPs [9] and allowing reuse in multiple instantiations. Complexity and intricacies of ontology semantics are encapsulated in the body of a GODP, shielding the user from errors (cf. also [14]).

The domain of robotics and AI should capitalize on the existence of such welcome tools improving the methodology of ontology development. With a separation of concerns, ontology experts try to encapsulate inherent complexity in the body of GODPs, providing a toolbox of GODPs for use by domain experts. In this paper we briefly recall the language used for writing GODPs in Sect. 2, and illustrate their use for robotics and AI with salient examples[1] in Sect. 3.

As a reader of this paper, you will classify yourself as an ontology expert, or a domain expert, or both. As a domain expert, you are encouraged to read only the headers in the GODP examples (up to and including explanatory comments after the "=" sign), skipping the details of the body thereafter; indeed, this is the whole point: the header gives information about how to use a GODP; this should be sufficient when accompanied by proper documentation, including an instantiation example.[2] Moreover, we have marked *in italics* those paragraphs that are meant to be read by ontology experts only.

2 Generic Ontology Design Patterns in Generic DOL

The *Distributed Ontology, Modeling and Specification Language* DOL [15] is a language for modular development of ontologies, independent of the formalism used at the basic level, e.g. OWL, first-order logic, Common logic (a standardized variant of SUO-KIF). It also allows establishing semantic relations between ontologies, such as theory interpretation or ontology alignment. Tool support is provided by *The Heterogeneous Tool Set*, Hets [16], that parses and analyzes DOL specifications and interfaces logic-specific tools, like provers or model finders.

DOL provides constructs for *uniting* ontologies, written O_1 **and** O_2, and for *extending* an ontology O_1 by new declarations and axioms, written O_1 **then** O_2 *(O_2 may be a fragment that is only well-formed in the context of O_1).* When all unstructured ontologies involved are written in OWL, this has the same expressivity as OWL imports.

[1] The complete set of this paper's examples can be found under https://ontohub.org/robot2019.

[2] Beyond the scope of this paper: GODPs should be well documented, including at least information along the lines of ontologydesignpatterns.org, also relating GODPs to each other.

```
pattern RelatedToS[Class: X; ObjectProperty: p Domain: X;
 Class: Y :: YS] = %% relates X via p to every Y in Y::YS
       Class: X SubClassOf: p some Y
  then RelatedToS[X;p;YS]
```

```
pattern NewHWDevice [Class: X; %% device to be introduced
  Class: AncestorOfX SubClassOf: HWDevice;
      %% ancestor of X in given HWDevice taxonomy
 {Class: OS SubClassOf: OSSystem}    :: OSSystemS;
      %% OSs or robotics frameworks X is supportedOn
 {Class: SW SubClassOf: SWComponent} :: SWComponentS;
      %% SW drivers that X is controlledBy
 {Class: R  SubClassOf: Regulation}  :: RegulationS ]
      %% regulations X compliesWith, e.g. some standard
given Device, %% HWDevice,OSSystem,SWComponent,supportedOn,controlledBy
      Regulation %% Regulation, compliesWith
= %% relates X to every OS, SW, R in the given sets
     Class: X SubClassOf: AncestorOfX
then RelatedToS[X; supportedOn;  OS :: OSSystemS]
 and RelatedToS[X; controlledBy; SW :: SWComponentS]
 and RelatedToS[X; compliesWith; R  :: RegulationS]
```

Fig. 1. NewHWDevice.

Generic DOL [13] is an extension of DOL with parameterized ontologies. The semantics of this extension is based on the semantics of generic specifications in CASL [17]. The most important aspect is that the parameter of an ontology is an ontology itself, and the properties specified by its axioms must be true for any argument in an instantiation. Moreover, a parameterized ontology may import ontologies, written after the list of parameters with the keyword **given**. *The semantics is that the symbols of these ontologies are visible in the parameters and the body, but will not be instantiated.*

Generic DOL has very recently [4] been further extended with list parameters, optional parameters, sequential semantics of parameters, local sub-patterns and convenient shorthand notations for arguments. We will make use of these features of Generic DOL for the application domain in the examples, and explain them as we go along.

Consider Fig. 1, an example for a GODP setting up a new hardware device: NewHWDevice defines X as a subclass of AncestorOfX, already defined in the Device ontology, with the intuition that each new hardware device is of a particular kind; using the auxiliary RelatedToS, it also relates it to every OS in OSSystemS (the operating systems or robotics framework where the new device is supported), SW in SWComponentS (the software controlling the new device), and R in RegulationS (the regulations that it complies with). The ontologies listed after the keyword **given** provide the necessary context for the GODP: the items defined in them are visible in the whole definition of the GODP, i.e.parameters

```
ontology NewHWDevice_Log = Regulation and Device
  %% gives Manipulator...HWDevice,Segments...HWDevice,Segment...HWDevice
  %% AdaptiveGripper-Gripper-EndEffector...HWDevice
then NewHWDevice[ UR3;     %% device to be introduced
     Manipulator;          %% ancestor of X in given HWDevice taxonomy
     [ROS];                %% OSs or robotics frameworks X is supportedOn
     [Ur_modern_driver];   %% SW drivers that X is controlledBy
     [IP64Rating,EN_ISO_13849_2008_PLd] ] %% regulations X compliesWith
and  NewHWDevice[Robotiq2F85; AdaptiveGripper;   [ROS];
     [RobotiqGripperAddOn,RobotiqQuickRelease,
      RobotiqFreeTurning,RobotiqTorqueTurning]; [IP40Rating] ]
  %% similarly for other grippers such as Robotiq3Finger, QBSofthand

ontology NewHWDevice_Log_Expanded = %% expansion for UR3 only:
 Class: UR3 SubClassOf: Manipulator
 Class: UR3 SubClassOf: supportedOn  some ROS
 Class: UR3 SubClassOf: controlledBy some Ur_modern_driver
 Class: UR3 SubClassOf: compliesWith some IP64Rating
 Class: UR3 SubClassOf: compliesWith some EN_ISO_13849_2008_PLd
```

Fig. 2. Instantiations of NewHWDevice in NewHWDevice_Log and sample expansion.

and body. For documentation purposes, a list of items provided by each ontology may be added as a comment, e.g. the class HWDevice of the ontology Device (which we assume to be defined, suitably populated).

The auxiliary pattern RelatedToS provides a very simple first example of a repetitive situation in ontological modelling: we want to state the all instances of a class must be related via a property on that class to some instance of another class, and this latter class is obtained by iterating over a list of classes. The constructor for lists is "::", with YS denoting the tail of the parameter list. When a list parameter is empty in an instantiation (or becomes empty at the end of a recursion), this instantiation is void, i.e. expands to the empty ontology, written {}.

NewHWDevice_Log in Fig. 2 then states some typical instantiations (the first one being verbose with comments on each argument, carried over from the corresponding parameter),[3] made in the context of previously declared taxonomies in Regulation and Device, explained in a comment (where we used "–" to mark direct subclasses and "..." to mark ancestors), and the beginning of an expansion: the Generic DOL ontology is replaced by the equivalent OWL one obtained by substituting the parameters with the corresponding arguments in the body of NewHWDevice.

This illustrates the typical workflow for developing ontologies using GODPs: the user will select an appropriate GODP for a particular development task from a collection provided by ontology experts, and provide a choice of arguments in instantiations; then the tool expands the instantiation to an OWL ontology. To

[3] We envisage a tool that generates such comments for arguments in instantiation templates.

choose suitable arguments, the user shall be assisted e.g. by code completion or some GUI elements.

In NewHWDevice, the parameter AncestorOfX is constrained to being a subclass of HWDevice: AncestorOfX must be a class somewhere in a "branch" of the taxonomical hierarchy emanating from HWDevice. Note that the property of being a subclass (the "is-a" property) is semantically a subset relation, leading to its transitivity along a "branch". This kind of constraint is very useful to express relationships in a taxonomy.

More generally, a subclass constraint for a parameter X of the form "p ***some*** *Y" would indicate that X must be related by p to the given Y (cf. RelatedToS). Thus a seemingly innocuous SubClassOf axiom may embody all the expressive power of OWL-DL to state that some semantic constraint must hold on an argument, in each instantiation. Moreover, since OWL-DL is decidable, checking such conditions for arguments with a suitable reasoner will always terminate.*

Such features of Generic DOL [4], demanding specific *semantic properties* for arguments via ontology parameters, are expected to greatly enhance consistency of modeling and safety against errors. In the absence of such semantic requirements, the only check performed is a syntactic one (e.g. is the argument a class or not) and this would not suffice to catch e.g. an erroneous instantiation of NewHWDevice, whose second argument is mistakenly not a subclass of HWDevice.

3 Qualitative Capabilities of Device Properties

In this section we illustrate our methodology of ontology development with GODPs by salient examples for the domain of robotics and AI. We have chosen examples from modeling device properties as this is an emerging area (see e.g. [18]), and nontrivial applications of GODPs for qualitative abstraction, graded relations, etc., can be demonstrated; the examples go beyond classical Ontology Design Patterns, making the need for a generative approach apparent.

Some of these patterns have already been introduced in [4]; we extend them further here in an attempt to devise a general toolbox, to be instantiated for this application domain. Thus we attempt to demonstrate on the one hand that the *development* of such a toolbox, embodying substantial complexity, *is no trivial matter* and should be delegated to *ontology experts*, and on the other hand that its *use* can be made *safe for domain experts* and, if well documented, rather easy.

In Sect. 3.3 we will show how the physical abilities of a robot, expressed by *quantitative data* in the technical documentation, can be turned into *qualitative semantic abilities*, and how the properties of components give rise to "higher" properties of composite objects. All such inference will be done automatically in OWL-DL extended with rules in the Semantic Web Rule Language, SWRL [11][4].

[4] The language definition is available at www.w3.org/Submission/SWRL/.

```
pattern ValSet [ Class: Val;     %% class of values to be defined
      Individual: v0 :: vS;      %% individuals to become members of Val
    ? ObjectProperty: greater ] %% optional order relation on Val
= %% all individuals vi from v0::vS become members of Val
let pattern OrderStep [Individual: vi; Individual: vj :: vS] =
           Individual: vj Types: Val Facts: greater vi
      then OrderStep[vj; vS]
in   Individual: v0 Types: Val
then SimpleOrder[greater;Val] and OrderStep[v0; vS]
then { DifferentIndividuals: v0 :: vS
       Class: Val EquivalentTo: {v0 :: vS} }
```

```
ontology ValSet_Payload = Device then %% gives HWDevice
 ValSet[Payload;[maxPayld1kg,maxPayld5kg,maxPayld10kg];bigger[Payload]]
```

Fig. 3. ValSet and instantiation in ValSet_Payload

3.1 Qualitative Values and Graded Relations

Qualitative values, corresponding to abstractions from quantitative data, occur quite often in practice. Cognitive science shows that they are related to the human need for doing away with irrelevant detail (precision in this case); they allow us to simplify abstract reasoning. In the context of grading, we introduce operations for combining qualitative values, allowing a kind of abstract calculation (cf. [5]).

In Fig. 3 we define sets of named individuals as values of Val [4]. As an example for an instantiation ValSet_Payload, various ranges of payloads are defined as abstract values (as fine-grained as desired). *Declarations for arguments such as Payload or maxPayld1kg are implicitly made by the instantiation, if they are not visible from the environment* [4]. Note that greater is an optional parameter in ValSet; thus Val may be an ordered set if desired, but it need not. The instantiation of SimpleOrder defines greater to be a transitive relation, but is void, if this argument is optional.

Grading with qualitative values occurs for human abilities (e.g. severely, moderately, slightly reduced), but makes analogous sense in robotics and AI. A general approach for grading in ontologies is the introduction of a sheaf of graded relations for a given relation [12]. This is needed because OWL only support binary relations, and the third argument of the relation, giving its qualitative value, is encoded in the name of a graded relation that is part of the sheaf. Consider GradedRels in Fig. 4 (cf. [4], shortened here): for a given relation p and set of qualitative values Val, it introduces a sheaf of graded relations p[v], one for each v in v::valS, and a relation has[Val] that recovers the grade from each member of the sheaf as stated by the SWRL rules introduced via the iteration of the Step sub-pattern.

In the instantiation GradedRels_Payld in Fig. 4, various subrelations of hasDevice, such as hasSegment, are graded according to their maximal Payload. The name p[v] in GradedRels is a *parameterized name*, depending on the name of the

```
pattern GradedRels [Class: S; Class: T;  %% domain and range
   ObjectProperty: p Domain: S Range: T; %% property to be graded
   Class: Val;                           %% class of values for grading
  {Individual: v Types: Val} :: valS ]   %% qualitative values vi
= %% a sheaf of graded relations p[vi], one for each vi in v::valS
  ObjectProperty: has[Val] Domain: T Range: Val
  Class: T SubClassOf: has[Val] some Val
  ObjectProperty: p[v] Domain: S Range: T SubPropertyOf: p
  Rule: S(?a) /\ T(?b) /\ p[v](?a, ?b) -> has[Val](?b, v)
  Rule: S(?a) /\ T(?b) /\ p(?a,?b) /\ has[Val](?b,v) -> p[v](?a,?b)}
then GradedRels[S;T;p;Val;valS]

ontology GradedRels_Payld = ValSet_Payload
then GradedRels[HWDevice; Segment; hasSegment; Payload;
                [maxPayld1kg,maxPayld5kg,maxPayld10kg]]
%% similarly for hasSegments,..., hasEndEffector, hasManipulator
```

Fig. 4. GradedRels and instantiations in GradedRels_Payld

```
pattern BoundedClasses [ Class: C;    %% to be separated into intervals
                         Class: Val; %% class of qualitative grades
  ObjectProperty: hasVal Domain: C Range: Val; %% => qualitative  grade
  DataProperty: p Range: xsd:integer;          %% => quantitative bound
 {Individual: v Types: Val} :: vS; %% qualitative  values vi for grades
  xsd:integer b          :: boundS] %% quantitative values bi for bounds
= %% interval subclasses C[vi] whose quantitative lower bound is bi
   Class: C[v] EquivalentTo: p some xsd:integer [>= b]
               SubClassOf: C, hasVal some {v}
then BoundedClasses[C; Val; hasVal; C; p; vS; boundS]

ontology ClassifyPayload = GradedRels_Payld
then BoundedClasses[HWDevice; Payload; has[Payload]; maxPayload;
           [maxPayld1kg,maxPayld5kg,maxPayld10kg]; [1,5,10]   ]
then {Individual: ur3  Types: UR3  Facts: maxPayload 3 }
```

Fig. 5. BoundedClasses and instantiation in ClassifyPayload

argument for v. During instantiation, e.g. in GradedRels_Payld, v is replaced with this name, e.g. maxPayld5kg, and v with hasSegment; during final expansion to OWL, hasSegment[maxPayld5kg] is *stratified* to hasSegment_maxPayld5kg (legal OWL).

The technical data in catalogues of manufacturers or resellers are largely quantitative or refer to adherence to standards and other regulations. A GODP will introduce a data property for each quantitative property of a device.

In Fig. 5 we introduce a pattern BoundedClasses for defining the subclass of all individuals of the class C, whose value of a data property p is beyond a certain bound b. Moreover, the relation hasVal will relate these individuals to the grade v corresponding to the bound. This is iterated for v and b over the lists vS and boundS, resp. The ontology ClassifyPayload shows how this may be used to classify hardware devices: the instance ur3 is stated as being able to carry at

```
pattern TabularComposition3 [ Class: D; %% domain of rx, domain of rz
                              Class: S; %%  range of rx, domain of ry
                              Class: T; %%  range of ry,  range of rz
  Class: Valx; Class: Valy; Class: Valz; %% grades in x, y, z coordinate
  ObjectProperty: rx Domain: D Range: S; %% graded relation, x coordinate
  ObjectProperty: ry Domain: S Range: T; %% graded relation, y coordinate
  ObjectProperty: rz Domain: D Range: T; %% graded relation, z coordinate
  %% below: n x 3 Composition Table; column headers xi, row headers yi
                             {Individual: x    Types: Valx} :: xChain;
 Individual: y0 Types: Valy; {Individual: x_y0 Types: Valz} :: x_y0Chain;
 Individual: y1 Types: Valy; {Individual: x_y1 Types: Valz} :: x_y1Chain;
 Individual: y2 Types: Valy; {Individual: x_y2 Types: Valz} :: x_y2Chain]
= %% composes sheafs of relations rx[xi] o ry[yj] -> rz[xi_yj]
  ...
```

```
ontology TabularComposition3_hasManipulator = ClassifyPayload then
TabularComposition3 [HWDevice;HWDevice;HWDevice;Payload;Payload;Payload;
  hasEndEffector;hasSegments;hasManipulator;
               [maxPayld1kg,maxPayld5kg,maxPayld10kg];
  maxPayld1kg; [maxPayld1kg,      {},          {}     ]; %% too heavy
  maxPayld5kg; [maxPayld1kg,maxPayld5kg,maxPayld5kg ]; %% downgraded
  maxPayld10kg;[maxPayld1kg,maxPayld5kg,maxPayld10kg]]
```

Fig. 6. TabularComposition3, instantiation in TabularComposition3_hasManipulator

most 3 kgs. Since 3 kgs is not among the bounds, we will need to classify it as an instance of one of the classes HWDevice[maxPayld1kg], HWDevice[maxPayld5kg] or HWDevice[maxPayld10kg]. ur3 will be classified as HWDevice[maxPayld1kg] using standard OWL reasoning, as it is able to carry all objects that weigh at most 1 kg, but not as HWDevice[maxPayld5kg] because it will not be able to carry, for example, an object that weighs 4 kg.

3.2 Compliance of Device Compositions

We will now attempt to show how *semantic* properties of devices can be deduced in ontologies: at first by deduction of the capabilities of a device from its components. In the example in Fig. 6, a sequence of segments is composed with an end-effector to form a manipulator. Instead of the objects involved, we show the composition of the relations hasEndEffector and hasSegments to hasManipulator, i.e. their graded versions, resp. This way, we can then deduce, that e.g. a device that hasManipulator[maxPayld5kg] is properly composed from components that are compliant with the Payload requirement.

For this we introduce a GODP that allows us to define the consistency of the composition in a tabular fashion. TabularComposition3 in Fig. 6 composes sheafs of relations rx[xi] o ry[yj] to rz[xi_yj], where xi_yj denotes the resp. entry for the grade of rz at position xi,yj of the table. The table is modeled by a fixed number of rows, where TabularComposition3 has 3 such rows (and a corresponding version for any other number of rows is easy to construct). Each row is modeled as a

```
ontology ValSet_VFType = ValSet[VFType; [VF2,VF3,VF4,VF5]; moreFingers]
```

... analogously for HandlingType, OppositionType, AbductionType, and GraspType

```
ontology GradedRels_Gripper = ClassifyPayload and ValSet_HandlingType
 and ValSet_OppositionType and ValSet_AbductionType and ValSet_VFType
then GradedRels[HWDevice; Gripper; hasEndEffector;
                                   HandlingType;      [Pow,Int,Prec]]
 and GradedRels[HWDevice; Gripper; hasEndEffector;
                                   OppositionType;   [Palm,Side,Pad]]
 and GradedRels[HWDevice; Gripper; hasEndEffector;
                                   AbductionType;          [Abd,Add]]
 and GradedRels[HWDevice; Gripper; hasEndEffector;
                                   VFType;         [VF2,VF3,VF4,VF5]]
```

```
ontology TabularAnd4_hasEndEffector = GradedRels_Gripper then
TabularAnd4[HWDevice;HWDevice;GraspForm;VFType;GraspType;
     hasEndEffector;hasEndEffector;hasEndEffector;
     [ Abd_Pow_Palm,   Abd_Pow_Palm,   Abd_Pow_Palm, Abd_Pow_Palm];
 VF2;[1LargeDiameter, 2SmallDiameter,  3MediumWrap,  10PowerDisk];
 VF3;[1LargeDiameter, 2SmallDiameter,  3MediumWrap,  10PowerDisk];
 VF4;[1LargeDiameter, 2SmallDiameter,  3MediumWrap,  10PowerDisk];
 VF5;[1LargeDiameter, 2SmallDiameter,  3MediumWrap,  10PowerDisk]]
and ...
```

Fig. 7. Gripper features and combination in TabularAnd4_hasEndEffector (abbreviated)

list. Note that these lists may have a length that is different from the number of rows, corresponding to rectangular but not necessarily square matrices.

In our instantiation example TabularComposition3_hasManipulator in Fig. 6, the matrix is square; the formatting allows easy readability. The matrix is sparse: some entries denote the empty ontology {}; for such an entry, the composition is void. This way we can model that an end-effector with a payload of more than 1 kg is considered to be too heavy for segments allowing a maximal payload of 1 kg. An end-effector with a maximal payload of 10 kg is allowed for segments allowing a maximal payload of 5 kg, but the resulting maximal payload is of course 5 kg and not 10 kg; it is "downgraded".

As it happens, the lower left triangle of the matrix is left-regular in terms of the downgrading. For such cases, specialized versions can be defined such as SymmetricDowngrading. In the full version of the examples in ontohub.org/robot2019, this simplified version is used for stating the compliance segments in a manipulator: only a segment with a lower or equal maximal payload can be added to the front of a sequence of segments.

3.3 Deduction of Grasping Capabilities

Now we would like to model the grasp types defined in the table in Fig. 4 of [6]. A GraspType is actually a combination of four "features": HandlingType, OppositionType, AbductionType, and VFType (cf. Fig. 7).

```
pattern GripperGraspFeatures
[Class: X SubClassOf: Gripper; %% gripper to add grasp features to
 Individual: htype Types: HandlingType;   %% admissible HandlingType
{Individual: otype Types: OppositionType} :: otypes; %% OppositionTypes
{Individual: atype Types: AbductionType}  :: atypes; %% AbductionTypes
 Individual: vf    Types: VFType                    ] %% VFType
given GradedRels_Gripper
= Class: X SubClassOf: has[HandlingType]   some  {htype},
                       has[OppositionType] only {otype::otypes},
                       has[AbductionType]  only {atype::atypes},
                       has[VirtualFingers] only  {vf}
```

```
ontology GripperGraspFeatures_Log = GradedRels_Gripper
then GripperGraspFeatures[Robotiq2F85;     Prec; [Pad]; [Abd];       VF2]
 and GripperGraspFeatures[Robotiq3Finger; Prec; [Pad]; [Abd];       VF3]
 and GripperGraspFeatures[QBSofthand;     Prec; [Pad, Side, Palm];
                                                        [Abd, Add]; VF5]
```

Fig. 8. GripperGraspFeatures and instantiations.

We assume here that the first three have already been combined to GraspForm, and are presented as stratified names; its values are used as column headers in TabularAnd4_hasEndEffector: thus e.g. Abd_Pow_Pad represents a combination of Abd for "abducted" of AbductionType, Pow for "power' of HandlingType, and Pad for the OppositionType; values of VFType are given as row headers. TabularAnd4 is analogous to TabularComposition3 in Fig. 6 in its structure: the entry in each combination of row and column defines an admissible result. Note that a column header may be repeated, if more than one resulting combination is admissible. In the instantiation TabularAnd4_hasEndEffector, e.g. the GraspForm Abd_Pow_Pad is introduced 3 times as a column header, giving rise to the admissible GraspTypes 31Ring, 28Sphere3Finger, or 18ExtensionType, resp., in various (but not all!) combinations with virtual finger VFTypes in row headers.

In Fig. 8 we introduce a pattern GripperGraspFeatures for modeling the grasping capabilities of grippers. Each new gripper X is related to the four feature types; the pattern then restricts the possible values of the features to those given in the arguments, using OWL nominals to turn individuals into classes. For HandlingType, we make the assumption that a morePrecise HandlingType implies all other less precise types as well. The ontology GripperGraspFeatures_Log presents three instantiations: to add a new gripper, one only needs to give values for the four features as arguments of the GripperGraspFeatures pattern. Using these we can now derive what kind of grasps are possible for each gripper G.

The reasoning is ensured by the following facts: the pattern for grippers gives the values of the four grasp features; the rules in the pattern GradedRels obtained via the instantiation GradedRels_Gripper ensure that if a device D has a gripper of type G as end-effector, the graded relations hasEndEffector[X], where X is each of the four values of the grasp features, hold for D and G; finally the composition table in TabularAnd4_hasEndEffector combines these to hasEndEf-

fector[GraspType], where GraspType is determined by the grasp features, encoding them all in this parameterized name.

To extend the example further for other kinds of grippers, one would probably need more information about the rigidity of the fingers or the kind of oppositions between fingers (in the sense of [6]), or analogous grasp forms for other kinds of end-effectors.

4 Conclusions

It is our belief that ontologies in the domain of robotics and AI (and elsewhere) can be developed in a safer way using GODPs, as supported by the following arguments. Development responsibility is divided between ontology experts as developers of GODPs and application domain experts as users. The complexity and intricacies of ontology semantics are encapsulated in the body of a GODP, shielding the users from errors. In instantiations, users only need to concentrate on appropriate arguments that satisfy the requirements specified in the parameters. The body of a GODP encapsulates a design decision for (re)use in various instantiations; this allows localization and confinement of revisions to this very body. Specialized GODPs for an application domain (e.g. NewHWDevice) ensure that a domain expert or end user, confined to a set of such GODPs suitable for a particular development or maintenance task, has no impact beyond. A log of instantiations (e.g. NewHWDevice_Log) documents the development process, including the major design decisions.

After devising a little toolbox of salient GODPs, we demonstrated its applicability for automatically deducing qualitative semantic capabilities of devices from quantitative data in technical documentations. We hope that such examples will motivate other domain experts in robotics and AI to join with ontology experts in order to further populate an emerging repository of general and application domain oriented GODPs.

One interesting future research enterprise is organizing the IEEE 1872-2015 standard CORA ontology [19] as a collection of GODPs. Benefits are increased modularity and thus easier maintenance, and that an ontology developed by systematically instantiating the GODPs from the CORA collection automatically adheres to the standard.

Some patterns presented here are very similar to those developed for the ontology in the EASE [2]: there, the pattern for objects states that every object is described by a certain description and has some quality (compare with NewHWDevice in Fig. 1).

Several GODPs presented in this paper, e.g. ValSet and GradedRels, originated from work on designing food ontologies to be used in the EU project CrowdHEALTH [7]. Conversely, TabularComposition3, TabularAnd4 etc. introduced here will now be used there, providing significant new ways for specifying composite properties of ingredients. This demonstrates how a pattern observed when modelling knowledge for a specific domain may be abstracted and then reused in completely different domains.

Acknowledgments. We are grateful to Till Mossakowski for his continuous cooperation in the development of Generic DOL, and Sebastian Bartsch, Daniel Bessler, José de Gea Fernandez, and Mihai Pomarlan for their advice and suggestions regarding robotics.

References

1. Barton, A., Seppälä, S., Porello, D. (eds.): Proceedings of the Joint Ontology Workshops, 23–25 September 2017, Graz, Austria. CEUR Workshop Proceedings, CEUR-WS.org (2019). http://ceur-ws.org/
2. Bateman, J., Beetz, M., Beßler, D., Bozcuoglu, A.K., Pomarlan, M.: Heterogeneous ontologies and hybrid reasoning for service robotics: the EASE framework. In: Ollero, A., Sanfeliu, A., Montano, L., Lau, N., Cardeira, C. (eds.) ROBOT 2017: Third Iberian Robotics Conference - Volume 1. Advances in Intelligent Systems and Computing, vol. 693, pp. 417–428. Springer (2017). https://doi.org/10.1007/978-3-319-70833-1_34
3. Blomqvist, E., Corcho, O., Horridge, M., Hoekstra, R., Carral, D. (eds.): Proceedings of the 8th Workshop on Ontology Design and Patterns (WOP 2017), CEUR Workshop Proceedings, vol. 2043. CEUR-WS.org (2017)
4. Codescu, M., Krieg-Brückner, B., Mossakowski, T.: Extensions of Generic DOL for Generic Ontology Design Patterns. In: Barton, et al. [1]. http://ceur-ws.org/
5. Cousot, P., Cousot, R.: Abstract interpretation: a unified lattice model for static analysis of programs by construction or approximation of fixpoints. In: Graham, R.M., Harrison, M.A., Sethi, R. (eds.) Fourth ACM Symposium on Principles of Programming Languages, Los Angeles, USA, January 1977, pp. 238–252. ACM (1977). https://doi.org/10.1145/512950.512973
6. Feix, T., Romero, J., Schmiedmayer, H., Dollar, A.M., Kragic, D.: The GRASP taxonomy of human grasp types. IEEE Trans. Hum. Mach. Syst. **46**(1), 66–77 (2016). https://doi.org/10.1109/THMS.2015.2470657
7. Gallos, P., Aso, S., Autexier, S., Brotons, A., Nigro, A.D., Jurak, G., Kiourtis, A., Kranas, P., Kyriazis, D., Lustrek, M., Magdalinou, A., Maglogiannis, I., Mantas, J., Martinez, A., Menychtas, A., Montandon, L., Picioroaga, P., Perez, M., Stanimirovic, D., Starc, G., Tomson, T., Vilar-Mateo, R., Vizitiu, A.M.: CrowdHEALTH: big data analytics and holistic health records. In: Shabo (Shvo), A., et al. (eds.) ICT for Health Science Research, Proc. EFMI 2019. Studies in Health Technology and Informatics, vol. 258, pp. 255–256. IOS Press, April 2019
8. Guizzardi, G.: Ontological foundations for structural conceptual models. Ph.D. thesis, University of Twente (2005)
9. Hitzler, P., Gangemi, A., Janowicz, K., Krisnadhi, A., Presutti, V. (eds.): Ontology Engineering with Ontology Design Patterns - Foundations and Applications, Studies on the Semantic Web, vol. 25. IOS Press (2016)
10. Hitzler, P., Gangemi, A., Janowicz, K., Krisnadhi, A.A., Presutti, V.: Towards a simple but useful ontology design pattern representation language. In: Blomqvist et al. [3]. http://ceur-ws.org/Vol-2043/paper-09.pdf
11. Horrocks, I., Patel-Schneider, P.F., Bechhofer, S., Tsarkov, D.: OWL rules: a proposal and prototype implementation. J. Web Semant. **3**(1), 23–40 (2005). https://doi.org/10.1016/j.websem.2005.05.003

12. Krieg-Brückner, B.: Generic ontology design patterns: qualitatively graded configuration. In: Lehner, F., Fteimi, N. (eds.) Proceedings of 9th International Conference KSEM 2016. Lecture Notes in Computer Science, vol. 9983, pp. 580–595 (2016). https://doi.org/10.1007/978-3-319-47650-6_46
13. Krieg-Brückner, B., Mossakowski, T.: Generic ontologies and generic ontology design patterns. In: Blomqvist et al. [3]. http://ontologydesignpatterns.org/wiki/images/0/0e/Paper-02.pdf
14. Krieg-Brückner, B., Mossakowski, T., Neuhaus, F.: Generic ontology design patterns at work. In: Barton et al. [1]. http://ceur-ws.org/
15. Mossakowski, T., Codescu, M., Neuhaus, F., Kutz, O.: The distributed ontology, modeling and specification language – DOL. In: Koslow, A., Buchsbaum, A. (eds.) The Road to Universal Logic, vol. 2, pp. 489–520. Birkhäuser (2015). http://www.springer.com/gp/book/9783319101927
16. Mossakowski, T., Maeder, C., Lüttich, K.: The heterogeneous tool set, hets. In: Grumberg, O., Huth, M. (eds.) TACAS. Lecture Notes in Computer Science, vol. 4424, pp. 519–522. Springer (2007). http://dblp.uni-trier.de/db/conf/tacas/tacas2007.html#MossakowskiML07
17. Mosses, P.D. (ed.): CASL Reference Manual, Lecture Notes in Computer Science, vol. 2960. Springer, Heidelberg (2004)
18. Ramos, F., Vázquez, A.S., Fernández, R., Alarcos, A.O.: Ontology based design, control and programming of modular robots. Integr. Comput. Aided Eng. **25**(2), 173–192 (2018). https://doi.org/10.3233/ICA-180569
19. Schlenoff, C., Prestes, E., Madhavan, R., Goncalves, P., Li, H., Balakirsky, S., Kramer, T., Migueláñez, E.: An IEEE standard ontology for robotics and automation. In: 2012 IEEE/RSJ International Conference on Intelligent Robots and Systems, pp. 1337–1342, October 2012
20. Skjæveland, M.G., Karlsen, L.H., Lupp, D.P.: Practical ontology pattern instantiation, discovery, and maintanence with reasonable ontology templates. In: van Erp, M., Atre, M., López, V., Srinivas, K., Fortuna, C. (eds.) Proceedings of ISWC 2018 Posters & Demonstrations. CEUR Workshop Proceedings, vol. 2180. CEUR-WS.org (2018). http://ceur-ws.org/Vol-2180/paper-60.pdf
21. Zambon, E., Guizzardi, G.: Formal definition of a general ontology pattern language using a graph grammar. In: Ganzha, M., Maciaszek, L.A., Paprzycki, M. (eds.) Proceedings of the 2017 Federated Conference on Computer Science and Information Systems, FedCSIS 2017, Prague, Czech Republic, 3–6 September 2017, pp. 1–10 (2017). https://doi.org/10.15439/2017F001

Meta-control and Self-Awareness for the UX-1 Autonomous Underwater Robot

Carlos Hernandez Corbato[1(✉)], Zorana Milosevic[2], Carmen Olivares[2], Gonzalo Rodriguez[2], and Claudio Rossi[2]

[1] Delft University of Technology, 2628 CD Delft, The Netherlands
c.h.corbato@tudelft.nl
[2] Centre for Automation and Robotics UPM-CSIC, Madrid, Spain

Abstract. Autonomous underwater robots, such as the UX-1 developed in the UNEXMIN project, need to maintain reliable autonomous operation in hazardous and unknown environments. Because of the lack of any kind of real-time communications with a human operated command and control station, the control architecture needs to be enhanced with mission-level self-diagnosis and self-adaptation properties an additional provided by some kind of supervisory or "metacontrol" component to ensure its reliability. In this paper, we propose an ontological implementation of such component based on Web Ontology Language (OWL) and the Semantic Web Rule Language (SWRL). The solution is based on an ontology of the functional architecture of autonomous robots, which allows inferring the effects of the performance of its constituents components in the functions required during the robot mission, and generate the reconfigurations needed to maintain operation reliably. The concept solution has been validated using a hypothetical set of scenarios implemented in an OWL ontology and an OWLAPI-based reasoner, which we aim at validating by integrating the metacontrol reasoning with a realistic simulation of the underwater robot.

1 Introduction

The objective of the European Project UNEXMIN[1] is to develop an underwater vehicle (see Fig. 1) capable of autonomously surveying old mine sites, that are nowadays flooded[2]. The information available regarding the structural layout of the tunnels of such mines is limited, imprecise or even totally lacking. Therefore, prior to any decision, a survey and prospecting of the mine tunnels network should be conducted. Since exploration by human divers is mostly ruled out due to the risks involved, the use of robotic systems appears the only possible solution.

Operating in such environments poses additional requirements, in addition to the "classical" Planning and GNC (Guidance, Navigation and Control) features

[1] H2020, Grant agreement No 690008.
[2] There is a high interest in re-opening some of these sites, since the European Union is largely dependent on raw materials imports.

M. F. Silva et al. (Eds.): ROBOT 2019, AISC 1092, pp. 404–415, 2020.
https://doi.org/10.1007/978-3-030-35990-4_33

Fig. 1. The UX-1 prototype during a dive in Kaatiala Mine (Finland). Image credits: UNEXMIN consortium, www.unexmin.eu

of autonomous robots. In fact, due to the lack of any kind of real-time communications with a human operated command and control station, the robot, besides of taking autonomous decisions regarding its mission, must be provided with enhanced fault-tolerance capabilities. Here, we decided to go beyond simple "fault tolerance". Our purpose is to augment the UX-1 control architecture with self-adaptation properties to ensure the reliability of its behavior, using the Metacontrol architectural framework by Hernandez et al. [9] and ontological reasoning. The UX-1 perception and motion systems are highly redundant, which allows for multiple (sub-optimal) configurations. Here, we focus on a subset of possibilities for its motion system (see Fig. 2), and demonstrate how, thanks to the self-diagnosis and self-adaptation capabilities provided by the Metacontrol, it can respond to thrusters failures with suitable reconfiguration of both its hardware and software.

We argue that ontological reasoning to drive Metacontrol operation fulfills the needs in the previous context. Ontologies are suitable to capture the system's architecture and capabilities with the appropriate level of abstraction. Ontological reasoning allows to separate the rules for metacontrol operation (i.e. diagnostics and reconfiguration) from the application specific knowledge, and the use of off-the-shelf reasoners, facilitating the validation of the architectural reconfigurations inferred.

This paper presents three novel contributions to the Metacontrol framework: (1) the extension of the Teleological and Ontological Metamodel for Autonomous Systems (TOMASys), to include models of quality attributes, concretely "performance", (2) the use of ontological reasoning for self-diagnostics and reconfiguration, and (3) its proof-of-concept application in the context of the UX-1 autonomous underwater robot.

The paper is organised as follows. Section 2 discusses related research on self-adaptation and ontologies in robotics. Section 3 presents the general Metacontrol framework and the extension of the TOMASys metamodel. Section 4 introduces the specific autonomous underwater robot and its functional architecture, and its ontological modelling with TOMASys. Finally, Sect. 5 discusses the benefits and limitations of solution proposed and addresses next steps in this research and Sect. 6 presents some concluding remarks.

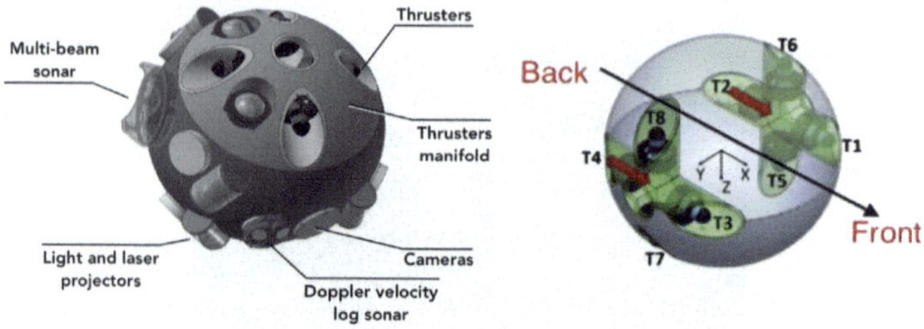

Fig. 2. Left: Main components if the UX-1 robot. Right: thrusters allocation.

2 Related Work

In robotics, ontologies and symbolic reasoning have proven effective for mission management and high-level representation of robot environments and tasks [1], thanks to their ability to represent heterogeneous knowledge. In particular, underwater autonomous robots are a very good testbed for this kind of systems, since their working environment is difficult and hazardous for humans, and hence the need for autonomy and resilience is much more compelling. Zhai *et al.* [18], developed a OWL and SRWL-based ontology to provide users of underwater robots with query services to know the status of the robotic systems and the underwater environment. We use the same technologies, but to implement automatic diagnosis and reconfiguration of an autonomous underwater robot. The Ontology for Autonomous Systems (OASys) [3] is meant to describe and drive the entire life-cycle of autonomous systems, from engineering to operation, base on metamodelling and ontologies.

Control systems of autonomous robots are software systems. The self-adaptation and the use of models at runtime has been extensively studied in software systems in the last decade [2,17]. Simple examples have shown that models of software properties, such as execution time, can be used to optimize the software design of a robot according to associated requirements, such as safety [4,15].

In a previous work [9] we demonstrated how the metamodelling approach in [3] can be used for self-adaptation at the mission level at runtime, going beyond component fault-tolerance. In this work, we advance in that roadmap with the use of ontologies to reason on the robot control architecture and its properties during the mission.

3 Metacontrol Framework

The work presented here extends the framework by Hernandez *et al.* [9] for self-adaptive autonomous systems. The core idea (Fig. 3) is to leverage the engineer knowledge of the system in the form of a runtime model, to drive its runtime self-adaptation capabilities. This knowledge is captured as a model of the functional architecture during system development [7], by using the Integrated Systems

Engineering and PPOOA (ISE&PPOOA) method for Model-Based Systems Engineering (MBSE) [6]. Following the functional approach in ISE&PPOOA allows to raise the level of abstraction in the system's representation and focus on representing the architectural properties that are particularly relevant for the system's capabilities required in the mission at hand. To create an application and domain independent solution, the Metacontrol framework departs from knowledge representation approaches in robotics, centered in application-specific models, by applying a metamodelling approach. The TOMASys metamodel [9] is designed to represent runtime models for metacontrol, based on other metamodels for MBSE (UML, SysML), so that eventually transformations can be defined to automatically generate the runtime model from the functional architecture model.

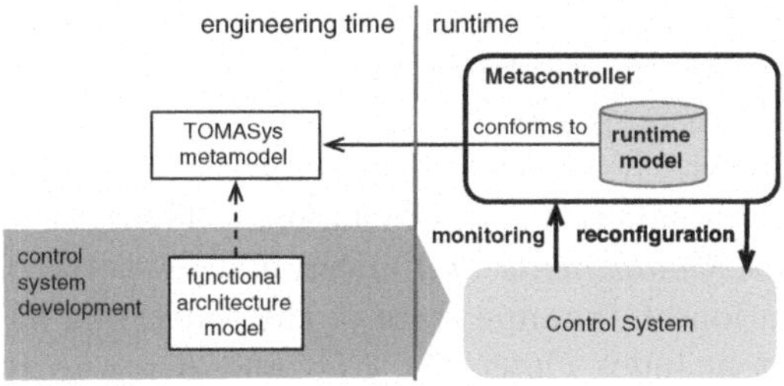

Fig. 3. The Metacontrol approach.

3.1 TOMASys

TOMASys is a metamodel to represent the functional architecture of an autonomous robot, and distinctively also its runtime state. TOMASys functional concepts are based on the theoretical framework for autonomous cognitive systems by Lopez [13], and TOMASys elements related to *structure* are inspired by models for component-based software [14]. Here we will only present briefly the functional elements in TOMASys, displayed in its ontological version in Fig. 4, a complete description of the original TOMASys metamodel can be found in [8].

A differentiating aspect in TOMASys is that it takes the functional approach from systems engineering and functional modelling. TOMASys explicitly captures the relation between the robot's objectives[3] and its control architecture through functional decomposition. At each instant, the system pursues a set of *Objectives*, or specific instances of the functional requirements of the system, i.e. the *Functions*. A *FunctionDesign (FD)* is selected in order to *Solve* each particular *Objective* using concrete system resources, i.e. *Components*. The *FunctionDesign* contains a set of *Roles* that define how certain *Components* in the system are configured to fulfill the Function. The instantiation or grounding of a FunctionDesign by the *Binding* of its defined *Roles* to actual *Components* is

[3] We use italics for the elements in TOMASys ontology.

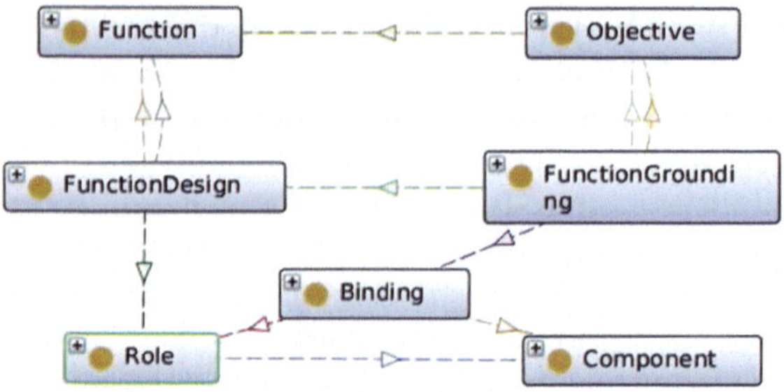

Fig. 4. Main classes in the OWL implementation of TOMASys

represented by a *FunctionGrounding (FG)*. The *FunctionDesign* may also *require* the realisation of other *Objectives*, which would in turn require the instantiation of other *FunctionDesign* that solve them, and so on. This is how TOMASys represents functional decomposition.

Summarily, a TOMASys model represents the control architecture of an autonomous robot through two sets of elements. The first set includes *Objectives*, *FGs*, *Bindings*, *Components* and additional property values and links to capture the information of dynamic state of the functional hierarchy of the system. The second set includes *Functions*, *FDs* and *Roles*, to represent the static knowledge about the system resources and architectural alternatives.

TOMASys originally supported simple fault-based reasoning for diagnosis and component fault-tolerance and functional reconfiguration [9]. This was implemented as fault-propagation in the functional hierarchy model (*FGs* and *Objectives*). In this work we address quality attributes of functions, using performance as an example. For this, we have extended TOMASys with three elements. *Performance* is a property of *Components*, *FunctionGroundings* and *Objectives*, and it is a real number [0.0, 1.0]. *Efficacy* is a property of *FunctionDesigns*, also [0.0, 1.0], that express their ability to solve the objective they address. The instantaneous performance achieved at runtime for each objective in the system depends on the performance of the *FunctionGrounding* solving it and the efficacy of the *FD* it grounds. To compute the performance of the *FG*, different models could be considered, and we have enabled the definition of different performance models for different *FDs* through SWRL rules. However, here we will consider only a default model for all *FDs*, which equals the performance of its grounded *FG* to the average of the performances of the components binded to the roles defined by the *FD*.

3.2 Ontological Reasoning for Metacontrol

One of the core contributions of this work is the implementation of TOMASys in an OWL ontology with SWRL rules. This way, the metacontrol operation for functional diagnosis and reconfiguration is implemented using an OWL reasoner, with the following benefits. First, it facilitates validation and verification by using standard reasoners for the metacontrol operation, and explicitly separating it from the model semantics. Finally, it opens the door to extend the runtime model with knowledge from other robot ontologies, and enables the use of the multiple ontological tools for metacontrol development.

The Semantic Web Rule Language (SWRL) extends OWL with the capability to specify Horn-like rules (i.e. statements of the form "if-then") to perform inferences over OWL individuals, so new knowledge about those individuals can be generated. An example of the SWRL rules developed to implement TOMASys semantics is shown in Table 1. Concrete, these rules allow to perform functional diagnosis: they identify which objectives are affected by an error in a component. R1 propagates an error detected in a component by the monitoring infrastructure, to the role it plays in a function, by setting the binding's status to ERROR. R2 scales that error to the function grounding, and R3 to the objective being realised by the function grounding. Rules 4 and 5 in Table 2 determine that a function design that uses a component that is not available because of being unique and in ERROR, is not realisable.

Table 1. Example of the SWRL rules implementing TOMASys semantics for error propagation.

R1	Binding(?b) ^binding_component(?b, ?c) ^c_status(?c, false) ->b_status(?b, false)
R2	Binding(?b) ^hasBindings(?fg, ?b) ^b_status(?b, false) ->fg_status(?fg, false)
R3	Objective(?o) ^fg_status(?fg, false) ^FunctionGrounding(?fg) ^realises(?fg, ?o) ->o_status(?o, false)

Table 2. SWRL rules implementing.

R4	Component(?c) ^typeC(?c, ?cc) ^c_status(?c, false) ^cc_unique(?cc, true) ->cc_availability(?cc, false)
R5	FunctionDesign(?fd) ^ComponentSpecification(?cs) ^cc_availability(?cc, false) ^ Role(?r) ^roleDef(?r, ?cs) ^typeC(?cs, ?cc) ^roles(?fd, ?r) ^ComponentClass(?cc) ->realisability(?fd, false)

We have used a rule-based reasoner that supports SWRL, concretely Pellet [16], and OWLAPI, to implement the metacontrol operation as symbolic inference, in a similar approach to that of [11]. The TOMASys metamodel assumes that the architectural alternatives (FDs) are finite and known, and the statuses of the system components are fully known through monitoring. We use OWLAPI constructs to implement our reasoning with the partial-closed world assumption. The metacontrol reasoning operation consists of functional diagnosis and reconfiguration.

Firstly, the **functional diagnosis** proceeds as follows. The monitoring observations about the system components are converted into assertions of OWL individuals. Then, the reasoner is executed on the updated ontology, and the status of *FGs* and *Objectives* is inferred using the TOMASys SWRL rules. In this work, in addition to the rules for functional diagnosis in Table 1, SWRL rules are defined for the *FDs* in the application ontology to implement the new TOMASys performance model (see Rule 5 in Table 3). These rules update the performance achieved in the fulfillment of each *Objective*, based on the instantaneous performance of the components involved in their realization[4] and the efficacy of the *FDs* they ground.

Secondly, the metacontrol reasoner infers the best reconfiguration to optimize *Objectives' performance*, based on the available resources, i.e. *Components*, and the design knowledge, i.e. the *Efficacies* of the *FDs* that can be implemented with them. This is done reasoning again with the application-specific rules for performance, in addition to the generic rules for TOMASys semantics.

4 Metacontrol of an Underwater Robot

In this section, we elaborate the application of the Metacontrol framework to implement reasoning for self-adaptation in the control architecture of the UX-1 robot. Concretely, the objective was to design the metacontroller to: (1) perform self-diagnosis of the navigation and motion subsystems, given the status of the thrusters, and (2) determine the best configuration of both subsystems based on the previous diagnosis.

The knowledge obtained during the engineering of both subsystems is that (i) according to the level of performance of the thrusters, different controllers shall be used for optimal performance, and similarly (ii) according to the controller used, one or the other navigation system is suitable. To account for this type of self-adaptation requirements, we have extended TOMASys to model performance as presented at the end of Sect. 3.1.

4.1 Control Architecture of the Underwater Robot

Figure 2 depicts the UX-1 prototype, highlighting both its perception and actuation means. As mentioned earlier, in this work we take into consideration only a subset of all the devices as a proof of concept. Concretely, we take into account only four of the eight thrusters (the ones dedicated to the forward and backward movement), and we consider only the subsystem to control the motions of the underwater robot, including two low-level possible controllers (one PID and one fuzzy) tuned for different thruster configurations. We use the ISE&PPOOA [5,7] to develop the hypothetical model of the control architecture for the motion subsystem and its alternative configurations. This is represented in Fig. 5, where only two of the multiple alternatives are displayed for each function, and the Navigation function is not detailed as it is not considered in the concept proof presented here.

[4] We assume here that this information is provided by the monitoring infrastructure or other specific observers.

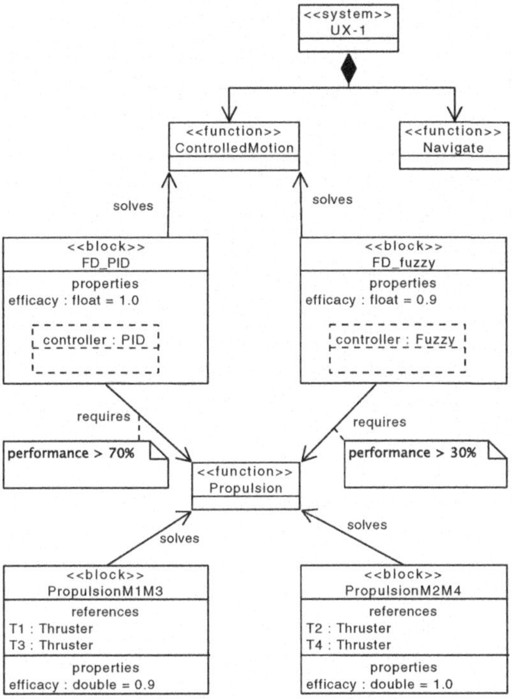

Fig. 5. Architectural alternatives for the UX-1 MotionControl and Propulsion functions, displayed using the SysML graphical notation.

For forward motion, the pair {T2, T4} (front thrusters) is preferred (`efficacy=1`), but any combination of T1/T2–T3/T4 is allowed, depending of the performance of the thrusters. For example, if T1 has bad performance, T2 can be employed. The rationale is that, according to the level of performance of the thrusters, different controllers shall be used for optimal performance. For the shake of our hypothetical scenario, it is assumed The PID controller works well when the performance of the thrusters results in a propulsion thrust above 70%, while the fuzzy controller performs better when the overall thrust is below 30% due to the instantaneous thruster's performance.

4.2 Ontology for the Underwater Robot

For the metacontroller to reason at runtime the optimal configuration of the control architecture of the US-1, the previous engineering knowledge is encoded in the ontology following the TOMASys metamodel.

The UX-1 ontology consist of two modules: the TOMASys ontology, and the application specific ontology for the underwater robot. The later contains the simplified version of the robot's control architecture described in Sect. 4.1.

TOMASys concepts have been implemented through OWL classes, complemented with SWRL rules for additional Metacontrol semantics, in the so called TBox (for terminological knowledge), which is used to represent the domain, i.e. the functional architecture of autonomous systems in this case. The model of the underwater robot's architecture is thus captured in the ABox (for assertions) by creating specific instances of the TOMASys classes for the different objectives, functions and components in the underwater robot's architecture.

In our simplified version of the UX-1, the system has the root *Objective* to `Navigate`, for which any *FunctionDesign* solving it requires the *Objective* `ControlledMotion`. Correspondingly, multiple *FDs* for `ControlledMotion` are available, each one corresponding to each of the different controller options presented in the previous section, and all of them requiring a specific objective for the function `Propulsion`. Finally, the different configurations of thrusters are modeled as different *FDs* for the *Function* `Propulsion`.

We have used the new TOMASys elements *performance* and *efficacy* already discussed in Sect. 3.1, and an associated set of SWRL rules to model the runtime performance of alternative architectures. For example (see Fig. 5), the *FunctionDesign* `FD_PID` solves the function `ControlledMotion` with an efficacy of 1.0 (optimal), and for this it has a role that contains a specification for the PID controller, and requires that an objective of *functionType* `Propulsion` is fulfilled with a performance higher than 0.7. At the lower level, multiple *FDs* are available to address objectives of *functionType* `Propulsion`. For example, `PropulsionM1M3` defines two roles, specifying the use of thruster T1 and T3 for propulsion, with an efficacy of 0.9 (they are not the optimal configuration for propulsion). During runtime operation, the performance achieved for each objective is given by the TOMASys model for performance, which has been implemented in application-specific SWRL rules such as R6 in Table 3. For example, the performance for the propulsion objective, when realized by a *FG* implementing the *FD* `PropulsionM1M3`, is given by multiplying the performance of both thrusters, weighted by the efficacy of the *FD*, which in this case is 0.9. Note that different performance models could have been specified for different *FDs*.

5 Discussion

The previous proof of concept shows the benefits of our metacontrol approach for the self-adaptation of robot control architectures. The ontological implementation of TOMASys allows to implement a general functional diagnosis for robot control architectures, novelly including performance considerations, by adding a model for this quality attribute in TOMASys. The approach followed for the ontological implementation of TOMSys has TOMASys elements implemented in the TBox, as OWL classes and SWRL rules complementing the metacontrol semantics, and the application-specific model represented through individuals in the ABox. This allows for a clear separation of application-specific knowledge and reuse of the TOMASys ontology file and reasoner across applications and domains.

Table 3. One of the SWRL rules that implement the TOMASys performance model for the UX-1.

R6	FunctionGrounding(?fg) ^typeF(?fg, ?fd) ^fd_efficay(?fd, ?eff) hasBindings(?fg, ?bA) ^hasBindings(?fg, ?bB) Binding(?bA) ^binding_role(?bA, role1-fd_move_fw_2m13) ^binding_component(?bA, ?motorA) ^ Binding(?bB) ^binding_role(?bB, role3-fd_move_fw_2m13) ^binding_component(?bB, ?motorB) ^ c_performance(?motorA, ?pA) ^c_performance(?motorB, ?pB) ^ swrlb:add(?aux1 , ?pA, ?pB) ^swrlb:divide(?aux2, ?aux1, 2.0) ^ swrlb:multiply(?eff, ?aux2, .0) -> fg_performance(?fg, ?aux)

The second contribution of this paper is the extension of the TOMASys metamodel to incorporate quality attributes in the model of the functional hierarchy. Previously, TOMASys only accounted for a "confidence" property of the functional design in the architecture, that allowed to propagate component's faults into the functional hierarchy, and diagnose their impact in the system's objectives. This approach was very limited and did not support considering more detailed quality attributes that are usually considered in the engineering of systems, e.g. performance, efficiency/power.

However, the proof-of-concept with the underwater robot have also manifested a difficulty with the current TOMASys metamodel to capture bottom-up design considerations. For example when the performance of some components (controllers) in the robot depends on which other components they are interacting with (thrusters). The solution we have applied is to make the dependency indirect through intermediate objectives (e.g. propulsion). An alternative solution is to create a "flatter" model, in which the interdependent components are captured as multiple roles in a FunctionDesign, having as many *FDs* as alternative configurations for those components.

OWL and SWRL present some limitations for our metacontrol, mainly related to their open-world assumption, that were exposed by the proof of concept. In practice, for TOMASys modelling one of the limitations of SWRL is that it cannot represent rules that require to iterate over individuals. In some cases, e.g. rule 5 in Table 1, we have been able overcome this limitation with the addition of local closed-world assumption by injecting facts with OWLAPI, implementing an application-independent SWRL rule, at the cost of ad-hoc metacontrol reasoning out of the standard reasoner, therefore reducing the original benefits obtained when implementing all the semantics explicitly in the ontology, e.g. verification and validation.

5.1 Future Work

Currently we are testing different ontology engineering approaches for the UX-1 runtime model and associated metacontrol reasoning designs. The representation limitations of TOMASys and OWL/SWRL need to be addressed. Then, we plan to implement the metacontrol operation with a realistic version of the UX-1 control architecture, and test it with a simulation of the underwater robot, analyzing the influence of the reasoner execution time in real reconfiguration scenarios.

In a next step, we plan to investigate the integration of TOMASys with the ontological standards in robotics CORA (IEEE 1872-2015) [10], which contains concepts to represent the complete triplet mission–system(architecture)–environment. This would allow to coordinate the metacontrol operation with the mission or task-planning. Finally, To address the ontological engineering burden (e.g. modelling of many alternative *FDs*, and their associated SWRL rules) we plan to explore ontology design patterns [12] and metamodelling transformations.

6 Concluding Remarks

In conclusion, we believe that component-level fault-tolerance is insufficient and self-diagnosis and self-adaptation capabilities are needed for autonomous robots, such as the underwater robot UX-1, that need to maintain reliable autonomous operation in hazardous, unknown environments.

The implementation of our Metacontrol framework for self-adaptation using OWL ontologies and SWRL rules for reasoning, extended with specific models of quality attributes of the control architecture, has demonstrated to be effective for this purpose, and the experience gained in this real-world application has raised a series of interesting questions and potential research lines.

Acknowledgements. This work was supported by the UNEXMIN (Grant Agreement No. 690008) and ROSIN (Grant Agreement No. 732287) projects with funding from the European Union's Horizon 2020 research and innovation programme, and has been co-funded by the RoboCity2030-DIH-CM Madrid Robotics Digital Innovation Hub ("Robotica aplicada a la mejora de la calidad de vida de los ciudadanos. fase IV"; S2018/NMT-4331), funded by "Programas de Actividades I+D en la Comunidad de Madrid" and cofunded by Structural Funds of the EU.

References

1. Beetz, M., Beßler, D., Haidu, A., Pomarlan, M., Kaan Bozcuoglu, A., Bartels, G.: KnowRob 2.0 — a 2nd generation knowledge processing framework for cognition-enabled robotic agents, pp. 512–519, May 2018
2. Bencomo, N., Götz, S., Song, H.: Models@run.time: a guided tour of the state of the art and research challenges. Softw. Syst. Model. **18**(5), 3049–3082 (2019)
3. Bermejo-Alonso, J., Hernández, C., Sanz, R.: Model-based engineering of autonomous systems using ontologies and metamodels. In: 2016 IEEE International Symposium on Systems Engineering (ISSE), pp. 1–8, October 2016
4. Brugali, D., Capilla, R., Mirandola, R., Trubiani, C.: Model-based development of QoS-aware reconfigurable autonomous robotic systems. In: 2018 Second IEEE International Conference on Robotic Computing (IRC), pp. 129–136, January 2018
5. Fernandez, J.L., Lopez, J., Gomez, J.P.: Feature article: reengineering the avionics of an unmanned aerial vehicle. IEEE Aerosp. Electron. Syst. Mag. **31**(4), 6–13 (2016)

6. Fernandez-Sánchez, J.L., Hernández, C.: Practical model based systems engineering. Artech House (2019)
7. Hernandez, C., Fernandez-Sanchez, J.L.: Model-based systems engineering to design collaborative robotics applications. In: 2017 IEEE International Systems Engineering Symposium (ISSE), pp. 1–6, October 2017
8. Hernández, C.: Model-based Self-awareness patterns for autonomy. Ph.D. thesis, Universidad Politécnica de Madrid, ETSII, Dpto. Automática, Ing. Electrónica e Informática Industrial, October 2013
9. Hernández, C., Bermejo-Alonso, J., Sanz, R.: A self-adaptation framework based on functional knowledge for augmented autonomy in robots. Integr. Comput. Aided Eng. **25**(2), 157–172 (2018)
10. IEEE Robotics and Automation Society: IEEE Standard Ontologies for Robotics and Automation. Technical report. IEEE Std 1872–2015, February 2015
11. Jajaga, E., Ahmedi, L.: C-SWRL: SWRL for reasoning over stream data. In: 2017 IEEE 11th International Conference on Semantic Computing (ICSC), pp. 395–400, January 2017
12. Krieg-Brückner, B., Mossakowski, T.: Generic ontologies and generic ontology design patterns. In: Workshop on Ontology Design and Patterns (WOP-2017), located at ISWC 2017, Wien, Austria, 21–25 October. CEUR (2017)
13. López, I., Sanz, R., Hernández, C., Hernando, A.: Perception in general autonomous systems. In: Grzech, A. (ed.) Proceedings of the 16th International Conference on Systems Science, vol. 1, pp. 204–210 (2007)
14. OMG: Robotic Technology Component Specification. Technical Report, formal/2008-04-04, Object Management Group, April 2008
15. Ramaswamy, A., Monsuez, B., Tapus, A.: Model-driven self-adaptation of robotics software using probabilistic approach. In: 2015 European Conference on Mobile Robots (ECMR), pp. 1–6, September 2015
16. Sirin, E., Parsia, B., Grau, B.C., Kalyanpur, A., Katz, Y.: Pellet: a practical OWL-DL reasoner. J. Web Semant. **5**(2), 51–53 (2007). Software Engineering and the Semantic Web
17. Weyns, D., et al.: Perpetual assurances for self-adaptive systems. In: de Lemos, R., et al. (eds.) Software Engineering for Self-Adaptive Systems III. Assurances, pp. 31–63. Springer, Cham (2017)
18. Zhai, Z., Martínez Ortega, J.F., Lucas Martínez, N., Castillejo, P.: A rule-based reasoner for underwater robots using OWL and SWRL. Sensors **18**, 3481 (2018)

An Apology for the "Self" Concept in Autonomous Robot Ontologies

Ricardo Sanz[1,2(✉)], Julita Bermejo-Alonso[3], Claudio Rossi[1], Miguel Hernando[1], Koro Irusta[2], and Esther Aguado[2]

[1] UPM-CSIC Centre for Automation and Robotics, Madrid, Spain
ricardo.sanz@upm.es
[2] Autonomous Systems Laboratory, Universidad Politécnica de Madrid, Madrid, Spain
[3] Facultad de Ciencias y Tecnología, Universidad Isabel I, Burgos, Spain
http://car.upm-csic.es, http://www.aslab.upm.es, http://www.ui1.es

Abstract. This paper focuses on the core idea that underlies all mechanisms for system self-awareness: "Self". Robot self awareness is a hot topic not only from a bioinspiration perspective but also from a more profound reflection-based strategy for increased autonomy and resilience. In this paper we address the uses and genealogy of the concept of "self", its value in the implementation of robots and the role it may play in autonomous robotic systems' architectures. We hence propose the inclusion of the "Self" concept in the future IEEE AuR standard ontology.

Keywords: Robots · Autonomy · Ontology · Engineering · Consciousness · Self

1 Introduction

The conference special session on *Core concepts for an Ontology for Autonomous Robotics. Genealogy and Engineering Practice* addresses the important issue of what are the *core concepts* for an ontology of autonomy in robots. During the last years, work of different authors, esp. in relation to the upcoming IEEE Standard for Autonomous Robotics (AuR) Ontology, has been addressing the identification of such concepts. However, to our understanding, there is one concept of major importance that may be included: the concept of *Self*.

Obviously the term has wide and complex meanings arising from different cognitive science domains —esp. psychology and sociology— but has gained relevance in the technical domains where self-x capabilities have been steadily deployed in adaptive and resilient systems.

The "Self" term appears repeatedly in the literature descriptions of robotic systems but it is not usually realised explicitly in implementations —e.g. as specific software subsystems or as a term in ontologies in the domain. Robots

M. F. Silva et al. (Eds.): ROBOT 2019, AISC 1092, pp. 416–428, 2020.
https://doi.org/10.1007/978-3-030-35990-4_34

can self-teach, self-repair or self-explain, but there is commonly no subsystem or entity named "Self".

In this paper we address the nature of a robotic concept of "self", its value and role in autonomous robotic systems, and propose its inclusion in the future IEEE AuR standard ontology.

2 Engineering Autonomous Robots

Autonomous robots are robots that can complete their missions without resorting to functions provided externally to them [1]. The most common use of the term "autonomy" in robotics is to refer to the *energy* and the *navigation* functions. For many people an autonomous robot is a self moving vehicle, that can decide on the fly the trajectory to follow to accomplish the movement. Going beyond movement, autonomy means capability for self-x, x being any class of function required by the system [27,37].

Precise definitions of autonomy shall always be specified in terms of environment, mission, system [38]. *Function*, with its dual nature [24], couples the mission —goals— with the system —capabilities. A fully autonomous robot is able to perform any needed function by itself (is able to achieve self-* in its environment, for its mission and with its resources). The variety of needed functions is enormous for complex missions or environments. From *energy* and *navigation* mentioned before, to *talking* with other agents, *repairing* broken parts or *negotiating* service contracts.

For example, the recently started European project ROBOMINERS[1] has the objective of developing an autonomous mining robot that is able to self *assemble*. The project will:

> *"develop a bio-inspired, modular and reconfigurable robot-miner for small and difficult to access deposits. The aim is to create a prototype robot that is capable of mining underground, underwater or above water, and can be delivered in modules to the deposit via a large diameter borehole.[...] This will then self-assemble and begin its [autonomous] operation."*

Fig. 1. A robot will have a "self" if it has any form of reflection—both percept or action.

In our Autonomous Systems (ASys) research program, we aim at improving systems resilience by adding self-control mechanisms based on self-awareness [15,35]. Self-awareness and self-action (see Fig. 1) are the two system-level, critical architectural aspects that are needed for reflective enhancement of autonomy. Self-awareness means

[1] https://robominers.eu/.

that the robot is aware of itself and self-action means that the robot is able to control itself. The deepness and integration of these awareness and action mechanisms are what determine the degree of selfhood. However, these concepts are still unclear and far for being formal enough to achieve systematic, model-based engineering. It is in the context of the engineering of these systems where the concept of "Self" needs a more precise definition.

3 Robot Selves

From time to time, an announcement hits the news headlines: a robot has passed a test for self-awareness *for the first time* [4,40]. A paradigmatic example of self-awareness test is the *mirror test* that is conventionally used to discern if an animal is self-aware or not [2]. However, these tests were created for human-like systems (*e.g.* great apes [11]) and its general validity (*i.e.* for robots) may be questionable.

Mere program-based mirror self recognition seems not enough for granting a self to a robot. For example, during the AAAI Fall Symposium Series of 2007, Haikonen performed a live experiment showing that a toy doll with some simple electronic components from Radio Shack was able to pass the mirror test [13]. Obviously this happened without any human-like significant "self" inside the doll (Fig. 2).

Fig. 2. A robot will have a "self" if it has any form of reflection, e.g. by observing itself in a mirror.

Does this imply that the mirror test is wrong or that some widely accepted "self" concept is wrong? The problem with biological-originated concepts lies in the simple fact that observation and experimentation of autonomous behaviour almost always happens with whole, opaque, alive systems. Experiments with parts are rare, limited and uninformative. We cannot observe, less modify, the inner mechanisms that generate these behaviours (as we can do with robots) (Fig. 3).

The perception that humans have of themselves —the experience of selfhood [18]— seems always serial and integrated in healthy conditions [44,46]. However, being *seriality* and *integration* considered necessary attributes of human consciousness, they are not necessary and maybe even counter-productive in artefacts [34].

For some, the self is an illusion created in a social context. Hood [16] says that the self emerges during childhood as a process of learning. Humans not only

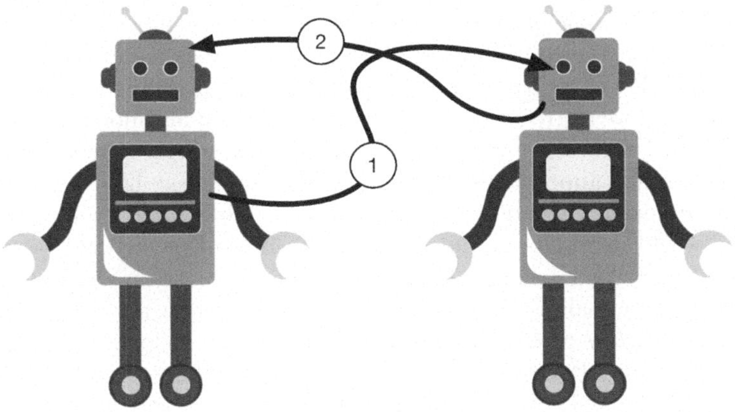

Fig. 3. A robot will have a "self" if it has any form of reflection, e.g. by interacting with another robot in a social context (*e.g.* through a dialogue).

learn from others, but also to learn to become like others. The architecture of the developing brain —its phenotypic expression— enables us to become social animals dependent on each other.

Robots are also social and their mental development can mimic the social development of humans. However, the social aspects of machines can go quite beyond human aspects. The personality of a group, the spirit of a nation are metaphors for shared ideas and attitudes, but in the case of machines it can be quite more in the form of deep integration of functions —e.g. in the domain of

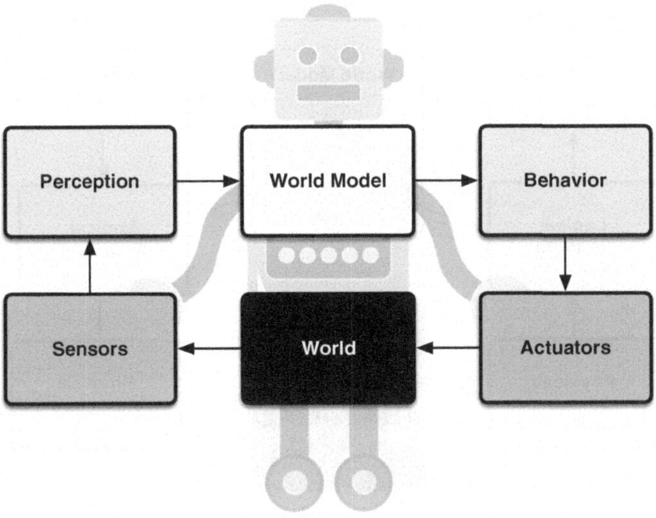

Fig. 4. An epistemic control loop uses knowledge about the world to perceive it and act meaningfully on it.

cloud robotics. The issue of hierarchical/scalable selves will become an important issue. For example, the ROBOMINERS project proposes the creation of a robot that self-assembles from parts. The robot shall have a self, but also the parts shall, because before aggregation they will be separated autonomous agents. The robot self shall be an aggregation of the subsystem selves. A federation-based, systems-of-systems, approach to its engineering may ensue [25].

From an external viewpoint we can always say that a robot has a self as the part of the world inside its boundary. From an internal viewpoint, a robot will have a "self" only if it has any form of reflection as system. Reflection is any process, with its sustaining mechanics, where the system turns to itself —this is pure *intentionality* in Brentano's sense [3]. Having reflection is a necessary condition but it may be not sufficient; many programming languages –like Java or Python— implement reflection but this does not necessarily imply that objects or classes do have a self.

We are interested in the concept of self and hence on the ontological/epistemic aspects of cognitive robot reflection. The epistemic control loop [36] (see Fig. 4) describes the basic operation of any cognitive system. In order to act on some part of the world the robot shall have a model of the world and an update mechanism for this model. Mental representation of the world is always necessary *for meaningful action* [7]. Anti-representationism is, however, a strong movement in cognitive science and robotics [5]. Some authors don't understand the fact that action and representation are two sides of the same coin [10].

When the object of perception/action of an epistemic control loop is the robot itself we will have *system reflection*. In this situation a self will necessarily appear implicitly or explicitly. The epistemological self —what the robot thinks about itself— will emerge in the representation of the ontic self —what the robot is.

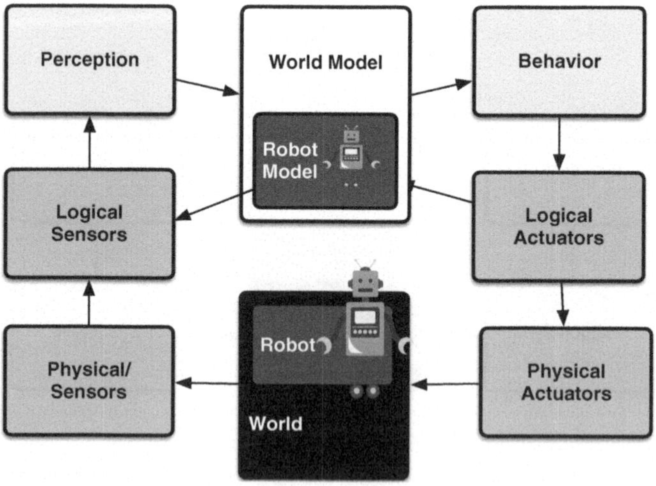

Fig. 5. The knowledge in the epistemic control loop model includes knowledge about the robot and the world to be perceived includes the robot itself.

The world to be perceived includes the robot and, hence, the knowledge in the epistemic control loop model includes both knowledge about the world and knowledge about the robot itself (see Fig. 5)[2]. In this process of the reflection, the object of observation/control can be physical parts of the robot —motors, wheels, cameras, etc.— but can also be mental parts of the robot —the controllers, filters, knowledge, memories, etc. The reflection on the mental parts seems strongly close to the current idea of self-awareness in humans, that is strongly tinted of cartesian dualism.

The basic epistemic control loop is completed with extra kinds of mechanisms: (i) inner thinking processes that are not bound to perception and action, (ii) modelling processes that update and learn representations, (iii) decision processes that are separated from action, and (iv) evaluation processes that estimate performance of the agent in terms of its objectives. These objectives can be internal to the agent or imposed externally —this last is of particular importance for artificial systems (see Fig. 6). The evaluation mechanisms can in fact act on the

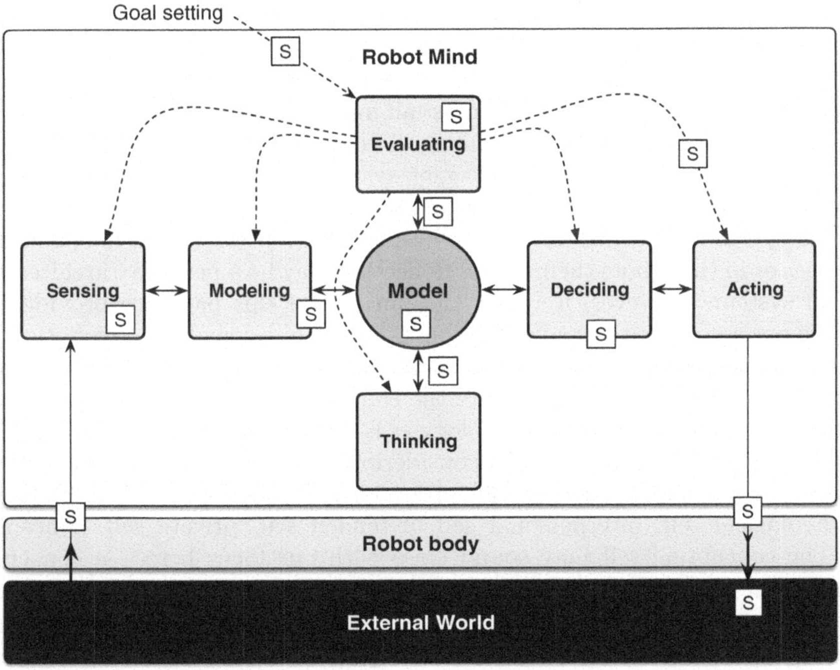

Fig. 6. The overall cognitive architecture is completed with inner thinking and evaluation processes. Note that being the robot an artificial being, its goals are set externally. The figure shows some specific places were a specific role of agent self has been described (marked with a boxed $\boxed{S}$).

[2] Note that even when we draw some boxes out of the world, everything is inside it.

inner behaviour of the agent to modify it according to circumstances. This meta-control can take the form of another cognitive control loop [8,14] or of emotional reactions [39].

The idea of self can appear in many forms across all these processes, being the self-image that forms part of the world model the most central to all of them.

4 The Concept of Self

The "Self" concept we are looking for is a concept that can be captured in an ontology to serve the purpose of supporting autonomous robot engineering processes and autonomous robot system operation. This concept emerges from tho major sources: (i) the transposition of humanly self concepts to the domain of robotics (esp. in social or cognitive robots), and (ii) the systems analysis of reflection.

4.1 Profiling the "Self" Concept

The second source —system reflection— is easier to identify and analyse, existing good, focused references on it [19–21,23,26,31]. Looking at the analysis of the previous section (see the roles of self in Fig. 6) we reach the conclusion that reflectional aspects can appear in almost all mental processes of cognitive agents. This may imply that "Self" is a mongrel concept because it addresses many entities and processes of cognitive agent systems that possess the property of reflection but that are substantially non-integrated. However, the(se) concept(s) requiere(s) a more detailed analysis to evaluate if they are needed in engineering processes or in the robots themselves. Reflection may be a pervasive architectural trait of systems, however, it is not the purpose of this paper to provide such analysis nor a systematic study of accounts of self in psychology or robotics, but to motivate its relevance as a concept for an ontology of autonomy. Further work will be necessary to properly capture the concept.

The first source —self in psychology— is harder and more confusing. However there are some sources worth considering for their connection to technical systems. In particular, Neisser [30] proposed five levels of analysis of human self: Ecological self, interpersonal self, extended self, private self, conceptual self. The conceptual self may sound close with our focus here —a concept of "Self"— but for Neisser it is not based on a systemic analysis but gets its content "from a network of socially-based assumptions and theories about human nature in general and ourselves in particular" [29]. Based on Neisser's categorisation, Lewis [22] proposed a series of levels of computational self-awareness and associated self-action: Stimulus awareness, interaction awareness, time awareness, goal awareness and Meta-self-awareness (this last strongly related to our interests on augmented resilience expressed here).

In our own laboratory, Gordillo [12] performed a systemic analysis of the self concept in cognitive science and robotics, reaching the conclusion that the many processes that underlie the concept of self can be reduced to four types of function:

Identity: ability of the system to distinguish itself with respect to the environment (esp. if there are other agents out there).
Proprioception: mechanisms that allow an agent to observe himself.
Self-awareness: being conscious of oneself.
Autobiographical memory: a set of experiences in which the agent himself is the protagonist; among which the planning of the future is included.

Another interesting work is that of Samsonovich [33] concerning the engineering of agency in a cognitive system. In these systems, esp. when situated in social environments, attribution of beliefs, values and/or goals represented in the system becomes essential. Using an explicit concept of "Self" these implicitly attributed content would be explicitly attributed to the self of the agent. In the words of Samsonovich:

> *"When the cognitive system becomes explicitly aware of this attribution, it acquires a self-regulation capacity allowing it to control, modify and develop its self-concept together with the attitudes attributed to the self, adjusting to dynamically changing contexts and personal experience."*

4.2 Genealogy of "Self"

The "Self" concept genealogy could, in principle, be traced to the two classes of origins: human mentality and technical system reflection. However, the history of the concept on the technical side is minor as it is mostly based on transposition from natural sciences.

The main genealogy of the concept lies in psychology and philosophy of mind. Cary [6] traces back the origin of the concept of inner self to St. Augustine in opposition to the self/soul conceived as the subject of external moral duties proposed by Plato in Phaedo and later by Aristotle. From then on, may scholars have addressed the issue, from Aquinas, Descartes and Spinoza [32] to Freud, James, Kierkegaard and Damasio [9].

A good account of the evolution of the concept of self in psychological and social sciences is the book *Sources of the Self* [42]. In this text Taylor explores the historical sources of the psychological understanding of "self" and, at the same time, it tries to clarify its very understanding.

However, being a concept of strong interest in human psychology, there is no consensus on a single definition of the concept of "Self" [12]. Many authors contribute their own definition, such as the five levels Neisser or the three levels of Damasio's theory of self. James [17] said that the problem of personal identity —what it is to be a self— was "the most puzzling puzzle with which psychology has to deal".

In general, given definitions understand the self as a multifaceted process, composed of numerous aspects, mostly in relation with the perception and modelling of the agent system own state and the associated self-awareness and sense of individuality.

4.3 The "Self" Concept in Computational Ontology

Many concepts in an engineering ontology have a dual ontological/ epistemological nature. Due to its origins in information technology ontologies are mostly considered as related to information structures; hence being epistemological entities. However, in engineering, sharing concepts related to physical things is also necessary; hence these concepts address ontological entities.

The analysis shown before in the context of self-awareness in cognitive architectures show that the concept of "Self" is of high relevance from both perspectives. However, it has not received much attention in formal ontology work. We were able to find some terms of the form self-x, but the only reference we were able to find to having a self was the concept "hasSelf" in OWL, which, unfortunately, was far from our interests.

The idea of reflection has been addressed in ontologies, but usually related to a characteristic of the ontology itself in the vein of computational reflection. We shall not confuse the issue of having a "Self" concept with the focus of reflexive ontologies [43] (an ontology with a set of self contained queries over instances in the domain of study).

The "Self" has been defined in many forms: a nucleus of identity, a subject of experience, a state of self-awareness, a body image, a narrative center. If we argue for its inclusion in an ontology for autonomous robots, What should be the focus? From a robot systems engineering point of view we see three major possibilities:

Self as system. The subject agent system performing the action. The object agent system receiving the action.
Self as locus of control. The source of authority in an action.
Self as knowledge. The system image represented in the system itself.

5 Cave Canem

Good, old cybernetics thought about intelligent systems within a unitary stance. There were no *a priori* distinctions between biological and technical systems in relation with the cybernetic processes they supported. Bioinspiration has always been a source of ideas for engineering. However, the anthropo-x trap lurks here. When thinking about robots, esp. concerning intelligent robots, we can easily fall equating robots and humans.

Much robotics is anthropocentric. It is anthroposcoped and anthropobiased, focusing only and specially on humans' lives, environments and activities. Worse enough, this discipline is also anthropomorphic: it shapes all theories, designs and realisations using the human form: from body shape and mechanics to abstract conceptualisations or personality. Protagoras would be satisfied: man is being used to measure all things robot.

Anthropocentrism can also affect the idea of self. The effect may be worse than in any other idea. We are in the risk of shaping our "Self" concept based on the human conceptions of self; with all their loads of nature and nurture. This shall be avoided by taking a strictly detached knowledge engineering viewpoint; the "Self" concept formulation shall be driven by the universality motive of classic cybernetics.

A sound concept of self shall be of applicability to humanoids interacting with humans, but also to non-humanoid robots —departing from both physical and mental human traits— and, by extension, to all classes of reflecting machines situated in all classes of environments and performing all classes of activities.

6 Conclusions

The link between autonomy and self-awareness is manifest from the most shallow analysis of architectures for adaptive resilience of robots. Traditionally, autonomy has been considered as a *mission-centric capability* but it is our understanding that it shall be instead considered as a *system-centric capability*.

The need for the explicit incorporation of a self into a robot is evident in any situation where it is useful for the robot to have knowledge about its internal state and its own functioning in a multiagent environment. The question is if it is necessary to have such a concept as explicit realisation or it is enough to have the mechanisms and associated concepts that animate the functions associated to the self (e.g. those identified by Neisser, Gordillo or Lewis).

In this brief apology of "Self" there are conspicuously missing pieces. The most important concerns the experience that the self-aware agent has of itself [28,41,45]. However, phenomenology has not yet much to say about robots. The question of machine qualia is still an open issue —as it is, indeed, in human psychology.

However, it is our opinion that having the "Self" concept as en explicit asset in the engineering and operation of such systems would help build better architectural descriptions and more robust and resilient self-management mechanisms. Robots need an explicit "Self" both at construction and runtime[3].

Acknowledgements. This work has received funding from the European Union's Horizon 2020 programme under grant agreement No 820971 — ROBOMINERS.

References

1. Antsaklis, P.J., Rahnama, A.: Control and machine intelligence for system autonomy. J. Intell. Robot. Syst. (2018). https://doi.org/10.1007/s10846-018-0832-6
2. Boyle, A.: Mirror self-recognition and self-identification. Philos. Phenomenol. Res. **97**(2), 284–303 (2018). https://doi.org/10.1111/phpr.12370
3. Brentano, F.: Psychology from an Empirical Standpoint. Routledge & Kegan Paul, London (1973). Trans. A.C. Rancurello, D.B. Terrell and L.L. McAlister. First published in 1874

[3] https://robominers.eu/.

4. Bringsjord, S., Licato, J., Govindarajulu, N.S., Ghosh, R., Sen, A.: Real robots that pass human tests of self-consciousness. In: 24th IEEE International Symposium on Robot and Human Interactive Communication, Kobe, Japan, 31 August–4 September 2015 (2015)
5. Brooks, R.A.: Intelligence without representation. Artif. Intell. **47**(1–3), 139–159 (1991)
6. Cary, P.: Augustine's Invention of the Inner Self. The Legacy of a Christian Platonist. Oxford University Press, Oxford (2000)
7. Conant, R.C., Ashby, W.R.: Every good regulator of a system must be a model of that system. Int. J. Syst. Sci. **1**(2), 89–97 (1970)
8. Cox, M.T., Raja, A., Horvitz, E.: Metareasoning: Thinking about Thinking. The MIT Press, Cambridge (2011)
9. Damasio, A.R.: Self Comes to Mind: Constructing the Conscious Brain. Willian Heinemann, London (2010)
10. Frisina, W.G.: The Unity of Knowledge and Action. SUNY Press, New York (2002)
11. Gallup, G.G.: Chimpanzees: self-recognition. Science **167**(3914), 86–87 (1970). https://science.sciencemag.org/content/167/3914/86
12. Gordillo Dagallier, L.: Extensión de la representación de un yo para una arquitectura de sistema autónomo. TFG en Ingeniería de las Tecnologías Industriales, July 2016. http://oa.upm.es/43482/
13. Haikonen, P.O.A.: Reflections of consciousness: the mirror test. In: Chella, A., Manzotti, R. (eds.) Proceedings of the 2007 AAAI Fall Symposium on AI and Consciousness: Theoretical Foundations and Current Approaches. Technical report FS-07-01, pp. 67–71 (2007)
14. Hernández, C., Bermejo-Alonso, J., López, I., Sanz, R.: Three patterns for autonomous robot control architecting. In: The Fifth International Conference on Pervasive Patterns and Applications - PATTERNS 2013, Valencia, 27 May–1 June 2013, pp. 44–51 (2013)
15. Hernández, C., Bermejo-Alonso, J., Sanz, R.: A self-adaptation framework based on functional knowledge for augmented autonomy in robots. Integr. Comput. Aided Eng. **25**, 157–172 (2018)
16. Hood, B.: The Self Illusion: How the Social Brain Creates Identity. Oxford University Press, Oxford (2012)
17. James, W.: The Principles of Psychology. Henry Holt and Co, New York (1890)
18. Jeannerod, M.: From self-recognition to self-consciousness. In: Zahavi, D., Grünbaum, T., Parnas, J. (eds.) The Structure and Development of Self-Consciousness. Interdisciplinary perspectives, Advances in Consciousness Research, vol. 59, pp. 65–88. John Benjamins Publishing Company (2004)
19. Kiczales, G., Des Rivières, J., Bobrow, D.G.: The Art of the Metaobject Protocol. MIT Press, Cambridge (1991)
20. Kounev, S., Kephart, J.O., Milenkoski, A., Zhu, X.: Self-Aware Computing Systems. Springer, Cham (2017)
21. Landauer, C., Bellman, K.L.: Meta-analysis and reflection as system development strategies. In: Metainformatics. International Symposium MIS 2003. LNCS, vol. 3002, pp. 178–196. Springer (2004)
22. Lewis, P.R., Chandra, A., Faniyi, F., Glette, K., Chen, T., Bahsoon, R., Torresen, J., Yao, X.: Architectural aspects of self-aware and self-expressive systems: from psychology to engineering. Computer **48**(8), 62–70 (2015)
23. Lewis, P.R., Platzner, M., Rinner, B., Tørresen, J., Yao, X.: Self-aware Computing Systems. Springer, An Engineering Approach (2016)

24. Lind, M.: Means and ends of control. In: 2004 IEEE International Conference on Systems, Man and Cybernetics, vol. 1, pp. 833–840 (2004)
25. Lluch, I., Golkar, A.: Architecting federations of systems: a framework forcapturing synergy. Syst. Eng. **22**, 295–312 (2019)
26. Maes, P.: Computational reflection. Technical report 87-2, AI Laboratory. Vrije Universiteit Brussel (1987)
27. Mann, D.: Ideality and "Self-X" - part 1: Things that do things for themselves. TRIZ J. (2003)
28. Nagel, T.: What is it like to be a bat? Philos. Rev. **83**(4), 435–450 (1974)
29. Neisser, U.: Five kinds of self-knowledge. Philos. Psychol. **1**(1), 35–59 (1988). https://doi.org/10.1080/09515088808572924
30. Neisser, U.: The roots of self-knowledge: perceiving self, it, and thou. Ann. N. Y. Acad. Sci. **818**(1), 19–33 (1997)
31. Perlis, D., Subrahmanian, V.S.: Meta-languages, reflection principles and self-reference. In: Handbook of Logic in Artificial Intelligence and Logic Programming, vol. 2, pp. 323–358. Oxford University Press (1994)
32. Remes, P., Sihvola, J. (eds.): Ancient Philosophy of the Self, The New Synthese Historical Library, vol. 64. Springer, Dordrecht (2008)
33. Samsonovich, A.V., Kitsantas, A., Dabbagh, N., Jong, K.A.D.: Self-awareness as metacognition about own self concept. In: Metareasoning: Thinking about Thinking. AAAI Workshop WS-08-07, pp. 159–162 (2008)
34. Sanz, R.: Consciousness, engineering and anthropomorphism. In: International Association for Computing and Philosophy – Annual Meeting, Warsaw, Poland (2018)
35. Sanz, R., Hernández, C., Hernando, A., Gómez, J., Bermejo, J.: Grounding robot autonomy in emotion and self-awareness. In: Kim, J.H., Ge, S.S., Vadakkepat, P., Jesse, N. (eds.) Advances in Robotics, pp. 23–43. Springer, Heidelberg (2009)
36. Sanz, R., Hernández, C., Rodriguez, M.: The epistemic control loop. In: Proceedings of CogSys 2010 - 4th International Conference on Cognitive Systems, Zurich, Switzerland, January 2010
37. Sanz, R., López, I., Bermejo-Alonso, J., Chinchilla, R., Conde, R.: Self-X: The control within. In: Proceedings of IFAC World Congress 2005, July 2005
38. Sanz, R., Matía, F., Galán, S.: Fridges, elephants and the meaning of autonomy and intelligence. In: IEEE International Symposium on Intelligent Control, ISIC 2000, Patras, Greece (2000)
39. Sanz, R., Sánchez-Escribano, M.G., Herrera, C.: A model of emotion as patterned metacontrol. Biol. Inspired Cogn. Arch. **4**, 79–97 (2013). http://www.sciencedirect.com/science/article/pii/S2212683X13000194
40. Takeno, J.: Creation of a Conscious Robot. Pan Stanford Publishing, Mirror Image Cognition and Self-Awareness (2013)
41. Tallis, R.: The Explicit Animal: A Defence of Human Consciousness. Palgrave Macmillan, London (1999)
42. Taylor, C.: Sources of the Self: The Making of Modern Identity. Harvard University Press, Cambridge (1989)
43. Toro, C., Sanín, C., Szczerbicki, E., Posada, J.: Reflexive ontologies: enhancing ontologies with self- contained queries. Cybern. Syst. **39**(2), 171–189 (2008). https://doi.org/10.1080/01969720701853467
44. Woźniak, M.: "I" and "Me": the self in the context of consciousness. Front. Psychol. **9**, 1656 (2018). https://doi.org/10.3389/fpsyg.2018.01656

45. Zahavi, D., Parnas, J.: Phenomenal consciousness and self-awareness: a phenomenological critique of representational theory. J. Conscious. Stud. **5**(5–6), 687–705 (1998)
46. Zahavi, D. (ed.): Exploring the Self. Philosophical and psychopathological perspectives on self-experience, Advances in Consciousness Research, vol. 23. John Benjamins Publishing Company (2000)

Knowledge and Capabilities Representation for Visually Guided Robotic Bin Picking

Paulo J. S. Gonçalves[1,2(✉)], J. R. Caldas Pinto[2], and Frederico Torres[2]

[1] Instituto Politécnico de Castelo Branco, Escola Superior de Tecnologia, Av. Empresário, 6000-767 Castelo Branco, Portugal
paulo.goncalves@ipcb.pt

[2] IDMEC, Instituto Superior Técnico, Universidade de Lisboa, Av. Rovisco Pais, 1049-001 Lisbon, Portugal
http://pessoas.ipcb.pt/paulo.goncalves/

Abstract. The paper presents an implementation of knowledge representation including the capabilities of the system, based on ontologies for a Visually Guided Bin Picking Task. The ontology based approach was used to define the work environment, the robot, the machine vision system, and the capabilities that are needed to be performed by the robotic system, to perform the bin-picking task. The work proposes a novel application framework that is able to locate the object to pick from the bin and place it in a cell from a kit. For that, the framework, delivers the task implementation (PDDL) files that should be executed by the robot. The method used to detect the objects is based on Chamfer Match (CM) and Oriented Chamfer Match (OCM) which take advantage of the image edge map. To complete the pose estimation problem the robot manipulator is equipped with a laser range finder that can measure the object height. The robotic system was validated experimentally with simulation. using the V-REP environment interfacing with ROS, where the knowledge representation and reasoning framework is implemented. The system showed its capability to correctly pick and place a specific object. Moreover, the ontology based approach was very useful to define the task, the actions to be performed by the robot, based on its capabilities.

Keywords: Knowledge representation · Ontologies · Robot capabilities · Industrial robots · Machine vision

1 Introduction

The present work is done in collaboration with Bitzer, a compressor company founded 80 years ago. It is specialized in the production of compressors mainly for acclimatization and refrigeration purposes.

In today's global economy, the competition level has never been higher. Countries such as China and India with low cost labour and huge industry investments have displaced or even closed many factories in the western countries [3]. A good

M. F. Silva et al. (Eds.): ROBOT 2019, AISC 1092, pp. 429–440, 2020.
https://doi.org/10.1007/978-3-030-35990-4_35

solution, nowadays, is to invest in flexible manufacturing production lines. A flexible manufacturing system is a system that can easily react to change. These type of production lines are computer based, and take advantage of latest technologies such as robotic manipulators, sensors, computer vision, artificial intelligence and are able to fully automate a manufacturing line.

The aim in this work is to develop a bin picking system that can improve the assembly of the valve mechanism of the compressor. Presently this task is manually done and involves small components that require precise operations. With this solution it is intended to develop a system that involves three main areas, namely: object detection using computer vision techniques, robotics to pick and place the detected objects, and ontologies. The later is used to model all knowledge, skills, and the visually guided bin-picking task.

The paper proposes an ontology framework to define the robotic system components, using CORA [6] (Core Ontologies for Robotics and Automation), and the robot capabilities and tasks (building from previous works, that include the skiROS [12] framework, and tasks representation [7,8]). Both the robot and the machine vision systems are defined using CORA definitions, which enables an ontological definition. Furthermore, the set of capabilities needed for the robot to operate with the objects, are defined, including the bin container and the cells in the kit of the final placing position. From the robot, machine vision system, environment, and capabilities identified, the knowledge model, i.e., the ontology, is obtained. The ontological system provides the PDDL [13] files with the actions to be executed by the robot to perform the task that validated the visually guided bin-picking.

2 Object Pose Recognition

2.1 Theoretical Model

Computer vision has evolved greatly in the last decades becoming an essential tool for industrial applications. One of the most challenging tasks are object detection, pose estimation and object recognition. There are two main approaches for the 3 main visual tasks (detection, recognition, and pose estimation): feature point base approach and shape matching base approach. The feature point technique has very effective algorithms such as the SIFT [2] or SURF [1] models. The main idea is to scan an object template through the image and find the best correspondence using a score function. The approach of the shape match model presented here derives from the Chamfer Matching [4], algorithm.

The algorithm used to perform the search is based on two main score processes, the first one, used in the first sweep is the Chamfer Match (CM) the second process used is a more robust derivation of the CM algorithm called Oriented Chamfer Match (OCM) [5]. The first method, Chamfer Match, uses edge maps of the query and template images, computing a score for each position on the query image. This score has pixel units and represents the mean pixel distance of the template edge map at some position in the query edge map.

It represents a mean error so, for smaller scores, there is higher probability of a detected object. Equation 1 describes the score calculations for a specific point in an image. Let $U = u_i$ and $V = v_j$ be the sets of template and query edge map respectively while $n = |U|$.

$$d_{CM}(U, V) = \frac{1}{n} \sum^{u_i \in U} min_{v_j \in V} ||u_i - v_j|| \tag{1}$$

This method is a good first approach to solve the matching problems. The second method is a derivation of the CM called Oriented Chamfer Match (OCM) and its approach is used in the final stages of the algorithm, once the first candidates have been selected. The final value of the OCM score is a more reliable correspondence criteria since the final value is a combination of pixel distance and edge orientation difference.

The OCM formula contains a constant, λ, which stands for a unit equalizer and determines the influence of the orientation score in the final value.

$$d_{OCM}(U, V) = \frac{1}{n} \sum^{u_i \in U} min_{v_j \in V}(||u_i - v_j|| + \lambda||\phi(u_i) - \phi(v_j)||) \tag{2}$$

2.2 Search Algorithm

The search algorithm used is a combination of two methods: the Chamfer Match and the Oriented Chamfer Match. There are three stages in which these methods are applied. After each stage, a selection procedure is done in order to determine the best candidates that proceed to the next stage. Each candidate is defined by a score, a coordinate, and an orientation. The first step consists of determining interesting areas to search for matches. Once the search points are defined, the program will apply the Chamfer Match, at each of these locations - this is the first stage. The second stage consists in updating the CM score to an OCM score for each candidate.

The third and final stage refines each match using the OCM method. Equation 3 formulates the refinement process. U and V stand for the query and templates edge maps respectively, The refinement area is the vicinity of the candidate location and orientation. The values u_x and u_y stand for the points around the candidate position while $V(\theta)$ is for templates close to the candidate orientation.

When a match is found, this is a local minimum in the OCM score map. In order to be a match a third threshold has to be defined. This one evaluates the final OCM score and is the one that defines if the candidate is a match or not.

$$d_{OCM_{std}}(U, V) = min(D_{ocm}(u_x, u_y, V(\theta))) \tag{3}$$

3 Knowledge and Capabilities Representation

In 2015, IEEE Robotics and Automation Society produced its first ever standard [9]. The standard is focused on a set of Core Ontologies for Robotics and

Automation (CORA), where ontologies were developed for Robot Definitions, Parts, and Position. From the represented knowledge, using ontologies, is then possible to describe robots, the environment and human co-workers, that jointly perform tasks. Moreover, the standard provides tools for a formal reference vocabulary to be used in the communication between humans and/or robots. However, the effort is not complete in terms of tasks and capabilities representations. Efforts are going on in recent works, e.g., [8] and [14], although theoretical efforts. Recently, specific ontologies were developed to incorporate skills definitions [12], in an ontological framework. This work, available on-line, https://github.com/frovida/skiros, was followed in this research work to implement the visually guided bin picking, the robot capabilities, as an extension of the CORA ontology standard.

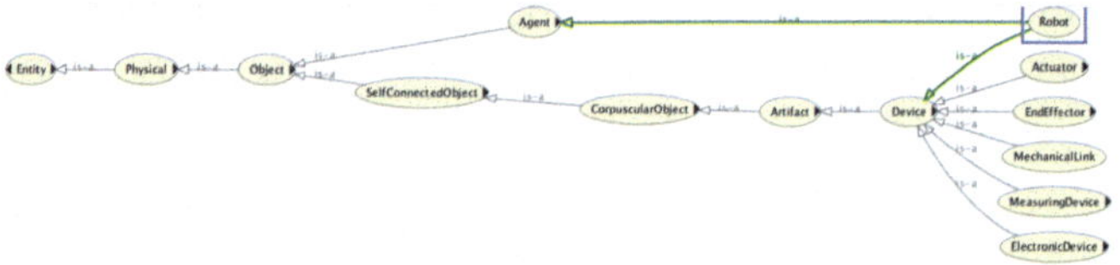

Fig. 1. The definition of Robot and related Devices, using SUMO and CORA.

In this paper was applied some standard definitions, and relations between those, where the visual bin picking knowledge is represented, in its levels, i.e., the machine components, capabilities, tasks and environment. The first part, it will be addressed in the current section. In the following sections the remaining levels will be addressed, and the overall system validated.

The Robot definition and its main parts are based on the IEEE standard [9], where the SUMO [10] upper ontology was used. In Fig. 1 is presented the definitions directly in the standard, where is depicted that a robot is a device and also, an agent that can reason, based on existing knowledge. Moreover, in Fig. 1, are also depicted several Objects, i.e., physical entities, that can be part of a robot, e.g., a measuring device, a mechanical link, etc.

Focusing on the developed ontology for this research work, and the knowledge representations of the robot components, in Fig. 2 is shown the definition of the large majority of them. The *Computer* class is defined as a *Device* – *PhysicalObject*, needed for the robotic system to computationally process the robot capabilities, and all the software that implements the framework. It is worth to mention the existence of the Concept 'MobileBase', not present in CORA, but is needed to represent mobile manipulators, a class of robots often used in industrial and/or service environments. In the present work, the Mitshubishi PA-10 robot used is an instance of the class *Arm*. The classes defined and its properties like *dependsOn, hasA, hasObject, hasParameter, spatiallyRelated, hasPostCondition* are defined in the ontology developed for this work, depicted in Fig. 3 and will be used in this section, and the following, to reason on the ontologies, and validate the proposed approach.

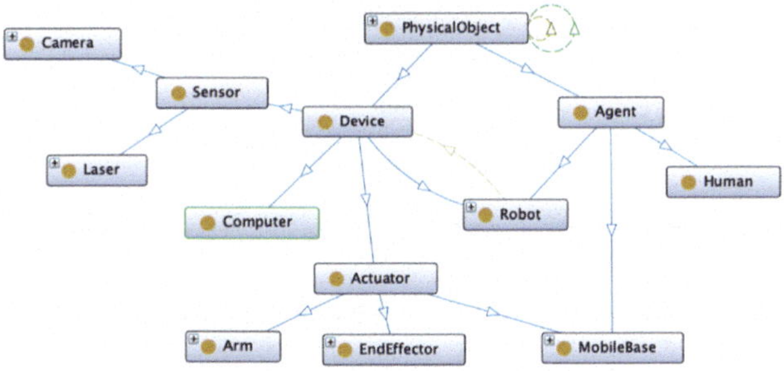

Fig. 2. Snapshot of the Ontology in which are defined the Robot and its components, within the current ontology.

Fig. 3. A snapshot of some of the properties defined in the ontology.

After the definition of the robotic system and all its components, follows the descriptions of the needed capabilities that the robot and the vision system needs to accomplish, the visually guided bin-picking task. The capabilities present in the ontology and implemented in the system are depicted in Fig. 4. The *Capability* class is an abstract *Concept*, who is related to conditions, pre and post, for its proper execution, along with parameters that completely define it. It is worth to mention that capabilities are often implemented with the use of hardware components, for example cameras. As such, they are dependent of software *drivers* and ROS nodes that implement them. This fact is also captured in the ontology.

The capabilities needed for the robotic system comprise, *Pick, Place*, specific to the *Arm*. For the vision capabilities, *Object_Recognition, Object_Pose_Calculus, Calibrate* were defined. Those were implemented as a module of the robotic system, because are not specific for the robot. The vision module is running continuously, publishing the next object pose *topic* to be picked by the *Arm*. The *Drive* capability is added in the ontology for driving the *MobileBase* of the mobile manipulator, also present in the laboratory.

From Fig. 4, it is clear that in the ontology are defined the concepts *Param* and *Condition*. These are important to completely define the *Capability* because it has pre and/or post conditions to be checked prior and/or after execution. Also, parameters must be sent to the *Capability* to precisely define it.

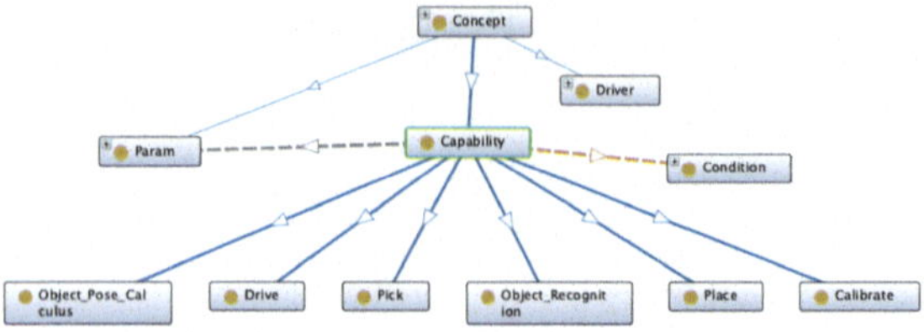

Fig. 4. Snapshot of the Ontology where are defined the needed Capabilities, within the current ontology.

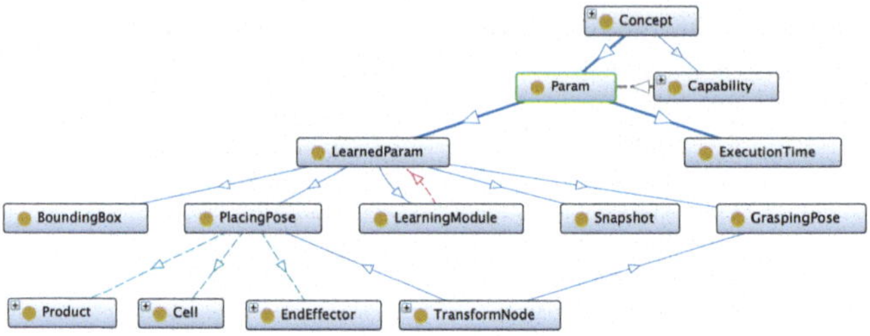

Fig. 5. Snapshot of the Ontology where are defined the Parameters needed to define the Capabilities, within the current ontology.

As of Fig. 5, one can define, for example, the minimum *ExecutionTime* and also $LearnedParam_{eters}$, via the learning modules that implement the vision related capabilities presented in Fig. 4. In the scope of this research work, it is required to obtain information of the *GraspingPose* from the *Snapshot* taken from the *Camera*. Also, for the *PlacingPose* that the robot must have to place the *Product* on the Kit *Cell*, using its *EndEfector*. Moreover, the *TransformNode* is used to obtain the *PlacingPose* via a spatial transformation. In this specific case, and since a *vacuum EndEfector* is used, the transformation is straightforward.

To implement the needed *Capabilities* and to define the *Task* to be performed by the robot, a set of *Conditions* must be taken into account, as presented in Fig. 6. These can also define *requirements* of the task to be performed by the robot. As such, are defined: *PropertyConditions, RelationConditions, PreConditions, PostConditions*. In the following section, examples of its implementation are presented. *RelationConditions* and *PropertyConditions* are related to the interaction of the physical objects that the robot must handle, and where to pick and place them in the current scenario, i.e., work environment. Pre and Post Conditions are important to describe the task, its flow, and to monitor its behaviour.

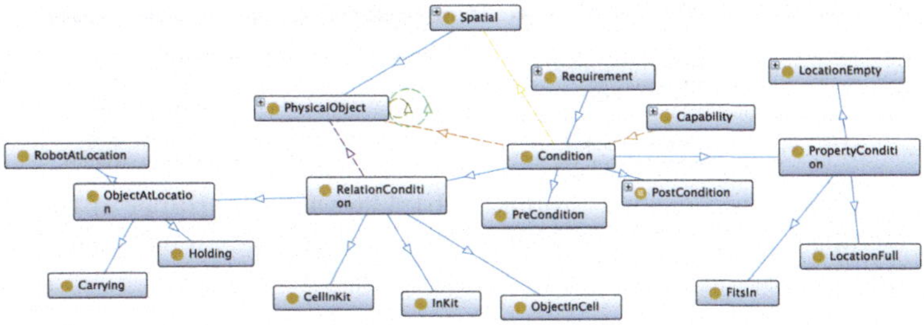

Fig. 6. Snapshot of the Ontology where are defined the set of Conditions that can be used to define the given Capabilities, within the current ontology.

4 Implementing the Framework

In this section is presented the main implementation steps of the visually guided bin-picking task. As such, the first step is to define the world model from the knowledge that exists in the framework ontology. For example, which robot to use, with which end-effector, camera, which objects to manipulate and so on. In Fig. 7 is depicted the implemented scenario, which contains a robot, and a large box that contains the objects to be picked. Moreover, is depicted in the figure the instances of each class, for example (Robot-3).

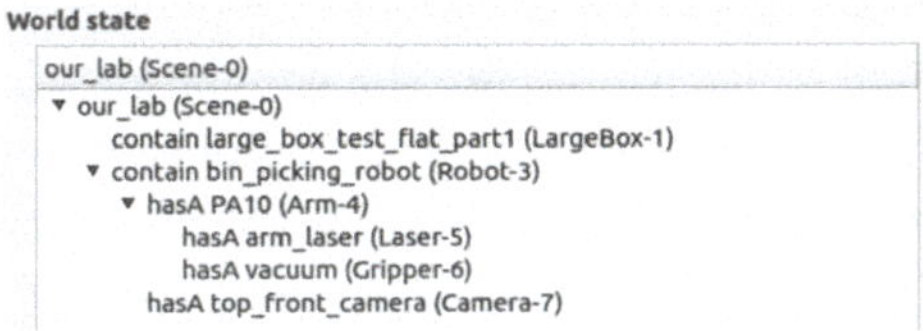

Fig. 7. Definition of the work environment, based on the ontology, including instances of the classes.

The robot definition */bin_picking_robot* is presented in Figs. 8a and b, which show all the properties set for these devices, which include the */tf* transformations of the ROS environment, the drivers, and so on.

In the following listing 4, is presented the definition of one of the main capability of the task, i.e., *place_PA10*. The definition has pre and post-conditions, which relate the objects to manipulate in the scene. Example of conditions use instance of the classes: *EmptyHanded, RobotAtLocation, ObjectAtLocation, Holding*.

(a) The bin picking robot instance, with a Mitsubishi PA10 *Arm*.

(b) The vacuum gripper of the PA10 robot, e.g., an instance of the *Gripper* class.

Fig. 8. The implemented Robot and the gripper instances.

In the next section are presented the results of the implementation of the ontological concepts defined in this framework.

Listing 1.1. Example of a Place instance with pre and post conditions.

```
<relation subject="Place-30" predicate="relatedTo" object="Scene-0" />
<relation subject="Place-30" predicate="hasPreCondition" object="Holding-31" />
<relation subject="Place-30" predicate="hasPreCondition" object="FitsIn-32" />
<relation subject="Place-30" predicate="hasPreCondition" object="LocationEmpty-33" />
<relation subject="Place-30" predicate="hasPreCondition" object="CellInKit-34" />
<relation subject="Place-30" predicate="hasPostCondition" object="LocationEmpty-35" />
<relation subject="Place-30" predicate="hasPostCondition" object="EmptyHanded-36" />
<relation subject="Place-30" predicate="hasPostCondition" object="Holding-37" />
<relation subject="Place-30" predicate="hasPostCondition" object="ObjectInCell-38" />
<relation subject="Place-30" predicate="hasPostCondition" object="InKit-39" />
```

5 Results and Discussion

This section is three fold, since it presents the simulation setup, the results from the object recognition step, presented in Sect. 2, and the results on the reasoning on the ontological framework.

5.1 Simulation Setup

The simulation was performed using the Virtual Robotic Experimentation Platform (V-Rep) [11] that is a 3D robotic simulation platform developed by Coppela Robotics. The simulation system is developed using the V-Rep software and the RosInterface, that is part of the V-REP API framework. The V-Rep environment allows to create a closer to reality scene, more specifically by using one of the available dynamic engines it is possible to recreate a cluttered scene.

In Fig. 9(a) is depicted the PA-10 robot and the boxes that have inside the *Flat_part1* of the compressor.

5.2 Object Detection

For the object detection needed for bin-picking, presented in Sect. 2, a two step procedure was performed. In the first step, real world images were captured and the results showed that the third method presented earlier in this paper, OCM std, obtains the best results, being consistent with several tries on cluttered

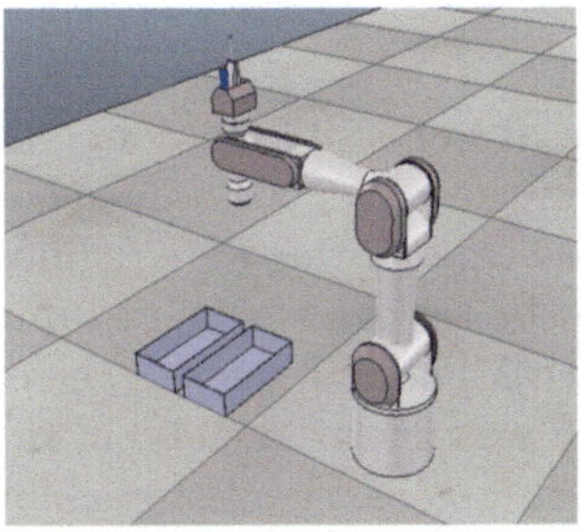

(a) The Robot and the Warehouse Box.

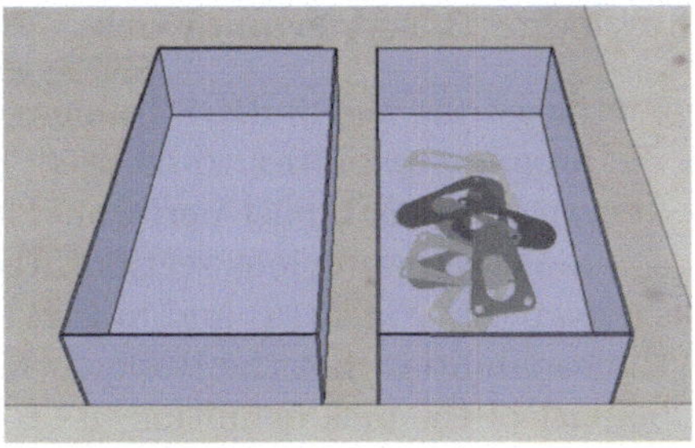

(b) The Box with the *Flat_part1*.

Fig. 9. The V-REP simulation setup.

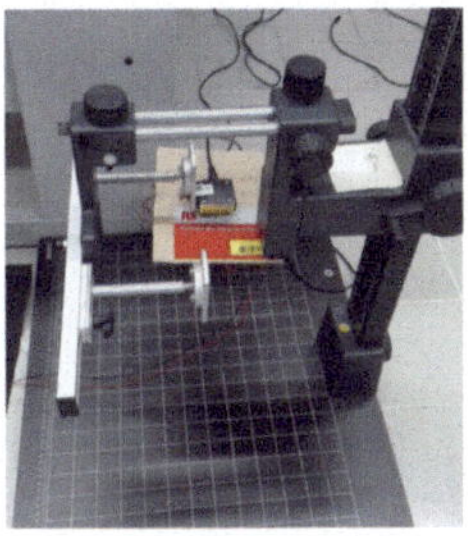

(a) The vision system setup.

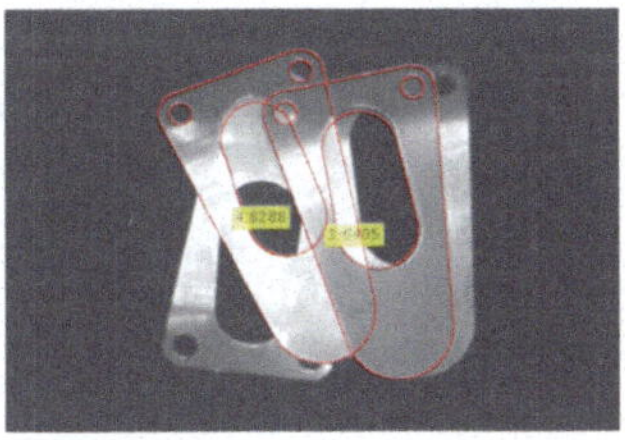

(b) Cluttered image captured, with scores for each object.

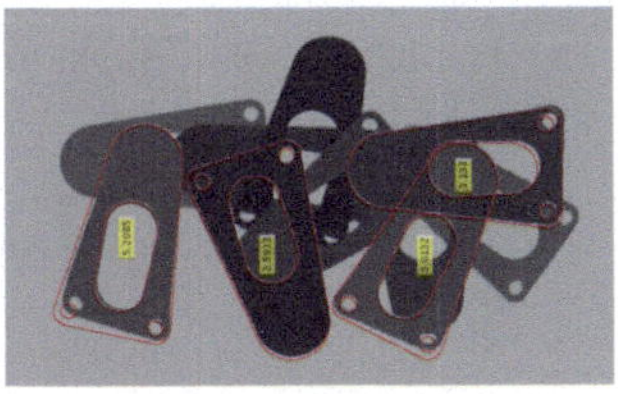

(c) Captured image using the V-REP simulator, with scores for each object.

Fig. 10. Object detection results.

images of the warehouse box. The object detection module provides a score for the several objects present in the image, and the one chosen for grasping is the one with lowest score, e.g., which is more similar to one of the trained images, i.e., not occluded with other objects, as depicted in Figs. 10(b), (c), using the vision setup depicted in Fig. 10(a). Table 1 shows the results of one hundred trials, of the vision system, with the object shown. Clearly validates the approach, with approximately 3 mm of error, perfectly suited for a vacuum gripper.

Table 1. OCM std results for cluttered scenes.

Mean score	Standard deviation	Include 98%
2.956	0.299	3.572

5.3 Knowledge Based Framework

Using the ontology and the SkiROS framework, it is possible to define tasks, by choosing the proper capabilities to be used, e.g. the *pick_PA10* and *place_PA10* that the */bin_picking_robot* must perform. These capabilities were defined in the ontology as presented in previous section. Also the tasks goals and example conditions, as depicted in Fig. 11(a) and in Fig. 11(b). The tool queries the ontology what are the capabilities that the Robot, which was declared to be in the scene, that can be part of the task definition, its conditions and goals to achieve.

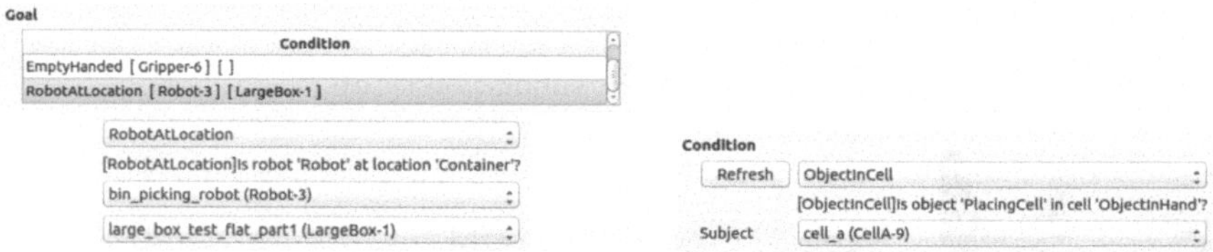

(a) Definition of a task goal. Robot at a given location.

(b) Definition of a Post condition.

Fig. 11. Goals and Conditions definitions in the framework.

The main result obtained from the SkiROS framework are the PDDL files, i.e, the domain and the problem files. Those are obtained directly from the framework, by reasoning the ontology and with the information given by the user, as presented in previous sections. The domain file, depicted in listing 5.3 presents the requirements, types, predicates and actions. There are presented the PA10 robot actions, e.g., pick and place, with start conditions and effects, along with end effects, that will enable the next capability to be used by the robot *Place_PA10* or to achieve the task goal, defined in the PDDL problem file.

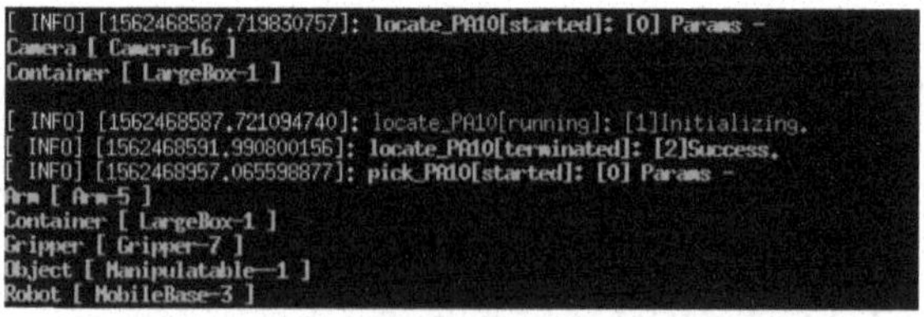

Fig. 12. Overall system running, displaying information to the user.

The problem file, presents the objects, initial state and goal specification. In this specific case ensures that the object can fit a specific cell in the Kit that must be filled with the parts. The framework can perform the planning of the task by querying the ontology and the defined world model.

Finally, in Fig. 12 is presented the start of the task execution, with information for the user, that is then capable to log information of all the components and workflow during the task execution. In this figure is also depicted the execution of the vision module outside the task plan, because of the design decision to keep the planning only for robot actions. The vision module operates continuously to locate the best object to pick, based on the score of the object detection algorithm. It is under development the inclusion of other actions in the task planner, related not only to the robot, but also to other modules that may exist is the scene, such as sensors, conveyors and so on.

Listing 1.2. The domain PDDL file generated by the framework with only the place_PA10 action.

```
(define (domain bin_picking)
(:requirements :typing)
(:types
        Agent Location Robot Arm Gripper Manipulatable Cell Kit - object
)
(:predicates
    (RobotAtLocation ?A - Agent ?L - Location)
    (EmptyHanded ?G - Gripper)
    (ObjectAtLocation ?L - Location ?M - Manipulatable)
    (Holding ?G - Gripper ?M - Manipulatable)
    (FitsIn ?C - Cell ?M - Manipulatable)
    (LocationEmpty ?C - Cell)
    (CellInKit ?K - Kit ?C - Cell)
    (ObjectInCell ?C - Cell ?M - Manipulatable)
    (InKit ?M - Manipulatable ?K - Kit)
)

(:durative-action place_PA10
    :parameters ( ?Arm - Arm ?Gripper - Gripper ?ObjectInHand - Manipulatable
                  ?PlacingCell - Cell ?PlacingKit - Kit ?Robot - Agent)
    :duration (= ?duration 1)
    :condition (and
        (at start (Holding ?Gripper ?ObjectInHand))
        (at start (FitsIn ?PlacingCell ?ObjectInHand))
        (at start (LocationEmpty ?PlacingCell))
        (at start (CellInKit ?PlacingKit ?PlacingCell))
        (at start (can_place_PA10 ?Robot))
    )
    :effect (and
        (at start (not (LocationEmpty ?PlacingCell)))
        (at start (not (Holding ?Gripper ?ObjectInHand)))
        (at end (EmptyHanded ?Gripper))
        (at end (ObjectInCell ?PlacingCell ?ObjectInHand))
        (at end (InKit ?ObjectInHand ?PlacingKit))
    )
)
)
```

6 Conclusions

The paper proposed an ontology driven framework to define capabilities and all the scenario knowledge for visual guided bin-picking tasks, capable to obtain the task goal to be achieved. The developed framework was successfully simulated and validated in a simulation scenario using V-REP and ROS. The developed framework was built based on the IEEE ORA standard and SUMO upper ontology, extending the skiROS tools. Specific contributions were developed to extend the previous ontologies to the presented application case, e.g., for interaction with environment, capabilities definition, and task planning.

The framework can perform the planning of the task by querying the ontology and the defined world model. Moreover, the user can follow the task execution, that is then capable to log information of all the components and workflow during the task execution. The execution of the vision module (object pose recognition) is performed outside the task plan, because of the design decision to keep the planning only for robot actions. As future work, it is under development the inclusion of other actions in the task planner, related not only to the robot, but also to other modules that may exist is the scene, such as sensors, conveyors.

Acknowledgments. This work was partially supported by FCT, through IDMEC, under LAETA, project UID/EMS/50022/2019. This work was partially supported by project 0043- EUROAGE-4-E (POCTEP Programa Interreg V-A Spain-Portugal).

References

1. Bay, H., Tuytelaars, T., Van Gool, L.: SURF: Speeded up robust features. In: ECCV, pp. 404–417 (2006)
2. Lowe, D.G.: Distinctive image features from scale-invariant key-points. IJCV **60**, 91–110 (2004)
3. Rapoza, K.: India: China's New Low-Cost Labor Hub? Forbes, September 2014
4. Barrow, H.G.: Parametric correspondence and chamfer matching: two new techniques for image matching. SRI Int. (1977)
5. Shotton, J.: Multiscale categorical object recognition using contour fragments. IEEE Trans. PAMI **30**(7), 1270–1281 (2008)
6. Prestes, E., Carbonera, J.L., Fiorini, S.R., et al.: Towards a core ontology for robotics and automation. Robot. Auton. Syst. **61**(11) (2013). https://doi.org/10.1016/j.robot.2013.04.005
7. Farinha, R., Gonćalves, P.J.S.: Knowledge based robotic system, towards ontology driven pick and place taks. Rom. Rev. Precis. Mech. Opt. Mechatron. **49**, 152–157 (2016). https://doi.org/10.17683/rrpmom.issue.49
8. Balakirsky, S., Schlenoff, C., et al.: Towards a robot task ontology standard. In: Proceedings of the 12th Manufacturing Science and Engineering Conference. Los Angeles, USA (2017)
9. IEEE Std. 1872-2015: IEEE Standard Ontologies for Robotics and Automation (2015). https://doi.org/10.1109/IEEESTD.2015.7084073
10. Pease, A., Niles, I., Li, J.: The suggested upper merged ontology: a large ontology for the semantic web and its applications. Working Notes of the AAAI 2002 Workshop on Ontologies and the Semantic Web, vol. 28 (2002)
11. Rohmer, E., Singh, S.P.N., Freese, M.: V-REP: a versatile and scalable robot simulation framework. In: 2013 IEEE IROS, 3–7 November 2013
12. Rovida, F., Crosby, M., Holz, D., Polydoros, A., Großmann, B., Petrick, R., Krüger, V.: SkiROS - A Skill-Based Robot Control Platform on Top of ROS. Springer Book on Robot Operating System, vol. 2. Springer, Cham (2017)
13. McDermott, D., Ghallab, M., Howe, A., Knoblock, C., Ram, A., Veloso, M., Weld, D., Wilkins, D.: PDDL—the planning domain definition language. Technical report CVC TR98003/DCS TR1165. Yale Center for Computational Vision and Control, New Haven, CT (1998)
14. Olszewska, J., Carbonera, J.L., Olivares-Alarcos, A., et al.: Ontology for autonomous robotics. In: 26th IEEE International Symposium on Robot and Human Interactive Communication (ROMAN), August 2017

Educational Robotics

Factors Influencing the Sustainability of Robot Supported Math Learning in Basic School

Janika Leoste(✉) and Mati Heidmets

Tallinn University, 10120 Tallinn, Estonia
leoste@tlu.ee

Abstract. Many countries are trying to enhance math learning by bringing technology into classroom. Educational robots are frequently considered as one of such technologies with the benefits of improving math learning motivation of students and helping them to acquire abstract math concepts. However, teachers often prefer traditional teaching methods, making innovation unsustainable. In this paper we used the feedback from 133 Estonian math teachers to study the factors that influence the sustainability of robot supported math teaching. Results indicate that there are two types of problems that teachers face when conducting robot supported math lessons. First, the problems that are caused by the initially inadequate method-related skills of teachers and students. These problems are fading over times, especially when teachers are able to use the help from a school-university partnership. Secondly, there are problems that cannot be eliminated by teachers or students by themselves, for example problems caused by shared use of robots. Our analysis of the experience of participating teachers indicates that making robot supported math teaching sustainable requires an additional systematic support from school management.

Keywords: Educational robots · Technology enhanced learning · Robot supported learning · School university partnership · Math

1 Introduction

The current revolution of digital technologies has had a tough impact on society, creating new kinds of necessities and products [1], but also exerting a lot of pressure on educational systems to provide labor market with people that have good STEM skills [2–6]. This problem is aggravated by the fact that students are losing interest towards learning STEM based domains at universities [2, 7–9]. Math, as one of the core STEM skills has been hit especially hard. For example, in Estonia, a country with an excellent PISA math score [10], the percentage of non-performers at basic school final math exam is above 20% (whereas the average percentage of non-performers at all final exams is much lower, about 10%) [11].

Many countries are trying to enhance math learning by bringing technology into classroom [4]. This approach is supported by numerous studies, demonstrating students' improved math learning motivation. For example, mobile apps have had positive results in making math learning more engaging for students in health schools [12], and inquiry based learning has improved students' math engagement [13]. One of the ways

M. F. Silva et al. (Eds.): ROBOT 2019, AISC 1092, pp. 443–454, 2020.
https://doi.org/10.1007/978-3-030-35990-4_36

for helping students to learn math is using educational robots as technological tools in math classrooms (robot supported math teaching) for visualizing math concepts [14], and supporting collaborative learning [15]. Studies indicate that while using educational robots in math classroom helps students to better understand math concepts, and encourage student collaboration, discussion, creativity and critical thinking, it will also increase their math learning motivation [16–20].

However, despite of teachers and school leaders being aware of various methods that can be used for enhancing teaching with technology, they are still inclined to use traditional lecture-based practices instead [6]. Several reasons are cited for this, starting from teachers' low STEM skills [21], immaturity of used technology [22, 23], lack of instructional materials, teacher training and support [6], to ideological differences between researchers and practicing teachers [24, 25]. School-university partnership (SUP) has been suggested as a suitable solution for overcoming these problems. Using SUP allows researchers and teachers to unify their contrasting worldviews, and to co-create a balanced and shared framework of values, knowledge and methods. SUP is also useful for establishing a scaffolding system that is relying on the peer support of both researchers and teachers [25–27]. Nevertheless, providing teachers with a framework that supports acquisition of new knowledge and teaching methods is not a guarantee for these knowledge and methods to become sustainable in practice. We felt it necessary to explore more deeply the factors that influence the sustainability of technology enhanced math learning.

In Estonia, more than 60% of primary schools have educational robots [28], and most of the schools also employ an educational technologist who is usually the first person helping teachers to implement new technologies [29]. Exploiting these available resources we designed a study that explores the sustainability problems of robot supported math teaching (Robomath).

This study is based on a SUP approach, providing teachers with a framework of Robomath method related training and ongoing support with the purpose of helping to transfer the ownership of the method from researchers to teachers and making the method sustainable in practice. The study involved 137 classes from the grades three and six. In each of these classes up to 20 Robomath lessons were conducted. These lessons were conducted by 133 math teachers who were aided by 56 educational technologists. At the end of the study the teachers' opinions about the factors influencing Robomath method's sustainability were collected and analyzed.

1.1 Research Questions

In this paper we are analyzing, based on the experience of participant teachers, the factors that influence the sustainability of Robomath method. For these purposes we are examining the problems teachers faced while using the method, the development of these problems over time, and the scaffolding that teachers considered necessary for the sustainability of Robomath method. We wanted to find out whether Robomath method is usable in real-life classroom context, if teachers are able to appropriate this method and if the method can be sustainable. For these purposes we defined following research questions:

1. What do teachers consider to be the main obstacles in applying Robomath method to the practice?
2. How do these obstacles change over time, in teachers' estimation?
3. What kind of support do teachers need in addition to learning Robomath method for this method to become a part of their practice?

We used teachers' lesson feedback diaries for answering the research questions 1 and 2. For answering the research question 3 we used semi-structured questionnaires.

2 Method

2.1 Study Design and Sample

In Estonia, the compulsory school education starts at age 7. The primary school grades 1 to 6 correspond to ISCED basic education stage [30]. We selected the grades 3 and 6 as the target of the study because for these grades the results of the national standardized math tests are available, allowing us to collect additional data about students' development.

The invitation to take part in the study was sent out by email to the principals of all Estonian primary schools in spring 2018 and the participation criteria were set as follows:

1. The school had to have at least one classroom set of certain educational robots[1].
2. The participation was allowed for the grades 3 and 6, provided that their math teachers were willing to conduct up to 20 Robomath lessons during the school year 2018/2019.
3. The school had to have an educational technologist to support participating math teacher(s), or the math teacher had to have previous educational robotics experience.

As a result of this invitation, 67 schools throughout Estonia, with 137 classes (98 in the 3rd grade and 39 in the 6th grade, with more than 2000 students) joined the study. 39% of participating schools had 101 to 500 students; 34% had 501 to 1000 students; 20% had up to 100 students; and 7% had more than 1000 students (the largest school in Estonia has 1395 students, the smallest school has 5 students [31]).

These classes were taught by 133 math teachers (96 third grade teachers and 37 sixth grade teachers) who were supported by 56 educational technologists. The statistical information about the participating teachers is following:

1. Sex: 97% women; 3% men.
2. Age distribution: 34% 50–59 years old; 33% 40–49 years old; 14% 30–39 years old; 10% 60 years or older; 9% younger than 30 years (the average age of math teacher in Estonia is 48 years [32]).
3. Previous contact with educational robotics at school: 73% of teachers had educational robotics training. However, only 32% of teachers had used educational robots at all (15% of teachers had used them more than 2 times).

[1] Description of the robotics platforms used in the study: http://bit.ly/2AvwZpB.

For participating teachers, following scaffolding options were available: (a) regional training days before the beginning of the study; (b) a special online teacher training environment eDidaktikum[2] for sharing experience with peers and seeking help from peers and researchers; (c) a co-creation focused training with six contact days that took place at the university, during which teachers co-created additional teaching materials and shared them with their peers.

The length of the study was one school year (from September 2018 to April 2019). During this period each participating class was given up to 20 Robomath lessons. The lessons were conducted by participating math teachers with the help of educational technologists of their schools when needed. For conducting Robomath lessons teachers used lesson plans that were created during the pilot study[3]. These lesson plans were available as interactive GeoGebra worksheets and as Word documents.

Each lesson plan was co-created by researchers and practicing math teachers during the pilot study and consisted of a math word problem and accompanying robotics exercises, focusing on a certain math concept. The math content of a lesson plan was based on the regular curriculum of the respective grade. The purpose of robotics exercises was to provide an environment for using math content in a real-life context. In order to help students to focus on math content, sample solution videos and coding examples with explanations were included into each lesson plan.

The scripting of lessons was relatively loose, allowing teachers to customize lesson plans and to leave out some of the robotics exercises, if needed. The lesson plans were designed to encourage student collaborative learning and student independence. Students were supposed to work in pairs. The role of the teacher was to counsel students and helping them to understand the math problem and the robotics exercises and their connections. The role of the educational technologist was to provide technical help before and during the lesson, to assist students with technical problems and to answer students' questions in case teacher was already occupied elsewhere.

2.2 Data Collection

For the purposes of answering our research questions we collected data using teachers' lesson feedback diaries and semi-structured teacher questionnaires. For analyzing the collected data we used content analysis method [33].

Teachers' Lesson Feedback Diaries. Teachers were asked to record their observations after each conducted lesson, using a Google Forms questionnaire. By the end of the study, in spring 2019, there were 1095 feedback entries. We sorted these entries by their time-stamps and analyzed the content, searching for information about the problems teachers experienced in the lessons and about the factors influencing the development of these problems. In total, 248 entries contained relevant information that was further coded and classified into 3 major categories, each having several significant clusters of meaning:

[2] https://edidaktikum.ee/en/home.

[3] Sample lesson plans (in English): http://bit.ly/2Q1n9AS.

1. Method related problems (Fig. 1), having 6 clusters of meaning.
2. Hardware related problems (Fig. 2), having 5 clusters of meaning.
3. Factors that influenced the problems (Fig. 3), having 5 clusters of meaning.

Semi-structured Teacher Questionnaires. In spring 2019, before the end of the school year, we asked participating teachers to express their opinions about the sustainability of Robomath method, using semi-structured teacher questionnaires via e-mail. Based on our research questions we formed questions for the questionnaire:

- What were the biggest problems for you in Robomath lessons?
- When comparing the first and last Robomath lessons, what changed the most?
- Do you consider it necessary and possible to continue with Robomath lesson in the next academic year?
- If you would like to continue with Robomath lessons in the new school year, what kind of support or help would you need?
- What would support the sustainable development and implementation of Robomath method in your school?

This questionnaire was filled by 30 teachers. For analysis of received information we used open coding, resulting in 8 clusters of meaning (Fig. 4).

3 Results

3.1 Question 1: What Do Teachers Consider to Be the Main Obstacles in Applying Robomath Method to the Practice?

We used teachers' lesson feedback diaries for answering this question. During the process of open coding we decided to distinguish method related problems from directly hardware related problems in order to have a more accurate understanding of particular problems.

We noticed that the greatest method related problems (Fig. 1) were students' inability to understand multiple sentence text tasks, their insufficient programming and robotics skills, their discomfort with method-related teaching practices, and a need for an assistant teacher. All of these issues were most pronounced in the first Robomath lessons and were somewhat interrelated, caused by new teaching practices that are relatively rare in regular math lessons: e.g. collaborative learning and peer support. It also introduced a learning object for students to apply their math knowledge to, requiring students to convert their math knowledge into the context of robotics and programming. The lack of students' experience of using their math skills outside of the regular math classroom context initiated thorough discussions and brought about a number of questions that teachers were not able to handle alone.

To a much lesser extent there were other problems present: it was sometimes difficult to form teams as students preferred to work individually, and it was not always easy for students and teachers alike to comprehend the connection between robotics exercises and underlying math constructs.

All of these problems were alleviated in later lessons as the confidence and skills of teachers and students grew. However, some teachers pointed out that students with special educational needs (SEN students) were unable to use the same teaching materials as other students and therefore got tired of using Robomath method.

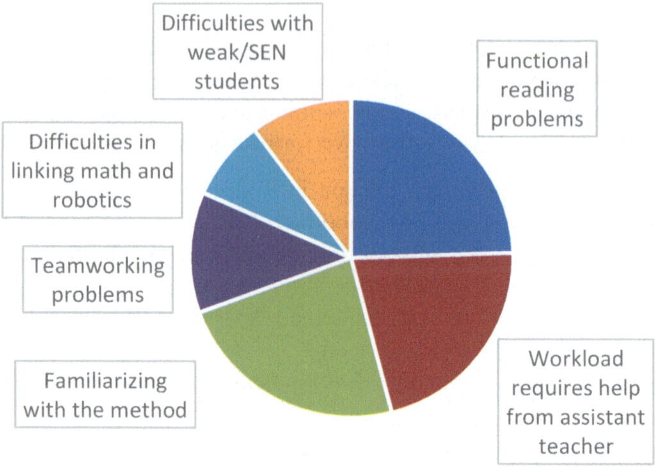

Fig. 1. Method related problems when using Robomath method.

Almost a half of the technical problems (Fig. 2) were robot platform specific. For example, robots did not drive straight, robot sensors functioned unreliably under certain conditions (poor lightning, empty battery, etc.), and situations where robot's erratic behavior could not be explained.

Still, more than half of the technical problems pointed at different issues. For example, in some schools the robots were shared with afterschool robotics clubs or were used by several classes, requiring robots to be reconstructed at the beginning of each Robomath lesson, or causing robots to malfunction due to empty batteries. There were also lesser problems related to computers and tablets that were used for programming robots, connection failures, and problems with technical infrastructure.

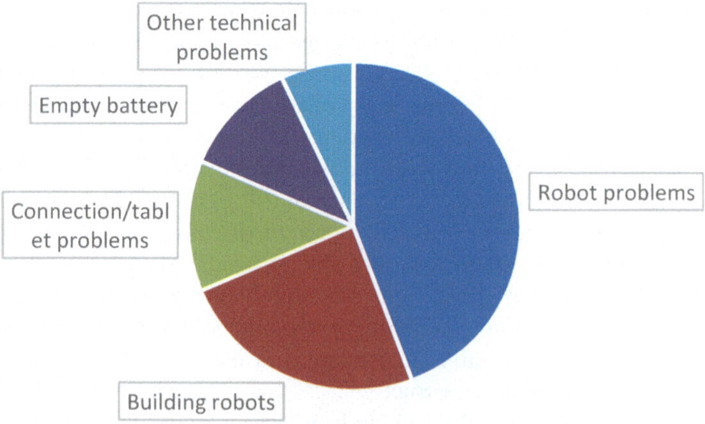

Fig. 2. Hardware related problems when using Robomath method.

3.2 Question 2: How Do These Obstacles Change Over Time, in Teachers' Estimation?

The data from teachers' feedback diaries shows that most of the problems teachers had in the first Robomath lessons were not present in the later lessons. Also, although the technical problems continued to emerge throughout the study period, they did not disturb the lesson flow anymore. There are several reasons for these developments (Fig. 3).

We feel that besides becoming accustomed to using the method, the major influencing factor was the growth of students' independence combined with their increased skills of programming, robotics and ability to connect math and robotics. Teachers pointed out that due to this growth of independence and skills the students became able to solve most of the technical problems on the fly and instead of asking teachers for help they were able to support and counsel each other.

Teachers started to see how the lesson plans meaningfully connected math and robotics, and began adapting teaching material to the needs of their students, making exercises easier to understand for weaker students and further reducing the need for an assistant teacher. In some occasions the adaptions made use of Robomath method in an entirely different context: i.e. they conducted Robomath days or weeks or Robomath demonstration days for their colleagues and parents.

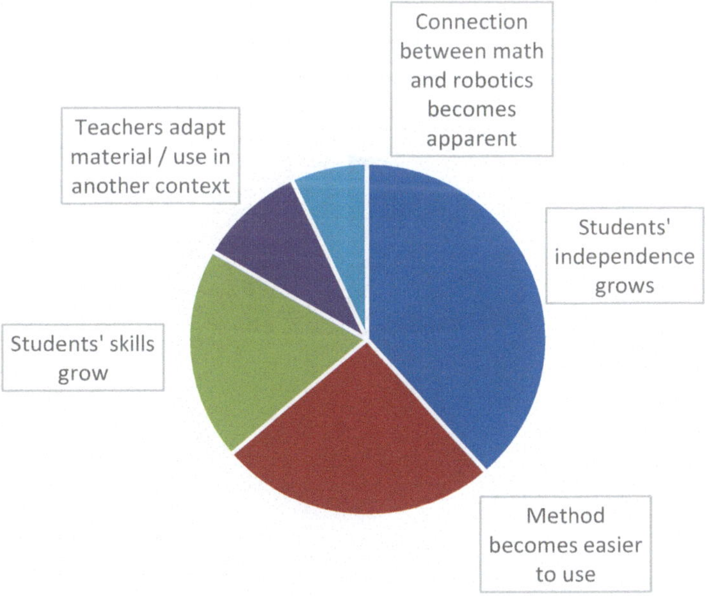

Fig. 3. Factors that influenced the obstacles when using Robomath method.

3.3 Question 3: What Kind of Support Do Teachers Need in Addition to Learning Robomath Method for This Method to Become a Part of Their Practice?

In order to answer this question we analyzed the data that was collected from teachers with semi-structured questionnaires. The data pointed out that the most important factors ensuring the sustainability of the method were having a support person and teaching materials (Fig. 4). The help that was expected from the support person, was to facilitate preparing lessons, to provide technical assistance, but also to take part in students' lesson discussions, answering their questions and giving advice.

The next critical element was having enough technical resources. In some of the participating schools the robots and tablets were used by several teachers and after-school clubs, causing tensions between colleagues, wasting time when re-building robots for math lessons, and, sometimes, ruining experiments when there was too little time between robot/tablet uses to recharge the batteries.

Teachers considered it also very important that their schools as whole would have positive attitudes towards technology enhanced learning (TEL). They regarded it essential to have the emotional support from their colleagues, to see more teachers engaged with similar teaching approaches, and to get full management support in ensuring Robomath method's sustainability. To a lesser extent, some teachers also found it necessary to get method-related training and allocating additional time for Robomath lessons – both of these factors are controlled by school's management.

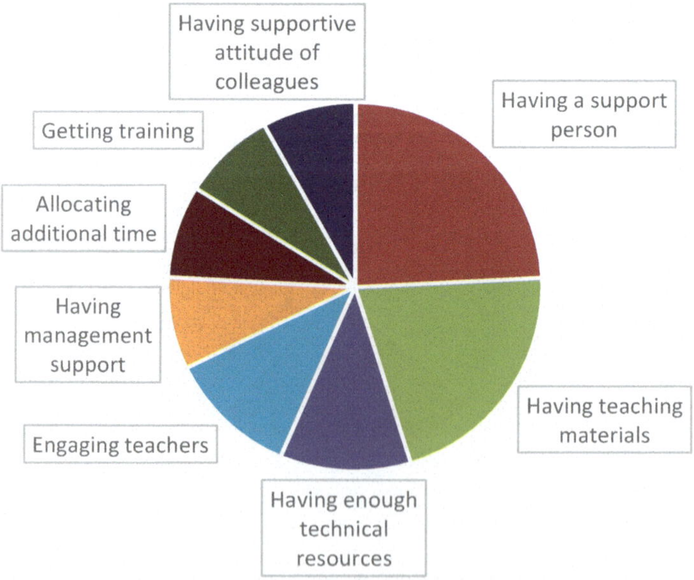

Fig. 4. Areas that need support when using Robomath method.

4 Conclusions and Discussion

In order to better understand the factors influencing sustainability of TEL we conducted a study in Estonia during the school year 2018/2019, examining the sustainability aspects of robot supported math teaching. In order to answer our research questions we used data from the lesson feedback diaries of 133 teachers and semi-structured questionnaires, completed by 30 teachers.

First we explored the problems that made conducting Robomath lessons difficult. We found that using educational robots was initially difficult for both teachers and students. As most of participating math teachers had no previous meaningful experience with educational robotics, they had to acquire basic programming and robotics skills, and to learn new methods suitable for technology enhanced teaching. Teachers also had to be able to counsel students, solve technical problems, and demonstrate the connections between math constructs and robotics exercises. All of these tasks required the help of educational technologists, at least in the beginning. However, this help could not alleviate the problems that were caused by shared use of robots and tablets, for example, empty batteries and incorrectly built robots.

Secondly we examined how and why these problems changed in time. While teachers in general recognized that using Robomath method became easier over time, we discovered that the growth of student skills and their independence allowed them to better understand exercises, to overcome most of the technical problems and to support each other, reducing thus teacher's workload. We could thus distinguish two general types of problems: those that faded over time (as the method related skills of students and teachers grew), and those that did not (the problems with SEN students and robot-related technical problems).

Thirdly we studied the factors teachers considered important for the sustainability of Robomath method. We found that helping teachers to get started had the greatest importance: they needed ready-made teaching materials, enough robots and tablets, and support persons who were able to assist teachers with both technical and methodological aspects. However, for continuous use of the method, it is necessary to have supportive attitudes of colleagues and managements, combined with an overall systematic approach that involves allocating extra paid time for conducting Robomath lessons, providing training, and engaging other teachers.

Based on the opinions of the teachers that participated in our SUP-based study we conclude that SUP approach can help implementing TEL methods into math classroom by providing initial training and teaching materials. However, providing teachers with a comprehensive method-related training does not guarantee the sustainability of method if teachers have to face method-induced problems alone (e.g. the problems of allocating time for preparing the lessons or problems caused by shared use of resources). Removing most of such obstacles requires conscious and systematic effort from the school that is trying to implement TEL methods. Besides providing enough technical resources, learning material and technical support persons, it is also necessary to engage a significant number of teachers, so that they could eventually scaffold each other as a community.

5 Limitations and Future Work

There are several areas of potential improvement in our study. First of all, our results rely on the opinions of math teachers only. For better understanding of the problem, also information from schools' management level should be collected. Secondly, for more reliable results, the study's length should be several years. Thirdly, the sample size could be enlarged by involving different grades, allowing thus getting more exact results, and by ensuring higher levels of feedback from participants.

This paper focused on using educational robots for supporting math teaching. It would be useful to study whether there are educational technologies more appropriate for implementing TEL approach in real life classrooms, and whether such an approach would yield better results if the technology would be used in lessons that integrate different subjects, including math.

In this study the teachers were provided with ready-made teaching materials. There are studies that recommend teacher co-creation as a way of ensuring better sustainability of education. It would be necessary to study whether such an approach would work if co-creation would be a result of hands-on learning of a technological domain such as robotics that is relatively unfamiliar to teachers.

We are currently launching a follow-up study that takes place from autumn 2019 to spring 2020. In this follow-up study we are examining the sustainability of TEL implementation when teachers conduct robot supported math lessons by using the materials they have co-created during monthly training sessions that are organized by researchers.

Acknowledgments. This project has received funding from the European Union's Horizon 2020 research and innovation programme under grant agreement No. 669074.

References

1. Schwab, K.: The Fourth Industrial Revolution: what it means, how to respond. World Economic Forum (2016)
2. The Commonwealth of Australia: Australia's National Science Statement. https://publications.industry.gov.au/publications/nationalsciencestatement/national-science-statement.pdf. Accessed 11 May 2019
3. European Commission: Employment and Social Developments in Europe Annual Review (2018). https://ec.europa.eu/social/BlobServlet?docId=19719&langId=en. Accessed 11 May 2019
4. Mullis, I.V.S., Martin, M.O., Loveless, T.: 20 Years of TIMSS: International Trends in Mathematics and Science Achievement, Curriculum, and Instruction. Boston College, Chestnut Hill (2016)
5. OECD: The Future of Education and Skills. Education 2030. OECD Publishing (2018). https://www.oecd.org/education/2030/E2030%20Position%20Paper%20(05.04.2018).pdf. Accessed 11 May 2019
6. UNESCO: School and Teaching Practices for Twenty-First Century Challenges. Lessons from the Asia-Pacific Region - Regional Synthesis Report. UNESCO Bangkok (2016)

7. Bacchus, A.: New Microsoft research points to the declining interest of girls in STEM, ways to close the gender gap. OnMSFT (2018). https://www.onmsft.com/news/new-microsoft-research-points-to-the-declining-interest-of-girls-in-stem-ways-to-close-the-gender-gap. Accessed 11 May 2019
8. Elliot, D.: STEM interest declining among teens. CBS Interactive Inc. (2013). https://www.cbsnews.com/news/stem-interest-declining-among-teens/. Accessed 11 June 2019
9. Ernst & Young Global: Research Reveals Boys' Interest in STEM Careers Declining; Girls' Interest Unchanged. Ernst & Young Global Ltd. (2018)
10. OECD: PISA 2015 Results (Volume I): Excellence and Equity in Education. PISA. OECD Publishing, Paris (2016). https://doi.org/10.1787/9789264266490-en
11. Palu, A., Kikas, E.: Matemaatikapädevus. Kikas, E. (Toim.), Õppimine ja õpetamine kolmandas kooliastmes. Üldpädevused ja nende arendamine. Eesti Ülikoolide Kirjastus OÜ (2015)
12. Willacy, H., Calder, N.: Making mathematics learning more engaging for students in health schools through the use of apps. Educ. Sci. **7**(2), 48 (2017)
13. Fielding-Wells, J., Makar, K.: Student (dis)engagement in mathematics. In: Conference Proceedings: Australian Association for Research in Education, At Brisbane, Australia (2008)
14. Savard, A., Highfield, K.: Teachers' talk about robotics: where is the mathematics? In: Proceedings of the 38th Annual Conference of the Mathematics Education Research Group of Australasia. Mathematics Education Research Group of Australasia (2015). https://www.merga.net.au/documents/RP2015-60.pdf. Accessed 11 June 2019
15. Gerretson, H., Howes, E., Campbell, S., Thompson, D.: Interdisciplinary mathematics and science education through robotics technology: its potential for education for sustainable development (a case study from the USA). J. Teach. Educ. Sustain. **10**(1), 32–41 (2008)
16. Han, I.: Embodiment: a new perspective for evaluating physicality in learning. J. Educ. Comput. Res. **49**(1), 41–59 (2013)
17. Kennedy, J., Baxter, P., Belpaeme, T.: Comparing robot embodiments in a guided discovery learning interaction with children. Int. J. Soc. Robot. **7**(2), 293–308 (2015)
18. Kopcha, T.J., McGregor, J., Shin, S., Qian, Y., Choi, J., Hill, R., Mativo, J., Choi, I.: Developing an integrative STEM curriculum for robotics education through educational design research. J. Form. Des. Learn. **1**(1), 31–44 (2017)
19. Werfel, J.: Embodied teachable agents: learning by teaching robots. In: Conference Proceedings (2014). http://people.seas.harvard.edu/~jkwerfel/nrfias14.pdf. Accessed 8 Aug 2018
20. Leoste, J., Heidmets, M.: Õpperobot matemaatikatunnis. MIKS.EE. Estonian Research Council (2019)
21. Rasinen, A., Virtanen, S., Endepohls-Ulpe, M., Ikonen, P., Judith Ebach, J., Stahl-von Zabern, J.: Technology education for children in primary schools in Finland and Germany: different school systems, similar problems and how to overcome them. Int. J. Technol. Des. Educ. **19**, 367 (2009)
22. Peters, V.: Preparing for Change and Uncertainty. The Oxford Handbook of Technology and Music Education. Oxford University Press, Oxford (2017)
23. Banke, J.: Technology Readiness Levels Demystified. NASA (2010)
24. Arhar, J., Niesz, T., Brossmann, J., Koebley, S., O'Brien, K., Loe, D., Black, F.: Creating a 'third space' in the context of a university–school partnership: supporting teacher action research and the research preparation of doctoral students. Educ. Action Res. **21**(2), 218–236 (2013)
25. Korthagen, F.: The gap between research and practice revisited. Educ. Res. Eval. **13**(3), 303–310 (2007)

26. Dimmock, C.: Conceptualising the research–practice–professional development nexus: mobilising schools as 'research-engaged' professional learning communities. Prof. Dev. Educ. **42**(1), 36–53 (2016)
27. Ley, T., Leoste, J., Poom-Valickis, K., Rodríguez-Triana, M. J., Gillet, D., Väljataga, T.: CEUR Workshop Proceedings (2018). http://ceur-ws.org/Vol-2190/CC-TEL_2018_paper_1.pdf. Accessed 11 May 2019
28. HITSA: ProgeTiiger programmis toetuse saanud haridusasutused 2014–2018. https://www.hitsa.ee/ikt-haridus/progetiiger. Accessed 11 May 2019
29. Lorenz, B., Kikkas, K., Laanpere, M.: The role of educational technologist in implementing new technologies at school. In: Zaphiris, P., Ioannou, A. (eds.) Learning and Collaboration Technologies. Technology-Rich Environments for Learning and Collaboration. LCT 2014. Lecture Notes in Computer Science, vol. 8524. Springer, Cham (2014)
30. Statistics Estonia: Mõisted ja metoodika (2018). http://pub.stat.ee/px-web.2001/Database/RAHVASTIK/01RAHVASTIKUNAITAJAD_JA_KOOSSEIS/04RAHVAARV_JA_RAHVASTIKU_KOOSSEIS/RV_0231.htm. Accessed 15 Mar 2019
31. Estonian Ministry of Education and Research: 2017/2018 õppeaasta arvudes. https://www.hm.ee/sites/default/files/2017-2018_oppeaasta_arvudes.pdf. Accessed 11 May 2019
32. Tõnisson, A.: Matemaatika – sidur, pidur, gaas. Õpetajate leht (2019)
33. Duriau, V.J., Reger, R.K., Pfarrer, M.D.: A content analysis of the content analysis literature in organization studies: research themes, data sources, and methodological refinements. Organ. Res. Methods **10**, 5–34 (2007)

Teaching Mobile Robotics Using the Autonomous Driving Simulator of the Portuguese Robotics Open

Valter Costa[1,2(✉)], Peter Cebola[1], Pedro Tavares[2], Vitor Morais[4], and Armando Sousa[2,3]

[1] INEGI - Institute of Science and Innovation in Mechanical and Industrial Engineering, Porto, Portugal
ee09115@gmail.com
[2] FEUP - Faculty of Engineering of the University of Porto, Porto, Portugal
[3] INESC TEC—INESC Technology and Science (formerly INESC Porto), Porto, Portugal
[4] Department of Electrical and Computers Engineering, Faculty of Engineering of the University of Porto, Porto, Portugal

Abstract. Teaching mobile robotics adequately is a complex task. Within the strategies found in the literature, the one used in this work includes the use of a simulator. This simulator represents the Autonomous Driving Competition of the Portuguese Robotics Open. Currently, the simulator supports two different robots and all challenges of the autonomous driving competition. This simulator was used at a Robotics course of the Integrated Master Degree in Informatics and Computing Engineering at the Faculty of Engineering of the University of Porto. In order to study the influence of the simulator in the college students learning process, a survey was conducted. The results and its corresponding analysis indicate that the simulator is suited to teach some of the mobile robotics challenges crossing several fields of study, including image processing, computer vision and control.

Keywords: Educational robotics · Autonomous driving simulator · Portuguese Robotics Open · Gazebo · ROS · Mobile robotics · Autonomous Driving Competition

1 Introduction

Teaching robotics to college students efficaciously represent is a daunting task. Within the strategies that may be employed to teach robotics, the one chosen in this work includes the use of a simulator. In previous work [8,9] an autonomous driving simulator inspired in the Autonomous Driving Competition—ADC—of the Portuguese Robotics Open [19]—PRO—was proposed. In that work, the simulator was capable to represent the track and some of the driving challenges of the ADC. As for the simulated robot, a 3D model was built, representing

M. F. Silva et al. (Eds.): ROBOT 2019, AISC 1092, pp. 455–466, 2020.
https://doi.org/10.1007/978-3-030-35990-4_37

the real robot named Conde, which has participated in the 2016, 2017 and 2018 editions of the PRO. Conde uses a differential drive steering locomotion system, in which its movement changes by varying the relative rate of rotation of its wheels. To balance the robot's structure, a castor wheel placed at the end of the robot. In order to detect and navigate along the track, two cameras (pointed down) were used; and to detect and recognize the signalling panels/traffic signs another one (pointed up) was also used. Currently, this simulator is used in the Robotics course of the Integrated Master Degree in Informatics and Computing Engineering at the Faculty of Engineering of the University of Porto since the 2017 edition. Although this simulator is able to replicate with rigour some of the ADC challenges of the PRO, it only had support for one robot - Conde. Moreover, in the educational context, it also lacks an analysis to infer the effects of using it in the task of learning mobile robotics. To overcome the aforementioned challenges, this work presents an update to this simulator, including the support of all challenges of the ADC of the PRO according to the rules of the 2019 edition, and a new robot called "Major Alvega" was also added. To also perform an analysis in the educational impact of the simulator, a survey was performed to the students that used this simulator. The survey results' and its corresponding analysis are also presented in this work.

The remainder of this paper is organized as follows: Sect. 2 presents the background work on simulators used to teach robotics; the design of the simulator and survey are described in Sect. 3; the results and its corresponding discussion are shown in Sect. 4; finally, Sect. 5 presents the conclusions and future work.

2 Background Work

Hardware-based solutions to teach robotics are usually costly and potentially difficult to distribute between students. To overcome these limitations, the use of simulators are a well known alternative. Simulators play a very import role in robotics, as concluded in [12], boosting the development process. In a pedagogical context, simulators may be used to introduce and teach some tools and challenges of robotics.

Blank et al. proposed "Pyro", which is a Python-based programming environment for teaching robotics [1]. According to their authors, Pyro was created to teach mobile robotics in a "coherent, abstract, and robot-independent manner". Pyro main goal was to reduce the learning curve to program robots by creating conceptualizations that are independent of specific robot platforms. By providing a tool that abstracts from the robot used, it was expected that the students had more time to work on the task of modelling robot "brains", discarding the specifics of robot's hardware.

RobLib [15] was released in 2000 and was designed to teach robot modelling and control to undergraduate students. No feedback from the students that used RobLib was collected.

In [14], a programming environment named Jago was created to teach problem solving, computing and information technology. This environment enables the students to code their programs in Java, and run their programs in a graphic simulator before deploying in a real robot (Lego Mindstorms). Assignments were created, and the students are challenged to solve simple tasks like solving a maze, mine fields, etc. in the simulator before testing their programs in Lego Mindstorms robot.

USARSim [5] was proposed in 2006 as a robot simulator for research and education. Albeit no evaluation was made regarding its teaching capabilities, this simulator is well known by the Robocup community.

Simbad [17] is a 3D robot simulator created for scientific and educational purposes. According to their authors, Simbad was designed for students and researchers that want to develop and test systems based on situated artificial intelligence and machine learning in the context of Autonomous Robotics and Autonomous Agents.

ERBPI—Easy Robot Behaviour Programming Interface was proposed in [2]. The main idea of this application is to substitute the imperative programming paradigm and use a behaviour-based approach. The behaviours are accomplished using a connectionist paradigm by establishing configurable connections between sensors and actuators. This application was designed for high school students.

The design of a robotics course was explored in [4], using Gazebo and JdeRobot middleware [3]. In that work, several exercises were created, including vision-based control and local navigation algorithm. The main conclusion that the authors of the proposed robotics teaching experience arrived is that the use of simulators like Gazebo are a good platforms for the students to learn robotics.

A robotics course targeting high school students using V-REP and LabVIEW was reported in [16]. The course's main goal was to introduce some of robotics challenges to high school students including the building of a robot. It is also reported that after taking the course, the students are capable of building simulations and control algorithms for their own robots with limited supervision.

Rviz was the tool chosen to teach robotics and ROS—Robotics Operating System in [7]. In this work, a collaborative game world was created, and the intrinsic competitive mindset of the game was explored, making the students able to solve the challenges through cooperation and collaboration.

RobWork [11] is a simulator created for robotics research and education, specially designed for working with manipulators and dextrous hands. Several add-ons packages, including a graphical user interface, dynamic simulation and hardware interfacing are also supported by this simulator. Despite their authors stated that received feedback from hundreds of students, no analysis was performed in the pedagogical context of teaching mobile robotics.

Another simulation environment for the ADC of the PRO based on MORSE—Modular Open Robots Simulation Engine—was proposed in [13]. This simulator was also designed with the aim of teaching some of the mobile robotics challenges, however no evaluation was made to the impact of the simulator in the robotic learning process.

Although several simulators and tools have been designed for robotics research and education, only a few works describe the learning experience.

In this work, both the simulator and its influence on the learning process is studied and presented.

3 Materials and Methods

This section presents the development details of the new robot model - Major Alvega - and the updates performed to the simulation world in order to support all challenges present in the ADC of the PRO. The survey performed to the students is also shown.

3.1 Autonomous Driving Simulator Design

The Autonomous Driving Competition of the Portuguese Robotics Open aims to replicate the real challenges of self-driving cars. It consists on an autonomous robot being able to perform various tasks without the input of human intervenients, such as: travelling along a closed track (approximately size of $17 \times 7\,\text{m}$), detecting signalling panels (traffic lights), avoiding obstacles, identifying vertical signs (traffic signs), parking in two different parking areas, and circumventing a work zone delimited by small traffic cones. To circumvent the building of a real track to test the robot, which can be impracticable due to logistics and/or financial cost, a simulator was developed. The simulator includes two different parts: design of the simulation world, and design of the robot models.

World Design - Autonomous Driving Competition of the Portuguese Robotics Open. As mentioned in the Introduction section, the first version of the simulation environment [8,9] was capable of replicate some of the challenges[1] (D1, D2, and P1) of the PRO's ADC. In this work, the simulation world was updated to support all of the challenges (see Footnote 1) defined in the competition: D1, D2, D3, D4, B1, B2, P1, P2, and V1. Figure 1 displays two different views of the simulation world, with various models to represent the signalling panels (the panels near the crosswalk), the obstacles (green blocks on the track), different traffic signs placed around the track, parking area (identified with a "P"), construction cones (the orange cones in the first half of the track), etc. By adding support to more challenges, it is expected to raise more interest and enthusiasm the students in the mobile robotics field of study. This simulator is open-source and freely available[2] for downloading to anyone.

[1] The challenges are defined in the competition ruling at: https://web.fe.up.pt/~robotica2019/images/fnr2019_Autonomous_Driving.pdf.
[2] Source code available at: https://github.com/ee09115/conde_simulator.

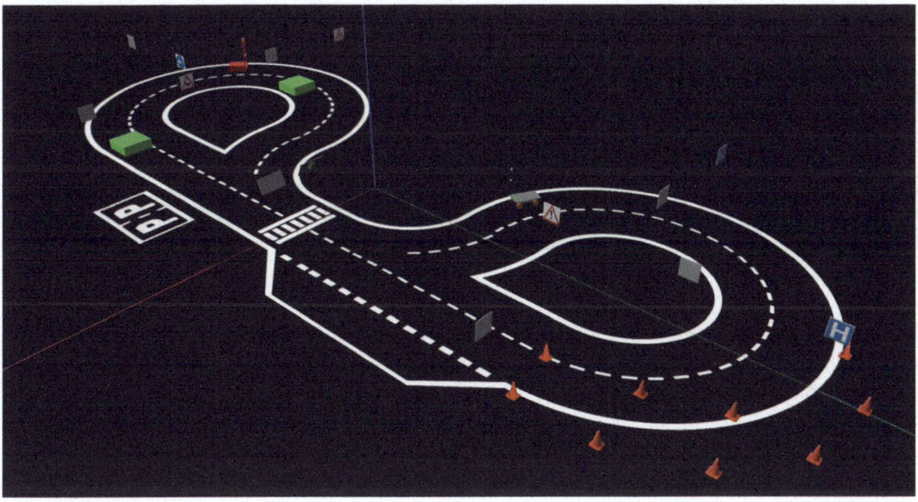

Fig. 1. Simulation world of the autonomous driving simulator for the Portuguese robotics open.

Robot Model Design - Major Alvega. Figures 2(a) and (b) present the robot Conde that is currently supported by the simulator. This simulator was used to boost the development process of the subsystems that form the Conde system architecture, and to overcome the necessity of physical space of the competition track. This simulator has proven to be a good approximation to the reality [10] and all of the software developed for the simulator was used in the real robot, with no need for software changes. Since Conde does not act as a real car due to its differential drive locomotion system and the underlying objective is to mimic a standard real-life car, another robot named Major Alvega was added.

Major Alvega is an Ackermann steering driving robot equipped with two cameras (one pointed up to see the signalling panels and traffic signs, and the other pointed down to see the track). Both cameras have fisheye lenses to increase the cameras Field Of View—FOV. Figure 2(c) shows the picture of the real robot. This robot has participated in the 2019 edition of the ADC of the PRO.

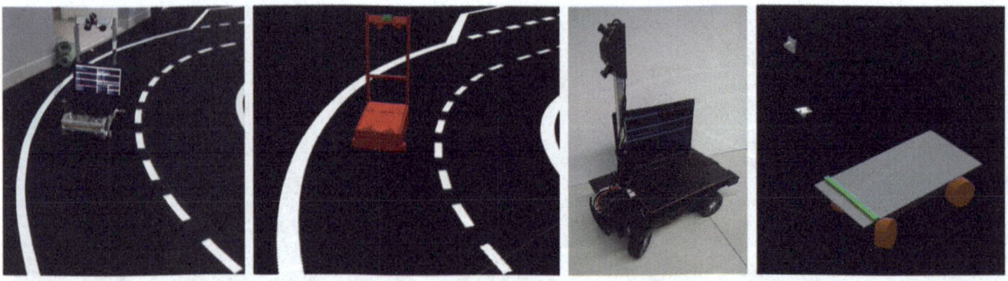

(a) (b) (c) (d)

Fig. 2. The robots supported by the simulator: (a) Conde robot on the competition track; (b) simulation of Conde robot on the competition track; (c) Major Alvega robot; (d) simulation of Major Alvega robot on the competition track.

Major Alvega System Architecture. All the subsystems that constitute this robot were developed under the ROS—Robotics Operating System—framework. The ROS system architecture for Major Alvega robot is represented in Fig. 3. Four of these nodes are of major importance to the operation of the robot, that warrants a detailed description. The first one is major_signaling_panel node that subscribes the topic /major_top_camera/image_raw and is responsible to detect and identify the signalling panels and/or vertical signs that appear around the track. The second one is the major_tracking node which is in charge of detecting and processing the images of the track published by the major_tracking_camera. The processed data from these two nodes are published in two topics (/signalling_panel_msg and /tracking_msg) that are subscribed by the /major_decision node. The decision node possesses the "intelligence" of the robot. From the processed information of the cameras, this node is responsible to calculate the velocity/angle references needed for the control node. Lastly, from the references subscribed in the /major_msg topic, the /major_control node is responsible to calculate and update the linear velocity and steering angle of the Major Alvega robot. Finally, this information is published in the /ackermann_cmd topic which updates the robot simulation in the Gazebo world.

The objective behind the division of these tasks in four main nodes is to simplify the solving of a complex problem by solving smaller problems. Moreover, this strategy facilitates the tasks division by each group of students (usually from 2 up to 4 students) that intend to use this simulator. By providing two different development platforms (Conde and Major Alvega robots) it is expected to: (i) introduce some of the challenges of mobile robotics area and, at the same time, cross several fields of study, including image processing, computer vision, control systems, artificial intelligence, etc.; (ii) captivate more students to the mobile robotics area.

The next section describes the target audience and details how the simulator was used in the 2017/2018 edition of the Robotics course.

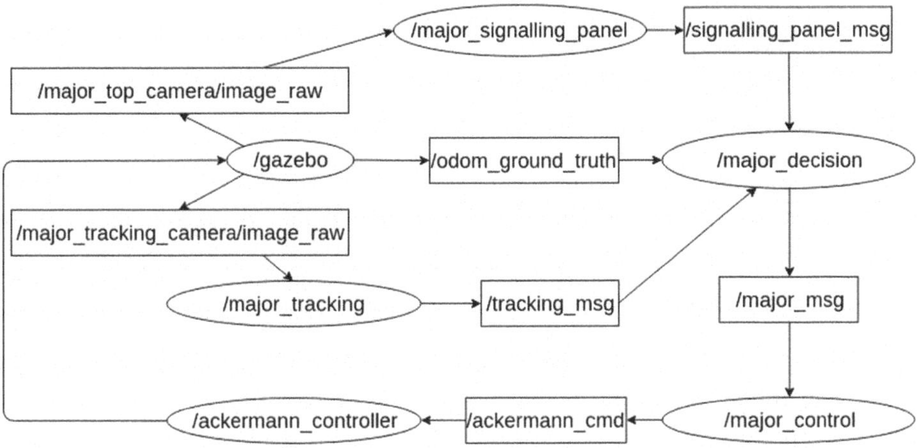

Fig. 3. Simplified ROS architecture for the Major Alvega simulator.

3.2 Sample Audience

The simulator was used in a course of the Master in Informatics and Computing Engineering at the Faculty of Engineering of the University of Porto in Portugal. The program is a 5 year integrated MSc. degree and the course where the tool was used is Robotics[3] and is optional of the 5th year.

The mentioned course starts with a general overview of robotics, then goes to a very quick generic introduction to Robotic Operating System (ROS) after which the students simulate a tiny reactive robot. Latter on, 5 weeks of the semester are mostly dedicated to a course project that involves making the simulator work, example course projects include simulated participation or augmenting features of the simulator (as mentioned earlier, the simulator is Open software). The course has 24 students but only 7 of them used the simulator, other elected other types of work. Each group has 2 or 3 students according to difficulty and work involved in the informal course work contract that each group must propose and get approved by course professors. Grading issues include development, working features, article and public presentation.

4 Results

Regarding the new robot model—Major Alvega—no tests were made to demonstrate how close the developed robot model is to the real robot. However, the authors can ensure that all the systems (ROS nodes) that were developed using this simulator were deployed in the real robot during the 2019 edition of the Autonomous Driving Competition of the Portuguese Robotics Open with no

[3] https://sigarra.up.pt/feup/en/UCURR_GERAL.FICHA_UC_VIEW?pv_ocorrencia_id=420040.

need to changes. This fact indicates that the robot model is capable of representing accurately the real robot.

4.1 Survey

To evaluate the influence and impact of the use of this simulator to teach mobile robotics, the a survey was performed to the students. This was done anonymously but in person, right after the public presentation and in paper format. With this approach, 100% of all possible responses were gathered. The survey was conducted in individually in Portuguese language, with the following questions:

Q1 - Which was the greatest difficulty felt during the installation/use of the simulator?
Q2 - Did the documentation provided proved to be useful? What would you add/change to improve it?
Q3 - What do you think of this simulator as a learning tool for mobile robotics?
Q4 - Which fields of study connected to mobile robotics did you learn?
Q5 - In a scale of 1 to 5, where 5 is strongly recommend and 1 is do not recommend, do you recommend this simulator to teach some of the challenges present in mobile robotics?
Q6 - Will you recommend to keep or remove the simulator in the next edition of the robotics course?
Q7 - Which are the weakest points of the simulator?

The first question, Q1, is to infer the quality of the instructions provided for the installation of this simulator. As in any software project, the documentation is of utmost importance, specially when using third-party created code. Question 2 aims to deduce how good is the documentation provided. It is noteworthy that the main goal of this simulator is to introduce and teach some of the challenges of mobile robotics, Question 3, Q3, intends to evaluate how adequate is this simulator to teach mobile robotics. Following on from the previous question, Q4 attempts to assess which areas of robotics did the students learn. Q5 has the purpose to evaluate quantitatively how good is the simulator to teach mobile robotics. Question 6 main goal is to objectively evaluate the necessity and the quality of the simulator to teach robotics. Finally, Q7 is an open question for the students to report which are the weakest points of this simulator, for future improvement.

This survey was performed to three groups of students that used this simulator, up to the total of 7 students. The responses to these questions are shown in Table 1.

Table 1. Responses on the survey performed to the students (Sx stands for Student x).

<table>
<tr><th rowspan="2">Question</th><th colspan="7">Responses</th></tr>
<tr><th>S1</th><th>S2</th><th>S3</th><th>S4</th><th>S5</th><th>S6</th><th>S7</th></tr>
<tr><td>Q1
Difficulty</td><td colspan="7">Easy to install.</td></tr>
<tr><td>Q2
Documentation</td><td colspan="7">The documentation was useful.
Add an "How to use it?";
Add short description to each ROS node.
Add information about IPM — Inverse Perspective Mapping.</td></tr>
<tr><td>Q3
Learning tool</td><td colspan="3">Simulator is relevant.</td><td colspan="2">Simulator is complete.</td><td colspan="2">Simulator is accessible.</td></tr>
<tr><td>Q4
Rel. Fields</td><td colspan="7">Image Processing, Computer Vision and Control Systems.</td></tr>
<tr><td>Q5
Recommend</td><td>4</td><td>4</td><td>4</td><td>4</td><td>4</td><td>5</td><td>5</td></tr>
<tr><td>Q6
Keep</td><td colspan="7">Keep the simulator.</td></tr>
<tr><td rowspan="2">Q7
Weak Point</td><td colspan="7">Documentation.</td></tr>
<tr><td colspan="2">The track is hard to solve.</td><td colspan="5"></td></tr>
</table>

The merged cells in questions Q1, Q2, Q3, Q4, Q6 and Q7 represent the number of students that answered the same response for those questions. For instance, in Q1 all students answered that the simulator is easy to install, in Q3 two students (S6 and S7) state that the simulator is accessible, etc.

The discussion of these results are presented in the following section.

4.2 Discussion

Analysing the answers presented in Table 1, it is possible to see that all of the students thought that the simulator is easy to install (Q1). Regarding the question about documentation, Q2, all of the students stated that the documentation provided was useful. Three of the students have made three suggestions: (i) add to the documentation an "how to use it file"; (ii) add a short description to each ROS node; and (iii) add information about IPM. With respect to suggestion (i), several videos were created and are publicly available that describe which challenges can be tackled when using this simulator. The links to those videos are included in the repository (see Footnote 2) README file. Also, concerning suggestion (ii), a section was added containing a short description of each ROS node in the README file. Lastly, suggestion (iii) was taken into consideration and two articles [6,18] that describe the IPM transformation were added to the repository. The responses at Q3 show that three of the students have considered the simulator relevant to teach mobile robotics, two of the students stated that the simulator is complete (this reveals that there are several mobile robotics challenges covered by this simulator), and two of the students described the simulator as accessible (which means that the learning curve is not steep, thus making achievable solving the challenges covered by this simulator). Image processing, computer vision and control systems were the areas pointed out of all students when asked which areas did they learn by using this simulator, Q4. The average value of ≈4.3 in a scale 1 to 5 was obtained in Q5. This value demonstrates that the students recommend this simulator as a teaching tool for some of the mobile robotics challenges. All of the students would keep the simulator in the next edition of the of the Mobile Robotics course, Q6. Regarding Q7, which are the weakest points of this simulator, all students explained that more documentation is needed, and the students S1 and S2 considered the track hard to solve. By proving more resources in the documentation, namely, the short description of ROS nodes, papers, and descriptive videos it is expected to fill the gap noticed in this simulator's documentation.

Considering all of the survey responses and its corresponding analysis, it is possible to conclude that this simulator is adequate to teach some of the mobile robotics challenges. Therefore, this simulator will be used in the next editions of the Mobile Robotics course of Integrated Master Degree in Informatics and Computing Engineering at the Faculty of Engineering of the University of Porto.

5 Conclusion

In this work, an update to the autonomous driving simulator [8,9] to support all challenges of the Autonomous Driving Competition of the Portuguese Robotics Open was performed. A new robot model—Major Alvega—was also added to this simulator. One of the main goals of this simulator is to provide a complete platform to teach some of the challenges of mobile robotics to college students. This simulator has been tested in the Mobile Robotics course of the Integrated Master Degree in Informatics and Computing Engineering at the Faculty of Engineering of the University of Porto. To evaluate how efficient is the proposed simulator to teach mobile robotics, a survey was performed to the students that used it. The survey analysis indicated that this simulator is adequate to teach mobile robotics areas, including image processing, computer vision and control systems. One of the main limitations identified in this simulator was the shortage of documentation. To fill this gap, descriptive videos, ROS node descriptions and papers were added as official documentation. By adding this information, it is expected that all the information needed was made available for the students, thus boosting the learning and developing process.

As future work, it is suggested the creation of an automatic referee. This will enable the possibility of organizing an in course simulated autonomous driving competition. The main goal is to exploit the competitive mindset as a tool for teaching mobile robotics as demonstrated in [7,8].

Acknowledgements. Authors gratefully acknowledge the funding of Project NORTE -01-0145-FEDER-000022 - SciTech - Science and Technology for Competitive and Sustainable Industries, co-financed by Programa Operacional Regional do Norte (NORTE2020), through Fundo Europeu de Desenvolvimento Regional (FEDER).

This work is partially financed by the ERDF European Regional Development Fund through the Operational Programme for Competitiveness and Internationalisation - COMPETE 2020 Programme within project POCI-01-0145-FEDER-006961, and by

National Funds through the FCT Fundao para a Cincia e a Tecnologia (Portuguese Foundation for Science and Technology) as part of project UID/EEA/50014/2013.

References

1. Blank, D., Kumar, D., Meeden, L., Yanco, H.: Pyro: a Python-based versatile programming environment for teaching robotics. J. Educ. Resour. Comput. **4**(3), 1–15 (2004). https://doi.org/10.1145/1083310.1047569. http://portal.acm.org/citation.cfm?doid=1083310.1047569
2. Caccavelli, J., Pedre, S., de Cristóforis, P., Katz, A., Bendersky, D.: A new programming interface for educational robotics. In: Research and Education in Robotics - EUROBOT 2011, pp. 68–77. Springer, Heidelberg (2011). https://doi.org/10.1007/978-3-642-21975-7_7. http://link.springer.com/10.1007/978-3-642-21975-7_7
3. Canas, J., González, M., Hernández, A., Rivas, F.: Recent advances in the JdeRobot framework for robot programming. In: Proceedings of the 12th RoboCity2030 Workshop, Madrid, pp. 1–21 (2013)
4. Canas, J., Martín, L., Vega, J.: Innovating in robotics education with Gazebo simulator and JdeRobot framework. In: CUIEET 2014. XXII Congreso Universitario de Innovación Educativa en las Enseñanzas Técnicas (2014)
5. Carpin, S., Lewis, M., Wang, J., Balakirsky, S., Scrapper, C.: USARSim: a robot simulator for research and education. In: Proceedings 2007 IEEE International Conference on Robotics and Automation, pp. 1400–1405. IEEE (2007). https://doi.org/10.1109/ROBOT.2007.363180. http://ieeexplore.ieee.org/document/4209284/
6. Costa, V., Cebola, P., Sousa, A., Reis, A.: Design hints for efficient robotic vision - lessons learned from a robotic platform. Lecture Notes in Computational Vision and Biomechanics, vol. 27, pp. 515–524. Springer, Cham (2018). https://doi.org/10.1007/978-3-319-68195-5_56. http://link.springer.com/10.1007/978-3-319-68195-5_56
7. Costa, V., Cunha, T., Oliveira, M., Sobreira, H., Sousa, A.: Robotics: using a competition mindset as a tool for learning ROS. In: Robot 2015: Second Iberian Robotics Conference, vol. 417, pp. 757–766. Springer, Cham (2016). https://doi.org/10.1007/978-3-319-27146-0_58. http://link.springer.com/10.1007/978-3-319-27146-0_58
8. Costa, V., Rossetti, R., Sousa, A.: Simulator for teaching robotics, ROS and autonomous driving in a competitive mindset. Int. J. Technol. Hum. Interact. **13**(4), 19–32 (2017). https://doi.org/10.4018/IJTHI.2017100102. http://services.igi-global.com/resolvedoi/resolve.aspx?doi=10.4018/IJTHI.2017100102
9. Costa, V., Rossetti, R.J., Sousa, A.: Autonomous driving simulator for educational purposes. In: 2016 11th Iberian Conference on Information Systems and Technologies (CISTI), pp. 1–5. IEEE (2016). https://doi.org/10.1109/CISTI.2016.7521461. http://ieeexplore.ieee.org/document/7521461/
10. Costa, V., Rossetti, R.J., Sousa, A.: Simulator for teaching robotics, ROS and autonomous driving in a competitive mindset. In: Rapid Automation, pp. 720–734. IGI Global (2019). https://doi.org/10.4018/978-1-5225-8060-7.ch033. http://services.igi-global.com/resolvedoi/resolve.aspx?doi=10.4018/978-1-5225-8060-7.ch033
11. Ellekilde, L.P., Jorgensen, J.A.: RobWork: a flexible toolbox for robotics research and education. In: ISR 2010 (41st International Symposium on Robotics) and ROBOTIK 2010 (6th German Conference on Robotics), pp. 1–7. VDE (2010)

12. Fagin, B., Merkle, L.: Measuring the effectiveness of robots in teaching computer science. In: Proceedings of the 34th SIGCSE Technical Symposium on Computer Science Education - SIGCSE 2003, p. 307. ACM Press, New York (2003). https://doi.org/10.1145/611892.611994. http://portal.acm.org/citation.cfm?doid=611892.611994
13. Fernandes, D., Pinheiro, F., Dias, A., Martins, A., Almeida, J., Silva, E.: Teaching robotics with a simulator environment developed for the autonomous driving competition. In: Merdan, M., Lepuschitz, W., Koppensteiner, G., Balogh, R., Obdržálek, D. (eds.) Robotics in Education, pp. 387–399. Springer, Cham (2020)
14. Flowers, T.R., Gossett, K.A.: Teaching problem solving, computing, and information technology with robots. J. Comput. Sci. Coll. **17**(6), 45–55 (2002). http://dl.acm.org/citation.cfm?id=775742.775755
15. Fonseca Ferreira, N., Tenreiro Machado, J.: ROBLIB: an educational program for robotics. IFAC Proc. Vol. **33**(27), 563–568 (2000). https://doi.org/10.1016/S1474-6670(17)37990-9. https://linkinghub.elsevier.com/retrieve/pii/S1474667017379909
16. Gawryszewski, M., Kmiecik, P., Granosik, G.: V-REP and LabVIEW in the service of education. In: Robotics in Education, pp. 15–27. Springer, Cham (2017). https://doi.org/10.1007/978-3-319-42975-5_2. http://link.springer.com/10.1007/978-3-319-42975-5_2
17. Hugues, L., Bredeche, N.: Simbad: an autonomous robot simulation package for education and research. In: From Animals to Animats 9, pp. 831–842. Springer, Berlin (2006). https://doi.org/10.1007/11840541_68. http://link.springer.com/10.1007/11840541_68
18. Oliveira, M., Santos, V., Sappa, A.D.: Multimodal inverse perspective mapping. Inf. Fusion **24**(1), 108–121 (2015). https://doi.org/10.1016/j.inffus.2014.09.003. http://www.sciencedirect.com/science/article/pii/S1566253514001031, http://linkinghub.elsevier.com/retrieve/pii/S1566253514001031
19. Portuguese Robotics Open: Robotica 2019 - Autonomous Driving. https://web.fe.up.pt/~robotica2019/index.php/en/conducao-autonoma-2

The Role of Educational Technologist in Robot Supported Math Lessons

Janika Leoste(✉) and Mati Heidmets

Tallinn University, 10120 Tallinn, Estonia
leoste@tlu.ee

Abstract. The decision makers of educational systems in different countries have started to realize the importance of technology enhanced learning (TEL) in order to prepare students for the world of 4th Industrial Revolution. However, in the grass root level, teachers are still reluctant to implement technology into their lessons. In this paper we investigate the feedback from 134 Estonian teachers, each of whom conducted with the help of educational technologists up to 15 robot supported math lessons, in order to find out which supportive roles did educational technologists have in these lessons. The results show that educational technologist's roles as a technical support person or a robotics teacher were more important during the first lessons, but the need for these roles faded fast. Instead, educational technologist's role as an assistant teacher, explaining tasks and answering students' questions, proved to have a greater importance, especially in the 3rd grade. Based on the results we suggest that in TEL lessons the subject teacher needs to be accompanied by an educational technologist who also has basic knowledge about the topic taught.

Keywords: Educational robots · Technology enhanced learning · Educational technologist · Math

1 Introduction

The explosion of digital technologies has impacted the world in two major waves: first, the new technologies enabled people to manage traditional needs differently, and now, the maturation and fusion of these technologies are creating entirely different needs and products [1] – changing thus the requirements that labor market places on workers, especially valuing STEM skills [2–5]. However, schools have not been successful in exploiting the possibilities of modern technologies, especially in STEM subjects where these technologies could be more relevant [6]. It is believed that this widening gap between technology rich everyday life and traditional teaching methods is making student interest towards STEM heavy fields of study to fade [2, 7–10]. Instead of creating and using new ways of teaching and learning that would make use of common technology, surrounding students in their everyday life, education has fallen into crisis where students suffer from sleep deprivation, stress and low performance [11, 12].

Schools' inertia towards changing their practices has several reasons. One of the challenges is overcoming teachers' skepticism. Teachers rely in their work on previously acquired comprehensive framework of knowledge that can only be altered if new

M. F. Silva et al. (Eds.): ROBOT 2019, AISC 1092, pp. 467–477, 2020.
https://doi.org/10.1007/978-3-030-35990-4_38

information is perceived as useful and applicable in real classroom settings [13–15]. Implementing new, technology-centered teaching practices is, however, a laborious and complex process [16], complicated by the fact that researchers and policy makers often fail to take into consideration teachers' real life requirements [17]. Poorly planned implementation attempts are destined to fail, resulting teachers to become distrustful towards any new innovation ideas [18]. Involving teachers into the design process of technology based educational innovation early on, or providing teachers with technical support during the implementation process are often seen as possible solutions to this situation [19].

The previous research proposes that educational robot as a learning tool for visualizing abstract math concepts and creating meaningful real-life like context for applying math knowledge (robot supported learning) is a suitable technology for changing math teaching practices and building student motivation to learn this subject [20–28]. Due to national programs, more than 60% of Estonian basic education schools have educational robots [29] while only 8% of teachers have tried to use these robots for teaching purposes [30], supposedly due to lack of their STEM skills [31]. Since Estonian schools generally also employ an educational technologist with the purpose of helping school to implement new technologies [32], we hypothesized that it in Estonian context math teachers could, with the help of educational technologists, use educational robots as technological teaching tools for making math lessons more engaging to students.

Based on these notions we designed a study with a duration of 1 school year (from September 2018 to April 2019), with a broader purpose of finding out how does the use of educational robots in math lessons shape student learning motivation and what would be the main challenges when conducting these lessons. In the context of this paper the aim was to clarify what roles educational technologists had when helping teachers to conduct robot supported math lessons.

The study was preceded by a pilot that took place in spring 2018, and involved 10 classes, 208 students and 9 math teachers. During the pilot we found out that math teachers in general had poor ICT skills and were, at the beginning, uncomfortable to accept teaching practices that accompanied robot supported learning (for example, collaborative learning, student autonomy, self-regulated learning, peer tutoring) [33]. We also discovered, that teachers perceived robot supported math lessons to be easier if they were accompanied by an educational technologist. These notions were taken into account when designing the main study. On the one hand we created a scaffolding structure for helping teachers to acquire elementary programming and educational robotics skills. On the other hand we requested participating schools to provide math teachers with the help from educational technologists when conducting robot supported math lessons.

1.1 Research Questions

This paper is about the study that was conducted in Estonia during the school year of 2018/2019. The study examined the possibilities of using educational robots as learning tools for helping students to learn math in real life like situations, making thus abstract math concepts more meaningful for students, and increasing student motivation towards learning math. During the study, regular math teachers, with the help of educational technologists of their respective schools, conducted up to 15 robot supported math lessons. The purpose of this paper is, based on the results of lesson feedback surveys of 134 math teachers, to analyze the roles of educational technologist when helping math teacher to conduct robot supported lessons, focusing on the following research questions:

1. How did the educational technologist's roles change in robot supported math lessons over time in the grades 3 and 6?
2. What similarities and differences were in the role of the educational technologist in the third and sixth grades?

2 Method

2.1 Study Design and Sample

The compulsory school start age in Estonia is 7. The ISCED basic education stage corresponds to the grades 1–6 in Estonian basic school [34]. We focused our study on the grades 3 and 6 because for these grades the national standardized tests for measuring math knowledge are available, making it possible to compare the students' development.

The full sample consisted of 67 schools with 137 classes (98 in the grade 3 and 39 in the grade 6) with more than 2000 students. The robot supported[1] math lessons[2] in these classes were conducted by 137 teachers, aided by 56 educational technologists. As 3 teachers withdrew from the study then the final sample size for math teachers was 134.

In the study we used 3 different educational robotics platforms: the Edison robot, the LEGO Mindstorms EV3 robot and the LEGO WeDo 2.0 robot. In the 3rd grade lessons all of these platforms were used, in the 6th grade lessons only LEGO Mindstorms EV3 robot was used. The major features of these robotics platforms are as follow [34]:

- The Edison educational robot was launched in mid-2014 by an Australian company Microbric. The base robot is small, self-contained and relatively robust. It has two individually controlled motors as actuators, a speaker, 2 LED lights, 2 IR transmitters, 3 buttons, and following sensors: IR receiver, line tracking sensor, 2 light sensors, and sound sensor. In the study the graphical programming environment EdBlocks was used for programming the robot. The Edison robot is relatively cheap and simple to use. However, its motor rotation sensors cannot be used with graphical programming languages, making robot movement somewhat inaccurate.
- The LEGO Mindstorms EV3 (EV3) robot belongs to the family of very popular LEGO robots. Compared to the Edison robot the EV3 is a constructor robot, meaning that students needs to build the robot before using it. In the educational set there are 3 motors, 1 color sensor, 2 touch sensors, 1 ultrasonic sensor, 1 gyro

[1] Full description of robots used in the study: http://bit.ly/2AvwZpB.

[2] Sample lesson plans (in English): http://bit.ly/2Q1n9AS.

sensor, robot's brain, necessary cables and LEGO bricks. For the purposes of the study the EV3 "driving base" model was used. The building of models had to take place outside the mathematics lessons (being built either by support personnel, teachers, or robotics club students). For the purposes of the study a tablet based LEGO Mindstorms Education EV3 Programming app was used. The EV3 robot is relatively accurate, intuitive to program and it is widely available in (Estonian) schools. However, its price is up to 7 times higher than that of the Edison robot.
- The LEGO WeDo 2.0 (WeDo) robot is also part of the LEGO educational robots family. This robot is designed for elementary students ages 7+ although it can also be successfully used with somewhat younger children. The WeDo robot is a constructor robot, and there are 1 motor, 1 tilt sensor, 1 motion sensor, robot's brain, necessary bricks and cables for building different models of robot. For the purposes of the current study the Milo science rover model was used. The robot was programmed with proprietary WeDo 2.0 LEGO Education app. The WeDo robot has child friendly appearance and is easy to use. However, it is relatively inaccurate, it lacks a second motor (making it impossible for it to turn) and it is relatively expensive.

The study took place from September 2018 to April 2019. During this period of time each participating teachers was asked to conduct together with an educational technologist of her school up to 15 robot supported math lessons. The lesson designs were previously co-created with teachers who participated the pilot study. Each lesson was presented as an interactive GeoGebra worksheet and consisted of math word problem and robotics exercises that used the data and concepts of the math word problem. Additionally, the lesson designs were also available as Word documents. The design of the exercises paid special attention to encouraging collaborative learning approach and student independence.

The mathematical content of the lesson designs was based on the regular curricula of the respective grade. The robotics exercises included sample solution videos and coding examples with explanations. The connection with real math curriculum made lessons progressively more difficult to solve.

The lessons were scripted relatively loosely, describing the order of solving exercises, approximate time cost per each task and different approaches for supporting students during the lesson. Teachers were encouraged to customize the lesson designs.

Students were supposed to work in these lessons autonomously, preferably in pairs. Teacher and educational technologist were advised to counsel students, help them in case of technical problems and to supervise overall classroom order, taking into consideration a relatively noisy nature of teams working out and testing their solutions on robots. Also, after each lesson teacher filled a lesson feedback survey (using Google Forms) that recorded various aspects of the lesson, including her estimation about the help she needed from educational technologist for conducting the lesson.

2.2 Data Collection

Based on the content of the pilot study teachers' lesson diaries we identified the major areas of problems that occurred in robot supported math lessons [24, 33]. For each of the problem areas we designed two questions that explored teacher's need for help from educational technologist in the robot supported math lesson. The questions were arranged into a semi-structured lesson feedback survey that was carried out using Google Forms. The questions of the survey, grouped by their corresponding problem areas were following:

- Technical support
  - "How much help did you need from educational technologist for setting up and maintaining robots?"
  - "How much help did you need from educational technologist for setting up the software in tablets?"
- Robotics teacher's tasks
  - "How much help did you need from educational technologist for consulting students when writing programs?"
  - "How much help did you need from educational technologist for checking the results of robotics experiments?"
- Pedagogical tasks
  - "How much help did you need from educational technologist for explaining the worksheet tasks?"
  - "How much help did you need from educational technologist for answering student questions?"

Each question had to be answered by using a 5-item Likert scale: "(1) did not need assistance at all", "(2) mostly did not need", "(3) was needed about half of the time", "(4) was mostly needed", and "(5) was needed all of the time".

3 Results

3.1 Question 1: How Did the Educational Technologist's Roles Change in Robot Supported Math Lessons Over Time in the Grades 3 and 6?

In order to find the answers to the research questions the following steps were carried out. Firstly, to reduce the impact of individual deviations on overall results we summarized the answers for individual lessons into 5 bigger categories (Periods): lessons 1 to 3 (Period 1), lessons 4 to 6 (Period 2), lessons 7 to 9 (Period 3), lessons 10 to 12 (Period 4), and lessons 13 to 15 (Period 5). Then we summarized the answers to the questions that described the same problem areas (Technical support, Robotics teacher's tasks, Pedagogical tasks).

Next we grouped the gathered answers according to the following three criteria: `[(grade); (Period); (problem area)]`. For example: (the grade 3) – (Period 1) – ("Technical support").

For each resulting group we counted the number of answers that indicated teacher's need for the help from educational technologist (Likert scale answers 3, 4 and 5). These resulting numbers were then compared to the overall number of answers in that specific group, allowing calculation of percentage values of teachers that needed help in each group. The final results are visualized on Figs. 1 and 2.

Educational Technologist's Roles in the 3rd Grade

According to the data presented in Fig. 1 the area where teachers needed the least support was that of the technical issues. During the first 3 lessons (Period 1) 30% of teachers needed educational technologist's assistance. However, this demand for help faded rapidly and settled on the level below 20% in Periods 3, 4 and 5.

Similarly fading was teachers' need for help with tasks of robotics teacher (advising students in programming, checking robotics exercises). In the first lessons (Period 1) 54% of teachers required educational technologist's help for performing these tasks while in the last lessons (Period 5) this number was 41%.

Somewhat surprisingly, the share of teachers who needed educational technologist's help was largest with regular pedagogical tasks, like explaining worksheet content and answering students' questions. Also, this was the only area where the necessity for help practically did not decline over time and instead went occasionally up (58% in Period 1, 56% in Period 5, but 63% in Period 4).

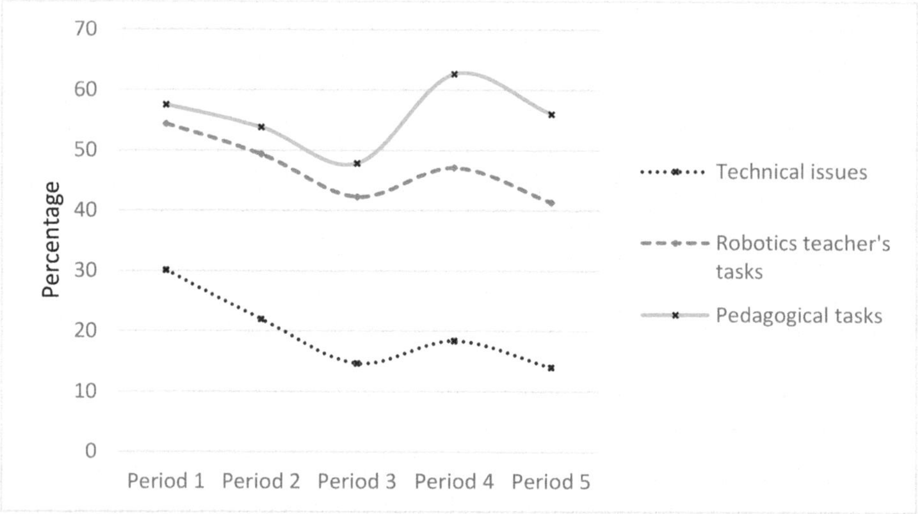

Fig. 1. Percentage of teachers, expressing the need for educational technologist's assistance during the robot-supported math lessons in the 3rd grade.

Educational Technologist's Roles in the 6th Grade

According to the data presented in Fig. 2 the area where 6th grade math teachers needed the least educational technologist's help was the one of technical issues, and the necessity was fading over time: in Period 1 the share of teachers needing help was 23% and in Period 5 it was 14%.

Somewhat striking is that teachers' need for help with robotics teacher's tasks and pedagogical tasks was practically the same over time, with initial (Period 1) values 52% and 51%, and declining to final values (Period 5) 38% and 36%, respectively.

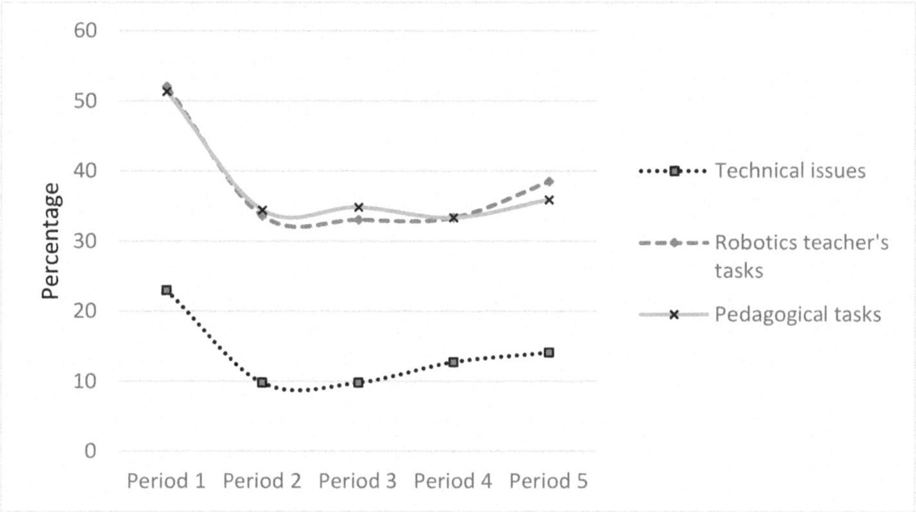

Fig. 2. Percentage of teachers, expressing the need for educational technologist's assistance during the robot-supported math lessons in the 6th grade.

3.2 Question 2: What Similarities and Differences Were in the Role of the Educational Technologist in the Third and Sixth Grades?

The data shows that in the both grades the teachers' demand for educational technologist's help was following the same trends: initially high demand for help faded during the first Period (in the 6th grade) or during the first two Periods (in the 3rd grade) and then remained on a relatively stable level. The area where least help was required was the one of technical issues. In the both grades only about 14% of teachers needed help here.

There were two major differences between the compared grades. First of all, the overall necessity for help with robotics teacher's tasks and with pedagogical tasks was higher in the 3rd grade. Secondly, in the 3rd grade, the need for help with pedagogical tasks was especially high and increased further when exercises became more difficult. We would like to point out that more than half (up to 63%) of the 3rd grade teachers were needing educational technologist's help with pedagogical tasks whereas in the 6th grade, after Period 1, the highest share of teachers needing help in any area was around one third of all teachers.

4 Conclusions and Discussion

Schools are inert in adopting TEL based innovative teaching methods although research shows that these methods may recover and increase student learning motivation in STEM subjects like math. It is often cited that teachers have skeptical views towards using modern technology based teaching tools in their subjects as they lack necessary technical knowledge for this [31]. As Estonia is in unique situation, having educational robots in more than half of the primary schools and educational technologists to support teachers [32] then we decided to test out our hypothesis that in Estonia math teachers could, with the help of educational technologists, use educational robots as technological teaching tools for making math lessons more engaging to students. For this purpose we designed a study that involved 134 math teachers who, with the help of 56 educational technologists, tried robot supported math teaching in their lessons over longer period of time (September 2018 to April 2019). In April 2019 we gathered teachers' feedback, using semi-structured surveys in Google Forms environment.

We found out, that in principle, math teachers were able to conduct technologically enhanced math lessons, using educational robots, when they were aided by educational technologists. However, in robot supported math lessons educational technologists' technical support related tasks had relatively little importance, and this importance faded over time even further. When considering the fact that the role of technical support had a little importance that was fading fast, we have to ask whether it is sustainable to support math teacher with educational technologist.

Educational technologist's role as a robotics teacher should be viewed cautiously. The fact that more than a third of math teachers required help in this area points to the possibility that a significant part of the lesson was not focused on teaching math.

We also discovered that educational technologists acted in these lessons, especially in the 3rd grade lessons, as assistant teachers, solving tasks of pedagogical nature. The importance of this role was high as more than half of the 3rd grade teachers relied on educational technologist's help in pedagogical tasks. We assume, based on the previous research [33], that using robots in math lessons will initiate change of teaching and learning practices in classroom. In robot supported math lessons students are actively testing out their existing, theoretical math knowledge with issues that resemble real life math problems, becoming thus intensely engaged with constructing their understanding of math. In this process it is natural for them to have a lot of questions.

However, the results show clearly, that math teacher needs assistance for conducting robot supported math lessons. The need for assistance could become lesser when using more mature technology [35] or if artificial intelligence would be able to assist both teacher and students [36] – both of these alternatives are, as of today, not feasible. Another way to approach this question would encourage more teacher teamwork [37] by having teachers of different subjects to co-creatively design and conduct joint lessons.

5 Limitations and Future Work

Our work has several areas of potential improvement. First of all, our data collection methods took into consideration only the viewpoint of math teachers. For having a more complete understanding, data from educational technologists and student learning outcomes should also be analyzed.

Our sample focused only on robot supported math teaching in the grades 3 and 6. We believe that including other primary school grades would result in bigger sample and more exact results. In this study teachers used only robotics platforms that were specified by the researchers. We feel that more math teachers would be encouraged to take part in study if they could choose the robotics platform by themselves, based on available robotics sets in their schools and on their personal preferences.

Further studies should also try to explore the ways that would make teachers' need for help in technical support and robotics teaching to fade more rapidly so that students could better focus on subject content.

In order to eliminate some of these shortcomings we are planning to conduct a follow-up study that will take place from autumn 2019 to spring 2020. In this follow-up study math teachers of any primary school grade will co-create teaching materials and are allowed to freely choose the robotics platform for their lessons. This process of co-creation is supported by regional monthly trainings, provided by researchers.

Acknowledgments. This project has received funding from the European Union's Horizon 2020 research and innovation programme under grant agreement No. 669074.

References

1. Schwab, K.: The Fourth Industrial Revolution: what it means, how to respond. World Economic Forum (2016)
2. The Commonwealth of Australia: Australia's National Science Statement. https://publications.industry.gov.au/publications/nationalsciencestatement/national-science-statement.pdf. Accessed 11 May 2019
3. European Commission: Employment and Social Developments in Europe Annual Review 2018. https://ec.europa.eu/social/BlobServlet?docId=19719&langId=en. Accessed 11 May 2019
4. Mullis, I.V.S., Martin, M.O., Loveless, T.: 20 Years of TIMSS: International Trends in Mathematics and Science Achievement, Curriculum, and Instruction. Boston College, Chestnut Hill (2016)
5. OECD: The Future of Education and Skills. Education 2030. OECD Publishing (2018). https://www.oecd.org/education/2030/E2030%20Position%20Paper%20(05.04.2018).pdf. Accessed 11 May 2019
6. Halonen, N., Hietajärvi, L., Lonka, K., Salmela-Aro, K.: Sixth graders' use of technologies in learning, technology attitudes and school well-being. EJSBS **XVIII** (2016)
7. Bacchus, A.: New Microsoft research points to the declining interest of girls in STEM, ways to close the gender gap. OnMSFT (2018). https://www.onmsft.com/news/new-microsoft-research-points-to-the-declining-interest-of-girls-in-stem-ways-to-close-the-gender-gap. Accessed 11 May 2019
8. Elliott, J.: Action Research for Educational Change. Open University Press, Bristol (1991)
9. Ernst & Young Global: Research Reveals Boys' Interest in STEM Careers Declining; Girls' Interest Unchanged. Ernst & Young Global Ltd (2018). https://www.ey.com/us/en/newsroom/news-releases/news-ey-research-reveals-boys-interest-in-stem-careers-declining-girls-interest-unchanged. Accessed 11 May 2019
10. The Star Online: Students taking up STEM subjects on decline last 10 years. Star Media Group Berhad (2018). https://www.thestar.com.my/news/nation/2017/07/16/students-taking-up-stem-subjects-on-decline-last-10-years-ratio-of-science-to-arts-classes-reversed/. Accessed 11 May 2019
11. Grudin, J.: Innovation and Inertia: Information Technology and Education in the United States. IEEE Computer Society (2018)

12. OECD: Teaching for the Future. Effective Classroom Practices to Transform Education. OECD Publishing (2018). http://www.oecd.org/education/teaching-for-the-future-9789264293243-en.htm. Accessed 11 May 2019
13. Aypay, A., Çelik, H.C., Sever, M.: Technology acceptance in education: a study of pre-service teachers in Turkey. Turk. Online J. Educ. Technol. **11**, 264–272 (2012)
14. Korthagen, F.: The gap between research and practice revisited. Educ. Res. Eval. **13**(3), 303–310 (2007)
15. Miller, M.D., Rainer, R.K., Corley, J.K.: Predictors of engagement and participation in an on-line course. Online J. Distance Learn. Adm. **6**, 1–13 (2003)
16. Laferrière, T., Montane, M., Gros, B., Alvarez, I., Bernaus, M., Breuleux, A., Allaire, S., Hamel, C., Lamon, M.: Partnerships for knowledge building: an emerging model. Can. J. Learn. Technol. **V36**(1) (2010)
17. Arhar, J., Niesz, T., Brossmann, J., Koebley, S., O'Brien, K., Loe, D., Black, F.: Creating a 'third space' in the context of a university–school partnership: supporting teacher action research and the research preparation of doctoral students. Educ. Action Res. **21**(2), 218–236 (2013)
18. Erss, M., Kalmus, V.: Discourses of teacher autonomy and the role of teachers in Estonian, Finnish and Bavarian teachers' newspapers in 1991-2010. Teach. Teach. Educ. **76**, 95–105 (2018)
19. Ley, T., Leoste, J., Poom-Valickis, K., Rodríguez-Triana, M.J., Gillet, D., Väljataga, T.: CEUR Workshop Proceedings (2018). http://ceur-ws.org/Vol-2190/CC-TEL_2018_paper_1.pdf. Accessed 11 May 2019
20. Acosta, A., Slotta, J.: CKBiology: an active learning curriculum design for secondary biology. Front. Educ. **3**, 52 (2018)
21. Barker, B., Ansorge, J.: Robotics as means to increase achievement scores in an informal learning environment. J. Res. Technol. Educ. **39**, 229–243 (2007)
22. Highfield, K., Mulligan, J., Hedberg, J.: Early mathematics learning through exploration with programmable toys. In: Figueras, O., Cortina, J.L., Alatorre, S., Rojano, T., Sepulveda, A., (eds.) Proceedings of the Joint Meeting of Pme 32 And Pme-Na Xxx, vol. 3, pp. 169–176. (PME Conference Proceedings). Cinvestav-UMSNH, Mexico (2008)
23. Kopcha, T.J., McGregor, J., Shin, S., Qian, Y., Choi, J., Hill, R., Mativo, J., Choi, I.: Developing an integrative STEM curriculum for robotics education through educational design research. J. Form. Des. Learn. **1**, 31–44 (2017)
24. Leoste, J., Heidmets, M.: Bringing an educational robot into a basic education math lesson. In: Robotics in Education - Current Research and Innovations. Springer (2019, in press)
25. Lindh, J., Holgersson, T.: Does lego training stimulate pupils' ability to solve logical problems? Comput. Educ. **49**, 1097–1111 (2007)
26. Papert, S.: Mindstorms: Children, Computers, and Powerful Ideas. Basic Books, New York (1980)
27. Samuels, P., Haapasalo, L.: Real and virtual robotics in mathematics education at the school–university transition. Int. J. Math. Educ. **43**, 285–301 (2012)
28. Werfel, J.: Embodied teachable agents: learning by teaching robots. In: Conference Proceedings (2014). http://people.seas.harvard.edu/~jkwerfel/nrfias14.pdf. Accessed 08 Nov 2018
29. HITSA: ProgeTiiger programmis toetuse saanud haridusasutused 2014–2018. https://www.hitsa.ee/ikt-haridus/progetiiger. Accessed 11 May 2019
30. Leppik, C., Haaristo, H.S., Mägi, E.: IKT-haridus: digioskuste õpetamine, hoiakud ja võimalused üldhariduskoolis ja lasteaias. Praxis (2017). http://www.praxis.ee/wp-content/uploads/2016/08/IKT-hariduse-uuring_aruanne_mai2017.pdf. Accessed 11 May 2019

31. Rasinen, A., Virtanen, S., Endepohls-Ulpe, M., Ikonen, P., Judith Ebach, J., Stahl-von Zabern, J.: Technology education for children in primary schools in Finland and Germany: different school systems, similar problems and how to overcome them. Int. J. Technol. Des. Educ. **19**, 367 (2009)
32. Lorenz, B., Kikkas, K., Laanpere, M.: The role of educational technologist in implementing new technologies at school. In: Zaphiris, P., Ioannou, A. (eds.) Learning and Collaboration Technologies. Technology-Rich Environments for Learning and Collaboration, LCT 2014. Lecture Notes in Computer Science, vol. 8524. Springer, Cham (2014)
33. Leoste, J., Heidmets, M.: Õpperobot matemaatikatunnis. Miks.ee. Estonian Research Council (2019). http://www.miks.ee/opetajale/uudised/opperobot-matemaatikatunnis. Accessed 11 May 2019
34. Leoste, J., Heidmets, M.: The impact of educational robots as learning tools on mathematics learning outcomes in basic education. In: Väljataga, T., Laanpere, M. (eds.) Digital Turn in Schools—Research, Policy, Practice. Lecture Notes in Educational Technology. Springer, Singapore (2019)
35. Banke, J.: Technology Readiness Levels Demystified. NASA (2010)
36. Smith, C.: Artificial intelligence that can teach? It's already happening. ABC Science (2018). http://www.abc.net.au/news/science/2018-06-16/artificial-intelligence-that-can-teach-is-already-happening/9863574. Accessed 11 May 2019
37. Polega, M., Neto, R.C., Brilowski, R., Baker, K.: Principals and teamwork among teachers: an exploratory study. Revista@mbienteeducação, vol. 12, no. 2, pp. 12–32 mai/ago. Universidade Cidade de São Paulo, São Paulo (2019)

Robot@Factory Lite: An Educational Approach for the Competition with Simulated and Real Environment

João Braun[1(✉)], Lucas A. Fernandes[1], Thiago Moya[1], Vitor Oliveira[1], Thadeu Brito[2], José Lima[2,3], and Paulo Costa[3,4]

[1] Federal University of Technology, Paraná, Curitiba, Brazil
jabraunneto1@gmail.com, lucas_azevedo33@hotmail.com, vfelipeoliveira@hotmail.com, thiagomoya@ymail.com

[2] Research Centre in Digitalization and Intelligent Robotics (CeDRI), Instituto Politécnico de Bragança, Bragança, Portugal
{brito,jllima}@ipb.pt

[3] Centre for Robotics in Industry and Intelligent Systems - INESC TEC, Porto, Portugal

[4] Faculty of Engineering of University of Porto (FEUP), Porto, Portugal
paco@fe.up.pt

Abstract. Teaching based on challenges and competitions is one of the most exciting and promising methods for students. In this paper, a competition of the Portuguese Robotics Open is addressed and a solution is proposed. The Robot@Factory Lite is a new challenge and accepts participants from secondary schools (Rookie) and universities. The concepts of simulation, hardware-in-the-loop and timed finite state machine are presented and validated in the real robot prototype. The aim of this paper is to disseminate the developed solution in order to attract more students to STEM educational program.

Keywords: Robotics competition · Simulation · Factory logistics

1 Introduction

Robotics competitions are one of the methodologies that drives technology development, as robotics competitions encourage students and researchers to develop new ways to solve a task. Examples such as robotic soccer, autonomous driving and among others have contributed to the advancement of algorithms that are later used in both industry and service robotics. Besides, it is well known that robotics competitions captivate the students attention, improve their intrinsic motivation and skills and also improves teamwork and social collaboration. An example of this incentive for the development of robotics through competition can be seen in [1], which analyzes the educational performance of engineering students in the First Lego League (FLL).

M. F. Silva et al. (Eds.): ROBOT 2019, AISC 1092, pp. 478–489, 2020.
https://doi.org/10.1007/978-3-030-35990-4_39

Having as base the Robot@Factory competition that started in 2011 Portuguese Robotics Open, the Robot@Factory Lite (R@FL) is a simplified version where the parts should be moved between warehouses and processing machines through a magnet bringing this competition more accessible for younger students. The organization provides a prototype and some libraries to deal with the robot I/O (magnet, RFID, motors, etc.). This paper presents a solution for this R@FL competition, based on the proposed prototype, that uses both Hardware-in-the-loop (HIL) and real robot approaches. The results show that the adopted solution solves the challenge. It is desired to present a solution that encourages further students to participate in robotics competition, namely the R@FL. This competition is aligned with the STEM topics, i.e., the proposed challenges encourage students to approach the topics of science, technology, engineering and mathematics.

During the competition, the students develop skills such as communication skills (they have to socialize and present the developed work), problem-solving skills (problems appear during the competition), teamwork skills (the students need to work together to develop solutions and compete), self-motivation (student motivation increases as they are facing real problems), and conflict resolution (if a problem appears, the students should work to solve it), among the others.

The paper as organized as follows: after this introduction, the related work is presented in Sect. 2. Section 3 addresses the rules and simulation environment. The adopted solutions for all rounds are addressed in Sect. 4 whereas results and conclusions with future work are stressed in Sects. 5 and 6 respectively.

2 Related Work

The challenges presented in robotic competitions provide the opportunity for researchers, students and enthusiasts to come up with creative solutions. Over the years, the methods and solutions found are vast, and therefore, it is important to have a benchmark of the methodologies developed [2]. By means of metric comparisons of the performances of the participating teams, [3] demonstrates the difficulty in judging the approaches of the competitors during the competition. Maritime robots can have various shapes, sizes, and application solutions, therefore the measurement during the challenges of the euRathlon 2015 competition was based on references from previous editions. Service robot challenges are made in RoboCup@Home based on household activities. For over seven years, the challenges of domestic robots have been influenced by the complexity and performance of tasks [4].

Benefiting of the challenge proposed in the Robot@Factory competition as a test, [5,6] demonstrates a platform for navigation, control, and localization of Automatic Guided Vehicle (AGV) robots. With the purpose of encouraging the development of human-robot cooperation applications, the RoCKIn@Work competition challenges its competitors to optimize small and medium factory processes [7]. Since the challenge is to simulate a real shop-floor situation, the developed system is able to avoid obstacles, determine the position of the mobile robot and indicate the paths it must take to get the product to be processed. By comparing Student Autonomous Underwater Vehicles Challenge - Europe (SAUC-E), An Outdoor Robotics Challenge for Land, Sea and Air (EURATHLON) in the 2014 and 2015 editions, and The European Robotics League (ERL) EMERGENCY 2017, [8] demonstrates the instability in the scoring and judging system of the teams during the challenges. While some systems demonstrate to favor the implementation of the application, other systems favor the development of the application project. On the other hand, participants are free to undertake any approaches as long as they respect the rules of each competition.

3 The Competition and Simulation

The R@FL competition has the objective to stimulate students and researchers to develop solutions to the challenges it presents [9]. The idea behind the challenge is an AGV to organize the materials in warehouses with processing machines. The layout of the competition can be seen in Fig. 1.

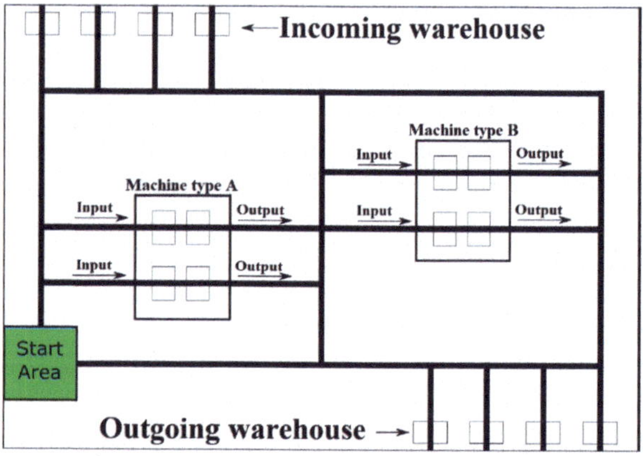

Fig. 1. Schematic of the competition environment [9].

As Fig. 1 illustrates, there are four incoming and outgoing warehouses alongside with two machines, A and B. In the incoming warehouses, there will be boxes that an AGV must deliver them to their correct locations simulating a real working warehouse. The outgoing warehouses represent the final destination to the processed products. In the same reasoning, machines A and B process the materials, and in this concept, they are pre-conditions for some of the materials before going to the outgoing warehouse, i.e., some materials need to be processed before being delivered to the outgoing warehouse. The start area can be chosen in the southwest or the northeast as the environment is symmetric in X and Y axes (machine A is always the machine near the start area). Thus, the competitors must implement an AGV capable of autonomously move, identify, manipulate and deliver the type of materials to their correct locations. There are three types of materials in the competition, which are represented by parts. The parts projected can be seen in Fig. 2.

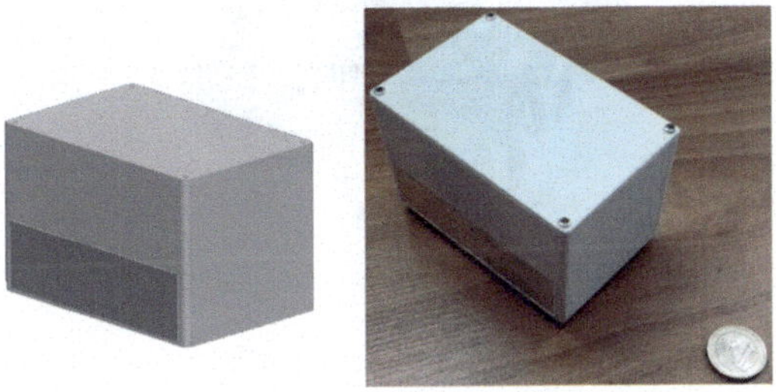

Fig. 2. Simulated and real parts with an RFID tag inside box em behind grey area [9].

The boxes dimensions can be seen in the competition rules [10]. As can be noted in Fig. 2, the boxes have a metal plate (grey area) and, this way the AGV can manipulate them by using electromagnets. Inside them, above the grey area, there are Radio Frequency Identification (RFID) tags. Thus the AGV, equipped with an RFID reader can identify the type of box that will be manipulated. In this sense, the types of boxes can be seen in the Table 1:

Table 1. Type of parts and its destinations.

Type of box	Destination
Raw	Machine A
Semi-processed	Machine B
Processed	Outgoing warehouse

As Table 1 displays, if the box identified is a raw one, this box will need to be processed by the two machines (A, B) before going to the warehouse. Therefore, if the box is a semi-processed material, it will be necessary just to process it through machine B before delivering to the outgoing warehouse. Finally, if the box is a processed one, the only task necessary is to deliver correctly to its final destination.

The competitors are free to implement the AGV they find suitable to the competition as long as the robot does not violate the dimensions rules [10]. However, the competition provides not only the full project (parts, bill of materials, project archives) for the recommended AGV but also a full manual that covers the implementation of the robot step by step. This was done to facilitate the integration of students to the competition. The recommended robot that was implemented can be seen in Fig. 3.

Fig. 3. The real AGV.

3.1 The Simulation

The competition also provides a simulation model. The SimTwo simulator provides a simulation considering the dynamic constraints that the real scenario has. Therefore, the simulator has a realistic 3D model of the robot and the competition scenario [11]. All those data can be freely modified, the graphic part in XML language and the simulator script in Pascal language. Therefore, the teams can adapt their robot model if they made a different one and, if they wish, modify the script provided by the competition staff. However, it is not necessary as all the tools needed are already coded. Thus, the competitors can validate their solutions easier and faster in the simulation before taking it to the real scenario. The simulation environment can be seen in Fig. 4.

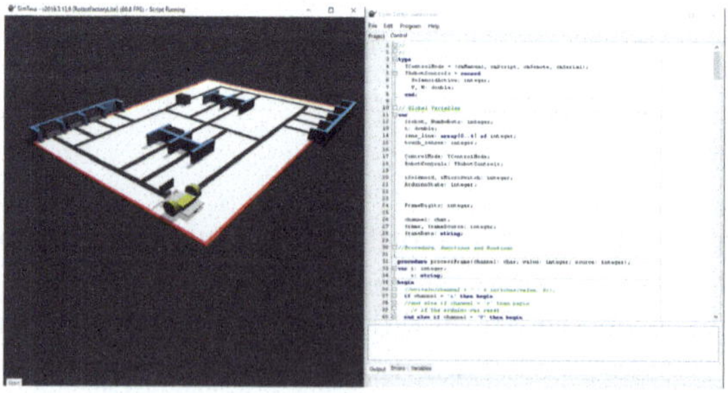

Fig. 4. Simulation environment. Left window is the graphic environment. Right window the code editor.

Although the simulator is realistic, it does not consider the microcontroller limitations such as available memory and processing speed. In this way, the competition staff provided a HIL tool coded in the simulator [9] as well. Therefore, the competitors can insert their microcontroller in the loop of the simulation, i.e., they program their solution to their microcontroller and then, by serial communication (USB), the microcontroller is inserted to the simulator loop. In this sense, the simulator sends the sensor data (line sensor, electromagnet and the micro switch) to the microcontroller and the information is processed. Soon after, the microcontroller sends the motors speeds to the simulator which is then processed dynamically and graphically. The main loop in the simulator run every 40 ms. In this way, the HIL loops roughly the same time. Figure 5 demonstrates this program loop.

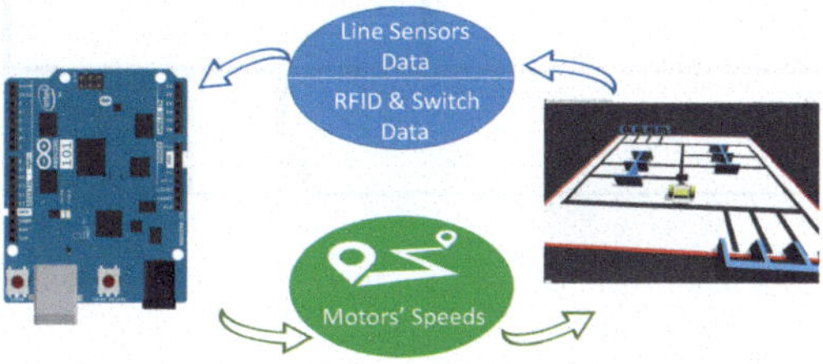

Fig. 5. HIL illustration within the code, provided by the organizers, it is possible to configure between the real movements or HIL mode. Adapted from [9].

4 Adopted Solutions

In this Section, the adopted solutions used in the three rounds of the R@FL competition are presented. As stated in the official rules [10], the boxes were identified through an RFID tag to differentiate the product type that they contain. However, in the first round this feature was not used because all boxes were processed materials and consequently had the same destination. The following subsections shows the logic of the developed code and the states that the robot performs through illustrative figures.

4.1 First Case

To perform the first case, it was adopted the Timed Finite-State Machine (TFSM) approach. This technique and the code was provided by the competition organizer. The TFSM idea is simple, each path that the robot execute compose a different state and each state is achieved through a selected action. As an example, the pick-and-place of the first box using the technique explained is shown in Fig. 6 and described below.

The initial state of the pick-and-place movement consists of going straight until the robot reaches the box, then, when the robot touches the box, the microswitch is triggered and the second state is activated. Thus, the box is coupled to the robot through activating the electromagnet. After that, the third state is started and the robot drives backwards. When the conditions for changing this state are true, the fourth state is activated and an 180° turn is performed.

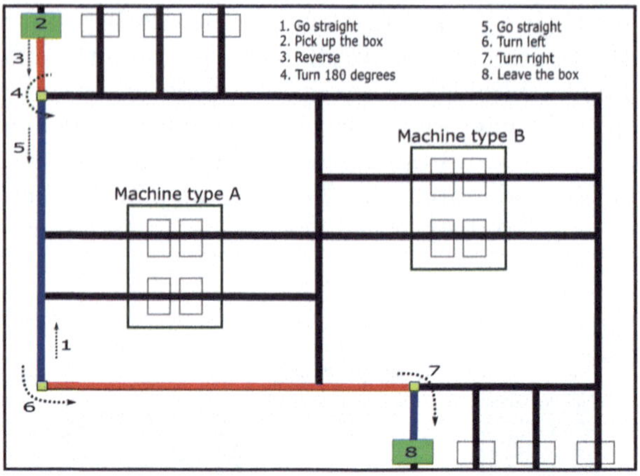

Fig. 6. Example of a path to deliver the first box.

Soon after, the fifth state is started, and the "go straight" command is given. When the requirements to activate the sixth and the seventh state are achieved the "turn to the left" and "turn to the right" commands are executed. Hereupon, the eighth and last state is accomplished leaving the box in the outgoing warehouse. The procedure for the other boxes is similar. Basically, a TFSM is based on the current state and the transitions between states when one or more conditions are satisfied.

4.2 Second Case

The second round includes the semi-processed materials and, consequently, new tasks to be performed. These boxes must be collected in the incoming warehouse and processed by either of the two machines, as can be seen in Fig. 8. From the different RFID tags, the robot must be able to identify the processed and semi-processed parts to correctly deliver the boxes. In this case the use of the TFSM is not recommended because the code would be large and inefficient resulting in a high memory consumption.

After studying the trajectories of the robot for all possibilities, it was noticed that several paths were repeated. For example, the path to go from the incoming warehouse to the outgoing warehouse is almost the same for all situations, as can be seen in Figs. 7 and 8. Thus, a generic path-travel function has been created. For each case the relevant information such as velocity and trajectory were transmitted to the function by parameters. In addition, the cases are determined according to the type of box, so the algorithm is able to identify and make the right decision in all situations.

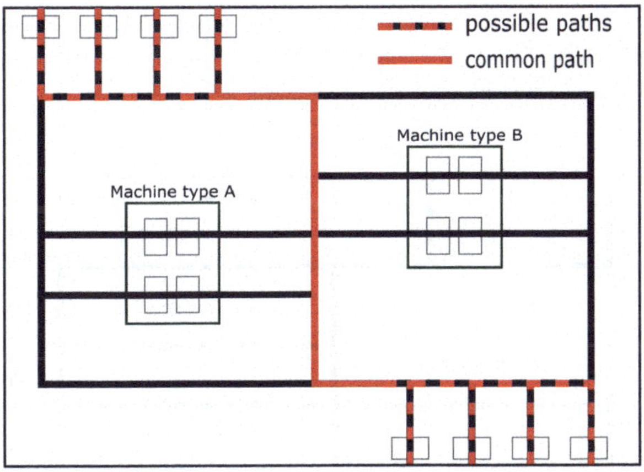

Fig. 7. Example of a path to deliver the processed parts.

The developed algorithm was ready to handle all position probabilities of all types of boxes for the first and second round. This means that if only the processed parts had been placed in the incoming warehouse, the robot could collect them and deliver them to the output.

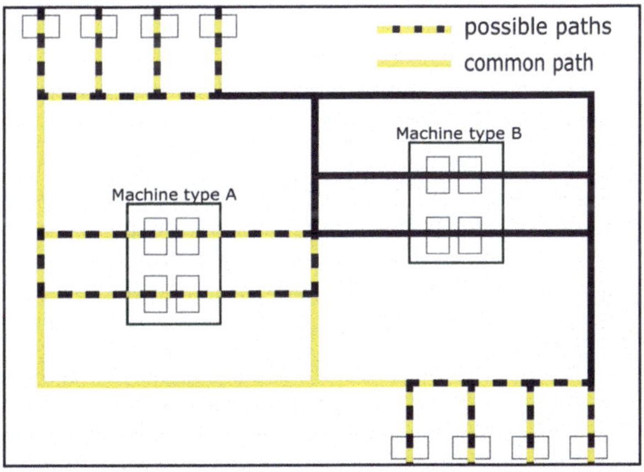

Fig. 8. Example of a path to the semi-processed parts.

4.3 Third Case

In the third round of the competition, the raw parts box was included. These boxes should be collected in the incoming warehouse, then be processed by Machine type A, after by Machine type B and finally delivered to the outgoing warehouse. From the logic used in the second round it was possible to develop the algorithm for the third round by including new paths and reusing the previously established paths. The new paths were necessary to allow passage through machine type B, as shown in Fig. 9.

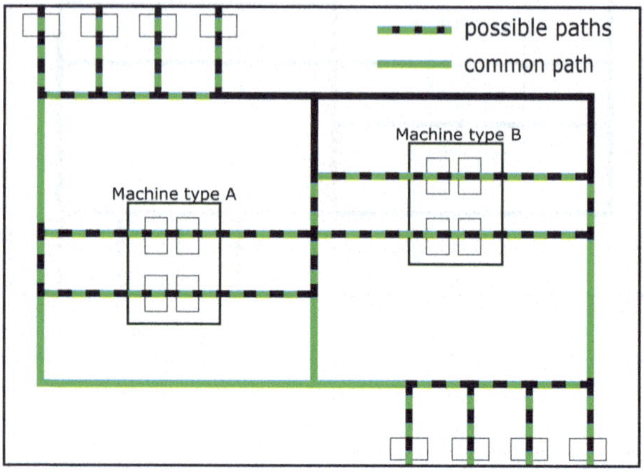

Fig. 9. Example of a path to the raw parts.

5 Results

The main objective of the experimental tests is to verify the performance of the robot and if the execution time of each round is in agreement with the time limit fixed by the rules. In this section will be presented the pseudo-codes of the implemented algorithms, the comparison between the applied methods, the memory consumption and the robot's performance in practice.

The first tests consisted of verifying the basic functions provided to competitors such as turn left or right, turn 180°, go straight, and drive backwards. Then the appropriate settings were set to adjust the movements of the robot. After that, the simulator was applied to become familiar with the competition scenario, as well as the possibility of performing the experiments without the physical robot. This resource was widely used in the initial tests. After the familiarization with the resources given by the competition, the development of the codes to solve the factory problems started. As previously mentioned, the technique used in the first case is different from the second and third cases. In the first one a TFSM technique was employed. To complete all steps, 61 states and 51% of the microcontroller's memory (SRAM) were used. This data is an estimated value extracted from Arduino IDE (Integrated Development Environment).

Algorithm 1. Main Function

```
function MAIN FUNCTION
    Go straight until touch the box
    Box ← Part_Type
    if Box == Processed Part then
        function ROUTE(1)
        end function
        function ROUTE(2)
        end function
    else if Box == Semi − Processed_Part then
        function ROUTE(3)
        end function
        function ROUTE(4)
        end function
        function ROUTE(5)
        end function
        function ROUTE(2)
        end function
    end if
end function
```

The second technique was based on a single function that covered all the necessary paths, reducing the size and complexity of the code. The developed function applied was the switch case statement. In this sense, each path is described as a case, and, since the same path can be traveled more than once, the same case can be called several times. In the main function of the program, exemplified in Algorithm 1, the detection of the product type is performed through an RFID tag reading function and the execution of the pick-and-place process through the Route function, shown in Algorithm 2. The source code consumed 58% of the microcontroller SRAM but if the TFSM approach was used, that value would certainly be bigger.

Algorithm 2. Function ROUTE

```
function ROUTE(int NumCase)
    switch NumCase do
        case 1
            Pick up the box and leave in outgoing warehouse
        case 2
            Go back to incoming warehouse
        case 3
            Pick up the box and leave in the machine
        case 4
            Pick up the box in machine
        case 5
            Leave the box in outgoing warehouse
end function
```

The third case was implemented based on the second round function, however, with some upgrades. For the robot to pick up the box and go through Machine A and Machine B, new paths were added. In the main function the novelty was the detection of a new RFID tag considering the raw parts. The source code consumed 67% of the microcontroller SRAM, a short increase of memory consumption in comparison with the second case.

Table 2. Performed times in the competition.

Round	Time	Boxes
1	2:00'57	4 blue boxes
2	3:48'80	2 blue boxes and 2 green boxes
3	2:58'00	1 red box, 1 green box and 1 blue box

The competition was divided into three days, which each day occurring a different round. In the first and second rounds the robot was able to pick-and-place all boxes required and, in the third round, the robot placed just 3 parts correctly. The performed times are presented in Table 2. Figure 10 shows the robot during the third round (real scenario).

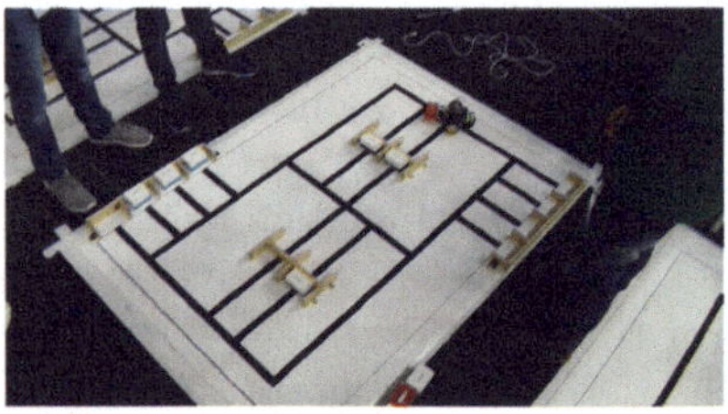

Fig. 10. Real robot in the third round.

6 Conclusion and Future Work

This paper presented a solution for the R@FL competition of the Portuguese Robotics Open. The simulation environment and tools provided by the organization were used to develop strategies to complete the three rounds. The importance of HIL in mobile robot applications to optimize the time of project implementation was evidenced.

The proposed solution in the first round is based in a TSFM technique, which was a suitable option because this turn required few number of states. However, as the remaining rounds had a higher number of paths implying in a larger set of states, the TSFM technique was not possible. Therefore, the solution was to simplify the code using functions with the common paths to deliver the parts.

The approach by function while compared with the TFSM method presented some advantages, principally, the development time required and the computational consumption. The robot could not deliver the four boxes in the last round due to non-systematic errors (irregular floor and sliding of the robot). These errors jeopardized the runs as our robot's decisions were time dependent. As future work, it is proposed to include encoders on the shaft of the motors to improve the control by using odometry. This will allow better precision of movements and potentially discard the time dependency in the robot decision making.

Acknowledgements. This work is financed by the ERDF – European Regional Development Fund through the Operational Programme for Competitiveness and Internationalisation - COMPETE 2020 Programme within project (POCI-01-0145-FEDER-006961), and by National Funds through the FCT – Fundação para a Ciência e a Tecnologia (Portuguese Foundation for Science and Technology) as part of project UID/EEA/50014/2013.

References

1. Oppliger, D.: Using first Lego league to enhance engineering education and to increase the pool of future engineering students (work in progress). In: 32nd Annual Frontiers in Education, vol. 3, p. S4D. IEEE (2002)
2. Holz, D., Iocchi, L., Van Der Zant, T.: Benchmarking intelligent service robots through scientific competitions: the RoboCup@ Home approach. In: 2013 AAAI Spring Symposium Series (2013)
3. Petillot, Y., Ferreira, F., Ferri, G.: Performance measures to improve evaluation of teams in the euRathlon 2014 sea robotics competition. IFAC-PapersOnLine **48**(2), 224–230 (2015)
4. Iocchi, L., Holz, D., Ruiz-del-Solar, J., Sugiura, K., Van Der Zant, T.: RoboCup@ Home: analysis and results of evolving competitions for domestic and service robots. Artif. Intell. **229**, 258–281 (2015)
5. Costa, P., Moreira, N., Campos, D., Gonçalves, J., Lima, J., Costa, P.: Localizaçao e navegacão de um robô môvel omnidirecional: caso de estudo da competição robot@ factory. VAEP-RITA **3**(1) (2015)
6. Costa, P.J., Moreira, N., Campos, D., Gonçalves, J., Lima, J., Costa, P.L.: Localization and navigation of an omnidirectional mobile robot: the robot@ factory case study. IEEE Revista Iberoamericana de Tec. del Aprendizaje **11**(1), 1–9 (2016)
7. Bischoff, R., Friedrich, T., Kraetzschmar, G.K., Schneider, S., Hochgeschwender, N.: RoCKIn@ work: industrial robot challenge. In: RoCKIn: Benchmarking Through Robot Competitions, vol. 47 (2017)
8. Ferreira, F., Ferri, G., Petillot, Y., Liu, X., Franco, M.P., Matteucci, M., Grau, F.J.P., Winfield, A.F.: Scoring robotic competitions: balancing judging promptness and meaningful performance evaluation. In: 2018 IEEE International Conference on Autonomous Robot Systems and Competitions (ICARSC), pp. 179–185. IEEE (2018)
9. Lima, J., Costa, P., Brito, T., Piardi, L.: Hardware-in-the-loop simulation approach for the Robot at Factory Lite competition proposal. In: 2019 IEEE International Conference on Autonomous Robot Systems and Competitions (ICARSC), pp. 1–6. IEEE (2019)
10. Robot at Factory Lite Competition files. https://github.com/P33a/RobotAtFactoryLite
11. Piardi, L., Eckert, L., Lima, J., Costat, P., Valente, A., Nakano, A.: 3D simulator with hardware-in-the-loop capability for the micromouse competition. In: 2019 IEEE International Conference on Autonomous Robot Systems and Competitions (ICARSC), pp. 1–6. IEEE (2019)

Web Based Robotic Simulator for Tactode Tangible Block Programming System

Márcia Alves[1(✉)], Armando Sousa[2], and Ângela Cardoso[3]

[1] FEUP, Porto, Portugal
msdsalves1@gmail.com
[2] INESC-TEC & FEUP, Porto, Portugal
[3] INEGI, Porto, Portugal

Abstract. Nowadays, with the increase of technology, it is important to adapt children and their education to this development. This article proposes programming blocks for young students to learn concepts related to math and technology in an easy and funny way, using a Web Application and a robot.

The students can build a puzzle, with tangible tiles, giving instructions for the robot execute. Then, it is possible to take a photograph of the puzzle and upload it on the application. This photograph is processed and converted in executable code for the robot that can be simulated in the app by the virtual robot or performed in the real robot.

Keywords: Education · Programming · Technology for education · Tangible system · Web application · ArUco · Simulator

1 Introduction

Since the birth of the internet, the development of technology has been increasing. With this growth, education also has to change and evolve [1]. Computational thinking (CT) is a set of thinking skills, habits, and approaches that are essential to solving problems using a computer [2] and has to be more present in education. However, students are reluctant to choose computer programming as a subject due to its perceived difficulty. But, it is well known that children that are introduced to computer programming will finish graduates in computer science in the future [1].

Tangible programming makes programming an activity that is accessible to the hands and minds by making it more direct, less abstract [2] and easy to understand when connected to robotics. Robotics apply and content knowledge in a meaningful and exciting way [1], so students are able to improve CT and think deeply, therefore they learn how the technology works.

Although it is difficult for teachers to include new things in the regular curriculum, because of the academic standards, the aim is to connect robotics with STEM (Science, Technology, Engineering, Mathematics) standards. Besides, it

M. F. Silva et al. (Eds.): ROBOT 2019, AISC 1092, pp. 490–501, 2020.
https://doi.org/10.1007/978-3-030-35990-4_40

is also difficult for schools, support the price of that kind of technology for all students [3].

Howsoever, the propose is Tactode as a Web Application, to program robots like a game in the classroom, based on programming blocks with tiles, similar to a puzzle that children need to solve. It is possible to create the puzzle, with tangible tiles, take a photograph and upload it in the application. In this way, children can see, in the app, a virtual robot simulating the code or send it to the real robot. This is useful to test the code created before execute in the real robot, or just to play in anywhere without having to carry the robot.

Hereupon, the suggestion is just a set of puzzle pieces and one computer, tablet or even a smartphone per group of students. In this way, the kids are able to learn about robotics and have a programming logic since a young age in a fun way and also improve team cooperation, working in a group to solve a problem [4]. Even for the teachers is easier to manage a small group of equipment than equipment for each student.

2 State of the Art

2.1 Programming Education

When teaching programming to children and young adolescents there are clear advantages in starting with an educational language [5], because it should have a simpler syntax, lower entry-level requirements and sometimes already provide activities aimed at grabbing the attention of the young. Also, it is getting easier everyday to have access to an educational language for every situation, particularly with development of educational programming languages by giant software companies such as Apple (Swift Playgrounds [6]), Google (Blockly [7–9]) and Microsoft (MakeCode [10]).

Block based languages are probably the most influential educational languages of today. As the name implies, they contain a set of predefined blocks of code, which the user typically drags and drops to form a program. Examples of this kind of languages are Scratch [5,11], Snap! [12], Stencyl [13], Blockly, MakeCode, Alice [14–16] and Etoys [17]. While Scratch drew inspiration from the precursors Alice and Etoys, it has since influenced most of the others.

2.2 Tangible Blocks Programming Language

A programming language is considered visual when it is mostly text independent, relying on icons and images to represent its elements. This is not, however, the consensus, as all the languages above use text in their blocks and yet they are typically considered visual. Examples of visual languages are ScratchJr [18], Kodu [19], the Lego block based visual programming language designed to program their EV3 [20] robot, Lightbot [21], the Fisher Price Think & Learn Code-a-Pillar [22] in its tablet application format, Coding Safari [23], SpriteBox [24] and codeSpark [25]. All these languages allow younger children, even before they can read, to learn programming concepts while playing.

A tangible programming language is a block-based language whose blocks can be physically grabbed and arranged by the programmer. In their 2015 study, Sapounidis et al. [26] established many advantages of using tangible interfaces to teach programming, particularly for children up to ten years old. The children completed the tasks faster, with fewer errors, more debugging of the errors they made, better collaboration, using a wider set of different blocks and, in some cases, achieving higher complexity [3,27]. Also, they considered the tangible interface more attractive, more enjoyable and, for the younger ones, easier to use.

As for disadvantages, tangible languages are more expensive and less portable, due to their physical aspect. However, these issues can be mitigated by using less expensive materials and make it possible for schools to manufacturing the tangible blocks themselves.

Examples of tangible languages are AlgoBlock [28], Electronic [29], Cubelets [30–32], the physical robot version of the Fisher Price Think & Learn Code-a-Pillar [33], TagTile [34], Quetzal and Tern [35], T-Maze [2,36], the Osmo Coding Family [37], and CodeBits [38].

The more recent tangible languages are replacing electronic components with image processing to execute their programs, unless they need the electronics because the language is simultaneously a robot, which is what happens in Cubelets and Code-a-Pillar.

2.3 Similar Projects

In order to understand the similar projects mentioned before, this section explains briefly some of them.

Lego Mindstorm EV3 is a set of programmable robotics construction for ages greater than 10 years. The aim is to build the own robot with Lego pieces, program and command it. The EV3 set includes bricks, motors, and sensors to build the robot and make it walk, talk, move. It also comes with the necessary software and App, where the robot can be easily programmed and controlled, with basic tasks, from PC, Mac, tablet or smartphone [39] (Fig. 1).

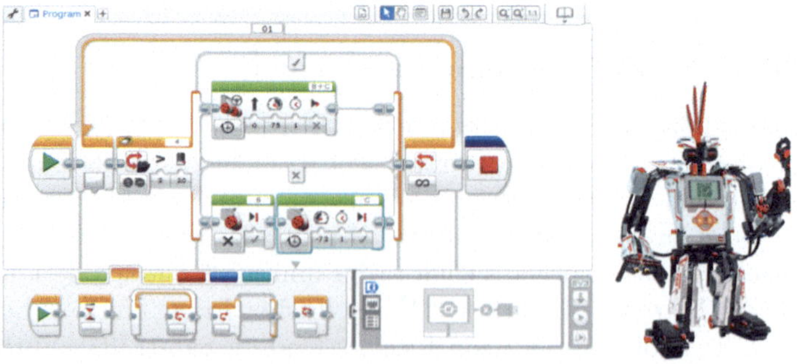

Fig. 1. The Lego Mindstorms EV3 programming interface and robot.

Ozobot has two ways to code their robots. They can be coded online with OzoBlockly or screen-free with Color Codes. The purpose is to inspire young minds to go from consuming technology to creating it. [40] OzoBlocky can be used with an application or in a Web browser. The application can be used in iOS or Android tablet, working with the Evo robot. The browser is also compatible with Evo robot, while used with a computer, or compatible with bit robot for a computer or tablet [41] (Fig. 2).

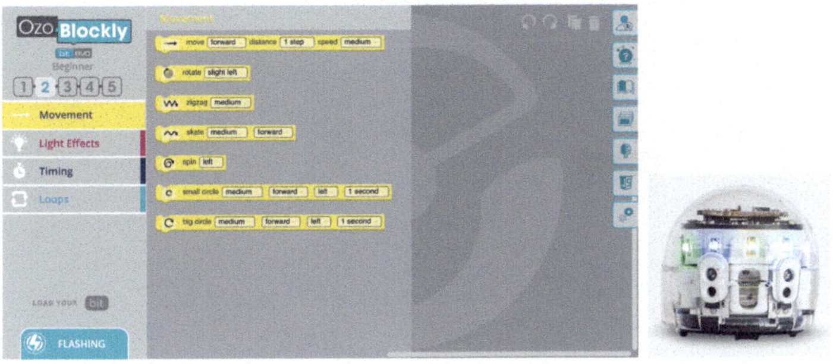

Fig. 2. The OzoBlockly Editor and Ozobot robot.

Open Roberta Lab is a free platform that makes learning programming easy from the first steps to program robots with multiple sensors and capabilities. It can be used at any time without installation by any devices, PC, Mac or tablet, with an Internet browser. Thanks to the programming language NEPO (a graphical programming language developed at Fraunhofer IAIS), simple programs can be created like puzzle pieces [42] (Fig. 3).

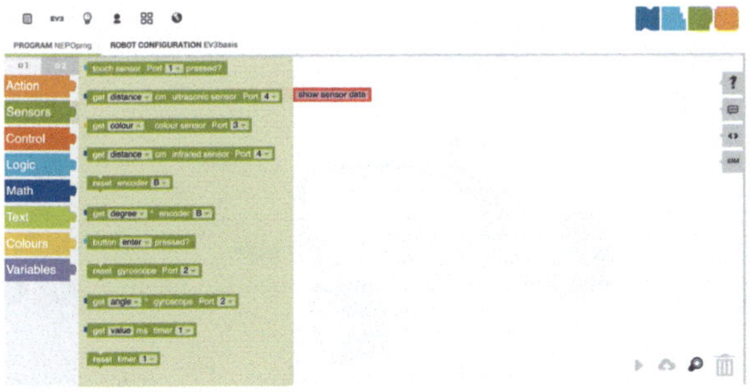

Fig. 3. Open Roberta Lab.

Blockly Games is a free Google project with a series of educational games that teach programming. There are different games, with different levels, so children who have not had prior experience are ready to use conventional text-based languages by the end of these games [43] (Fig. 4).

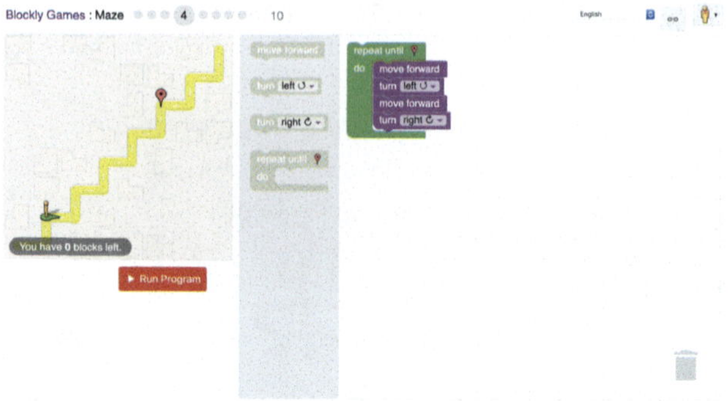

Fig. 4. Blockly Games.

Scratch is made for children between six and eight years old learn how to code and important strategies for solving problems, designing projects, and communicating ideas [44]. The activity is mixing graphics, animations, photos, music, and sound, supporting different types of projects like stories, games, animations, simulations, so people are all able to work on projects they care about [5] (Fig. 5).

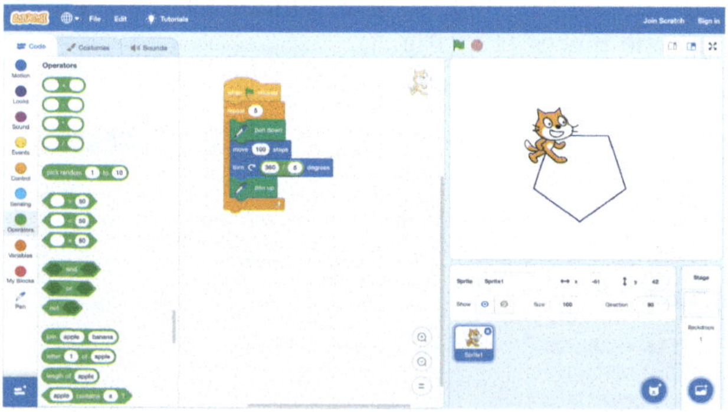

Fig. 5. Scratch.

2.4 Development Tools

The Tactode application aims to simulate the real robot into the application. In order to be accessible for all users, the Tactode application should be compatible with all platforms [45]. Nowadays, exist several options to create that.

It was used Ionic Framework, that generates applications for multiple systems from a single source code, using AngularJs and TypeScript [46]. Besides, it was used Threejs to develop the simulator. Ionic uses Cordova to have access to host operating systems features such as Camera, GPS, Flashlight, etc. It includes mobile components, typography, interactive paradigms, and an extensible base theme [47]. In Tactode, the principal feature used is Camera. Threejs is a high-level JavaScript library and Application Programming Interface used to create and display animated 3D graphics in web browser. The use of WebGL allows complex animations to be created without having to use plugins [48].

3 Tactode Programming System

Tactode Programming System is made for elementary school children with the aim of teaching them how to program in a fun and interactive way, by building puzzles that represent a chain of commands which a robot will follow, so they can see how it reacts to the different puzzle compositions. The mobile device will capture a photograph of the tangible puzzle, process it, create the commands for the robot and show the results in a simulator. The focus is the simulation of the robot and, in this section, will be addressed the requirements for this simulator, its architecture and possible challenges that children can create.

3.1 Requirements

The purpose of Tactode Programming System is to be easy to understand for the users, so they can immediately see how Tactode works and easy to use and install. It has a Web App prepared to process the tangible puzzle directly in the browser.

There are two ways to upload a puzzle on the app:

- Upload a previous photo by searching an image on the device.
- Take a photo directly, using a hand-held camera device where the application is running.

After uploading the photo, children have two options to see the execution.

- Simulating in the application, seeing a virtual robot execution.
- Real execution, using a real robot.

3.2 Architecture

This section will explain how Tactode pieces are defined and built in the simulator. Every uploaded puzzle is processed as Abstract Syntax Tree Structure (AST). Each original piece has a corresponding Block and each of them has an array of other Block objects (children) and also a parent Block object. In this way, each object knows the parent and the children. However, there are some extra elements in the AST that do not have a piece in Tactode, such as:

- RootBlock: special Block with no parent that serves as the root of the AST. Its children are the command blocks that are not inside of any control flow block, which means that have no indentation in the tangible language.
- BodyBlock: child of control flow blocks RepeatBlock, ForeverBlock, IfBlock, ElseBlock or WhileBlock.
- ConditionBlock: child of RepeatBlock, IfBlock or WhileBlock and as the name suggests, it contains the condition to be verified by these control flow blocks;

Figure 6 shows an example of a square - a loop, running four times, that inside move forward, with a fixed distance, and turn right/left 90°.

In this case, Root Block has one child - Repeat Block - and this block has three more children:

- Condition Block that usually has a child the number of repeats of the cycle.
- Body Block where are introduced the main instructions. It has two children: Forward Block and Turn Left Block. Forward Block has two more children - Distance and Speed - each with Number block as children. Turn Left Block is similar but instead of distance, it has Angle Block.
- End Repeat Block only ends the repeat, as the name indicates.

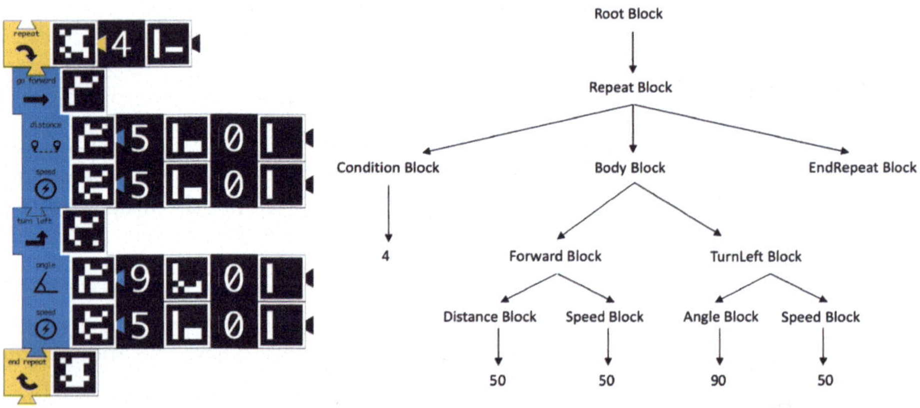

Fig. 6. AST of a square.

4 Challenges

For each target platform, a set of challenges is designed for experiments. These challenges are detailed in this section. There are many possibilities to program each target. Challenges were designed to improve educational value. Kids will improve math concepts by using operators such as addition, subtraction, division, multiplication, and tact to move using velocity and sensors. They can also program flow control, such as repeat, forever, while, if, and else. There are four main challenges: regular polygon construction, two types of obstacle reaction and line follow.

4.1 Regular Polygon

This challenge is intuitive for children understand. They need to know some concepts about regular polygons. Regular Polygons have all sides with the same length and their internal and external angles have the same amplitude. There are n regular polygons with n sides where the amplitude of each angle the robot needs to turn is 360°/n.

In Fig. 7 it is possible to see a program for building a pentagon and its simulation. Note that, the tag pen down is only available for platform scratch, so this program only works when the platform scratch is selected. Another exception is that Ozobot does not implement events like the tag Flag. So, this program can be built without pen down tag in Cozmo, Shero, Robobo, and Ozobot (without flag). In Scratch and Robobo, it is also feasible to ask the user a question like how many sides the user wants. After the question, the program waits to the answer and uses that to build the polygon, like Fig. 7 shows.

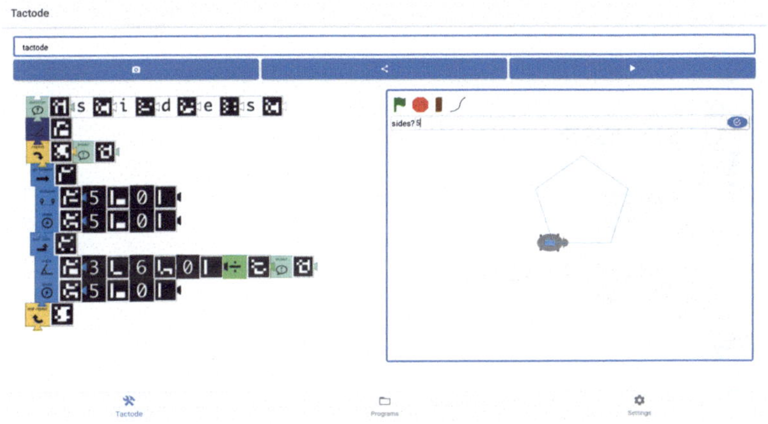

Fig. 7. Puzzle and simulation of a pentagon.

4.2 Obstacle Reaction

The obstacle reaction and sensors are for Ozobot. In these examples, the robot can stop or running away when facing an obstacle. The obstacle can be placed anywhere and moved around the scene. In Figs. 8 and 9 can be seen different robot reactions to that situations. In both cases, it is told to the robot to build a square. The robot tries to complete the orders but in Fig. 8 it stops when the obstacle is in front of it and, in Fig. 9, the robot pick back direction, running away from the obstacle.

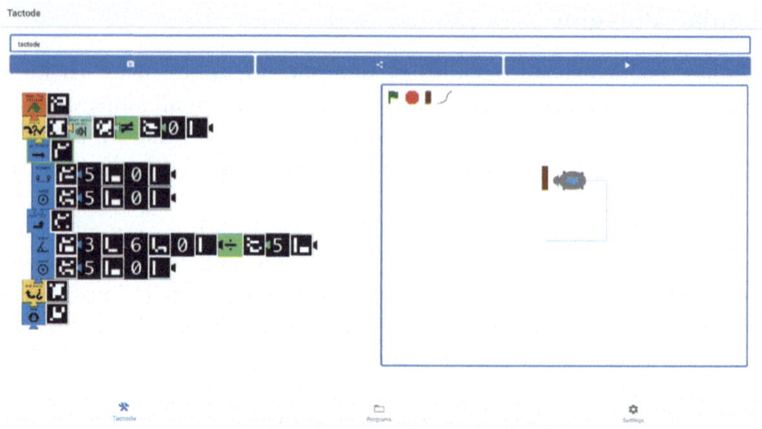

Fig. 8. Puzzle and simulation of stopping in front of an obstacle.

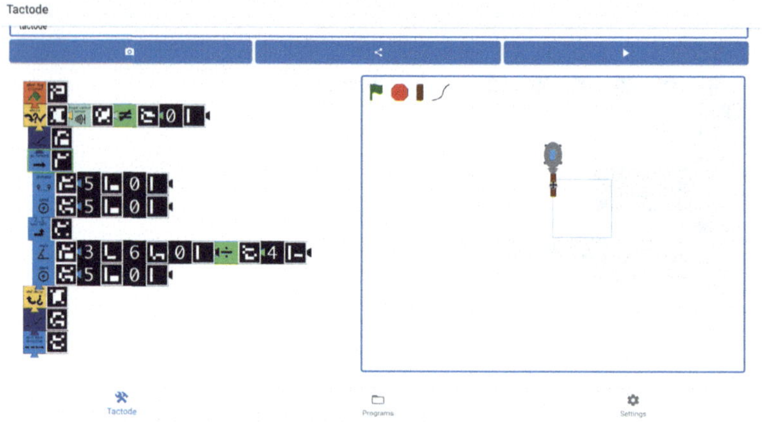

Fig. 9. Puzzle and simulation of running away of an obstacle.

4.3 Follow Line

This functionality only works in Ozobot because it is the only one with line detection capabilities. The line is generated randomly when the button line is clicked. The Fig. 10 shows that while the robot sees a black line, it follows it. When the line ends, the robot no longer sees the line and the while loop ends, stopping the movement.

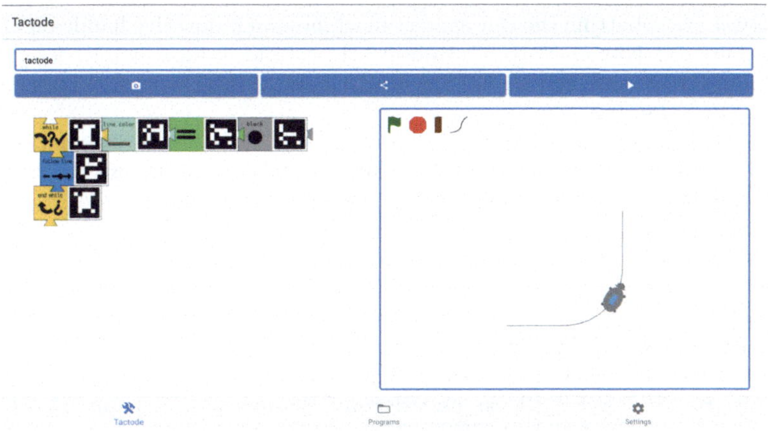

Fig. 10. Puzzle and simulation of a line follow.

5 Conclusion

This project focuses on motivating children, attending elementary education, engineering, and programming, using an application and robotics, so they can understand how computer science works. Children can give orders to a virtual robot, run it on the simulator or a real robot. To be accessible to more people, it would be interesting if the applications could be a Progressive Web App (PWA), which runs on any browser with Internet access. The PWA only needs internet to open the application and then it can run offline. In this way, Tactode can be more general and accessible to society anytime, anywhere. More challenges could be created, a labyrinth, for example, and the current ones could be improved. An interesting improvement would be if it were possible to paint to be followed, with different colors. Another could be to choose what kind of reactions the robot might have when facing an obstacle, such as approaching the obstacle instead of running away and being available on all sides (/ sensors) of the robot (front, left frontal, right frontal and back).

References

1. Eguchi, A.: Bringing robotics in classrooms. In: Khine, M. (ed.) Robotics in STEM Education, pp. 3–31. Springer, Cham (2017). https://doi.org/10.1007/978-3-319-57786-9_1
2. Wang, D., Wang, T., Liu, Z.: A tangible programming tool for children to cultivate computational thinking. Sci. World J. **2014** (2014). https://doi.org/10.1155/2014/428080
3. Cardoso, A., Sousa, A., Ferreira, H.: Programming for young children using tangible tiles and camera-enable handheld devices. In: 11th Annual International Conference of Education, Research and Innovation, pp. 6389–6394 (2018). https://doi.org/10.21125/iceri.2018.2504

4. Chetty, J.: Combatting the war against machines: an innovative hands-on approach to coding. In: Khine, M. (ed.) Robotics in STEM Education: Redesigning the Learning Experience, pp. 59–83. Springer, Cham (2017). https://doi.org/10.1007/978-3-319-57786-9_3
5. Resnick, M., Maloney, J., Monroy-Hernández, A., Rusk, N., Eastmond, E., Brennan, K., Millner, A., Rosenbaum, E., Silver, J., Silverman, B., Kafai, Y.: Scratch: programming for all. Commun. ACM, 60–67 (2009). https://doi.org/10.1145/1592761.1592779
6. Apple: Swift Playgrounds. https://www.apple.com/swift/playgrounds
7. Fraser, N.: Ten things we've learned from Blockly. In: 2015 IEEE Blocks and Beyond Workshop (Blocks and Beyond), pp. 49–50 (2015). https://doi.org/10.1109/BLOCKS.2015.7369000
8. Pasternak, E., Fenichel, R., Marshall, A.N.: Tips for creating a block language with Blockly. In: 2017 IEEE Blocks and Beyond Workshop (B B), pp. 21–24 (2017). https://doi.org/10.1109/BLOCKS.2017.8120404
9. Google for Education: Blockly. https://developers.google.com/blockly
10. Microsoft: MakeCode. https://makecode.com
11. Lifelong Kindergarten Group at the MIT Media Lab: Scratch (2005). https://scratch.mit.edu
12. Mönig, J.: Snap! http://snap.berkeley.edu/about.html
13. Chung, J.: Stencyl. http://stencyl.com
14. Pausch, R., Burnette, T., Capehart, A.C., Conway, M., Cosgrove, D., DeLine, R., Durbin, J., Gossweiler, R., Koga, S., White, J.: Alice: rapid prototyping for virtual reality. IEEE Comput. Graph. Appl. **15**, 8–11 (1995). https://doi.org/10.1109/38.376600
15. Cooper, S., Dann, W., Pausch, R.: Alice: a 3-D tool for introductory programming concepts. J. Comput. Sci. Coll. **15**, 107–116 (2000). http://dl.acm.org/citation.cfm?id=364133.364161
16. Carnegie Mellon University: Alice. https://www.alice.org
17. Kay, A., et al.: Squeakland. http://www.squeakland.org
18. Lifelong Kindergarten Group at the MIT Media Lab: ScratchJr. https://www.scratchjr.org
19. Microsoft Research: Koduv (2009). https://www.kodugamelab.com
20. Lego: Mindstorms: Learn To Program (2013). https://www.lego.com/en-us/mindstorms/learn-to-program
21. Yaroslavski, D.: LightBot (2017). http://lightbot.com
22. Fisher Price: Think & Learn Code-a-Pillar Application. https://www.fisher-price.com/en_US/brands/think-and-learn/learning-apps/index.html
23. Hopster: Coding Safari. https://www.hopster.tv/coding-safari/
24. SpriteBox LLC: SpriteBox. http://spritebox.com/hour.html
25. codeSpark: codeSpark Academy: Kids Coding. https://codespark.com
26. Sapounidis, T., Demetriadis, S., Stamelos, I.: Evaluating children performance with graphical and tangible robot programming tools. Pers. Ubiquitous Comput. **19**, 225–237 (2015). https://doi.org/10.1007/s00779-014-0774-3
27. Cardoso, A., Sousa, A., Ferreira, H.: Easy robotics with camera devices and tangible tiles. In: 11th Annual International Conference of Education, Research and Innovation, pp. 6400–6406 (2018). https://doi.org/10.21125/iceri.2018.2506
28. Suzuki, H., Kato, H: AlgoBlock: a tangible programming language, a tool for collaborative learning. In: Proceedings of the 4th European Logo Conference, pp. 297–393 (1993)

29. Wyeth, P., Purchase, H.: Designing technology for children: moving from the computer into the physical world with electronic blocks. In: Information Technology in Childhood Education Annual, vol. 2002, pp. 219–244 (2002). http://eprints.gla.ac.uk/14107/
30. Modular Robotics: Cubelets (2012). https://www.modrobotics.com/cubelets/
31. Correll, N., Wailes, C., Slaby, S.: A one-hour curriculum to engage middle school students in robotics and computer science using cubelets. In: Ani Hsieh, M., Chirikjian, G. (eds.) Distributed Autonomous Robotic Systems, pp. 165–176. Springer, Berlin (2014). https://doi.org/10.1007/978-3-642-55146-8
32. Wohl, B., Porter, B., Clinch, S.: Teaching computer science to 5–7 year-olds: an initial study with scratch, cubelets and unplugged computing. In: Proceedings of the Workshop in Primary and Secondary Computing Education, pp. 55–60. ACM, New York. https://doi.org/10.1145/2818314.2818340
33. Fisher Price: Think & Learn Code-a-Pillar. https://fisher-price.mattel.com/shop/en-us/fp/think-learn/think-learn-code-a-pillar-dkt39
34. KUBO Robotics: KUBO (2017). https://kubo-robot.com
35. Horn, M.S., Jacob, R.J.K.: Designing tangible programming languages for classroom use. In: Proceedings of the 1st International Conference on Tangible and Embedded Interaction, pp. 159–162. ACM, New York (2007). https://doi.org/10.1145/1226969.1227003
36. Wang, D., Zhang, C., Wang, H.: T-Maze: a tangible programming tool for children. In: Proceedings of the 10th International Conference on Interaction Design and Children, pp. 127–135. ACM, New York (2011). https://doi.org/10.1145/1999030.1999045
37. Osmo: Osmo Coding Family. https://www.playosmo.com/en/coding-family/
38. Goyal, S., Vijay, R.S., Monga, C., Kalita, P.: Code bits: an inexpensive tangible computational thinking toolkit for K-12 curriculum. In: Proceedings of the TEI 2016: Tenth International Conference on Tangible, Embedded, and Embodied Interaction. pp. 441–447. ACM, New York (2016). https://doi.org/10.1145/2839462.2856541
39. LEGO group: Support - Mindstorms (2018). https://www.lego.com/en-us/mindstorms/support
40. Ozobot: About us. https://ozobot.com/about-us
41. Ozobot: Getting Started Guide. https://files.ozobot.com/stem-education/ozoblockly-getting-started.pdf
42. Roberta: Learning to program intuitively in the Open Roberta Lab. https://www.roberta-home.de/en/lab/
43. Google for Education: Blockly Games: About. https://blockly-games.appspot.com/about?lang=en
44. Scratch: About Scratch. https://scratch.mit.edu/about
45. Ionic: Browser Support (2019). https://ionicframework.com/docs/intro/browser-support/
46. Ionic: What is Ionic Framework? (2019). https://ionicframework.com/docs/intro
47. Ionic creator: Custom Code Editing (2019). https://docs.usecreator.com/docs/custom-code-editing
48. Threejs: Three.js (2019). https://threejs.org/

Development of an AlphaBot2 Simulator for RPi Camera and Infrared Sensors

Ana Rafael[1], Cássio Santos[2], Diogo Duque[1], Sara Fernandes[1(✉)], Armando Sousa[1,3], and Luís Paulo Reis[4,5]

[1] FEUP - Faculty of Engineering, University of Porto, Porto, Portugal
{up201405377,up201406274,up201405955,asousa}@fe.up.pt
[2] State University of Feira de Santana, Feira de Santana, Bahia, Brazil
sss.cassio@gmail.com
[3] INESC TEC - INESC Technology and Science, Porto, Portugal
[4] DEI/FEUP, Informatics Engineering Department, Faculty of Engineering of the University of Porto, Porto, Portugal
lpreis@fe.up.pt
[5] LIACC/UP, Artificial Intelligence and Computer Science Laboratory of the University of Porto, Porto, Portugal

Abstract. In recent years robots have been used as a tool for teaching purposes, motivating the development of fully virtual environments for combined real/simulated robotics teaching. The AlphaBot2 Raspberry Pi (RPi), a robot used for education, has no currently available simulator. A Gazebo simulator was produced and a ROS framework was implemented for hardware abstraction and control of low-level modules facilitating students control of the robot's physical behaviours in the real and simulated robot, simultaneously. For the demonstration of the basic model operation, an algorithm for detection of obstacles and lines was implemented for the IR sensors, however, some discrepancies in a line track timed test were detected justifying the need for further work in modelling and performance assessment. Despite that, the implemented ROS structure was verified to be functional in the simulation and the real AlphaBot2 for its motion control, through the input sensors and camera.

Keywords: AlphaBot2 RPi · Educational robotics · Gazebo simulator · Line following · Obstacle avoidance · Robotics Operating System (ROS)

1 Introduction

The use of robots for teaching purposes has proven to be a motivational tool for students to learn science, technology, engineering and mathematics (STEM) [1]. However, due to low budget for resources, which limit the number of available robots in a university and the necessity of repair in case of damage, the reduced access to these materials may be a constraint at the development and

M. F. Silva et al. (Eds.): ROBOT 2019, AISC 1092, pp. 502–514, 2020.
https://doi.org/10.1007/978-3-030-35990-4_41

experimenting phases of a project. Robot simulations allow for an inexpensive and fast debugging process, experimental repeatability and easy alteration in the environment and robot dynamics, giving the possibility to test changes in numerous parameters without compromising the real robot [2]. Such platforms provide a stand-alone environment, resulting in no restrictions from lab hours or power management inherent to all physical systems.

Computer simulations of mobile robots require the modelling of kinematic characteristics imposed by the robot's wheels [3]. Physics inaccuracies (e.g., friction, gravity, mass, force, etc.) often result in strange and inaccurate behaviours in the simulation resulting in the impossibility of evaluating robot's performance with faulty models [4].

The AlphaBot2 (represented in Fig. 1) is a compact two-wheeled robot with infrared (IR) sensors, 5 lower ones for line tracking and 2 superior for obstacle avoiding, a RPi camera (oriented by two servo motors), a micro SC card, a 5 V/2.5 V USB adaptor, IR remote, and some additional components. This robot uses a Raspberry Pi 3 which is a series of small single-board computers with the Raspbian operating system so that it is possible to move the respective robot. It is easily configured and has demo codes for its basic behaviours, described and available in [5]. It has a low price (minimum 82$) and is a great kit for teaching programming and control strategies for the performance of various services.

The motivation for the creation of a simulation model of the AlphaBot2 is related to the non-existence of a model for this robot. A simulation model of it allows the better organisation of classes combining real/simulated robots, facilitates its behaviour control eliminating noise and uncertainty introduced by the real world and facilitates its use in classes since there is no need for the real robot at all time for all the students.

Fig. 1. Assembled AlphaBot2 RPi front view [5].

The present work aims to implement the camera image acquisition, the control of the pan and tilt motors, the reading of the IR sensors and the control of the wheels' motors in the real and simulated robot, through the Gazebo simulator [6]. The final purpose is to test and finely tune the virtual model for verification of similar performances between the real and simulated robot specifically for the RPi camera and the IR sensors. All the simulations were created using Gazebo and its plugins, namely the laser sensor or camera sensor plugins as well as ROS commands [7]. Gazebo is an open source simulator which allows the creation/recreation of complex 3D environments encountered by the next generation of mobile robots.

The current paper starts by presenting the most relevant work produced up to date in Sect. 2, then proceeds to a description of the developed robot and the implemented simplifications to it, in Sect. 3. Additionally, the ROS message-passing architecture is represented in order to have a fair understanding of the developed tasks, based in the one developed in the Conde project [8]. Section 4 presents the results for the tested parameters and developed control algorithm. Finally, Sect. 5 outlines the main conclusions and future work.

2 Line Following and Obstacle Avoidance Related Work

Focusing the following analysis in line following and obstacle avoidance robots reviewed in the literature the most distinctive studies and experimental activities are referred below.

In [9], a robot achieves high speeds while following lines independently. This robot uses the novel square-topology IR sensor matrix that allows the anticipation of a turn by sensing a curve ahead. Study [10] uses image preprocessing for line following robots, through the Track-Before-Detect algorithm. In addition, this robot and its algorithms use deep learning to estimate the line and the area beyond the line.

Article [11] presents an algorithm that allows a wheeled robot to follow a path and avoid obstacles that cross in front of him. Additionally, [12] accepts user input for line selection.

Finally, paper [13] describes the implementation of a method that efficiently detects and avoids obstacles through a 2D LiDAR of an autonomous robot. This method extracts spatial information from a point-cloud laser using segmentation and clustering methods. In order to detect the geometric structure of the obstacle, the Convex hull algorithm was used.

3 Methodologies

This section yields a description of the developed robot's model in a 3D simulator, the implemented world, a simplification of the Portuguese autonomous driving competition and the ROS architecture.

3.1 Design of the Robot Model

Since many robot components did not need to be simulated, only a robot with the chassis, sensors, wheels, casters and camera was created, where the positions of these components are the same as the positions they occupy in the real robot. The most significant measurements for the several components are detailed in Table 1.

Table 1. AlphaBot2 RPi measurements.

Components	Weight (g)	LWD[a] or DT[b] or D[c] (cm)	Height (cm)
Base board	150	10.9×0.15	1.6
Pi board	45		
Wheels	–	4.1×1.7	–
Balance wheel	–	1	1.5
Camera	55	$3.7 \times 3.3 \times 7.5$	–
Pillars	–	–	3.8
Raspberry Pi	40	$8.4 \times 5.5 \times 1.7$	-
Total			**15.5**

[a]Length × Width × Depth.
[b]Diameter × Thickness.
[c]Sphere × Diameter.

For the design of the robot in the Gazebo simulator a URDF (*.xacro*) file was written. In it, the links define the inertial values, collisions and visuals for each of the AlphaBot2 components. For simplification reasons, the robot was designed resorting to simple geometries like cylinders and boxes. In a more advanced representation, the robot's chassis is represented through a mesh, the final model is represented in Fig. 2.

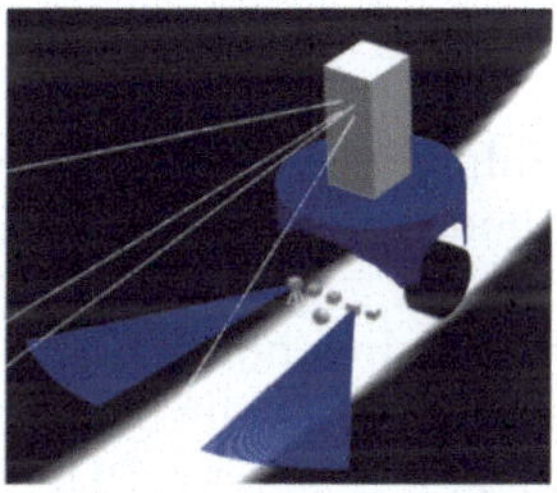

Fig. 2. Model of the AlphaBot2 RPi simulated in Gazebo.

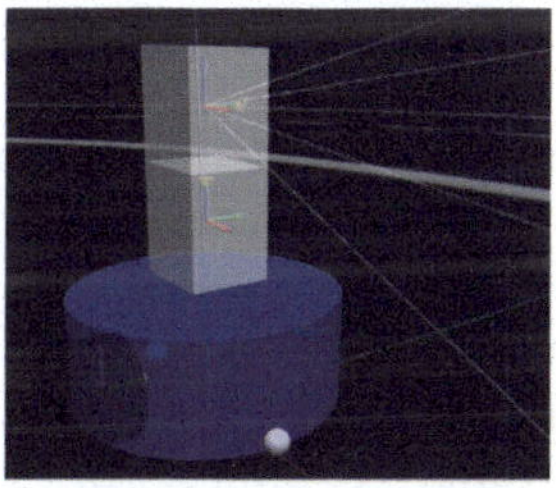

Fig. 3. Visualisation of the model joints in the Gazebo simulation.

The inertial values are calculated for these shapes having in consideration each of the components' weight. The joints locate the points of articulation between elements and the plugins set the needed connections between the Gazebo simulation environment and ROS. The camera orientation is controlled through two motors for pan-tilt action interfaced with the Gazebo plugin *libgazebo_ros_control.so* and the camera is activated with the *libgazebo_ros_camera.so*. As to the wheels these are controlled with the *libgazebo_ros_diff_drive.so*. A representation of the model joints is presented in Fig. 3, the axis of rotation for each one has a circular arrow indicating the allowed movement.

With regards to the IR sensors, first, the 5 lower sensors, with a vertical scan to detect lines, used the Gazebo plugin for a camera type sensor. Secondly, the 2 upper sensors with a range between -15° and 15°, making a total of 30° and a minimum distance of 1 cm and maximum of 10 cm for obstacle detection, use the plugin that creates laser sensors.

3.2 System Architecture

The system architecture is based on the ROS meta-operating system. Through the implementation of nodes which publish and subscribe to specific topics, message-passing between processes is accomplished. Two levels of packages (pkg) are responsible for the control of the robot's behaviour in the Gazebo simulation and in the real robot. Hence, on the side of the Gazebo simulation, two nodes are implemented for the readout of the camera sensor, which publishes a raw image, and a pkg for the control of the joints present in the camera stand (motions of pan and tilt) which subscribes to the angle positions for the two motors.

Considering first the modules for the camera related components control, the connections between nodes using ROS for the real are presented in Fig. 4, and in Fig. 5 is shown the simulator architecture. With regards to the IR sensors the node architecture used is delineated in Fig. 6.

Using these packages, students and developers can program and test their own applications subscribing or publishing to the topics that will be described next.

Description of ROS Nodes

alphabot2_control_node_real. Subscribes to "/alphabot2/control" and translate it to drive control the Alphabot2. (This node only runs on a RaspberryPI);

alphabot2_pantilt_node_real. Subscribes to "/alphabot2/vertical" and "/alphabot2/horizontal" topics and control the Pan and Tilt position of the RaspCam on Alphabot2. (This node only runs on a RaspberryPI);

alphabot2_pantilt_node_gazebo. Same as "alphabot2_pantilt_node_real" but controls Gazebo model virtual joints instead of real robot Pan-Tilt;

alphabot2_top_sensors_middleman. Receives info from Gazebo top sensors and re-transmits in standardised format to "/alphabot2/top_sensors";

alphabot2_bottom_sensors_middleman. Receives info from Gazebo bottom sensors and re-transmits in standardised format to "/alphabot2/bottom_sensors";

alphabot2_handler. Reads information provide by Alphabot2's Infrared sensors and re-transmits in standardised format to "/alphabot2/top_sensors" and "/alphabot2/bottom_sensors". Also subscribe to "/alphabot2/control" and translate it to drive control the Alphabot2. (This node only runs on a RaspberryPI)

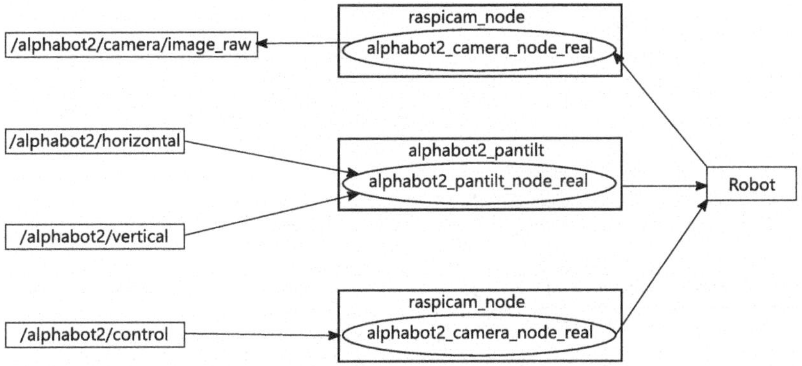

Fig. 4. ROS architecture for the camera related nodes in the real robot.

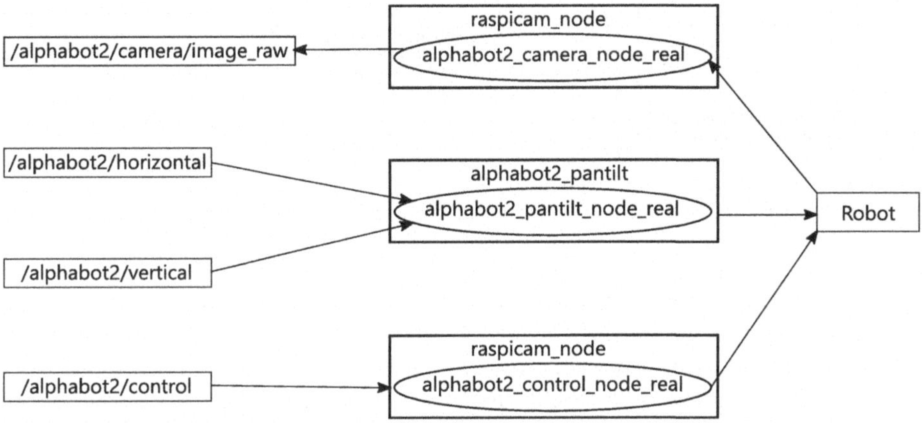

Fig. 5. ROS architecture for the camera related nodes in the simulator.

Description of ROS Topics

/alphabot2/control. Publish a message of type *geometry_msgs/Twist* setting *linear velocity* x and *angular velocity* z to control the movement of the AlphaBot2;

/alphabot2/vertical. Publish a message setting a value between -90 and 90 to control the tilt angle of the camera (Message type: *std_msgs/Float64*);

/alphabot2/horizontal. Publish a message setting a value between -90 and 90 to control the pan angle of the camera (Message type: *std_msgs/Float64*);

/alphabot2/camera/image_raw. Subscribe to this topic to read a message of type *sensor_msgs/Image*;

/alphabot2/top_sensors. Receives the result of the top sensors to be used by the real robot and the simulated robot. (Message type: *std_msgs /Int32MultiArray*);

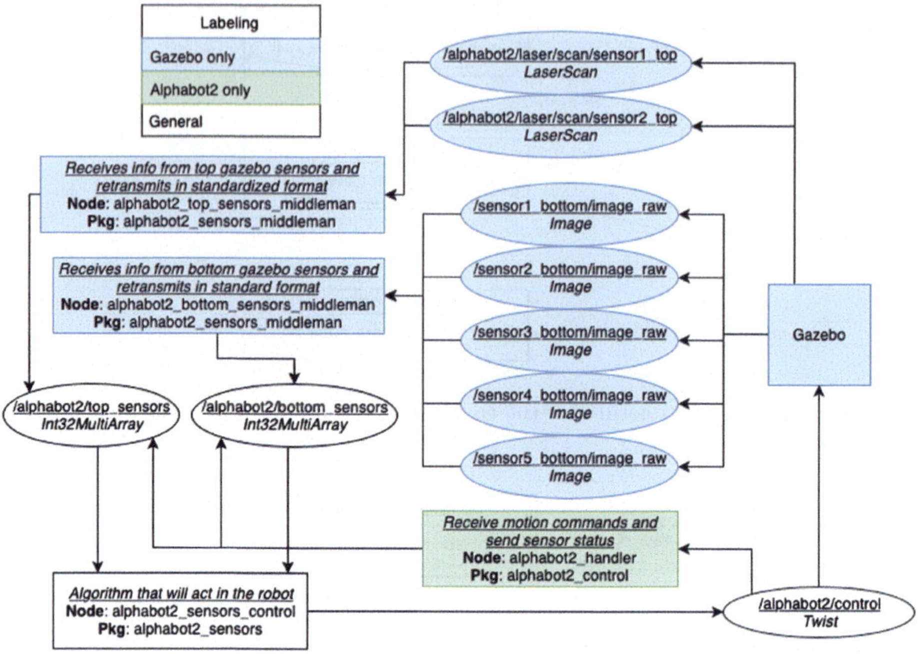

Fig. 6. ROS architecture for the IR sensor's related components.

/alphabot2/laser/scan/sensorx_top. Subscribes to the two upper sensors, represented by laser scanners to detect obstacles. x can take the value of 1 or 2.

/alphabot2/bottom_sensors. Receives the result of the bottom sensors to be used by the real robot and the simulated robot. (Message type: *std_msgs/Int32MultiArray*).

/sensorx_bottom/image_raw. Subscribes to the five bottom camera sensors to detect the lines brightness. x can take intriguer values from 1 to 5.

3.3 Robot Behaviour Control

This robot follows a subsumption based architecture like the one shown in Fig. 7 for demonstration of the basic operation of the model, regarding the main purpose of this research. According to the architecture, the robot presents only 4 possible behaviours. Initially, the robot moves itself around the map searching for a line (initial movement). As soon as the robot detects a line, it follows the respective line. When an object is detected the robot avoids the collision by deviating its path to the opposite direction of the obstacle. If the obstacle is in front of the robot it can deviate either to the left or to the right. Most of the code used by the real robot was taken from the demo code on the AlphaBot2 website, so one can program both its software and its hardware. Even so, it was still necessary to change some details, namely the linear and angular velocity translated values in the real AlphaBot so that both the real and the simulated robot walked at equal speeds when searching for the line to follow or avoid obstacles.

Obstacle Avoidance. To detect obstacles, it is verified if the values read by the two upper sensors are infinite. If they are not infinite, then the robot is in the presence of an obstacle. In this stage, an array with two boolean values representing the detection of obstacles by each of the sensors publishes to topic "/alphabot2/top_sensors".

After this initial analysis, the robot subscribes to the topic "/alphabot2/top_sensors". If one of the values is high it means there is an obstacle and the robot moves in the opposite direction with a minimal linear velocity of 0.1 m/s and an angular velocity of ±0.6 rad/s^2.

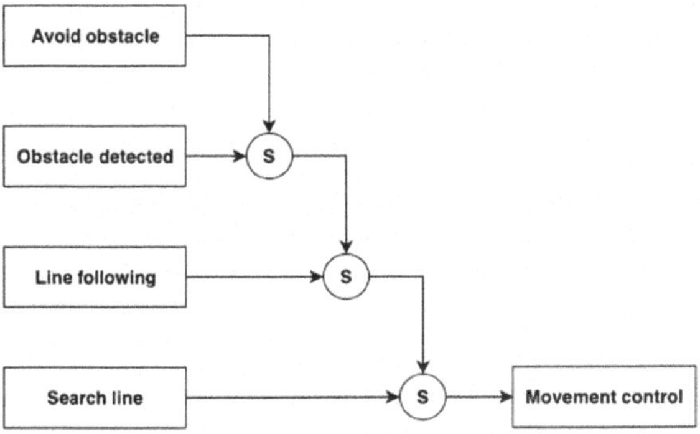

Fig. 7. Subsumption based architecture

Line Following. To follow lines, first, the brightness percentage scanned by the five bottom sensors is calculated, publishing them on the topic "/alphabot2/bottom_sensors". Then, the robot subscribes to this topic and the minimum and maximum brightness captured by the robot and the deviation of brightness values for the lateral sensors and the central sensors are calculated. Then the below conditions are assessed.

1. If the line has not yet been found and the maximum brightness is less than 50%, the robot will search the line, with a linear velocity of 0.4 m/s and an angular velocity of −0.1 rad/s^2.
2. Otherwise:
   (a) If the brightness of the middle sensor is greater than 50% the line was found.
   (b) If the absolute difference between the brightness of the two intermediate sensors is less than 50 and either of these two has brightness above 50, the robot is on top of the line and will have to move forward with a linear velocity of 0.4 m/s.

(c) If the maximum brightness (among all sensors) is above 50 and the absolute difference between the brightness of the 2 edge sensors is less than 50, the robot will have to slightly correct its trajectory, having a linear velocity of 0.4 m/s and an angular velocity of 0.1 rad/s^2 multiplied by a 1 or −1, depending on the turning direction.

(d) If none of the previously described cases occurs, the robot will have to completely change its trajectory, adopting a linear velocity of −0 m/s and an angular velocity of 0.1 rad/s^2 multiplied by a 1 or −1, depending on the turning direction.

4 Simulations and Experiments

In order to verify the correct implementation of the model and system architecture, nodes and topics, the robot was launched is an empty Gazebo's word (Fig. 8), and messages were published to the topics /alphabot2/vertical and /alphabot2/horizontal through the *rostopic command-line tool.*

Using the command: *rostopic pub /alphabot2/vertical std_msgs/Float64 "data: -45"* the camera starts to face the ground, represented in Fig. 9 where it is possible to see the new position of the Tilt and its vertical axis as the dashed red line which is tilted −45°. To control the Pan a std_msgs/Float64 is published to the /alphabot2/horizontal topic using the command: *rostopic pub /alphabot2/horizontal std_msgs/Float64 "data: -45"* (Fig. 10) and *rostopic pub /alphabot2/horizontal std_msgs/Float64 "data: 45"* (Fig. 11). It is possible to see that the camera is positioned to the right of its initial position when a negative value is submitted to the horizontal topic and it is positioned in the left if a positive value is submitted.

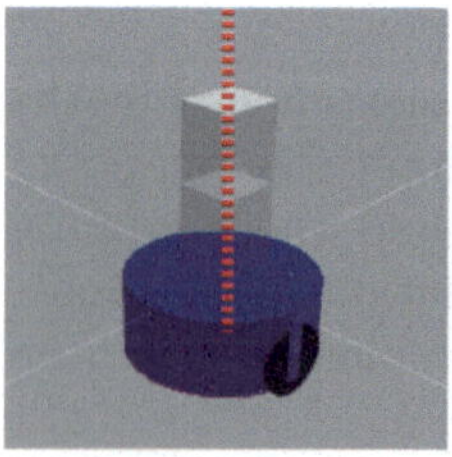

Fig. 8. Initial pose

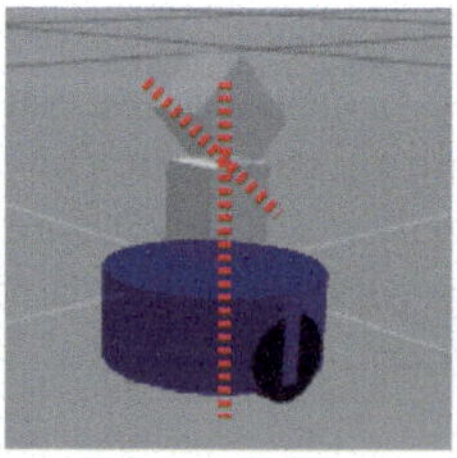

Fig. 9. −45° tilt

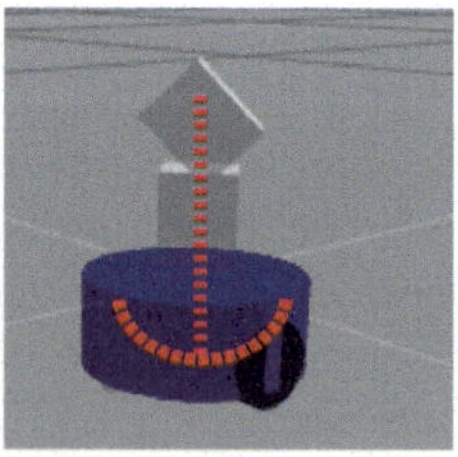

Fig. 10. −45° pan

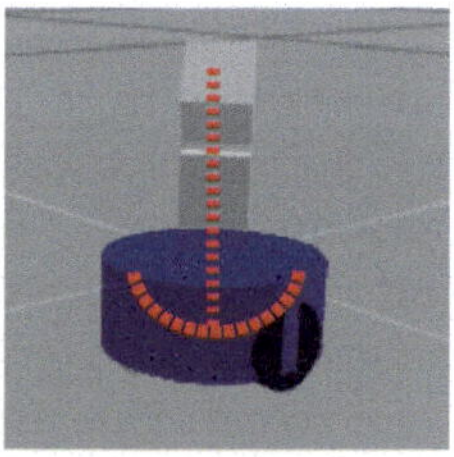

Fig. 11. 45° pan

For the verification of the camera functioning in the Gazebo simulation, a package called alphabot2_tracking was created to subscribe to the /alphabot2/camera/image_raw topic. With the robot spawned on the world with simple objects (Fig. 12) it is possible to visualize both the RGB image captured from the camera (Fig. 13) showing that the gazebo library is working properly for the camera sensor. On top of this node the user has the freedom to operate to what best fits its needs. Figure 14 exemplifies these transformations by applying a Canny Edge Detection for future line detection using Hough Transform. This information can later be used for the control of the robot's behaviour through, for example, a PID control for the adjustment of the robot's orientation with regards to that line or recognition of traffic signs, and treadmills.

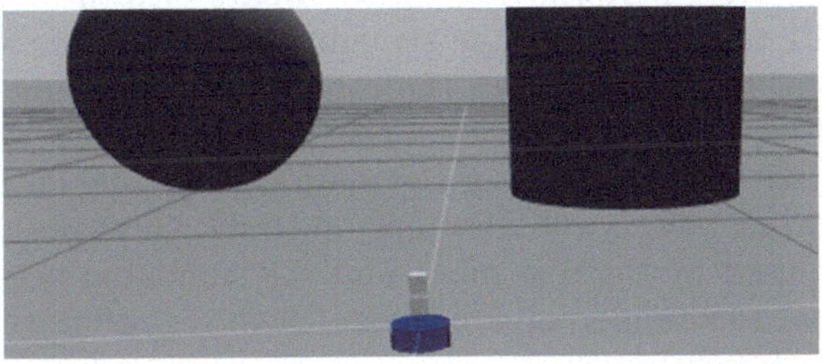

Fig. 12. Gazebo simulation with simple objects.

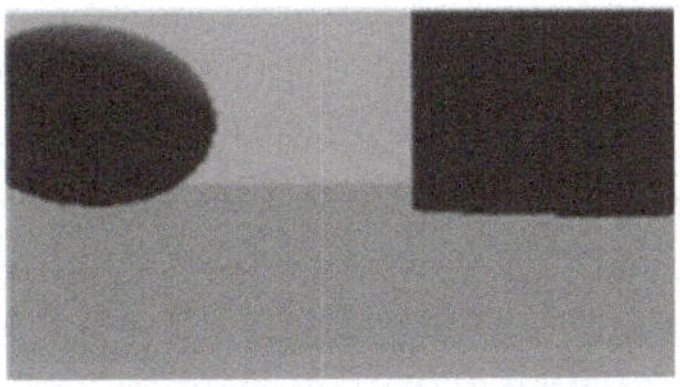

Fig. 13. Camera RGB image.

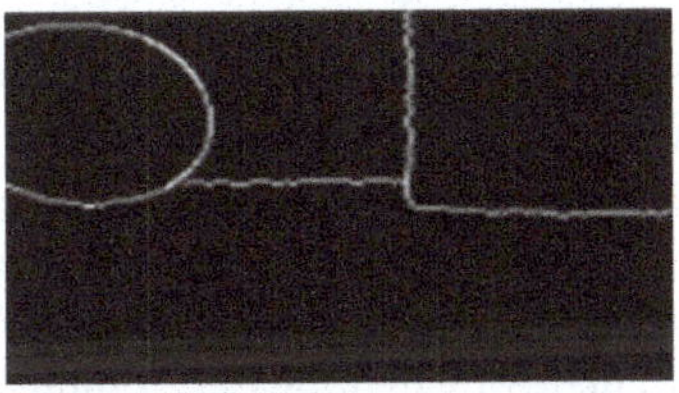

Fig. 14. Canny edge detection.

Fig. 15. Straight and hexagonal tracks for testing.

Table 2. Time results captured and absolute and relative errors.

	Simulated robot	Real robot	Absolute error	Relative error
T1	25 s 08 ms	18 s 02 ms	6,66	36,96%
T2	33 s 02 ms	16 s 53 ms	16,09	97,34%
T3	3 m 22 s 35 ms	24 s 13 ms	138,62	584,16%

With the purpose of testing the implementation of the line following and obstacle avoidance correspondence between the real and simulated roots, some tests were conducted. The two robots were tested in two different lanes, one of 0.89 m in a straight line and one of hexagonal shape with 0.30 m of side, in order to test its movement and its execution time. In addition, the robots were also tested with obstacles to be avoided.

3 different tests were performed on either robot, with the tracks represented in Fig. 15. The T1 test represents the straight line track test, with no obstacle in the course. The T2 test evaluates the robots in the straight line track with an obstacle in the course. The last one, the T3 test, verifies the robots in the hexagonal track, with no obstacles in the main course. The measured times are shown in Table 2, as well as the absolute and relative errors of the measurements. The code used by the real robot and the simulated one is very similar, although there are some differences between them regarding the programming of the AlphaBot's hardware.

5 Discussion and Conclusion

In this paper a simulator and several ROS packages were presented for the AlphaBot2 hardware abstraction and control of its low-level modules, allowing for an easy introduction of students to the field of robotics. The AlphaBot2's simulator was produced in Gazebo, a 3D simulator, and tested by modeling the AlphaBot2 robot. The simulator's components communicate with each other using the ROS framework which facilitates the user's work by suppressing concerns with low-level issues and hardware, making him only concerned with the logic behind the task. The user can then easily use the drive control of the AlphaBot2, its pan-tilt control, as well as viewing information obtained by the camera and sensors.

Moreover, the present implementation allows the real AlphaBot2 RPi and the simulated one to follow lines and avoid obstacles. The real robot performed better than the simulated one, regardless of the track used for testing. This may be due to errors in either the AlphaBot size measurements or the construction of the actual and simulated Gazebo tracks, as well as errors due to the positioning of the two robots in the different tracks. In addition, the fact that the AlphaBot2 uses some code already on the vendor's website may be causing it to perform better. With all the results obtained, it is verified that the two robots' implementation are not perfectly equal and their use together has some limitations. Even though in simpler tests such as T1 and T2 the performance was, though different, similar enough to demonstrate the concept works, in T3 we can see how much work still needs to be done.

In conclusion, the current work successfully achieved a simulator of the AlphaBot2, despite its limitations and room for further improvement. With relation to future developments, enhancements to the *.xacro* model can be made by providing a more accurate representations of the robot design in the Gazebo simulation, which can be done, for example, with *.stl* files for each of the components.

However, these would yield superior computational requirements. Additionally, it is of paramount importance the enhancement of the AlphaBot's simulation model, in terms of joint kinematic and dynamic properties for better calibration with the real robot, so that it does not oscillate so much when making turns and changing speeds. Furthermore, only differences in time in the tasks between the simulator and the real hardware were evaluated, future studies ought to evaluate other values such as robot's trajectory, sensors reading, among other parameters. The use of the robot with an adapted track from the Conde's original one would further drive the development of special algorithms.

Supplementary Materials: The developed simulator's code can be found at https://github.com/ssscassio/alphabot2-simulator.

Acknowledgements. This work is partially financed by the ERDF – European Regional Development Fund through the Operational Programme for Competitiveness and Internationalisation - COMPETE 2020 Programme within project POCI-01-0145-FEDER-006961, and by National Funds through the FCT – Fundação para a Ciência e a Tecnologia (Portuguese Foundation for Science and Technology) as part of project UID/EEA/50014/2013. This research was partially supported by LIACC - Artificial Intelligence and Computer Science Laboratory of the University of Porto (FCT/UID/CEC/00027/2019).

References

1. Davcev, K., Koceska, N., Koceski, S.: A review of robotic kits used for education purposes. In: International Conference on Information Technology and Development of Education – ITRO, Zrenjanin, Serbia, pp. 152–155, June 2019
2. Yusof, Y., Hassan, M.A., Mohd Saroni, N.J., Che Wan Azizan, W.M.F.: Development of an educational virtual mobile robot simulation (2011)
3. Siciliano, B., Sciavicco, L., Villani, L., Oriolo, G.: Robotics: Modelling, Planning and Control. Springer, London (2010)
4. Pepper, C.T., Balakirsky, S.B., Scrapper Jr., C.J.: Robot simulation physics validation — NIST. In: Performance Metrics for Intelligent Systems (PerMIS 2007), December 2007
5. AlphaBot2 - Waveshare Wiki. https://www.waveshare.com/wiki/AlphaBot2
6. Gazebo. http://gazebosim.org/
7. ROS.org — Powering the world's robots. http://www.ros.org/
8. Costa, V., Rossetti, R.J.F., Sousa, A.: Autonomous driving simulator for educational purposes. In: 2016 11th Iberian Conference on Information Systems and Technologies (CISTI), pp. 1–5, June 2016
9. Roy, A., Noel, M.M.: Design of a high-speed line following robot that smoothly follows tight curves. Comput. Electr. Eng. **56**, 732–747 (2016)
10. Matczak, G., Mazurek, P.: Dim line tracking using deep learning for autonomous line following robot. In: Silhavy, R., Senkerik, R., Oplatkova, Z.K., Prokopova, Z., Silhavy, P. (eds.) Artificial Intelligence Trends in Intelligent Systems. Advances in Intelligent Systems and Computing, pp. 414–423. Springer, Cham (2017)

11. Hassan Tanveer, M., Recchiuto, C.T., Sgorbissa, A.: Analysis of path following and obstacle avoidance for multiple wheeled robots in a shared workspace. Robotica **37**(1), 80–108 (2019)
12. Javed, M., Hamid, S., Talha, M., Ahmad, Z., Wahab, F., Ali, H.: Input based multiple destination, multiple lines following robot with obstacle bypassing. ICST Trans. Scalable Inf. Syst. **5**, 154472 (2018)
13. Ghorpade, D., Thakare, A.D., Doiphode, S.: Obstacle detection and avoidance algorithm for autonomous mobile robot using 2D LiDAR. In: 2017 International Conference on Computing, Communication, Control and Automation (ICCUBEA), pp. 1–6, August 2017

Artificial Intelligence Teaching Through Embedded Systems: A Smartphone-Based Robot Approach

Luis F. Llamas[1], Alejandro Paz-Lopez[2], Abraham Prieto[2], Felix Orjales[1], and Francisco Bellas[2(✉)]

[1] Integrated Group for Engineering Research, Universidade da Coruña, A Coruña, Spain
{luis.llamas, felix.orjales}@udc.es

[2] CITIC Research Center, Universidade da Coruña, A Coruña, Spain
{alpaz, abprieto, francisco.bellas}@udc.es

Abstract. Following the recommendations of the European Commission, with the aim of positioning the EU as a leader in the technological revolution that is yet to come, Artificial Intelligence (AI) teaching at University degrees should be updated. Current AI subjects should move from theoretical and virtual applications towards what is called "specific AI", focused on real embedded devices, using data from real sensors and interacting with their environment to solve problems in the real world. These real devices must have the computing power to process all the information that comes from their sensors and also full network connectivity, to allow the connection with other intelligent devices. This work belongs to an Erasmus Plus proposal in such direction, called TAIREMA, which aims to provide a set of tools to include low-cost embedded devices at classes to support AI teaching. One of these tools is a smartphone-based robot called Robobo, which is the main topic of this paper. We will present its main features, mainly in software aspects, and we will describe some specific teaching units that have been developed in classes during the last year in AI subjects.

Keywords: Artificial intelligence education · Educational robots · ROS · Embedded devices

1 Introduction

The European Commission, published in early December 2018 the first edition of its Artificial Intelligence Plan for 2019 and 2020, under the title "Coordinated Plan on the development and use of Artificial Intelligence Made in Europe - 2018" [1]. This plan aims to ensure synergies between actions at the national and EU levels to maximize the impact and spread the benefits of AI across Europe, in order to achieve a leader role of the member countries in the AI revolution. This document clearly establishes that education is one of the key areas where the European AI plan will focus. In the short time, by increasing the preparation of specialized students in AI topics at university degrees and, in the future, by introducing such topics in the secondary and high school levels.

M. F. Silva et al. (Eds.): ROBOT 2019, AISC 1092, pp. 515–527, 2020.
https://doi.org/10.1007/978-3-030-35990-4_42

Recently, the Spanish Government has published a document which describes the AI Spanish strategy [2], encouraging that current AI university studies should move from a general approach to AI, more theoretical, towards what they call "specific AI", more applied and embedded in real devices. This way, students will be able to realize how AI can be really used in real systems, automating day-to-day processes in the industry and in many social application fields.

The subject of AI was introduced many years ago as part of many computer science degrees at universities. Its study has been increasing during the last years and currently almost all degrees and masters teach general AI or AI related subjects [3]. Most of the AI courses one can find nowadays in European Universities follow, with different depth, the topics proposed by Russell and Norvig in their classical book [4]. Such topics are tested by coding the different methods and algorithms and running them over mathematical optimization benchmarks, game play learning, and some simplified simulations of real problems, like simulated robotics [5–11].

The market revolution of AI that is starting relies on transferring such solutions of virtual problems to real embedded systems, like robots, smart home devices, smartphones or wearables. In addition, AI teaching must include concepts on distributed intelligence, to deal with all the concepts of the internet of things that will create a network of interconnected intelligent devices. Therefore, including real devices from the initial steps of learning of AI allows to understand the differences among the use of synthetic data sets or the use of real time data obtained from the sensors of these devices and to apply real time actions based on them. On the other hand, it is not easy to include long-term devices in AI teaching, because this is a field in continuous improvement, which implies to update the devices frequently.

This paper belongs to the EU Erasmus Plus proposal TAIREMA (Teaching Artificial Intelligence with REal MAchines), which intends to transfer this tendency towards using real embodied and embedded systems in the teaching of AI in classrooms of European Universities and make that affordable. The project is based on the use of ROS and ROS2 libraries and different compatible low-cost devices. Specifically, three main types of technologies are considered:

1. *Low cost robots*: they represent the first level of embodied technology. They will provide real data in real time and execute actions that affect their actuators also in real time. Several possible solutions can be considered, but they have to be equipped with high-level sensors in order to be useful for AI teaching, like cameras, microphones, tactile screens, speakers, and so on. Moreover, they must support different types of wireless connections, and they must be connected to the internet.
2. *Personal activity trackers and wearables:* in a second level we consider embedded technology. Based on the information provided by humans carrying some devices we can extract and analyze different data to find different patterns. Health conditions, sport activity, stress level, location, pace, sleeping/sitting/standing time and not only for one person based patterns but also for collective trends or states. Again, they are intended to provide local, real time, real data to train the algorithms. The level of actuation will be weaker in this case but still available as notifications to warn or inform the human (or humans in the same region, condition or environment). Again, it maintains a high quality real data for a low cost philosophy of the project.

3. Finally, in the least reactive level we include *online services*: this represents any type of available online server which produces real time local information. There are many of those services nowadays and with the popularity of the IoT everyday more and more will be created. They can offer information of several kinds: weather, different pollutant levels, traffic conditions, real time webcams, presence control or some specific servers implemented in facilities or particular homes to control ambient parameters as temperature, light, noise, etc. Again, it will produce an adequate quality/quantity information per invested cost. The reactive level is very low or null and its isolated usage will be focus in data analysis. However, as it will be part of an interconnected architecture it can provide extra information to one of the two previous approaches and affect their behavior. For instance, to analyze the relation among pollution/weather or traffic to personal parameters of activity or health or to affect the behavior of different robots working in some specific environments.

This work presented in this paper is focused in the first type of embedded device, low-cost robots, to show the kind of project we have been solving with robots at classes in this new teaching approach. Specifically, we will describe a smartphone-based robot, called Robobo, we have been developing since 2016 for AI teaching at different educational levels, and which was presented previously in [12]. In the next section, the robot will be described, focusing the explanation in the improvements carried out into the TAIREMA project, most of them related to the software architecture and the ROS/ROS2 compatibility.

Fig. 1. The Robobo robot

2 The Robotic Platform: Robobo

Robobo (see Fig. 1) is an educational robot that combines the computing power and capabilities of a modern smartphone with the versatility of a mobile robot base. The use of a smartphone as the brain of the robot not only provides the necessary processing power but also offers a variety of sensors and high-level capabilities, like the camera, internal inertial unit, microphone, touch screen interaction, or plenty of connectivity possibilities via Wi-Fi, Bluetooth, 4G or NFC. This approach provides us with an affordable and adaptable environment to support AI teaching, with a focus on dealing with the complexity of real devices and interaction with physical environments from the early steps. Both aspects are necessary when we aim to cope with more realistic tasks using AI tools and methods.

In this section, new aspects of the platform are presented, like an extended architecture with integrated support for ROS2 [13], Python programming and robot simulation, together with support for Android and iOS devices.

2.1 Sensing and Actuation Capabilities

The robot, as previously stated, is composed by the combination of a smartphone with a robotic base. This base is what gives the robot its mobility and obstacle avoidance capabilities. It is equipped with four motors, two of them used by the wheels and another two on the Pan/Tilt unit where the phone is mounted (Fig. 2). All of the motors have encoders attached to them, which allows the user to control them by position or speed, also giving the option to calculate the odometry of the robot. There are seven user programmable RGB LEDs, five of them located on a ring in the front of the robot, and the other two in the tail. These LEDs can be used to indicate the status of the task being performed by the robot. The base also has eight infrared distance sensors, that can be used to detect objects that are close to the robot and to program different strategies to avoid them. Some of the sensors are pointed to the ground, allowing to detect potential falls and stopping the robot before falling. Figure 2 shows a layout of the Robobo base, including all these elements.

The connection between the smartphone and the base is implemented using a Bluetooth 2.0 interface in the Android platform and a Bluetooth 4.0 Low Energy interface in the iOS platform. The use of both technologies is transparent to the user and no configuration is needed in the robot base to switch between different mobile operating systems. The communication between the smartphone and the base is

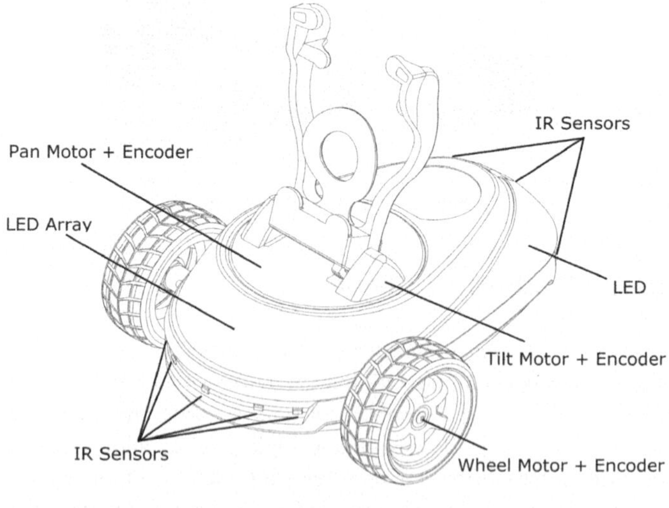

Fig. 2. Main sensors and actuators of the Robobo base.

encoded using a scalable byte-level protocol, allowing the expansion of the base capabilities without breaking the compatibility with previous versions. This is achieved with a protocol built on top of two types of messages: commands, that are used to execute actions in the base, and status, that carry sensor information from the base to the smartphone. Thus, new commands and status can be incrementally added when new sensors or actuators are implemented in the base without affect compatibility with old versions of the smartphone software. Furthermore, from the end-user point of view, programming the sensors/actuators of the base and the sensors/actuators of the smartphone is presented through the same API and conceptual model, facilitating their joint use in high-level AI applications.

2.2 Software Platform and Programming Capabilities

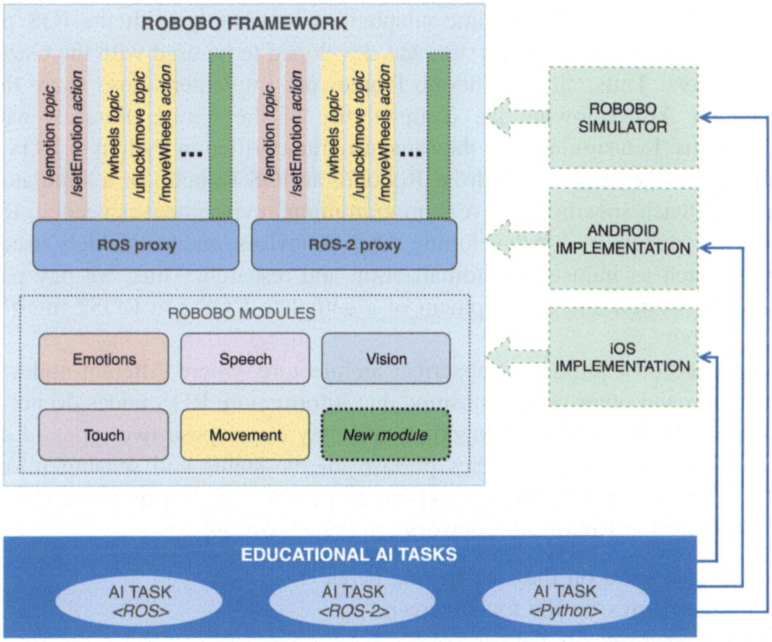

Fig. 3. Schematic view of the Robobo software environment.

In order to obtain a flexible software platform that allows the robot to grow with new capabilities, a scalable and modular software architecture was designed and implemented (Fig. 3).

This architecture provides two different ways to program the robot: native programming using the Robobo Framework in Java for Android devices and Swift for iOS ones, and remotely, using the remote-control feature that exposes actuation commands and sensing status. These remote-control capabilities allow the programming of the robot with any language that implements the Robobo remote protocol, a simple JSON based protocol, that can be interfaced through different network technologies. Currently websocket and ROS/ROS2 are supported, and also libraries in Python and Javascript programming languages are available.

The software framework provided by the Robobo platform is composed by its core support modules and a series of modules that implement the specific functionalities. The detail of this modular architecture was previously described in [12]. In this work, we have expanded the Robobo development environment with a series of tools and programming capabilities, with the main objective of providing a richer and more versatile platform for the purpose of AI teaching. Specifically, the platform pursues the objective of allowing the students to easily take advantage of tools and algorithms that are available in commonly used development environments in the context of AI and robotics, such as ROS and Python. Following this approach, a new implementation of the Robobo Framework, in this case for the iOS platform, was developed. This implementation follows the same design of the previous Android version, but it has been developed in Swift, with some adaptations to comply with the iOS platform requirements. Also, a simulated version was developed to be used with the Gazebo and V-Rep simulators. Thus, all the Robobo Framework implementations share the same communication API, allowing the compatibility of the remote libraries with both mobile platforms. In a similar way, the same program written in Python or ROS may be executed indistinctly using an Android Robobo, an iOS Robobo or a simulator.

In order to teach specific AI, a real programming environment is needed. ROS is a state-of-the-art framework for developing robot behaviors, and it is widely used in real applications, such as industrial automatization and research. Thus, we pay particular attention to the design and development of a complete ROS and ROS2 interfaces for the Robobo robot.

ROS is based on a publisher/subscriber architecture, where different nodes publish data to a stream and other nodes consume that information. ROS nodes do not need to be in the same machine to work together, and they can expose two types of communication: topics and services. Topics publish all the status and sensing information gathered by the robot while services expose all possible actuations of the device. The topics provide by the Robobo framework are the following:

- */accel:* Acceleration on the smartphone IMU.
- */fling:* Fast gestures on the touch screen.
- */irs:* IR distance sensors data.
- /leds: Current led color.
- /orientation: Orientation (yaw, pitch and roll) on the smartphone IMU.
- /tap: Taps on the touch screen.
- /emotion: Current emotion displayed on the robot screen.
- /unlock/move: Notification when a movement ends (synchronize movements).
- /unlock/talk: Notification when text-to-speech end, used to synchronize speech.
- /wheels: Position and speed of the wheels.
- /image/compressed: Live feed of the smartphone camera.
- /camera_info: Information of the camera.
- /String: Raw robot status, a string encoded in robobo remote protocol JSON.
- /int8/level: Battery level.

- /int16/panPos,/int16/tiltPos: Pan and Tilt position.
- /int32/level: Ambient light level.

On the other hand, the services provided by the Robobo framework are:

- /movePanTilt: Moves the Pan/Tilt unit of the robot.
- /moveWheels: Moves the wheels of the robot.
- /playSound: Plays a predefined sound from a sound library.
- /resetWheels: Resets the count of the wheel encoders.
- /setCamera: Selects between the front and back camera of the smartphone.
- /setEmotion: Sets the current emotion displayed on the robot screen.
- /setSensorFrequency: Sets the period of the sensors.
- /setLed: Sets the color and intensity of the LEDs.
- /talk: Text-to-speech call.

This set of topics and services can be further expanded thanks to the modular architecture. New modules implementing new features can define new Status and new Commands that will be exposed through the ROS interface of the robot (topics and services).

Moreover, on top of the modular architecture of the Robobo Framework, a new module implementing a remote interface that supports ROS2 (the new version of ROS that is under development) was built. The new ROS2 network architecture is based on UDP multicast, allowing the automatic discovery of the nodes as they join the network and consequently avoiding the need of centralized management [13]. The ROS2 module supports all the topics and services present in the ROS module. Furthermore, we implement a configurable and modular solution, based on the dependency injection pattern, that makes easy to change between the ROS and the ROS2 module. Such design will facilitate the experimentation and transition to the new ROS2 version.

The next section illustrates, through examples, how this set of functionalities and tools, combined with other common libraries (i.e. OpenCV or TensorFlow), has been effectively used to teach the application of AI to real problems in real environments.

3 Examples of AI Projects

This section describes two particular teaching units we have solved with Robobo robots in the context of specific AI during the year 2019. The first unit is focused on introducing students in the application of reactive architectures for intelligent robotics operating in real environments. In the second one, a task for human attendance monitoring is proposed to be collectively solved by a group of Robobo robots using an evolutionary algorithm as the optimization strategy.

3.1 Reactive Architecture

As a first example of the type of teaching unit that could be carried out in the scope of the TAIREMA proposal using Robobo, we will describe one focused on the practical implementation of an architecture for autonomous robots. In this case, the subsumption architecture proposed by Brooks [14] was used, which is very interesting for introducing students in the reactive approach to intelligent robotics [4]. Due to its simplicity, it can be implemented with any type of robot, even those with poor sensing and acting capabilities [15]. But in this case, we aim to show how such a simple architecture could be applied in a realistic and challenging task from an AI perspective, thanks to the technological features provided by Robobo.

Fig. 4. Experimental setup for the "autonomous recycling" teaching unit.

Specifically, in this teaching unit, students had to solve a robotics challenge called "autonomous recycling". The experimental setup is shown in Fig. 4, where we can see colored cylinders (red, green and blue) placed on a table. In front of the table there is a wall with 3 recycling areas, marked again with red, green and blue labels, and containing a text corresponding to three types of waste (organic, paper and glass). The goal of the students was to design a controller for Robobo so it can place each cylinder in its corresponding recycling area by pushing it with a gripper accessory attached to its front (see Fig. 4). Both, the objects and the recycling areas, were randomly placed in the environment at the beginning of the execution.

Students had to implement the software architecture displayed in Fig. 5, where five basic behaviors, that make up the global solution, were provided by the teachers. Following the subsumption architecture principles, each of the behaviors should be implemented using an ASFM (augmented finite state machines), although any other approach could be allowed if desired, like a rule-based system, a neural network or a controller obtained through reinforcement learning. From top to bottom, the proposed behaviors were the following:

- *Command detection:* it takes, as sensorial input, the output provided by the Robobo speech recognition module, which uses the PocketSphinx library. In this case, students had to detect simple voice commands like STOP, BACK or START, which allow the teacher to interact with the robot in a simple and natural way. The output of this behavior can perform a subsumption over all the others, as shown in the diagram, controlling the wheel motors.

- *Avoid*: the infrared sensors of the base are the input to this behavior which simply avoids colliding with cylinders and walls by controlling the motor wheels.
- *Pick up*: it uses, as input, the output of a camera library that provides real-time object detection, in this case used just to detect the nearest cylinder position and color. This library was selected by students in a previous teaching unit from the set of available ones compatible with ROS and ROS2, like those present in OpenCV [16] or TensorFlow [17]. With this information, the "pick up" behavior was devoted with moving Robobo wheels towards the nearest cylinder until it was caught with the gripper.
- *Deliver*: this behavior uses a label recognition library, which provides the text written in the label (OCR) and its color [18]. Again, such a library was selected by students in a previous teaching unit, from those compatible with ROS, which in this case implied combining an OCR process with color detection. The output of this behavior controlled Robobo movement towards the proper recycling area.
- *Cruise*: finally, the cruise behavior was simply an asynchronous motor wheel movement, and all the remaining behaviors perform a subsumption over it.

Each of these behaviors could control other Robobo actuators if decided by the students, like the Pan or Tilt motors, the facial expressions displayed on the LCD screen, or the speech production to communicate any robot state, but they are not shown in the scheme of Fig. 5 for the sake of clarity.

Figure 4 contains four snapshots of a typical solution obtained by the students. Top left image corresponds to the initial step, where the robot detects the first cylinder (the green small one). Top right and bottom left images display how the robot carries a cylinder towards the corresponding areas. Finally, bottom right image shows the final step, with all the cylinders correctly delivered. What must be highlighted here is that students not only develop a reactive architecture, but they do it in a real embedded system which uses real sensors (vision, speech).

3.2 Onboard Online Evolution

As a second example of teaching unit we have tested in the scope of the TAIREMA project with Robobo, we focus on the topic of distributed artificial intelligence. Specifically, a collaborative task was proposed to be performed by a group of Robobo robots. The automatic optimization of the collective behavior was carried out by means of evolutionary techniques which, in this case, had to run online on the real robots, avoiding the use of a simulator as much as possible.

As a first step, students had to implement an Embodied Evolution algorithm [19] to be run by each Robobo. This way, the high computing capabilities and communication power of Robobo were exploited to provide autonomy to each of the robots and to perform the optimization in a distributed manner. In particular, we proposed the use of the Distributed Differential Embodied Evolution [20], a distributed and embodied variation of the original Differential Evolution algorithm [21]. Figure 6 shows a pseudocode of the algorithm that students had to implement on each robot using ROS.

The specific task proposed in this teaching unit was an attendance monitoring task, in which the set of five Robobo robots were used to surveil a closed space, locate any present person, and collectively create a dynamic attendance map. The task requires each robot to be able to perform two activities: a reliable navigation and self-location, and the tracking of the attendees. These activities rely on the use of two physical sensors: the encoders of the motors and the smartphone's camera. Associated to them, a set of libraries that perform complex operations over the data were provided to students, so they had to understand their operation, but they did not have to implement them:

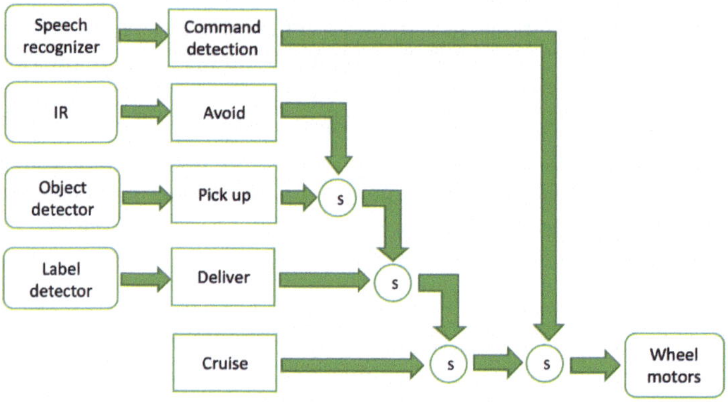

Fig. 5. Diagram of the subsumption architecture used in the "autonomous recycling".

- *Apriltag*: this library detects artificial beacons, called Apriltag [22], to provide the location and orientation of the robot.
- *Human tracking*: this is an OpenCV and TensorFlow based library that use the image coming from the camera to track human bodies. This information, together with the location and orientation of the robot, sets the position of the human.
- *Odometry*: it provides the relative position and orientation of the platform and pan-tilt unit of the Robobo based on the information provided by the encoders, and on the initial position and orientation. It produces a small but accumulative error which on the long term introduces a drift in the pose estimation.
- *Navigation*: this contains a set of functions that use the estimated location of the robot, its goal position, and the map of the public space to define the trajectory that the robot has to follow from one place to the next. It, also, automatically visits the location of Apriltags when the estimated accuracy of the self-location becomes too low and then the navigation is not reliable.
- *Communication*: it sets a communication channel to exchange information among robots to create a common map which includes the position of the detected humans, the time of the detection, and the position of the Robobos. This map produces the inputs for the control system.

The students also took advantage of the remote control libraries available for Robobo. This control system uses the information coming from the processed sensors and sets the behavior of the robot, this is, the movement of the wheels and the position of the pan unit to orientate the camera in the most convenient direction. We proposed the use of a simple multilayer perceptron Artificial Neural Network (ANN) as the control system, since it has been widely proved its adequacy to model this type of behaviors. Consequently, the genotype of the individuals that make up the population of the evolutionary algorithm contained the weights of the ANN. Students had to adjust the ANN size and algorithm parameterization in order to obtain a successful solution to the task.

Algorithm 1: Distributed Diff. Embodied Evolution

```
initialize p robots
while evolution is active do
    simulate one step
    map individual robot fitness→f_r
    for robot ← 1 to p do
        if age_r == T_maturity then
            if f_r < f_parent,r then
                active ← parent
            end
        end
        if age_r > T_maturity then
            if rand(0, 1) < P_matingandReplacement
              then
                active ← recombination.DE operator
                f_parent,r ← f_r
                age ← 0
            end
        end
        age_r++
    end
end
```

Algorithm 2: Distributed Diff. Embodied Evolution

```
Operator recombination.DE() is
    pbest ← 0.15;
    create a pool with affine genotypes within the
      population→ P_elig;
    best ← random individual from top pbest from
      P_elig;
    random1 ← random individual from P_elig;
    random2 ← random individual from P_elig;
    randomGen ← randi[0, genotype.length];
    for i ← 0 to genotype.length do
        if rand(0, 1) < crossRate or
          i == randomGen then
            active.g[i] ← active.g[i] + F*(best.g[i] -
              active.g[i]) + F*(random1.g[i] -
              random2.g[i]);
        else
            active.g[i] ← current.genotype[i];
        end
        return active
    end
end
```

Fig. 6. Pseudocode of the distributed differential evolution algorithm.

Finally, to evaluate each of the controllers of the robots, and therefore feed the evolutionary algorithm that will optimize the value of their parameters, the students had to implement a fitness function which was periodically calculated for each robot. To make things simpler, the used fitness was just the normalized number of humans detected during a specific period. The overall conclusion of this second teaching unit was good, and students realized the complexity of optimizing a distributed intelligent system online, although the control of such a high number of robots was slow, which implied a number of working hours higher than expected.

4 Conclusions

Introducing specific Artificial Intelligence in University degrees requires of low-cost tools with high computing power, top-end sensing capabilities, and a permanent communication to the network. In addition, new teaching units must be developed that cover the AI topics considering the limitations imposed by real-time operation and embedded devices. Here, we have described one possible approach towards this new AI teaching, based on the use of the Robobo robot, which exploits all the capabilities provided by current smartphones. Robobo programming is based on ROS, enabling students to import many available libraries in computer vision, machine learning or robotics, and facilitating the communication between devices. Two examples of teaching units have been described, providing a clear idea of the type of challenges that should be faced.

Acknowledgments. This work has been partially funded by ROSIN project (Agreement 732287), Ministerio de Ciencia, Innovación y Universidades of Spain/FEDER (RTI2018-101114-B-I00), Xunta de Galicia and FEDER (ED431C 2017/12).

References

1. European Comission, Coordinated Plan on Artificial Intelligence (COM(2018) 795 final). https://ec.europa.eu/digital-single-market/en/news/coordinated-plan-artificial-intelligence
2. Spanish Ministry of Science, Innovation and Universities: Spanish RDI Strategy in Artificial Intelligence (2019). http://www.ciencia.gob.es/stfls/MICINN/Ciencia/Ficheros/Estrategia_Inteligencia_Artificial_EN.PDF
3. Universities with AI programs (2019). http://www.aiinternational.org/universities.html
4. Russell, S., Norvig, P.: Artificial Intelligence: A Modern Approach. Prentice Hall, Upper Saddle River (2010)
5. McArthur, D., Lewis, M., Bishary, M.: The roles of artificial intelligence in education: current progress and future prospects. Ournal Educ. Technol. **1**(4), 42–80 (2005)
6. DeNero, J., Klein, D.: Teaching introductory artificial intelligence with pac-man. In: Proceedings First AAAI Symposium on Educational Advances in Artificial Intelligence (2010)
7. Langley, P.: An integrative framework for artificial intelligence education. In: Proceedings of the 9th Symposium on Educational Advances in Artificial Intelligence. AAAI Press (2019)
8. Blank, D., Kumar, D., Meeden, L., Yanco, H.: The Pyro toolkit for AI and robotics. AI Mag. **27**, 39–50 (2006)
9. Hugues, L., Bredeche, N.: Simbad: an autonomous robot simulation package for education and research. In: LNCS, vol. 4095. Springer, Heidelberg (2006)
10. Parsons, S., Sklar, E.: Teaching AI using LEGO mindstorms. In: AAAI Spring Symposium 2004 on Accessible Hands-on Artificial Intelligence and Robotics Education (2004)
11. Miller, D.P., Nourbakhsh, I.: Robotics for education. In: Siciliano, B., Khatib, O. (eds.) Springer Handbook of Robotics. Springer Handbooks. Springer, Cham (2016)
12. Bellas, F., et al.: Robobo: the next generation of educational robot. In: ROBOT 2017: Advances in Intelligent Systems and Computing, vol. 694. pp. 359–369 (2018)
13. ROS and ROS2 web page. http://www.ros.org
14. Brooks, R.A.: A robust layered control system for a mobile robot. IEEE J. Robot. Autom. **2**, 14–23 (1986)
15. Murphy, R.: Introduction to AI Robotics. MIT Press, Cambridge (2000)
16. OpenCV web page – Image processing. https://docs.opencv.org/3.4/d7/da8/tutorial_table_of_content_imgproc.html
17. TensorFlow web page – Learn and use ML. https://www.tensorflow.org/tutorials/keras
18. Tesseract OCR web page. https://opensource.google.com/projects/tesseract
19. Bredeche, N., Haasdijk, E., Prieto, A.: Embodied evolution in collective robotics: a review. Front. Robot. AI **5**, 12 (2018)

20. Trueba, P., Prieto, A.: Improving performance in distributed embodied evolution: distributed differential embodied evolution. In: ALIFE Proceedings, pp. 222–223. MIT Press (2018)
21. Storn, R., Price, K.: Differential evolution - a simple and efficient heuristic for global optimization over continuous spaces. J. Glob. Optim. **11**(4), 341–359 (1997)
22. April Tag website. https://april.eecs.umich.edu/software/apriltag

Human-Robot Scaffolding, an Architecture to Support the Learning Process

Enrique González[1], John Páez[2], Fernando Luis-Ferreira[3(✉)], João Sarraipa[3], and Ricardo Gonçalves[3]

[1] Facultad de Ingeniería, Pontificia Universidad Javeriana, Bogotá, Colombia
egonzal@javeriana.edu.co
[2] Facultad de Ciencias y Educación, Universidad Distrital Francisco José de Caldas, Bogotá, Colombia
johnpa@live.com
[3] CTS, UNINOVA, Dep.º de Eng.ª Electrotécnica Faculdade de Ciências e Tecnologia, Universidade Nova de Lisboa, Caparica, Portugal
{flf, jfss, rg}@uninova.pt

Abstract. Recognizing and diagnosing learner's cognitive and emotional state to intervene assertively is an important aspect to improve learning processes. This mission that can be supported by social robots in educational contexts. A cognitive architecture to manage the robot's social behavior with handling capacity is presented. The human-robot scaffolding architecture is composed of three systems: multimodal fusion, believes, and scaffolding. Those recognize verbal and nonverbal data from user and from the mechanical assembly task, acknowledges the user's cognitive and emotional state according to the learning task and configure the actions of the robot based on the Flow Theory. It establishes relations between challenges and skills during the learning process, presenting also the theoretical analysis and explorative actions with children to build each subsystem of architecture. The present research contributes to the field of human-robot interaction by suggesting an architecture that seeks the robot's proactive behavior according to learner's needs.

Keywords: Social robots · Scaffolding · Constructionism · Human-robot interaction

1 Introduction

During the learning process, the social robots can give physical, cognitive and emotional support based on the learner's characteristics. In the physical aspect, robots can assist the user during the mechanical assembly process by manipulating and reorganizing the blocks. For example, robots can change the blocks to foster new ways of thinking about the problem. In the cognitive aspect, robots can give support through three strategies: focus lessons, guided instructions, and collaborative work. For example, if the user is so confused then robots teach specific lessons through verbal cues or prompts. And finally, they can give emotional support to encourage the user during the learning process. For example, when the user is bored, a robot shows a

M. F. Silva et al. (Eds.): ROBOT 2019, AISC 1092, pp. 528–541, 2020.
https://doi.org/10.1007/978-3-030-35990-4_43

happy face to foster learners to continue the learning process. In overall, the assertive robot intervention permits to stablish, maintain, change, and finish learning events through cognitive, emotional, and physical strategies.

In the field of education, there is research interest in how to use the artificial cognitive systems to support learning processes which have been historically supported by humans. Some advancements are related to topics such as: relation between physical presence and perceived support effect [6, 29]; mechanisms of emotional communication [28]; non-verbal language [35] and non-verbal user behavior recognized by robots [46]. However, cognitive conditions of current anthropomorphic robots allow to use them as collaborative agents in collaborative learning tasks. This document describes the model of the Human-Robot Scaffolding Architecture. By the time, some semiotic explorations about the cognitive and emotional behavior were analyzed. According to the findings, the body language and emotions are used as tools to express the learning process. Traditionally, the emotions are related to social interactions but in the learning environments the emotions as boredom, anxiety and flow have not been explored.

2 Architecture Development

The basic idea of the architecture is in accordance to Beliefs, Desires, ands Intentions Theory. The architecture has been designed to foster the learning process through the agent's beliefs which are grounded by the cognitive and emotional learner's behavior. When learners are solving the problem then the robot creates a cognitive and emotional model. The first step to create the model is to recognize the task state and the body language. The second step consists in evaluating the learner's knowledge about the problem and based on this value evaluate the learner's skills to solve the problem. The third step defines what action could be carry out by the robot. The actions are based on the scaffolding psychological model. Finally, but not least important, the robot's actions are performed according to the robot's skills such as show emotions, make movements, and give clues to the learner.

System I: Multimodal Fusion

As mentioned above, the proposed artificial cognitive architecture in three modules is presented in Fig. 1. The first module acknowledges the student's behavior (multimodal fusion). The proposed method recognizes the tradition of the work developed at the Technische Universität München around the JAST project. Having said this; The architecture recognizes actions of verbal and non-verbal communication of the subject [2]. The robot interprets integrates the information of the different communication channels and estimates the cognitive state of the user [3, 4]. The context for interpreting information is the mechanical task, which is related to the solution of a transformation problem. That is a problem with a defined space [30, 32, 33, 48]. The actions developed by the robot contribute to foster the learner's cognitive development [31].

Sensorial Processing Subsystem

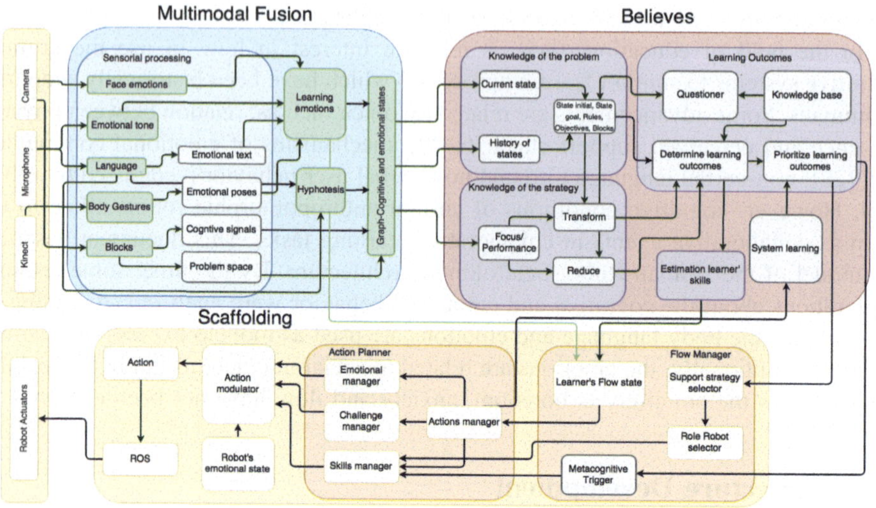

Fig. 1. Human-robot scaffolding architecture

The observation of student's behavior is estimated by modules as blocks' positions, blocks' movements, emotional state, verbal judgments and cognitive gestures, as can be seen in the same figure. The positions of the blocks allow to determine two aspects: the knowledge of the problem and the knowledge of the strategy to solve the problem. The knowledge problem is linked to the use of the operator's (an action to change a state) problem, application of rules, and the knowledge of the different states (initial, intermediate, goal) during the process of solving a problem. The knowledge to develop the strategy involves three objectives: to transform, to reduce and to apply. They are part of the Mean-End Analysis strategy. On the other hand, blocks' positions are described through three-dimensional coordinates according to mechanical restrictions. With each movement, a new node is created in the user`s problem space. Some tests have been carried out with a camera and a Kinect device, although an accelerometer device into each block could be another option.

The kinematic description of the blocks is measured during each change of the state of the problem developed by the learner. This contributes to acknowledge the learner's cognitive process within the problem's solution space. For example, the variation of block's trajectory indicates changes on the user's reasoning [8], the object manipulation discovers relationships between concepts and spatial thinking, and recognition of spatial patterns of the problem and promotes the choice of operators even by omitting deliberative processes [36].

In order to develop the kinematic descriptions each coordinate is taken during the trajectory between two nodes is measured. Based on these data three quantitative descriptors are determined: block's velocity, change of trajectory, and blocks' dropping. Figure 2 presents the analysis of four trajectories generated by some students. In task context, the trajectories generated by the movement of the blocks could indicate cognitive actions such as hypothesis evaluation, goal change, safety and insecurity in knowledge. A continuous block's trajectory means self-confidence; a change of direction or bounce indicates a change of objective. To contrast the theoretical issues, the code has been implemented using OpenCv and Python. The technological tools can be the same as proposed above.

The emotional state indicates the user's cognitive mood to undertake, develop and complete a learning task. The cognitive arrangement depends on two aspects: The user skills and the challenge of the problem [16]. The combination of both variables generates three emotional states: anxiety, flow, and boredom [21]. These can be divided into eight areas: anxiety, concern, comfort and optimism, curiosity, interest, indifference, and boredom [27].

In order to determine the user emotional learning's states different data from face movement are taken. Some alternative software as Emotient and Affective have been tested during the design of this architecture. The verbal judgments allow to give sense to the cognitive processes and the knowledge of the user during the process of problems solution. Cognitive processes are related to aspects such as thought, attention, strategies, knowledge, and hypothesis. The interpretation of verbal judgments has two background aspects: the context of the task and the cognitive processes [12]. According to the information processing theory, the context of the task determines the operators, the rules, and the knowledge to solve the problem [37]. With this data, the emotional, cognitive and meta-cognitive states are recognized. According to the Intelligent Tutors Systems, the most use characteristics to model the student profile are: knowledge level (52.8%), cognitive features (40.75%), affective features (16.85%), misconceptions (15.75%), and meta-cognitive features (6.74%) [10]. Alternatives as CMU Sphinx to recognize emotions and Synesketch to recognize textual emotion recognition have been tested to design this architecture.

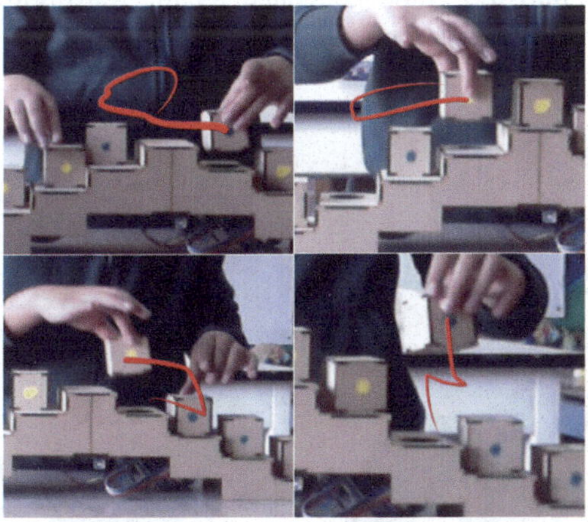

Fig. 2. Kinematic characteristics

Body gestures expose the actions of thought in the problem-solving process. Characteristics such as body posture, facial expressions, eye movement and hand movement have been studied [17], based on the degree of abstraction, during the problem-solving process. The study of gestures involves three aspects: recognition, cognitive contribution, and estimation of mental models. First, the recognition involves coding gestures according to the characteristics of the problem, and segmenting them according to their occurrence, size, and quantity [36]. Second, cognitive contribution involves recognizing how its occurrence evokes implicit knowledge, promotes spatial representation, and introduces information to solve the problem [1]. Finally, but not least, is the estimation of mental models to acknowledge aspects such as knowledge of strategy, strategy change, problem difficulty and solver's expertise [1, 11].

Different body movements during the problem-solving process have been analyzed. This kind of movements with or without blocks are useful to carry out assertive robot`s intervention. Figure 3 presents four typical cognitive gestures presented in experimental sessions during problem-solving process: confused, spatial reasoning, iconic movements, and the intention to apply an operator and change the problem state. The optional devices to get data from fingers, wrist, arm, shoulders and head movements are cameras, Kinect, and particularly RealSense by Intel.

In general, the main methods of perception used in robots like Kismet, Cog, iCub, GRACE, Robox, Reckman, Robovie, RUBI, AMARIII, Papero, Huggable, MEXI, ROMAN, BARTHOC, BIRON, Fritz, ASIMO, iCat, AIBO, Albert Einstein and YouBot are presented. Three actions are highlighted. First, extraction of characteristics of signals like video, audio, tactile and sensors. Second, reduction of dimensionality through techniques such as analysis of main components, analysis of linear discriminants and preservation of local projections. Third, semantic comprehension through the recognition, tracking and segmentation of objects [20].

As a conclusion of System I, the information obtained from the sensorial processing module is organized into the graph generated by the transition of states during of solution of the transformation problem.

Fig. 3. Cognitive gestures

According to Fig. 4, the nodes indicate the changes of the state in the problem space, and the arcs represent the relation between cognitive gestures, verbal judgments, and emotional states.

System II: Believes

As mentioned above, Fig. 1 shows the proposed artificial cognitive architecture in three modules. The second module diagnoses cognitive and emotional states (believes). The main objective is prioritizing the learning goals through four aspects: the knowledge of the problem, the knowledge of the strategy, the learning objectives and questioner module [10] (See Fig. 1).

Subsystem: Knowledge of the Problem

The problem knowledge level is determined through the student's behavior observation. Characteristics as the initial state, goal state, intermediate states, operators, errors, misconceptions, and rules are analyzed. In the transformation problems as Hanoi Tower, errors are detected by infringing rules and misconceptions are detected through the development of the strategy [14].

The current state module takes information from the previously constructed graph. Figure 4 presents a bit of the graph. The module evaluates the learner's actions related to the problem. The actions describe the learner's procedural knowledge related to the use of rules and operators. The evaluation process means the user's knowledge of the problem. With this information, the objectives for procedural support are estimated.

The history states module takes information from the previously constructed graph. The module evaluates the learner's actions related to use of means-ends analysis strategy. The learner's actions describe their cognitive knowledge related to two objectives of this strategy. First, to transform (to find the difference between the current state and the target state). Second, to reduce (to find the operator that reduces the difference). As a result, the evaluation indicates the knowledge of the strategy. With the above information, the goals for cognitive support are estimated.

Subsystem: Knowledge of the Strategy

The strategy knowledge level is acknowledged through data of the task. The cognitive characterization process involves four steps: labeling data, distilling data features, developing detector and validating. Data as time invested in the transition of each problem state and previous interactions are evaluated [13]. Also, the metacognitive characteristics as self-regulation, self-evaluation, self-explanation, and self-efficacy could be assessed. The memory state sequence shows aspects as the learner's thinking states and the doubts and pauses which correspond to changes in the direction of thought. Each memory state is joined to a knowledge about the solution process. This kind of information is useful to determine the cognitive student needs (transforming, reducing) which are necessary to learn the mean-end analysis strategy. Also, the information of memory states is useful to estimate the learner's space problem, which is a dynamic structure and it is defined by sub-goals (transforming, reducing, applying), storage goals, connection table, structure goals, and production system.

Different methods of cognitive task analysis as think-aloud protocols, content analysis, process isolation, situated studies, hierarchical task analysis, link analysis, operational diagram sequence, timeline analysis and GOMS has been analyzed to determine the learning outcomes. The learning outcomes have two categories: procedural and cognitive.

Subsystem: Learning Outcomes

The questioner module is a useful methodology to know the hypotheses generated by the users during the solution of the problem. Whit this kind of answers, the robot can estimate aspects such as student's beliefs, self-evaluation, biases, and heuristics used for reasoning, even the robot's effect [19].

As conclusion, the learning objectives are grouped into two categories. First, the objectives related to the knowledge of the problem. For example, the use of rules or operators to change states. Second, the objectives related to the knowledge of the strategy to solve the problem. For example, the recognition of the objectives to transform and reduce the problem, which is useful to implement the strategy of means-ends analysis.

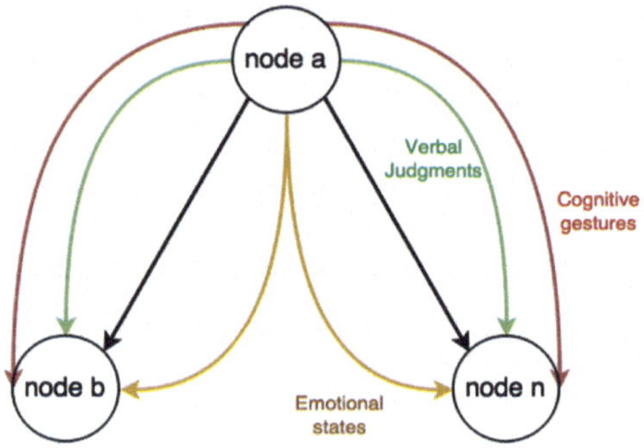

Fig. 4. Graph characteristics

System III: Scaffolding

As mentioned above, Fig. 1 shows the proposed artificial cognitive architecture in three modules. The third module plans and creates the intervention according to two conditions: learning curve (scaffolding) and Flow theory. Such theory shows a correlation between Challenge and Skills. The relation describes emotional states as anxiety when learners have not skills to solve the problem, but the challenge is high. Elsewhere, the relation describes boredom when learners are more skillful than the challenge requirements. After that, learning outcomes are prioritized and the intervention strategy is defined (See Fig. 1). Figure 5 describes the Flow theory which is described as a balance between skills and task demands. The brown line represents the flow's evolution from a low level to a high level. The brown line divides the flow area into two equal parts. The axis y represents the level of the challenge of the problem. For example, in the Hanoi tower, the number of the disks involved in the problem. The axis x represents the user's skills. For example, the knowledge of the strategy.

The *Ci* represents the current challenge level of the problem. The *Si* represents the learner`s skills which have been estimated by the previous subsystem. When the system determines the learner's flow level, three actions (c, e, s) are developed to bring closer the learner to the ideal flow zone.

Subsystem: Flow Manager

The flow manager is composed by four modules. First, learner's flow state. The learner's flow state acknowledges the balance between the learner's skills and the emotional state to change the cognitive load through different cognitive, emotional or physical actions. Second, support strategy selector. According to Fig. 1, the robot has four strategies to support the learning process: focus lesson, guided instruction, collaborative work, and independent work. Third, role robot selector. The robot can take three different roles to support assertively the learning process: peer, tutor, and learner. Four, metacognitive trigger. The function is generated thinking process based on the graph evolution. The learning process is foster through four strategies: focus lesson, guided instruction, cooperative work and independent work. By each strategy, a role is assigned. There are three alternatives: tutor, peer, and learner. As tutor, the robot interventions to explain and guide the learner's actions. As peer, the robot intervenes through verbal suggestions and physical interventions negotiated with the learner. As learner, the robot carries out questions related to the development of the task and through erroneous physical actions. In the three roles mentioned above, five aspects are implemented in robot: emotion, gestures, verbal judgments, kinematic description, positioning the blocks.

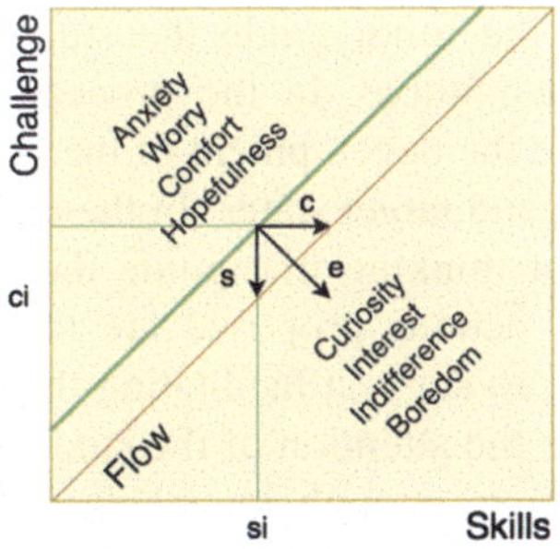

Fig. 5. Flow theory as decision robot's rule

According to Fig. 5, Ci and Si show a relation which the robot needs either change the challenge or reduce the level of skills to foster the flow state. The green line is build based on the learner's emotional state, the challenge is related to the problem and the skills are related to the learner. The action C is related to modify the Challenge of the problem. For example, reduce the amount of block in the Jumper game. The action E means to perform some action in order to give assertive emotional support while the learning process. The action S are related to the learner's Skills because of the robot can select action to adjust the problem based on learner's skills.

The robot's morphology affects alternatives of gestural behavior and therefore its process of communication with the user. The robot can perform different kind of movements as emblematic, descriptive, rhythmic, deictic, symbolic, expressive and regulated. In humanoid robots, the non-verbal behavior of the robot affects the perception of the subjects [52]. In non-humanoid robots and without emotions: the behavior is related to the movement associated with the task.

Subsystem: Action Manager

According to the architecture and the scaffolding theory, robot decides which the best option to support the learning process. There are three ways. First, the robot gives emotions support. Second, the robot presents new information or gives support based on hits and missed done by the learner. The support's strategy could consider physical intervention as moving the blocks. Third, the robot changes the task's challenge. In this option, the robot changes the task's complexity and the number of blocks is reduced or augmented.

Emotional expression done by social robots during the learning process contribute to the feedback of task performance. Emotional manifestations expressed by the robot are: seeking information, attention, and interest, inviting and controlling interaction, influencing others and presenting their emotional state according to the conditions of a learning activity [50]. As a collaborative agent, robot has physical and cognitive artificial infrastructure (reason and emotion) to foster thought's actions during learning. One way is through emotional responses that allow increasing, refine and restructure the mental models of the learning situation. The way of expressing emotions, which is part of the personality of the robot, facilitates the understanding of the actions of the subject and affects his learning process.

The verbal judgments of the robot guide the student's learning process in two aspects: procedural and metacognitive. In the procedural aspect, verbal judgments made by the robot aim to guide the development of the problem posed using operators, recognition of rules and initial and target states. In the metacognitive aspect, the verbal judgments stimulate actions of thinking that guide the knowledge of the strategy of analysis of means-ends. Four actions proposed are to ask to evaluate the student's knowledge and understanding, to suggest facilitating the cognitive and meta-cognitive process, to point out to change the attention of the students, to explain in the moments where the student does not have enough knowledge of the problem or the implementation of the strategy to solve it [15].

As was mentioned above, the main mechanism of the scaffolding of the robot is the movement of the blocks to contribute to the process of solving the problem of transformation such as the Hanoi towers, the stair set and stacking blocks. The position of the blocks and the kinematic conditions for during positioning influence the user's thought processes [7].

In order to effectively support the robot, is necessary to consider aspects such as: adaptive behavior [39], use of skills learned from other tasks for use in new tasks [25], dynamic transition of responsibility between robot and learner [41], recognition of the cognitive state of the user [34], increase of dialogue resources through nonverbal behaviors [5], assertive suggestions appropriate to the needs of the learner [40, 43, 44, 47].

In addition, support during decision-making requires consideration of aspects such as the dialogue system for the understanding of robot interventions [33, 34], motor interaction requirements [23], The development of interactive behaviors through actions such as pointing objects [24], non-verbal actions such as gaze, proximity and development of iconic, metaphoric, deictic and vocal gestures robot [9].

The scaffolding process not only depends on student characteristics, also depends on the strategy and the task's characteristics. The above aspects open possibilities of generating levels of granularity in the scaffolding process. Figure 6 elicits a process of interaction with the robot. The physical appearance of the robot generates expectations of behavior, cognitive, emotional support, evaluation mechanisms and continuous feedback.

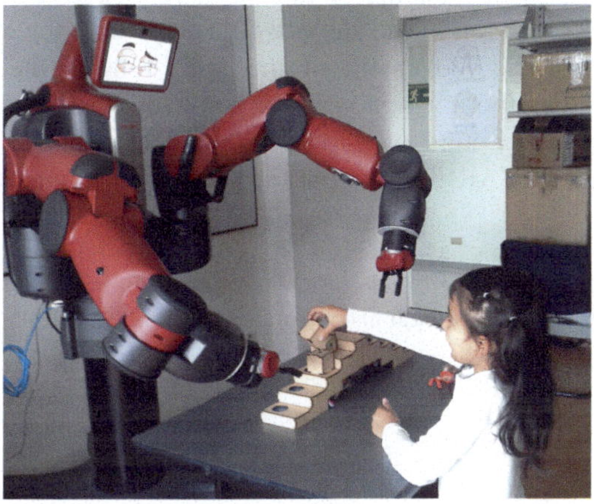

Fig. 6. Baxter robot

3 Discussion and Conclusions

The technologies available these days are useful for countless applications. That is de case for robotics that can be adapted to different cases where the robot can be another participant in an environment created for a specific goal. That is the case of the present work where robots are used to establish Human-Robot interaction for the improvement of the learning environment. The same case could be used for the healthcare domain where tests are being designed for patients with mild dementia that are experiencing cognitive decline. Especially in older patients, the permanent availability of the robot without time constraints, can be useful as the older patients can use it when they are willing to make cognitive rehabilitation any time without the need of a therapist for those designed training exercises.

As conclusion, in education, robots have been considered as learning tools. They have evolved from simple tools that only followed students' instructions to complex cognitive artificial systems which allow robots to behave as tutor, peer, or learner. Each proposed behavior has been inspired by different pedagogical and psychological theories despite having been constrained by the contemporary technical conditions. Nowadays, embodiment, emotions, and physical interaction are topics of interest in the cognitive convergence challenge in the Human-Robot Interaction. For convergence to be effective and to contribute to the learning process, it is necessary for the cognitive architecture of the robot to develop three actions: to observe student behavior, to diagnose their cognitive and emotional states, and to intervene assertively according to their learning curve.

The robot as a collaborative agent recognizes several characteristics of the student and average during the learning process. The concept of mediation suggests an assertive intervention of the robot in a way that promotes cognitive effort during the development of the task and the student does not lose interest.

The shape of the robot influences the human-robot interaction and the behavior, which affects the user's understanding. In humanoid robots, the non-verbal behavior of the robot affects the perception of the subjects. In robots with the ability to manipulate and express emotions, the movement of the arms reflects the behavior. In non-humanoid robots and without emotions: the behavior is related to the movement associated with the task. Future work will promote inclusion of the results so far achieved in the scope of CARELINK AAL Project for Dementia patients.

Acknowledgements. The work has also been promoted under the project CARELINK, AAL-CALL-2016- 049 funded by AAL JP, and co-funded by the European Commission and National Funding Authorities FCT from Portugal and the national institutions from Ireland, Belgium and Switzerland.

References

1. Alibali, M.W., Spencer, R.C., Knox, L., Kita, S.: Spontaneous gestures influence strategy choices in problem solving. Psychol. Sci. **22**(9), 1138–1144 (2011)
2. Knoll, A., Hildenbrandt, B., Zhang, J.: Instructing cooperating assembly robots through situated dialogues in natural language. In: 1997 Proceedings of the IEEE International Conference on Robotics and Automation, vol. 1, pp. 888–894. IEEE (1997)
3. Knoll, A.C.: Distributed contract networks of sensor agents with adaptive reconfiguration: modelling, simulation, implementation and experiments. J. Frankl. Inst. **338**(6), 669–705 (2001)
4. Knoll, A.: A basic system for multimodal robot instruction. In: Pragmatics and Beyond New Series, pp. 215–228 (2003)
5. Alves-Oliveira, P., Janarthanam, S., Candeias, A., Deshmukh, A., Ribeiro, T., Hastie, H., Paiva, A., Aylett, R.: Towards dialogue dimensions for a robotic tutor in collaborative learning scenarios, pp. 862–867 (2014). https://doi.org/10.1109/ROMAN.2014.6926361
6. Bainbridge, W.A., Hart, J.W., Kim, E.S., Scassellati, B.: The benefits of interactions with physically present robots over video-displayed agents. Int. J. Soc. Robot. **3**(1), 41–52 (2011)
7. Baxter, G.D., Ritter, F.E.: Designing abstract visual perceptual and motor action capabilities for use by cognitive models. Technical report 36, ERSC Center for Research and Development, Instruction and Training, Department of Psychology, University of Nottingham, (1996)

8. Blauvelt, G.R., Eisenberg, M.: Machineshop: A Design Environment for Supporting Children's Construction of Mechanical Reasoning and Spatial Cognition. University of Colorado at Boulder, Boulder (2006)
9. Chandra, S., Alves-Oliveira, P., Lemaignan, S., Sequeira, P., Paiva, A., Dillenbourg, P.: Can a child feel responsible for another in the presence of a robot in a collaborative learning activity, pp. 167–172 (2015). https://doi.org/10.1109/ROMAN.2015.7333678
10. Chrysafiadi, K., Virvou, M.: Student modeling for personalized education: a review of the literature. In: Advances in Personalized Web-Based Education, pp. 1–24. Springer (2015)
11. Chu, M., Kita, S.: The nature of gestures' beneficial role in spatial problem solving. J. Exp. Psychol. Gen. **140**(1), 102 (2011)
12. Crandall, B., Klein, G.A., Hoffman, R.R.: Working Minds: A Practitioner's Guide to Cognitive Task Analysis. MIT Press, Cambridge (2006)
13. d Baker, R.S., Corbett, A.T., Roll, I., Koedinger, K.R., Aleven, V., Cocea, M., Hershkovitz, A., de Caravalho, A.M.J.B., Mitrovic, A., Mathews, M.: Modeling and studying gaming the system with educational data mining. In: International Handbook of Metacognition and Learning Technologies, pp. 97–115. Springer, New York (2013)
14. Essa, A.: A possible future for next generation adaptive learning systems. Smart Learn. Environ. **3**(1), 16 (2016)
15. Fisher, D., Frey, N.: Guided Instruction: How to Develop Confident and Successful Learners. ASCD, Chicago (2010)
16. Freire, T., Tavares, D., Silva, E., Teixeira, A.: Flow, leisure, and positive youth development. In: Flow Experience, pp. 163–178. Springer (2016)
17. Goldin-Meadow, S.: Talking and thinking with our hands. Curr. Dir. Psychol. Sci. **15**(1), 34–39 (2006)
18. Granados, L.F.M., Londoño, E.A.A.: Análisis de Protocolos: Posibilidad metodológica para el estudio de procesos cognitivos. Universidad Pedagógica Nacional (2001)
19. Hacker, D.J., Dunlosky, J., Graesser, A.C. (eds.): Handbook of Metacognition in Education. Routledge, Abingdon (2009)
20. Yan, H., Ang Jr., M.H., Poo, A.N.: A survey on perception methods for human–robot interaction in social robots. Int. J. Soc. Robot. **6**(1), 85–119 (2014)
21. Harmat, L., Andersen, F.Ø., Ullén, F., Wright, J., Sadlo, G. (eds.): Flow Experience: Empirical Research and Applications. Springer, Berlin (2016)
22. Hayes, B., Scassellati, B.: Challenges in shared-environment human-robot collaboration. In: Learning, vol. 8, p. 9 (2013)
23. Jarrassé, N., Sanguineti, V., Burdet, E.: Slaves no longer: review on role assignment for human-robot joint motor action. Adapt. Behav. **22**(1), 70–82 (2014). https://doi.org/10.1177/1059712313481044. Cited by 4
24. Kanda, T., Miyashita, T., Osada, T., Haikawa, Y., Ishiguro, H.: Analysis of humanoid appearances in human–robot interaction. IEEE Trans. Robot. **24**(3), 725–735 (2008)
25. Guerin, K.R., Riedel, S.D., Bohren, J., Hager, G.D.: Adjutant: a framework for flexible human-machine collaborative systems. In: 2014 IEEE/RSJ International Conference on Intelligent Robots and Systems, pp. 1392–1399. IEEE (2014)
26. Kim, M.C., Hannafin, M.J.: Scaffolding problem solving in technology-enhanced learning environments (TELEs): bridging research and theory with practice. Comput. Educ. **56**(2), 403–417 (2011)
27. Kort, B., Reilly, R.: Analytical models of emotions, learning and relationships: towards an affect-sensitive cognitive machine. In: Conference on Virtual Worlds and Simulation (VWSim 2002) (2002)

28. Kwak, S.S., Kim, Y., Kim, E., Shin, C., Cho, K.: What makes people empathize with an emotional robot? The impact of agency and physical embodiment on human empathy for a robot. In: 2013 IEEE RO-MAN, pp. 180–185. IEEE, August 2013
29. Mann, J.A., MacDonald, B.A., Kuo, I.H., Li, X., Broadbent, E.: People respond better to robots than computer tablets delivering healthcare instructions. Comput. Hum. Behav. **43**, 112–117 (2015)
30. Giuliani, M., Foster, M.E., Isard, A., Matheson, C., Oberlander, J., Knoll, A.: Situated reference in a hybrid human-robot interaction system. In: Proceedings of the 6th International Natural Language Generation Conference, pp. 67–75. Association for Computational Linguistics (2010)
31. Giuliani, M., Knoll, A.: Using embodied multimodal fusion to perform supportive and instructive robot roles in human-robot interaction. Int. J. Soc. Robot. **5**(3), 345–356 (2013)
32. Rickert, M., Foster, M.E., Giuliani, M., By, T., Panin, G., Knoll, A.: Integrating language, vision and action for human robot dialog systems. In: Universal Access in Human-Computer Interaction. Ambient Interaction, pp. 987–995. Springer (2007)
33. Foster, M.E., Bard, E.G., Guhe, M., Hill, R.L., Oberlander, J., Knoll, A.: The roles of haptic-ostensive referring expressions in cooperative, task-based human-robot dialogue. In: Proceedings of the 3rd ACM/IEEE International Conference on Human Robot Interaction, pp. 295– 302. ACM (2008)
34. Foster, M.E., Giuliani, M., Isard, A., Matheson, C., Oberlander, J., Knoll, A.: Evaluating description and reference strategies in a cooperative human-robot dialogue system. In: IJCAI, pp. 1818–1823 (2009)
35. Tielman, M., Neerincx, M., Meyer, J.-J., Looije, R.: Adaptive emotional expression in robot-child interaction. In: Proceedings of the 2014 ACM/IEEE International Conference on Human-Robot Interaction, pp. 407–414. ACM (2014)
36. National Research Council: Learning to think spatially: GIS as a support system in the K-12 curriculum. National Academies Press (2005)
37. Brooks, N.B., Barner, D., Frank, M., Goldin-Meadow, S.: The role of gesture in supporting mental representations: the case of mental abacus arithmetic. University of Chicago (2015)
38. Newell, A., Simon, H.A.: Human Problem Solving, vol. 104, no. 9. Prentice-Hall, Englewood Cliffs (1972)
39. Pea, R.D.: The social and technological dimensions of scaffolding and related theoretical concepts for learning, education, and human activity. J. Learn. Sci. **13**(3), 423–451 (2004)
40. Fournier-Viger, P., Nkambou, R., Nguifo, E.M., Mayers, A., Faghihi, U.: A multiparadigm intelligent tutoring system for robotic arm training. IEEE Trans. Learn. Technol. **6**(4), 364–377 (2013)
41. Ramacliandran, A., Scassellati, B.: Adapting difficulty levels in personalized robot-child tutoring interactions, vol. WS-14-07, pp. 56–59 (2014)
42. Reardon, C., Zhang, H., Wright, R., Parker, L.E.: Response prompting for intelligent robot instruction of students with intellectual disabilities, pp. 784– 790 (2015). https://doi.org/10.1109/ROMAN.2015.7333651
43. Reardon, C., Zhang, H., Wright, R., Parker, L.E.: Response prompting for intelligent robot instruction of students with intellectual disabilities. In: 2015 24th IEEE International Symposium on Robot and Human Interactive Communication (RO-MAN), pp. 784–790. IEEE, August 2015
44. Reidsma, D.: The EASEL project: towards educational human-robot symbiotic interaction. In: Lecture Notes in Computer Science (Including Subseries Lecture Notes in Artificial Intelligence and Lecture Notes in Bioinformatics, vol. 9793, pp. 297–306 (2016). https://doi.org/10.1007/978-3-319-42417-0_27

45. Serholt, S., Basedow, C.A., Barendregt, W., Obaid, M.: Comparing a humanoid tutor to a human tutor delivering an instructional task to children, pp. 1134–1141 (2015). https://doi.org/10.1109/HUMANOIDS.2014.7041511
46. Tabak, I.: Synergy: a complement to emerging patterns of distributed scaffolding. J. Learn. Sci. **13**(3), 305–335 (2004)
47. Kanda, T., Hirano, T., Eaton, D., Ishiguro, H.: Interactive robots as social partners and peer tutors for children: a field trial. Hum.-Comput. Interact. **19**(1), 61–84 (2004)
48. Thien, N.D., Terracina, A., Iocchi, L., Mecella, M.: Robotic teaching assistance for the "tower of hanoi" problem. Int. J. Dist. Educ. Technol. **14**(1), 64–76 (2016). https://doi.org/10.4018/IJDET.2016010104
49. Müller, T., Ziaie, P., Knoll, A.: A wait-free real-time system for optimal distribution of vision tasks on multi-core architectures. In: ICINCO-RA, no. 1, pp. 301–306 (2008)
50. Turner, J.E., Waugh, R.M., Summers, J.J., Grove, C.M.: Implementing high-quality educational reform efforts: an interpersonal circumplex model bridging social and personal aspects of teachers' motivation. In: Advances in Teacher Emotion Research, pp. 253–271. Springer (2009)
51. Van De Sande, B.: Properties of the Bayesian knowledge tracing model. JEDM-J. Educ. Data Min. **5**(2), 1–10 (2013)
52. Salem, M., Eyssel, F., Rohlfing, K., Kopp, S., Joublin, F.: To err is human (-like): effects of robot gesture on perceived anthropomorphism and likability. Int. J. Soc. Robot. **5**(3), 313–323 (2013)

Azoresbot: An Arduino Based Robot for Robocup Competitions

José Cascalho[1,2,3,6(✉)], Armando Mendes[1,2,4,6], Alberto Ramos[6], Francisco Pedro[6,8], Nuno Bonito[6], Domingos Almeida[8], Pedro Augusto[7], Paulo Leite[6,9], Matthias Funk[1,6], and Arturo Garcia[5,6]

[1] Faculdade de Ciências e Tecnologia, Universidade dos Açores, Ponta Delgada, Portugal
{jose.mv.cascalho,armando.b.mendes,gunther.ma.funk}@uac.pt
[2] NIDeS - Núcleo de Desenvolvimento em e-Saúde, Universidade dos Açores, Ponta Delgada, Portugal
[3] BioISI - Biosystems and Integrative Sciences Institute, FCUL -Universidade de Lisboa, Lisboa, Portugal
[4] Algoritmi, Universidade do Minho, Braga, Portugal
[5] Instituto de Investigação em Vulcanologia e Avaliação de Riscos, Universidade dos Açores, Ponta Delgada, Portugal
arturo.m.garcia@uac.pt
[6] GRIA - Grupo de Robótica e Inteligência Artifical, Universidade dos Açores, Ponta Delgada, Portugal
[7] Escola Secundária Domingos Rebelo, Ponta Delgada, São Miguel, Portugal
[8] Escola Básica Integrada de Rabo de Peixe, Ribeira Grande, São Miguel, Portugal
[9] GLOBALEDA, Ponta Delgada, Portugal

Abstract. Robotic competitions in the context of the Robocup festival or similar, obey to a certain set of challenges that demands for each robot to have a set of sensors and actuators adapted to different environments. In this paper we present the Azoresbot robot that was developed in the context of a robotic regional competition which intends to foster the learning of robotics and programming in Azorean schools. In the context of this festival, the robot was presented as a kit that teams had to build, to test and, then, to program for different competitions. This robot was created thinking in the modularity and adaptability needed to the different challenges in the competitions.

Keywords: Educational robotics · Arduino · Robocup

1 Introduction

Robotics is considered a key subject in Science Technology Engineering and Mathematics (STEM) education by several authors. Robotics activity usually implies not only a mechanical design but also electronic skills and programming, all these connected to mathematics and physics as a background knowledge [3]. Some authors considerer educational robotics as positive by the fact that they

M. F. Silva et al. (Eds.): ROBOT 2019, AISC 1092, pp. 542–552, 2020.
https://doi.org/10.1007/978-3-030-35990-4_44

are anchored in projects implying a set of competences related to the conception, implementation and construction and control of robotic devices, which foster the STEM interest [14].

Nonetheless, some critics are also mentioned. The robots used in schools tend to be limited to a narrow range of possible applications by the fact that schools usually have restriction on hardware budget spending [3] and that this set of applications are far from the robotics solutions applied in real environments [8].

The use of robotics in schools are not just mentioned as a way to provide knowledge in STEM areas but also to foster learning by facilitating new kind of practices in classes. In particular and with respect to the design engineering processes, i.e the capability of using resources available to construct a product, Kaloti-Hallak [7] refers how robotics subject helps the learning process by changing the mentoring style to a student-centered activity and supporting the absence of a textbook. An extra-curricular competition-oriented nature of the activities is also a factor that produce effects on the learning results.

In the project presented in this paper there was the commitment to organize a first regional robotics competition in Azores behind the intentionality of creating a new didactic mobile robot. The fact that many schools didn't have any robot and students were expected to have few programming skills, triggered the necessity to provide a robot that could be build up by teams and, at the same time, to be adjusted to the simplest challenges, where they have to follow a line, avoiding obstacles or to detect colors.

Although the robotic kit described in this paper was prepared to be used in the simplest challenges in robotics competitions, their design was thought to provide a final product that could be upgraded and enhanced to more complex challenges.

The first regional robotics competition was organized by the Association of Programming and Robotics of Azores and a group of professors from Azores University with collaboration of teachers and students from robotic clubs in two different schools of Ponta Delgada.

2 Robots for Education

There is a large number of robots for educational proposals. The recent surge of interest by schools was due to the acknowledgement that robotics activity has a role on fostering learning in STEM areas [1,9][1]. It is difficult to cover all the types of educational robots that nowadays can be used at schools. Nevertheless, a distinction of types can be made if some target features are considered.

Concerning the adopted strategies to provide a robot to be used by different groups of students (e.g. elementary and high school), a set of robots are being equipped with a resourceful set of sensors. Example of these type of robots are the *Robobo* [3] or the *Sparki* and *E-puck* mobile robots [15].

[1] Some authors argue that this idea must be further studied as it seams that there is a different perception depending on students' gender [11].

Other solutions point out to the hands on activity of building a robot and to its adaptability to different challenges or goals. The paradigmatic robot of this type is the Lego Mindstorms, which is also used in specific competitions [7]. This robot uses specially conceived actuators and sensors connected to a specific micro-controller/computer brick. Using a robot which can be used by children of a wide age range, supported by adapted interactive tools for different learning goals [13] or a robot that can be built based on preconceived modules [12], are two other approaches to the concept of modularity in educational robotics.

There are, however, some robots that use Arduino and invite students to use a 3D printer to print robot pieces. These are usually intentional low cost solutions. One of these solutions is provided by the Portuguese Informatics Teachers Association (ANPRI) with a set of Arduino robots adapted to specific tasks using a suitable set of sensors, accordingly[2].

Another perspective is to adopt robotic competitions as a motivation for learning STEM areas in an engineering class [10]. Some robots are even prepared to be used in a specific task, such as the robot used in the *Robot@FactoryLite* competition in the Robocup initiative in Portugal[3] where a fully adapted robot to that specific competition is used in a controlled environment, where the main challenge is focused on programming the best planning strategies.

Nowadays, there is a diversity of prefabricated robots as "black box" for children who "are invited to play or interact without understanding "what's inside" and how it works" [2]. This scenario contrasts with the "maker movement" in educational robotics that stresses the importance of designing and building activity, instead of just "using them".

In the robotic competitions, robots are usually built and programmed by participant teams which adapt them to specific challengers. Instead of off the shelf robots, students that apply to these competitions must provide a robotic solution that is singular and usually built entirely using their own resources. In the same line of thought, kits area provided for initiatives where the main goal is to address robotics at an initial level[4]). In these contexts, the learning target is not focused on problem solving but mainly to learn about the different components and how to program the robot to execute simple tasks.

Recently, a taxonomy was proposed by Catlin et al. [4] which identifies three main types of educational robots, the Build Bots, the Use Bots and the Social Bots. The taxonomy is also provided on-line[5] where for each type different classes and subclasses provide a way to characterize each robot based on its locomotion, power, commands and control, etc. In this taxonomy, the robot described in this paper would fit in the Build Bots type, in the class *Class Build Systems* or *Robot Kits*, as it will be clear in the next section.

---

2 http://www.anpri.pt/anprino/.

3 https://web.fe.up.pt/~robotica2019/index.php/en/robot-factory-lite-2.

4 See RoboOeste (http://www.robooeste.educacaotorresvedras.com/) and RoboParty (https://www.roboparty.org/).

5 https://robots-for-education.com/.

3 The Modular Concept

The idea of making a modular design come up with the necessity of having a low cost two wheel differential drive robot that could be adapted to problems with different degrees of complexity. To support that modularity it was decided to have the control component on the top layer and to put motoring, drivers and power source, on the bottom layer. Moreover, a breadboard was used to the wired connections between the Arduino and sensors. With this strategy it was expected to support additional control modules and other sensors.

Table 1 depict the main components select to AZORESBOT robot with the ultrasonic, line follower and color sensors, the motors and the RGB led to signal a detected color. The total cost of this solution was about 100 euro.

Table 1. List of the components used in the AZORESBOT robot.

Component	Reference
Micro-controller	Arduino Mega 2560
Motor driver	L298N
Motor	Micro motor DC 140 rpm
Ultrasonic sensor	HC-SR04
Color sensor	TC3200-WS
Line follower	TCRT5000
Led	LED RGB
Battery	Li-ion 18650

The other conceptual idea behind this robot was to provide it as a kit that could be built without using a soldering iron.

Fig. 1. The bottom layer with the motors.

Finally, all the different parts of the robot should be built using a 3D printer, following the idea presented by ANPRI.

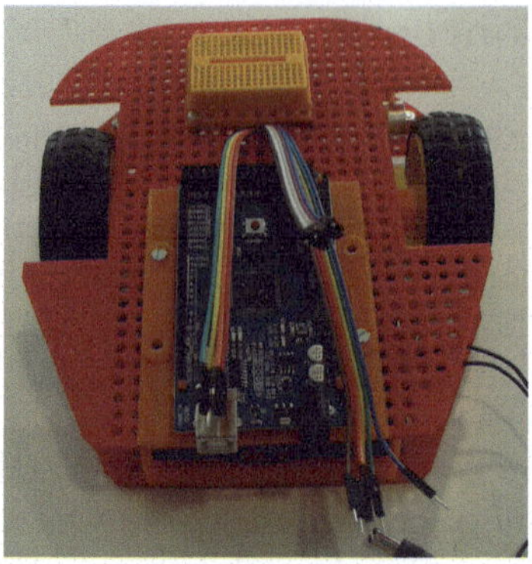

Fig. 2. The top layer with the Arduino micro-controller.

The adopted micro-controller was the ATmega2560 based Arduino Mega 2560 firstly because it had a number of connectors to all the sensors planned to be used. And, secondly, because of the ease of use of Arduino IDE, the programming environment and the libraries support for the different Arduino boards, sensors and actuators.

Figures 1 and 2 show the bottom and the top boards of the robot, where the bottom board supports motors and motor driver as well as proximity and ground sensors and the top board have the Arduino Mega connected. Figure 3 shows a detail of the ground sensor attached to the bottom layer.

Fig. 3. The sensors in the robot front bottom layer.

Figure 4 shows the small breadboard that was used to connect wires from the bottom layer to Arduino Mega whereas in Fig. 5 it is presented a robot in competition with all the connected wires. In this figure is shown the front with three ultrasonic sensors and a line tracking infrared sensor. The top layer only has the Arduino board and the breadboard connectors. It is on the top of this layer that other modules can be added, as an upgrade to improve robot functionalities. The color sensor was placed in one of the sides of robot to detect ground colors. This sensor is used in the *First Challenger* competition (see Sect. 5).

Fig. 4. The top board with the Arduino micro-controller and the breadboard.

Fig. 5. The AZORESBOT robot.

The type of boards used (i.e. pegboards) allowed to place ultrasonic sensors on the top layer, instead of on the bottom layer as the Fig. 5 shows, or to adjust them to different position. Some teams have decided not to use all the three ultrasonic sensors distributed in the kit, because they knew that not all were necessary for the competition.

4 Programming the Robot

The robot was programmed using the Arduino IDE. Along the testing of the hardware, before the regional robocup festival has started, students from robotic clubs were invited to build, to program and to test the robot. This activity was a first test for the robot and had a positive feedback. Notwithstanding it was decided to create libraries that could provide a higher level of programming for each actuator and sensor, presuming that most of the participants in the robotic competitions didn't know the C programming language syntax. Moreover it was decided to add a Ardublockly interface because it was expected previous practice in Scratch by some of the teams.

```
Motor::Motor(int enpin, int inlpin, int inrpin)
{
  pinMode(enpin, OUTPUT);
  _enpin = enpin;
  pinMode(inlpin, OUTPUT);
  _inlpin = inlpin;
  pinMode(inrpin, OUTPUT);
  _inrpin = inrpin;
}
void Motor::avancar(int vel){
  digitalWrite(_inlpin,HIGH);
  digitalWrite(_inrpin,LOW);
  analogWrite(_enpin,vel);
}
void Motor::recuar(int vel){
  digitalWrite(_inlpin,LOW);
     digitalWrite(_inrpin,HIGH);
  analogWrite(_enpin,vel);
}
void Motor::parar(){
  digitalWrite(_inlpin,HIGH);
  digitalWrite(_inrpin,HIGH);
  analogWrite(_enpin,0);
}
```

Fig. 6. The interface that supports the execution of higher level functions to run forward, backward and to stop a motor in the robot. The name of the functions are in Portuguese, *avancar* means *forward*, *recuar* means *backward* and *parar* means *to stop*.

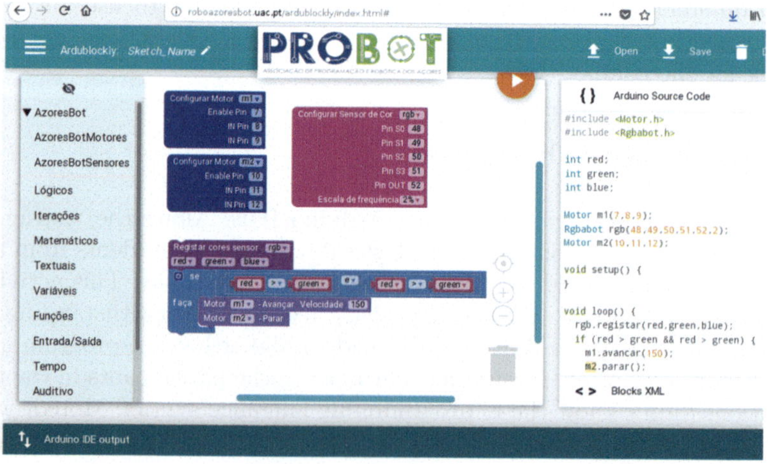

Fig. 7. Programming interface for the AZORESBOT robot.

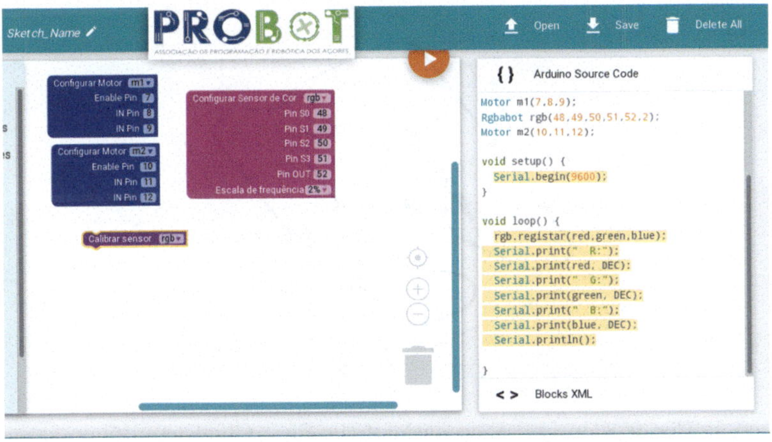

Fig. 8. Ardublockly calibration block for RGB color sensor.

4.1 Library in the Arduino IDE

An example of the code that supports the higher level interface in the set of libraries created is shown in the Fig. 6. It depicts one example of code created for powering the motors. Teams have just to use this higher level functions, *avancar*, *recuar* and *parar* (i.e. forward, backward and stop) for each motor, to move the robot.

4.2 The Role of Ardublockly

An extension using Ardublockly libraries was used to support programming using blocks. The extension were prepared to be used with motors and sensors for the AZORESBOT robot. Figure 7 shows the blocks for the motors and the color sensor are used in the interface. On the right of the figure, the code in C (or C++) corresponds to the code generated using the blocks.

One of the concerns related to the use of the robot in the competitions, was the need to calibrate the RGB sensor. A *calibration block* was created and teams could test the value of output for different colors in the real environment using the monitor in Arduino IDE (Fig. 8).

The software was added to a web server to be downloaded by the teams. The Ardublockly interface was accessed using a browser.

5 Testing the Robot

The final test of the robot AZORESBOT was the First Challenger junior competition in the regional festival in Azores. This challenge was created for the

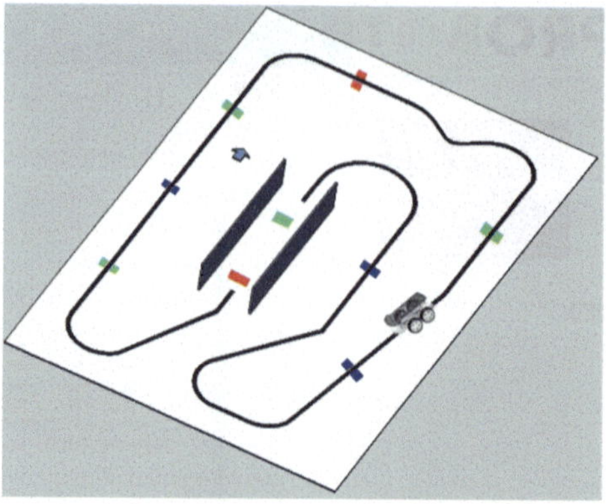

Fig. 9. Conceptual scenario of the First Challenger [5].

portuguese robocup competition and held for the first time in Porto[6]. The challenge [5] is divided into three tasks with increasing complexity allowing the teams to successively test the robot using only the line tracking sensor, then the line tracking and the RGB color detection sensor and, finally, both sensors used in the previous challenges with the ultrasonic sensor helping the robot to pass a tunnel where there is no line to follow (see Fig. 9).

The robot was distributed as a kit for each team. All the pieces in the kit were made by a 3D printer except the wheels for a small group of robots which used the ones that usually came with the motors package[7]. Teams were guided by a document where they had the instructions to build the robot step-by-step. Simple pre-program on-line examples for using the different sensors and the motors were also available to all teams.

6 Discussion

The robotic regional festival received 24 teams along three days and all of them used the AZORESBOT robot in the competition. None of the teams were succeeded in the last task of the *First Challenger*. Four teams completed the second task. But, most interestingly, in the final day, all the teams could program the AZORESBOT robot following the line.

Some teams had difficulties to make the necessary wired connections mainly because they were a little bit complex when all the sensors were used. The library that supported the interaction with the Arduino IDE software was a fundamental tool for the teams success. Teams aware of the Scratch software, were the same that used the Ardublockly interface.

[6] This challenge was created by André Dias and Vitor Cerqueira from Academia de Robótica/ISEP (https://web.fe.up.pt/~robotica2019/index.php/en/first-challenger3.).

[7] In Figs. 3 and 5 it is possible to see different wheels, used by robots due to lack of time to print all the kits pieces.

In the end of the festival, all teams were very enthusiastic about the robot and their achievements. A post-event survey were conducted to professors group tutors, one for each team that participated in the festival, with excellent results. For instance 86%, belief that all the objectives were completely fulfilled and the overall assessment was 83% "very good" and the rest "good". In spite of these results, some comments appoint improvements for next events. One of the most relevant was more programming workshops (17%) and 15% of the respondents asked for more time to mount the robot but mostly to program.

Although satisfied by the success of the robot, two different goals were delineated after the end of the robotic festival:

(a) Increasing the processing capacity of the robot to support more complex challenges. This goal can be achieved by adding a shield that connects Arduino to Raspberry Pi as it usually addressed in projects focused on first semester activities at several universities (e.g. see [16], where the concept of modularity is also considered an important feature).
(b) To evaluate the use of this educational robot. A possible framework to be used is described in [6], where an evaluation framework was created to be applied to a wide range of educational robotic tools, based on the concept of *educational robotics systems* which has three components i.e. *Robot*, *Interface* and *Tasks*.

7 Conclusions

This paper describes the construction and the use of the AZORESBOT robot created in the context of the first regional robotic competition in Azores organized by the Association of Programming and Robotics of Açores and the Group of Robotics and Artificial Intelligence of Universidade dos Açores (GRIA) with the collaboration of professors and students from robotic clubs of two different schools. This robot was tested and proved to be adapted to different challenges in the competition.

During the competition, the distributed kit was built by the teams and then programmed and tested, using the Arduino IDE with the help of libraries that could support a higher level of programming. The use of Ardublockly interface was available because it was expected that some teams in the festival have previous programming practice in Scratch software.

The modular design of the robot come up with the necessity to build a robot that could have, additionally, attached new modules to increment its capacity to solve increasingly more complex problems, as necessary. The robot behaved very well along the competition, with all the teams in the end, showing their enthusiasm about using the robot and satisfied with their own achievements.

Acknowledgements. This project was supported by Luso-American Foundation, project 119/2019 and by the Azores Government by the Regional Department of Science and Technology, PRO-SCIENTIA financial program. The authors wish to thanks to PROBOT - Associação de Programação e Robótica dos Açores and to the other organizations and volunteers that supported the event and the robot design.

References

1. Alimisis, D.: Educational robotics: open questions and new challenges. Themes Sci. Technol. Educ. **6**(1), 63–71 (2013)
2. Alimisis, D., Alimisi, R., Loukatos, D., Zoulias, E.: Introducing maker movement in educational robotics: beyond prefabricated robots and "black boxes", pp. 93–115. Springer (2019)
3. Bellas, F., Naya, M., Varela, G., Llamas, L., Bautista, M., Prieto, A., Duro, R.J.: Robobo: the next generation of educational robot. In: Ollero, A., Sanfeliu, A., Montano, L., Lau, N., Cardeira, C. (eds.) ROBOT 2017: Third Iberian Robotics Conference, pp. 359–369. Springer (2018)
4. Catlin, D., Holmquist, S., Kandlhofer, M., Angel-Fernandez, J., Cabibihan, J.J., Csizmadia, A.: EduRobot taxonomy, pp. 333–338. Springer (2019)
5. Dias, A., Cerqueira, V.: Regras: first challenger. Academia de Robótica, Sociedade Portuguesa de Robótica (2019). https://web.fe.up.pt/~robotica2019/
6. Giang, C., Piatti, A., Mondada, F.: Heuristics for the development and evaluation of educational robotics systems. IEEE Trans. Educ. 1–10 (2019)
7. Kaloti-Hallak, F., Armoni, M., Ben-Ari, M.: The effect of robotics activities on learning the engineering design process. Inform. Educ. **18**, 105–129 (2019)
8. Kandlhofer, M., Steinbauer, G.: Evaluating the impact of educational robotics on pupils technical-and social-skills and science related attitudes. Robot. Auton. Syst. **75**, 679–685 (2016)
9. Levy, R.B.B., Ben-Ari, M.M.: Robotics activities–is the investment worthwhile? In: Brodnik, A., Vahrenhold, J. (eds.) Informatics in Schools: Curricula, Competences, and Competitions, pp. 22–31. Springer (2015)
10. Muramatsu, S., Chugo, D., Yokota, S., Hashimoto, H.: Student education utilizing the development of autonomous mobile robot for robot competition. J. Robot. Mechatron. **29**(6), 1025–1036 (2017)
11. Negrini, L., Giang, C.: How do pupils perceive educational robotics as a tool to improve their 21st century skills? J. E-Learn. Knowl. Soc. **15**, 77–87 (2019)
12. Pacheco, M., Fogh, R., Lund, H.H., Christensen, D.J.: Fable II: design of a modular robot for creative learning. In: 2015 IEEE International Conference on Robotics and Automation (ICRA). IEEE (2015)
13. Riedo, F., Chevalier, M., Magnenat, S., Mondada, F.: Thymio II, a robot that grows wiser with children. In: 2013 IEEE Workshop on Advanced Robotics and Its Social Impacts, pp. 187–193 (2013)
14. Viegas, J.A., Villalba, K.: Education and educative robotics. Rev. de Educacion a Distancia (RED) (2017)
15. Vrochidou, E., Manios, M., Papakostas, G.A., Aitsidis, C.N., Panagiotopoulos, F.: Open-source robotics: investigation on existing platforms and their application in education. In: Proceedings of the 26th International Conference on Software, Telecommunications and Computer Networks (SoftCOM 2018), Symposium on: Robotic and ICT Assisted Wellbeing, Croatia (2018). https://doi.org/10.23919/SOFTCOM.2018.8555860
16. Wei, Z., Berry, C.A.: Design of a modular educational robotics platform for multidisciplinary education. In: 2018 ASEE Annual Conference & Exposition, Salt Lake City, Utah (2018)

BulbRobot – Inexpensive Open Hardware and Software Robot Featuring Catadioptric Vision and Virtual Sonars

João Ferreira[1(✉)], Filipe Coelho[1], Armando Sousa[1,2], and Luís Paulo Reis[3,4]

[1] FEUP - Faculty of Engineering, UP - University of Porto, Porto, Portugal
joao.martins.ferreira@fe.up.pt

[2] INESC TEC - INESC Technology and Science, Porto, Portugal

[3] LIACC/UP - Artificial Intelligence and Computer Science Laboratory, UP, Porto, Portugal

[4] DEI/FEUP - Informatics Engineering Department, FEUP, Porto, Portugal

Abstract. This article proposes a feature-rich, open hardware, open software inexpensive robot based on a Waveshare AlphaBot 2. The proposal uses a Raspberry Pi and a chrome plated light bulb as a mirror to produce a robot with an omnidirectional vision (catadioptric) system. The system also tackles boot and network issues to allow for monitor-less programming and usage, thus further reducing usage costs. The OpenCV library is used for image processing and obstacles are identified based on their brightness and saturation in contrast to the ground. Our solution achieved acceptable framerates and near perfect object detection up to 1.5-m distances. The robot is usable for simple robotic demonstrations and educational purposes for its simplicity and flexibility.

1 Introduction

Autonomous navigation in robots is not a new subject. Since robots became mobile, navigation has been a main field of study [8]. However, the hardware is very different from the early times, not only have sensors become faster and more precise, have batteries grown and materials become more accessible, but also computing units are made smaller, more efficient and faster [14] every day. These innovations provide greater autonomy for robots; they can endure longer run times without the need to recharge, can detect obstacles easier, predict and calculate new paths faster and solve more complicated problems.

Using computer vision to create autonomy in robots has been implemented with success in a lot of products, most publicly in autonomous cars. Despite that, many computer vision algorithms used are still too heavy to implement in lower-end hardware, like the Raspberry Pi 3, while having real-time frame processing, especially considering the number of variables that systems like the ones implemented in autonomous cars are subjected to. However, if we have a controlled environment, where the robot navigates through a semi-static background and

M. F. Silva et al. (Eds.): ROBOT 2019, AISC 1092, pp. 553–564, 2020.
https://doi.org/10.1007/978-3-030-35990-4_45

the obstacles are marked with very distinct colors, simpler algorithms can be used to detect obstacles.

In this project, an AlphaBot2 kit was used as a base for the autonomous robot. For the computing unit, a Raspberry Pi 3 Model B+ was used, with a Raspberry Pi Camera Module (B) Rev 2.0 installed. The team designed a support for the camera on top of the unit, and the supports for a mirror, where a light bulb with a half-spherical mirror surface was mounted (Fig. 2d).

2 Related Work

Newer computing units provide a lot more computing power in a small form factor, allowing for tasks such as object detection based on computer vision to be possible. Object detection is one of the most important features in intelligent robotics [2], and there is a great deal of research and scientific documentation on this subject. Several types of vehicles are running computer vision algorithms with real-time performance [2] to aid navigation and provide a greater deal of security [11], such as unmanned aerial vehicles (UAVs) or automated guided vehicles (AGVs) [4–6,12].

Obstacle detection is a subset of object detection, and the ability to detect objects allows robots to perform more elaborated tasks than only obstacle avoidance, like object/vehicle tracking [11], human detection and action recognition [9] or object searching and identification. These tasks became more intuitive if the sensor(s) used to detect the objects are cameras. This allows using computer vision algorithms to accomplish the given challenges. Since this project is only about detecting obstacles that have a distinct color from a semi-uniform background, a lightweight approach was taken, based only on image color processing.

These contributions to autonomous robots are one of the many pieces of evidence that this field is in a big expansion. And for the robot to be fully autonomous, it has to actuate (e.g. arms, wheels, etc.) based only on its perceptions (e.g. sensors), without any human intervention. This can make educational kits either limited or expensive. As can be seen by [7,18] and [17], there are already several implementations with various development platforms available that allow for a satisfactory environment for investigating simple navigation tasks (e.g. visual programming, integrated development environments, etc.). However, these platforms tend to provide lower sensor support, restricting the possibilities to study and develop true autonomous behavior.

Due to this, a project entitled LIDAR for Scribble 2 was set with the goal of providing an alternative implementation for a project already established in education. The idea was to implement a low-cost LIDAR, an IOIO-OTG board and an inexpensive Android device [16].

Another project is the Pi-puck extension board, with open-source hardware design and supporting software infrastructure, which promises to provide a low-cost alternative to existing e-puck extension boards. The main goal is to use an already successful platform like e-puck and improve on its limitations by designing an interface with the Raspberry Pi, taking advantage of the Pi's computation, memory, networking, storage, and image processing capabilities and using the e-puck microcontroller for the low level actions like motor and sensor interfacing [15].

A third interesting project was developed by Csaba Kertész, aimed at using an inexpensive Raspberry Pi camera to take pictures of the sky and processing them using a CNN to detect clear sky on pictures. This project was trying to solve some existing problems with previous implementations based on thresholds, like uncertainties derived by image interpolation or difficulty in distinguishing dark from the sky. The proposal achieved an accuracy of 95.75% [10].

Besides these projects, there are many others including robots using catadioptric systems to enhance their vision [1,3,13]. While these produced successful results, they rely on custom shaped mirrors that are very expensive to produce. This demonstrates that there is space for improvement in the field of inexpensive omnidirectional vision systems.

3 System Description

Our system uses a Waveshare AlphaBot2 as the robot basis with a Raspberry Pi 3B+ for its brain and an off-the-shelf light-bulb secured to 3D printed supports. It can be replicated through the following steps:

3.1 Initial Configuration

For the OS, we used an image of the Raspbian Operating System, where we installed and configured a VNC server for remote access to the Raspberry and built the OpenCV library to use for image processing. With everything on the software side working, we mounted the Raspberry Pi on the AlphaBot2, with an additional script to turn off the buzzer at every boot.

3.2 Utilities

After this we decided to solve the networking issues. Being that we were working on the Raspberry Pi through VNC, if we were to change locations we would have no way to set the new network credentials and, even if we did, we would have to guess its IP address to connect with it. To address these problems we designed a script that looks for any USB flash drive connected at boot and looks for a text file with a specific name. Inside that text file, the first line would correspond to the network SSID and the second line to the password. To find out the IP address, this script would also write the addresses on a text file inside the USB flash drive, if the USB was found.

For remote operations convenience, we also wanted to see the IP address of the RPi without having to physically access it. So we tagged each instance of the AlphaBot2 with a unique identifier and created a dedicated repository, where branches would get automatically named after every identifier. Then, a second script would write to a text file the RPi IP address and commit that file to the remote repository, in the respective branch. While this method would only work when the network had internet connection, this was a valuable convenience, with the additional advantage that it was easy to use and that we had all the AlphaBots listed in a single location, always up to date.

With these utility scripts implemented, we could bring our system anywhere and work on it, knowing we would always be able to remote access it, even if the network was not previously configured. We also added a quick response (QR) code and a near-field communication (NFC) tag to the robot that takes us to the URL of the branch of our specific AlphaBot, making it very quick and easy to access its IP address.

3.3 Support Structures

To start the development of the support structures, every important detail from the AlphaBot's top board, such as its dimensions, cutouts and screw locations and sizes, was measured and replicated inside a 3D modelling software to later use as reference (Fig. 1A).

A camera mount was designed with the purpose of being screwed in using the existing holes and holding the camera centered on the board (Fig. 1C). The mount has two clips that allow the camera to slide in, be held in place, and be taken out at will (Fig. 1B).

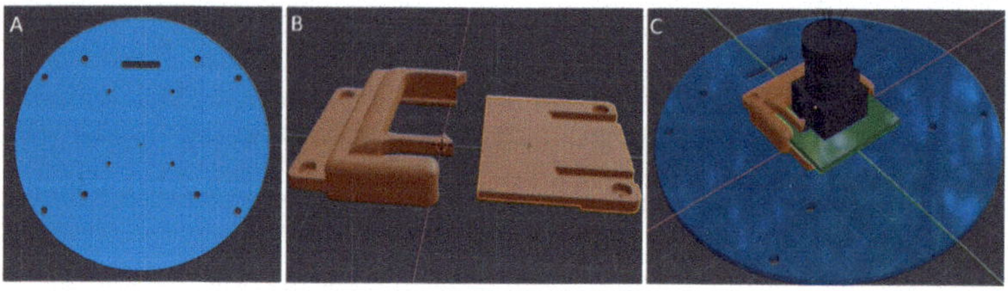

Fig. 1. A: AlphaBot2 top board 3D model. B: The two pieces of the camera mount. C: Render of the camera mount installation on the board.

A more complex task was to design the mirror mount. The mirror must stay centered with the base, elevated from the camera, and sturdy enough so it does not wobble with the AlphaBot's movement. After several iterations, we ended up with an armature around the AlphaBot board and a four-legged design bringing the mirror up as seen in Fig. 2a.

The legs are split into two parts, where the bottom parts are screw shafts allowing for fine adjustments of the leg height, and consequently the distance between the mirror surface and the camera, in increments of 1 mm (Fig. 2b). The splitting of the legs into two parts also allows us to have legs longer than the maximum build volume for our 3D printer. Some measurements of the final results can be seen in Table 1.

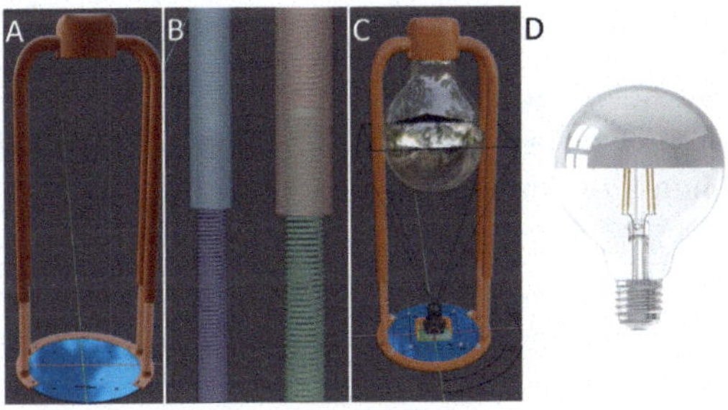

Fig. 2. (a) The light bulb support design. (b) The design of the adjustable legs. (c) Render of the full mounted upper part of the robot. (d) The light bulb with half spherical mirror surface.

Table 1. Mirror support structure properties

Upper leg height	205 mm	
Lower leg height	84 mm	
Number of threads	75	
Thread pitch	1 mm	
Total leg height variance	Min	Max
	214 mm	284 mm
Distance from lamp socket to light bulb base	127 mm	
Distance from AlphaBot board to camera lens	27 mm	
Distance between camera lens and light bulb base	Min	Max
	60 mm	130 mm

The reason for having two pairs of two legs close to each other instead of only three legs was to unclutter the front and back of the robot as much as possible as these are the directions most important to the robot's movement. For three legs to provide a sturdy mount, forced at least one of them to obstruct the view to the front or back. Finally, the lamp socket was a simple E27 lamp socket with 4 slots for the legs to slide in. The final assembly can be seen in Fig. 2c.

4 Working Solution

The final steps for the success of our solution were to: Firstly place the mirror at an ideal height; Then capture a reflection from the mirror and use a polar coordinate system to transform the circular reflection captured into a linear frame where the coordinates represent the angle and the distance; Mask out the parts of the image that correspond to the robot's supports; And finally detect when there is an object near the robot and return its information as a bounding box in polar coordinates.

4.1 Mirror Ideal Height

To calculate the ideal height for the mirror surface, we created a mathematical model for the system as can be seen in Fig. 3. This helped us better understand the different components of the system and their interaction.

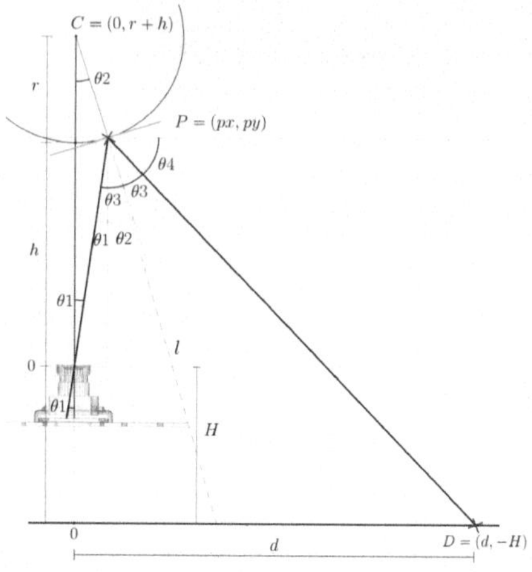

Fig. 3. Distance relation diagram.

From there we started by calculating the coordinates of the point P relative to the center of the camera lens, which will be placed at the origin O. P is the point that light coming from an object on the ground (point D) would have to hit to be reflected to the camera of our robot. The first step to figure out the coordinates of this point is to understand that a point captured by the sensor on the camera, if in focus, has a linear relationship between its position on the sensor and its position in 3D space that can be given by tracing a line $\overleftrightarrow{OP}$ that intersects the point in the sensor and the center of the camera lens, extending to find P. This means that by using the angle that a point and the center of the lens make with the normal of the sensor, marked in the diagram as θ_1, one can easily construct the equation for the line that intersects them.

The value of θ_1, for now, is the FOV of the sensor, but later can also be understood as the "partial FOV" of the point in the sensor and easily calculated trigonometrically by seeing how much it deviates from the center of the sensor (in a ratio) and the sensor's maximum FOV value. This will allow us to use this formula for other applications explained later.

We intersect this line OP with the equation for the sphere that is our mirror (Eq. 1) and get the coordinates of point $P(px, py)$ dependent on the values of: the height of the mirror to the lens (h), the radius of the mirror (r), and the partial FOV of the point in the sensor (θ_1) (Eq. 2).

$$\begin{cases} py = cot(\theta_1)px \\ px^2 + (py - (h + r))^2 = r^2 \end{cases} \tag{1}$$

$$\left[px = \frac{(r+h)cot(\theta_1) - \sqrt{r^2 cot^2(\theta_1) - 2hr - h^2}}{cot^2(\theta_1)+1}, py = cot(\theta_1)px \right] \tag{2}$$

After knowing the position of point P, we can then proceed to calculate θ_2 which is the angle of the normal of the reflection. This is the angle at which the angle θ_3 is reflected as can be seen in the diagram. This angle is easily calculated knowing the coordinates of point C and point P (Eq. 3). With this, we also get to know the value of θ_3 which, as can be seen in the diagram is simply given by Eq. 4.

$$\theta_2 = atan\left(\frac{-r + py + h}{px}\right) + \frac{\pi}{2} \tag{3}$$

$$\theta_3 = \theta_1 + \theta_2 \tag{4}$$

Finally, we can calculate θ_4 (Eq. 5) and by using it as the slope for the line that starts on P and intersects with the ground, we can find the x value of the point D, knowing it has a y value of $-H$, where H is the height of the robot measured from the camera lens to the floor. Knowing point D, we know its distance to the center of the robot, as it is its x component. Put simply, the distance (d) can be given by Eq. 6.

$$\theta_4 = \theta_2 + \theta_3 - \frac{\pi}{2} \tag{5}$$

$$d = px - \frac{py + H}{tan(\theta_4)} \tag{6}$$

From all of this, we can find out the ideal mirror height using the known parameters and the intended distance, and solving for h.

Notice that all this is done unidimensionally as we only intend to find the ideal mirror height suited for our purposes. Because of this, only the FOV of the direction of the camera with the least FOV is important, that is, in this case, being that the camera's aspect ratio is 4:3, the vertical FOV. However, we can use this similarly bidimensionally to, using a point on the image and calculating its "partial FOV" coordinates (as mentioned earlier), calculate the distance an object is from the robot.

4.2 Capturing a Frame

When capturing a frame from the mirror, we implemented a zoom function that makes use of the fact that the picamera python library can crop the frame in the GPU at the time it is taken before down-sampling the image. This makes it so that image quality is preserved and performance is not lost.

This function is important because the mirror surface in the light bulb is not perfectly round, and the light bulb screw is always expected to be slightly lob-sided, which means that the center of the polar coordinate system could be not centered with the robot reflection, causing distortion problems. To overcome that, we marked the ring around the camera lens with a fluorescent color and captured a frame with $zoom = 5$ and $resolution = (512; 512)px$ (Fig. 4A left), binarized it by finding the pixels belonging to a predefined color range (Fig. 4A center) and used a circle detection algorithm present in OpenCV to detect the circle and its center (Fig. 4A right).

Now that we have the center of our polar coordinate system, we capture the new frames with $zoom = 1.1$ and $resolution = (360; 360)px$ (Fig. 4B left). The choice of 1.1 for the $zoom$ is done to compensate for the loss of information in the edges when re-centering the image and was taken into account when choosing the height of the mirror.

With the center of the polar coordinate system found, we proceed by applying a linear polar transformation to the frames, obtaining an image where the coordinate system denominates the angle and distance of a point relative to the center of the robot (Fig. 4B right).

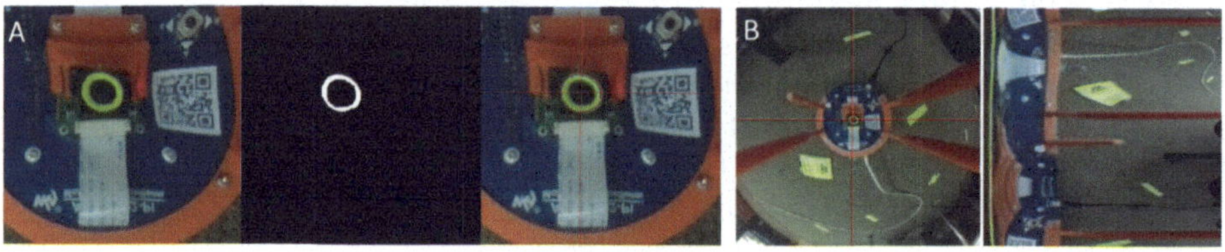

Fig. 4. A: Center of polar coordinate system detection. B: Transforming the spherical capture into a polar coordinate system.

We then detect objects by analyzing the brightness and saturation differences between each pixel and the perceived average background, and using this to binarize the image. After that, we find the bounding boxes that fully contain every perceived connected object. This process can be seen in Fig. 5.

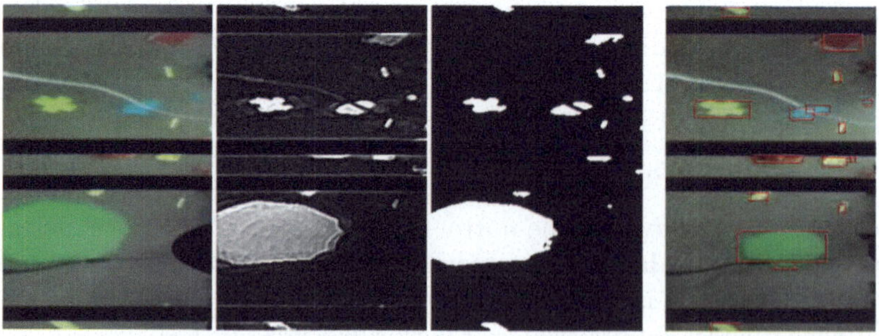

Fig. 5. The several steps of the image processing.

For ease of use, we abstract this information into what we call virtual sonars, which we setup by detailing their position around the robot, their cone of vision, and max and min ranges in a configuration file. We are then able to get the perceived distance readings by polling each of the virtual sonars.

5 Experiments and Results

Initially, when calculating the ideal height for the mirror, we worked under the assumption that for this, we simply needed to be at a distance that allowed us to see the whole mirror. We used the System of Eq. 1 to find the maximum real value solutions for the height, using the horizontal and vertical FOVs of the camera. We found that for the camera to see the whole mirror, horizontally ($FOV = 43.85$), the mirror needed to be at 77.9 mm from the lens, and vertically ($FOV = 33.66$), more importantly, the mirror needed to be at a height of 116.6 mm. After testing this height empirically (116 mm), we found out that this was indeed not the intended solution. In fact, only about half the image was usable, as most of the outer image presented reflections non-convergent to the ground.

With this in mind, we reworked our calculations, this time solving for a known max view distance. This made the problem more complex, but solvable, using Eq. 2 through 6. By using the known values of $r = 95$ mm, $H = 81$ mm and the intended max view distance of $d = 2$ m, we solved for the height of the mirror which we found to be approximately $h = 80$ mm. This value is well within the range of heights of the legs of our support, as can be seen in the last line of Table 1. Despite that, and to account for the fact that the mirror is not perfect, we decided to choose a slightly bigger height of $h = 100$ mm, which gives us a margin of error in case the lamp is too lob-sided and a big portion of the frame needs to be cut to re-center it. We can see the effects that changing the height of the mirror has on the reflection distance captured by the camera in Fig. 6a.

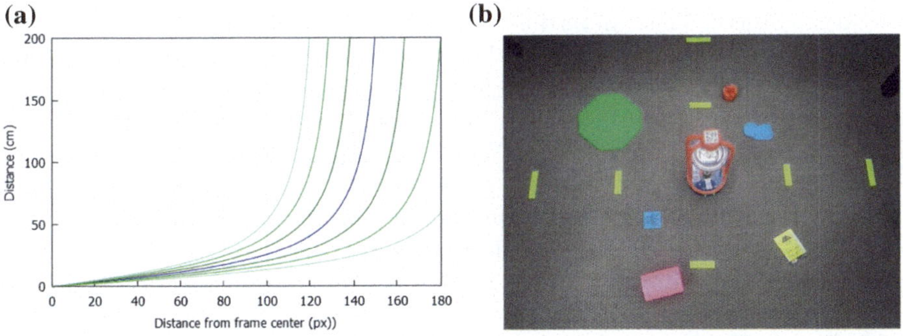

Fig. 6. (a) Reflection distance for different mirror heights. (From 130 mm on the leftmost line to 70 mm on the rightmost one, in increments of 10 mm). Blue was the chosen - 100 mm. (b) AlphaBot surrounded by obstacles with distinct colors and marks every 25 cm.

To test the system, we placed the AlphaBot in a location where the background had a uniform grayish color, surrounded by objects with very distinct colors and markings every 25 cm (Fig. 6b). This simple setup was chosen to prevent the shortcomings of any object detection algorithms we would develop from affecting the evaluation of the omnidirectional viewing system in itself, which was the focus of our work.

The system managed to detect successfully the objects in the surrounding area up to a distance of 1 m, and correctly identified their direction and distance from the robot. After 1 m, although the robot could still detect big enough objects, the identification of their distance started to suffer because of the way the distance per pixel grows at that point (Fig. 7).

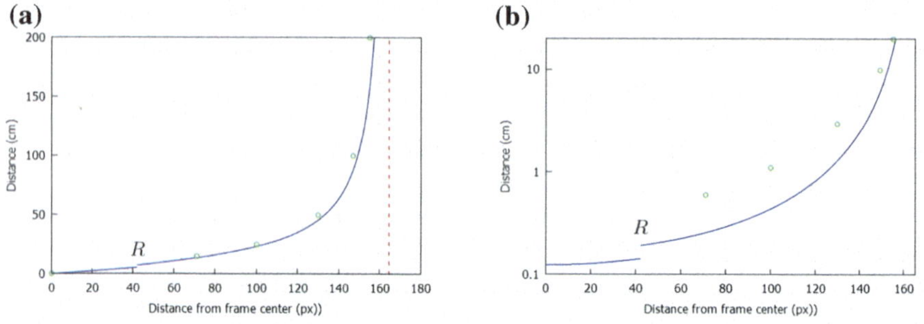

Fig. 7. (a) Final reflection distance across the image, for a 360 × 360 resolution - predicted (blue) vs real average measurements (green). (b) The expected error in distance prediction across the camera sensor (blue) vs the average measured real error (green).

Figure 7 shows the results of tests done to assess the system's distance reporting and compares it to the predicted values had the mirror been a perfect sphere. As we can see on Fig. 7a, the system constantly underestimates the expected distance for an object. This can be due to several factors, like the bulb being slightly smaller than what it advertises, the detection algorithm giving oversized bounding boxes, or the lens not being perfectly in focus smearing the object's color over several pixels. Nonetheless, we find these results acceptable and consider the system to be very promising.

Point R on both Figs. (7a and b) represents the transition between the robot looking at itself and the ground, thus the discontinuity in both functions. This point is also illustrated in Fig. 3 with the line l.

We were able to obtain a stable 15 fps with full object detection and obstacle avoidance working. To achieve this, the frame had to be no bigger than (360;360) px and their processing had to be multi-threaded. At a resolution of (800;800) px the framerate lowered to 8 fps and the robot started to overshoot when avoiding obstacles. Lowering the resolution further than (360;360) px yielded no performance gain.

Some oscillation of the mirror could be seen, although very little, which translated to a minor motion blur in the captured frames while in motion. This did not greatly affect the robot's performance.

The implementation of the obstacle avoidance was rather simple and was not a target of too many tests. This system was only a proof of concept to demonstrate that the obstacle detection could work and it was not part of the main focus of this project. Still, it performed acceptably, identifying objects quickly and accurately, serving as further proof that the system was viable.

6 Conclusions and Future Work

Implementing a 360° obstacle detection system on a low-budget system is a challenging task. The system's clear bottleneck is the computing power. Even so, we consider this implementation to be a success, as it only affects the detection speed and, consequently, the maximum movement speed of the robot.

Overall, this paper presented a very inexpensive robot adequate for education and small demonstrations. The robot is based on an AlphaBot 2 RPi platform that is, in turn, based in a common Raspberry Pi hardware. A 3B+ model was used for tests. A catadioptric vision system is used to get rid of moving parts and get 360 virtual sensor coverage. The mirror is "Off the Shelf", a chrome-mirrored light bulb Fig. 2d. 3D printed mechanical adaptations are presented in order to allow vision near the robot. Center calibration is easy and allows for variation of parameters. Objects with vivid colors are considered obstacles and sample software is provided for future improvements.

Future work on the mechanical part should include mechanical protection of the light bulb. Other future work includes further functionalities on the software part (transform into a library), standard ROS support, etc.

Acknowledgements. This research was partially supported by LIACC - Artificial Intelligence and Computer Science Laboratory of the University of Porto (FCT/UID/CEC/00027/2019).

References

1. Benosman, R., Deforas, E., Devars, J.: A new catadioptric sensor for the panoramic vision of mobile robots. In: Proceedings of the IEEE Workshop on Omnidirectional Vision (Cat. No. PR00704), pp. 112–116, June 2000. https://doi.org/10.1109/OMNVIS.2000.853816
2. Chen, P., Dang, Y., Liang, R., Zhu, W., He, X.: Real-time object tracking on a drone with multi-inertial sensing data. IEEE Trans. Intell. Transp. Syst. **19**(1), 131–139 (2018). https://doi.org/10.1109/TITS.2017.2750091
3. Cho, D., Park, J., Tai, Y., Kweon, I.: Asymmetric stereo with catadioptric lens: high quality image generation for intelligent robot. In: 2016 13th International Conference on Ubiquitous Robots and Ambient Intelligence (URAI), pp. 240–242, August 2016. https://doi.org/10.1109/URAI.2016.7625745

4. Colomina, I., Molina, P.: Unmanned aerial systems for photogrammetry and remote sensing: a review. ISPRS J. Photogramm. Remote. Sens. **92**, 79–97 (2014). https://doi.org/10.1016/j.isprsjprs.2014.02.013
5. Darma, S., Buessler, J.L., Hermann, G., Urban, J.P., Kusumoputro, B.: Visual servoing quadrotor control in autonomous target search. In: 2013 IEEE 3rd International Conference on System Engineering and Technology, pp. 319–324, August 2013. https://doi.org/10.1109/ICSEngT.2013.6650192
6. Ergezer, H., Leblebicioglu, K.: Path planning for UAVs for maximum information collection. IEEE Trans. Aerosp. Electron. Syst. **49**(1), 502–520 (2013). https://doi.org/10.1109/TAES.2013.6404117
7. Eubanks, A.M., Strader, R.G., Dunn, D.L.: A comparison of compact robotics platforms for model teaching. J. Comput. Sci. Coll **26**(4), 35–40 (2011)
8. Gomez, C., Hernandez, A.C., Crespo, J., Barber, R.: Integration of multiple events in a topological autonomous navigation system. In: 2016 International Conference on Autonomous Robot Systems and Competitions (ICARSC), pp. 41–46, May 2016. https://doi.org/10.1109/ICARSC.2016.47
9. Hoshino, S., Niimura, K.: Robot vision system for real-time human detection and action recognition. In: Intelligent Autonomous Systems 15, pp. 507–519. Springer, Cham (2019)
10. Kertész, C.: Clear sky detection with deep learning and an inexpensive infrared camera for robot telescopes. In: 2018 15th International Conference on Control, Automation, Robotics and Vision (ICARCV), pp. 1698–1702, November 2018. https://doi.org/10.1109/ICARCV.2018.8581095
11. Kho, Y.H., Abdulla, A.E., Yan, J.C.Z.: A vision-based autonomous vehicle tracking robot platform. In: 2014 IEEE Symposium on Industrial Electronics Applications (ISIEA), pp. 173–176, September 2014. https://doi.org/10.1109/ISIEA.2014.8049893
12. Lim, H., Sinha, S.N.: Monocular localization of a moving person onboard a quadrotor MAV. In: 2015 IEEE International Conference on Robotics and Automation (ICRA), pp. 2182–2189, May 2015. https://doi.org/10.1109/ICRA.2015.7139487
13. Lopes, G., Ribeiro, F., Pereira, N.: Catadioptric system optimisation for omnidirectional RoboCup MSL robots. In: Röfer, T., Mayer, N.M., Savage, J., Saranlı, U. (eds.) RoboCup 2011: Robot Soccer World Cup XV, pp. 318–328. Springer, Heidelberg (2012)
14. Michael, N., Mellinger, D., Lindsey, Q., Kumar, V.: The GRASP multiple micro-UAV testbed. IEEE Robot. Autom. Mag. **17**(3), 56–65 (2010). https://doi.org/10.1109/MRA.2010.937855
15. Millard, A.G., Joyce, R., Hilder, J.A., Fleşeriu, C., Newbrook, L., Li, W., McDaid, L.J., Halliday, D.M.: The Pi-puck extension board: a Raspberry Pi interface for the e-puck robot platform. In: 2017 IEEE/RSJ International Conference on Intelligent Robots and Systems (IROS), pp. 741–748, September 2017. https://doi.org/10.1109/IROS.2017.8202233
16. Miller, K.S., Robila, S.A.: LIDAR for Scribbler 2. In: 2017 IEEE Long Island Systems, Applications and Technology Conference (LISAT), pp. 1–6, May 2017. https://doi.org/10.1109/LISAT.2017.8001957
17. Rubenstein, M., Cimino, B., Nagpal, R., Werfel, J.: AERobot: an affordable one-robot-per-student system for early robotics education. In: 2015 IEEE International Conference on Robotics and Automation (ICRA), pp. 6107–6113. IEEE (2015)
18. Weiss, R., Overcast, I.: Finding your bot-mate: criteria for evaluating robot kits for use in undergraduate computer science education. J. Comput. Sci. Coll. **24**(2), 43–49 (2008)

Field Robotics In Challenging Environments

Trajectory Planning for Time-Constrained Agent Synchronization

Yaroslav Marchukov(✉) and Luis Montano(✉)

Instituto de Investigación en Ingeniería de Aragón (I3A), University of Zaragoza, Zaragoza, Spain
{yamar,montano}@unizar.es

Abstract. In the present paper we focus on the problem of synchronizing two agents in movement. An agent, knowing the trajectory of a teammate who it must exchange data with, has to obtain a trajectory to synchronize with this mate before going to its own goal. We develop the trajectory planner for the agent that it is constrained by the time to synchronize with the mate. Firstly, we define the dynamic communication area, produced by a teammate agent in movement, as well as the different parts of this area, used by the proposed planner. Then, we develop a method to obtain trajectories for an agent in order to be able to synchronize with a teammate in movement, whose trajectory is known. Simulated results show that the proposed approach is able to provide the solution according to two chosen criteria: distance or time.

Keywords: Connectivity constraints · Trajectory planning

1 Introduction

Consider a scenario where a team of agents is used to gather data from some environment that lacks a communication infrastructure. The mission of the agents is to reach some locations of interest, or goals, take measurements and deliver the gathered data to a static Base Station (BS). To do this, the agents have to establish a connectivity link with the BS, directly or using other agents in role of relay, retransmitting the information. However, the establishment of the link with the BS it is not always possible, due to: the distance to the goals, the dimensions of the scenario, the limited ranges of the wireless devices, and the number of the available agents in the team. Therefore, the information of the measurements taken at the goals must be delivered going directly to the BS. At the same time, from the BS, new information from new locations of interest may be requested, generated them automatically or by some human operator.

Another approach may be devoting only some agents to go to the BS in order to retransmit the information of their teammates, that are only being used to reach the goals. The last ones, are called *worker* agents, that are used only for

M. F. Silva et al. (Eds.): ROBOT 2019, AISC 1092, pp. 567–579, 2020.
https://doi.org/10.1007/978-3-030-35990-4_46

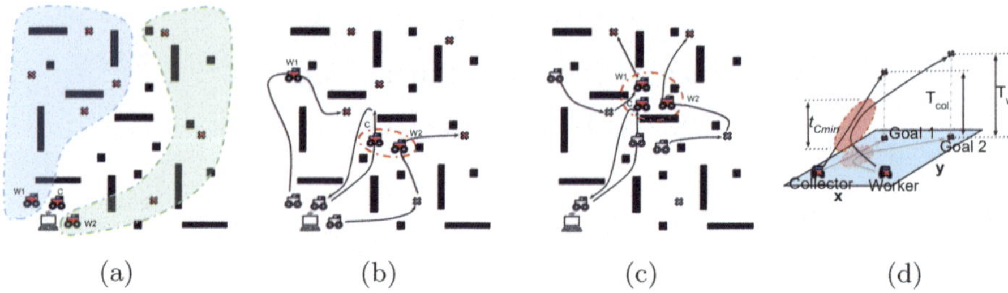

Fig. 1. Illustrative description of application of the proposed method. In (a) the tasks are allocated. C and W denote collector and worker, respectively. The trajectories for data gathering and synchronization are illustrated in (b)–(c). (d) depicts the worker-collector synchronization, during the communication time $t_{c_{min}}$, in the *x-y-time* space.

working purposes. That is, they are only visiting the goals and taking measurements. The first ones are what we call *collector* agents, which are travelling in the scenario a constant trajectory in a persistent way, receiving the data of the workers in order to retransmit it to the BS. Hence, the workers know about the collectors position along the time and can decide where and when to share the data with a collector agent. An illustrative example of the application of the proposed trajectory planner is depicted in Fig. 1.

The tasks of the mission are allocated to the workers in Fig. 1(a). The trajectory of the collector agent is shared between the team. Then, the data gathering mission starts and the workers visit their goals, taking measures and deciding where and when to share the data with the collector Fig. 1(b)–(c). They obtain the best trajectory according to the time to transmit all the gathered data and constrained to the trajectory of the collector. That is, before it returns to the BS for uploading, Fig. 1(d).

The planning of the data gathering mission was developed in [10]. In the present paper, we develop a planning algorithm to compute a trajectory of a worker agent for synchronization, in movement, with a collector teammate with a known trajectory. We evaluate the proposed approach for a single collector case and in scenarios where the worker should choose among multiple collectors in the environment to transmit the data, based on two criteria: distance or time.

2 Related Works

In this work we consider a solution to the trajectory planning for synchronization in data gathering missions. The developed methods for exploration missions [2,9,11] cannot solve our scenario either, since in their scenarios there no exist time constraints for synchronization between agents. Our scenario shares more similarities with patrolling works [3,4,12]. The agents patrol a predefined path, obtained from an environment graph, making observations and synchronizing with their teammates just during the time to share data. However, the predefined paths are the main drawback of these methods. It must guarantee that at some moment the agents will meet each other. However, all the robots never deviate from their paths, that could be inefficient in data gathering missions. In [6], the agents periodically meet each other in order to share information between them. But, the problem of planning paths to make concur large teams of agents may become intractable. In our approach, we propose the usage only of some members of the team that will travel a known path, the collector agents. So, the worker agents plan their trajectories, without disrupting the travelled paths of the collectors.

The trajectory computation involves the time in which the path is traversed, so the planner explores different temporary options to reach the goal. The runtime of optimal methods such as Dijkstra or A* scale with the number of dimensions. The randomized multiple-query algorithms such as Probabilistic Roadmaps (PRM) [1], are still computationally heavy for our problem and do not really provide better results than other sampling-based techniques. The Rapidly-exploring Random Trees (RRT) [8], due to the single-query and randomized nature, are faster than PRMs. However, they still obtain very poor quality solutions in terms of trajectory costs.

The proposed time-constrained planner is based on the Fast Marching Method (FMM) [13], due to its advantageous properties. On the contrary to the sampling-based techniques, FMM is slower, but it computes the optimal paths. So, if there exists a possible path that accomplishes the time constraints, the solution is found. Regarding optimal solvers, the Dijkstra's algorithm, as FMM, evaluates all the possibilities from the starting point to every position of the grid and it obtains the best path to the goal. However, the FMM is more efficient in terms of precision. The interpolation of the distance is more accurate to the real distance. Another possibility is the A* algorithm, which rapidly guides the search of the path to the goal using heuristics. In the scenarios considered in our work, the goals to be reached by every robot are known, but the position where the mate is intercepted is ignored a priori. Thus, the needed subgoals for communication are not explicitly computed to be used for an A* path planning algorithm. This makes difficult to find out a good heuristic. Therefore, in the case of evaluating many options for exchanging information positions, it is necessary to execute the A* algorithm the same number of times as positions of all the connectivity areas, which is computationally inefficient.

3 Problem Setup

The problem solved here is the computation of a trajectory to simultaneously reach some goal location and synchronize with a collector agent in movement. In other words, to find a trajectory to a goal location that traverses the dynamic communication area of a collector robot. This involves a spatio-temporal planning. Thus, let us define some location of the scenario as x. Each position is visited at some time t, so we can define a node $n = [x\ t]^T$, as the position and time in the scenario. Throughout this paper $x(n)$ and $t(n)$ will denote position and time of some node n. Since the space is three-dimensional, the distance between a pair of nodes includes the difference between times, besides the Euclidean distance.

We denote τ as a trajectory travelled by an agent, which can be defined as a sequence of contiguous nodes $\tau = [n_0, n_1, \ldots n_N]$, where N denotes the number of nodes in the trajectory. The expression $x(\tau)$ refers to each position of the trajectory and $t(\tau)$ to each of their respective times. We define $T(\tau)$ as the total time to travel the trajectory τ. The worker must communicate with a collector mate, which is assigned for relaying the information to the BS. The trajectory of the collector mate is expressed as τ_c, and it generates a dynamic communication area $A(\tau_c)$. This area is also composed by nodes n formed by positions and times. The details of computation of this area are explained in Sect. 5. An example of this area in a scenario with obstacles is depicted in Fig. 2(a).

Knowing the size of the packages to transmit, the worker determines the minimum time to fulfill the transmission, $t_{c_{min}}$. So, the time that the trajectory of the worker must remain within $A(\tau_c)$ to fulfill the data transmission, is denoted with $t_c(\tau)$. In the case of having a goal to reach, the worker must traverse the communication dynamic area $A(\tau_c)$ and then reach the goal. In absence of a goal, the obtained trajectory only must remain within $A(\tau_c)$ during $t_{c_{min}}$. Thus, the optimal trajectory for synchronization τ^* is defined as:

$$\begin{aligned} \tau^* &= \underset{\tau}{\operatorname{argmin}}(J(\tau)) \\ &\text{subject to} \quad t_c(\tau) \geq t_{c_{min}} \end{aligned} \tag{1}$$

where the cost J is computed using the normalized times and distances of all the possible trajectories τ, expressed as:

$$J(\tau) = w_t\overline{t(\tau)} + w_d\overline{d(\tau)} \qquad \overline{(\cdot)} = \frac{(\cdot)}{max(\cdot)} \tag{2}$$

where the parameters $w = (w_t, w_d)$, with $w_t + w_d = 1$, represent the weighting factors of the time and distance, and $max(\cdot)$ is the maximum value of the vector.

4 Obtaining the Trajectories with FMM

The proposed approach uses the Fast Marching Method (FMM) [13] to obtain the trajectories. The FMM can be easily described as a propagation of a wavefront from some source location along all the map, computing the distance from the source to each point of the map. Thus, the wavefront is initialized at some position x and it is extended uniformly in all the directions, solving the equation:

$$|\nabla D(x)|F = 1 \tag{3}$$

where $\nabla D(x)$ expresses the distance gradient from the source x. F represents the velocity of propagation of the wavefront, so it is propagated faster for higher values and slower for lower values of F. In our scenario, $F = 0$ when the position contains an obstacle and $F = 1$ denoting free space. After propagating the wavefront over all the scenario, a path to any position x_g from the source can be obtained by descending the gradient $\nabla D(x)$ from x_g. This is the first advantage of FMM. It only requires a unique gradient computation to be able to obtain the paths to all the desired positions. The second advantage is an accurate distance approximation. FMM uses several neighboring nodes to interpolate the distance, instead of one used by Dijkstra, offering more precise distances. Therefore, the paths obtained with FMM are the shortest and fastest ones.

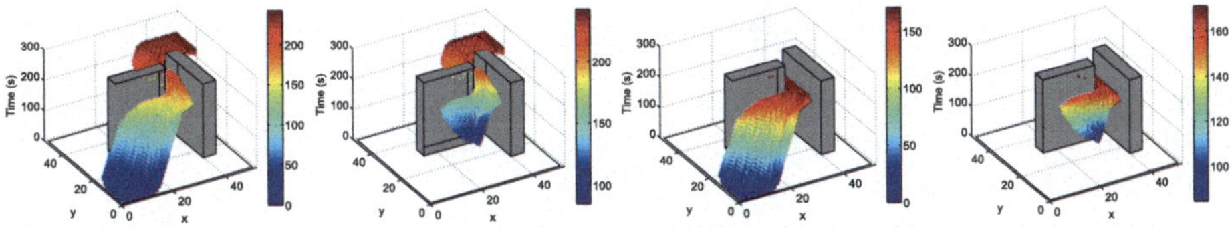

(a) $A(\tau_c)$ (b) A_{reach} (c) A_{ca} (d) A_{feas}

Fig. 2. Communication area decomposition. Gray objects are obstacles.

5 Dynamic Communication Area

In order to synchronize with the collector, the worker has to know where and when to do it. The collector may be in movement or not. For instance, the latter case, it is the particular case of a static Base Station. So instead of defining different strategies for communication, we obtain its communication area. This way, it does not matter if the collector is moving or not. The planner only obtains a trajectory that traverses this area during $t_{c_{min}}$. In the case of moving, the collector drags its communication area along its trajectory, becoming dynamic. We consider that both, worker and collector, are equipped the same wireless antennas, with a limited signal range. Therefore, we define the communication area as all the nodes n of the scenario that are within the signal range of the antenna and non-obstructed *line-of-sight*. Formally expressed as:

$$A(\tau_c) : \{n \mid dist(\tau_c, n) \wedge LoS(\tau_c, n)\} \tag{4}$$

where $LoS(\tau_c, n)$ is a boolean function that checks if exists *line-of-sight* between the trajectory τ_c and the node n, using the Bresenham algorithm [7]. $dist(\tau_c, n)$ function is the condition of distance between τ_c and n, formulated as:

$$dist(\tau_c, n) : \begin{bmatrix} \|x(\tau_c) - x(n)\| \leq d_{th} \\ |t(\tau_c) - t(n)| \leq \epsilon \end{bmatrix} \tag{5}$$

where ϵ is a small value fixed by the user and the distance threshold d_{th} is established based on the propagation parameters that assure communication, according to the communication model from [5]:

$$P_{RX} = P_{TX} - 10\gamma log_{10}(d_{th}); \qquad d_{th} = 10^{\frac{P_{TX} - P_{RX}}{10\gamma}} \tag{6}$$

where P_{TX}, P_{RX} represent the transmitted/received power and γ is the path-loss exponent. The example of dynamic communication area is depicted in Fig. 2(a).

Clearly not all the area is necessary, since the worker and the collector start their trajectories from different locations. So, some nodes of $A(\tau_c)$ are not reachable by the worker. Which means that including all the nodes of $A(\tau_c)$ will only increase the computation time to analyze all the possible trajectories, some of which are unfeasible for synchronization. Thus, we obtain the reachable nodes where and when the worker could reach the collector at its trajectory:

$$A_{reach} : \{n \in A(\tau_c) \mid d_{grad}(x_w, x(n))/v_w \leq t(n)\} \tag{7}$$

where x_w is the initial position of the worker, v_w is its maximum attainable speed and $d_{grad}(x_w, x(n))$ is the value of the gradient of $x(n)$, for the source x_w. The reachable area is illustrated in Fig. 2(b).

The amount of information which the worker must share requires a minimum time for transmission ($t_{c_{min}}$). So only those nodes of the communication area, which will guarantee the transmission of the data, are considered by the planner. Thus, we define the communication assurance area (Fig. 2(c)), which guarantees the complete information transmission as:

$$A_{ca} : \{n \in A(\tau_c) \mid t(n) \leq T(\tau_c) - t_{c_{min}}\} \tag{8}$$

where $T(\tau_c)$ denotes the time of the trajectory of the collector. In other words, the last moment when the collector can exchange data.

In conclusion, the velocity of the worker and the required communication time constrain the area. Thus, we define the feasible area, as the area that must be reached in order to ensure the communication, depicted in Fig. 2(d):

$$A_{feas} : A_{reach} \cap A_{ca} \tag{9}$$

If $\exists n_{feas} \in A_{feas}$, it means that the mission may be accomplished if the location of $x(n_{feas})$ is reached no later than $t(n_{feas})$. However, if $A_{feas} = \emptyset$, there is no solution. Therefore, the worker guides the search to the goal through this communication area, to synchronize with the collector, explained in Sect. 6.

6 Trajectory Planner

The trajectory planner obtains trajectories that traverse the dynamic communication area of the collector. This means that the suitable trajectories traverse A_{reach}. As described in the previous chapters, a unique gradient computation using FMM provides all the possible paths to all the points of the scenario. Therefore, we compute two gradients: one from the initial position of the worker expressed as $\nabla D(x_{ini})$, and another from the goal position defined by $\nabla D(x_{goal})$. This way, descending both gradients from any position that belongs to A_{reach}, allows building a path which traverses the communication area.

The communication area A_{reach} described in the Sect. 5, allows multiple communication possibilities. We consider three ways of actuation, according to the minimal communication time ($t_{c_{min}}$), the distance to A_{reach}, and the relative velocities between the worker and the collector.

- If the worker is faster and $t_{c_{min}}$ is short, one of the options is to *intercept* the collector, reaching the area as fast as possible. The worker traverses it, exchanging the data, and proceeds to reach the goal. This procedure is illustrated in Fig. 3(a) and it attempts to minimize the time of the mission.
- If the priority of the mission is to reduce the travelled distance, the worker adopts a *lazy* approach. Reaching the area, *waiting* in some position transmitting the data, meanwhile the collector goes towards its goal. The example of this kind of trajectory is depicted in Fig. 3(b). This situation and the previous one occur when exist some points where the interval, between the lower and upper bounds of the communication area, is greater than $t_{c_{min}}$.
- When $t_{c_{min}}$ is a significant time, it is necessary to *follow* the collector, as illustrated in Fig. 3(c). Therefore, the worker computes the optimal position where to intercept to the collector, follows it until finishing the data transmission, and then it continues to its goal.

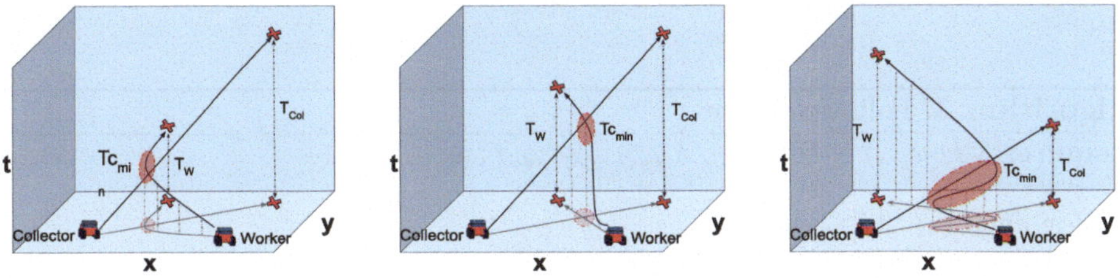

(a) Interception (b) Waiting (c) Following

Fig. 3. Possible trajectories for synchronization, based on transmission time.

Algorithm 1. Intercept routine

Require: $\nabla D(x_{ini}), \nabla D(x_{goal}), A_{reach}$
$\tau_{tot} \leftarrow \emptyset$
for each $n \in A_{reach}$ **do**
  $\tau_{ini} \leftarrow build\_traj(n, \nabla D(x_{ini}))$
  $n_{inter} \leftarrow \{n \mid x(\tau_{ini}) \cap x(A_{reach})\}$
  $n_i \leftarrow n_{inter}[1]$ ▷ First position
  while $t(n_i) < t_{min}(A_{reach})$ **do**
    $\tau_{ini} \leftarrow [x(n_i)\ \Delta t]$ ▷ Wait
  end while
  $\tau_{goal} \leftarrow build\_traj(n, \nabla D(x_{goal}))$
  $\tau_{tot} \leftarrow \tau_{ini} \cup \tau_{goal}$ ▷ Concatenate
end for
return τ_{tot}

Algorithm 2. Wait routine

Require: $\nabla D(x_{ini}), \nabla D(x_{goal}), A_{reach}$
$\tau_{tot} \leftarrow \emptyset$
for each $n \in A_{reach}$ **do**
  $\tau_{ini} \leftarrow build\_traj(n, \nabla D(x_{ini}))$
  $n_e \leftarrow \tau_{ini}[end]$ ▷ Last position
  while $t(n_e) < min(t_{max}(n), t(n_e) + t_{c_{min}})$ **do**
    $\tau_{ini} \leftarrow [x(n_e)\ \Delta t]$ ▷ Wait
  end while
  $\tau_{goal} \leftarrow build\_traj(n, \nabla D(x_{goal}))$
  $\tau_{tot} \leftarrow \tau_{ini} \cup \tau_{goal}$ ▷ Concatenate
end for
return τ_{tot}

The three situations are analyzed using three routines: *Intercept*, *Wait* and *Follow*. Each of them computes all the possible trajectories that accomplish the connectivity time constraint defined in Eq. 1. Then, the optimal trajectory is chosen based on the optimality criteria of Eq. 2.

The first one, the *Intercept* routine, is described in Algorithm 1. The algorithm evaluates all the positions of A_{reach} in l.2. The trajectories are built by descending the gradient $\nabla D(x_{ini})$, from each position of A_{reach}, in l.3. Then the method, obtains the first point where the worker achieves the area in l.4-5. Here, the worker waits to the first communication with the collector defined as $t_{min}(A_{reach})$, l.6-8. Δt represents a temporal or vertical slice of the communication area. Finally, the remaining trajectory to the goal, which traverses the area A_{reach} to communicate with the collector, is built, descending the gradient to the goal $\nabla D(x_{goal})$ in l.9. Finally, the complete trajectory is formed, concatenating τ_{ini} and τ_{goal}, in l.10.

The *Wait* procedure, Algorithm 2, follows the same principle as the interception. The difference is that the trajectory remains in the same position until the collector leaves, dragging the communication area or until the communication time is high enough to fulfill the data transmission $t_{c_{min}}$, in l.4-7. In l.4 *end* stands for the last position of the trajectory. The rest of the procedure is the same.

Algorithm 3. Follow routine

Require: $\nabla D(x_{ini}), \nabla D(x_{goal}), A_{reach}, A_{feas}, F, w$
1: $\tau_{tot} \leftarrow \emptyset$
2: **for each** $n \in A_{feas}$ **do**
3: $\quad \tau_{ini} \leftarrow build\_traj(n, \nabla D(x_{ini}))$
4: $\quad$ **while** $t(\tau_{ini}[end]) < t_{min}(A_{reach})$ **do**
5: $\quad\quad \tau_{ini} \leftarrow [x(\tau_{ini}[end])\ \Delta t]$ ▷ Wait until area arrives
6: $\quad$ **end while**
7: **end for**
8: $\tau^*_{ini} \leftarrow choose\_traj(\tau_{ini}, w)$ ▷ Using eq.1
9: $F_{area} \leftarrow F(!A_{reach}) = 0$ ▷ Only the points of the area
10: $\mathbf{x}_{inter} \leftarrow \tau^*_{ini}[end]$ ▷ Last positions of τ^*_{ini}
11: $\nabla D(\mathbf{x}_{inter}) \leftarrow gradient(\mathbf{x}_{inter}, F_{area})$
12: $\mathbf{n}_{cont} \leftarrow contour(A_{reach})$
13: **for each** $n \in \mathbf{n}_{cont}$ **do**
14: $\quad \tau_{inter} \leftarrow build\_traj(n, \nabla D(\mathbf{x}_{inter}))$
15: $\quad \tau_{goal} \leftarrow build\_traj(n, \nabla D(x_{goal}))$
16: $\quad \tau_{tot} \leftarrow \tau^*_{ini} \cup \tau_{inter} \cup \tau_{goal}$
17: **end for**
18: **return** τ_{tot}

The *Follow* routine is summarized in Algorithm 3 and is somewhat more complex. As described above, in this case the minimum communication time for exchanging requires to follow the collector. Thus, the trajectory necessarily requires to enter into A_{feas}, whose points assure a complete data exchange. So, the trajectories are obtained, remaining in the same point until the first communication with the collector, as described in l.2-7. From here, the worker needs to move within the communication area A_{reach}, so it is necessary to compute a new gradient for this purpose, and obtaining new trajectories inside this area. Since A_{feas} may contain a large number of points, it is computationally expensive to calculate $|A_{feas}|$ gradients. Thus, we choose the optimal trajectories according to Eq. 1 in l.8, for values of w selected by the user. The gradient has to cover only the required area A_{reach}, thus the rest of the points are discarded, in l.9, in order to reduce the time of computation. The point where the worker will leave the communication area A_{reach} will be some point of its contour. So the gradient $\nabla D(\mathbf{x}_{inter})$ is computed only within A_{reach} in l.10-11, in order to obtain all the possible trajectories to the contour points $\mathbf{n}_{cont}$ obtained in l.12. Finally, the stretch of the trajectories for synchronization are obtained descending this gradient in l.14. Here, the worker, in the case of being faster than the collector and getting out of A_{reach}, waits in order to remain within it. The remaining stretch to the goal is obtained in l.15.

The entire procedure to obtain the optimal trajectory is shown in Algorithm 4. The chosen trajectory is the one that minimizes the constrained criteria defined in Eq. 1. It is important to highlight that the expensive procedures as area computation (l.1) and segmentation (l.2) are executed once. This also occurs with the gradient computation, it is only executed three times: one for initial position, one for the goal position, and another one in Algorithm 3 to obtain the stretch within A_{reach} for *Follow* procedure. The time to descend the gradient with *build_traj* routine is much less time-consuming. This makes it possible to evaluate a large amount of possible trajectories without a drastic increase of the computation time.

Algorithm 4. Complete routine

Require: $x_{ini}, x_{goal}, \tau_c, F, w$
1: $A(\tau_c) \leftarrow com\_area(\tau_c)$ ▷ Using eq.4
2: $[A_{reach}, A_{feas}] \leftarrow area\_parts(A(\tau_c))$ ▷ Using eq.7-9
3: $\nabla D(x_{ini}) \leftarrow gradient(x_{ini}, F)$ ▷ Using FMM
4: $\nabla D(x_{goal}) \leftarrow gradient(x_{goal}, F)$ ▷ Using FMM
5: $\tau_i \leftarrow intercept(\nabla D(x_{ini}), \nabla D(x_{goal}), A_{reach})$ ▷ Alg.1
6: $\tau_w \leftarrow wait(\nabla D(x_{ini}), \nabla D(x_{goal}), A_{reach})$ ▷ Alg.2
7: $\tau_f \leftarrow follow(\nabla D(x_{ini}), \nabla D(x_{goal}), A_{reach}, A_{feas}, F, w)$ ▷ Alg.3
8: $\tau^* \leftarrow choose\_traj(\{\tau_i, \tau_w, \tau_f\}, w)$ ▷ Using eq.1
9: **return** τ^*

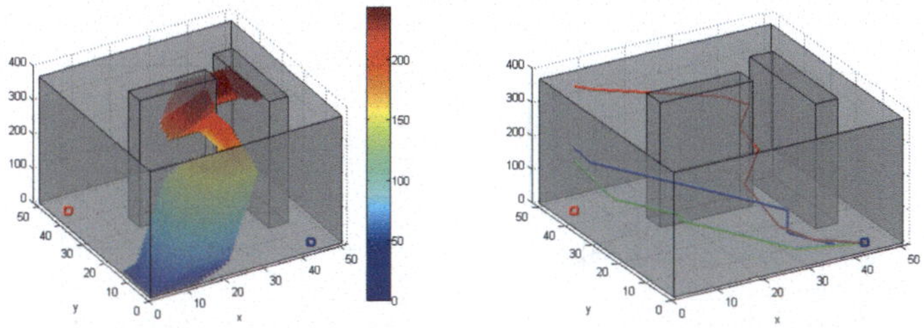

(a) Communication area (b) Obtained trajectories

Fig. 4. Obtained trajectories in presence of one collector. In (a) the collector goes from [5, 5] to [45, 45], creating a dynamic communication area $A(\tau_c)$. The worker must go from [45, 5] to [5, 45], but previously synchronizing with the collector. In (b) the green, blue, and red lines represent the trajectories obtained by *Intercept*, *Wait*, and *Follow* routines, respectively.

7 Results

In this section we present the simulated results of the proposed planner. The simulator was implemented in MatLab and tested on a computer with i7-4770 processor clocked at 3.4 GHz with 8 GB of RAM. We consider two cases: one worker to synchronize with a single collector in the environment and one worker to synchronize with one collector from multiple present in the environment. In this work we have not considered the collision avoidance problem, but it can be solved with a reactive navigator such as Dynamic Window.

Single Collector Case. We evaluate two situations: when the worker is faster and the amount of data is small, thus, $t_{c_{min}}$ is a little time; and when both agents have the same speed and $t_{c_{min}}$ is a long time. The obtained trajectories are depicted in Fig. 4(b), and the results are shown in Table 1. The times are expressed in seconds and distances in meters. The Algorithm 4 is executed varying the weighting factor w of Eq. 2. Setting $w_t = 1$, the time must be minimized and the optimal trajectory (green line) is obtained by interception routine of Algorithm 1. However, establishing $w_d = 1$, the optimal trajectory (blue line) is provided by waiting routine of Algorithm 2, reducing the travelled distance, but increasing the time spent. In the case of increasing $t_{c_{min}}$, the optimal trajectory (red line) is obtained by *Follow* method of Algorithm 3, because the others do not provide a solution, since they do not accomplish the constraint of Eq. 1. We select $t_{c_{min}} = 165$ s, which represents almost 100% of time of A_{reach}, so that the worker deviates from its goal, following the communication area trace generated by the collector in movement.

Table 1. Obtained results for scenario of Fig. 4.

	Worker/collector speed	$t_{c_{min}}$	$t(\tau)$	$d(\tau)$	$t_c(\tau)$
Intercept	2	30	147.68	65.36	33.94
Waiting	2	30	186.05	60.08	31.11
Following	1	165	369.14	92.28	166.23

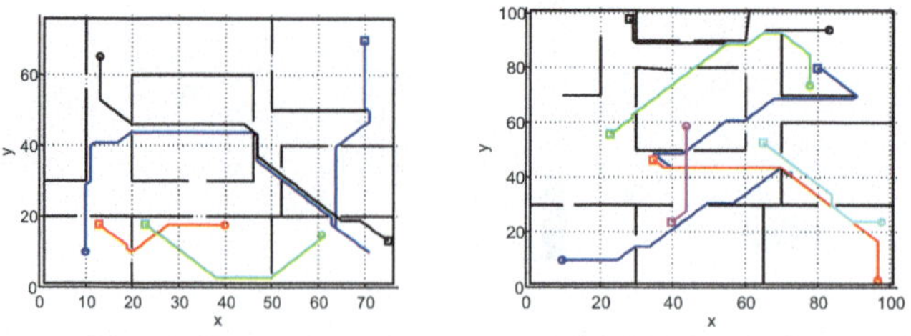

(a) 3 collectors. (b) 5 collectors.

Fig. 5. Circles and squares represent initial and goal positions, respectively. Different color lines are trajectories of the collectors. The worker (blue circle) must reach a location (blue square). In (a) the worker follows the black collector trajectory, in (b) the red one.

Multiple Collectors Case. We also test our method for multi-robot scenarios, where there are multiple collectors in the environment. The simulations are performed in two scenarios, depicted in Fig. 5. We consider all the collectors present in the environment as possible candidates to transmit the data. The worker has to obtain the trajectory, choosing the best collector to share data with, according to Eqs. 1 and 2. We present the mean results based in 100 random collectors distributions for each scenario. That is, randomly generating the initial and goal locations of each collector. The trajectories of the collectors are sampled each cell to obtain $A(\tau_c)$, in a grid with squared cells of 1 m. We use 3 and 5 collectors in scenarios of Fig. 5(a) and (b), respectively.

We evaluate the method performance for different communication necessities, so that, varying $t_{c_{min}}$. We fix the speed of the worker to be twice the speed of the collectors. The results for *Wait* and *Intercept* routines, setting $t_{c_{min}}$ to 10 and 30 s, are depicted in Table 3. The results of *Follow* procedure appear in Table 4, with $t_{c_{min}}$ being 75% and 100% of possible communication time, in other words, of A_{reach}.

As described in Sect. 6, some procedures require a unique execution. Table 2 shows the times of these procedures for both scenarios. The time of computation of the gradients from start and goal position of the worker is around 4 s. We also show the computation time to obtain the trajectories of the collectors (τ_c), form the communication area ($A(\tau_c)$) and decompose it, expressed in seconds. Obviously, for the bigger scenario with more collectors in the team, the values are bigger.

In order to assess the optimality of the trajectories with time-constrained communication, we provide the values of distance and time of the direct trajectories, without connectivity constraints, in Table 2. The distance and time of the direct path are $d(\tau_{dir})$ and $t(\tau_{dir})$, respectively.

When we set $w_d = 1$, the distance of the trajectory must be minimized, so the optimal trajectories are provided by *Wait* routine. By contrast, if the chosen weight is $w_t = 1$, the trajectories come from *Intercept* routine, at the expense of being longer. The mean results appear in Table 3. The results are based on: the number of analyzed trajectories ($|\tau|$), computation time (t_{comp}), the time $t(\tau^*)$, the distance $d(\tau^*)$ and the communication time $t_c(\tau^*)$ of the optimal trajectory. The times are expressed in seconds and the distance in meters.

The trajectories provided by *Wait* procedure have practically the same distance of the direct path $d(\tau_{dir})$ in Table 2. The trajectories of lower times are obtained by *Intercept* routine, but it is difficult to obtain similar times to $t(\tau_{dir})$, because the worker considerably deviates from τ_{dir}. It is remarkable that the computational time of *Wait* is smaller than for *Intercept*. This is because *Intercept* method requires an extra time to intersect the trajectories with the area A_{reach}.

When the amount of data to transmit is large, so that $t_{c_{min}}$ is high, the worker is forced to follow some collector until fulfilling the data transmission. Obviously, the worker is drastically deviated from its original goal, but it accomplishes the information exchange, as shown in Table 4. It is noteworthy that the computational time of *Follow* is considerably lower than the other procedures. On the

Table 2. Computation times and direct paths values

Scenario	Computation times (s)			Direct paths values	
	∇D	$\tau_c + A(\tau_c)$	Parts of $A(\tau_c)$	$d(\tau_{dir})$ (m)	$t(\tau_{dir})$ (s)
Figure 5(a)	3.81	18.29	15.42	99.5	79.5
Figure 5(b)	3.91	29.84	26.09	122.75	98.2

Table 3. Results of *Intercept* and *Wait* procedures for scenarios of Fig. 5.

$t_{c_{min}}$	Routine	Scenario of Fig. 5(a)					Scenario of Fig. 5(b)				
		$\|\tau\|$	t_{comp}	$t(\tau^*)$	$d(\tau^*)$	$t_c(\tau^*)$	$\|\tau\|$	t_{comp}	$t(\tau^*)$	$d(\tau^*)$	$t_c(\tau^*)$
10	Intercept	1677	24.01	105.82	100.9	17.57	3661	73.79	105.4	126.67	12.89
	Wait	1677	12.7	115.13	99.7	13.58	3661	38.67	111.86	122.75	12.27
30	Intercept	1677	23.94	121.57	102.75	32.4	3661	73.87	129.32	127.96	31.07
	Wait	1677	12.7	134.86	102.75	32.09	3661	38.71	136.89	127.98	31.11

Table 4. Results of *Follow* routine for scenario of Fig. 5.

Scenario	$t_{c_{min}}$	$\|\tau\|$	t_{comp}	$t(\tau^*)$	$d(\tau^*)$	$t_c(\tau^*)$
Figure 5(a)	75%	945	5.82	148.14	117.22	73.76
	100%	234	1.8	149.06	121.28	77.84
Figure 5(b)	75%	1378	12.31	199.13	154.76	110.27
	100%	238	2.73	208.2	167.5	110.08

one hand, this is because the number of analyzed trajectories ($|\tau|$) is much lower, since the worker only must navigate within the communication area of some collector (A_{reach}), as described in Algorithm 3. On the other hand, increasing the minimal communication time, leaves fewer trajectory possibilities. So, this fact also reduces the number of trajectories to analyze.

8 Conclusions

In the present paper we have developed a trajectory planning method for a worker agent to synchronize, in order to exchange data in movement, with a collector agent with a known trajectory. Using this planner, the worker is able to obtain the best trajectory based on the amount of data to share. We have formally defined the dynamic communication area of a collector agent in movement and the different parts of this area are used in order to compute the trajectories for synchronization. The method uses the Fast Marching Method to compute the optimal trajectory, by analyzing multiple possible solutions, based on two different metrics: distance and time. As future work, we want to include dynamic obstacles that can be present in the environment. For example, in scenarios with human presence, they must be included in the planning process, in order to plan collision free trajectories and to preserve the connectivity.

References

1. Amato, N.M., Wu, Y.: A randomized roadmap method for path and manipulation planning. In: Proceedings of IEEE International Conference on Robotics and Automation, vol. 1, pp. 113–120, April 1996
2. Banfi, J., Li, A.Q., Basilico, N., Rekleitis, I., Amigoni, F.: Asynchronous multirobot exploration under recurrent connectivity constraints. In: 2016 IEEE International Conference on Robotics and Automation (ICRA), pp. 5491–5498, May 2016
3. Díaz-Bánez, J.M., Caraballo, L.E., Lopez, M.A., Bereg, S., Maza, I., Ollero, A.: A general framework for synchronizing a team of robots under communication constraints. IEEE Trans. Robot. **33**(3), 748–755 (2017)
4. Farinelli, A., Iocchi, L., Nardi, D.: Distributed on-line dynamic task assignment for multi-robot patrolling. Auton. Robot. **41**(6), 1321–1345 (2017)
5. Goldhirsh, J., Vogel, W.: Handbook of propagation effects for vehicular and personal mobile satellite systems, December 1998
6. Hollinger, G.A., Singh, S.: Multirobot coordination with periodic connectivity: theory and experiments. IEEE Trans. Robot. **28**(4), 967–973 (2012)
7. Joy, K.I.: Breshenham's algorithm. In: Visualization and Graphics Research Group. Department of Computer Science, University of Carolina, December 1999
8. Lavalle, S.M.: Rapidly-exploring random trees: a new tool for path planning. Technical report, Department of Computer Science, Iowa State University (1998)
9. Majcherczyk, N., Jayabalan, A., Beltrame, G., Pinciroli, C.: Decentralized connectivity-preserving deployment of large-scale robot swarms. In: 2018 IEEE/RSJ International Conference on Intelligent Robots and Systems (IROS), pp. 4295–4302 (2018)
10. Marchukov, Y., Montano, L.: Multi-agent coordination for on-demand data gathering with periodic information upload. In: Advances in Practical Applications of Survivable Agents and Multi-Agent Systems: The PAAMS Collection, pp. 153–167. Springer, Cham (2019)
11. Mukhija, P., Krishna, K.M., Krishna, V.: A two phase recursive tree propagation based multi-robotic exploration framework with fixed base station constraint. In: 2010 IEEE/RSJ International Conference on Intelligent Robots and Systems, pp. 4806–4811, October 2010
12. Portugal, D., Rocha, R.: MSP algorithm: multi-robot patrolling based on territory allocation using balanced graph partitioning. In: Proceedings of the 2010 ACM Symposium on Applied Computing, SAC 2010, pp. 1271–1276. ACM (2010)
13. Sethian, J.A.: A fast marching level set method for monotonically advancing fronts. Proc. Natl. Acad. Sci. USA **93**(4), 1591–1595 (1996)

Graph-Based Robot Localization in Tunnels Using RF Fadings

Teresa Seco[1(✉)], María Teresa Lázaro[1], Carlos Rizzo[2], Jesús Espelosín[1], and José Luis Villarroel[3]

[1] Instituto Tecnológico de Aragón (ITAINNOVA), Zaragoza, Spain
{tseco,mtlazaro,jespelosin}@itainnova.es

[2] Eurecat, Centre Tecnològic de Catalunya, Robotics and Automation Unit, Barcelona, Spain
carlos.rizzo@eurecat.org

[3] Aragón Institute for Engineering Research (I3A), University of Zaragoza, Zaragoza, Spain
jlvilla@unizar.es

Abstract. Robot localization inside tunnels is a challenging task due to the hostile conditions of the environment. The GPS-denied nature of the scenario together with the low visibility, slippery surfaces, and the lack of distinguishable features, make traditional robotics methods based on cameras or laser unreliable. In this paper, we address the robot localization problem with an alternative graph-based localization approach, taking advantage of the periodic nature of the RF signal fadings that appears inside tunnels under certain transmitter-receiver settings. Experimental results in a real scenario demonstrate the validity of the proposed method for inspection applications.

Keywords: RF fadings · Tunnel-like environments · Graph localization

1 Introduction

Inspection tasks in tunnel-like environments are crucial in order to detect and identify critical characteristics during the construction of the tunnel, rescue missions or regular service routines. In recent years, robots seem to be the best candidates to perform these tasks mainly due to their flexibility and the harsh and even dangerous conditions of the environment that makes the human intervention risky.

However, accurate robot localization in tunnel-like scenarios represents a challenge due to the darkness and absence of distinguishable features in their longitudinal direction that makes traditional methods, based on cameras or laser sensors, inefficient. Moreover, GPS sensors cannot be used in underground environments. [1] presents an autonomous platform for exploration and navigation in mines where the localization is based on the detection and matching of natural

M. F. Silva et al. (Eds.): ROBOT 2019, AISC 1092, pp. 580–592, 2020.
https://doi.org/10.1007/978-3-030-35990-4_47

landmarks over a 2D survey map using a laser sensor. In the case of tunnels, these natural features are almost non-existent.

Recent promising works explore the use of Radio Frequency (RF) signal for indoor localization. In [2] the authors propose the use of an Ultra Wide-Band (UWB) ranging sensor in combination with a LiDAR to obtain the localization in a tunnel fusing the information with a Gaussian Particle Filter (GPF). Nevertheless, the use of RF-based indoor localization implies a previous commissioning step to place at least three anchor nodes with high precision in the infrastructure to calculate the position by trilateration algorithms.

In [3,4] the authors present several intensive studies about the RF signal propagation inside tunnels. Those works show that, on the one hand, tunnel-like environments behave as waveguides extending the communication range, but on the other hand, the signal suffers from strong attenuation (fadings). The authors also demonstrate that it is possible, under certain transmitter-receiver setups, to obtain predictable periodic fadings. The periodic nature of the RF signal is exploited in [5] to design a discrete robot localization system based on the identification of the RF signal minima and matching them with the known signal propagation model acting as an RF map.

Recent advances in the field of graph-SLAM result in new localization approaches that model the localization problem as a pose-graph optimization [6] with the advantage of easily incorporating measurements from different sources of information to the graph, not only local (wheel odometry) but also global measurements (GPS, IMU).

Taking into account the aforementioned works, in this paper we address the robot localization problem in tunnels as an online pose-graph localization problem, where we originally introduce the results of our RF signal minima detection method into the graph optimization taking advantage of the periodic nature of the RF signal inside tunnels.

Our approach consists of identifying the minima of the signal which are related to a global position provided by a previously obtained RF map (corresponding to the signal propagation model). The absolute position of each minimum is added as a constraint to the pose-graph that is being generated with the information provided by the odometry during the displacement of the robot. Each time new information is incorporated into the graph, it is optimized and the position of the robot is corrected allowing to locate the main characteristics to be inspected more accurately. The main advantages of approaching the robot localization problem using a graph-based representation are twofold: it allows to easily incorporate delayed measurements into the estimation process, and to recover (undo) from wrong decisions such as the inclusion of incorrect measurements.

The paper is structured as follows. The next section describes the related work about the fundamental aspects of the electromagnetic propagation in tunnel-like scenarios. The proposed method to identify the RF minima signal is presented in Sect. 3. The formulation of the graph-based localization problem together with a detailed description of the strategy followed to incorporate the minima to the graph is explained in Sect. 4. Section 5 presents the results obtained in the real scenario. Finally, the conclusions are set out in Sect. 6.

2 Related Work: Fundamentals of Electromagnetic Propagation in Tunnels

As stated before, previous works [3,4,7] demonstrate two different behaviors of the RF signal in tunnels: on the one hand, the tunnel acts as an oversized waveguide, extending the communication range if the wavelength of the signal is much smaller than the tunnel cross-section dimensions. On the other hand, strong fadings appear due to the interaction between the propagation modes present in the waveguide. It is important to highlight that we refer to fadings in a spatial domain, as a consequence of the constructive and destructive interference between propagating modes (using a modal theory approach) or propagating rays (using a raytracing approach), unlike the well-known small-scale fadings, which are understood as temporal variations of the channel.

Depending on the distance from the transmitter, due to the different attenuation rate of the propagation modes, two regions can be distinguished in the signal. In the *near sector*, all the propagation nodes are present provoking fast fluctuation on the signal (*fast-fadings*). Once the higher order modes (which have higher attenuation rate) are mitigated with the distance the lower modes survive, giving rise to the *far sector*, where the *slow-fadings* dominate [8]. A specific periodic signal is obtained under the most adequate transmitter-receiver configuration (Fig. 1(a)). These studies also demonstrate that the period of the fadings depends on the operating frequency and the tunnel dimensions.

Lastly, the authors adopt the Modal Theory approach, modeling the tunnel as a rectangular dielectric waveguide. We encourage the reader to see [9] for a complete 3-D fadings structure analysis in tunnels. With this approximation, the obtained theoretical propagation model matches closely the experimental data. The similarity between both signals (Fig. 1(b)) are enough to make us consider them useful for localization purposes, using the propagation model as a position reference.

3 RF Signal Minima Detection

As stated before, the agreement between the signal propagation model and the real RF signal let us consider the first one as an RF map, which relates the RSSI values to the distance along the tunnel. Due to the noisy nature of the RF signal, the most distinguishable features of the RF waveform are the valleys (fadings). The goal of the presented method is to identify the minima of the real signal during the robot displacement and to extract the reference position associated to each valley from the RF Map. The information provided by the virtual minima detector will be added to the pose-graph as it will be explained in Sect. 4.

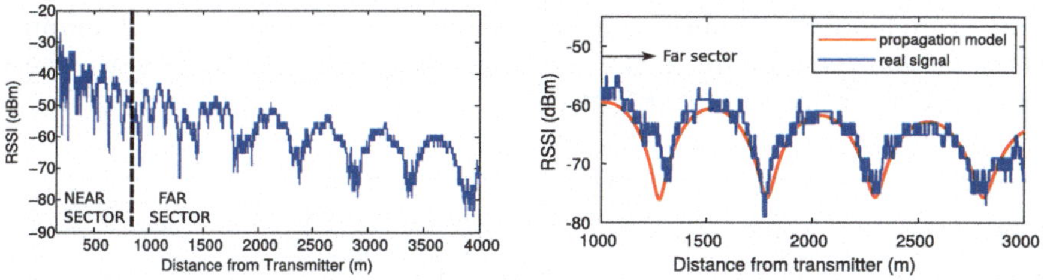

(a) Periodic Fadings

(b) Propagation model vs real data

Fig. 1. Measured received power at 2.4 GHz inside the Somport tunnel, from [4]. The transmitter was kept fixed and the receiver was displaced along 4 km from the transmitter. The signals were sampled with a spatial period of 0.1 m. In (b), the red line represent the modal theory simulations, and the blue the experimental results for the far sector.

The first step of the proposed method consists of extracting a discrete model representing the theoretical minimum model from the RF Map. Using the propagation model described in Chaps. 2.1 and 3.2 of [10], it is possible to know the position of each valley along the tunnel and then, the theoretical minimum model can be obtained in advance. During the displacement of the vehicle, the algorithm tries to match the discrete real model, generated during the movement, with the theoretical model. When the two models match, a minimum is found and the information about the estimated position of the minimum together with its corresponding position in the map is available.

Figure 2 shows the steps of the proposed strategy. The discrete theoretical model is extracted from the RF signal model in advance (Fig. 2(a)). The theoretical model consists of a set of points (x, y) where x is the position corresponding to each theoretical RSSI value y. Both values are provided by the RF map. The real model is obtained by accumulating points (x_t, y_t) during a certain period of time T corresponding to a fixed distance D (Fig. 2(b)). x_t is the position estimated by the odometry and y_t corresponds to the actual RSSI value provided by an RF sensor. Once the real model is available, the matching process starts using the previously recorded theoretical model. The points enclosed in the B blue area represent the real model used to describe the matching procedure that involves the following steps:

- Relate the theoretical model to the reference system of the real model (Fig. 2(c)). Both models have in common the minimum value.
- For each real point, calculate the Mahalanobis distance d_m between the real point and the closest neighbors from the theoretical model Fig. 2(d).
- Classify each point as inlier or outlier based on the Mahalanobis distance.
- If the number of inliers is greater than a certain threshold and the ratio between left and right inliers is balanced, we can conclude that a minimum has been found Fig. 2(e).

- The Mahalanobis distance is again calculated between the real minimum detected and the minimums of the theoretical model, selecting the theoretical one with the least distance Fig. 2(f).

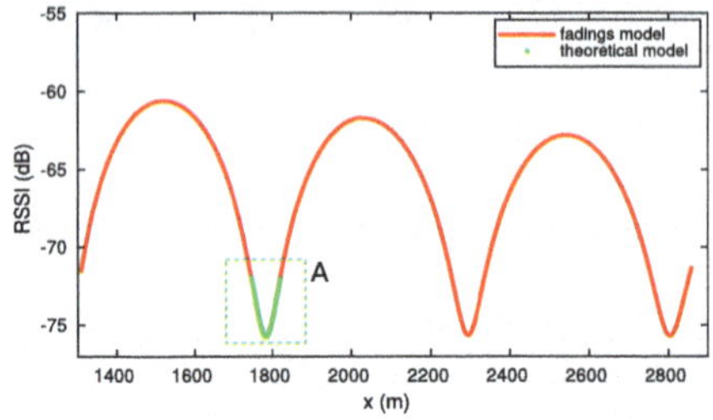

(a) Theoretical model extraction from RF signal model

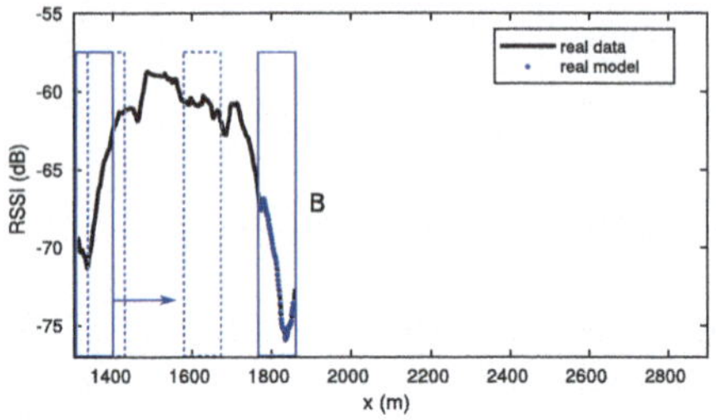

(b) Real model extraction

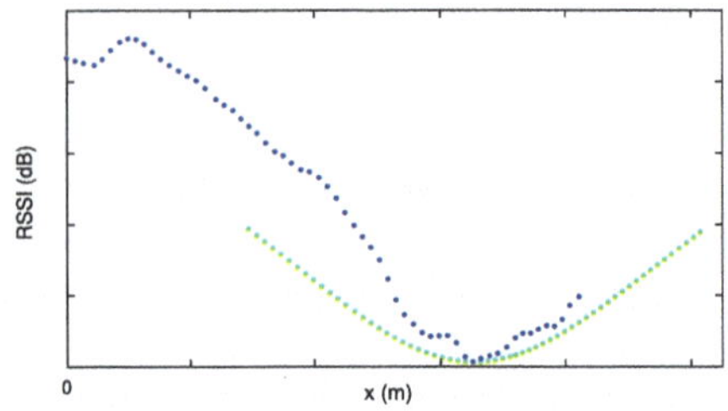

(c) Zero referenced models

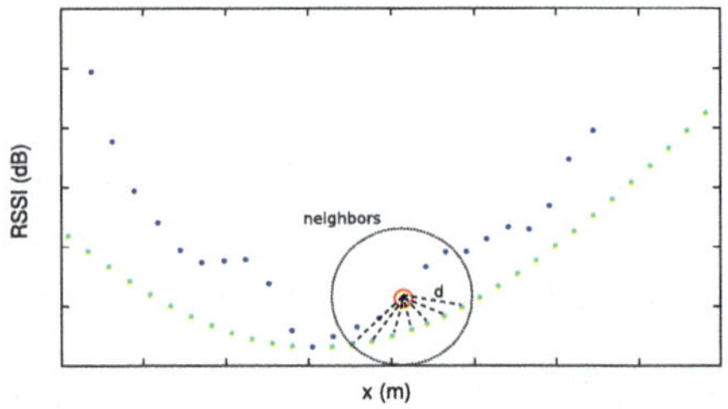

(d) Inlier/outlier classification

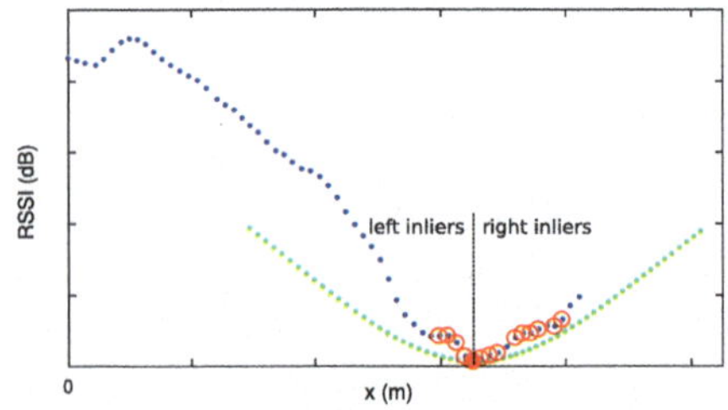

(e) Minimum detection

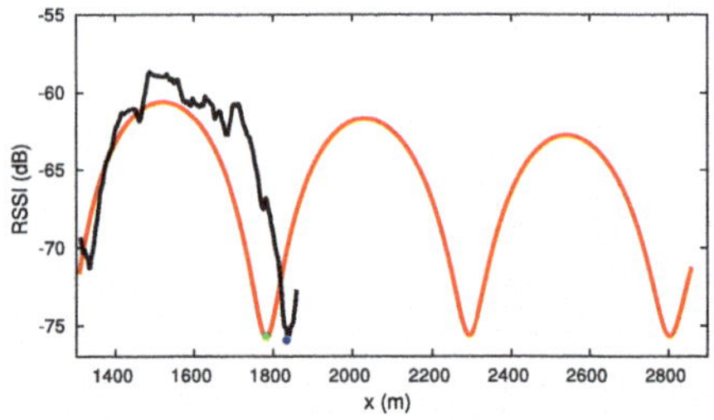

(f) Estimated and reference minimum pose

Fig. 2. RF signal minima detection steps: (a) Theoretical model (green points inside the dashed green square A) extracted from the RF signal model (red). (b) Real model generation during the displacement of the vehicle from the real RF signal. (c) Both models referenced to the same system coordinates. (d) Point classification depending on the Mahalanobis distance between the real data and the closest neighbors from the theoretical model (e) Minimum detection if the number and proportion of inliers satisfy the threshold. (f) Estimated position by the odometry (black point) and position reference from the RF Map (green point) of the detected minimum

The resulting data are the estimated position provided by the odometry (x_{T-k}) and the position reference of the RF map (z_i), both corresponding to a minimum of the RF signal. The uncertainty of the position reference (δ) is a measure of the RF signal model fidelity with respect to the ground truth. This process is repeated iteratively using a sliding window to generate the discrete real model (dashed blue area in Fig. 2(b)). It is worthy to notice that the information provided by the virtual sensor corresponds to delayed measurements, i.e., the position of the minimum is detected at a timestamp after its appearance. The strategy followed to add these measurements to the pose-graph is explained in the following section.

4 Graph-Based Localization Using RF Fadings

In our approach we model the robot localization problem as a graph-based pose optimization problem. The trajectory of the robot $\mathbf{x}_{0:T} = \{\mathbf{x}_0, \ldots, \mathbf{x}_T\}$ is represented as a graph where nodes symbolize discrete robot positions $\mathbf{x}_t$ at time step t. Nodes in the graph are related by *binary* measurements encoding relative position constraints between two nodes $(\mathbf{x}_i, \mathbf{x}_j)$ characterized by a mean $\mathbf{z}_{ij}$ and information matrix $\boldsymbol{\Omega}_{ij}$. These relative measurements are typically obtained through odometry or scan matching. Furthermore, it is possible to incorporate into the graph global or prior information associated only to one robot position $\mathbf{x}_i$ by means of *unary* measurements $\mathbf{z}_i$ with information matrix $\boldsymbol{\Omega}_i$. These unary measurements typically come from sensors providing direct absolute information about the robot pose such as GPS or IMU. Let $\hat{\mathbf{z}}_i = \mathbf{h}(\mathbf{x}_i)$ and $\hat{\mathbf{z}}_{ij} = \mathbf{h}(\mathbf{x}_i, \mathbf{x}_j)$ be the expected unary and binary measurements given the current estimation of the nodes. The errors committed in the estimation can be obtained as:

$$\mathbf{e}_i = \mathbf{z}_i - \hat{\mathbf{z}}_i, \quad \mathbf{e}_{ij} = \mathbf{z}_{ij} - \hat{\mathbf{z}}_{ij} \tag{1}$$

The goal of a graph-based approach is to find the configuration of nodes that minimizes the sum of the errors introduced by the measurements, formulated as:

$$\mathbf{x}^* = \underset{\mathbf{x}}{\operatorname{argmin}} \sum_{i,j} \mathbf{e}_{ij}^T \boldsymbol{\Omega}_{ij} \mathbf{e}_{ij} + \sum_i \mathbf{e}_i^T \boldsymbol{\Omega}_i \mathbf{e}_i \tag{2}$$

The above Eq. 2 poses a non-linear least-squares problem that can be solved iteratively using the Gauss-Newton algorithm.

Our approach for localization inside tunnels considers measurements coming from two sources of information: odometry data and RF signal minima detection using the procedure described in previous Sect. 3. Odometry measurements are straightforwardly introduced into the graph as binary constraints encoding relative displacement between consecutive nodes $(\mathbf{x}_{t-1}, \mathbf{x}_t)$. Additionally, the output provided by the minima detection mechanism can be considered as an absolute positioning system inside the tunnel which can be used as a unary measurement during the graph optimization process.

As previously mentioned, RF signal minima detection is obtained on a posterior time T in which it actually occurred. This implies incorporating in the estimation process information referred to a past position $\mathbf{x}_{T-k}$. This can be handled thanks to the use of a graph representation, having an impact on the current pose estimation after the optimization process. Next Subsect. 4.1 describes the proposed mechanism to incorporate the RF signal minima measurements into the graph.

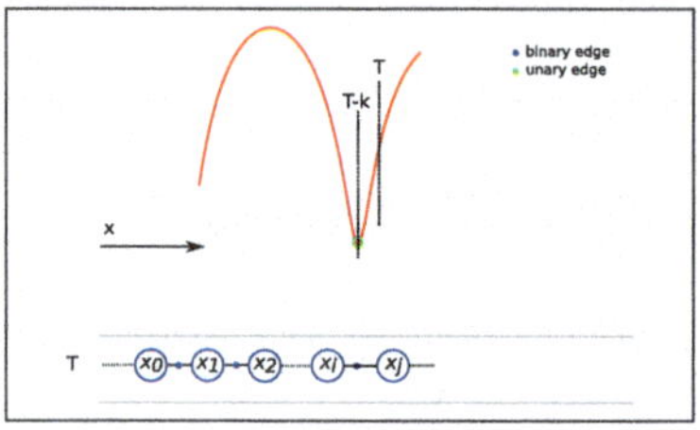

(a) Minimum detection

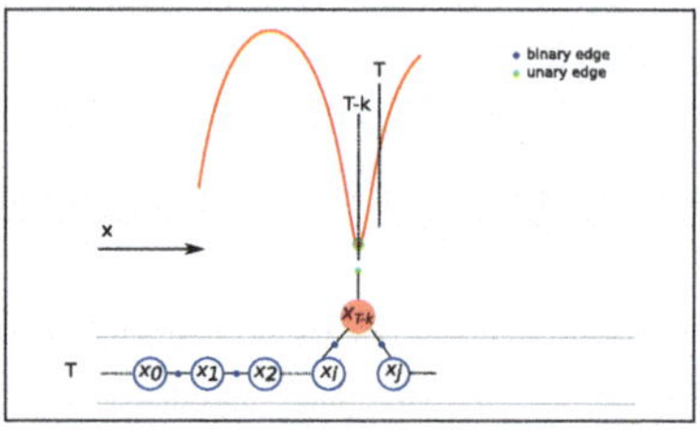

(b) Minimum node insertion

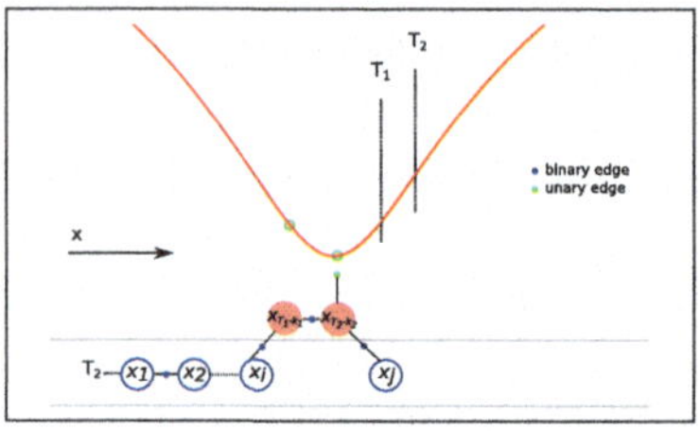

(c) False positive case

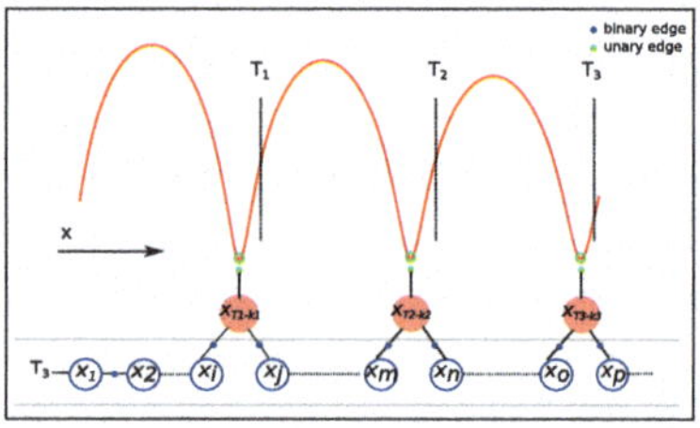

(d) Final graph

Fig. 3. Pose-graph creation steps: (a) Minimum identification at time T. (b) Insertion of the node and the unary constraint corresponding to the detected minimum. (c) False positive case detail, deactivation of the previous unary edge. (d) Resulting pose-graph after three minimums.

4.1 Management of RF Fadings Minima Detection in the Pose-Graph

As introduced in the previous section, the graph based localization approach consists in the representation of a set of discretized robot poses from the robot trajectory as nodes in the graph. Once the constraints derived from the measurements are introduced into the graph the error minimization process takes place, where the optimization time depends directly on the number of nodes.

Graph-based localization and mapping systems usually perform a rich discretization of the robot trajectory, where the separation between nodes ranges from few centimeters to few meters. This type of dense discretization would be intractable in a tunnel-like environment with few distinguishable features where the length of the robot trajectory is measured in the order of magnitude of kilometers. It is therefore necessary to maintain a greater distance between nodes to manage a sparser and more efficient graph.

Under an RF signal minimum detection event we need to associate a unary constraint to the past robot position where the minimum occurred. In view of the need to maintain a sparse graph, it can happen that the referred robot position is not represented in the graph as a node, having to modify the current graph structure to include it.

The procedure to include the unary measurement corresponding to a past robot position $\mathbf{x}_{T-k}$, is illustrated in Fig. 3 and described in the following:

- At timestamp T, a RF signal minimum corresponding to timestamp $T-k$ is identified. Since robot position $\mathbf{x}_{T-k}$ is not present in the graph, we determine between which two nodes $\mathbf{x}_i$ and $\mathbf{x}_j$ it should be included, based on the timestamps stored in each node. We also maintain a buffer containing the odometry information associated to each timestamp (Fig. 3(a)).
- Once the two nodes are identified, the new node $\mathbf{x}_{T-k}$ is inserted into the graph connected to nodes $\mathbf{x}_i$ and $\mathbf{x}_j$ by taking into account their original relative odometry information and the unary edge is associated to the node $\mathbf{x}_{T-k}$. Previous odometry measurement connecting $\mathbf{x}_i$ and $\mathbf{x}_j$ is removed to prevent double-counting of information (Fig. 3(b)).
- In the event of detecting another minimum corresponding to the same minimum in the RF map, the unary constraint of the previous minimum is deactivated and same procedure is followed (Fig. 3(c)). This can be the case of false positives or improved detections after the accumulation of more data.

5 Experimental Results

In order to validate the proposed graph-based localization approach, all the algorithms involved in the process were implemented in $MATLAB^{TM}$ and tested with real data collected during an experiment developed in the Somport tunnel.

5.1 Scenario and Experimental Setup

The old out-of-service Somport railway tunnel was selected to carry out the experiments. It is a 7.7 km long tunnel connecting Spain with France with a change in slope at approximately 4 km from the Spanish entrance. It has a horseshoe-shape cross section, around 5 m high and 4.65 m wide.

An all-terrain vehicle was used as the mobile platform simulating a service routine. It was equipped with two SICK DSF60 0.036 deg resolution encoders and a SICK LMS200 LIDAR. Due to the specific characteristics of this tunnel, with lateral galleries and emergency shelters, it is possible to obtain the real localization of the platform (ground truth) along the tunnel fusing all the data sensor using the algorithm described in [11] with a previously built map. Without

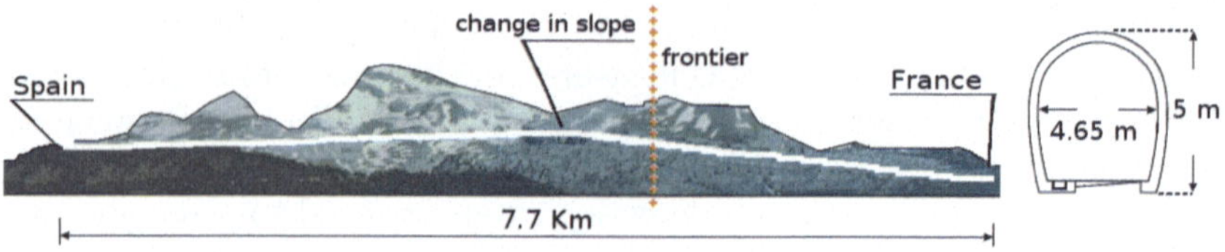

(a) The Somport tunnel

(b) Encoder

(c) Laser sensor

(d) RF receivers

Fig. 4. Experimental setup

these landmarks, it would not be feasible to apply this method because of the lack of relevant features along the tunnel. The ground truth is only used for comparison purposes.

The platform was also equipped with two RF Alpha receivers placed at 2.25 m in height from the ground and with the antennas spaced 1.40 m apart. The transmitter, a TPLINK tl-wn7200md wireless adapter with Ralink chipset, was placed at approximately 850 m from the entrance of the tunnel, 3.50 m above the ground and 1.50 m from the right wall. Using a 2.412 GHz working frequency and under this receiver-transmitter setup, the expected fadings period is around 512 m. Figure 4 shows the experimental setup.

The mobile platform moved up to about 3000 m from the transmitter position along the center of the tunnel in straight line with almost negligible heading variations. This behaviour during the experiment makes feasible the simplification of the general formulation of our approach, where $\mathbf{x}$ refers to (x, y, θ), to a one dimension problem where $\mathbf{x}$ corresponds to the longitudinal distance from the transmitter. During the displacement of the vehicle, the data provided by the sensors were streaming and logging with a laptop running Robot Operating System (ROS) [12] on Ubuntu.

The RF data used to validate the proposed method are the RSSI values provided by the rightmost antenna. It should be noted that the proposed graph-based approach is intended to solve the localization problem in the area of the tunnel where the periodic fadings are observable (far sector). For this reason, the data corresponding to the near sector have been removed from the data set.

5.2 Algorithm Implementation

As stated before, the nodes are added to the graph each time the platform travels a certain distance, which in this case is 40 m. The selected value provides sufficient discretization of the total distance travelled avoiding the complexity of a more dense graph guaranteeing enough resolution between minima. The binary edges $\mathbf{e}_{ij}$ model the constraints between two consecutive nodes $(\mathbf{x}_i, \mathbf{x}_j)$ with the relative position between them calculated using the odometry data: $\mathbf{z}_{ij} = (x_j^{odom} - x_i^{odom})$, $\boldsymbol{\Omega}_{ij} = [\epsilon]^{-1}$, where x^{odom} is the position estimated by the odometry and ϵ represents the uncertainty of the odometry with a value of 0.02.

If a minimum is detected at time T, the estimated position $\mathbf{x}_m$ of this minimum provided by the odometry (x_m^{odom}) is added as a new node in the pose-graph. The position reference of the minimum $(x_m^{RF_{map}})$ provided by the RF map is considered as the measurement $\mathbf{z}_m$ and it is included as a global information with a unary edge $\mathbf{e}_m$ associated to this new minimum node, being $\boldsymbol{\Omega}_m = [\delta]^{-1}$ the information matrix. δ corresponds to the uncertainty of the measurement and, due to the fact that the positions provided by the RF map closely represent the ground truth, it has a very low value (10^{-4}).

The strategy explained in Sect. 4.1 is used to introduce this delayed measurement into the graph. Each time a new node or measurement is added to the graph, the optimization process takes place. Even if nodes separation is large in the graph, our approach guarantees continuous robot localization by accumulating the odometry data to the last estimated robot position in the graph.

5.3 Results

Minima Detection. Figure 5 shows the results of the minima detection method. The number of points accumulated to create the real model corresponds to a distance D of 80 m. This value is selected based on the distance corresponding to the theoretical minimum model. The RSSI data provided by the RF receiver is represented related to the position estimated by the odometry and the RF signal model related to the ground truth (Fig. 5(a)). The results shows the ability of the proposed algorithm to identify the minima of the signal although the real signal waveform does not exactly match with the RF signal model due to the noisy nature of the real signal and the odometry errors. As can be seen in Fig. 5(b), two different values have been identified corresponding to the same RF map minimum (second and third). The mechanism explained in Sect. 4.1 is used to handle this situation.

Graph-Based Localization Results. Although the pose-graph generation and optimization take place online during the displacement of the vehicle, the results presented in this section correspond not only to the vehicle localization along the tunnel but also to the position correction after the service routine.

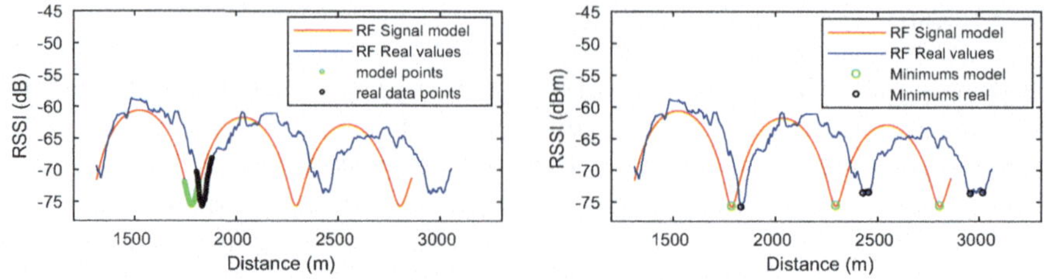

(a) Theoretical and real model (b) Minimums detected

Fig. 5. Results of the minima detection process

Figure 6(a) shows the initial graph with the odometry and the minimum nodes, and the resultant graph after the optimization process. It is clearly seen how the vehicle positions represented by the nodes are corrected after the optimization.

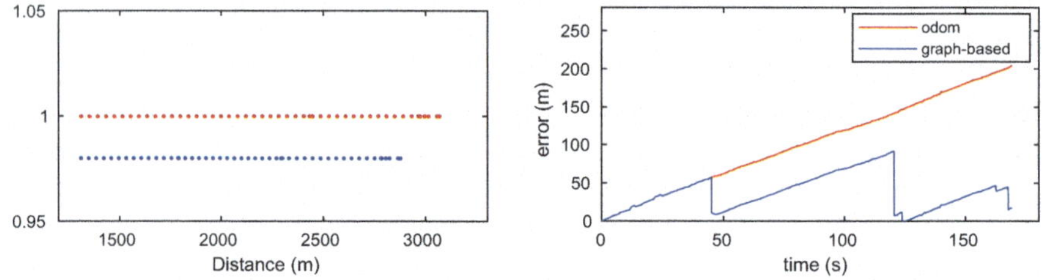

(a) Graph before and after optimization (b) Error pose

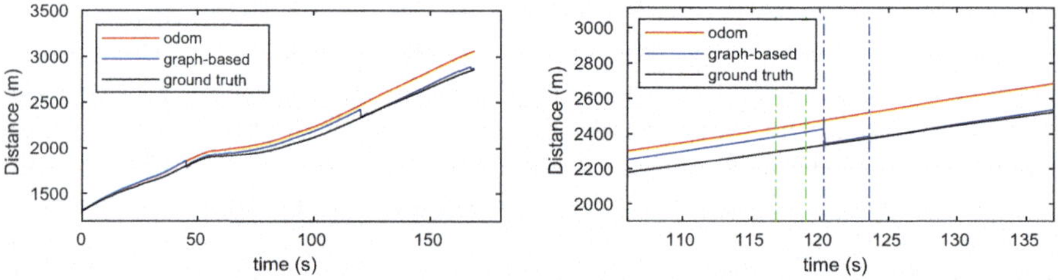

(c) Pose estimation (d) Minimum detection

Fig. 6. Results of the online pose-graph localization approach. (a) Node graph before (red) and after (blue) the graph-optimization process. The y axis values have been set only for visualization purposes to avoid overlapping of the trajectories. (b) Pose error during the displacement of the vehicle. (c) Estimated position along the tunnel provided by the odometry and our proposed method. (d) Detail corresponding to a detected minimum. The dashed blue line represents the minimum detection instant and the dashed green line the previous time when the minimum occurs

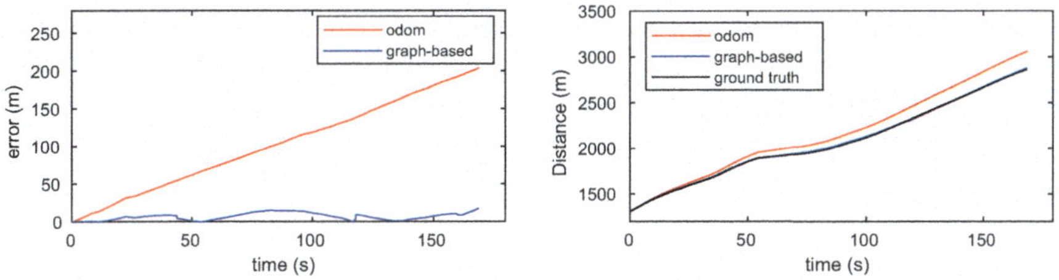

(a) Error pose (b) Pose estimation

Fig. 7. Results of the pose-graph approach after the service routine of the vehicle

The position error during the movement of the robot is shown in Fig. 6(b). Each time a minimum is detected, the position of the vehicle is corrected and therefore, the error is reset. As previously mentioned, the detection of the minimum is delayed with respect to the instant at which the minimum appears. The effect in the position correction can also be observed in Fig. 6(c) and in Fig. 6(d) in detail.

As stated before, one of the main benefits of the proposed approach is the ability to modify the location of some features observed during the route of the vehicle. Figure 7 shows the results when the position and the error are calculated once the tunnel has been traversed. The position error along the tunnel remains limited under acceptable values in comparison with the error using only the odometry which increases along the time as shown (Fig. 7(a)). The estimated position obtained through our proposed method follows closely the real position of the vehicle as can be seen in Fig. 7(b).

6 Conclusions

This paper have presented a graph-based localization approach for tunnel-like environments using two main sources of information: the odometry data and the absolute positions provided by an RF signal minima detector based on a theoretical fadings model that acts as an RF map. The feasibility of the proposed approach has been validated with the data collected during experiments developed in a real tunnel scenario.

The empirical results demonstrate the validity of the proposed minima detection method even when the RF actual signal and the RF signal model differs due mainly to odometry uncertainty and amplitude differences in the RSSI signal values. Additionally we prove the robustness of the method against scale differences in the signal. The results also show that the localization error is greatly reduced after the graph optimization. As a consequence, it is possible to locate features of interest observed during the inspection task more accurately.

Future work will be aimed at improving the continuous localization in the tunnel by incorporating into the graph additional sources of information such as galleries detection or the results of a scan matching process.

Acknowledgments. This work has been supported by the Spanish Project *"Robot navigation and deployment in challenging environments - Robochallenge"* (DPI2016-76676-R-AEI/FEDER-UE)

References

1. Bakambu, J.N.: Integrated autonomous system for exploration and navigation in underground mines. In: 2006 IEEE/RSJ International Conference on Intelligent Robots and Systems, pp. 2308–2313, October 2006
2. Zhen, W., Scherer, S.: Achieving robust localization in geometrically degenerated tunnels. In: Workshop on Challenges and Opportunities for Resilient Collective Intelligence in Subterranean Environments, June 2018
3. Rizzo, C., Lera, F., Villarroel, J.L.: UHF and SHF fading analysis using wavelets in tunnel environments. In: IEEE 78th Vehicular Technology Conference, pp. 1–6, September 2013
4. Rizzo, C., Lera, F., Villarroel, J.: Transversal fading analysis in straight tunnels at 2.4 GHz. In: 13th International Conference on ITS Telecommunications, pp. 313–318, November 2013
5. Seco, T., Rizzo, C., Espelosín, J., Villarroel, J.L.: Discrete robot localization in tunnels. In: ROBOT 2017: Third Iberian Robotics Conference, pp. 823–834. Springer (2018)
6. Imperoli, M., Potena, C., Nardi, D., Grisetti, G., Pretto, A.: An effective multi-cue positioning system for agricultural robotics. IEEE Robot. Autom. Lett. **3**(4), 3685–3692 (2018)
7. Rizzo, C., Lera, F., Villarroel, J.: A methodology for localization in tunnels based on periodic RF signal fadings. In: 2014 IEEE Military Communications Conference (MILCOM), pp. 317–324, October 2014
8. Dudley, D., Lienard, M., Mahmoud, S., Degauque, P.: Wireless propagation in tunnels. IEEE Antennas Propag. Mag. **49**(2), 11–26 (2007)
9. Rizzo, C., Lera, F., Villarroel, J.L.: 3-D fadings structure analysis in straight tunnels toward communication, localization, and navigation. IEEE Trans. Antennas Propag. **67**(9), 6123–6137 (2019)
10. Rizzo, C.: Propagation, localization and navigation in tunnel-like environments. Ph.D. thesis, University of Zaragoza, July 2015
11. Lazaro, M., Castellanos, J.: Localization of probabilistic robot formations in SLAM. In: IEEE International Conference on Robotics and Automation, pp. 3179–3184, May 2010
12. Quigley, M., Conley, K., Gerkey, B.P., Faust, J., Foote, T., Leibs, J., Wheeler, R., Ng, A.Y.: ROS: an open-source robot operating system. In: ICRA Workshop on Open Source Software (2009)

A RGBD-Based System for Real-Time Robotic Defects Detection on Sewer Networks

Luis Merino(✉), David Alejo, Simón Martinez-Rozas, and Fernando Caballero

Service Robotics Laboratory, Universidad Pablo de Olavide, Seville, Spain
{lmercab,daletei,fcaballero}@upo.es, s.martinez.r@ieee.org

Abstract. In this paper we summarize the automatic defect inspection onboard the sewer inspection ground platform SIAR. We include a general overview of the software and hardware characteristics of our platform, making a special emphasis on the sensing devices and software systems that are used for defect inspection. The main detection algorithm makes use of the a priori knowledge of ideal sections of the sewers that can be found in the Geographic Information Systems (GIS), and uses a variant of the Iterative Closest Point (ICP) algorithm for finding structural and serviceability defects. Then, we describe the software modules that are in charge of storing the alerts found by the detection system and of displaying them to the operator. The whole system has been tested in two field scenarios on different locations of the real sewer network of Barcelona, Spain.

Keywords: Sewer inspection · Defect detection · Field robotics

1 Introduction

Sewers represent a very important infrastructure of cities. The state of the sewer network has to be assessed in order to intervene if damages, blockages and other hazards are discovered. This is a labour intensive task. For instance, Barcelona has a 1532 km long network, in which 50% can be visited by operators. Furthermore, sewer inspections require many people to work in risky and unhealthy conditions. Sewers are classified as confined spaces which require special health and safety measures, in addition to other risks present like slippery sections, obstacles or biological risks from the potential contact with waste water.

Including a robotic solution for sewer inspection could thus provide many advantages. First, it would reduce labor risks, as it prevents operators to enter in such spaces, in which health hazards are common. Second, robots can acquire precise data and process it in real-time to perform a 3D-mapping of the environment or to automatically detect structural defects on the environment. Last, the use of such systems can reduce the costs involved in the inspection of sewers.

And thus, there are a wide variety of robotic platforms for sewer inspection already in the market, mainly for pipe inspection (like Alligator, Minigator,

M. F. Silva et al. (Eds.): ROBOT 2019, AISC 1092, pp. 593–605, 2020.
https://doi.org/10.1007/978-3-030-35990-4_48

Multigator and Flexigator wheeled robots from IBAK or Geolyn's tracked robots to name a few), but also for larger ones (like ServiceRoboter wheeled solution from Fraunhofer IFF). In this paper, we consider the SIAR robotic solution developed in the frame of the EU Project ECHORD++.

Most of the previous systems are manually operated and relay information for the operator, who performs the inspection. Providing automatic inspection capabilities can reduce inspection times.

Again, most of the available research has been designed for inspecting conventional sewer pipe networks. For example, the KARO system is aimed at analyzing the rings of light projected onto the internal pipe wall so as to detect pipe deformations and obstacles, and detecting leakage using microwave sensors [1]. The PIRAT system classifies defects using a feed-forward neural network classifier, which is trained off-line [2]. Unfortunately, these kind of experimental platforms usually lack of testing in real scenarios: their experimentation is limited to a laboratory replica of the real sewers or a very small part of the real sewer network.

There is also work on automatic problem detection which uses data collected by closed-circuit television (CCTV) systems, such as the early framework proposed in [3]. More recently, neural network and deep learning approaches are becoming more and more common. For example, an automated recognition process for infiltration defects in sewer pipes can be found in [4]. In [5] an optical flow based approach for CCTV camera motion analysis to automatically identify, locate and extract frames of inspection video which likely include defects is presented. Also, the work in [6] uses Hidden Markov models to generate consistent anomaly detection using CCTV. However, the inspection based on CCTV systems is not able to get metric measures to the detected anomalies, they have to infer them by taking into account the cross section of the pipe. Moreover, they lack the flexibility offered by a mobile platform.

To fill this gap, this paper presents a method that is able to automatically detect defects on the sewer network. To this end, our system uses range data available from multiple RGB-D sensors onboard of our SIAR robotic platform, which is presented in Sect. 2. Furthermore, it is able to distinguish two main classes of defects: structural defects and serviceability defects, as presented in Sect. 3. In addition, the tools developed to make this information available to the operators in real-time are presented in Sect. 4. Finally, we tested the system in the real sewer network of Barcelona (see Sect. 5.2). Section 6 analyzes the results of the system and lists the future challenges and research directions.

2 Overview of the SIAR Platform

A new ground robot, the SIAR platform[1], has been developed to tackle the requirements of the sewer inspection application [7]. In this section, we will briefly describe the robotic platform, paying special attention to the sensor payload that is used for the detection of defects on the sewer network.

[1] http://siar.idmind.pt/.

(a) (b) (c) (d)

Fig. 1. (a) Detail of the SIAR platform. (b–c) the robot being introduced in the manhole. (d) SIAR platform inside the sewer network.

(a) (b)

Fig. 2. (a) The robot carries seven RGB-D cameras. (b) These cameras provide high-resolution 3D point clouds of the environment, used for automatic defect detection.

The SIAR robot is a six-wheeled differential ground platform. The wheel traction system is composed by six sets of independent motors and 260 mm off-road wheels. Each of these sets has an independent suspension arm that connects to the central robot frame. Due to the specifications of the problem, the robot must be introduced through a manhole, which is noticeably smaller than the width of the tunnels (see Fig. 1b, c). Furthermore, the robot should negotiate sewer sections and gutters of different widths. For this reason, a width adjustment mechanism has been included so that the robot can adapt to these different structural changes in the working environment.

2.1 Sensor Payload for Structural Inspection

The robot is equipped with a set of navigation and environmental perception sensors. The main element of the sensor payload is a set of seven Orbbec Astra RGB-D cameras[2] that provides visual and depth information. The cameras are placed symmetrically as presented in Fig. 2a. There are three RGB-D cameras looking forward, three backwards and one upwards. For each direction, the robot has a camera parallel to the ground to detect damage in the tunnel and to illustrate the operator controlling the robot. Additionally, we disposed two cameras facing downwards and sideways to visualize closer range obstacles and defects.

[2] https://orbbec3d.com/product-astra-pro/.

The upward facing camera is placed over the center of the robot. This camera is used for detecting flaws in the tunnel dome and for identifying sewer elements such as manholes. Figure 2b represents a combined point cloud with the depth information of all cameras. Additionally, we placed a camera on the end-effector of a robotic arm with five degrees of freedom. This camera can be controlled by the operator to obtain closer looks of areas of interest.

For navigation, the robot also uses encoders to control the velocity of the traction motors and an encoder plus a potentiometer to control the position of the platform width actuator. The robot is also equipped with an inertial sensor. The information of these sensors and the RGB-D cameras is used for estimating the position of the robot according to GIS data, as presented in [8].

3 Serviceability and Structural Defects Inspection System

There are two types of problems that are considered by the current inspection system:

- Sewer serviceability inspection: determine when there is debris in the gutters and floor that may obstruct the sewer.
- Structural defects inspection: determine the presence of cracks, fractures, breaks and collapses in the structure of the sewers.

This system uses the 3D input data given by the onboard set of RGB-D cameras of the robot to generate, in real-time, serviceability and/or structural defects alarms that can be displayed in the Control Station for the operator. Furthermore, those alarms are time-stamped and geo-located (by considering the robot localization system), and a report is generated when the mission finishes for mission de-briefing and post-processing.

The processing pipeline of the system is summarized in Fig. 3. We describe the main modules of the system in the following:

1. **Automatic detection of sewer type:** a point cloud is generated by combining information from all cameras. This model is aligned with a database of 3D virtual models of the different section types stored in the robot, according to the drawings of the different sections present in the zone to inspect. The virtual point cloud which contain labels for the different parts of the sewer: gutter, curbs, walls and roof, see Fig. 4b. The alignment is carried out using the Iterative Closest Point (ICP) algorithm [9]. ICP is initialized by generating the model point clouds in the frame of the cameras, which approximately aligned with the gathered point cloud. The model with the lowest ICP error is selected as the current section type, Fig. 4a.
2. **Segmentation of sewer elements:** once the virtual model is aligned with the 3D data, the current point cloud is segmented into the different parts of the sewer; that is, each 3D point is classified as either gutter, curbs, walls or roof. Each point of the cloud is labeled according to the label of the closest point in the virtual 3D model of the sewer, which contains this information. Figure 5 shows the results of the segmentation.

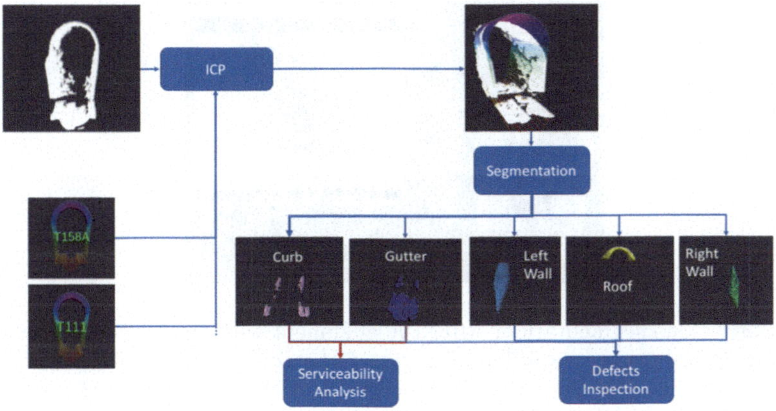

Fig. 3. Main processing pipeline. ICP algorithm is used to estimate (and align) the current section type by comparing the 3D data (in white) with virtual models from the database. Then, the parts related to serviceability (curb and gutter, sill, etc) are segmented from the input point cloud. These parts are analyzed to estimate potential serviceability and structural alarms.

Fig. 4. (a) Measures for Sections T111 and T158A. (b) Center and Right: Virtual ideal 3D models for section types T111 and T158A. Right: the current 3D data (white) is aligned using ICP to all the models in the database searching for the one that best fit the data.

3. **Serviceability inspection:** with the information of the two above processes, the serviceability analysis begins. Only the parts corresponding to the gutter, curbs and sill are considered in such analysis. Then, on one hand, the method extracts those points that separate from the model further than a minimum distance, as these points may indicate deviations from the ideal section model, and thus a potential serviceability problem. At the same time, the absolute maximum, minimum and mean height of the 3D points segmented as points of the gutter are computed. This way it is possible to estimate if the gutter could be blocked by comparing with the ideal minimum, maximum and mean height of the gutter. The same is carried out with the curb. As a result, potential alarms are generated (see Fig. 6). A temporal consistency filter is

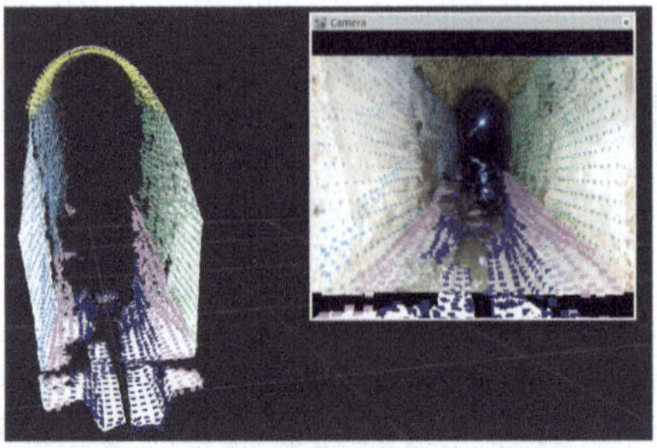

Fig. 5. Point-cloud segmentation. After the alignment with the section, the points are segmented according to the different parts of the sewer. Left: points segmented as gutter (purple), curb (pink), left wall (blue), right wall (green) and roof (yellow). The points projected back on the frontal camera of the robot can be also seen.

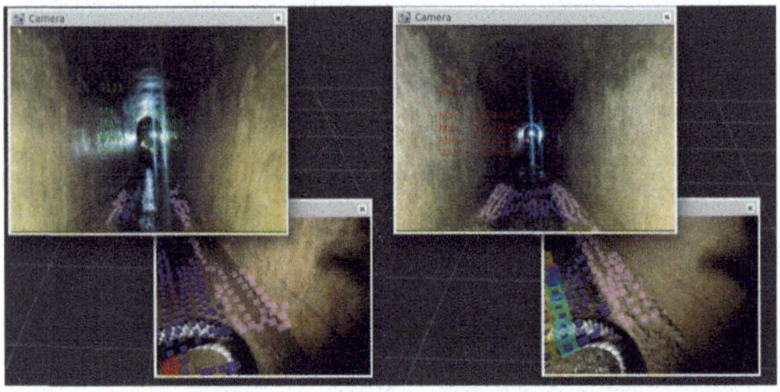

Fig. 6. Left: the serviceability of the gallery is correct. The frontal and down right cameras are shown (there is a down left camera that is not shown for clarity). In green, the absolute values for the height of the gutter are displayed in the frontal image for the operator. In purple and pink, the 3D points corresponding to the gutter and curb respectively. Right: serviceability alarm. The values are displayed in red. Many of the 3D points in the down-right camera (marked as colored points) are detected as departing from the ideal section (the gutter is blocked by debris).

employed to filter out false alarms due to brief misreadings or misalignment of the sensor data.

4. **Automatic structural defects inspection:** structural defects alarms can be raised by estimating the error between the ideal section model and the 3D data gathered by the sensor. This error is estimated by nearest neighbour search between each point in the cloud and the virtual model of the section. The points for which the error is above a threshold are candidates for potential defects. The threshold is user defined, and can be used to balance the size of the defects and the rate of false alarms, but the system is able to detect defects of the order of centimetres. The arm onboard the SIAR platform is equipped with a camera that can be used for close inspection in order to confirm the potential defects highlighted by the module.

4 Graphical User Interface

The information provided by the online Serviceability and Structural Defects Inspection System (see Sect. 3) can be displayed to the operator in real-time by using a custom made Graphical User Interface (GUI) at the base station. This GUI can operate in the following modes:

- The exploration mode allows the operator to have localization, real-time images, depth information and SIAR proceptive information (see Fig. 7). This mode is recommended for taking metric measurements.
- The inspection mode (see Fig. 8a) includes images from the inspection arm camera. It is recommended for navigation and for obtaining details of defects.
- In the mission execution mode the operator can easily get information about the alerts generated during the experiment (see Fig. 8b).

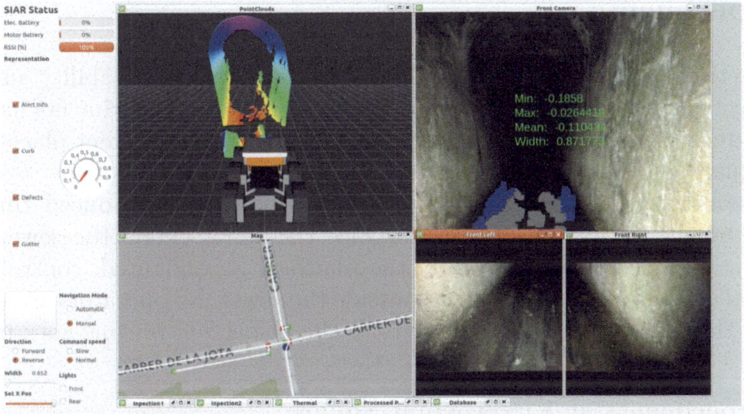

Fig. 7. Exploration view of the GUI. On the leftmost panel, proceptive information of the platform can be found. On the middle panel, we can see the depth cloud information obtained from the onboard cameras (top) and the localization of the platform with the GIS data (bottom). The rightmost panel shows the composed views of the RGB images from the onboard cameras. The serviceability information is overlayed in green.

The software has been developed as a C++ application for Ubuntu 16.04 and uses the Robotic Operating System (ROS) Kinetic. We make extensive use of the convenient RViz tool [10] for 3D representation and augmented reality purposes. We used the convenient rviz-satellite plugin[3] for GIS representation.

[3] https://github.com/gareth-cross/rviz_satellite.

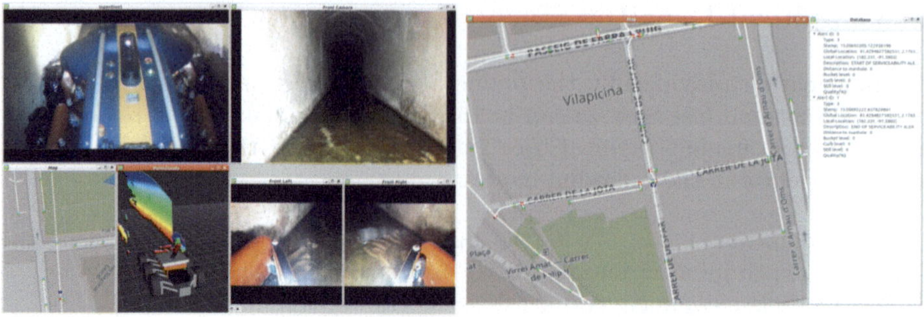

(a) (b)

Fig. 8. (a) Inspection view of the GUI. There is an additional image shown from the camera on the arm (up-left). (b) Mission execution view of the GUI. The generated alerts are shown in the rightmost panel and displayed over the GIS.

5 Experiments

In this section, we present the results obtained from the serviceability and structural defects inspection system in two experiments. These experiments have been carried out in the real sewer network of Barcelona in two different places: Virrei Amat Square and Passeig Garcia Faria.

To perform each experiment, the platform must be introduced through a manhole by the aid of the operators. Once the robot enters the sewer and is left in a stable position to navigate, the platform is configured, controlled and monitored by the base station that is outside the sewer. No operators are required to follow the platform inside the sewers as the inspection routine goes by.

5.1 Experiments at Virrei Amat Square

In this experiment, more than 500 m of sewer galleries were inspected (see Fig. 9). We deployed three repeaters in different manholes along to ensure network connectivity between the SIAR platform and the base station.

During the inspection, four serviceability defects were automatically detected and localized by the platform. In particular, our algorithm detected two main zones where there are very noticeable defects where human intervention would be needed for cleaning purposes.

We will now detail the contents of the four automatically generated defects. For each defect, the detection system generated two alerts: one at the beginning of the problem and another when the serviceability is restored. For each alert, we provide the GPS coordinates, time and the distance to the closest manhole.
Serviceability Defect 1: Alerts 1 and 2. Figure 10a shows a composition of the three images from the frontal cameras. In this case, there are noticeable sediments on the gutter and in the leftmost curb.

Fig. 9. Inspection report of the experiments at Virrei Amat square. The path followed by the SIAR platform is marked in a red line. The generated alerts by the system are shown in red exclamation marks.

- **Begins:** 41.429 959°N, 2.176 372°E. Local Time: 2018-07-04-10:03:07. Distance to closest manhole: 17.2 m. Closest manhole: MH 30.
- **Ends:** 41.429 964°N, 2.176 373°E. Local Time: 2018-07-04-10:03:12. Distance to closest manhole: 17.6 m. Closest manhole: MH 30.

Serviceability Defect 2: Alerts 3 and 4. Figure 10b presents a composition of the three images from the frontal cameras. In this case, sediments are accumulated in the gutter and they prevent the water from flowing.

- **Begins:** 41.429 397°N, 2.176 346°E. Local Time: 2018-07-04-10:32:44. Distance to closest manhole: 10.5 m. Closest manhole: MH 31.
- **Ends:** 41.429 392°N, 2.176 345°E. Local Time: 2018-07-04-10:32:48. Distance to closest manhole: 10.8 m. Closest manhole: MH 31.

Serviceability Defect 3: Alerts 5 and 6. Figure 10c presents a composition of the three images from the frontal cameras. Sediments are accumulated in the gutter and prevent the water from flowing. The accumulated water at the end of the serviceability defect can be found in the video[4].

- **Begins:** 41.429 374°N, 2.176 344°E. Local Time: 04-07-2018-10:33:02. Distance to closest manhole: 12.8 m. Closest manhole: MH 31.

[4] https://robotics.upo.es/papers/robot_2019_alert.mp4.

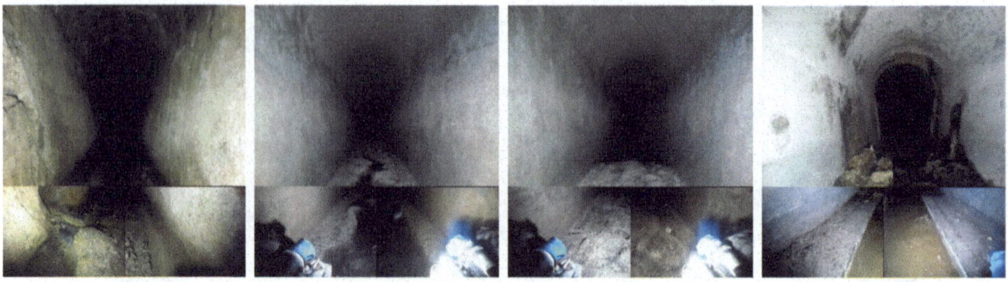

(a) (b) (c) (d)

Fig. 10. (a–d): Composed snapshot in Alerts 1–4 in Virrei Amat, respectively.

Fig. 11. Inspection report of the experiments at Passeig de Garcia Faria. The localization of the SIAR platform is plotted in a red line. The alerts generated by the system are marked in red exclamation marks.

- **Ends:** 41.429 333°N, 2.176 344°E. Local Time: 2018-07-04-10:33:22. Distance to closest manhole: 16.7 m. Closest manhole: MH 31.

Serviceability Defect 4: Alerts 7 and 8. Figure 10d presents a composition of the three images from the rear cameras. We found a noticeable sediment over the rightmost curb and some sediments at the gutter.

- **Begins:** 41.429 209°N, 2.175 506°E. Local Time: 2018-07-04-11:56:21. Distance to closest manhole: 5.3 m. Closest manhole: MH 35.
- **Ends:** 41.429 097°N, 2.175 630°E. Local Time: 2018-07-04-11:58:07. Distance to closest manhole: 9.4 m. Closest manhole: MH 37.

5.2 Experiments at Passeig de Garcia Faria - Barcelona

The SIAR platform inspected a 200 m long straight-line sewer during this experiment. Next we detail the defects found by the proposed system. In particular, it found three defects and one structural element that are marked in Fig. 11:

Serviceability Defect 1: Alerts 1 and 2. The system detected an anomaly in the serviceability in the first part of the inspected area. In this case, the water covered the complete surface of the floor and thus the anomaly was reported. The obtained data indicates an anomaly in the gutter levels as shown in Fig. 12a.

- **Begins:** 41.407 281°N, 2.218 564°E. Local Time: 2018-12-13-11:57:47. Distance to closest manhole: 7.4 m. Closest manhole: MH 161.

- **Ends:** 41.407 144°N, 2.218 369°E. Local Time: 2018-12-13-12:02:47. Distance to closest manhole: 19.3 m. Closest manhole: MH 102.

Structural Defect 1: Alert 3. Structural defect 1. Figure 12b and b.1 show the visual and 3D representations, respectively. In this case, a noticeable obstacle of 14.4 cm can be found in the rightmost curb.

- Location: 41.407 116°N, 2.218 330°E. Local Time: 2018-12-13-12:08:55. Distance to closest manhole: 12.2 m. Closest manhole: MH 102.

Structural Element 1: Alert 4. The system detected an inlet in one of the sides of the gallery. Figure 12c and c.1 show visual and 3D representations of the surroundings of the inlet, respectively. We estimated its height on 19.1 cm, which can be of use for determining if it has been built according to the laws.

- Location: 41.406 434°N, 2.217 364°E. Local Time: 2018-12-13-12:08:55. Distance to closest manhole: 1.1 m. Closest manhole: MH 105.

Structural Defect 2: Alert 5. Figure 12d and d.1 show visual and 3D representations, respectively. We observe a 7.2 cm tall inlet with its sediments.

- Location: 41.406 327°N, 2.217 217°E. Local Time: 2018-12-13-12:20:01. Distance to closest manhole: 8.9 m. Closest manhole: MH 109.

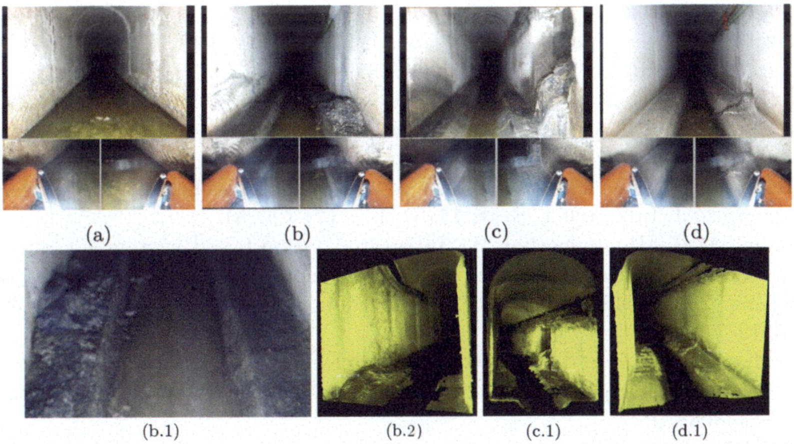

Fig. 12. (a) Composed snapshot in Serviceability Defect 1 of the experiment in Passeig Garcia Faria. (b) Composed snapshot in Structural Defect 1, with an obstacle in the rightmost curb. (b.1) Detail of the obstacle obtained with the arm camera (b.2) Obtained 3D mesh in Structural Defect 1 (the obstacle is on the left). (c) Composed snapshot in Structural Element 1. There are some sediments under an inlet on the rightmost curb. (c.1) 3D mesh in Structural Element 1. (d) Composed snapshot in Structural Defect 1. It is on the rightmost curb and is composed by sediments that come from an inlet. (d.1) 3D mesh in Structural Defect 2.

6 Conclusion

In this paper we have presented a system for automatic defect report generation on the sewer networks. This detection is performed online, and the obtained defects are shown to the operator immediately, which allows to allocate more resources to validate the defect or just gathering more information of interest. The complete system includes the defect detection system that fuses the information of up to 7 RGB-D cameras and a GUI for representing the report in real-time. The system has been demonstrated during the field experiments on the real sewer networks of Barcelona.

The system currently detects automatically potential alarms that are then confirmed by the operator. Future work includes the automatic classification of the detected anomalies and defects as cracks, collapses, structural elements, etc, by using learning approaches. The information of the structural elements can be included as features in the localization system in order to increase the precision of both the localization of the alerts and of the localization of the platform.

References

1. Kuntze, H., Haffner, H.: Experiences with the development of a robot for smart multisensoric pipe inspection. In: IEEE International Conference on Robotics and Automation, vol. 2, pp. 1773–1778 (1998). https://doi.org/10.1109/ROBOT.1998.677423
2. Kirkham, R., Kearney, P.D., Rogers, K.J., Mashford, J.: PIRAT - a system for quantitative sewer pipe assessment. Int. J. Robot. Res. **19**(11), 1033–1053 (2000). https://doi.org/10.1177/02783640022067959
3. Moselhi, O., Shehab, T.: Automated detection of surface defects in water and sewer pipes. Autom. Constr. **8**(5), 581–588 (1999). https://doi.org/10.1016/S0926-5805(99)00007-2
4. Shehab, T., Moselhi, O.: Automated detection and classification of infiltration in sewer pipes. J. Infrastruct. Syst. **11**(3), 165–171 (2005)
5. Halfawy, M.R., Hengmeechai, J.: Integrated vision-based system for automated defect detection in sewer closed circuit television inspection videos. J. Comput. Civ. Eng. **29**(1), 04014024 (2015). https://doi.org/10.1061/(ASCE)CP.1943-5487.0000312
6. Moradi, S. and Zayed, T.: Real-time defect detection in sewer closed circuit television inspection videos, pp. 295-307 (2017). https://doi.org/10.1061/9780784480885.027
7. ECHORD++: Utility infrastructures and condition monitoring for sewer network. Robots for the inspection and the clearance of the sewer network in cities (2014). http://echord.eu/public/wp-content/uploads/2015/11/20141218_Challenge-Brief_Urban_Robotics.pdf
8. Alejo, D., Caballero, F. and Merino, L.: RGBD-based robot localization in sewer networks. In: IEEE/RSJ International Conference on Intelligent Robots and Systems (IROS), pp. 4070–4076, September 2017. https://doi.org/10.1109/IROS.2017.8206263

9. Besl, P.J., McKay, N.D.: A method for registration of 3-D shapes. IEEE Trans. Pattern Anal. Mach. Intell. **14**(2), 239–256 (1992). https://doi.org/10.1109/34.121791
10. Kam, H., Lee, S.-H., Park, T., Kim, C.-H.: RViz: a toolkit for real domain data visualization. Telecommun. Syst. **60**, 337–345 (2015)

Detecting Indoor Smoldering Fires with a Mobile Robot

Carolina Soares da Conceição[1], João Macedo[1,2](✉), and Lino Marques[1]

[1] Institute of Systems and Robotics, University of Coimbra, 3030-290 Coimbra, Portugal
engcarolinasoares@gmail.com, {jmacedo,lino}@isr.uc.pt
[2] Centre for Informatics and Systems, University of Coimbra, 3030-290 Coimbra, Portugal

Abstract. This paper presents a mobile robot to detect indoor smoldering fires. The robot uses a custom multisensory system able to measure a set of environmental parameters including CO, CO2, NO2, O3 and airborne particles. The paper proposes a sensory fusion method to detect fires with this system and evaluates that method in realistic environments. The obtained results show that the proposed method is able to detect fires which are undetected by a commercial system and that it is able to discriminate between burning materials.

Keywords: Fire detection · Mobile robot · Random forest · Sensory fusion

1 Introduction

Urban fires cause enormous financial losses, environmental impacts and deaths [1]. The financial losses are related to material damages and operational interruptions, while environmental impacts are associated with the gases from the combustion, the water used in the combat and the waste from the burned materials. The loss of human lives is often caused by smoke inhalation, with the victims being both the occupants of the building or fire and rescue team members. Tragic events of building fires still occur nowadays, even with the advances in fire safety standards regarding automatic fire detection systems. The most commonly used detector in these systems is the smoke sensor (ionization or optical technology). The majority of the systems are only able to identify a single combustion product and thus are unable to discriminate between actual fires or the presence of aerosols or non-fire particles, generating high rates of false alarms and consequently reducing the occupant confidence in the system [2]. In recent years, there has been a growing interest in developing new equipment and methods to promote a more effective detection. These studies seek technological alternatives capable of identifying, alerting and fighting fires more effectively. One of the proposed solutions is the implementation of autonomous robots to actively patrol indoor spaces to search for and declare the presence of fire based on various techniques of sensory perception [3]. Moreover, researchers have also been interested in developing approaches for autonomous multi-robot exploration and fire detection in large warehouse environments [4].

M. F. Silva et al. (Eds.): ROBOT 2019, AISC 1092, pp. 606–616, 2020.
https://doi.org/10.1007/978-3-030-35990-4_49

The application of the technologies with two or more fire detection criteria is important to avoid false alarms. Scorsone et al. [2] reports the development of an electronic nose based on a set of eight conductive polymer sensors for fire detection. The device may be used, for example, in an aspiration detection system where the air in a given environment is continuously aspirated and sent to a fire detection panel. Luo and Su [5] propose an adaptive fusion algorithm for fire detection, using a smoke sensor, flame sensor, and temperature sensor to detect a fire. Sucuoglo et al. [6] designed, manufactured and tested a mobile robot platform equipped with early fire detection unit to search through prescribed paths to detect fire events.

Another important limitation of traditional fire detection systems is related with the time required for detecting the combustion products. The positioning of the sensors is static at the ceiling, and the detection of fires is done passively. Unlike a fixed fire detection system, a robot can move inside a building and search for abnormal situations. The robot may act autonomously or be remotely controlled by an operator [7]. Many of the existing robots for dealing with fires are tele-operated. While they already are a step in the right direction, as they allow the operator to be at a safe distance [8], there are limitations regarding wireless communications and the effectiveness in the simultaneous control of more than one robot. For those reasons, efforts should be made to increase the autonomy of the robots. Cabrita et al. [9] developed a prototype of a mobile robot that could perform programmed indoor patrolling through predefined waypoints while continuously obtaining environment parameters (e.g., CO2, CO, NOx, SO2, VOCs). This also has the advanced of not needing an operator to be constantly controlling the robot.

This paper reports a method for detecting initial signs of fires which are undetected by commercial systems. This is done by using a mobile robot that patrols office-like environments. It uses a SpreadNose [10] module to perceive the air conditions and feeds that information to a Machine Learning Classifier, which shall decide whether there is a fire and, ideally, what material is burning.

This paper is organized as follows: Sect. 2 presents the materials and methods used in this work; Sect. 3 presents the experimental results, both training the system and validating it in the real world; finally, Sect. 4 presents the conclusions drawn and gives suggestions for future endeavors.

2 Materials and Methods

This section presents the materials used to develop and test the proposed fire detection system, composed by a mobile robot equipped with a sensor module for perceiving the air quality. It moves on to describe the methods used to create knowledge from the information of the air conditions and, finally, presents the test environment.

2.1 Mobile Robot

This work used an Erratic mobile robot (manufactured by Videre Design LLC) used as patrolling agent. This robot, shown in Fig. 1(a), is a differential-driven unit with two drive wheels mounted on a central axis. A caster wheel is mounted at the rear of the robot, ensuring its stability. The robot is equipped with an Asus Eee PC for onboard computation, a Hokuyo UBG-04LX-F01 laser range finder (LRF) to support obstacle avoidance and Simultaneous Localization And Mapping (SLAM), and a SpreadNose sensor module (Fig. 1b), which provides the information regarding various airborne substances. This sensor module is further described in Sect. 2.2. Due to the limited capabilities of the onboard computer, a distributed system was devised. In this system, the onboard laptop collects the information from the sensors and sends it to a remote server, which performs all computations related with SLAM and classification of the sampled air. This system is based on the Robot Operating System (ROS) [11], which serves as the backbone for distributing the computation, and provides the packages for interfacing with the robot, the LRF and for performing SLAM. Table 1 presents the versions of the software used in this work.

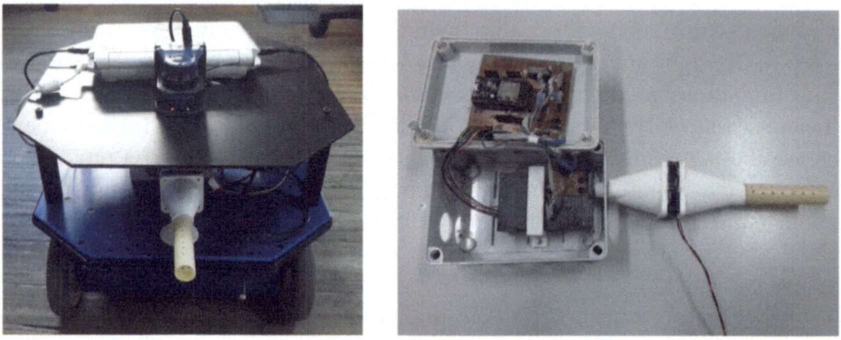

(a) (b)

Fig. 1. (Left) Videre Erratic robot equipped with a Hokuyo Laser Range Finder, an Asus Eee PC and a modified version of a SpreadNose sensor module. (Right) Open view of the modified SpreadNose module featuring an external fan.

Table 1. Version specification of relevant software tools.

System	Software tool
Server	Ubuntu 16.04
	ROS Kinetic
	RVIz 1.12.17
Onboard PC	Ubuntu 12.04
	Groovy 1.9.55

2.2 SpreadNose

The SpreadNose module [10] consists of a set of sensors devised for monitoring the air quality in outdoor urban spaces. Each module uses various sensors for providing information regarding the concentrations of various airborne substances (Table 2). In this work, a sensor module is used to equip the robot, transferring the information to its onboard computer through an USB interface. As SpreadNose was initially intended to monitor urban pollution, it was necessary to improve its ability for capturing small concentrations of smoke. This was done by fitting an add-on to the sensor, which contains a fan to improve its aspiration ability. The resulting module is depicted in Fig. 1.

Table 2. Technical information of the sensors in the SpreadNose module.

Environmental variable	Sensor	Range	Technology
CO	Sensortech MiCS 4514	1–100 ppm	Metal oxide semiconductor
NO2		0.05–10 ppm	
CO2	Amphenol Telaire T6713	0–5000 ppm	Non-dispersive infrared
O3	Sensortech MiCS 2614	10–1000 ppb	Metal oxide semiconductor
PM1	Plantower PMS5003	0.3–1.0 μm	Laser scattering principle
PM2,5		1.0–2.5 μm	
PM10		2.5–10.0 μm	
Temperature	Sensirion SHTC3	−40–125 °C	Bandgap temperature sensor
Relative humidity		0–100%	Capacitive humidity sensor

2.3 Classification

According to Wald [12], data fusion can be defined as a formal structure in which means and tools are expressed for combining data from different sources. It is intended to obtain higher quality information, where the exact definition of "higher quality" will depend on the application. In this work, information provided by various sensors is used as input to a machine learning classifier that decides whether the robot is in the presence of a smoldering fire. The goal of this classifier is twofold: firstly, it aims to be able to distinguish between smoldering fire and non-fire situations to generate alarms prior to what a commercial alarm can, whilst reducing the amount of false alarms; secondly, it aims to classify the type of burning material, to enable the response team to identify the fire source more easily. Paper and electrical cables are used to create the fires, as they are two common causes of fires in office-like environments.

Prior to any classification, data must be collected. This is done by using the mobile robot to sample the air in diverse situations and, for each situation a dataset was built containing the same amount of randomly drawn samples. The information collected is used as input to a Random Forest classifier [13]. The choice of classifier was based on empirical experimentation, during which the Random Forests showed to adequately and quickly perform the intended task. A Random Forest Classifier consists of an ensemble of Decision Trees. Each tree is fit to a subset of the training data and provides an output for each input sample. The output of the Random Forest Classifier is the mode of the classes outputted by its Decision Trees. Thus, Random Forests may be considered to be white box classifiers, as for each input sample it is possible to inspect the decision path along the trees that compose each classifier.

2.4 Test Environment and Fire Materials

The proposed approach was tested in a laboratory environment, where a commercial fire detection system was mounted to serve as baseline for comparing the proposed approach (Fig. 2a). This is a class A system, manufactured by GE Security, model 2X-F1. Two optical DP2061N sensors are connected to the control panel, one being mounted 2.3 m off the ground and another at the same height as the robot. The system was set up using the parameters recommended by the manufacturer. Its output is qualitative, simply generating an alarm once the output of one of its sensors goes beyond the predefined threshold. A goal of the present work is to detect a smoldering fire prior to the commercial system and, ideally, correctly classifying the combustible material. In order to give an advantage to the commercial system, the fire is ignited directly beneath its sensor, in a metal cup placed over the X mark (3 on Fig. 2b).

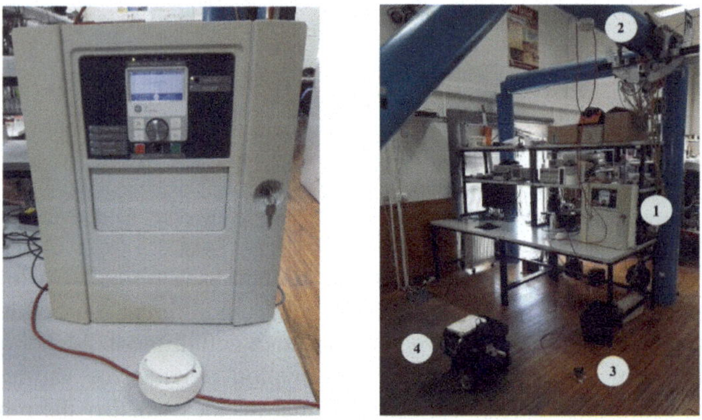

(a) (b)

Fig. 2. (Left) Fire Detection Panel 2X-F1 Ge Security and optical detector DP2061N. (Right) View of the experimental environment. A commercial fire alarm system is placed on top of the desk (1), connected to a sensor that is mounted on the blue structure, 2.3 m off the ground (2). The fires are ignited in the metal cup, directly beneath the sensor and on top of the X mark (3). The robot patrols the environment (4).

According to data provided by the Government of the United Kingdom [1], in the 2017/2018 financial year 11.141 urban fires took place, 57% of which were caused by electrical malfunctions. Moreover, in these environments there are usually significant amounts of paper material, which are a prime candidate for fueling fires. For those reasons, the proposed approach is tested with fires created by short-circuiting sections of a PVC isolated electrical wire with 0.4 mm^2 of section and approximately 10 cm length and by firing small paper strips with approximately 8 cm in length.

Mapping

In most robotic experiments, the robot must be able to localize itself in the environment. This work is not an exception, as the robot must be able to efficiently patrol the environment and properly signal the fires detected. The map of the test environment was not available *a priori* and, as it is a laboratory, the objects within it change place frequently. For those reasons, it was chosen to integrate Simultaneous Localization and Mapping (SLAM) in the robot's navigation, for which open-source ROS packages (e.g. gmapping) were used. The resulting map is shown on Fig. 3. Using RViz, a graphical interface contained in ROS, it is possible to manually insert waypoints in the map for defining the patrol routes.

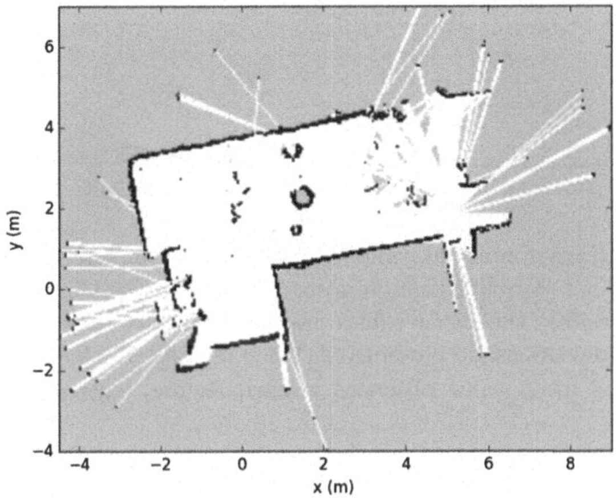

Fig. 3. Map of the laboratory created by gmapping.

3 Experimental Results

This section presents the experiments made for implementing and testing the proposed approach. It starts by presenting the creation of the Machine Learning model for detecting fires and classifying their source. Afterwards, the real-world validation is presented.

3.1 Model Training

The first step of this work consisted on using the SpreadNose module to collect measurements of the normal conditions of the test environment over the course of several days (NSL on Fig. 4). Afterwards, the module was used to collect measurements from 0.25 m to 2 m away from paper (FP on Fig. 4) and burning electrical wires (FW on Fig. 4), each experiment lasting for approximately 7 min. Analyzing the collected data, it was possible to verify that there were significant variations in the values of the various features from day to day. For that reason, it was necessary to find a way to assess only the difference between the different scenarios, rather than the absolute values of the features. A normalization procedure was employed, which consists of gathering data of a non-contaminated environment and computing the mean value of each feature. Afterwards, the data of the fire scenarios is collected, and each sample is normalized using the mean value computed for the non-contaminated environment. The normalized mean values of each environmental variable in the three situations are presented in Fig. 4, for both the training (Fig. 4a) and validation data (Fig. 4b).

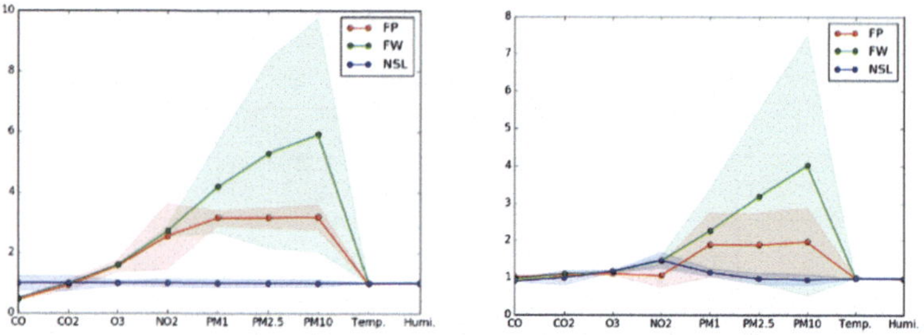

Fig. 4. Values of each environmental variable for each class (i.e., Fire with Paper (FP), Fire with electrical Wire (FW) and Normal Situation in the Laboratory, (NSL)) in the training data (left) and validation data (right). The mean values are presented as circles and connected by lines, whereas the standard deviations are presented as faded regions. The values presented in this plot are normalized by the mean value observed in each feature, when there are no fires in the environment.

Initial empirical tests showed that Random Forests are able to adequately distinguish samples collected from a normal environment, i.e. a non-fire situation, from samples collected in the presence of various types of burning substances. Having set on the classifier to use, a representative dataset was built for each situation (i.e., baseline environment, burning paper, and burning electrical wire). All datasets have the same size, and the samples are drawn randomly. As the goal of this work is to detect fires at an early stage, it is not likely that the air temperature and relative humidity are affected by it in a meaningful manner. For that reason, these two environmental variables are not used in the remaining experiments.

Table 3 presents the accuracy of Random Forests when classifying samples of various situations using as input the signal of each sensor separately, as well as the signals of all sensors. These values were obtained from 50 independent experiments, using different partitions of the collected datasets. In this work, Python 2.7.12 and `scikit-learn` version 0.18rc2 were used to implement the Random Forest Classifier, and the default parameters for this classifier were employed.

As can be seen from Table 3, it was possible to achieve very good accuracy values using any feature independently. However, using the 7 features available improves the performance of the system, producing a perfect mean accuracy value. It should be noted that, to reduce the minimum of cross-contamination, the data of each scenario was collected quite far apart. So, while the accuracy values may seem too good, it is possible that the classifier is distinguishing not only between the three scenarios, but also between time of day. This hypothesis is supported by the fact that the train data and validation data were collected in distinct days and their plots (Fig. 4) are quite different. Also, the datasets related to fire situations are likely to contain instances where the robot is sensing a clean environment but are classified differently than a clean environment.

Table 3. Accuracy of the random forest classifier using each sensor separately and together.

Feature	Min. score	Max. score	Mean score	Std. dev. score
CO	0.905	0.933	0.921	0.006
CO2	0.725	0.766	0.749	0.009
O3	0.974	0.986	0.980	0.003
NO2	0.945	0.961	0.951	0.004
PM1	0.943	0.964	0.956	0.004
PM2.5	0.961	0.976	0.969	0.003
PM10	0.967	0.976	0.969	0.003
All features	0.998	1.00	1.00	0.000

A Mann-Whitney test was used to perform pairwise comparisons of the accuracy values obtained using the different inputs. At a 95% confidence interval, there are statistically significant differences between the accuracies obtained with most inputs. The exceptions are when comparing the performance obtained using only PM10 to that of only using CO2, PM1 or PM2.5. In all other situations, this test showed that using different inputs lead to significantly different performances and, proved that using all features results in the best accuracy. For that reason, all 7 features shall be used in the remaining experiments.

3.2 Real-World Validation

In real world validation, the robot was controlled by a remote control connected to the ROS Master and the commands were transmitted through a WiFi network to the robot. Initially, three rounds were conducted in the environment without smoke interference to evaluate and record the values under normal conditions (Fig. 5a). Subsequently, smoldering paper fire was produced with the use of a lighter and data recorded (Fig. 5b). Finally, a power supply was used to generate 10 A through a thin electrical wire isolated with PVC. This current was sufficient to burn the wire isolation and generate some smoke. Figure 5 presents the trajectories followed by the robot in each situation as well as the concentration of airborne particles sensed.

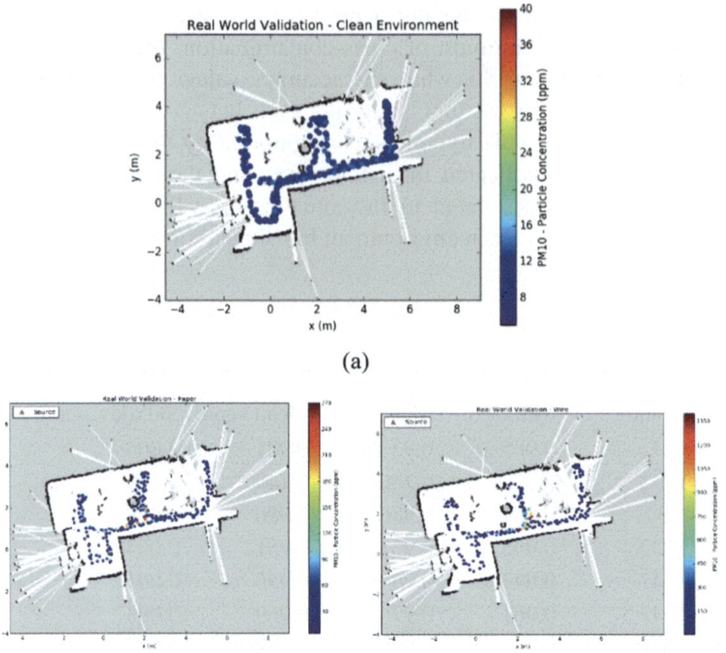

(a)

(b) (c)

Fig. 5. Particle concentration (PM10) measurement chart of the three scenarios. (A) Normal environment. (B) Paper burning. (C) Electrical wire burning.

The data collected from the validation process was fed into the previously trained classifier for deciding whether there were fires at each moment. In the scenario with no fires, the system behaved flawlessly, generating no false alarms.

In the scenario where paper was burnt, the system considered that 46 out of the 341 (i.e. 13%) of the data points corresponded to fires, more specifically, to paper-fueled fires. The first detection took place 3 min and 12 s after the experiment began, approximately 0.89 m away from the source. Conversely, the last detection took place 3.29 m away from the source, at the end of the experiment. This indicates that during the course of the experiment, the environment was being continuously filled with products resulting from burning paper, making it easier for the robot to detect the fire from farther away.

Finally, in the scenario with electrical fires, the system considered that 142 out of 318 (44.65%) data points correspond to fire situations, but only 79 of which were considered as electric-type fires. The first detection took place after 2 min and 41 s, approximately 0.59 m away from the source.

In the two fire-related environments, it is to be expected that some samples are classified as no-fire situations, as the range of the SpreadNose sensor is limited. The misclassifications between fire types are likely to be due to the classifier having learned to distinguish between more than the three scenarios, but also between the air conditions exhibited during the collection of each training dataset. Looking further into the data, it seems that these misclassifications are related to the concentrations of the airborne particles, being the samples with higher concentrations classified as electrical-generated fires, which is consistent with the plots exhibited in Fig. 4.

During the validation process, the commercial fire detection system was connected, but yielded no alarms. As a result, the proposed system can be considered to be a significant improvement over commercial units, generating no false alarms and detecting fires that would have gone unnoticed by conventional detection systems for much longer times. Moreover, it was somewhat possible to identify the material being burnt.

4 Conclusions and Future Work

This work presented a method for detecting fires at an early phase using a mobile robot. The mobile robot was equipped with a SpreadNose module, i.e., a sensor module designed for assessing air quality, and was used to patrol an indoor environment. The real world validation showed that the proposed method was able to detect fires that go undetected by a commercial system, whilst generating no false alarms. Moreover, the proposed approach is able to accurately identify when a fire is being fueled by burning paper, whilst still making some errors classifying electrical-generated fires.

In the future, further efforts should be made to improve the classification ability between fire types. This could be done by collecting more data in different conditions (e.g. temperature, relative humidity and pollutants). Another possible improvement can be the addition of other sensors to the robot both for detecting different chemical compounds or for providing visual abilities (e.g. thermal cameras). The robot could also learn the common patterns in each environment to improve the classification performance. For instance, a scent of burning metal could be classified as a fire in a clean room but the same should not be true in a workshop. Moreover, the robot could be modified to further assess the toxicity of the environment, by analyzing the synergistic effects of other gases such as deficient levels of O2 and the increase of HCl.

Acknowledgements. This work has been supported by OE - national funds of FCT/MCTES (PIDDAC) under project UID/EEA/00048/2019. J. Macedo also acknowledges the Portuguese Foundation for Science and Technology for Ph.D. studentship SFRH/BD/129673/2017.

References

1. Home Office: Fire statistics. GOV.UK Homepage. https://www.gov.uk/government/collections/fire-statistics. Accessed 07 Feb 2019
2. Scorsone, E., Pisanelli, A.M., Persaud, K.C.: Development of an electronic nose for fire detection. Sens. Actuators B: Chem. **116**, 55–61 (2006)
3. ur Rehman, A., Necsulescu, D.S., Sasiadek, J.: Robotic based fire detection in smart manufacturing facilities. IFAC-PapersOnLine **48**(3), 1640–1645 (2015)
4. Marjovi, A., et al.: Multi-robot exploration and fire searching. In: 2009 IEEE/RSJ International Conference on Intelligent Robots and Systems. IEEE (2009)

5. Luo, R.C., Su, K.L.: Autonomous fire-detection system using adaptive sensory fusion for intelligent security robot. IEEE/ASME Trans. Mechatron. **12**(3), 274–281 (2007)
6. Sucuoglu, H.S., Bogrekci, I., Demircioglu, P.: Development of mobile robot with sensor fusion fire detection unit. IFAC-PapersOnLine **51**(30), 430–435 (2018)
7. Hong, J.H., Min, B.C., Taylor, J.M., Raskin, V., Matson, E.T.: NL-based communication with firefighting robots. In: IEEE International Conference on Systems, Man, and Cybernetics, pp. 1461–1466 (2012)
8. Chang, P.H., Kang, Y.H., Cho, G.R., Kim, J.H., Jin, M., Lee, J., Kim, Y.B.: Control architecture design for a fire searching robot using task oriented design methodology. In: SICE-ICASE International Joint Conference, pp. 3126–3131 (2006)
9. Cabrita, G., Sousa, P., Marques, L., Almeida, A.T.: Infrastructure monitoring with multi-robot teams. In: Proceedings of the 2010 IEEE/RSJ International Conference on Intelligent Robots and Systems, IROS (2010)
10. Palhinha, J.P.P.: SpreadNose Sistema de Monitorização de Poluição em Ambientes Urbanos. Universidade de Coimbra (2018)
11. Quigley, M., Gerkey, B., Smart, W.D.: Programming Robots with ROS: A Practical Introduction to the Robot Operating System, 1st edn. O'Reilly Media Inc., California (2015)
12. Wald, L.: Definitions and terms of reference in data fusion. In: Joint EARSeL/ISPRS Workshop, Fusion of Sensor Data, Knowledge Sources and Algorithms for Extraction and Classification of Topographic Objects, vol. 32, pp. 2–6 (2015)
13. Breiman, L.: Random Forests - Random Features. Technical report 567, Statistics Department, University of California, Berkeley (1999). ftp://stat.berkeley.edu/pub/users/breiman

Future Industrial Robotics

Perception of Entangled Tubes for Automated Bin Picking

Gonçalo Leão[1(✉)], Carlos M. Costa[1,2], Armando Sousa[1,2], and Germano Veiga[1,2]

[1] FEUP - Faculty of Engineering, University of Porto, Porto, Portugal
{goncalo.leao,asousa}@fe.up.pt
[2] INESC TEC - INESC Technology and Science, Porto, Portugal
{carlos.m.costa,germano.veiga}@inesctec.pt

Abstract. Bin picking is a challenging problem common to many industries, whose automation will lead to great economic benefits. This paper presents a method for estimating the pose of a set of randomly arranged bent tubes, highly subject to occlusions and entanglement. The approach involves using a depth sensor to obtain a point cloud of the bin. The algorithm begins by filtering the point cloud to remove noise and segmenting it using the surface normals. Tube sections are then modeled as cylinders that are fitted into each segment using RANSAC. Finally, the sections are combined into complete tubes by adopting a greedy heuristic based on the distance between their endpoints. Experimental results with a dataset created with a Zivid sensor show that this method is able to provide estimates with high accuracy for bins with up to ten tubes. Therefore, this solution has the potential of being integrated into fully automated bin picking systems.

Keywords: Bin picking · Industrial robots · Linear objects · Pose estimation · Robot vision

1 Introduction

Bin picking corresponds to the task of taking an object out of a box with an open lid for subsequent manipulation [1]. It is usually decomposed into a set of sub-tasks, including object recognition, pose estimation and computation of an appropriate grasping position for a robot gripper to pick up a piece.

Since this task is present in many industrial processes, its automation will lead to substantial increases in productivity in a wide array of businesses. However, this problem is quite challenging as the items are often arranged in a random fashion, and are thus subject to occlusion and entanglement. One example of an item that is highly prone to entanglement are the wire-harnesses in the automotive industry. Due to its complexity, currently, there is no general solution to this problem.

This paper focuses on tubes and on the bin picking sub-task of pose estimation. The tubes are placed randomly in a bin, making them especially vulnerable

M. F. Silva et al. (Eds.): ROBOT 2019, AISC 1092, pp. 619–631, 2020.
https://doi.org/10.1007/978-3-030-35990-4_50

to entanglement. The goal is to develop a computationally efficient algorithm to process point clouds from a depth sensor, with the assumption that all the tubes have a common well-known radius, with a small variability. In most industrial scenarios, this assumption is realistic and acceptable. The algorithm makes no other assumptions regarding the tubes' geometric properties, namely their curvature, so that the solution can be applied to a varied set of industrial contexts.

Section 2 consists of a brief review of the literature regarding pose estimation for bin picking. Section 3 presents the approach to determine the tubes' geometry, detailing the experimental setup, the algorithm and how the solution was evaluated. Some results are also reported regarding the algorithm's computational efficiency and accuracy. Lastly, Sect. 4 concludes with some discussions of the contributions and lines for future work.

2 Literature Review

Research for bin picking dates back 50 years, but much of the work that was published was not dedicated to bin picking itself but rather to one or more of its sub-tasks. The emergence of new technologies for depth sensing, namely active stereo, led to the creation of many vision-based techniques with increased accuracy and robustness.

Many approaches for pose estimation resorted to 3D models of the objects to be perceived. Bolles and Horaud [2] presented one of the earliest approaches to locate objects in bins using 3D sensors. They used the edges from the depth image to extract local features, such as circular arcs, and followed a hypothesize-and-verify paradigm to find matches between the image and the object model.

One recent project on object localization is DoraPicker [3], which participated in the 2015 Amazon Picking Challenge. This system begins by filtering its input cloud by downsampling with a voxel grid and a statistics-based outlier removal. It then computes an initial estimate of the object's pose using LINEMOD and refines it using the well-known Iterative Closest Point (ICP) algorithm.

Kita and Kawai [4] proposed a method to estimate the pose of twisted tubes for bin picking by applying a region growing-based segmentation using the surface normals and then determining the tube's principal axis by examining cross-sections of the segmented region. The estimate is then improved by using the tube's model and a modified version of ICP that uses a skeleton of the tube.

The model-less approaches have the advantage of handling intra-class shape variations, such as tubes with varying curvatures. However, this also requires a clean segmentation of the target objects. One very common method used to obtain the pose of a bin's objects without having a model apriori is by matching with shape primitives.

Cylinders are sometimes employed for the pose estimation of tubes. Taylor and Kleeman [5] describe a split-and-merge segmentation algorithm that begins by identifying smoothly connected regions with the surface normals. Surface elements, such as ridges and valleys, are then found to check if a given region's shape is consistent with a cylinder. If so, fitting is performed with a least squares solver, and regions are merged iteratively when the model for the combined region has a lower residue. If the region is not consistent, it is split into smaller parts and the algorithm is applied recursively.

Qiu et al. [6] focused on reconstructing industrial site pipe-runs and performed a statistical analysis of the surface normals for global similarity acquisition, which they claim is more robust to noise than local features. This analysis involves using Random Sample Consensus (RANSAC) to obtain the cylinders' principal direction. Cylinder positions are then extracted via mean-shift clustering to detect circle centers on the orthogonal plane that contains the projection of the points belonging to cylinders with a given direction. Finally, pipe sections are joined using several criteria such as the distance and skew between cylinders.

An alternative to cylinders is to fit splines to model curved tubes. Bauer and Polthier [7] used a moving least squares technique to compute a spine curve, using a partial 3D view of the tube. The spine is then approximated to an arc-line spline using heuristics.

3 Algorithm and Experimental Validation

3.1 Experimental Setup

A Zivid One Plus L depth sensor (Fig. 1b) was chosen due to its high accuracy, having a spatial resolution of 0.45 mm (on the plane that is perpendicular to the sensor's optical axis) and a depth precision of 0.3 mm at 1.2 m. This high-end commercial device uses active stereo with structured light to acquire depth information and is also able to capture color, which was not used to show that this method is agnostic to the object's color.

The depth sensor was positioned looking downwards towards a bin filled with tubes for electric installations, with a vertical distance of 85 cm relative to the bin's bottom (Fig. 1a). They were made out of Polyvinyl chloride (PVC), with a diameter of 2.5 cm and a length of 50 cm. The tubes were bent with arbitrary

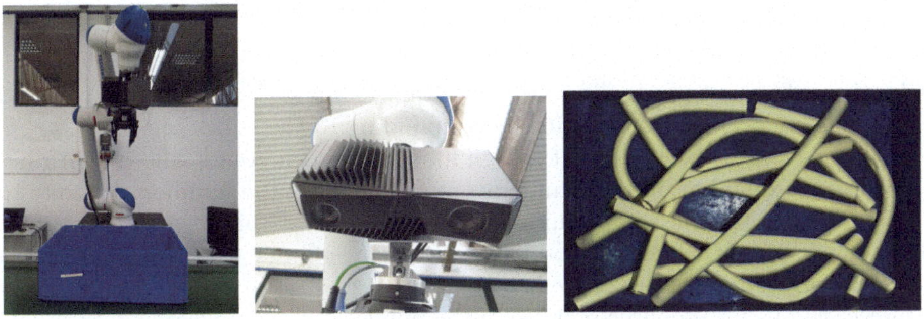

(a) Overview (b) Zivid sensor (c) Bin with bent PVC tubes

Fig. 1. Experimental setup

angles (Fig. 1c). The bin's dimensions are 55 cm (length) × 37 cm (width) × 20 cm (height).

3.2 Point Cloud Processing

The program responsible for obtaining a model for the tubes from the sensor's point cloud was developed using Point Cloud Library (PCL), which contains a wide array of useful algorithms for point cloud processing.

The solution is divided into four main steps: filtering, segmentation, cylinder fitting and tube joining. Algorithm 1 presents an overview of the solution. Figure 3 shows an example of the proposed steps.

Algorithm 1. Overview of the tube perception algorithm

Input: *Cloud* - Point cloud
Output: *Tubes* - Set of tubes

1: $Cloud \leftarrow applyFilters(Cloud)$
2: $Segments \leftarrow regionGrowingSegmentation(Cloud)$
3: $Tubes \leftarrow \emptyset$
4: **for each** $segment \in Segments$ **do**
5: $\quad tube, inliers, outliers \leftarrow fitCylinder(segment)$
6: $\quad$ **while** $size(inliers) >= min_{inliers}$ **do**
7: $\quad\quad Tubes \leftarrow Tubes \cup tube$
8: $\quad\quad tube, inliers, outliers \leftarrow fitCylinder(outliers)$
9: $EndPointPairs \leftarrow \emptyset$
10: **for each** $pair \in EndPoints(Tubes) \times EndPoints(Tubes)$ **do**
11: $\quad$ **if** $isNearby(pair)$ **and** $areDifferentTubes(pair)$ **then**
12: $\quad\quad EndPointPairs \leftarrow EndPointPairs \cup pair$
13: $EndPointPairs \leftarrow sortByDistance(EndPointPairs)$
14: **for each** $(endPoint1, endPoint2) \in EndPointPairs$ **do**
15: $\quad$ **if** $areCompatible(endPoint1, endPoint2)$ **then**
16: $\quad\quad (tube1, tube2) \leftarrow (tubeOf(endPoint1), tubeOf(endPoint2))$
17: $\quad\quad Tubes \leftarrow Tubes \cup unite(tube1, tube2) \setminus \{tube1, tube2\}$

Filtering. The first phase (line 1 in Algorithm 1) filters the point cloud received from the sensor so that only points corresponding to the tubes remain.

Filtering begins with a pass-through filter to remove points whose depth lies outside a reasonable range. A random sampler reduces the number of points to a fixed amount by removing points with a uniform probability. A radius outlier filter erases points which do not have enough neighbors within a sphere of a given radius. These filters are efficient at removing noise since the regions of interest tend to form denser regions on the point cloud, comparatively to points due to noise.

This first set of three filters is intended to reduce the number of points where the surface normals will be computed. The normals are estimated using a least-square method to fit a tangent plane to each point using the covariance matrix formed by the neighboring points from the raw point cloud [8]. The curvature of each point is estimated by using the eigenvalues of this matrix. The unfiltered point cloud was used for the search surface so that there are more points available around each neighborhood, thus increasing this operation's accuracy. Noisy points are not considered to have a significant impact in these estimates as they are greatly outnumbered by the relevant data.

The points from the plane belonging to the bin's bottom are then removed with RANSAC, an iterative, non-deterministic method that determines a model's coefficients by finding a sufficiently large subset of points (inliers) which are within a given distance from it. The distances from the points to the plane take into account both the point's coordinates and its normal vector. The minimum distance for a point to be considered as a inlier must be carefully chosen since if it is too small, the filter will not be able to remove points from the bin's bottom, and if it is too large, points belonging to the tubes may also be removed (the points whose normals are facing upwards are quite vulnerable to this problem), which will degrade the performance of the following steps of the algorithm. The advantages of using RANSAC include its simplicity and its robustness to outliers, even when in high proportion. The biggest drawback is that, due to its stochastic nature, it has no upper bound for the number of iterations needed to find a model that fits well the data. This leads to a trade-off between the execution time and the probability of computing an accurate model.

A second random sampler and radius outlier filter are applied to reduce the size of the resulting cloud to make the following processing steps more computationally efficient and remove points that may have remained from the bin's bottom.

The filtering process ends with a statistical outlier filter. This filter begins by computing the average μ and standard deviation σ of the distances of all the points to their k nearest neighbors [9]. The cloud's points are then scanned a second time and a point is considered as an outlier (and thus removed) if the average distance $\overline{d}$ follows inequality (1), where $mult$ is a multiplier that helps to regulate how restrictive the filter is. In this case, k and $mult$ were set to 100 and 3.0, respectively.

$$\overline{d} > \mu + mult * \sigma \qquad (1)$$

Segmentation. The second phase (line 2 in Algorithm 1) aims to divide the filtered point cloud (which should only contain points from the tubes) into regions where the surface normal's direction varies smoothly. Each region corresponds to a continuous and non-occluded portion of a tube. Therefore, the points should be clustered so that each visible or partially visible tube is associated with at least one segment, and each segment pertains to exactly one tube.

To perform this segmentation, a region-growing algorithm is used where each region starts with a seed point assigned to it and is progressively expanded by adding nearby points for which the angle between their normal and the one for the seed point is below the smoothness threshold [10]. Additional points may be added as seeds for the same region if their curvature is sufficiently low. When there are no more seed points to grow a given region, the algorithm creates a new cluster and sets as the initial seed the unlabeled point with the lowest curvature. The thresholds used by this algorithm must be properly defined to achieve a balance between over and under-segmentation.

Cylinder Fitting. The third phase (lines 3–8 in Algorithm 1) processes each segment independently. For each segment, the RANSAC algorithm repeatedly constructs cylinders, with the outliers of each cylinder serving as input for the next instance of RANSAC, until the number of remaining points is too low (below 100) for a new cylinder to be fit. Each call to the RANSAC algorithm runs up to 1000 iterations and attempts to find a cylinder with a radius between 0.7 cm and 1.5 cm. Associating multiple cylinders to each segment allows the algorithm to model curved tube sections.

The seven parameters for the cylinders' model are the 3D coordinates of a point and the three components of a unit vector to describe the axis, and its radius. The cylinder's length is not used as a parameter for RANSAC to reduce the number of iterations needed to produce good estimates for the other parameters. Instead, the length is computed after each run of the RANSAC algorithm by applying a rotation to the model's inliers so that the cylinder's axis is aligned with the z axis, and finding the inliers with minimum and maximum z coordinate. This method also allows the computation of both of the cylinder's endpoints, at center of the bases, which are used on the tube joining phase.

In order to avoid the overlapping of cylinders created from the same segment, after each run of RANSAC, the outliers within a slightly wider and longer cylinder with the same axis and center as the newly-created cylinder are removed. This bigger cylinder was set to be extended by 1 cm in length at both bases and to have a radius of 2 cm, which will always be larger than any reasonable cylinder fitted by RANSAC (since the tubes have a radius of 1.25 cm). The overlapping degrades significantly the performance of the tube joining phase since the algorithm may be unable to recombine both tubes into one.

Tube Joining. The fourth and final phase (lines 9–17 in Algorithm 1) consists in combining the cylinders created in the previous phase to form complete tubes. Each tube is modeled as a linked list of cylinders.

Initially, it is assumed that each cylinder corresponds to one and exactly one tube, even for cylinders that were generated by the same segment. One advantage of not considering all of the cylinders of the same segment to belong to the same tube is that the algorithm becomes more robust to occurrences of under-segmentation during the second phase.

This phase starts by considering all pairs of endpoints of distinct tubes and computing two distance metrics: the euclidean distance that separates both endpoints, and an angular distance, which describes how curved a junction would need to be to link both tubes. The angular distance, measured in degrees, is the

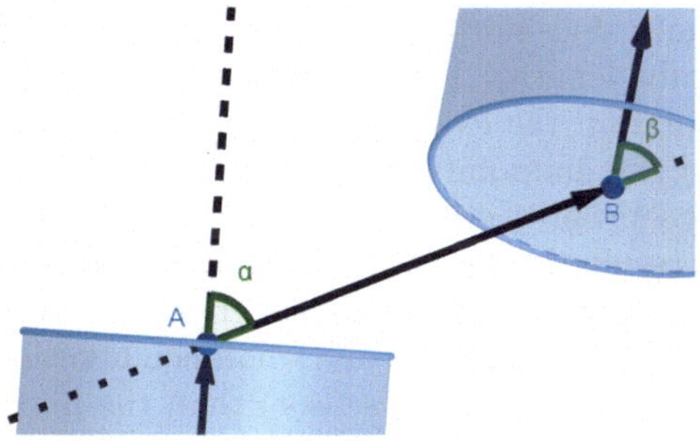

Fig. 2. Visual representation of the angular distance, sum of the angles α and β

sum of two angles formed by the unit vectors of the cylinder axes associated with each endpoint with a vector that links both endpoints, as depicted in Fig. 2.

The endpoint pairs with an euclidean distance above 10 cm or angular distance above 90° are discarded since they are unlikely to belong to the same tube. The remaining endpoint pairs are ordered using a distance metric that combines the euclidean and angular distance, as presented in Eq. (2). A coefficient was multiplied to the angular distance in an attempt to make both distances comparable (since one is expressed in meters and the other in degrees). The coefficient was set to 0.001 since the maximum acceptable distance for both metrics (10 cm and 90°) was assumed to be 'equivalent' (with $\frac{0.1}{90} \approx 0.001$). In the exceptional case where the euclidean distance is below 1.25 cm (the radius of a tube), it is very likely that the cylinders belong to the same tube, so the angular distance is set to 0 to increase the chances of the cylinders being combined in case there are not perfectly aligned (due to RANSAC's stochastic factors).

$$dist(A, B) = euclideanDist(A, B) + 0.001 * angularDist(A, B) \tag{2}$$

By adopting a greedy approach, the remaining endpoint pairs are processed in ascending order of distance, based on the rationale that the closer the cylinders are, the more likely they belong to the same tube. This is akin to the 'Joint Reconstruction' stage proposed by Qiu et al. [6], which also joins tubes based on their euclidean distance (which they call 'gap distance') and relative angles.

As each pair is processed, the endpoints' respective tubes are combined into a single tube that is modeled by the union of cylinders that belonged to both of the original cylinders. The resulting tube's length is estimated as the combined length of its cylinders lengths in addition to the euclidean distance between the endpoint connections. It should be noted that since it is common for the junctions between cylinders to be curved, this estimate is slightly lower than the actual length. If the cylinder overlapping problem was not solved in the previous phase, then these estimates would overshoot the actual length.

When processing the pairs, three constraints are applied to reduce the probability of a wrong junction of tubes. Firstly, the endpoints must belong to different tubes. Secondly, a 'length constraint' imposes that two tubes can only be joined if their combined length does not exceed 60 cm (a margin of error of 10 cm was added to the 50 cm tube length). This heuristic can be applied for other sorts of tubes if an upper bound for the length is known. Finally, a 'visibility constraint' prevents the union of two endpoints when there is an empty gap between them. This was implemented by projecting both endpoints onto a 2D range image of the point cloud that resulted from the filtering phase from the sensor's point of view. Afterwards, the midpoint between both endpoint projections is computed and it is determined whether any pixel in a small neighborhood around this midpoint has less depth (i.e. closer to the sensor) than the maximum depth among both endpoints. If there is no such pixel, then the tubes cannot be combined. To the best of the authors' knowledge, this is the first publication that presents these last two constraints for the problem of tube reconstruction.

3.3 Results

A dataset[1] with 50 point clouds of the bin with different amounts of tubes in various arrangements was constructed to evaluate the performance of the solution. There are five different test cases for each value for the number of tubes, ranging from 1 to 10. The tubes, bin and sensor properties are the same as those described in Sect. 3.1. Tables 1 and 2 present the average value of different performance metrics with respect to the number of tubes.

The results in Table 1 were obtained with an Intel Core i7-8750H processor, with 2.20 GHz. Each test case was evaluated five times to ensure more reliable execution times. The phase that takes the longest is the filtering, where the slowest operation was the plane fitting, as a large number of iterations was used for RANSAC. The increase of the filtering time with the number of tubes is likely due to a shift between the proportions of the points belonging to the bin's bottom and to the tubes: as there are less points on the bin's plane, RANSAC needs more iterations to converge to an acceptable model. It can be speculated that this increase in execution time should not increase indefinitely with the number of tubes and will stabilize once the amount of tubes is high enough for the bin's bottom to not be visible. The tube joining phase has a remarkably low execution time, with an order of magnitude of 1 ms.

Overall, the execution time of the perception algorithm has an order of magnitude of 1 s, which is rather reasonable for industrial applications.

To evaluate the performance of the segmentation phase, two annotators counted the number of visible continuous tube sections, using color images captured by the Zivid sensor for the 50 test cases. Ideally, there would be a one-to-one mapping between clusters and tube sections. The 'segmentation error' metric in Table 2 is the relative error between the number of clusters produced by the segmentation phase and the number of tube sections. This error is the greatest for cases with one tube. This is likely caused by a moderate amount of leftover noise, due to imperfect removal of the bin's bottom plane. The increase of this

[1] The 'Entangled Tubes Bin Picking' dataset is available at https://github.com/GoncaloLeao/Entangled-Tubes-Bin-Picking-Dataset.

Table 1. Average execution time for the algorithm's four phases

No. of tubes	Filtering time (s)	Segmentation time (s)	Fitting time (s)	Joining time (s)	Total time (s)
1	0.49	0.08	0.13	0.0036	0.70
2	0.60	0.14	0.16	0.0043	0.90
3	0.60	0.14	0.17	0.0050	0.92
4	0.60	0.14	0.21	0.0056	0.95
5	0.63	0.15	0.19	0.0069	0.97
6	0.62	0.14	0.20	0.0083	0.97
7	0.64	0.14	0.19	0.0089	0.98
8	0.69	0.14	0.19	0.0094	1.03
9	0.83	0.14	0.19	0.0095	1.17
10	0.94	0.14	0.20	0.0098	1.28

error in cases with more tubes is due to some tube sections being too small (as a result of a growing number of occlusions), and thus not having enough points for the region growing algorithm to be able to create clusters for them.

One metric used to assess the performance of the tube joining phase in a given test case are the average and standard deviation for the lengths of the tubes. Ideally, the average length should be 50 cm and the standard deviation should be minimal. According to Table 2, the tube joining phase appears to have a good performance overall. As the number of tubes increases, it is natural for these metrics to worsen since the visible surface area of the tubes decrease, giving less information about each individual tube for the algorithm to work with.

The tube length metrics are not sufficient to assess with great confidence the performance of the perception algorithm since the lengths are estimates and do not account for incorrect matchings between tube sections of similar length. Ergo, along with the number of tubes produced by the solution ('Number of joint tubes'), the annotators counted how many tubes were correctly and partially correctly modeled. A tube (in real-life) is considered to be 'correct' if it is associated with one and exactly one virtual tube which has a 'sufficient' amount of cylinders to cover its visible sections and does not have cylinders in sections of the bin where the tube is not present. A 'partially correct' tube only differs in the fact that multiple virtual tubes can be assigned to it. As seen in Table 2, the algorithm performs well even in cases where the bin is fuller. It should be noted that, in a bin picking system, the bin is scanned after removing each tube, so it is acceptable if there are few partially correct tubes, as long as at least a few tubes are correctly modeled. As more tubes are removed, the remaining ones have more chances of having a correct model.

Figure 3 illustrates the full process for one of the test cases with ten tubes (test case '10_bin_picking2' from Fig. 1c). The number of remaining points after each filter is also shown. In this case, the algorithm performed quite well since

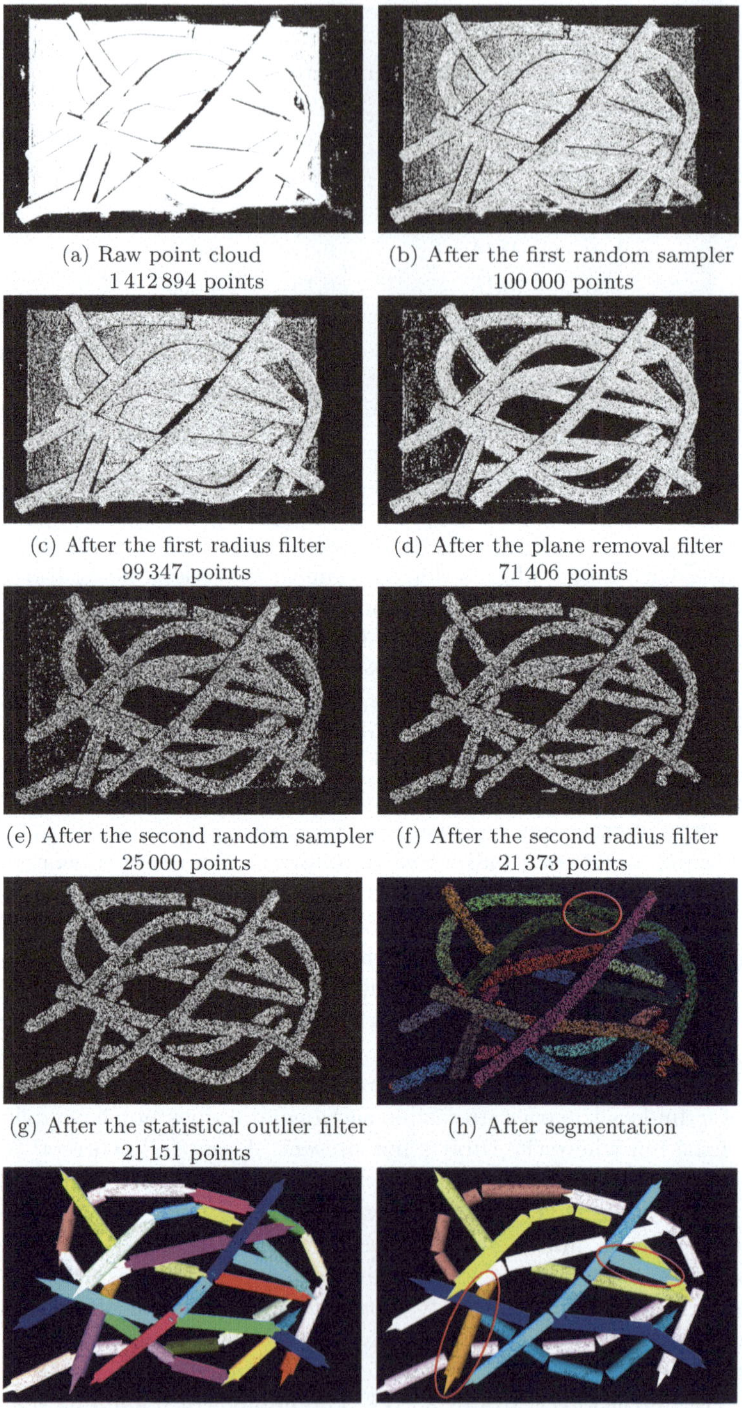

(a) Raw point cloud
1 412 894 points

(b) After the first random sampler
100 000 points

(c) After the first radius filter
99 347 points

(d) After the plane removal filter
71 406 points

(e) After the second random sampler
25 000 points

(f) After the second radius filter
21 373 points

(g) After the statistical outlier filter
21 151 points

(h) After segmentation

(i) After fitting cylinders

(j) After joining tubes

Fig. 3. Results of the algorithm's steps for the case shown in Fig. 1c

Table 2. Accuracy metrics for the segmentation, fitting and joining phases

No. of tubes	Segmentation error	Avg. length (m)	Std. deviation length (m)	No. of joint tubes	No. of correct tubes	No. of partially correct tubes
1	2.00	0.439	0.00105	1.2	0.8	0.2
2	0.83	0.439	0.08911	2.4	2	0
3	0.20	0.502	0.00824	3	3	0
4	0.00	0.468	0.03406	4.4	3.6	0.4
5	0.04	0.480	0.03661	5.2	4.8	0.2
6	0.00	0.481	0.03135	6.2	5.6	0.4
7	0.06	0.454	0.08529	7.6	6.4	0.6
8	0.06	0.414	0.12530	9.2	7	1
9	0.13	0.386	0.13101	10.6	6.6	2.2
10	0.12	0.347	0.15357	12.6	7	3

nine of the tubes are correctly modeled and the remaining one is partially correct (the two tube models corresponding to the partially correct one are marked with a red ellipse in Fig. 3j). It is interesting to notice that in Fig. 3h there was an occurrence of under-segmentation (marked with a red ellipse) that the cylinder fitting phase was able to recover from.

4 Conclusions and Future Work

The algorithm presented in this paper processes a point cloud from a depth sensor and provides a model for a set of tubes of equal radius but varying curvatures in a bin where they are arranged randomly. Using a primitive fitting approach for pose estimation, rather than starting from an initial 3D model of the tubes, allows the solution to handle intra-class variability. Some distance-based heuristics presented by Qiu et al. [6] alongside some novel constraints, such as checking for a gap between the cylinders, also enable it to deal with occlusions and entanglement. This is proven by the experimental results, where the solution was able to accurately describe the shape of most tubes in bins with up to ten tubes.

Using this solution's output, many heuristics can be used to select which tube to pick up next, such as choosing the one with the least occlusions. This solution can thus be integrated into bin picking systems of a vast variety of industries. Despite the tubes that were used in the experiments being rigid, this method has the potential of being compatible with flexible tubes, for which there are very few bin picking solutions.

Another relevant contribution is the 'Entangled Tubes Bin Picking' dataset, which can be used by the robotics community to benchmark other solutions to this challenging problem.

This work opens several lines for future research. Firstly, the solution's performance can be measured using different kinds of tubes, with other lengths and radii, possibly made of a more flexible material. Tests can also be conducted using other sensors, ideally those with less precision. These experiments can lead to an enrichment of the dataset. Secondly, the performance of the algorithm's fitting phase (execution time and accuracy) can be compared to an alternative where a spline is fitted to each segment, as suggested by Bauer and Polthier [7]. Little to no modifications will need to be done to the other three phases of the solution. Lastly, to decrease the algorithm's execution time, a decision procedure can be devised to decide if the plane filtering should be applied to the cloud.

Acknowledgments. The research leading to these results has received funding from the European Union's Horizon 2020 - The EU Framework Programme for Research and Innovation 2014–2020, under grant agreement No. 723658.

References

1. Alonso, M., Izaguirre, A., Graña, M.: Current research trends in robot grasping and bin picking. In: International Joint Conference SOCO 2018-CISIS 2018-ICEUTE 2018, San Sebastian, Spain, vol. 771, pp. 367–376. Springer, Cham, June 2019
2. Bolles, R.C., Horaud, R.P.: 3DPO: a three dimensional part orientation system. Int. J. Robot. Res. **5**(3), 3–26 (1986)
3. Zhang, H., Long, P., Zhou, D., Qian, Z., Wang, Z., Wan, W., Manocha, D., Park, C., Hu, T., Cao, C., Chen, Y., Chow, M., Pan, J.: DoraPicker: an autonomous picking system for general objects. In: IEEE International Conference on Automation Science and Engineering, Fort Worth, TX, USA, pp. 721–726. IEEE, August 2016
4. Kita, Y., Kawai, Y.: Localization of freely curved pipes for bin picking. In: IEEE International Conference on Emerging Technologies and Factory Automation, ETFA, Luxembourg, Luxembourg, pp. 1–8. IEEE, September 2015
5. Taylor, G., Kleeman, L.: Robust range data segmentation using geometric primitives for robotic applications. In: Proceedings of the Fifth IASTED International Conference on Signal and Image Processing, Honolulu, HI, USA, pp. 467–472. Acta Press (2003)
6. Qiu, R., Zhou, Q.Y., Neumann, U.: Pipe-run extraction and reconstruction from point clouds. In: Proceedings of the 13th European Conference on Computer Vision - ECCV 2014, Zurich, Switzerland. LNCS, vol. 8691, pp. 17–30. Springer, Cham (2014)
7. Bauer, U., Polthier, K.: Parametric reconstruction of bent tube surfaces. In: Proceedings - 2007 International Conference on Cyberworlds, CW 2007, Hannover, Germany, pp. 465–474. IEEE, October 2007
8. Rusu, R.B.: Semantic 3D object maps for everyday manipulation in human living environments. KI - Künstliche Intelligenz **24**(4), 345–348 (2010)

9. Rusu, R.B., Blodow, N., Marton, Z., Soos, A., Beetz, M.: Towards 3D object maps for autonomous household robots. In: IEEE International Conference on Intelligent Robots and Systems, San Diego, CA, USA, pp. 3191–3198. IEEE, October 2007
10. Rabbani, T., van den Heuvel, F., Vosselmann, G.: Segmentation of point clouds using smoothness constraint. In: Maas, H.G.R., Schneider, D. (eds.) ISPRS 2006: Proceedings of the ISPRS Commission V Symposium, vol. 35, pp. 248–253. International Society for Photogrammetry and Remote Sensing (ISPRS), Dresden, Germany (2006)

Applying Software Static Analysis to ROS: The Case Study of the FASTEN European Project

Tiago Neto[1,2], Rafael Arrais[1,2(✉)], Armando Sousa[1,2], André Santos[2,3], and Germano Veiga[1,2]

[1] Faculty of Engineering, University of Porto, Porto, Portugal
[2] INESC TEC - INESC Technology and Science, Porto, Portugal
rafael.l.arrais@inesctec.pt
[3] Universidade do Minho, Braga, Portugal

Abstract. Modern industry is shifting towards flexible, advanced robotic systems to meet the increasing demand for custom-made products with low manufacturing costs and to promote a collaborative environment for humans and robots. As a consequence of this industrial revolution, some traditional, mechanical- and hardware-based safety mechanisms are discarded in favour of a safer, more dependable robot software. This work presents a case study of assessing and improving the internal quality of a European research mobile manipulator, operating in a real industrial environment, using modern static analysis tools geared for robotic software. Following an iterative approach, we managed to fix about 90% of the reported issues, resulting in code that is easier to use and maintain.

Keywords: Software static analysis · Safety · Mobile manipulator · ROS

1 Introduction

The shifting of paradigm imposed by the ongoing Fourth Industrial Revolution is introducing a new set of constraints and opportunities for industrial enterprises. These constraints and opportunities are catalyzing the introduction of flexible, adaptable and collaborative human-robot hybrid systems which can enable even small and medium enterprises to adapt to paradigm changes in market demand, often characterized by increasing customization [2]. These systems are materializing as collaborative robotic solutions in industrial applications and as autonomous mobile robotics in sectors ranging from agriculture to intralogistics, operating in dynamic and unstructured environments shared with humans.

Such advanced robotic systems, operating in cross-sectorial domains of activity, sensing and interacting with complex and unstructured environments require

M. F. Silva et al. (Eds.): ROBOT 2019, AISC 1092, pp. 632–644, 2020.
https://doi.org/10.1007/978-3-030-35990-4_51

the integration and support of the technologies, models, and functional components that enable robotic operations. In this context, the safety of humans operating and interacting with potentially dangerous equipment is a core scientific and technological challenge. Thus, and to cope with market demand for product customization or demanding field applications, contemporary robotics must drastically alter the safety assurance paradigm.

Traditionally, roboticists majorly relied on mechanical-based methodologies, such as physical barriers, to account for safety behaviour. However, as modern systems need to be flexible, adaptive and collaborative to adhere to the ongoing industrial revolution, software-based safety assurance mechanisms are emerging as a complement to traditional safety procedures. Also, software-based safety assurance can play important social and psychological roles to foster the acceptance of robots in human-populated environments and to promote collaboration between humans and robots. This change affects the robotics ecosystem and calls for techniques to promote best software engineering practice guidelines for the development of safety-critical software, suitable for the robotics development environment.

In a clear contrast with the current necessities, particularly in the cutting edge of innovation efforts, this meticulous attention to software engineering guidelines and safety assurance of software-based components is often overlooked [4] due to the experimental nature of developments, the complexity of the systems, and the difficulties associated with validating the software-based safety mechanisms in physical hardware.

Over the last decade, frameworks such as the Robot Operating System (ROS) [3] have emerged as de facto standards for robotic software development, with an increasing presence in the industrial environment. ROS provides roboticists with abstractions and a vast amount of libraries that widely simplifies and speeds up the development of advanced robotic systems. However, these benefits come with a price, in particular, the intrinsic difficulty to fully assess and validate ROS-based software and external libraries in what regards their compliance with safety protocols or even guidelines for software engineering best practices.

The Safety Verification for Robotic Software (SAFER) project, in which this work is integrated, brings together the expertise of computer scientists, with a background on software system design and analysis, and experienced robot engineers, to overcome the aforementioned shortcomings of ROS-based software development. One of the project's main output is the High Assurance ROS (HAROS) tool [5], a static analyzer of ROS-based software, that can extract valuable information from the source code without the need for executing it (or even compiling it, in many cases). The application of this tool during the development process promotes compliance with software engineering best practices and can be a valuable tool to allow developers to assess the safety compliance of their software. Furthermore, by promoting the creation of better-structured source code, its readability, maintainability, and scalability are deeply improved, potentially resulting not only in increased safety compliance but also in long-term financial gains, as the produced source code is easier to work with.

In this paper, the application of the HAROS tool to a complete stack of ROS-based software powering a mobile manipulator operating in an industrial environment is explored, with the objective of assessing and iteratively improve the code quality. This is attained with the fixing of dangerous sections of code that could lead to errors and failures. The remainder of the paper is organized as follows: Sect. 2 presents a conceptual overview of some of the discussed domains, as well as a brief state of the art of the subject; Sect. 3 presents a detailed description of the industrial utilization of the developed mobile manipulator, its hardware composition, and its software architecture; Sect. 4 highlights the principal scientific contribution of this research work, by presenting the methodology and results obtained from the application of the HAROS tool to guide ROS-based software development; and, finally, Sect. 5 draws some conclusions and points out some future work roadmap.

2 Related Work

A deciding factor in the adoption of robotic systems in real-world scenarios is related to the trust levels that humans have in their utilization. To fully promote the mass adoption of robotic systems in manufacturing, complying with the ongoing industrial revolution, users need to be fully confident in their operation. In what concerns these systems in a broader sense, trust can be defined as a combination of reliability, safety, security, privacy, and usability [7].

Static analysis techniques are one of many software engineering techniques that can elevate the quality of code, and thus also increase trustability in the developed system. This conceptually simple and time-efficient technique allows, since an early phase of development, the extraction of precious information from a program without running or even compiling it. Among the collected information, compliance of the code with given specifications, internal quality metrics and conformity with coding standards are amongst the most valuable metrics [5]. Static analysis tools evolved to be able to deal with industrial applications, containing millions of lines of code. In [1], the authors provide a comparative analysis of three of the most powerful and popular static analysis tools for industrial purposes, namely *PolySpace*, *Coverity* and *Klocwork*, yet none of it seems to be directly applicable to the ROS ecosystem.

In the domain of robotics, ROS, an open-source tool-based framework that provides developers with a large set of libraries and abstractions to ease the difficult task of developing robotic software [3]. Since its introduction, ROS is increasingly being introduced in industrial applications. However, ROS does not impose strict development rules to ensure its safety. Due to the great diversity of ROS applications, there is no solution to completely analyse and verify ROS programs in a formal way and certify their safety to guarantee correct behaviour of robots.

As an alternative to the lack of intrinsic safety compliance mechanisms in ROS and the underlying difficulty to validate such compliance, software static analysis can yield valuable information about the behaviour of each of its subsystems and the interactions between them, thus allowing developers to preemptively verify if the source code is according to the requirements and, consequently and implicitly, improving its safety compliance capabilities [4].

Despite the potential of this technique, applying it to ROS is not so straightforward. As previously mentioned, ROS is very customizable, has a large number of primitives and can be written in several programming languages. This diversity leads to an extremely complex and unfeasible ad hoc solution for an arbitrary ROS system. Nevertheless, for a more restricted set of ROS subsystems, and a bounded set of constraints, it could be achievable [4].

An example of a static analyser for ROS-based code is HAROS. HAROS is being developed having two fundamental ideas in mind: one is the integration with ROS specific settings, and the other is that it should not be restrictive, thus allowing the use of a wide range of static analysis techniques. The latter notion leads to HAROS allowing the integration and use of third-party analysis tools, as plug-ins [5]. This tool allows the fetching of ROS source code, its analysis and the compilation of a report in an automatic way. Therefore, it can be easily used, even by developers without extensive knowledge of ROS or static analysis techniques.

With HAROS, the user first chooses which packages should be analysed, and according to the required analysis, HAROS will dynamically load the adequate plug-ins. The properties that are analysed can be of two categories: rules or metrics. Rules report violations as individual issues, while metrics return a quantitative value, which can, in turn, result in a set of issues [5]. Once the configuration and analysis steps are concluded, the results are portrayed to the user in both a graphical form and by a list of issues, which can be filtered by their type. In its graphical form, the results are portrayed to the user in both a graphical form and by a list of issues, which can be filtered by their type. In its graphical form, the results visually display the analyzed metrics, and, most importantly, the system-wide and intra-node architecture and properties.

On [4], the authors focused on interpreting the outputs of applying a static analysis provided by ROS on a set of popular and publicly available ROS packages. Collecting this kind of information is important to elucidate about less used or even misused features and is also useful for developers of static analysis tools to determinate which features are more relevant to be supported [4]. HAROS was also used by the authors of [6], to extract and analyze the architecture of a field robotic system for the agriculture domain at static time. This verification provides valuable information during the development phase, which was used to ensure that safety design rules were well implemented in the architecture of the studied robot, validating and improving the safety of the system [6].

In this work, HAROS is applied on an industrial robotic system not only with the purpose of validating this tool, but also, and more critically, to attempt to verify and improve the safety of the system and, indirectly, the maintainability of the source code, as it will be demonstrated in Sect. 4.

Fig. 1. FASTEN Mobile manipulator developed for application in an Embraer industrial plant.

3 Case Study Description

The case study for the work was the H2020 Flexible and Autonomous Manufacturing Systems for Custom-Designed Products (FASTEN) project. This project aims to develop, demonstrate, validate, and disseminate a modular and integrated framework able to efficiently produce custom-designed products. In order to achieve that it integrates digital service/products manufacturing processes, decentralized decision-making and data interchange tools. Thus, to achieve a fully connected and responsive manufacturing system, several technologies are being developed, as is the case of sophisticated self-learning, self-optimizing, flexible and collaborative advanced robotic systems. As proof of concept, a mobile manipulator, capable of assembling and transporting kits of aerospace parts is being developed. Currently, at the scenario, Embraer Portugal S.A. (Embraer), the industrial end-user of the project, stores the parts used for wing assembly in a Automated Warehouse System (AWS). The kitting operation, composed by the retrieval of components from the AWS is a repetitive, non-ergonomic and non-added-value task which can be automatized to improve performance and working conditions. Furthermore, by relying on an automatic solution to assemble kits, Embraer can further enhance the traceability of its intralogistics process.

For this, an automated solution is being developed (Fig. 1). It is composed by an Automated guided vehicle (AGV) with an omnidirectional traction configuration, fitted with a collaborative robotic manipulator. So, this mobile manipulator is capable of traversing the logistic warehouse in any direction and cooperate with human operators in the assembly of kits, increasing the automation level and freeing human operators for more added-value tasks.

The software architecture of this system is being developed with three main objectives in mind, that lead to three structural ideas. The first objective is to reduce the cost of adapting robot applications by promoting code re-usability.

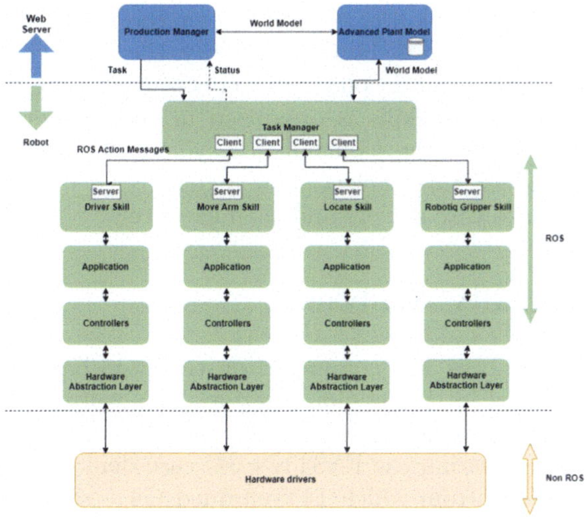

Fig. 2. High-level software architecture of the FASTEN robot system.

To achieve this, a skill-based robot programming approach was used. The second objective is to promote intuitive and flexible robot programming, achieved by task-level orchestration. The third objective is to support generic interoperability with manufacturing management systems and industrial equipment.

As depicted in Fig. 2, this robotic system has a distributed architecture. In the server-side implementation, there are two components, the Production Manager (PM) and the Advanced Plant Model (APM) [8], while on the robot side of the architecture, there are the skills and the Task Manager (TM). The APM keeps a near real-time model of the production environment. The PM is responsible to manage the production resources, control the execution of the production schedules and it is also responsible for monitoring the ongoing performance of the different production tasks.

On the robot, one of the most important components is the TM. The TM has two primary functions: it (i) provides integration between the robot and other modules on the system, like the APM or the PM, and (ii) is responsible for the orchestration of tasks, using the skills of the robot. On the TM there is a ROS Action Client for each skill and on each skill, there is a ROS Action server. This is due to the fact that skills are implemented using ROS Actions. The TM uses skills by defining a goal and sending it to the respective Action server. When the execution is completed it receives, from the skill Action server, the result and additional information about the outcome of the performed action.

For the H2020 FASTEN demonstrator, the robotic system has been instantiated with four different skills: (i) Move Arm Skill, (ii) Gripper Skill, (iii) Locate Skill, and (iv) Drive Skill. The Move Arm Skill is responsible for the movement of the robotic manipulator. The Gripper Skill is responsible for the actuation of the gripper. The Locate Skill is responsible for the recognition and localization of the parts that need to be handled. Finally, the Drive Skill is responsible for the movement of the robotic platform and ensuring that the movement is collision-free. Each of these skills is organized in three different parts, which are the Application Layer, the Controllers Layer, and, finally, the Hardware Abstraction Layer. These three layers allow a goal received from the TM to be transmitted to the hardware drivers and then executed.

4 Software Quality Analysis

A software quality analysis was conducted on the ROS-based mobile manipulator software presented in the previous section. This software stack comprised the set of functional components, in the form of ROS source code and launch files, responsible for powering the FASTEN use case demonstrator. In total, 22 packages were analysed, from which 14 contained C++ source code, while the remaining contained Python source code or only ROS launch files. The C++ source code amounted to approximately 200,000 lines of code.

To conduct this analysis, the HAROS tool was used. After an initial overview analysis of the complete system, its source code issues were listed and grouped by category for each ROS package. The remainder of the analysis was iterative. This means that the source code issues and model inconsistencies discovered were addressed in several iterations. After each iteration, the obtained results were re-evaluated with the HAROS tool and the strategy for the next iteration was drawn. This iterative approach was elected due to the intrinsic difficulty to address all software issues in a single run, allowing developers to assess, in each iteration, if the proposed changes do not impose constraints on the integrity of the system. In addition, addressing all software problems in a single passage would most likely originate novel issues that would be hard to trace the origin of. Moreover, an iterative methodology was employed in order to promote the continuous integration paradigm.

The conducted analysis can be divided into two distinct phases. The Architecture Analysis, presented in Subsect. 4.1, allows developers to have the full-scale system-wide and intra-node overview of the system and assess if the developed architecture is according to the specifications. The Static Code Analysis, presented in Subsect. 4.2 refers to the reasoning on the source code of each software application that composes the system. This analysis allows developers to catch safety-critical issues, and assess if the code complies with normative standards and guidelines, thus empowering not only the safety of the whole robotic system but also the underlying code maintainability and scalability.

4.1 Architectural Analysis

The architecture analysis is the differentiator feature that separates the HAROS tool from the remaining static analysis tools. For this feature, it is necessary to inform HAROS which ROS launch files should be analysed. Then, with that

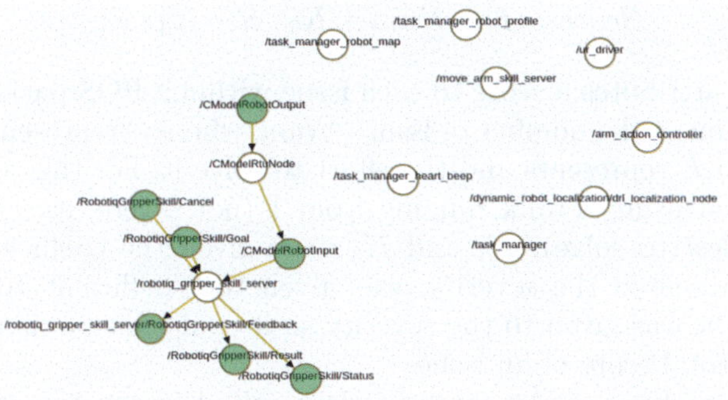

Fig. 3. Architectural analysis of the robotic system as displayed by the HAROS web-based visualization tool.

information, HAROS extracts the ROS nodes that are being launched by that file and the arguments that are being passed during the launch. However, in its current version, HAROS is not capable of finding a node that is being launched conditionally.

As the FASTEN mobile manipulator development is adopting a methodology where the ROS launch file of each sub-system is conditional it was necessary to provide hints via a YAML configuration file required by HAROS. These hints provide HAROS with additional information about which ROS topics are subscribed or published by each ROS node that composes the system.

The visualization of the output of this architectural analysis in the HAROS user interface is depicted in Fig. 3. This visualization component provides a good insight into what is to be expected from the application ROS nodes. Nevertheless, since this extraction could not be automated and had to be provided by hints, the model extraction tool validity and correctness is questionable for the purposes of this case study.

4.2 Static Code Analysis

Initial Analysis. This initial analysis contains the raw data collected using the HAROS tool. The issues were divided into 3 categories: Formatting, Code Standards, and Metrics. The first category, Formatting, encloses issues related to indention, whitespaces and the placement of braces. The second, the Code Standards, encloses issues related to the compliance with code standards, i.e. adhering to a specific style of programming or restricting oneself to a subset of the programming language. Finally, the Metrics encloses issues related to internal quality code metrics, such as cyclomatic complexity or the maintainability index.

Since it was impossible and impractical to solve every issue with one run, the intervention process, guided by the issues reported by HAROS, was divided into several iterative steps. Furthermore, it was necessary to determine which issues would be tackled first. To elect the first issues to be tackled, a model, described by Eq. 1 is proposed.

$$Score = K_1 \cdot Num + K_2 \cdot S + K_3 \cdot E; \quad (1)$$

This model attributes a score to each issue within a ROS package. The score is a weighted sum of the number of issues, *Num*, where S represents the severity of the issue and E represents and the effort to solve it. For this analysis, S and E were classified using a rank ranging from 1 (not severe, easiest to solve) to 3 (severe, hardest to solve). K_1 and K_3 were given the coefficient 1 while to K_2, which represented the severity, was given the coefficient 10. The biggest coefficient weight was given to the severity so it could have a more pronounced impact on the total score of an issue.

The initial analysis of the source code resulted in the report of a total of 28,040 issues, as can be seen in detail on Table 1.

First Iteration. For this first iteration, it was assumed that the code did not follow any code standard format since the code was developed by various development teams, and it also simplified the code format standard uniformization to be conducted. Analysing the results of the initial analysis, it is pretty clear that most of the issues are of the formatting type, as can be seen in Table 1, which means that they should be the first ones to be tackled. Since the code is vast it would be impractical and extremely time-consuming to correct all the formatting issues by hand. So, to tackle this kind of issues an automatic approach was taken. The chosen tool was the *Clang-Format* along with Visual Studio Code.

The *Clang-Format* was used to format the code accordingly to the Google C++ style guide. The decision to chose Google C++ style instead of ROS C++ Style was based on the fact that the portability of the majority of the source code to this style guide would be more straightforward. After the use of this tool, some additional adjustments had to be done by hand. This was necessary to ensure that the code still compiled. The adjustment done by hand were mostly related with include orders since the automatic tool rearranged the header files in such a way that compilation was not possible.

This first iteration allowed to eliminate 14 types of issues, from 66 in the initial analysis to 52 at the end of the first iteration. This was mostly because of the reduction of the Formatting issues from 24 to 10. On the total number of issues, it was registered a decrease of 22,686 issues. Even though the Formatting and Code Standard issues decreased, the Metric issues increased. The cause of this was the changes made to respect the line length that triggered an increase in the use of vertical lines. This increase originated a spike in the number of functions to have more than 40 lines of code, which, in its turn, triggered more Metric issues.

Second Iteration. For the second iteration, one of the issues with a higher score was the *line length*. Since the automatic formatting did not solve this, the source code was manually analysed to understand the root of this issue. There were two explanations: (i) functions with long names could not be solved, and (ii) comments with section markers could not be automatically processed.

Regardless, this could be solved by reducing the number of repeated characters without removing the code separation.

Another issue with a high count of occurrences was the *Non-const Reference Parameters*. This issue was caused by variables being passed by reference, but not using the keyword *const* as recommended by the Google C++ style guide. This issue has two possible solutions. The first is to use the keyword *const* if the variable does not need to be changed inside the function and the other, which requires more effort, is to pass by a pointer and to change the code according to this demand. However, since the second solution was the one that needed to be applied more often, it was opted to leave the code as-is, to avoid cross-package errors that could be hard to track. Also, this type of issue did not represent a safety threat.

The issues of the type *Integer types* were also among the issues with a higher count. These issues were mostly triggered by the use of the type *size_t*, but also by the use of the type *short* or *long*. The usage of the type *size_t* is allowed by Google C++ style guide when it is appropriate, which was the case for the totality of occurrences, and for that reason, it was not changed. When types such as *short* were being used, they were replaced by size-specific types, such as *int16_t*.

In this iteration, issues with *whitespaces*, *copyright*, and *contructors* were tackled. The *copyright* issues were solved by adding a copyright statement to each file, while the *constructors* issues were addressed by making constructors with single argument explicit. Furthermore, issues related to *casting* were also solved during this iteration. However, at the end of the iteration, HAROS still identified 2 casting issues. Yet, while manually inspecting the code, it was found that these were not casting issues, but were, in fact, false positives.

Finally, in this iteration, the issues with the *floating point* were solved. These issues were caused by float point expressions that were expecting exact equality, which is not compliant with the MISRA C++ guidelines, deeming it unsafe. The solution for these issues was to rewrite the expressions in a way that did not test equality directly and that was compliant with the guidelines.

Overall, 2,498 issues were solved in this iteration, which reduced the total of issues to solve to 2,859 at the end of this iteration. The formatting issues decreased from 10 to 5 and the code standard issues from 34 to 32. However, the average severity and average effort to solve increased from 1.85 to 1.95 and from 1.68 to 1.83, respectively. This is justified by the fixing of more issues with lower severity and lower effort to solve. Nevertheless, this was also a successful iteration, since it led to a reduction of around 53% of issues reported in the previous iteration.

Third Iteration. This third and final iteration focused on solving issues related to cyclomatic complexity, functions that were not considered safe and also analysed other issues to understand their causes.

Table 1. Static code analysis results of the initial analysis and subsequent iterations.

		Types of issues	Issues	Average severity	Average effort to solve	Total score
Initial	Formatting	24	24511	1.61	1.47	34414
	Code standard	34	3175			
	Metric	8	356			
	Total	66	28043			
First	Formatting	10	2327	1.85	1.68	10545
	Code standard	34	2288			
	Metric	8	478			
	Total	52	5357			
Second	Formatting	5	253	1.95	1.83	7267
	Code standard	32	2126			
	Metric	8	480			
	Total	45	2859			
Third	Formatting	5	253	1.93	1.90	6485
	Code standard	30	1883			
	Metric	8	467			
	Total	43	2603			

Among the metrics, the cyclomatic complexity is the easiest to change and improve. Despite that, it does not mean that it is a simple issue to fix. Some functions with high cyclomatic complexity are impossible to do in a less complex way, as their purpose is to verify a set of conditions that can not be easily changed. Others are simply just too complex, and it is therefore very risky to change them without incurring in drastic changes to the behaviour of the software, as this code belongs to robotic software that is responsible for the implementation of very specialized and complex features, such as computer vision algorithms. Areas like this require some specialized expertise to alter those algorithms, which complicate the task of changing these algorithms. However, for some of these functions, it is possible to understand their purpose without deep knowledge of the area. For some of those, it is possible to achieve the same result using less complex ways. Thus, during this iteration, it was possible to reduce the cyclomatic complexity of functions with a cyclomatic complexity score as high as 17. Above that value, it was opted not to change them due to the high probability to introduce errors. For these more complex functions, it is recommended intervention from a development team with higher expertise in the domain.

In spite of this last iteration not being able to solve as many issues as the previous ones, most of the issues solved in this iteration were harder to solve. Most of the issues solved in this iteration were also more severe, which reflected on the decrease in the average severity. On this iteration, 256 issues were solved, which led to a decrease in the total number of issues from 2,859 to 2,603 at the end of these iterations (around 9%). In this iteration, 2 Code Standard issues were also eliminated, reducing the total type of issues to 43 and the Code Standard issues to 30.

5 Conclusion

Overall, the source code analysis allowed to solve 25,440 issues, which represents a reduction of 90% of issues from the initial analysis. Some of the fixed issues were deemed to be dangerous and could potentially compromise the run-time functioning of the mobile manipulator. As such, the alterations performed by this work undoubtedly allowed the improvement of the safety and maintainability of the source code, and, correspondingly, the FASTEN mobile manipulator operation in an industrial environment.

With this analysis, it was also clear that the introduced improvements could benefit the development process in the long run. Thus, the methodology described in the paper is being applied during nominal development procedures. As such, the FASTEN mobile manipulator development teams are actively using the proposed methodology and applying the HAROS tool in a continuous integration fashion, as to check for potential issues prior to any source code commit.

In the future, this methodology will be applied to other use cases, as an attempt to replicate the improvements in the domains of code maintainability and safety to other robotic systems.

Acknowledgments. This work is financed by the ERDF – European Regional Development Fund through the Operational Programme for Competitiveness and Internationalisation - COMPETE 2020 Programme and by National Funds through the Portuguese funding agency, FCT - Fundação para a Ciência e a Tecnologia within project POCI-01-0145-FEDER-029583. The research leading to these results has also received funding from the European Union's Horizon 2020 - The EU Framework Programme for Research and Innovation 2014–2020, under grant agreement No. 777096.

References

1. Emanuelsson, P., Nilsson, U.: A comparative study of industrial static analysis tools. Electron. Notes Theor. Comput. Sci. **217**, 5–21 (2008)
2. Lasi, H., Fettke, P., Kemper, H.G., Feld, T., Hoffmann, M.: Industry 4.0. Bus. Inf. Syst. Eng. **6**(4), 239–242 (2014)
3. Quigley, M., Conley, K., Gerkey, B., Faust, J., Foote, T., Leibs, J., Wheeler, R., Ng, A.Y.: ROS: an open-source robot operating system. In: ICRA Workshop on Open Source Software, Kobe, Japan, vol. 3, p. 5 (2009)
4. Santos, A., Cunha, A., Macedo, N., Arrais, R., dos Santos, F.N.: Mining the usage patterns of ROS primitives. In: 2017 IEEE/RSJ International Conference on Intelligent Robots and Systems (IROS), pp. 3855–3860, September 2017
5. Santos, A., Cunha, A., Macedo, N., Lourenço, C.: A framework for quality assessment of ROS repositories. In: 2016 IEEE/RSJ International Conference on Intelligent Robots and Systems (IROS), pp. 4491–4496, October 2016

6. Santos, A., Cunha, A., Macedo, N.: Static-time extraction and analysis of the ROS computation graph. In: 2019 Third IEEE International Conference on Robotic Computing (IRC), pp. 62–69. IEEE (2019)
7. Sha, L., Gopalakrishnan, S., Liu, X., Wang, Q.: Cyber-physical systems: a new frontier. In: 2008 IEEE International Conference on Sensor Networks, Ubiquitous, and Trustworthy Computing (SUTC 2008), pp. 1–9 (2008). http://ieeexplore.ieee.org/document/4545732/
8. Toscano, C., Arrais, R., Veiga, G.: Enhancement of industrial logistic systems with semantic 3D representations for mobile manipulators. In: Ollero, A., Sanfeliu, A., Montano, L., Lau, N., Cardeira, C. (eds.) ROBOT 2017: Third Iberian Robotics Conference, pp. 617–628. Springer, Cham (2018)

Autonomous Robot Navigation for Automotive Assembly Task: An Industry Use-Case

Héber Sobreira[1], Luís Rocha[1], José Lima[2(✉)], Francisco Rodrigues[1], A. Paulo Moreira[3], and Germano Veiga[3]

[1] INESC TEC - INESC Technology and Science, Porto, Portugal
{heber.m.sobreira,luis.f.rocha,francisco.a.rodrigues}@inesctec.pt
[2] INESC TEC - INESC Technology and Science and CeDRI - Research Centre in Digitalization and Intelligent Robotics, Polytechnic Institute of Bragança, Bragança, Portugal
jllima@ipb.pt
[3] INESC TEC - INESC Technology and Science and Faculty of Engineering of University of Porto, Porto, Portugal
amoreira@fe.up.pt, germano.veiga@inesctec.pt

Abstract. Automobile industry faces one of the most flexible productivity caused by the number of customized models variants due to the buyers needs. This fact requires the production system to introduce flexible, adaptable and cooperative with humans solutions. In the present work, a panel that should be mounted inside a van is addressed. For that purpose, a mobile manipulator is suggested that could share the same space with workers helping each other. This paper presents the navigation system for the robot that enters the van from the rear door after a ramp, operates and exits. The localization system is based on 3DOF methodologies that allow the robot to operate autonomously. Real tests scenarios prove the precision and repeatability of the navigation system outside, inside and during the ramp access of the van.

Keywords: Autonomous robot · Navigation · Human cooperation

1 Introduction

Automobile industry presents one of the most flexible productivity caused by the number of models variants due to the customer needs. Moreover, the workers' ergonomics should be attended. This demanding pushes the manufacturers to look for new solutions that increase the flexibility, reduce the production costs and also finds a better ergonomic posture to the worker. A particular inconvenient for the worker is the assembly operations inside the vehicle, where the human worker has to go inside/outside the vehicle several times per shift and many assembly operations are near the floor of the vehicle. The posture inside

M. F. Silva et al. (Eds.): ROBOT 2019, AISC 1092, pp. 645–656, 2020.
https://doi.org/10.1007/978-3-030-35990-4_52

the vehicle (crouch) promotes injuries. These issues encourage researchers to propose solutions that solve the problems in a collaborative way with workers. This paper presents a solution, based on a mobile manipulator, that navigates a mobile robot, in an autonomous way, to enter the van from the rear door. It also addresses the localization system and validates it in a real scenario. It demonstrates the integration of advanced methodologies of localization, navigation and control for mobile robots. The development of this heterogeneous ecosystem of navigation and localization solutions. This paper is organized as follows: after a brief introduction, the state of the art of the industrial localization systems is presented. Then, Sect. 2 addresses the use-case scenario and describes the motivation for this work. In Sect. 3, the adopted mobile robot platform is presented as well as its hardware and software components. Sections 4 and 5 address the localization and navigation methodologies for the developed system. Section 6 evaluates the results through the precision and the repeatability of the navigation system. Finally, Sect. 7 concludes the paper and points out some future work direction.

1.1 State of the Art

Actually, Industrial mobile robots (AGVs, Automatic Guided Vehicle), can self-localize and move autonomously without human intervention. They are used to transport materials between work stations in warehouses and production lines.

AGVs are used in industrial environments for more than 50 years and both the algorithms and hardware used has been evolving in order to increase the accuracy, robustness and flexibility while decreasing costs of the overall system. Although, we are focusing in industrial autonomous robots (AGVs), the localization problem is transversal to all indoor autonomous robots application areas. Regarding the localization systems applied to the industrial mobile robots, it is common to use several solutions such as [1,2]:

- Wire Guidance (following a buried cable in the floor)
- Strip Guidance (magnetic or colored strips arranged on the floor, line detection is performed by hall effect sensors or optical sensors)
- Based on Marker (embedded in the ground, markers can be magnetic labels, reflectors, passive RF, geometric shapes or bar codes)
- Trilateration and Triangulation (detecting the localization of the robot through beacons usually arranged in high parts of the walls with a laser)

Meanwhile, in the last decade localization based on natural marks has been increasing [3,6]. These natural marks are composed by a set of distances and angles to the detected objects (such as doors, walls, furniture, etc.) that can be acquired through an on-board laser range finder. This method has the main advantage of not requiring the installation of dedicated reflectors in the environment, which in some factories might not be a viable option. On the other hand, it is expected that even without special markers and straight corridors, the localization system remains robust. Besides these advantages, this approach needs to process a significant amount of sensor data efficiently in order to provide real-time localization. Therefore, the map-matching algorithms must be optimized in terms of accuracy, processing time, convergence speed and also sensor noise robustness. The map-matching is a method of self-localization for mobile robots in which the local environment map (actual data acquired by the robot) is matched with and already stored map. Authors have worked with several industry applications based on the perfect match algorithm [7,8,10]. With these topics in mind, the paper addresses the localization and navigation of a mobile platform (able to perform assembly tasks) that allow the robot to move inside the van through a ramp and position itself to operate autonomously.

2 Use-Case Description

A high dynamic environment characterizes this use case scenario where the presence of Human operators is constant. Here, the main requirements for the navigation system are to be reliable, dynamic and adaptable to the real environment. The proposed solution, based on a mobile manipulator, is composed by the localization system that allows the robot, in an autonomous way, to navigate and enter the van from the rear door. Previously, the mobile manipulator carries a kit of screws that will be delivered to the worker which is inside the van. Once inside, the wheeled robot should create the path and positioning itself where its manipulator reachability accesses the screw position. The localization can be done resorting to different approaches such as 3DoF and 6 DoF. The 3DoF localization were done with a real robot in real scenario (although a simplified one with planar surface). Navigation problem can be addressed through four different approaches:

- Navigation in the factory (outside the van)
- Navigation in the ramp
- Navigation in the ramp-van transfer
- Navigation inside the van

3 Mobile Platform Description

In this section we make a short introduction to the mobile platform used, detailing the hardware and software configuration used to test the developed navigation and localization algorithms.

3.1 Hardware Description

For our test on a real scenario we used a commercial mobile platform, which was built on top of a Husky UGV, an outdoor research robot from Clearpath Robotics [4]. It has a size of $990 \times 670 \times 390$ mm and a maximum speed of 1,0 m/s.

For the purpose of our localization and navigation tests we assembled a Sick laser LSM151 in Husky's robot front. This laser has an aperture angle of 270° and an operating range of 50 m with a scanning frequency of 25/50 Hz and an angular resolution of 0.25/0.5° (Figs. 1 and 2).

Fig. 1. ColRobot platform, based on a husky platform with UR10 arm attached [4,5].

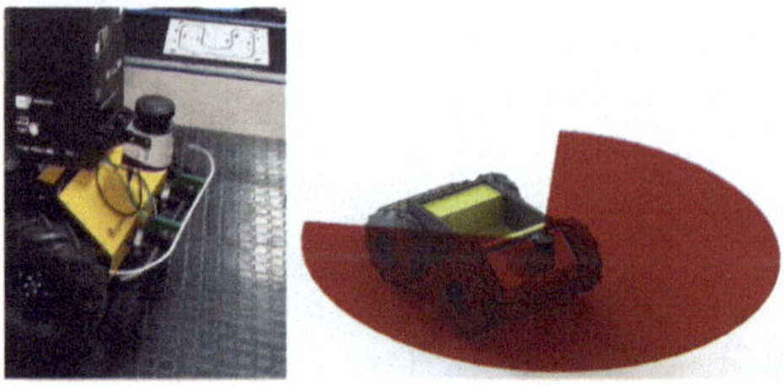

Fig. 2. Attached laser Sensor (LSM151 Sick) used in the experiment.

3.2 Software Description

Concerning the software architecture, we can divide it into several modules: (i) the localization, (ii) the decision, (iii) the controller, and (iv) the ground truth (Fig. 3).

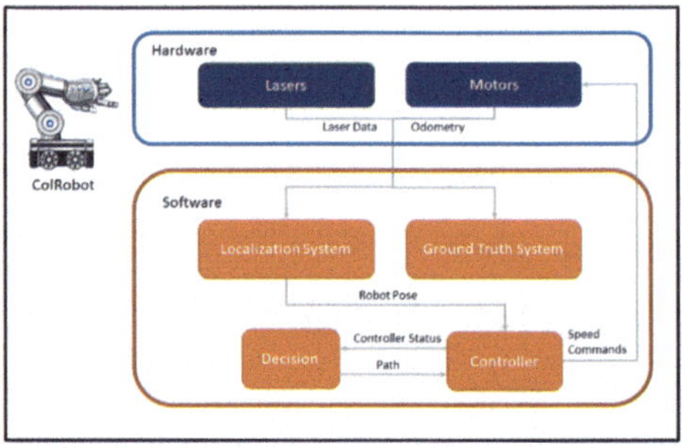

Fig. 3. The orange rectangle represents the software whereas the blue rectangle the hardware. Different modules and their interaction.

The localization system, (i), has the responsibility to determine the pose of the robot in the environment. It uses as input both the data from the Sick laser range finder and the odometry from the vehicle wheels. In this module we use a map matching algorithm, the Augmented Perfect Match, along with a map switch approach. When the robot is outside the van, we use a 2D map of the environment, built using SLAM, to determine its position. When inside the van, we use only the inner counters of the van interior, also pre acquired using SLAM, for the self-localization of the robot. Restricting the map to the van, allows the system to increase the positioning accuracy, which is important for the subsequent arm screwing operation. The map transition determined by the position of the robot path.

The decision module, (ii), is responsible for the definition of the robot trajectory, which is computed based on a fixed graph built on top of the environment map.

The controller, (iii), is responsible for guiding the robot in order to perform a trajectory. It uses as input the pose of the robot determined by the localization system.

The ground truth system (iv), has as purpose to determine the real pose of the robot with high precision. With it, we can estimate the error related to the navigation system, composed by the localization system and the controller.

4 Localization System

To solve the robot localization problem, for the use-case presented earlier, we analyzed several algorithms namely the: Augmented Perfect Match (APM), the Iterative Closest Point (ICP) and the Normal Distribution Transform (NDT). These algorithms were compared using different metrics and we have concluded that the APM is lighter in terms of computational weight, and also presents higher tolerance to orientation errors, making it a very interesting approach for the problem at hands. For more detail about this comparison please refer to [11].

Based on this conclusions we decided to use the APM for the navigation outside the van, since in this scenario the precision requirements are not so hard. Inside the van, and only if higher precision is needed, we propose to use a 6 DoF localization algorithm, LUT-ICP [11], allowing in this way to achieve a lower positional error of the robotic arm, important for some operations.

In more detail, the 3DoF localization system uses the result of APM, as a laser observation measurement, and fuses it with the vehicle's odometry data using an Extended Kalman Filter (EKF). The algorithm of Matching is based on the light computational Perfect Match algorithm, described by Lauer et al. in [12]. In this algorithm the vehicle pose is computed using 2D distance points from the surrounding environment. These points are acquired with a laser range finder and are matched with the map of the building previously computed. Therefore, the vehicle pose is calculated by trying to minimize the fitting error between the data acquired and the environment map. For details, see [9].

Despite of the solution presented, during the tests in real scenario, and in what concerns the localization, we verified that the 3 DoF localization system achieved sufficiently good results (as presented in Results section), guaranteeing the minimum requirements so that the robotic arm can carry out the remaining operations.

5 Parametric Trajectory Controller

The trajectory controller block determines the speed of wheels that allow to follow the desired trajectory, in a closed loop way, as presented in Fig. 4, based on the robot pose. During the phase where the robot is climbing to the inside of the van, it will perform a fixed trajectory. Such will avoid potential hazards for both the robot and human operators in the area. Therefore, we will use a path-following controller.

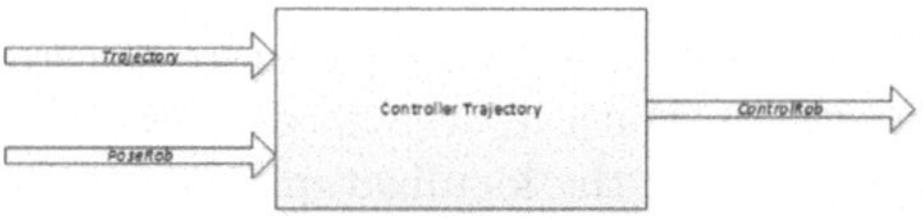

Fig. 4. Controller inputs and outputs.

The *PoseRob* represents the pose and orientation of the robot related to the absolute referential of the navigation:

$$PoseRob = \begin{bmatrix} x \\ y \\ \theta \end{bmatrix} \tag{1}$$

The *Trajectory* is composed by two parametric equations (Fx,Fy) which define a set of reference positions related to the absolute referential of the navigation (see Fig. 5). Fx and Fy are two n-order polynomials, which define such positions through the parameter *t*. The starting point of the trajectory corresponds to t = 0 and the end point corresponds to t = 1.

$$Trajectory\,(t) = \begin{bmatrix} F_x\,(t) \\ F_y\,(t) \end{bmatrix} : t \in \begin{bmatrix} 0 \ 1 \end{bmatrix} \tag{2}$$

$$F_x = \sum_{i=0}^{n-1} A_i * t^i \tag{3}$$

$$F_y = \sum_{i=0}^{n-1} B_i * t^i \tag{4}$$

The *ControlRob* is the control command sent to the hardware and is composed by linear (V) and angular velocity (W).

$$ControlRob = \begin{bmatrix} V \\ W \end{bmatrix} \tag{5}$$

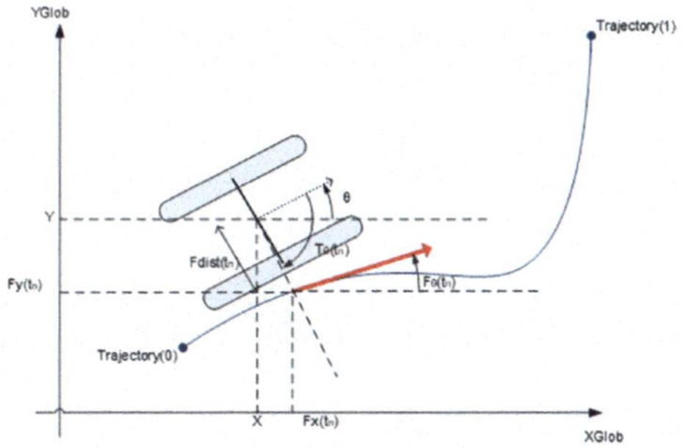

Fig. 5. Controller trajectory.

A path-following controller tends to minimize two types of errors. The first one that represents the distance between the robot and the path, and the second one related to the difference between the orientation of the robot and the orientation of the path.

The variable t_n represents the parameter t which minimizes the distance between a position on the trajectory and the pose of the robot.

$F_{Dist}(t_n)$ is the distance error between the robot and the trajectory:

$$F_{Dist}\left(t_n\right) = \sqrt{\left(F_x\left(t_n\right) - x\right)^2 + \left(F_y\left(t_n\right) - y\right)^2} \tag{6}$$

$F_{\theta}(t_n)$ is the reference orientation for the robot defined by *Trajectory*(t_n)

$$F_{\theta}\left(t\right) = Atan2\left(\frac{\partial F_x\left(t\right)}{\partial t}, \frac{\partial F_y\left(t\right)}{\partial t}\right) \tag{7}$$

where

$$\frac{\partial F_x\left(t\right)}{\partial t} = \sum_{i=1}^{n} i * A_i * t^{i-1} \tag{8}$$

$$\frac{\partial F_y\left(t\right)}{\partial t} = \sum_{i=1}^{n} i * B_i * t^{i-1} \tag{9}$$

$Err_{\theta}(t_n)$ is the orientation error between the robot and the trajectory and can be calculated as follows:

$$Err_{\theta}\left(t_n\right) = NormAng\left(F_{\theta}\left(t_n\right) - \theta\right) \tag{10}$$

$T_{\theta}(t_n)$ is the angle defined by the pose of the robot and the closest point on the trajectory.

$$\begin{aligned} T_{\theta}\left(t_n\right) = NormAng(&Atan2(F_x\left(t_n\right) - x, \\ &F_y\left(t_n\right) - y) - \theta) \end{aligned} \tag{11}$$

The linear velocity (V) is constant and one of the arguments passed with the trajectory.

The angular velocity (W) is defined as a function of the distance and orientation error and the value of the feedforward (*feedForward(t_n)*).

if $T_\theta(t_n) > 0$:

$$\begin{aligned} w = &K_{P,D} * Err_D(t_n) + K_{I,D} * \int Err_D(t_n) \\ &+ K_{P,\theta} * Err_\theta(t_n) + K_{I,\theta} * \int Err_\theta(t_n) \\ &+ FeedForward(t_n) \end{aligned} \tag{12}$$

else:

$$\begin{aligned} w = &-K_{P,D} * Err_D(t_n) - K_{I,D} * \int Err_D(t_n) \\ &+ K_{P,\theta} * Err_\theta(t_n) + K_{I,\theta} * \int Err_\theta(t_n) \\ &+ FeedForward(t_n) \end{aligned} \tag{13}$$

$K_{P,D}$ and $K_{I,D}$ are parameters regarding the proportional and integral of the controller related to the distance error, while $K_{P,\theta}$ and $K_{I,\theta}$ are related to the orientation error.

Regarding the $feedForward(t_n)$, this is determined by applying the derivative to the following equations:

$$R = \frac{V}{W} \tag{14}$$

where, R is the radius of the trajectory determined by the linear and angular velocity. The radius of the trajectory $Trajectory(t)$ is defined by the following equations:

$$T_{Radius}(t) = \frac{\sqrt{\left(\frac{\partial F_x(t)}{\partial t}\right)^2 + \left(\frac{\partial F_y(t)}{\partial t}\right)^2}}{\frac{\partial F_\theta(t)}{\partial t}} \tag{15}$$

where:

$$\frac{\partial F_\theta(t)}{\partial t} = \frac{\frac{\partial^2 F_y(t)}{\partial t^2} * \frac{\partial F_x(t)}{\partial t} - \frac{\partial^2 F_x(t)}{\partial t^2} * \frac{\partial F_y(t)}{\partial t}}{\left(\frac{\partial F_x(t)}{\partial t}\right)^2 + \left(\frac{\partial F_y(t)}{\partial t}\right)^2} \tag{16}$$

Combining Eqs. 14, 15 and 16 and assuming R is $T_{Radius}(t)$ we have the following definition of $feedForward(t_n)$:

$$\begin{aligned} &feedForward(t) \\ &= \frac{V * \left(\frac{\partial^2 F_y(t)}{\partial t^2} * \frac{\partial F_x(t)}{\partial t} - \frac{\partial^2 F_x(t)}{\partial t^2} * \frac{\partial F_y(t)}{\partial t}\right)}{\sqrt{\left(\frac{\partial F_x(t)}{\partial t}\right)^2 + \left(\frac{\partial F_y(t)}{\partial t}\right)^2}\left(\left(\frac{\partial F_x(t)}{\partial t}\right)^2 + \left(\frac{\partial F_y(t)}{\partial t}\right)^2\right)} \end{aligned} \tag{17}$$

6 Results

As a way to evaluate the precision and repeatability of the navigation system inside a van, we conducted a set of experiments, where, the robot executed two trajectories autonomously, allowing it to climb into the van and out of it. When it was inside the van, we activated the ground truth system to determine and evaluate its final pose.

6.1 Ground Truth System Description and Characterization

The ground truth system relied on a beacon-based localization algorithm installed in the van, which resorts to a method implemented by Sobreira et al. [10]. Such system estimates a pose of a robot through an Extended Kalman Filter while using cylindrical beacons and it has a position error of 0.005 m and an orientation error of 0.2°. On our application, we installed four beacons inside the van.

As the presence of elements with high reflectivity may affect the detection of dark objects, we covered the beacons while the robot was climbing into the van. When the robot was inside of it, we uncovered the beacons and determined its final pose.

6.2 Precision and Repeatability Results

The main goal of tests presented in this section was to evaluate the precision and repeatability of the navigation system inside a van. The results from 20 experiments are presented in the following table and figures.

The Table 1 represents the position and orientation maximum errors, the standard deviations and the average of the absolute errors.

Table 1. Position and orientation maximum errors

	X(m)	Y(m)	θ(deg)
Maximum error	0,019	0,014	2,216
Standard deviation	0,005	0,005	0,598
Average of absolute errors	0,004	0,004	0,351

The Fig. 6 represents a histogram of the position error of the X axis, while the Fig. 7 demonstrates the normal distribution of the errors we observed on our experiments.

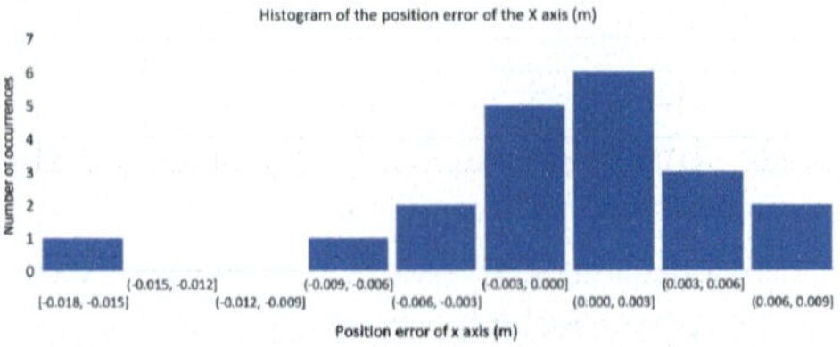

Fig. 6. Histogram of the position error of the X axis.

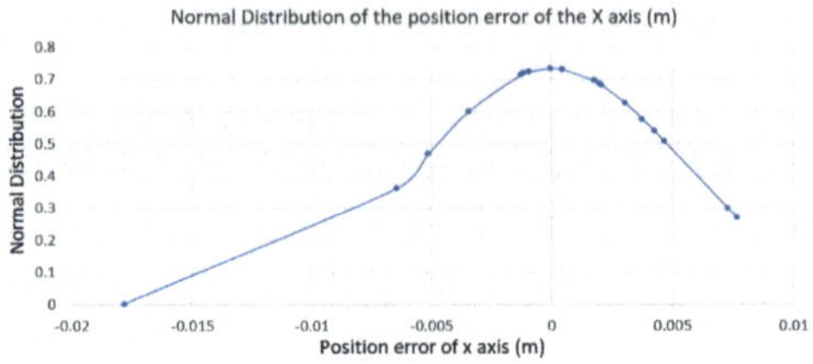

Fig. 7. Normal distribution of the position error of the X axis.

The Fig. 8 represents a histogram of the position error of the Y axis, while the Fig. 9 demonstrates the normal distribution of the errors we observed on our experiments.

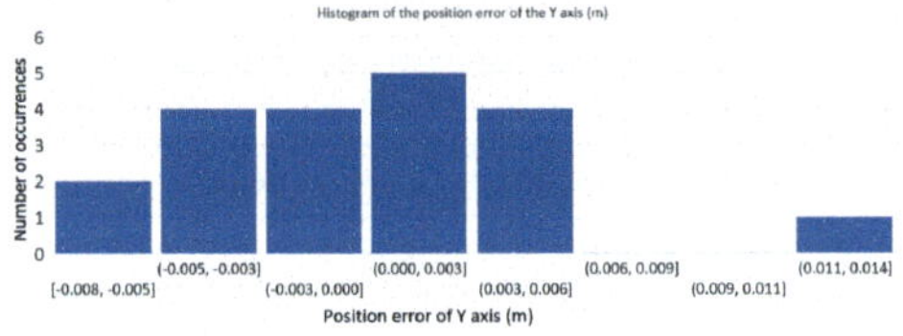

Fig. 8. Histogram of the position error of the Y axis.

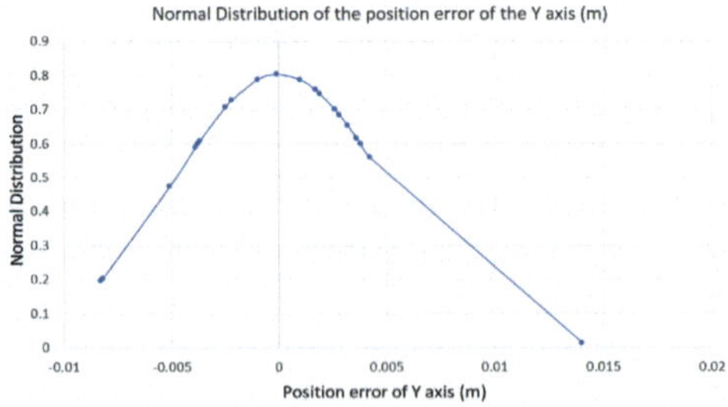

Fig. 9. Normal distribution of the position error of the Y axis.

The Fig. 10 represents a histogram of the orientation errors, while the Fig. 11 demonstrates the normal distribution of the errors we observed on our experiments.

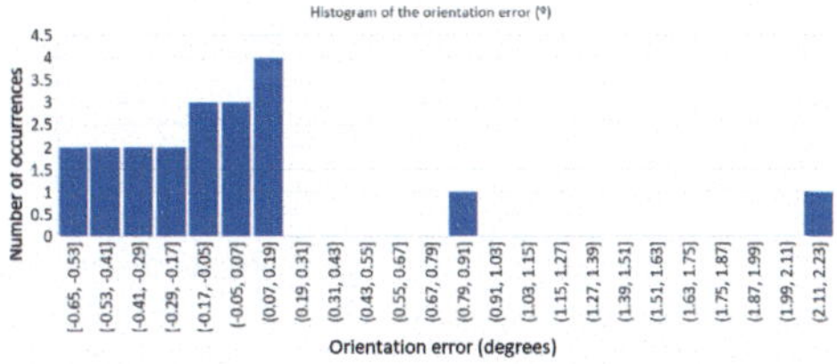

Fig. 10. Histogram of the orientation error.

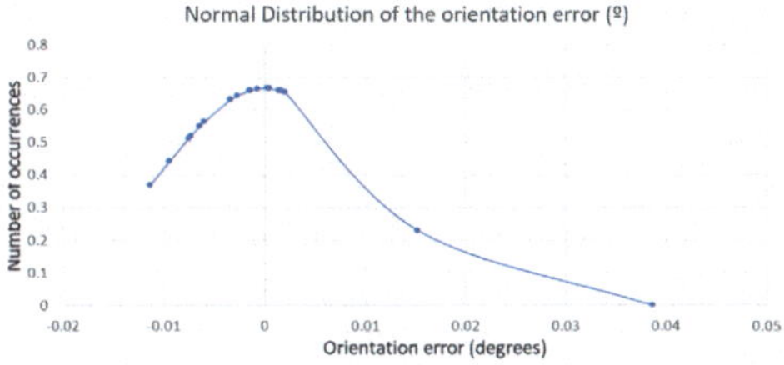

Fig. 11. Normal distribution of the orientation error.

Analyzing Table 1, it is possible to verify that our system had 0.019 m as maximum error in X, 0.014 m in Y and 2.216° in the orientation of the robot. However, observing each figure, we can notice a presence of an outlier, which influenced the maximum error observed for each variable. This outlier was due to the irregularity of the van's floor. This could have caused the robot to be tilted to one side during the trajectory, which induced in error the localization algorithm. Even though the maximum error was influenced by the presence of each outlier, such errors can be corrected by a vision system with a camera close to the robotic arm end-effector.

7 Conclusion and Future Work

The present work addresses the navigation system for the robot that enters the van from the rear door after a ramp, operates and exits. It is intended that should mounted a panel inside the van. It is suggested a mobile manipulator that could collaborate with workers helping each other. The presented localization system is based on 3DOF methodologies that allow the robot to localize, navigate and operate autonomously. Real tests scenarios prove the precision and repeatability of the navigation system outside, during the access ramp of the van and presents a maximum error of 0.019 m in X, 0.014 m in Y and 2.216° in the orientation.

Acknowledgment. This work is financed by the ERDF – European Regional Development Fund through the Operational Programme for Competitiveness and Internationalisation - COMPETE 2020 Programme, and by National Funds through the Portuguese funding agency, FCT - Fundação para a Ciência e a Tecnologia, within project SAICTPAC/0034/2015- POCI-01- 0145-FEDER-016418.

References

1. Schulze, L., Wullner, A.: The approach of automated guided vehicle systems. In: 2006 IEEE International Conference on Service Operations and Logistics, and Informatics, pp. 522–527 (2006)
2. Schulze, L., Behling, S., Buhrs, S.: Automated guided vehicle systems: a driver for increased business performance. In: Proceedings of the International MultiConference of Engineers and Computer Scientists, pp. 19–21 (2008)
3. Pinto, M., Sobreira, H., Moreira, A., Mendonça, H., Matos, A.: Self-localisation of indoor mobile robots using multi-hypotheses and a matching algorithm. Mechatronics **23**(6), 727–737 (2013)
4. Clearpath Robotics manufacturer site. https://www.clearpathrobotics.com/. Accessed 01 Sept 2019
5. ColRobot project site. https://www.colrobot.eu/. Accessed 01 Sept 2019
6. Tomatis, N.: BlueBotics: navigation for the clever robot [entrepreneur]. IEEE Robot. Autom. Mag. **18**(2), 14–16 (2011)
7. Sobreira, H., Moreira, A., Costa, P., Lima, J.: Mobile robot localization based on a security laser: an industry scene implementation. In: ROBOT 2015 - Second Iberian Robotics Conference, November 2015
8. Sobreira, H., Moreira, A., Costa, P., Lima, J.: Robust mobile robot localization based on security laser scanner. In: 2015 IEEE International Conference on Autonomous Robot Systems and Competitions (2015)
9. Sobreira, H., Pinto, M., Moreira, A., Costa, P., Lima, J.: Robust robot localization based on the perfect match algorithm. In: 11th Portuguese Conference on Automatic Control CONTROLO 2014. Lecture Notes in Electrical Engineering (2014)
10. Sobreira, H., Moreira, A., Costa, P., Lima, J.: Robust mobile robot localization based on a security laser: an industry case study. Ind. Robot: Int. J. **43**(6), 596–606 (2016). https://doi.org/10.1108/IR-01-2016-0026
11. Sobreira, H., Costa, C., Sousa, I., Rocha, L., Lima, J., Farias, A., Costa, P., Moreira, A.: Map-matching algorithms for robot self-localization: a comparison between perfect match, iterative closest point and normal distributions transform. J. Intell. Robot. Syst. **93**(3–4), 533–546 (2018)
12. Lauer, M., Lange, S., Riedmiller, M.: Calculating the perfect match: an efficient and accurate approach for robot self-localization. In: RoboCup Symposium, Osaka, Japan, 13–19 July 2005, pp. 142–153 (2005)

Smart Data Visualisation as a Stepping Stone for Industry 4.0 - a Case Study in Investment Casting Industry

Ana Beatriz Cruz[1(✉)], Armando Sousa[2], Ângela Cardoso[3], Bernardo Valente[4], and Ana Reis[5]

[1] FEUP, Porto, Portugal
up201403564@fe.up.pt
[2] INESC-TEC and FEUP, Porto, Portugal
[3] INEGI, Porto, Portugal
[4] Zollern & Comandita, Maia, Portugal
[5] INEGI and FEUP, Porto, Portugal

Abstract. With present day industries pressing for retrofitting of current machinery into Industry 4.0 ideas, a large effort is put into data production, storage and analysis. To be able to use such data, it is fundamental to create intelligent software for analysis and visualisation of a growing but frequently faulty amount of data, without the quality and quantity adequate for full blown data mining techniques. This article case studies a foundry company that uses the lost wax method to produce metal parts. As retrofitting is underway, modelling, simulation and smart data visualisation are proposed as methods to overcome data shortage in quantity and quality. The developed data visualisation system is demonstrated to be adapted to the requirements and needs of this company in order to approach full automation ideas. Such data visualisation system allow workers and supervisors to know in real time what is happening in the factory, or study the passage of manufacturing orders for a specific area. Data Analysts can also predict machinery problems, correct issues with slow changing deviations and gather additional knowledge on the implementation of the process itself.

1 Introduction

Nowadays, we are going through the fourth industrial revolution, also known as Industry 4.0. This is a digital revolution, whose main objectives are increasing the efficiency of operation and productivity, as well as increasing the level of automation, thus making the companies more competitive, as concluded in [1].

This digital revolution is driving the implementation of tools and intelligent platforms that produce a greater amount of data and information for analysis, as explained in [2]. The storage of large amounts of data allows an analysis of the conditions in which a product was created, as well as the use of machine learning techniques, to make an early prediction of the occurrence of product defects or machine failures.

M. F. Silva et al. (Eds.): ROBOT 2019, AISC 1092, pp. 657–668, 2020.
https://doi.org/10.1007/978-3-030-35990-4_53

However, not all companies are prepared for this type of revolution, because many have at least some very primitive processes, where human work and control predominate. An example are investment casting companies that use the lost wax casting method, where at least the last part of the process is mostly manual. In this type of manufacturing, the occurrence of failures in some sections, more specific in the casting department, is quite frequent and difficult to control. Generally, data acquisition and monitoring are also very scarce, due to the existing manufacturing processes.

For this project, the investment casting company *Zollern & Comandita*, was used as a case study, with the objective of developing a data visualisation system. The project began with a study of the manufacturing process. Then, because it is the section where the managers have less access to the information necessary to make their decisions, the focus was placed in the foundry department. Since data collection is at its early stages, it was necessary to develop a simulator, which produces relevant data, very close to what would be real.

Subsequently, an expert data visualisation system was developed, to allow an intuitive comprehension of the information. This analysis can be done in real time, in fact the system is capable of recognising deviations from what is expected and emits alerts when something goes wrong. The tool also allows to view data from a finished production order, which makes it possible to understand the conditions of its production, as well as to make the cause-effect study of its defects or qualities.

2 Industry Context, Objectives and Requirements

Lost wax casting is a method of producing metal parts with high precision. This process has eight stages of development until the final product is obtained, as shown in Fig. 1 and demonstrated on the web page [3].

First, wax forms of the product to be manufactured are injected. Then one or more of these pieces are welded to a common trunk of wax, which is called the tree. Subsequently, several layers of ceramic are made around this tree. After the ceramic layers are thoroughly dried, the inner wax is removed so that the metal alloy can be poured in a liquid state into the tree. After the metal solidification, the ceramic is broken and the final product is obtained by separating the parts of the common trunk.

A study was done at *Zollern & Comandita* to understand which section has the most issues and the foundry was the one that stood out in the high number of incidences. This section consists of the phases of wax removal, tree sinterization, metal alloy preparation and pouring.

At the foundry, the trees are placed inside a furnace (autoclave), and are subjected to a pressure of 9 bar for 15 min. This high pressure removes about 95% of the wax inside the tree. After that, the trees are stored until there is room to move to a second furnace. This is a rotary furnace, where the trees are placed in small sections, called buds, with at most three trees per bud, and run through the five different zones within it until they reach the exit. Figure 2 represents this furnace and its structure.

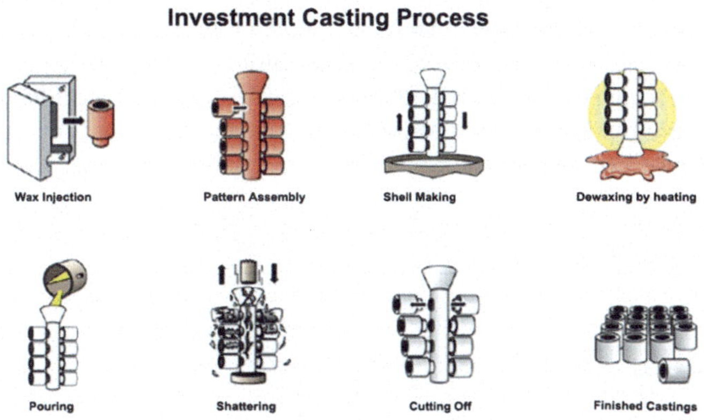

Fig. 1. Investment casting process, as shown in [5].

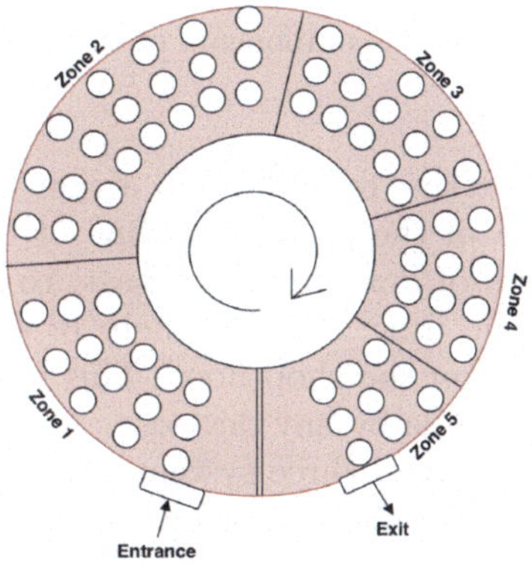

Fig. 2. Scheme of the rotary furnace.

The first zone of the furnace aims to burn the remaining wax inside of the tree. Trees must stay in this area for 15 min or more, depending on the type of metal part being produced. Zones 2, 3, 4 and 5 aim to increase the temperature and to sinter the trees to be cast. This phase is important to avoid thermal shocks and prevent the metal alloy from starting to solidify before reaching all areas of the interior of the tree.

Whilst the trees are in the rotary furnace, the metal alloy to be poured is being prepared in an induction oven. At the moment, there is very little data being collected from the induction ovens, only the temperatures of the alloy are measured and even that is done very sparsely. Once the trees have gone through all zones of the rotary furnace, which takes approximately 2 h, and the metal alloy is prepared, pouring begins, but no data is collected.

Although there are ideas to eventually collect more data from alloy preparation and pouring, those phases are not included in this project, because the specific way in which the data will be collected is yet to be decided.

Figure 3 explains the scheme sequence of the foundry section. At *Zollern & Comandita* there are two autoclaves, two rotary furnaces and three induction furnaces.

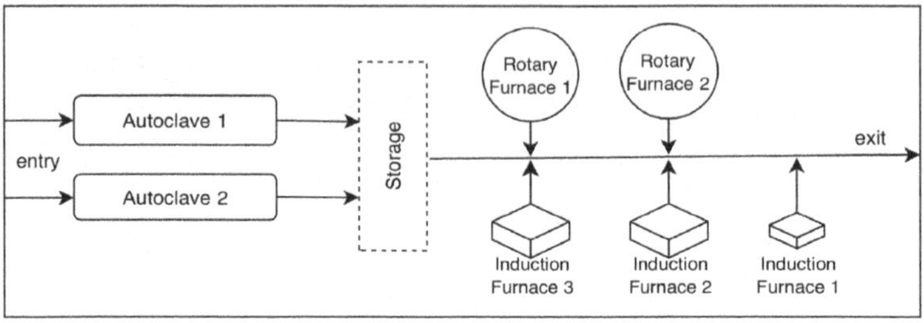

Fig. 3. Scheme of the main components of the foundry department.

This project's main objective is to develop a system where it is possible to visualise the collected data of the production line of the company. It must be able to filter large amounts of data from the past, to something more specific and easy to interpret by the user. This will help in understanding and better tracking what is happening on the shop floor. This system must also have the characteristic of being accessible by any device.

Based on the above objectives, three system requirements were devised:

R1. View data from a finished production order: Viewing data from a production order that has already been completed allows the managers to establish cause-and-effect relationships. These relationships will identify the characteristics in the manufacturing process that led to defects in the final product.

R2. Get real-time insight into what is happening on the shop floor: Understanding what is happening on the shop floor in real time is important for section managers. That way, they can quickly perceive what is being produced and if everything is working within the expected parameters.

R3. Emit alerts if certain variables are outside the expected values: Real-time warnings when important manufacturing conditions are not verified may allow the managers to quickly correct the problem, or at least to interrupt production.

3 State of the Art

With the high amount of data coming from Industry 4.0, some systems were developed to allow its visualisation. These systems have been created with in general purpose, so that they can be applied in any context, giving the user the possibility to choose the way of presenting the data. Following are two well known examples of this kind of data viewers.

The ***Q-DAS*** [6] software is specialised in the computerisation of statistical procedures with a focus on quality management applications including Statistical Processes. This software has several tools, such as *QS-STAT*, which allows the user to make statistical analyses of the collected data by producing analysis reports, which refer to the values collected by sensors at certain time intervals.

Another platform that allows visualisation of data is ***Grafana*** [7], which is a dashboard that allows the user to query, visualise, alert and understand data metrics, regardless of where they are stored.

Despite their wide usage, these two platforms are for local use, that is they can only be used in a computer that has the software installed. As such, these tools are not always the best for companies that are in the initial stages of implementing Industry 4.0 ideas.

4 Data Structure

Data is one of the main elements for the development of this work. As such, an architecture was created for the database, facilitating its access and interpretation.

4.1 Production Data

To study and analyse what is being produced, it is necessary to keep all the information from production orders. The tables shown in Fig. 4 were created to store only informative/static data. A production order has associated a type of part to be produced and several trees. Once the trees enter the foundry department they are organised in shelves, hence there is another table that associates the trees with their respective shelves.

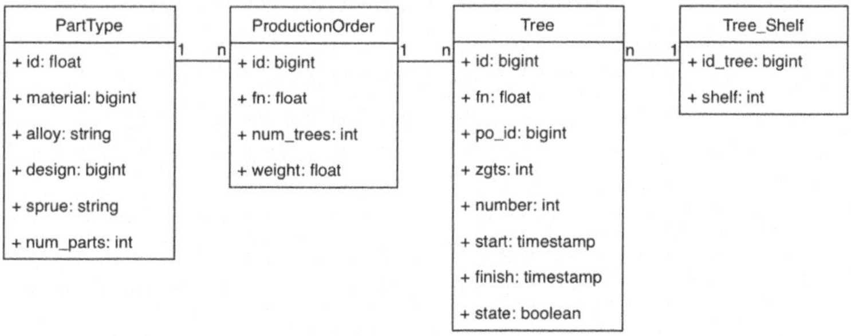

Fig. 4. Database scheme for the main production information.

4.2 Autoclave Data

The foundry department starts with the processing of shelves of trees through the autoclave. The data of this part of the process is stored in three tables. The first table stores general information about the autoclave cycles, while the second and third tables store temperature and pressure data, for each second that the cycle lasts, in the main chamber and the steam generator. There is also a table which keeps information about the production orders that have already exited the autoclave and are ready for the rotary furnace. The data in this table is inserted through a trigger that runs when a shelf exits the autoclave.

Figure 5 demonstrates the architecture of autoclave part of the database and the trigger (red arrow) between the *Autoclave* table and the *Storage* table.

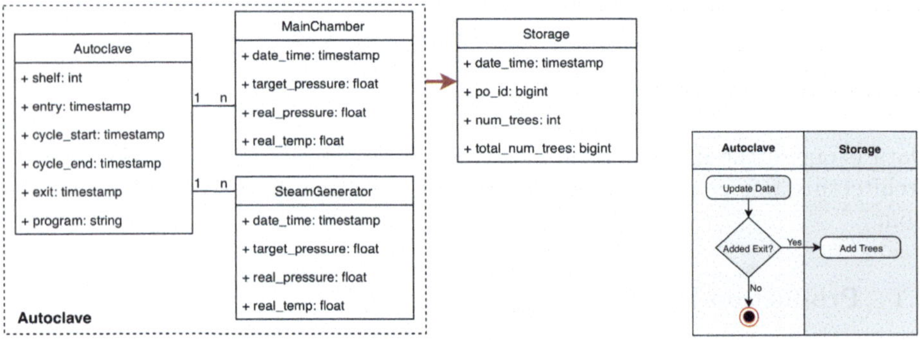

(a) Autoclave tables (b) Trigger

Fig. 5. Database scheme for the autoclave section with the trigger from the *Autoclave* table to the *Storage* table.

4.3 Rotary Furnace Data

In the case of the rotary furnace, it is necessary to record the location of each bud over time. Just knowing the bud that is in the furnace entrance at each moment, it is possible to know the ids of the remaining first buds of each zone, using Eq. 1, where e is the id of the entrance bud and n_k is the number of buds from the entrance to the start of zone k and b_k is the id of the bud at the start of zone k.

$$\begin{aligned} b_k &= ((((e + n_k - 1) \bmod 24) + 24) \bmod 24) + 1, \\ k &= \{2, 3, 4, 5\}, \\ n_k &= \{5, 12, 17, 21\} \end{aligned} \tag{1}$$

The tables for the rotary furnace are depicted in Fig. 6. In particular, the table *Bud_Zone*, stores the bud at the start of each zone over time. This information is generated with a trigger on each rotary furnace advance, which is represented by the green arrow in Fig. 6 and the scheme of left side of Fig. 7. Thus, it is easier to know how much time each tree spent in each zone, given that we also know the bud in which each tree is loaded.

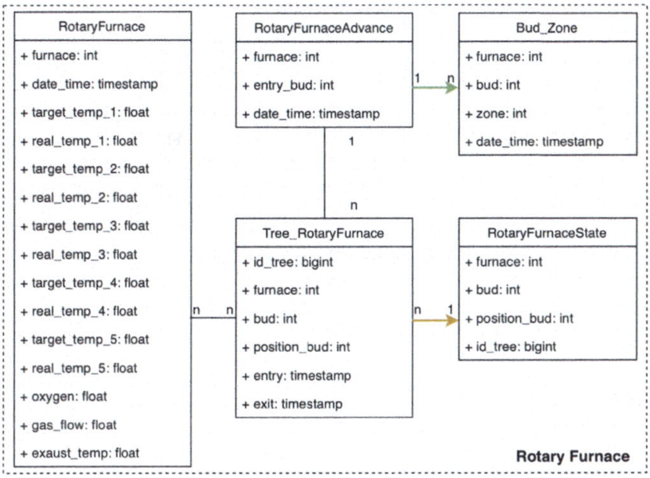

Fig. 6. Database scheme for the rotary furnace section.

To make reading the current state of the rotary furnace simpler, each time a new tree is inserted into the rotary furnace, the *RotaryFurnaceState* table is updated by a trigger, which is represented by the yellow arrow in Fig. 6 and the scheme on the right side of Fig. 7. The *RotaryFurnaceState* table always has the current state of each bud, saving the *id* of the trees it has at that moment, without maintaining history.

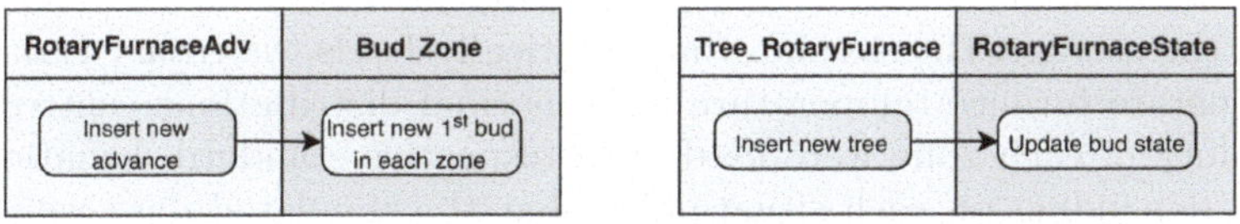

Fig. 7. Triggers on the rotary furnace database.

4.4 Alerts Data

In order for the managers to have instant alerts when an issue arises, the database also includes an alerts table, whose diagram is shown on the left of Fig. 8. The entries of this table are entirely generated by triggers. For example, when a new autoclave cycle starts, if the steam generator pressure is not approximately 12 bar, a new alert is added to the table as shown by the scheme on the right of Fig. 8.

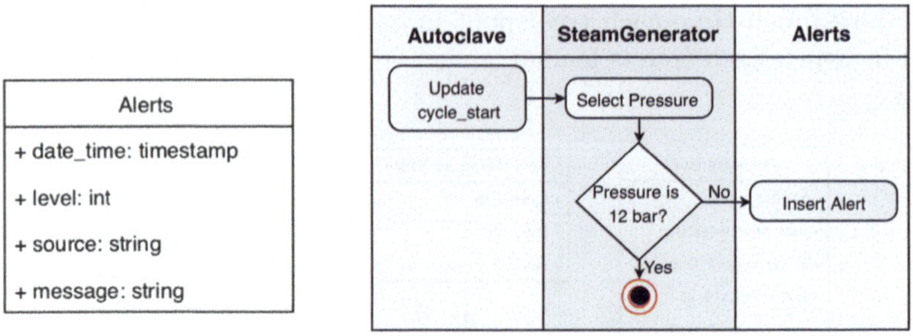

(a) Table of alerts (b) Example of a trigger of an alert

Fig. 8. Alerts data.

5 Simulation Tool

At the start of this project, *Zollern & Comandita* did not have all the data necessary for the visualisation tool to be built. As such, it was necessary to develop a simulator to generate the required data. This tool is designed to automatically populate all tables in the database, with values very close to the expected ones.

For the tables with the information of the pieces to be produced, some real samples were used while others where simulated to look like the real samples.

After information of the manufacturing orders is inserted in the database, the system begins to simulate the entire manufacturing process of the foundry department in real time. The autoclave and the rotary furnace are simulated by different components, which operate independently and are only connected by the data in the database. The simulator generates the sporadic event of the arrival of a new shelf in the autoclave. As soon as a shelf exits the autoclave, it is added to the *Storage* table by a trigger. With this table, the simulator of the rotary furnace always knows the shelves it has available to use. Every time the rotary furnace has room for a new production order and there are completed orders in the storage, the simulator automatically loads the rotary furnace.

To generate furnace temperature data, normal distributions centred on the target values of each zone were used. The respective standard deviations were adjusted, depending on each situation so that the simulated data would be as similar as possible to real samples of the rotary furnace.

The simulation tool generates data and stores it in the database. This data can be used in developing any tool for the company, as well as studying what is more relevant to monitor. When real data is available, the company can store it in the database, replacing the simulator, and the remaining components of this project will continue to work.

6 Visualisation Tool

To be able to visualise the data, a graphical interface has been developed. The idea is to add features to this interface in due time, as the managers decide what is more relevant for them to see, but also as data becomes available and is integrated in the database, throughout the whole factory. A web interface was chosen so that it is easily accessed from any computer or mobile device that has a network connection. It was taken into account that the interface should be responsive to adapt to the size of the screen of the device being used.

The visualisation system was designed as a typical web application, with a front end component, responsible for presenting the information to the users, and a back end component, which manages the access to the database, as shown in Fig. 9. Communication between these two components is achieved through HTTP (Hypertext Transfer Protocol) requests, which comply with the REST (Representational State Transfer) architectural style.

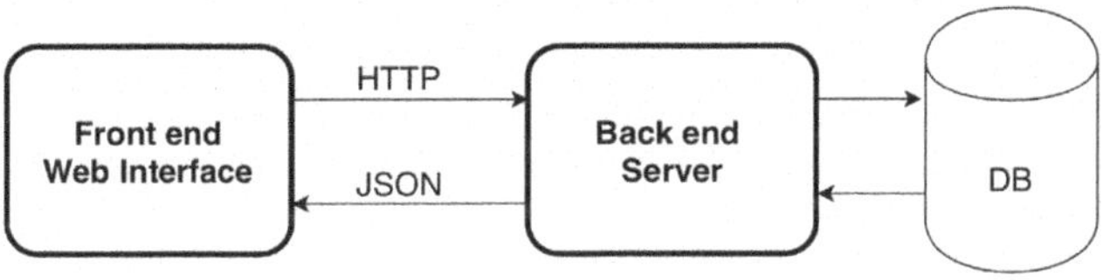

Fig. 9. Architecture of the visualisation tool.

For the development of front end component, the web framework *Angular* was used, because it offers a wide variety of tools that help with the development of such interfaces. In particular, the plugin *Chart.js* was used for the graphs, and the framework *Bootstrap* was used in order to easily obtain a responsive user interface.

Due to the fact that *Angular* is an MVC (Model View Controller) based framework, but also because this architectural style is well suited for web applications, the architecture of the front end follows the MVC pattern. As such, there is a representation of the data structure in the Model, which is simultaneously used as source of the information displayed in the View and queried by the Controller upon user interaction.

As for the server, the *DJango* framework was used, which follows the Model Template View (MTV) architectural style. MTV is similar to MVC, but depending on the source, there are some differences. In any case, the core ideas of code separation according to its purpose are the same. For the back end, the Model was used to represent the data and communicate with the database, while the View is responsible for replying to the HTTP requests from the front end using the JSON format. There is also a layer of business logic, that is responsible for the necessary processing of data before sending it to the front end. Typically, the Template layer is responsible for presenting the information to the user, not the content itself, but the way it is presented. For the back end component, this layer was not developed, because that responsibility belongs to the front end component.

7 Results

In this section, the results of the database, simulator and visualisation tool developed up to this point for the foundry department of *Zollern & Comandita* are presented, while validating the requirements elicited in Sect. 2.

R1. View data from a finished production order

With the developed tool it is possible to visualise data generated by the simulator that represents the passage of a certain production order by the existing furnaces. For example, one can see the data simulated for the autoclave and the resulting charts. Figure 10 contains a graph of the pressure variation in the main chamber during a full autoclave cycle, corresponding to a given production order. This information is preprocessed in the back end, which determines in which shelves that order was stored, before entering the autoclave, and then obtains the cycle data for those shelves. In this case, only one graph is shown because the whole order was stored in a single shelf.

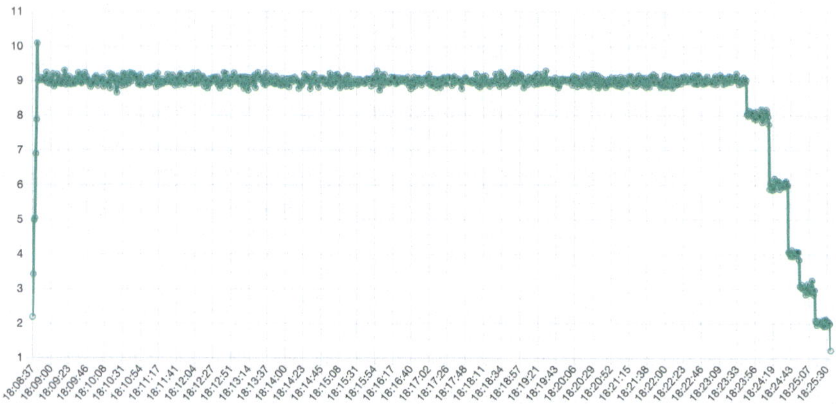

Fig. 10. Variation of the pressure in the main chamber of the autoclave during the cycle of the shelf containing production order 8127337.

R2. Get real-time insight into what is happening on the shop floor

One of the features requested by the managers of the foundry department was the ability to quickly check the state of the rotary furnace, which was implemented through the scheme in Fig. 11. This diagram shows how the furnace is loaded in real time, but it can also be used to check the progress of a given production order within the furnace over time. For each position of each bud one can see, if it is loaded and the number of the tree in that position. The chart is also interactive, allowing the user to scroll the mouse over to obtain a quick overview of the contents, or to click in a bud and obtain more detailed information.

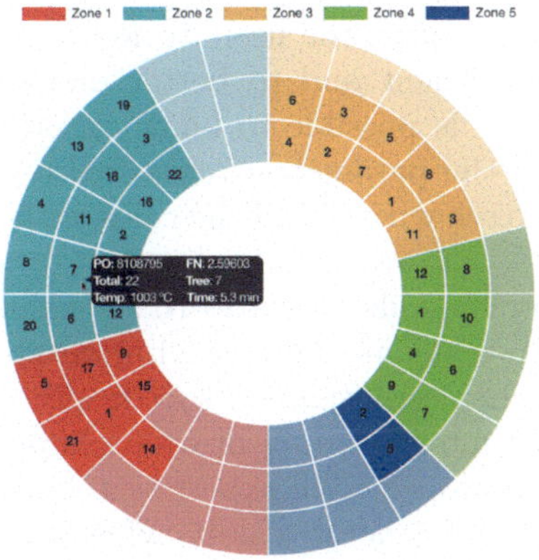

Fig. 11. Representation of the rotary furnace in the data visualisation system.

R3. Emit alerts if certain variables are outside the expected values

To verify this requirement, specific failures were introduced in the simulator, activating the triggers that generate alerts in the database. These alerts are received in the visualisation tool and shown both in the form of a pop-up and in a drop-down notifications menu, as depicted in Fig. 12.

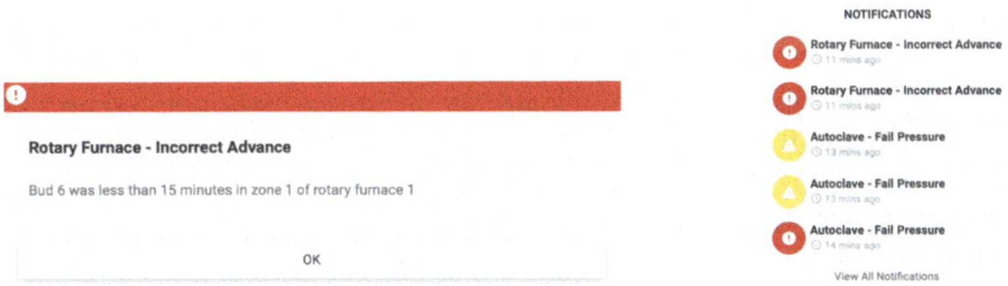

(a) Pop-up notification (b) Drop-down notifications

Fig. 12. Notifications on visualisation tool

8 Conclusions

As has been shown, it is essential that data visualisation systems are developed to facilitate access and interpretation of the large amounts of data acquired.

This work proposes a data visualisation system that allows the company *Zollern & Comandita* the possibility to understand what happens on the shop floor in real time in a simpler and more intuitive way. Wherever the users are, they can know what is being produced at that moment and verify if the whole process is following its normal operation.

The developed tool also allows the users to follow a given production order through the foundry department and to better understand the probable origins of high defect rates. It is possible to know exactly when an order passed through each furnace and the parameters registered during their processing. It will, therefore, be easier to see whether all the rules of the manufacturing process have been met or not.

Existing tools have some similar features to the visualisation tool developed in this work, like seeing in real time data saved in database. However, this visualisation tool has some schemes developed specifically to be applied at *Zollern & Comandita*, to help the interpretation of whats happening in the shop floor. In addition, the system developed is an expert system, it compares process data with expected parameters and generates alerts when something is wrong, so that the managers can take immediate action.

With this tool, the company may also conduct a cause-and-effect study in the foundry department. This will improve its manufacturing process, making it more efficient and able to decrease the percentage of defective parts.

Acknowledgements. Authors gratefully acknowledge the funding of Project MAGIC_4.0: Afinação de grão por correntes eletromagnéticas rotativas e tecnologias da indústria 4.0 para a fundição por cera perdida (POCI-01-0247-FEDER-038128), co-financed by Programa Operacional Competitividade e Internacionalização (COMPETE 2020) through Fundo Europeu de Desenvolvimento Regional (FEDER).

References

1. Lu, Y.: Industry 4.0: a survey on technologies, applications and open research issues. J. Industr. Inf. Integr. **6**, 1–10 (2017). https://doi.org/10.1016/j.jii.2017.04.005
2. Zysman, J., Kenney, M.: The next phase in the digital revolution: Intelligent tools, platforms, growth, employment. Commun. ACM **61**, 54–63 (2018). https://doi.org/10.1145/3173550
3. Pattnaik, S., Karunakar, D.B., Jha, P.K.: Developments in investment casting process–a review. J. Mater. Process. Technol. **212**(11), 2332–2348 (2012). https://doi.org/10.1016/j.jmatprotec.2012.06.003
4. Morgan, R., Grossmann, G., Schrefl, M., Stumptner, M.: A model-driven approach for visualisation processes. In: ACM International Conference Proceeding Series, art. no. a55 (2019). https://doi.org/10.1145/3290688.3290698
5. Sand Casting, Investment Casting and Die Casting in China. www.castingquality.com/casting-technology/investment-casting-tech/investment-casting-process.html
6. Q-DAS. https://www.q-das.com/en
7. Grafana Labs. https://grafana.com

Development of an Autonomous Mobile Towing Vehicle for Logistic Tasks

Cláudia Rocha[1], Ivo Sousa[1], Francisco Ferreira[1], Héber Sobreira[1], José Lima[1,2(✉)], Germano Veiga[1,3], and A. Paulo Moreira[1,3]

[1] INESC TEC - INESC Technology and Science, Porto, Portugal
{claudia.d.rocha,ivo.e.sousa,francisco.a.rodrigues,heber.m.sobreira, germano.veiga}@inesctec.pt

[2] CeDRI - Research Centre in Digitalization and Intelligent Robotics, Polytechnic Institute of Bragança, Bragança, Portugal
jllima@ipb.pt

[3] Faculty of Engineering of University of Porto, Porto, Portugal
amoreira@fe.up.pt

Abstract. Frequently carrying high loads and performing repetitive tasks compromises the ergonomics of individuals, a recurrent scenario in hospital environments. In this paper, we design a logistic planner of a fleet of autonomous mobile robots for the automation of transporting trolleys around the hospital, which is independent of the space configuration, and robust to loss of network and deadlocks. Our robotic solution has an innovative gripping system capable of grasping and pulling non-modified standard trolleys just by coupling a plate. Robots are able to navigate autonomously, to avoid obstacles assuring the safety of operators, to identify and dock a trolley, to access charging stations and elevators, and to communicate with the latter. An interface was built allowing users to command the robots through a web server. It is shown how the proposed methodology behaves in experiments conducted at the Faculty of Engineering of the University of Porto and Braga's Hospital.

Keywords: Mobile robot · Autonomous driving · Trolley docking · Ergonomics

1 Introduction

Even nowadays, carrying high loads and performing repetitive tasks is majorly done manually by employees. In the particular case of hospital environments, humans are responsible for the transportation of meal trolleys, laundry carts and other logistic tasks. These procedures tend to be recurrent and performed during long distances. Adding this to the high weight of the cars, which could sum up to almost three hundred kg, makes the health integrity of the employees become at risk.

The routine of carrying heavy objects, performing repetitive forceful tasks and inadequate postures tends to cause musculoskeletal disorders. These are

M. F. Silva et al. (Eds.): ROBOT 2019, AISC 1092, pp. 669–681, 2020.
https://doi.org/10.1007/978-3-030-35990-4_54

injuries or disorders of the muscles, nerves, tendons, joints, cartilage, and spinal discs, e.g. sprains, back pain, carpal tunnel syndrome, hernias that could get worse or even persist longer due to the work environment or performing continuously these tasks. Therefore, the goal of ergonomics is to diminish stress and eradicate disorders associated with the excessive use of muscles, bad postures and recurring tasks [1,5]. Thus, advancing solutions based on autonomous robotic systems capable of performing these tasks, is an important step towards improving the overall health of human resources and also achieving more efficient distributions of the personnel. However, these systems must adapt well to the environments where they shall make the difference, especially concerning safety issues. In crowded and chaotic locations such as hospitals, this is not trivial to address.

In this paper, we describe the development of an autonomous robotic system which may be used in different indoor environments without modifying them. Moreover, the system is able to use standard trailers without significantly changing them, as it only requires them to have a simple low-cost plate attached. It has a fleet administrator that supervises, coordinates and creates a list of missions ordered by an employee through a graphical interface. The specific use case of meal transporting hospital trolleys in a medical environment is here demonstrated but the concept can be applied to several areas. Experiments were conducted in both controlled and real environments, in a collaboration with TRIVALOR/GERTAL/ITAU in Braga's Hospital. We believe that such system will allow to reduce the injuries previously mentioned, to automate processes and to conduct operators to specific tasks which cannot be performed by a robot.

The structure of this paper is the following: Sect. 2 presents the work regarding the application of mobile robots in hospital environments, more precisely in the context of transporting trolleys. The architecture overview of the presented robotic system is described in Sect. 3. Then, the modules referred in the System Architecture will have a more detailed explanation in each corresponding Subsection. Section 4 describes the experimental tests conducted to validate the robotic system and presents the results achieved. Finally, Sect. 5 concludes the paper and point out some directions regarding the future work.

2 Related Literature

The development of smart logistics systems means the development of independent and flexible external and internal logistics solutions. The forthcoming Industry 4.0 means an accelerating growth of the efficiency of production systems [15]. There are a huge number of application of AGVs (Automated guided vehicles) in industries. They are revolutionising the way manufacturers move goods between places, increasing efficiency. Hospitals can also be benefited when moving trolleys with meals and linens.

Robotics can help hospitals to maintain care workflows and give staff more time for patient care. There are several robotic approaches applied to hospital logistics, but most of them are used for edutainment activities [10] or to transport medication in specific conveyors. This research has been carried out for some years, being an example the HelpMate robot, that carries late meal trays, sterile supplies, medications, medical records, reports, samples, specimens and mail while exhibiting humanlike behavior as it navigates [7]. According to our knowledge, there are no available approaches that transport, grasp and pull non-modified heavy standard trolleys. Most approaches transport the goods inside the robot and do not attach to an external trolley, as they are expecting a very particular kind of car model. Some commercial examples are the Moxi robot that is an hospital robot assistant that helps clinical staff with non-patient-facing tasks like gathering supplies and brings them to patient rooms and delivers lab samples. The proposed low cost solution allows to tow a trolley without modifying it, by simply coupling a plate into the latter.

Well known industrial mobile robot manufacturers with the capability of transportation and manipulation, such as MIR [11] and Swisslog [20], offer a lot of products and add-ons for AGVs. Inspired on the solutions referred previously and in the state of the art from automotive and industrial environment (such as BMW [6] and kuka) and hospital mobile robotics such as SAVANT "Target-free" AGV Hospital Cart Transportation System [14] and EvoCart [2], it is desired in this paper to present a new developed solution oriented to hospital facilities to transport food and linens that uses a low cost gripping system that handles the non-modified heavy standard trolley and transports it.

3 System Architecture

The modular architecture presented in Fig. 1 shows the underlying combination of software and hardware modules, and also the way how the users are able to command the robots. The developed system is composed by the integration of independent subsystems, allowing to add or remove any component without affecting the rest of the system, using ROS [9]. Given the importance of keeping hospital management tasks beyond reach, the architecture was designed to allow one or more users to communicate with the robots via the hospital private network.

Users are going to interact with the robotic system through an user-friendly and intuitive graphical interface designed to adapt their work needs. This may be done using different hardware, such as laptops and mobile phones. To streamline this process, a web server was created as the user's central command distribution unit. Regarding the software modules included in the robots, Fig. 1 shows a subset of them, the most relevant ones, in order to facilitate comprehension and analysis of the system. Our goal is to develop a fleet of trolley-transporting robots for an hospital. Each robot is independent and has its own state. In the next subsections is described more precisely.

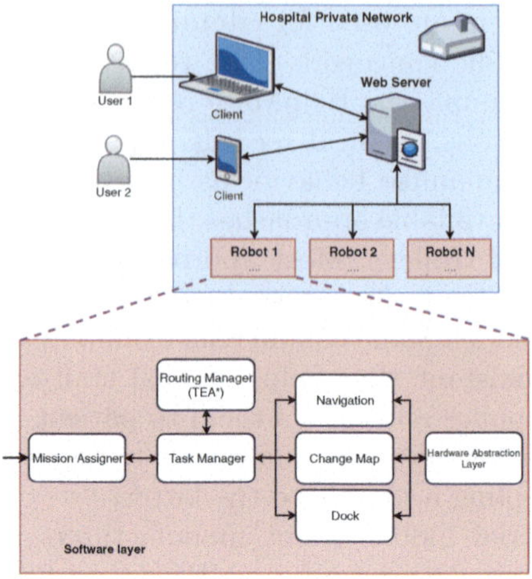

Fig. 1. System architecture diagram.

3.1 Hardware System

The robotic system comprises a mobile platform and a gripping system. The robot has a differential traction system with two spring-mounted drive-wheels with one motor each and four support wheels (Fig. 2a). It also has two Laser Range Finder SICK S300 which confer a 360° of view point and a bumper which incorporates a switch that is triggered on impact, allowing to detect obstacles and change behavior without sustaining damage. We assembled a tow arm on top of the robotic platform to act as a "hand" for grasping and pulling the trolleys (Fig. 2b). It consists of a structure which is just controlled by a linear actuator allowing two distinct movements. The orientation of the tow arm is known by a rotary absolute encoder of 10 bits. Arduino Mega was the microcontroller used to manage the trolley docking system. We considered a ROS node for monitoring connection and disconnection events, through Linux kernel event logs, allowing the system to be more robust to fails. Moreover, lighting and buzzing mechanisms were integrated to give feedback to the operator about the robot's state. To deal with a wide range of hospital car models, we have designed a simple, low-cost and easily integrable prototype plate for the docking task. This plate allows the robot to be attached rigidly to a trolley to be transported to its destination.

3.2 Human Robot Interaction

This module is responsible for enabling communications between users and the fleet of robots, in the sense that it allows users to define the operational behavior of the robots.

To control the robots, the user sends missions to them through a web application. As shown in Fig. 3, the user is allowed to send the robot from a given location to another one, to cancel one or all of the assigned displacements, and

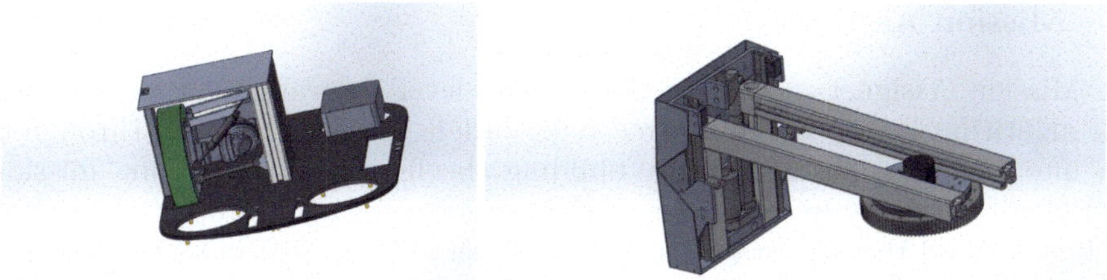

(a) (b)

Fig. 2. Hardware components: (a) traction system; (b) gripping system.

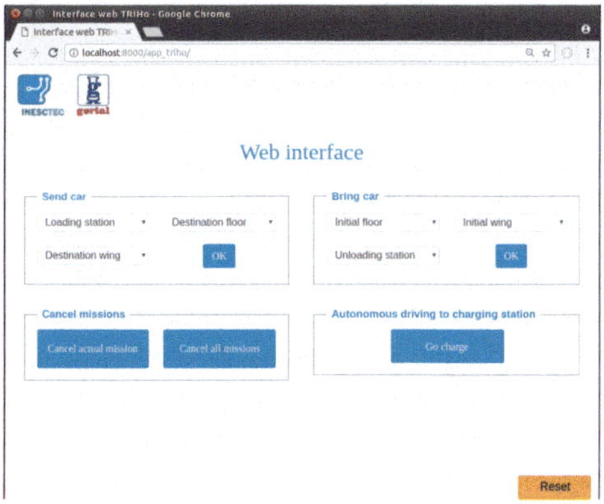

Fig. 3. Web interface prototype showing the fields where the user defines and cancels the missions.

to reset the state of the robot (by means of a joystick, the robot is sent to a predefined location, and reset is manually performed by pushing a button).

If any problem affects the server, disabling its proper operation, the robots may also be locally controlled without recurring to the external server, a state that is initiated after pushing specific physical buttons located on top of the robots. We are developing the web server using Django, a set of tools in python for streamline web development[1]. The web server provides the web page made in HTML and CSS in the browser of the device and the user responds to the forms. Then, the server receives the intended mission and sends it to the mission assigner. In reverse, the mission assigner gives feedback to the web server about the status of the mission. As currently implemented in our prototype, both communications are bidirectional using the HTTP protocol, differing in the type of data exchanged. In the deployment phase, we are switching HTTP for HTTPS, which is an extension of the first, where encryption allows the communications to be protected.

[1] https://www.djangoproject.com/.

3.3 Mission Assigner

The Mission Assigner module is the highest level software layer in the robot. This algorithm is responsible for receiving orders given by the user through the web interface and, autonomously, building the list of corresponding missions. For that, it is aware, in real time, of the current pose of the robot, the state of the battery and the arm state (actuated or not). Then, it decides, for example, if the robot needs to go charge itself or if new missions should be assigned. It also allows to cancel a single or all the missions stored in the robot. It is able to prioritize certain types of missions and to reset if any unexpected failure happens. This module communicates with the Task Manager through ROS Bridge Suite by using services and topics.

3.4 Action Planning: Time Enhanced A* (TEA*) and Token Manager

In a transportation system composed of a fleet of robots, it is important to choose the most optimized route for each one in order to prevent deadlocks between them. Nevertheless, unexpected events, such as the appearance of obstacles or delays during the execution of transporting tasks, need to be taken into consideration. Having this in mind, the Time Enhanced A* was chosen for the management of a fleet of robots. The TEA* Algorithm is used to determine a trajectory that a robot will have to perform for the coordination of multiple robots. It calculates each trajectory constantly, therefore it can be recognized as an on-line method. It is a path planning algorithm which resorts to a fixed graph to determine a trajectory based on the information of the 2D graph and the symbolic pose of each AGV (the edge each robot is occupying at the moment). A third dimension, *Time*, was added to enhance the information of the graph [12,13]. In other words, a temporal layer is created where each vertex can be considered as occupied or free. The robots perform each transportation task along edges, where the final point in each trajectory corresponds to a vertex in the graph. Therefore, this method searches by vertexes, where the cost function is associated to the length of the edges. As in the A* algorithm, that function results of the sum of the covered distance and the distance until the final point. The former represents the length of the edges and the latter the euclidean distance to the destination point. Figure 4 shows an example on how the TEA* works. From a fixed graph, where the red lines represent the edges, the blue circles represent the vertexes and the green arrow represents the desired orientation on each vertex, the algorithm determines the best route in order to reach a final point. In fact, that is represented by the yellow line over the chosen edges.

The token manager's module is a high-level supervision software, running in a central station, where it defines blocking and non-blocking areas in the trajectory to avoid deadlocks. It supervises, through UDP connections, the blocking areas' state (occupied or free), the last robot occupying that area and the real-time pose of each robot in the graph. Based on this information, the system is, additionally, able to deal with network fails.

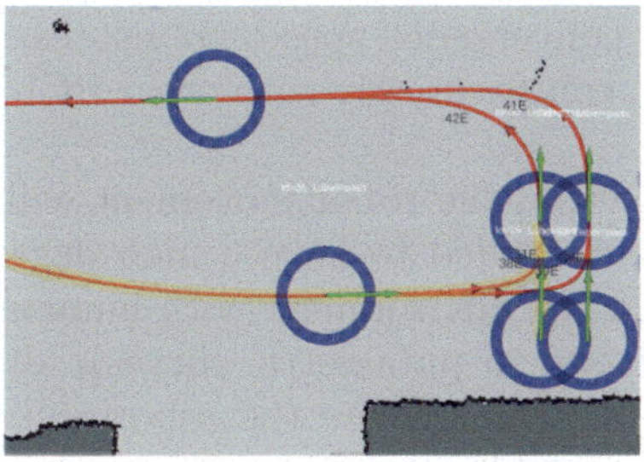

Fig. 4. Chosen trajectory by the TEA* (yellow) over a fixed graph. The red line represents the edges between vertexes (blue circle). The green arrow states the pretended orientation in each vertex.

3.5 Task Manager

At the robot level, Task Manager acts as an orchestrator, receiving top level missions from Mission Assigner and then breaking them into a sequence of specific tasks. These tasks are executed as ROS actions[2], in which the Task Manager interacts as the action client. For each action, there is a corresponding lower level action server node responsible for ensuring its execution, e.g., drive, dock and change map. Although each action has its own input parameters in order to execute the intended objective in a specific way, they are all executed similarly by the Task Manager, with an unified interface. This improves the modularity of the system, as it is easy to add or remove new functionalities (action servers).

As an exception, due to the higher behavioral complexity, the drive action is treated differently by the Task Manager. Before each one, the Task Manager interacts with the TEA* module, so it can receive the full path between the robot's current pose and the destination vertex. Then, the path is sent to the drive action which synchronizes its execution with the specific trajectory controller.

3.6 Navigation

The navigation system is composed by several modules, e.g. trajectory editor, differential/tricycle trajectory controllers, localization, map changer, which cooperate with each other so the robot can move around the environment.

For the localization process, the robot uses an extended version [4,19] of the Perfect Match algorithm [8]. It is a localization algorithm that does not require to change the target operation environment, as it uses natural features to estimate the robot's pose, e.g., walls, doors, cornerstones. It uses outlier rejection to be able to operate even in dynamic environments, when the input sensor data may be different than the previously created map. As seen in [16,17], the Perfect Match is a computationally light algorithm which performs better (considering

[2] http://wiki.ros.org/actionlib.

a 2D space) in terms of convergence speed, robustness to initialization errors, and has similar precision, comparatively with other PCL based localization algorithms.

In this specific application, the robot movement will be based in predefined fixed trajectories, as it is intended to have a high degree of predictability and repeatability, when sharing an environment with humans. So, by using a trajectory drawing software (based in parametric curves), all the possible paths will be defined, in a form of a graph, where the robot will be moving. Graphs are constituted by vertexes (possible stopping poses) and edges (parametric curves that connect two vertexes), as seen in Fig. 5. Each edge has its own forward and backwards linear velocities associated, which is constant when the robot is moving along that specific path. On the other hand, each vertex may have some actions associated with it, e.g., exchange map, interact with elevator or order to dock a trolley. The specific trajectory performed by the robot for each objective is calculated and optimized by the routing TEA* algorithm, better described in Sect. 3.4.

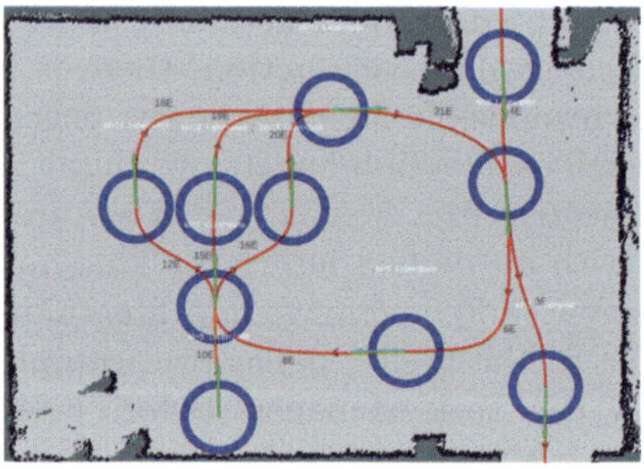

Fig. 5. Graph example. The red lines represent the edges and the blue circles represent the vertexes, with orientation given by the green arrows.

This robot base, as previously mentioned, has two traction wheels and the drive control is based on the differential traction model. When the robot is pulling the previously docked car, the robotic system exchanges to tricycle traction model, with the robot base acting as the direction wheel. This assures more precision and repeatability in the path following task and when some maneuvers are performed (more critical when driving backwards), e.g., parking the trailer and entering elevator.

This robot system ensures multi-floor navigation. As such, after mapping the robot's target environment (e.g., multi floor building), all maps are saved to be later used by the robot. There is a specific module responsible for coordinating the current navigation map with the robot's tasks, using all previously saved maps and the transformations between them.

3.7 Docking

In this section, a general overview on the different docking modes used on our application is presented. The robot performs a docking task to the elevator in order to change between floors, to reach the cars for their transportation and to move to the charger for the autonomous charging of the robot.

Elevator. The process responsible for the robot entering and exiting an elevator is controlled by a specific module, where the Perfect Match algorithm is used by the robot itself to find its location regarding the object of interest. A small map of the elevator contours is used, in order for the robot to estimate its pose in relation to the elevator. Then, when the clear space in the elevator is identified, the robot can generate an on-line trajectory in order to reach a predefined pose inside the elevator. However, it can also use an already existing path of the graph. This is done both when the robot is operating in differential and tricycle modes (with a trailer attached to its arm). It is possible for the robot to enter and exit the elevator either forward or backwards.

Car and Charger. We resorted to a Beacon-based Localization Algorithm [3,18], for the detection of the trolley or the charger. This algorithm is divided in several modules, for example the first one is responsible for the Kalman filter which handles the sensor fusion with the odometry. Other modules try to use the data from a laser range-finder to determine a possible position of cylindrical beacons and try to identify them. However, the usage of cylindrical beacons is unsuitable for our application due to the fact of their three dimensional shape, they can be easily damaged or moved and are voluminous, which led us to choose planar beacons. For this system to work, it needs as input: (a) the pose of the car we want the robot to dock, (b) the pose of the robot and (c) the detected beacons by the laser range finder. As the measured distance between the beacons attached to the car or the charger are known, we can group the detected ones in pairs. Each group may represent a possible pose of the object we want the robot to dock. This will allow to discard false beacons as well as give a more precise pose estimation of the car or charger to the Kalman Filter that processes the Beacons Localization Algorithm (Fig. 6a). With the poses as an input, we compute the theoretical pose of the car referenced to the robot. If this matches with a pose estimated by the detected beacons, the docking process is initialized. During the docking process, the robot will follow a line to approach the docking pose while testing if both beacons are visible through the whole trajectory (Fig. 6b).

When the robot reaches the final pose, a final verification is performed. In fact, it is determined if there is any deviation of the pose of the robot towards the pose of the car. If this verification is successful, a message is emitted to the upper level, which sends the signal for the arm to descend.

During the docking process, certain situations may lead to a failure: (a) the pose estimated by the detected beacons does not match the expected pose, (b) one or both beacons are not detected while the robot is approaching the trolley

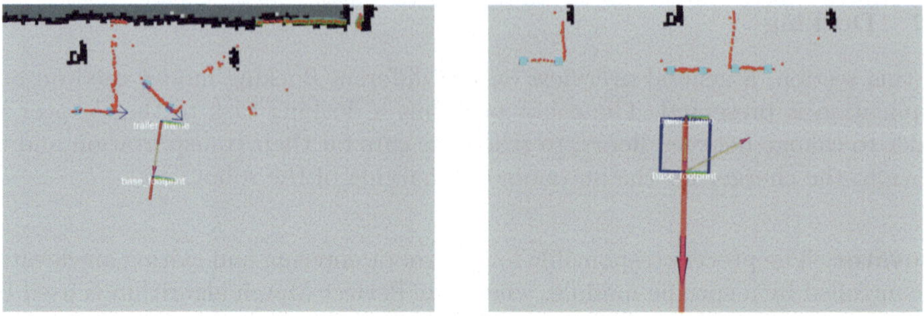

(a) (b)

Fig. 6. Docking process: (a) detection of the beacons (cyan squares) and pairing (blue arrow); (b) robot following a line in the docking process.

or the charger and (c) the failure of the final verification. When any of these scenarios occurs, an error message is emitted to the upper level, starting a retry process.

4 Experiments and Results

We conducted a set of experiments to evaluate our system, which can be seen in the following link: https://youtu.be/TMnIOADs-dM. We tested each module separately, e.g. we instructed the robot to dock a trolley (blue and orange block) or an elevator, to drive to a specific destination with the car docked (violet and red) and also without it. Upon the conclusion of these preliminary tests, the system was tested as a whole. In this last experiment (Fig. 7), the robot docked the trolley and carried it from a floor to another, reached its destination, undocked the car and returned to the starting point (green block). When the robot arrived, a mission to return the car to its original position was sent.

Fig. 7. Map of the ground floor of Braga's Hospital. The blue and orange blocks represent loading areas, the violet and the red blocks represent the unloading areas and the green block represent the charging station's area and the base position.

In order to evaluate the robotic system performance, several experiments were made keeping the same sequence of the operations. Previously, we set up a multi-

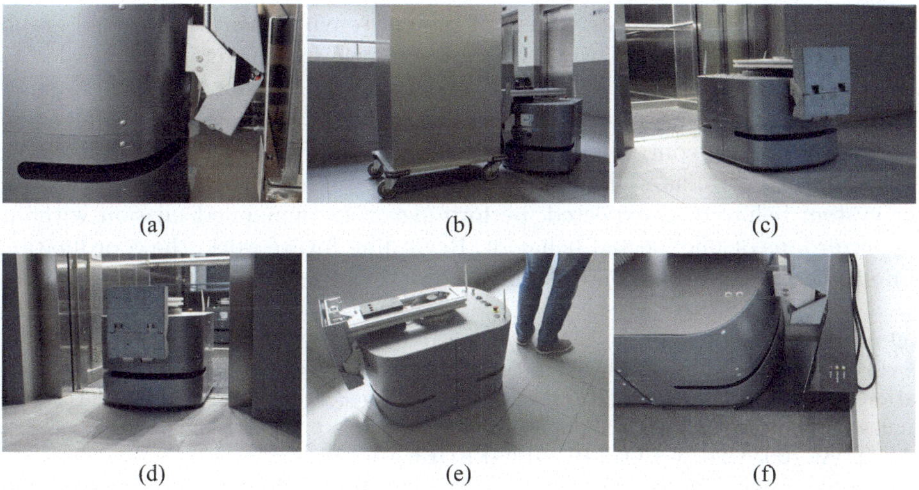

Fig. 8. Snapshots taken during the experiments: (a) robot docks the trolley; (b) robot navigating to a specific vertex; (c) robot docking the elevator; (d) robot undocking the elevator; (e) robot stops if an object or a person is detected in its security area; (f) robot charging.

floor trajectory involving communications with an elevator. Then, we provide a list of missions through a web interface. The robot fetched the trolley by docking it (Fig. 8a), went to the vertex before the elevator (Fig. 8b) and sent an order to the elevator to move to that floor. When it arrives, the robot starts the docking process to the elevator (Fig. 8c) and the elevator goes to the desired floor. After, the door opens and the robot leaves the elevator (Fig. 8d). It follows its path and, if any obstacle appears, the robot detects it and stops (Fig. 8e). The robot concludes its mission by un-docking the car at a predefined location. As soon as the mission finishes, it charges by docking to the charging station (Fig. 8f). All the experiments were conducted in the Faculty of Engineering of the University of Porto and in Braga's Hospital. To evaluate our system, we performed various tests in distinct routes and we used diverse obstacles to assure its flexibility. The outcome of these experiments was successful.

5 Conclusions and Future Work

This paper described the development of an autonomous robotic system for the transportation of hospital trolleys in medical environment. It presented the software architecture of the system, which is divided in several modules. A human machine interface was developed to simplify the interaction with the system. The graphical interface, where a user can outline a sequence of missions to be executed was presented. This system can be used in other types of environments, e.g. inside factories, without major changes to the environment itself. To deploy this system in other facilities, a map of the new environment and a fixed graph are required. We preferred fixed trajectories over free navigation in order to achieve a more predictable system. This is important in our application, since the robot may transport heavy loads in crowded environments. The whole system is being tested in a controlled environment in the Faculty of Engineering of the University of Porto, as well as in an hospital environment. As stated in Sect. 4, our system behaved as expected, performing every demanded mission without any hazardous or unexpected behavior. Regarding future work, the coordination of multiple mobile robots on different environments will be addressed.

Acknowledgements. This work is financed by the ERDF – European Regional Development Fund through the Operational Programme for Competitiveness and Internationalisation - COMPETE 2020 Programme, and by National Funds through the Portuguese funding agency, FCT-Fundação para a Ciência e a Tecnologia, within project SAICTPAC/0034/2015- POCI-01- 0145-FEDER-016418.

Authors would like to acknowledge to Trivalor, Itau and Gertal for the support of the project RDH.

References

1. Center for Disease Control and Prevention: Work-related musculoskeletal disorders and ergonomics (2018). https://www.cdc.gov/workplacehealthpromotion/health-strategies/musculoskeletal-disorders/index.html
2. EvoCart: Evocart - advanced motion technology (2019). https://www.oppent-evo.com/en/home-en/
3. Ferreira, F., Sobreira, H., Veiga, G., Moreira, A.: Landmark detection for docking tasks, pp. 3–13 (2018)
4. Gouveia, M., Moreira, A., Costa, P., Reis, L., Ferreira, M.: Robustness and precision analysis in map-matching based mobile robot self-localization (2009)
5. Iowa State University Department Environment Health and Safety: Ergonomics (2016). http://publications.ehs.iastate.edu/ergo/
6. Kochan, A.: BMW uses even more robots for both flexibility and quality. J. Ind. Robots (2005). https://doi.org/10.1108/01439910510600173
7. Krishnamurthy, B., Evans, J.: HelpMate: a robotic courier for hospital use. In: IEEE International Conference on Systems, Man, and Cybernetics (1992)
8. Lauer, M., Lange, S., Riedmiller, M.: Calculating the perfect match: an efficient and accurate approach for robot self-localization. In: Bredenfeld, A., Jacoff, A., Noda, I., Takahashi, Y. (eds.) RoboCup 2005: Robot Soccer World Cup IX, pp. 142–153. Springer, Heidelberg (2006)
9. Martinez, A., Fernández, E.: Learning ROS for Robotics Programming. Packt Publishing, Birmingham (2013)
10. Messia, J., Ventura, R., Lima, P., Sequeira, J., Alvito, P., Marques, C., Carriço, P.: A robotic platform for edutainment activities in a pediatric hospital. In: 2014 IEEE International Conference on Autonomous Robot Systems and Competitions (ICARSC) (2014)
11. MiR: Mobile industrial robots (2019). https://www.mobile-industrial-robots.com/en

12. Santos, J., Costa, P., Rocha, L., Moreira, A., Veiga, G.: Time enhanced A*: towards to the development of a new approach for multi-robot coordination. In: IEEE International Conference on Industrial Technology (ICIT) (2015)
13. Santos, J., Costa, P., Rocha, L., Vivaldini, K., Moreira, A.P., Veiga, G.: Validation of a time based routing algorithm using a realistic automatic warehouse scenario. In: Reis, L.P., Moreira, A.P., Lima, P.U., Montano, L., Muñoz-Martinez, V. (eds.) Robot 2015: Second Iberian Robotics Conference, pp. 81–92. Springer, Cham (2016)
14. Savant Automation: Savant automation - AGV systems (2019). http://www.agvsystems.com/
15. Skapinyecz, R., Illés, B., Bányai, A.: Logistic aspects of industry 4.0. In: IOP Conference Series: Materials Science and Engineering, vol. 448, no. 1 (2018)
16. Sobreira, H., Rocha, L., Costa, C., Lima, J., Costa, P., Moreira, A.P.: 2D cloud template matching-a comparison between iterative closest point and perfect match. In: 2016 International Conference on Autonomous Robot Systems and Competitions (ICARSC), pp. 53–59 (2016)
17. Sobreira, H., Costa, C.M., Sousa, I., Rocha, L., Lima, J., Farias, P.C.M.A., Costa, P., Moreira, A.P.: Map-matching algorithms for robot self-localization: a comparison between perfect match, iterative closest point and normal distributions transform. J. Intell. Robot. Syst. (2018). https://doi.org/10.1007/s10846-017-0765-5
18. Sobreira, H., Moreira, A., Costa, P., Lima, J.: Robust mobile robot localization based on a security laser: an industry case study. Ind. Robot: Int. J. **43**, 596–606 (2016)
19. Sobreira, H., Pinto, M., Moreira, A.P., Costa, P.G., Lima, J.: Robust robot localization based on the perfect match algorithm. In: Moreira, A.P., Matos, A., Veiga, G. (eds.) CONTROLO'2014 – Proceedings of the 11th Portuguese Conference on Automatic Control, pp. 607–616. Springer, Cham (2015)
20. Swisslog: Swisslog - healthcare and logistics automation (2019). https://www.swisslog.com/

Author Index

M. F. Silva et al. (Eds.): ROBOT 2019, AISC 1092, pp. 683–685, 2020.
https://doi.org/10.1007/978-3-030-35990-4